AF743865

Prof. Dr. Oliver C. Ruppel |
Prof. Dr. Christian Roschmann |
Dr. Katharina Ruppel-Schlichting [Eds]

# Climate Change: International Law and Global Governance

## Volume II: Policy, Diplomacy and Governance in a Changing Environment

Nomos

This publication was produced in cooperation with the

Konrad
Adenauer
Stiftung

Die Deutsche Nationalbibliothek lists this publication in the
Deutsche Nationalbibliografie; detailed bibliographic data
is available in the Internet at http://dnb.d-nb.de.

ISBN 978-3-8329-7796-2

Language Editing: Julie Streicher & The Word Factory

The content of the articles, including any final errors or omissions that remain, is the sole responsibility of the individual contributors. The editors have made every effort to acknowledge the use of copyright material. Should any infringement of copyright have occurred, please contact the editors, and every effort will be made to rectify omissions or errors, in the event of a reprint or new edition.

1. Edition 2013
© Nomos Verlagsgesellschaft, Baden-Baden 2013. Printed in Germany.

This work is subject to copyright. All rights are reserved, whether the whole or part of the material is concerned, specifically those of translation, reprinting, re-use of illustrations, broadcasting, reproduction by photocopying machine or similar means, and storage in data banks. Under § 54 of the German Copyright Law where copies are made for other than private use a fee is payable to »Verwertungsgesellschaft Wort«, Munich.

Contents

FOREWORD 9

PREFACE 11

ACKNOWLEDGEMENTS 13

THE EDITORS 15

THE CONTRIBUTORS 17

**PART I: GLOBAL CLIMATE GOVERNANCE – DEVELOPMENTS AND CHALLENGES**

1 Intersections of Law and Cooperative Global Climate Governance – Challenges in the Anthropocene 35
*Oliver C. Ruppel*

2 International Climate Change Policy: Where do we Stand? 101
*Nadia von Bassewitz*

3 Climate Change Lawfare 171
*Siri Gloppen & Asuncion Lera St. Clair*

4 Greening Permanent Sovereignty through the Common Concern in the Climate Change Regime: Awake Custodial Sovereignty! 201
*Werner Scholtz*

**PART II: GLOBAL CLIMATE GOVERNANCE AND DIPLOMACY**

5 Power in Global Climate Governance 217
*Babette Never*

6 CBDR as a Principle of Inspiring Actions rather than Justifying Inaction in the Global Climate Change Regime 235
*Achala C. Abeysinghe & Gilberto Arias*

7 Beyond International Climate Negotiations: Climate Diplomacy from a Foreign Policy Perspective 259
*Dennis Tänzler & Alexander Carius*

8 Confronting Complex Global Challenges: Comparing the Climate Change and Law of the Sea Negotiations 275
*Gregg B. Walker*

9 UNEP plus X? A Critical Assessment of Reform: Proposals and Implications for the International Climate Regime 313
*Dirk Hanschel*

**PART III: A TRANSDISCIPLINARY MIX: CLIMATE CHANGE, POLITICS AND FINANCE**

10 Transdisciplinarity: Theory and Visions on Global Transdisciplinary Processes for Adapting to Climate Change 329
*Roland W. Scholz*

11 The Politics of Climate Change: Review and Future Challenges 357
*Holger Haibach & Kathrin Schneider*

12 International Climate Finance: Policies, Structures and Challenges 375
*Cord Lüdemann & Oliver C. Ruppel*

**PART IV: CLIMATE CHANGE POLICY, COOPERATION AND PROTECTION EFFORTS**

13 Climate Change Law and Policy Positions in the African Union and Related Developments in Selected African Countries 411
*Oliver C. Ruppel*

14 Council of Europe Climate Law Standards and Perspectives 449
*Agata Rogalska-Piechota*

15 Renewable Energy Policy in the European Union: A Contribution to Meeting International Climate Protection Goals? 477
*Christian Calliess & Christian Hey*

16 China-Africa Cooperation: Joint Engagement in Adaptation to Climate Change 529
*Wanxin Li, Sven Grimm & Harrie Esterhuyse*

17 The BRICS Partnership: Development and Climate Change Policy from an African Perspective 549
*Oliver C. Ruppel & Katharina Ruppel-Schlichting*

**PART V: CLIMATE CHANGE RESPONSES, EQUITY AND SUSTAINABLE DEVELOPMENT**

18 Sustainable and Inclusive Adaptation to Climate Change and Public Policy Challenges in Central America 573
*Julie G. Lennox*

19 Mainstreaming Sustainable Development into National Climate Change Responses: Assessing the Legal Options to Reinforce Equity 601
*Robert Kibugi*

20 Agenda 21 and Climate Protection: The Development of Global and Local Governance for Environment and Development – Observations from Research in Namibia 621
*Manfred O. Hinz*

21 Ethics for Climate Justice and Sustainability through Value-based Approaches: A New Tanzanian Model and Paradigm Shift 661
*Aidan G. Msafiri*

**PART VI: CLIMATE CHANGE, SECURITY AND THE MIGRATION DEBATE**

22 Science, Facts and Fears: The Debate on Climate Change and Security 685
*Marcel Leroy & Fana Gebresenbet*

23 Dangerous Anthropogenic Climate Change from the Perspective of Adaptation 713
*Kristie L. Ebi & Ian Burton*

24 Climate Change, Disasters and Migration: Current Challenges to International Law 739
*Erika Pires Ramos*

25 Enhancing Adaptation Options and Managing Human Mobility in the Context of Climate Change: Role of the United Nations Framework Convention on Climate Change 761
*Koko Warner*

26 Adaptation to Climate Change under Changing Urban Patterns: The Climatic Perspective of Migration 785
*María Máñez Costa, Kathleen Schwerdtner Máñez & Sainab Husain Paragay*

27 Climate-change-induced Movement of Persons in Africa: Human Rights Responses to Aspects of Human Security 799
*Oliver C. Ruppel & Sanita van Wyk*

**PART VII: CLIMATE CHANGE AND THE LOSS AND DAMAGE DEBATE**

28 Framing the Loss and Damage Debate: A Thought Starter by the Loss and Damage in Vulnerable Countries Initiative 829
*Sönke Kreft, Koko Warner, Sven Harmeling & Erin Roberts*

29 Legal and Policy Responses to Loss and Damage Associated with Climate Change 843
*M. Hafijul Islam Khan*

30 Insurance Solutions in the Context of Climate-Change-Related Loss and Damage: Needs, Gaps and Roles of the UNFCCC in Addressing Loss and Damage 877
*Koko Warner, Sönke Kreft, Michael Zissener, Peter Höppe, Christoph Bals, Thomas Loster, Joanne Linnerooth-Bayer, Silvio Tschudi, Eugene Gurenko, Armin Haas, Simon Young, Paul Kovacs, Andrew Dlugolecki & Aaron Oxley*

Annex
Contents Volume I
Climate Change: International Law and Global Governance
Volume I: Legal Responses and Global Responsibility 923

## FOREWORD

It is with great pleasure and honour that I am writing a foreword for this eminent work, which seeks to promote the international rule of law, contribute to durable global peace, avoid conflict, lead to more effective protection of human rights, as well as sustain economic progress and development.

The two volumes of *Climate Change: International Law and Global Governance* describe important topics in respect to mankind and the future that lies ahead. Perhaps, the most important topics are the regulatory and diplomatic aspects of climate change.

In June 2013, UN Secretary-General Ban Ki-moon said "the scientific community plays a key role in finding new ways to combat climate change." When looking at this publication I must add that the same issue, most likely, also applies to the legal community! It serves as a valuable tool in harnessing the full strength of the global community, catalysing ambitious action, persuading the reduction of emissions, and strengthening climate resilience.

In 1979, Pope John Paul II named St. Francis of Assisi the patron saint "of those who promote ecology". Interestingly, the recently elected Pope Francis chose his name in honour of the historic preacher, who similarly conducted and was famous for his rather unconventional way of life. Born into wealth, St. Francis of Assisi eventually renounced all of his belongings, aspiring to live a life of wilful poverty in the quest for increased social justice. With respect to this quest, this publication also addresses the promotion of ecology, (un-) conventionalism, distribution of wealth, alleviation of poverty, and the promotion of global social justice.

In light of the impacts of climate change, international regimes face serious concerns with issues such as human rights, global trade, territorial sovereignty, or migration. Legal responses and global responsibilities, therefore, gain an increased political meaning as they encompass legal and policy responses of climate change (e.g. via liability or jurisdiction, and litigation).

In March 2013, the EU Commissioner for Climate Action, Connie Hedegaard, presented a speech at a Conference at Harvard University where she commented inter alia on the "pattern of more frequent and more severe extreme weather worldwide." She said: "What we see fits with the scientific community's projections of what a warming world will be like - except that their projections are actually becoming reality even faster than they themselves expected. As President Obama has said, we can either believe that these events were just a coincidence, or we can choose to believe in the overwhelming judgment of science and act before it's too late."

The challenge of strengthening national and international climate change policy, sustainable development, and increasing equity around the world are above the capacity of national governments. Thus, international climate change cooperation and protection efforts are crucial not only in the context of national but also global security.

Only recently German Chancellor Angela Merkel called for an internationally binding climate pact to be completed by 2015. "Waiting is not an option," she said. In addition, German Environmental Minister Peter Altmaier mentioned that 2015 will be an important year in climate negotiations. He said: "The international awareness that we need to reach, as a milestone by 2015, is growing," adding that progress "in many areas is still too slow and not enough."

In light of the aforementioned, I commend the editors of this significant work! This publication is not merely an inventory but, furthermore, one of the first academic attempts to systematically address both international climate change law and global climate change governance from a variety of doctrinal, transdisciplinary and thematic perspectives. As a political foundation the Konrad-Adenauer-Stiftung is committed to fostering democracy and the rule of law, implementing social and market-economic structures, and promoting human rights. In this respect, the protection of the environment, as well as issues of climate change and sustainable development are major concerns to this foundation. It is, thus, a privilege for the Konrad-Adenauer-Stiftung to support this important publication. This is a remarkable reflection of the commitment and expertise displayed by the editors and contributors to whom we are very grateful.

*Hans-Gert Poettering was born on 15 September of 1945 in Bersenbrueck (Lower Saxony, Germany). Since 1979 he has been a Member of the European Parliament whose President he was from 2007 to 2009. Since 2010 he is President of the Konrad-Adenauer-Stiftung. Hans-Gert Poettering belongs to the Group of the European People's Party (EPP) and is a member of the Christian Democratic Union (CDU, Germany).*

*Berlin, 17 June 2013*

# PREFACE

International climate change law is not only a new and emerging legal discipline. In fact, climate change in many ways permeates public and private law, as well as national and international law, creating intersections of law in its diverse procedural and substantive fields. This two-volume publication on ***Climate Change: International Law and Global Governance*** deals with international law and the multiple regulatory regimes which presently reflect fragmentation in the absence of a universal climate change regime. International climate change law, global climate governance and diplomacy are interrelated and extremely complex: the publication explores these areas from a variety of doctrinal, transdisciplinary and thematic perspectives.

**Volume I:** ***Legal Responses and Global Responsibility*** attempts to assess the most pressing impacts of climate change on various international law regimes and their responses thereto. In doing so, the volume inter alia reflects on international climate change law as a new international law discipline; climate change and human rights; climate change, international trade and investment law; the law of the sea and sea-level rise; judicial review and international climate change litigation; and multiple crosscutting issues such as mitigation regulation, natural resource management and climate-engineering.

As a point of departure, **Volume II:** ***Policy, Diplomacy and Governance in a Changing Environment*** reflects on the United Nations Convention on Climate Change (UNFCCC) and the most pressing impacts of climate change on international diplomacy and global governance. This is highlighted from various transdisciplinary and geopolitical perspectives with a special focus on the challenges of strengthening national and international climate change policy, promoting sustainable development and increasing equity around the world, which go beyond the capacity of national governments. Various international climate change cooperation and protection efforts are analysed, also in the context of global security, climate-induced migration movements, adaptation, and the loss and damage debate.

The effectiveness of the international response to climate change depends upon the legal tools available and the political will to ensure effective implementation. An enabling legal environment, underpinned by good governance and respect for the rule of law, is a prerequisite for greater international climate change equity. In this spirit, it is hoped that this publication can make a humble contribution towards ensuring more global justice, human security and international peace.

***The Editors***

***Oliver C. Ruppel   Christian Roschmann   Katharina Ruppel-Schlichting***

# ACKNOWLEDGEMENTS

A multi-authored publication such as this is an enormous team effort. With so many contributors from all continents of the world, our thanks go first and foremost to all the distinguished authors.

Moreover, we are very grateful to the Konrad-Adenauer-Stiftung which generously supported this enormous project in its various phases including the publication itself. In September 2012 the Konrad-Adenauer-Stiftung Rule of Law Program for Sub-Saharan Africa in collaboration with the Faculty of Law of the University of Stellenbosch, South Africa, held two consecutive conferences, one on *Climate Change and Governance* and another on *Climate Change, Legal Responses and Global Responsibility* in Stellenbosch, South Africa. The conferences deliberated on the effects and the legal aspects of climate change on governance and other pertinent issues. The participants – many of them also represented in this publication – comprised international experts drawn from around the world.

Our special thanks go to Professor Dr. Hans-Gert Poettering (Member of the European Parliament and President of the Konrad-Adenauer-Stiftung); Dr. Gerhard Wahlers (Deputy Secretary-General of the Konrad-Adenauer-Stiftung); and Dr. Christian Hübner (Coordinator for Environmental, Climate and Energy Affairs of the Konrad-Adenauer-Stiftung) for their continuous support.

We also thank NOMOS Publishers Baden-Baden and particularly Professor Dr. Johannes Rux and Dr. Phillipp Küsgens for their professional support.

***The Editors***

***Oliver C. Ruppel   Christian Roschmann   Katharina Ruppel-Schlichting***

## THE EDITORS

**Prof. Dr. Oliver C. Ruppel** is Professor of Public and International Law at the Faculty of Law of the University of Stellenbosch, South Africa; Professor Extraordinaire at several universities around the world and Fellow of the Ciucci Centre for Law and Social Development, an initiative of the Stonehage Charitable Trust. Until 2010 he held one of the worldwide 14 academic chairs of the World Trade Organisation (WTO), Geneva, Switzerland, which he established at the University of Namibia, Windhoek, where he had previously also served as the Director of the national Human Rights and Documentation Centre. He currently serves as Coordinating Lead Author for the Chapter on Africa in the United Nations Intergovernmental Panel on Climate Change (IPCC), Working Group II. He is an international arbitrator with the Association of Arbitrators of Southern Africa, member of the South African and Swiss Branches of the International Law Association, and member of various editorial boards, including the *NIALS Journal of Environmental Law*, and the legal series *Law and Constitution in Africa*, Nomos Publishers, Germany. His latest publications include: Ruppel, Oliver C. & Katharina Ruppel-Schlichting (Eds), 2013, *Environmental Law and Policy in Namibia. Towards Making Africa the Tree of Life*, Windhoek/Essen, Orumbode Press/ Welwitschia Verlag Dr. A. Eckl; Ruppel, Oliver C. & Gerd Winter (Eds), 2011, *Justice from Within: Legal Pluralism in Africa and beyond*, Hamburg, Kovač.

**Prof. Dr. Christian Roschmann** is the Director of the Rule-of-Law-Programme for Sub-Saharan Africa of the Konrad-Adenauer-Foundation, based in Nairobi, Kenya. He is Professor at the Hochschule Harz, University of Applied Sciences, in Saxony-Anhalt, Germany and was visiting Professor at Strathmore Law School, Nairobi, Kenya in 2012 -2013. He was previously Director of the Legal and Human Rights Department of the Ministry of Foreign Affairs of Timor Leste. His latest publication is: Roschmann, Christian et al., *Kenya's Constitutional Review. Experience and Lessons to be Learned for the Constitution-making Process in Tanzania,* to be published in the near future in: Series 3, Tanzanian-German Centre for East African Legal Studies, Dar-es-Salaam.

**Dr. Katharina Ruppel-Schlichting** is the Director of the Legal Research and Development Trust for Africa, Windhoek, Namibia and Stellenbosch, South Africa. She is an international a legal consultant and author of a variety of legal academic writings. Her latest publications include the following: Ruppel, Oliver C. & Katharina Ruppel-Schlichting (Eds), 2013, *Environmental Law and Policy in Namibia. Towards Making Africa the Tree of Life*, Windhoek/Essen, Orumbode Press/Welwitschia Verlag Dr. A. Eckl; Ruppel, Oliver C. & Katharina Ruppel-Schlichting, 2012, Climate Change and Human Security: Relevant for Regional Integration in SADC?, in: Hartzenberg, Trudi, Gerhard Erasmus & Andre Du Pisani (Eds), *Monitoring Regional Integration in Southern Africa – 2011 Yearbook for Regional Integration*, Stellenbosch, tralac, 32–71.

# THE CONTRIBUTORS

**Dr. Achala C. Abeysinghe** is Senior Researcher and Team Leader in the Global Climate Change Law, Policy and Governance Group at the International Institute for Environment and Development (IIED), London, United Kingdom. Her interests are in equity and fairness issues in international climate change negotiations, adaptation to climate change and finance for climate change adaptation. She is the team leader of the Global Climate Governance Programme and Head of the European Capacity Building Initiative (ecbi) workshops programme at IIED. She is an adviser to the Chairman of the Least Developed Country (LDC) Group in the UNFCCC negotiations and a Lead Author of the Chapter on Climate Resilient Pathways: Adaptation, Mitigation and Sustainable Development for the Fifth Assessment Report of the Intergovernmental Panel on Climate Change (IPCC). One of her latetest publications is: Horstmann, Britta & Achala C. Abeysinghe, 2011, The Adaptation Fund of the Kyoto Protocol: A Model for Financing Adaptation to Climate Change?, *Climate Law* 2 (3), 415–437.

**Ambassador Gilberto Arias** is Senior Policy Advisor on Climate Change and Sustainable Development Initiatives. For the period 2009–2011 he was the Ambassador of the Republic of Panama to the United Kingdom in a bilateral capacity; Permanent Representative to the International Maritime Organisation (IMO); and Principal Negotiator for Panama to the UNFCCC in multilateral work.

**Christoph Bals** is Policy Director at Germanwatch, Bonn, Germany. He is a Board Member of the *Stiftung Zukunftsfähigkeit* (Foundation for Sustainability), the Munich Climate Insurance Initiative (MCII) and the Renewables Grid Initiative (RGI). He is a Member of the Nerve Centre of the Global Campaign for Climate Action (GCCA) and has co-initiated many initiatives in the area of climate and development as well as climate and economics – for example, e5, e-mission 55, and atmosfair. One of his latest publications is: Bals, Christoph, 2013, Die menschliche Gesellschaft bewegt sich auf eine Zeit steigenden Risikos zu. Hat sie die Entwicklung bereits wahrgenommen? Nur wenig spricht dafür, *Die Furche* 02/2013.

**Prof. Dr. Ian Burton** is Emeritus Professor at the University of Toronto, Canada and an independent scholar and consultant. He is Scientist Emeritus Environment Canada and a Lead Author in the Fifth Assessment Report of the Intergovernmental Panel on Climate Change (IPCC). He is a Visiting Fellow at the International Institute for Environment and Development (IIED) and a Member of the Board of Trustees of the World Academy of Arts and Science, and of the Advisory Boards of the Collaborative Adaptation Research Initiative in Africa and Asia (CARIAA) and the International Centre for Climate Change and Development (ICCCAD). His latest publications include the following: Fankhauser, Samuel & Ian Burton, 2013, Spending Adaptation Money Wisely, in: Haites, Erik (Ed.) *International Climate Finance*, London, Earthscan, 178–193; Ian Burton et al., 2011, *Integrated Research on Disaster Risk*, Forensic Investigations of Disasters: The FORIN Project, Beijing, IRDR FORIN Publication No. 1.

**Prof. Dr. Christian Calliess** is Professor of Constitutional, European and Environmental Law at the Faculty of Law of the Freie Universität Berlin, Germany. Since 2008 he is a Member of the Advisory Council on the Environment of the Federal Government of Germany (Sachverständigenrat für Umweltfragen – SRU), and a Member of the International Law Association (ILA) Committee *The Legal Principles Relating to Climate Change*. Until the end of 2012 he was a Judge at the Higher Administrative Court (Oberverwaltungsgericht) of the Federal States of Berlin and Brandenburg, Germany. He is a former Member of the Federal Government´s Comission on Nanotechnology (Nanokommission der deutschen Bundesregierung). His latest publications include: Calliess, Christian, 2013, The Dynamics of European Citizenship: From Bourgeois to Citoyen, in: Court of Justice of the European Union (Ed.), *The Court of Justice and the Construction of Europe: Analyses and Perspectives on Sixty Years of Case-Law – La Cour de Justice et la Construction de l'Europe: Analyses et Perspectives de Soixante Ans de Jurisprudence*, Berlin, Springer, 425–441.

**Alexander Carius** is a Co-founder and Managing Director of adelphi, Berlin, Germany. The core topics of his research and consultancy work are: resources and governance, climate and energy, and development and security. He also provides advice on institutional matters with respect to environmental and development policy, policy integration and the development regional and multilateral cooperation frameworks. He advises federal ministries, aid agencies, the European Commission and various international organisations. He has been an evaluator for the European Commission's

Research Framework Programmes since the 1990s and evaluates programmes and projects for aid agencies and international organisations. He is a Board Member of various projects and journals and sat on the German Government's Advisory Panel on Civilian Crisis Prevention, Conflict Resolution and Peace Consolidation from 2005 to 2009. One of his latest publications is: Tänzler, Dennis & Alexander Carius (Eds), 2012, *Climate Diplomacy in Perspective. From Early Warning to Early Action*, Berlin, Berliner Wissenschafts-Verlag.

**Dr. Andrew Dlugolecki** is Visiting Research Fellow, Climatic Research Unit, University of East Anglia, United Kingdom. One of his latest publications is: Leurig, Sharlene & Andrew Dlugolecki, 2013, *Insurer Climate Risk Disclosure Survey: 2012 Findings & Recommendations*, Boston, CERES.

**Dr. Kristie L. Ebi** is an independent consultant who conducts research on the impacts of and adaptation to climate change, including on extreme events, thermal stress, foodborne safety and security, and vectorborne diseases. She also is a Guest Professor at Umeå University, Sweden and a Consulting Professor at Stanford University, United States of America. She was the Executive Director of the Intergovernmental Panel on Climate Change (IPCC), Working Group II, Technical Support Unit (TSU) from 2009–2012. One of her latest publications is: Ebi, Kristie L., Stéphane Hallegatte, Tom Kram, Nigel W. Arnell, Timothy R. Carter, Jae Edmonds, Elmar Kriegler, Ritu Mathur, Brian C. O'Neill, Kewyan Riahi, Harald Winkler, Detlef P. van Vuuren & Timm Zwickel, 2013, A New Scenario Framework for Climate Change Research: Background, Process, and Future Directions, *Climatic Change*.

**Harrie Esterhuyse** is Research Analyst at the Centre for Chinese Studies (CCS) at the University of Stellenbosch, South Africa. One of his latest publications is: Esterhuyse, Harrie, 2012, Climate Change Risk and Response: Droughts as Extreme Weather Events in China, Centre for Chinese Studies Discussion Paper 5/2012.

**Fana Gebresenbet** finished his MSc in Physical Land Resources from Gent University, Belgium, in September 2008, and has worked as a Researcher at the Africa Programme of the University for Peace on issues of climate security. He has now joined the Institute for Peace and Security Studies of Addis Ababa University (AAU), Ethiopia, and is a PhD candidate at the joint PhD programme in Global and Area Studies offered by the AAU and the

University of Leipzig, Germany. His latest publications include the following: Kefale, Asnake & Fana Gebresenbet, 2012, Discrepancies between Traditional Coping Mechanisms to Climate Change and Government Intervention in South Omo, Ethiopia, in: Berhe, Mulugeta G. & Jean-Bosco Butera (Eds), *Climate Change and Pastoralism: Traditional Coping Mechanisms and Conflict in the Horn of Africa*, IPSS and UPEACE.

**Prof. Dr. Siri Gloppen** is Professor of Political Science with a research focus in the intersection between law and politics at the Department of Comparative Politics at the University of Bergen, Norway. She is Senior Researcher and former Research Director of the Chr. Michelsen Insitute (CMI), an independent development research institute in Bergen, Norway, where she is the Head of the Centre for Law and Social Transformation. From April 2013 she is also Research Coordinator at the PluriCourts Centre of Excellence at the University of Oslo, Norway. One of her latest publications is: Gloppen, Siri & Varun Gauri, 2012, Human Rights Based Approaches to Development: Concepts, Evidence, and Policy, *Polity* 44 (4), 485–550.

**Dr. Sven Grimm** is the Director of the Centre for Chinese Studies (CCS), a leading African research institution for innovative and policy relevant analysis of the relations between China and Africa, at the University of Stellenbosch, South Africa. Previously he was a Research Fellow at the German Development Institute (DIE) in Bonn, Germany and the Overseas Development Institute (ODI) in London, United Kingdom. His latest publications include: Grimm, Sven & Meryl Burgess, 2013, Greening Chinese-African Relations?, in: Li, Xing & Abdulkadir O. Farah (Eds), *China-Africa Relations in an Era of Great Transformations*, Farnham, Ashgate Publishers, 71–86; Grimm, Sven & He Wenping, 2012, Emerging Partners and Their Impact on African Development, in: Lundsgaarde, Erik (Ed.), *Africa Toward 2030 Challenges for Development Policy*, Hampshire, Palgrave Macmillan, 196–223.

**Dr. Eugene Gurenko** is Lead Insurance Specialist at the World Bank Insurance Practice, where he is actively involved in developing catastrophe risk management solutions for World Bank client countries. He is furthermore Executive Board Member of the Munich Climate Insurance Initiative (MCII) and author of numerous professional publications on catastrophe insurance and reinsurance. One of his publications is: Gurenko, Eugene, 2007, *Climate Change and Insurance*, London, Earthscan.

**Dr. Armin Haas** is Fellow of the Institute for Advanced Sustainability Studies, Potsdam, Germany. One of his latest publications is: Lass, Wiebke, Armin Haas, Jochen Hinkel & Carlo C. Jaeger, 2013, Avoiding the Avoidable: Towards a European Heat Waves Risk Governance, in: Shi, Peijun & Carlo C. Jaeger, *Integrated Risk Governance, IHDP-Integrated Risk Governance Project Series*, Volume 1, Beijing and Berlin, Springer.

**Holger Haibach** is Country Director of the Konrad-Adenauer-Stiftung for Namibia and Angola, based in Windhoek, Namibia. From 2002 to 2011 he was a Member of the German Parliament (Deutscher Bundestag) and inter alia functioned as Spokesperson of the CDU/CSU Group on Economic Cooperation and Development.

**Dr. Dirk Hanschel** is Reader in Energy Law at the School of Law, University of Aberdeen, Scotland. From 1999–2005 he participated in an interdisciplinary research project on the *Institutionalization of International Negotiations Systems* funded by the German Research Foundation. In parallel, he was working as a Senior Research Assistant at the University of Mannheim, Germany. One of his latest publications is: Hanschel, Dirk, 2011, Developing a Legal Toolkit – Institutional Options to Remove Stumbling Blocks in the Climate Negotiations, in: Sjöstedt, Gunnar & Ariel M. Penetrante (Eds), *Climate Change Negotiations: A Guide for Resolving Disputes and Facilitating Multilateral Cooperation*, London, Routledge Publications.

**Sven Harmeling** is Team Leader for International Climate Policy at Germanwatch, Bonn, Germany. One of his latest publications is: Harmeling, Sven (with support from Anja Esch, Linde Griesshaber, David Eckstein, Lisa Junghans, Smita Nakhooda & Taryn Fransen), 2013, *The German Fast-Start Finance Contribution*, Bonn, Germanwatch, ODI and WRI.

**Dr. Christian Hey** is Secretary General of the German Advisory Council on the Environment (Sachverstänigenrat für Umweltfragen – SRU). One of his latest publications is: Hey, Christian, 2012, Low-carbon and Energy Strategies for the EU – The European Commission's Roadmaps: A Sound Agenda for a Green Economy? *Gaia* 1/2012, 43–47.

**Prof. Dr. Manfred O. Hinz** is Professor of Law at the Faculty of Law of the University of Bremen, Germany and the Centre for African and Migration Studies. He is also Research Professor at the Faculty of Law of the University of Namibia, where he previously served as the Dean and held the UNESCO Chair. He published widely in the fields of Legal and Political

Anthropology, Constitutional and International (Economic) Law. His recent publications include: Hinz, Manfred O., 2013, Customary Law and the Environment, in: Ruppel, Oliver C. & Katharina Ruppel-Schlichting (Eds), *Environmental Law and Policy in Namibia. Towards Making Africa the Tree of Life*, Windhoek/Essen, Orumbode Press/Welwitschia Verlag Dr. A. Eckl, 359–384; Hinz, Manfred O., Oliver C. Ruppel & Clever Mapaure (Eds), 2012, *Knowledge Lives in the Lake. Case Studies in Environmental and Customary Law from Southern Africa*, Windhoek, Namibia Scientific Society.

**Prof. Dr. Peter Höppe** is the Head of Munich Re's *Geo Risks Research/ Corporate Climate Centre*, Munich, Germany. Before joining Munich Re, he has worked in different university institutes with a research focus on effects of atmospheric processes and air pollutants on humans and the general assessment of environmental risks. Currently his research focus is on trends of natural catastrophes and their drivers and on climate change effects on the insurance industry in general. He is a Scientific Member and Advisor in several associations. Furthermore, he is the Founder and Chairperson of the Munich Climate Insurance Initiative (MCII) and one of the initiators of the Desertec Industrial Initiative (Dii). One of his recent publications is: Warner, Koko, Sönke Kreft, Michel Zissener, Peter Höppe, Christoph Bals, Thomas Loster, Joanne Linnerooth-Bayer, Silvio Tschudi, Eugene Gurenko, Armin Haas, Simon Young, Paul Kovacs, Andrew Dlugolecki & Aaron Oxley, 2012, Insurance Solutions in the Context of Climate Change-related Loss and Damage: Needs, Gaps, and Roles of the Convention in Addressing Loss and Damage, Munich Climate Insurance Initiative (MCII) submission to the SBI Work Programme on Loss and Damage, October 2012, Policy Brief No. 6, Bonn, United Nations University Institute for Environment and Human Security (UNU-EHS).

**Sainab Husain Paragay** is a Scientist at the University of Makassar, Indonesia. One of her recent publications is: Schwerdtner Máñez, Kathleen, Sainab Husain, Sebastian C.A. Ferse & Maria Máñez Costa, 2012, Water Scarcity in the Spermonde Archipelago, Sulawesi, Indonesia: Past, Present and Future, *Environmental Science and Policy* 23, 74–84.

**M. Hafijul Islam Khan** is an Environmental Lawyer, working on legal issues related to environment and climate change. He is Executive Director at the Centre for Climate Justice-Bangladesh (CCJ-B) and has worked with: Germanwatch, Bonn, Germany; Bangladesh Environmental Lawyers Association (BELA), Dhaka, Bangladesh; and has been working as the Core Team Member, with LDCs climate negotiators group on loss and damage associated with climate change to assist the LDCs Chair and other group members. One of his latest publications is: Khan, Hafijul I., 2008, Policy Reforms in Response to Climate Change, IUCN-BD.

**Dr. Robert M. Kibugi** is Lecturer in Environmental Law at the Centre for Advanced Studies in Environmental Law and Policy (CASELAP) at the University of Nairobi, Kenya. He is a Member to Board of IUCN Academy of Environmental Law (IUCNAEL), Ottawa, Canada for the African Continent. One of his latest publications is: Kibugi, Robert, 2012, Implementing Stewardship in Kenyan Land Use Law: The Case for a Sustainability Extension, in: Martin, Paul, Anel Du Plessiss, Yves LeBouthilier, Li Zhiping & Qin Tianbao (Eds), *Environmental Governance and Sustainability*, Cheltenham, Edward Elgar.

**Prof. Paul Kovacs** is Executive Director, Institute for Catastrophic Loss Reduction at the Western University, Ontario, Canada. He is the President and CEO of the Property and Casualty Insurance Compensation Corporation (PACICC) and Adjunct Research Professor of Economics at the Western University. He is a Lead Author for the North America Chapter of the Intergovernmental Panel on Climate Change (IPCC), Working Group II, in the Third, Fourth and Fifth Assessment Reports.

**Sönke Kreft** is Policy Officer, Climate, Development and Insurance for the NGO Germanwatch e.V., Bonn, Germany and Project Leader of the *Loss and Damage in Vulnerable Country Initiative*. He is furthermore Research Associate at the United Nations University, Institute for Environment and Human Security (UNU-EHS), Bonn, Germany. One of his latest publications is: Kreft, Sönke, 2013, Loss and Damage: Unpacking the Doha Decision 3/CP.18, Germanwatch discussion briefing prepared in the context of the Loss and Damage in Vulnerable Country Initiative.

**Julie G. Lennox** is the Climate Change Focal Point, Agricultural Development Unit Chief and Coordinator of the Economics of Climate Change in Central America Initiative, Subregional Headquarters in Mexico of the Economic Commission for Latin America and the Caribbean of the United Nations. She is the Coordinator of the institutional publication *The Economy of Climate Change in Central America: Summary 2012*, ECLAC, COSEFIN, CCAD/SICA, UKAID & DANIDA.

**Prof. Dr. Marcel Leroy** is Senior Researcher at the UN-mandated University for Peace, Africa Programme, Addis Ababa, Ethiopia, directing projects on the impact of environmental factors and climate change on security in Africa. After having served as Head of Mulitilateral and Regional Affairs at the NATO Secretariat in Brussels, Belgium, he worked as Special Advisor for Multilateral Issues and Political Advisor to the European Commission and the Council of the European Union from 1997 until 2008, completing various assignments in Africa, including as representative to the Organisation of African Unity and with the EU Special Representative for Sudan. His recent publications include the following: Leroy, Marcel & Fana Gebresenbet, 2011, Climate Conflicts in the Horn of Africa?, *Conflict Trends*; Leroy, Marcel (Ed.), 2009, *Environment and Conflict in Africa: Reflections on Darfur*, Addis Ababa, University for Peace Africa Programme.

**Prof. Dr. Wanxin Li** is Assistant Professor at the Department of Public Policy at the City University of Hong Kong, China. The topics of her publications include environmental information disclosure, environmental compliance and enforcement, eco-innovation policies, and subjective well-being in China. One of her latest publications is: Li, Wanxin & Paul Higgins, forthcoming 2013, Controlling Local Environmental Performance: An Analysis of Three National Environmental Management Programs in the Context of Regional Disparities in China, *Journal of Contemporary China* 22 (81).

**Dr. Joanne Linnerooth-Bayer** is Program Leader in the Risk, Policy and Vulnerability Program at the International Institute for Applied Systems Analysis (IIASA) in Laxenburg, Austria. One of her latest publications is: Burton, Ian, O. Pauline Dube, Diarmid Campbell-Lendrum, Ian Davis, Richard J.T. Klein, Joanne Linnerooth-Bayer, Apurva Sanghi, & Ferenc Toth, 2012, Managing the Risks: International Level and Integration Across Scales, in: IPCC/Intergovernmental Panel on Climate Change, 2012, *Managing the Risks of Extreme Events and Disasters to Advance Climate Change Adaptation*, A Special Report of Working Groups I and II of the Intergovernmental Panel on Climate Change, Field, Christopher B., Vicente Barros, Thomas F. Stocker, Dahe Qin, David J. Dokken, Kristie L. Ebi, Michael D. Mastrandrea, Katharine J. Mach, Gian-Kasper Plattner, Simon K. Allen, Melinda Tignor & Pauline M. Midgley (Eds), Cambridge, Cambridge University Press, 393–435.

**Thomas Loster** is the Chairperson of the Munich Re Foundation, Munich, Germany since 2004. He is a geographer and was a member of the Geoscience Research Group at Munich Reinsurance Company, Munich, for 16 years. He was in charge of issues relating to Natcat trends, weather perils, climate change and climate policy. The Munich Re Foundation addresses major global challenges including environmental and climate change, population growth and disaster prevention. Building resilience through disaster prevention as well as through microinsurance and other tailor-made insurance solutions is one key pillar of the work of Munich Re Foundation.

**Cord Lüdemann** is a LLD Candidate at the Faculty of Law of the University of Stellenbosch, South Africa. He is a Lawyer at the German Federal Network Agency (Bundesnetzagentur), Bonn, Germany. One of his latest publications is: Lüdemann, Cord, 2013, The Role of Renewable Energies in Industrialised and Developing Countries, in: Ruppel, Oliver C. & Katharina Ruppel-Schlichting (Eds), 2013, *Environmental Law and Policy in Namibia. Towards Making Africa the Tree of Life* (Second Edition), Windhoek/Essen, Orumbonde Press/Welwitschia Verlag Dr. A. Eckl, 235–244.

**Prof. Dr. María Máñez Costa** is the Head of the Economics and Policy Department at the Climate Service Center (CSC) in Hamburg, Germany. She works on political consulting, focused on international climate policy, climate economics, climate and security as well as participation approaches. She is also Visiting Professor at the University of Barcelona, Spain. One of her latest publications is: Schwerdtner Máñez, Kathleen, Sainab Husain, Sebastian C.A. Ferse & Maria Máñez Costa, 2012, Water Scarcity in the Spermonde Archipelago, Sulawesi, Indonesia: Past, Present and Future, *Environmental Science and Policy* 23, 74–84.

**Prof. Dr. Aidan G. Msafiri** is Professor of Climate Ethics, Business Ethics, Social Ethics and Ethics of Sustainability at the Stella Maris Mtwara University College, a Constituent College of St. Augustine University of Tanzania. He serves on the Advisory Board of Globethics.net East Africa based in Nairobi, Kenya. He has published four books and a number of book chapters and articles in national and international academic journals. One of his latest publications is: Msafiri, Aidan G., 2012, *Globalisation of Concern II, Essays on Education, Health, Climate Change and Cyberspace*, Geneva, Globethics.net.

**Dr. Babette Never** is Research Fellow at the German Institute of Global and Area Studies (GIGA) in Hamburg, Germany. One of her latest publications is: Never, Babette, 2012, Who Drives Change? Comparing the Evolution of

Domestic Climate Governance in India and South Africa, *Journal of Environment and Development* 21 (3), 362–387.

**Aaron Oxley** is Executive Director, RESULTS UK, London, United Kingdom. Until 2009 he was Chief Information Officer, MicroEnsure, UK.

**Dr. Erika Pires Ramos** is Federal Attorney at Brazilian Institute for Environment and Natural Renewable Resources (IBAMA) in São Paulo, Brazil since 2007. She is the Founder of the South American Network for Environmental Migrations (RESAMA); Researcher at various projects such as the project on *Les catastrophes et les droits de l'homme* (CADHOM), France, the Research Group on Environmentally Displaced Persons (NEPDA), Paraíba, Brazil, and the International Law and International Relations Studies Centre (ORBIS), São Paulo, Brazil. She is furthermore a Scientific Reviewer in the Global Environment Outlook – GEO 5 UNEP, published in 2012. Her latest publications include the following: Pires Ramos, Erika & Isis Akemi Morimoto, 2012, The Strengthening of the National Environment System in the Framework of the Treaty of Environmental Education for Sustainable Societies and Global Responsibility, in: Weiss Bliacheris, Marcos & Maria Augusta Soares de Oliveira Ferreira (Eds), *Sustainability in Public Administration: Values and Practices of Socio-environmental Management*, Belo Horizonte; Pires Ramos, Erika & Andrea Pacheco Pacífico, forthcoming 2013, Humanitarian Asylum for Forced Migrants: The Case of Haitians' Arrival in Brazil.

**Erin Roberts** is Coordinator, Loss and Damage in Bangladesh Study at the International Centre for Climate Change and Development, Bangladesh. She contributed to the recent publication by Asaduzzaman, Mohammed et al., 2013, Assessing the Risk of Loss and Damage Associated with the Adverse Effects of Climate Change in Bangladesh, Dhaka, ICCCAD.

**Agata Rogalska-Piechota** is Legal Counsel at the Circuit Chamber of Legal Counsel in Warsaw, Poland. She is Chief Specialist at the Department of Proceedings before International Human Rights Protection Bodies, Ministry of Foreign Affairs, Poland. One of her latest publications is: Rogalska-Piechota, Agata, 2011, Prawny charakter zarządzeń tymczasowych (interim measures) w orzecznictwie Europejskiego Trybunału Praw Człowieka, in: Balcerzak, Michal, Tadeusza Jasudowicz & Julii Kapelańska-Pręgowska (Eds), *Europejska konwencja praw człowieka i jej system kontrolny – perspektywa systemowa i orzecznicza*, Toruń, Katedra Praw Człowieka UMK, 359–372.

**Prof. Dr. Oliver C. Ruppel** is Professor of Public and International Law at the Faculty of Law of the University of Stellenbosch, South Africa; Professor Extraordinaire at several universities around the world and Fellow of the Ciucci Centre for Law and Social Development, an initiative of the Stonehage Charitable Trust. Until 2010 he held one of the worldwide 14 academic chairs of the World Trade Organisation (WTO), Geneva, Switzerland, which he established at the University of Namibia, Windhoek, where he had previously also served as the Director of the national Human Rights and Documentation Centre. He currently serves as Coordinating Lead Author for the Chapter on Africa in the United Nations Intergovernmental Panel on Climate Change (IPCC), Working Group II. He is an international arbitrator with the Association of Arbitrators of Southern Africa, member of the South African and Swiss Branches of the International Law Association, and member of various editorial boards, including the *NIALS Journal of Environmental Law*, and the legal series *Law and Constitution in Africa*, Nomos Publishers, Germany. His latest publications include: Ruppel, Oliver C. & Katharina Ruppel-Schlichting (Eds), 2013, *Environmental Law and Policy in Namibia. Towards Making Africa the Tree of Life*, Windhoek/Essen, Orumbode Press/ Welwitschia Verlag Dr. A. Eckl; Ruppel, Oliver C. & Gerd Winter (Eds), 2011, *Justice from Within: Legal Pluralism in Africa and beyond*, Hamburg, Kovač.

**Dr. Katharina Ruppel-Schlichting** is the Director of the Legal Research and Development Trust for Africa, Windhoek, Namibia and Stellenbosch, South Africa. She is an international a legal consultant and author of a variety of legal academic writings. Her latest publications include the following: Ruppel, Oliver C. & Katharina Ruppel-Schlichting (Eds), 2013, *Environmental Law and Policy in Namibia. Towards Making Africa the Tree of Life*, Windhoek/Essen, Orumbode Press/Welwitschia Verlag Dr. A. Eckl; Ruppel, Oliver C. & Katharina Ruppel-Schlichting, 2012, Climate Change and Human Security: Relevant for Regional Integration in SADC?, in: Hartzenberg, Trudi, Gerhard Erasmus & Andre Du Pisani (Eds), *Monitoring Regional Integration in Southern Africa – 2011 Yearbook for Regional Integration*, Stellenbosch, tralac, 32–71.

**Kathrin Schneider** currently works as management assistant for the foreign location of the EDAG Group, an engineering partner of the International Mobility Industry, in Puebla, México. She completed an internship at the Konrad-Adenauer-Stiftung, Namibia office from September to December 2012.

**Prof. Dr. Werner Scholtz** is Professor of Law in the Department of Public Law and Jurisprudence of the University of the Western Cape, South Africa. He is a Research Associate at the South African Institute for Advanced Constitutional, Public, Human Rights and International Law (SAIFAC) and has established the Regional African Law and Human Security Initiative during 2010. He is inter alia a Member of the International Law Association (ILA) Committee on the Legal Principles of Climate Change; the ILA Committee on Recognition/Non-recognition and the European Society of International Law (ESIL) Interest Group on International Environmental Law. One of his latest publications is: Scholtz, Werner & Gerrit Ferreira, 2011, Much Ado About Nothing? The SADC Tribunal's Quest for the Rule of Law Pursuant to Regional Integration, *Zeitschrift für ausländisches öffentliches Recht und Völkerrecht / Heidelberg International Law Journal*, 331–358.

**Prof. Dr. Roland W. Scholz** chaired for two decades the Natural and Social Science Interface in the Department of Environmental Sciences at the ETH (Swiss Federal Institute of Technology), Zurich, Switzerland. A mathematician, psychologist and decision theorist by training, he is particularly interested in environmental systems analysis, human-environment interactions, environmental decisions, and risk assessment. He has led numerous large-scale transdisciplinary processes to foster sustainable transitions of urban and regional systems. Currently, he is a Member of the Fraunhofer Project Group for Materials Recycling and Resource Strategies (IWKS), an institute at the engineering-society interface in resources management. One of his latest publications is: Scholz, Roland W., 2011, *Environmental Literacy in Science and Society: From Knowledge to Decisions*, Cambridge, Cambridge University Press.

**Dr. Kathleen Schwerdtner Máñez** is Senior Scientist at the Leibniz Center for Marine Tropical Ecology in Bremen, Germany. One of her latest publications is: Schwerdtner Máñez, Kathleen, Sainab Husain, Sebastian C.A. Ferse & Maria Máñez Costa, 2012, Water Scarcity in the Spermonde Archipelago, Sulawesi, Indonesia: Past, Present and Future, *Environmental Science and Policy* 23, 74–84.

**Prof. Dr. Asuncion Lera St. Clair** is Research Director at the International Centre for Climate and Environmental Research-Oslo (CICERO), Oslo, Norway and Associated Senior Researcher at the Chr. Michelsens Institute (CMI), Bergen, Norway. The philosopher and sociologist is former Professor of Sociology at the University of Bergen. She is a Lead Author of the

Fifth Assessment Report of the Intergovernmental Panel on Climate Change (IPCC), Working Group II. Furthermore, she is a Member of the Joint Programming Initiative Connecting Climate Knowledge for Europe (JPI Climate); and Member of several Editorial Boards of international journals such as GEC, Global Social Policy, J Global Ethics, and Global Governance. One of her latest publications is: Gasper, Des, Ana V. Portocarrero & Asuncion Lera St. Clair, 2013, Climate Change and Development Framings: A Comparative Analysis of the Human Development Report 2007/8 and the World Development Report 2010, *Journal of Global Environmental Change* 23 (1), 28–39.

**Dennis Tänzler** is the Head of Climate and Energy Policies at adelphi, Berlin, Germany. His research focuses on climate and energy policies as well as on peace and conflict studies. He has published widely on international affairs and environmental policy and is the Programme Director of the Environment, Conflict and Cooperation Platform. One of his latest publications is: Tänzler, Dennis & Alexander Carius (Eds), 2012, *Climate Diplomacy in Perspective. From Early Warning to Early Action*, Berlin, Berliner Wissenschafts-Verlag.

**Dr. Silvio Tschudi** is Retrocession Manager at Allianz SE Reinsurance, Munich, Germany and Executive Board Member of the Munich Climate Insurance Initiative (MCII), Munich, Germany. One of his latest publications is: Warner, Koko, Sönke Kreft, Michel Zissener, Peter Höppe, Christoph Bals, Thomas Loster, Joanne Linnerooth-Bayer, Silvio Tschudi, Eugene Gurenko, Armin Haas, Simon Young, Paul Kovacs, Andrew Dlugolecki & Aaron Oxley, 2012, Insurance Solutions in the Context of Climate Change-related Loss and Damage: Needs, Gaps, and Roles of the Convention in Addressing Loss and Damage, Munich Climate Insurance Initiative (MCII) submission to the SBI Work Programme on Loss and Damage, October 2012, Policy Brief No. 6, Bonn, United Nations University Institute for Environment and Human Security (UNU-EHS).

**Sanita van Wyk** is a LLD Candidate in International Environmental Law at the Faculty of Law at the University of Stellenbosch, South Africa, where she was a Member of the Overarching Strategic Research and Outreach Project on Combating Poverty, Homelessness and Socio-Economic Vulnerability under the Constitution and served as an Administrative and Research Assistant in the Department of Public Law. Currently she is an Erasmus Mundus Scholar at Freie Universität Berlin, Germany. One of her latest publications is: Ruppel, Oliver C. & Sanita van Wyk, 2011, Climate Change Induced Movement of Persons in Africa: Human Rights Responses to Aspects of Human Security?, *South African Journal of Environmental Law and Policy* 18, 49–64.

**Nadia von Bassewitz** works as a Consultant in the field of Competition Law for the International Institute for Advanced Economic Studies in Ghana, drafting regulatory instruments, undertaking studies, as well as organising training for various national and regional competition authorities in Africa. Previously, she was a Lecturer for International Business Law, WTO Law and Competition Law at the Polytechnic of Namibia as well as at the Law Faculty of the Addis Ababa University (Ethiopia). From 1998 to 2004 she served as a Consultant and Temporary Agent at the European Commission, General-Directorate Competition in Brussels, Belgium. One of her latest publications is: von Bassewitz, Nadia, 2013, International Climate Change Policy and Legislation: Where do we stand?, in: Ruppel, Oliver C. & Katharina Ruppel-Schlichting (Eds), *Environmental Law and Policy in Namibia. Towards Making Africa the Tree of Life* (Second Edition), Windhoek/Essen Orumbonde Press/Welwitschia Verlag Dr. A. Eckl, 311–358.

**Prof. Dr. Gregg B. Walker** is Professor of Communication and Member of the Faculties of Environmental Sciences, Forest Ecosystems and Society, Marine Resource Management, Public Policy, and Water Resource Management, and the Natural Resources Leadership Academy at the Oregon State University, United States of America. He is Co-director of the Climate Change Project for Mediators Beyond Borders International and Advisor and Facilitator with the National Collaboration Cadre of the US Forest Service. His latest publications include: Daniels, Steven E., Gregg B. Walker & Jens Emborg, 2012, The Unifying Negotiation Framework: A Model of Policy Discourse, *Conflict Resolution Quarterly* 30 (1), 1–14; Walker, Gregg B. & Steven E. Daniels, 2012, The Nature and Role of Agency Leadership: Building and Sustaining Collaboration in Natural Resource Management and Environmental Policy Decision-Making, in: Rigling Gallagher, Deborah (Ed.), *Environmental Leadership*, Thousand Oaks, CA, Sage.

**Dr. Koko Warner** is the Head of the Environmental Migration, Social Vulnerability, and Adaptation Section at the United Nations University, Institute for Environment and Human Security (UNU-EHS), Bonn, Germany. She is a Lead Author for the Fifth Assessment Report of the Intergovernmental Panel on Climate Change (IPCC), Working Group II and Founding Member and Executive Board Member of the Munich Climate Insurance Initiative (MCII). Furthermore, she is the UNU focal point to the HLCP and the UN-FCCC, and the climate adaptation and the Nairobi Work Programme. Her recent publications include: Warner, Koko, 2012, Human Migration and Displacement in the Context of Adaptation to Climate Change: The Cancun Adaptation Framework and Potential for Future Action, *Environment and Planning C: Government and Policy* 30 (6), 1061–1077; Hill, Harvey, John Wiener & Koko Warner, 2012, From Fatalism to Resilience: Reducing Disaster Impacts Through Systematic Investments, *Disaster* 36 (2), 175–194.

**Dr. Simon Young** is the CEO of the Caribbean Risk Managers Ltd. and a Risk Management Consultant. One of his latest publications is: Young, Simon & Ekhosuehi Iyahen, 2009, Innovative Risk Transfer Options as Adaptation Strategies to Growing Hydro-Meteorological Risks in the Caribbean Basin, in: United Nations World Meteorological Organisation, *Climate Sense*, Leicester, Tudor Rose.

**Michael Zissener** is Research Associate at the United Nations University Institute for Environment and Human Security (UNU-EHS), Bonn, Germany with a focus on the Munich Climate Insurance Initiative (MCII). He helps managing and coordinating an implementation-focused project on disaster risk management, risk reduction (DRR), and risk transfer in the Caribbean. One of his latest publications is: Warner, Koko, Sönke Kreft, Michel Zissener, Peter Höppe, Christoph Bals, Thomas Loster, Joanne Linnerooth-Bayer, Silvio Tschudi, Eugene Gurenko, Armin Haas, Simon Young, Paul Kovacs, Andrew Dlugolecki & Aaron Oxley, 2012, Insurance Solutions in the Context of Climate Change-related Loss and Damage: Needs, Gaps, and Roles of the Convention in Addressing Loss and Damage, Munich Climate Insurance Initiative (MCII) submission to the SBI Work Programme on Loss and Damage, October 2012, Policy Brief No. 6, Bonn, United Nations University Institute for Environment and Human Security (UNU-EHS).

# PART I: GLOBAL CLIMATE GOVERNANCE – DEVELOPMENTS AND CHALLENGES

1

# Intersections of Law and Cooperative Global Climate Governance – Challenges in the Anthropocene*

*Oliver C. Ruppel*

*Abstract*

In an age primarily shaped by people, the so-called *Anthropecene*, mankind is faced with enormous challenges posed by the effects of climate change, *de facto* and *de iure*. This article explores the various intersections of law related to climate change. The discussion of such intersections, suggesting an interdisciplinary approach to climate change, is particularly important as there is no clearly demarcated field of climate change law. Without doubt, the endless ramifications of climate change preclude any claim to exhaustiveness. However, many of the major legal issues that have emerged, are being sketched in this article. Intersections can be found between environmental law, human rights law, the law of the sea and world trade law among others. It is argued here that more coherence in the intersections of law and increased cooperative global climate governance should lead the way to cope with the challenges ahead, i.e. the challenges in the Anthropocene.

## *A. Introduction*

When recalling the recent United Nations climate process at the eighteenth Conference of the Parties (COP) to the United Nations Framework Convention on Climate Change (UNFCCC) and the eighth Conference of Parties serving as the Meeting of the Parties to the Kyoto Protocol (CMP) in Doha, Qatar last December, one may wonder what the diplomatic value of such massive negotiations really is. In the last days of the conference, many had already seen the talks close to collapse and were wondering whether COP18

---

* This article was the basis for the author's inaugural lecture held at the University of Stellenbosch, Faculty of Law, on 19 March 2013.

would need to be reconvened in 2013. Only last-minute decisions lead to a finalisation of the rules for the Kyoto Protocol's second commitment period and agreement on a work programme for the new negotiation track to deliver a new agreement by 2015.

Unfortunately, climate change is apparently not waiting for the slow timetables of diplomats. The Doha meeting took place at the end of a year (2012) of increasingly stark warnings both on paper and delivered by Mother Nature herself. The United States (US) suffered from a record drought, foreshadowing the permanent dust bowl the US Midwest is probably going to be turned into by climate change. Hurricane Sandy submerged vast swaths of the US East Coast including New York. Arctic sea ice reached a new record low, 50% below the long-term average. Shortly before the Doha conference the World Bank published a report warning of "cataclysmic consequences" if climate change was not reined in.[1] And while the Doha conference was underway the Philippines were battered by Bopha, a typhoon of near-unprecedented strength that caused hundreds of deaths.

The 'diagnosis' of planet earth seems rather clear in that constantly growing human and industrial activities have caused dramatically increased emissions of greenhouse gases, which in turn cause the global climate to change rapidly and probably irreversibly. The 'symptoms' of climate change are likely to cause more and more natural disasters, extreme weather events and climate induced migration movements. All of these undesirable happenings can be considered as a threat against all aspects of human security with a potential to cause national and cross-boundary conflict and thus endanger global peace and security. The 'therapy' against the symptoms of climate change is much less clear and will be discussed in this article. It is argued here that more coherence in the intersections of law and increased cooperative global climate governance should lead the way to cope with the challenges ahead, i.e. the challenges in the Anthropocene.

## *B. Anthropocene – The Age of Man*

The famous atmospheric chemist and Dutch Nobel Prize winner Paul Crutzen initially coined the term *anthropocene*. The term has ancient Greek roots: *anthropo* meaning *human* and *cene* meaning *new*. In 2000 Crutzen

1 World Bank (2012).

realised that we live in an age primarily shaped by people. From their trawlers scraping the floors of the seas to their dams impounding sediment by the gigatonne, from their stripping of forests to their irrigation of farms, from their mile-deep mines to their melting of glaciers, humans were bringing about an age of planetary change. Crutzen suggested this age be called Anthropocene – "the age of man".[2]

Mankind has now inhabited or visited almost all places on earth; even set foot on the moon – and the exploration continues. The expansion of mankind, both in numbers and per capita exploitation of the earth's resources, has been astounding. During the past three centuries the world's population increased tenfold to 7 billion, accompanied e.g. by a growth in cattle population to 1,500 billion. Urbanisation has increased tenfold in the past century. In only a little while we are deemed to exhaust the fossil fuels that were generated over millions of years. Thirty to fifty per cent of the land surface has been transformed by human action, and mankind uses more than half of all accessible fresh water. Considering these and many other major and still growing impacts of human activities on earth and atmosphere, it has become more than appropriate to emphasise the central role of mankind in geology, ecology and law by proposing the term *Anthropocene* for the current historical epoch as we already know that the impact of human activities has and will have severe consequences for present and future generations.[3]

For the purpose of this article the human being is seen as the root of the problem, the subject of vulnerability that requires protection, the nucleus of the law and the target of cooperative global climate governance aiming at maintaining peace and security at the same time. The predominant challenges in the Anthropocene, especially in regard of climate change, will be briefly sketched below. Typologically significant of the Anthropocene these challenges must be seen related to the level of complexity, the degree of uncertainty and the novelty that actually surrounds climate change in a process that involves ever-changing circumstances that can hardly be fully controlled. As a combination of legal and policy analysis this article shall also examine selected aspects of the framework of international law and governance in the field of climate change.

2 Crutzen & Stoermer (2000); *The Economist* (2011).

3 *The Economist* (2011).

In 2011, Pope Benedict XVI addressed the German Bundestag illustrating the sources of law in nature and reason by making reference to the popular interest in ecology as a means of respecting nature:[4]

> Yet I would like to underline a point that seems to me to be neglected, today as in the past: there is also an ecology of man. Man too has a nature that he must respect and that he cannot manipulate at will. Man is not merely self-creating freedom. Man does not create himself. He is intellect and will, but he is also nature, and his will is rightly ordered if he respects nature, listens to it and accepts himself for who he is, as one who did not create himself. In this way, and in no other, is true human freedom fulfilled.

In 2012, the Club of Rome launched a Report entitled *2052 – A Global Forecast for the Next Forty Years*.[5] In it, author Jorgen Randers tries to answer the question of what our world will look like in forty years' time. Some of the findings include the following:[6]

> Humanity is in overshoot (mainly climate-related) and the landing will not be soft .... Humanity has a forty-year window to avoid the most serious negative consequences of its decades-long overconsumption splurge. The process of adapting humanity to the planet's limitations may be too slow to stop planetary decline. Global population will grow, peaking at 8.1 billion people in 2042 because of rapid decline in urban fertility. CO2 emissions will peak in 2030, because of a shift toward low-carbon sources of power and heat. Nevertheless, CO2 concentrations will grow, and the global average temperature will pass the danger threshold of +2 C by 2050, and peak at 2.8 C in 2080, which could trigger self-reinforcing "run-away" warming with a possible collapse in the second half of the 21st century.

Translating the aforementioned statements into the context of the Anthropocene raises the following questions, among others: How many people will the planet be able to support in future? Will runaway climate change take hold? Where will the quality of life improve, and where will it decline? While the process of adapting humanity to the planet's limitations has started, Randers rightfully holds that the "human response could be too slow".[7]

4 Benedict XVI (2011).
5 Club of Rome (2012).
6 (ibid.).
7 (ibid.).

### *I. Climate Change and Natural Disasters*

Natural disasters are on the increase in the Anthropocene and in this context climate change cannot be viewed in isolation. "Disaster" means a calamitous event or series of events resulting in widespread loss of life, great human suffering or distress, or large-scale material or environmental damage, thereby seriously disrupting the functioning of society.[8] There is wide scientific consensus that the increased number and intensity of climate change induced natural disasters, such as earthquakes, volcano eruptions, tsunamis and hurricanes, is of alarming concern.[9] Recent incidents include among others the Indian Ocean tsunami (2004), Hurricanes Katrina (2005) and Sandy (2012), Typhoon Bopha in the Philippines (2012), and the earthquakes in Pakistan (2005), Haiti (2010) and Fukushima (2011). The World Bank in a report published in 2012 warned of "cataclysmic consequences" if climate change was not reined in.[10]

The 2012 Special Report of the Intergovernmental Panel on Climate Change (IPCC) titled *Managing the Risks of Extreme Events and Disasters to Advance Climate Change Adaptation (SREX)*[11] demonstrates shockingly that the severity of the impacts of extreme and non-extreme weather and climate events depends strongly on the level of vulnerability and exposure to these events. Basic risks to which people are subjected by displacement include landlessness, joblessness, homelessness, marginalisation, food insecurity, increased morbidity, loss of access to common property resources, and social disarticulation. Particular groups and conditions have been identified as having differential exposure or vulnerability to extreme events; for example race/ethnicity, socioeconomic class and caste, gender, age (both the elderly and children), migration, and housing tenure (whether renter or owner) are among the most common social vulnerability characteristics.[12] "During the period from 1970 to 2008, over 95% of deaths from natural disasters occurred in developing countries."[13]

8 International Law Commission (ILC) Draft Article 3 on the protection of persons in the event of disasters of the International Law Commission A/CN.4/L.758, available at http://daccess-dds-ny.un.org/doc/UNDOC/LTD/G09/626/84/PDF/G0962684.pdf?OpenElement, last accessed 16 February 2013.

9 IPCC (2012).

10 World Bank (2012).

11 IPCC (2012).

12 (ibid.).

13 (ibid.).

The increase of natural disasters also poses challenges for international law and the international governance framework, especially when it comes to coordination, disaster relief and international cooperation. The international community, even if willing, is not easily able to provide relief to disaster victims. The duty to provide relief is largely incumbent upon the state within whose territory and jurisdiction the disaster occurs. This problem is rooted in the notion of state sovereignty, one of the most defining principles of international law.[14]

An increase in the concentration of greenhouse gases in the atmosphere heightens the possibility that mechanisms that could lead to catastrophic or extreme climate change will be triggered, notwithstanding with the fact that there is uncertainty as to when and how exactly such mechanisms will be triggered. Not reducing GHG emissions, however, means subjecting future generations to the risk of severe harm.[15] Considering the dangers related to natural disasters and the extremity of the risks involved for future generations, there is in fact no right to presuppose that the effects of climate change will be far from catastrophic.[16] In other words, "postponing emissions cuts is in some ways like putting a revolver to future people's heads and hoping that there is no bullet in the chamber".[17] From the point of view of justice, it has been stated that –[18]

> the nature of [climate change catastrophes] requires us to take drastic precautions against further [climate change] that could lead us to pass the tipping points that cause them. This is the case notwithstanding the fact that we are in a state of strong uncertainty with respect to these events; indeed, our strong uncertainty with respect to them – given their nature – makes the case for action to prevent them even more persuasive.

To develop global strategies leading to sustainability of ecosystems against human induced impacts will be one of the greatest tasks of mankind, requiring new and intensive research efforts that will pose many challenges to international law and global governance. Dealing with a global problem like climate change will require a strong legal framework embedded in more effective global institutions in future. International law and global governance – traditionally viewed as separate academic disciplines, i.e. law, po-

14 Evans (2004).
15 See World Bank (2010); Gardiner (2004:576).
16 Macer et al. (2011:13).
17 Macer et al. (2011).
18 McKinnon (2009:200).

litics and social sciences – need to become part of a more integrated, coherent, interdisciplinary and holistic interplay, where international law and global governance eventually manage to get a grip on the arguably most significant challenge of our time – climate change.

## *II. Climate Change and Human Security*

The protection of the vital core of human lives in ways that enhance human freedoms and human fulfilment is at the core of the concept of human security. Providing human security means protecting individuals and the community from violent conflicts and from denial of civil liberties and to ensure freedom of expression and belief. It also encompasses the idea of satisfying the basic needs of individuals for food, shelter and clothing.[19]

Climate change has the potential to impose additional pressures on the various aspects of human security. Interrelating issues between climate change and human security include water stress, land use and food security, health security, and environmentally induced migration amongst others. Adverse climate events not only deepen poverty vulnerability in developing countries,[20] they impact on all aspects of human security, either directly or indirectly. The impacts of climate change on the agricultural sector are probably of most direct and profound nature. Impacts of climate change, droughts and floods in particular, will have an impact on food availability, food access and nutrient access.[21]

The ultimate damages of climate change may significantly affect economic growth.[22] Climate extremes exert substantial stress on low-income populations in particular. The poor are most vulnerable to multiple dimensions of climate change such as heat waves, sea level rise, the destruction of coastal zones and water shortages due to drought.[23] Health security is another important aspect of human security endangered by the impacts of climate change and the effects on health will exacerbate inequities between rich and poor.[24] Africa is particularly vulnerable in this regard as threats to health

19 UNDP (1994).
20 Ahmed et al. (2009).
21 Kotir (2010).
22 Lecocq & Shalizi (2007).
23 Hope (2009).
24 Costello et al. (2009).

security are usually greater for poor people in rural areas, particularly children, due to malnutrition and insufficient access to health services, clean water and other basic necessities. Major killer diseases such as malaria expand their coverage as a result of global warming. Global and regional climatic variability enhances the risk of a further spread of other infectious diseases such cholera,[25] dengue fever,[26] and meningitis.[27]

### *III. Climate Change, Conflict and Migration*

The impacts of climate change on violent conflicts and changing migration patterns are further aspects related to the aforementioned concept of human security, and again with particular relevance on the African continent. While violent conflict can be seen as a driver of vulnerability to climate change, migration is a stressor that increases vulnerability to climate change. The linkage between climate related environmental variability and conflict has attracted much attention and debate.[28] Yet, in 2011 Achim Steiner, Executive Director of the United Nations Environment Programme (UNEP), asserted that climate change is a "threat multiplier" that has fundamental implications for weather, settlements, infrastructure, food insecurity, livelihoods and development. Competition over scarce water and land, exacerbated by regional changes in climate, was already a key factor in local conflicts in Darfur, the Central African Republic, northern Kenya and Chad.[29]

Climate induced migration[30] is an aspect closely related to the concept of human security.[31] Notwithstanding the fact that there is no certainty as to what exactly climate change will mean for migration patterns, there seems to be consensus that climate change will over time lead to population movements. Migration can be an adaptation strategy and can enhance adaptive

25 De Magny et al. (2007).

26 Jansen & Beebe (2010).

27 Cuevas et al. (2007).

28 See for example Scheffran & Battaglini (2011); Barnett & Adger (2007); Nordås & Gleditsch (2007); Raleigh (2010); Raleigh & Urdal (2007); Theisen (2008).

29 United Nations Security Council (2011).

30 The terminology with regard to environmentally induced migration is varying and inconsistent and creates conflicts of a legal nature when it comes to the question as to whether or not a person can be classified as a refugee with the legal consequences of international refugee law. See Warner et al. (2010); Kälin & Schrepfer (2012:28).

31 Foresight (2011).

capacity[32] People migrate either temporarily or permanently, within their country or across borders, and many have an environmental signal in their reason for migration. The African continent[33] and small island nations around the globe are most likely to be among those who will produce the most climate migrants in future. The total number of displaced people in Africa increased almost 700,000 in 2008 to 1.7 million in 2010.[34]

The causes for displacement and migration are manifold; however, climate change is one of the interlinking issues. Potential drivers of migration are push and pull factors related to the region or country of origin or destination respectively, and intervening factors that facilitate or restrict migration, all of which may interact in different ways.[35] The available evidence suggests that, globally, the large majority of people displaced by disasters caused by sudden-onset hazards (hurricanes, floods, earthquakes, etc.) remain temporarily and internally displaced with people returning home to rebuild their homes and lives.[36] This might be different in the case of slow-onset disasters such as droughts and sea level rise with increasing cross-border movement of a permanent nature.[37]

## C. Intersections of Law

The aforementioned scenarios have surely attracted the reader's concern. In order to address this concern, it is necessary to call for effective regulation in order to prevent the worst case. In this context the law comes in: "Law is the major instrument by which mature societies consolidate their internal and external relationships" and "without legal rules, the life of a society becomes unpredictable and aleatory".[38] For good reason, there is no clearly defined term, nor a marked branch of the law, which would cover all legal implications of climate change. Subsuming climate change under any legal structure is a challenging task due to the endless ramifications of climate change and particularly due to the interdisciplinary nature of climate change

32 Barnett & Webber (2010).
33 For a focus on climate-induced migration from Africa to Europe see White (2011).
34 IDMC (2011).
35 Black et al. (2011).
36 Tschakert & Tutu (2010); IDMC (2011).
37 US National Intelligence Council (2010).
38 Tomuschat (2012:1283).

and its impacts on various segments of our planet. Climate change can therefore only be tackled through a combination of political, legal and natural science tools. Climate change, biodiversity loss, the marine environment, ozone depletion, genetic resources, intellectual property issues, international trade and human rights – among others – are strongly interrelated. There are numerous intersections of law that occur when climate change is looked at from a legal perspective. Efforts to curb climate change have given rise to the evolution of some new principles and concepts of international law, including among others the principle of common but differentiated responsibilities, the notion of common concern of humankind and the need for protection of the most vulnerable.[39]

Climate change permeates the law in many ways, creating intersections of law in its diverse fields. If one would brand a new discipline *climate change law*, this would be both international and domestic in nature and include (at least) two complementary dimensions: *procedural* and *substantive*.

The *procedural* dimension is related to the right to information, the right to participate in decision-making, and the right of access to justice. Climate change opens a multitude of challenges of a procedural nature. To what extent these challenges are relevant depends on the following aspects, among others: The question of whether and under what conditions an individual, organisation or state has the right to commence action needs to be addressed. The issue of *locus standi* is of great relevance in respect of judicial enforcement, which still needs specific attention. So far public interest litigation is scarce. Yet it seems to be most suitable in the context of climate change. Another focal point deals with the question of who would be the proper addressee of claims relating to climate change damages, and whether a right to environment is to be enforced vertically between individuals and/or horizontally between individuals and states. Moreover, the question of enforcement at the national or international level is of particular interest in the globalising world, where the climate knows no boundaries. In the ICJ judgment in the so-called Pulp Mills case the Court for instance held as follows:[40]

> [T]here are situations in which the parties' intent upon conclusion of the treaty was, or may be presumed to have been, to give the terms used – or some of them – a meaning or content capable of evolving, not one fixed once and for all, so

39 Schrijver (2011:1285).

40 Case concerning Pulp Mills on the River Uruguay (*Argentina v Uruguay*) International Court of Justice, 20 April 2010, General List No. 135.

> as to make allowance for, among other things, developments in international law.

The *substantive* dimension of climate change law is far reaching and incorporates among others constitutional law, administrative law, environmental law, water law, criminal law, the law of nuisance, the law of delict, insurance law and even tax law. On a vertical level, intersections of law occur on a very broad scale of the different but interrelated branches of the law with the underlying assumption that climate change law consists of the sum of legal provisions protecting the climate itself and those that protect the climate from the negative effects of climate change. This scale ranges from environmental law (with its multiple sub-branches such as biodiversity law, environmentally relevant provisions within the law of the sea, outer space law, energy and mining law, and specific legal instruments relating to climate change, etc.) to human rights law, humanitarian law, trade and investment law, the law on the use of force, criminal law, and liability law among others.[41]

On a horizontal level, climate change law intersections can be found at the different levels of international and national law. The horizontal level entails international law[42] with multilateral agreements on the global, regional and sub-regional level, bilateral (and unilateral) agreements, general principles of law, customary international law, case law, and other instruments such as declarations, agendas among others. National law may consist of constitutional law, statutory law, common law, case law, customary law, policies, strategies and action plans and other relevant instruments. Climate related –[43]

> policies are for instance central to the development of sustainable energy generation and markets. Laws governing sustainable energy development and supply cut across many sectors such as mining, forestry, agriculture, environment, water, industry, electricity, and petroleum, and hence require coordination – a complex challenge that is not easily overcome.

41 For an overview of legal issues relevant to climate change see for example Brunnée et al. (2012).

42 For further details see Rayfuse & Scott (2012).

43 Ruppel & Ruppel-Schlichting (2012:46).

***Figure 1:[44] Intersections of Law and Cooperative Global Climate Governance: Challenges in the Anthropocene***

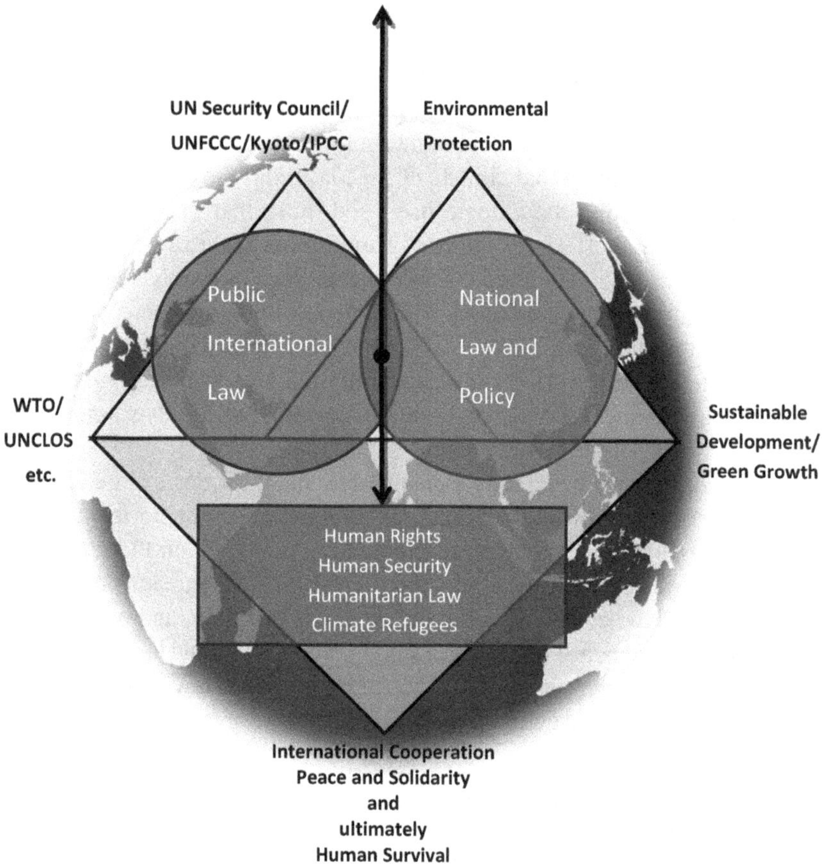


As Figure 1 demonstrates, intersections not only occur with regard to the question whether it is national or international law that applies, or both, but also within the categories of national or international law themselves. A further problem is the demarcation between 'hard' and 'soft' law. Some of the sources of national and international law are obligatory; others are of a non-binding nature. In the climate change context, the lack of globally applicable enforceable legal obligations is without doubt one of the major de-

44 Figure realised by Cord Lüdemann.

ficiencies[45] and one of the major subjects of and challenges for current climate change negotiations.

Furthermore, there has been an emergence of global administrative law forming trans-governmental regulation and administration in such fields as –[46]

> security, the conditions on development and financial assistance, environmental protection, banking and financial regulation, law enforcement, telecommunications, trade in products and services, intellectual property, labour standards, and cross-border movement of populations, including refugees. Increasingly, these consequences cannot be addressed effectively by isolated national regulatory and administrative measures.

Summarising it can be stated that cross-cutting themes thus include, among others, the relationship between international environmental law and general principles of international law; conflicts among differing legal regimes; the range of approaches to the regulation of activities within and beyond areas under national jurisdiction; the role and impact of competing state interests in the negotiation and enforcement of international regimes; the challenge of regulating in the face of scientific uncertainty; the role of both 'soft' and 'hard' law in addressing the global problem; and the potential contribution of the judiciary and international tribunals in the further development of climate change law.

The intersections of international climate change law and multiple overlapping regulatory bodies reflect the fragmentation of global climate change governance in the absence of a universal climate change regime. This makes international climate change law extremely complex and global climate governance not very orchestrated. This overlapping complexity in the different climate change (related) regimes can be observed in various United Nations conventions, the international human rights regime, the world trade order under the World Trade Organization (WTO), multilateral environmental agreements (MEAs) and other international legal instruments that (directly or indirectly) deal with climate change, such as the Vienna Con-

45 Spier (2012:49).
46 Kingsbury et al. (2005:16).

vention on Ozone Depletion, the Montreal Protocol,[47] the Convention on Biodiversity, the London Dumping Convention, the Convention on International Trade in Endangered Species of Wild Fauna and Flora (CITES), the RAMSAR Convention on Wetlands of International Importance and the Convention on the Conservation of Migratory Species of Wild Animals, among others. Same applies for geo-engineering, nuclear technology, intellectual property, international investment and finance regimes.

For the purpose of this article, the following sections shall only reflect on the climate regimes around the UNFCCC, the work of the IPCC, the role of the United Nations Security Council, the international human rights regime, international refugee law, the law of the sea regime (UNCLOS) and the world trade order (WTO).

### *I. The UN Framework Convention on Climate Change and the Kyoto Protocol*[48]

The international legal climate change regime is a product of international law, which has developed rapidly over the past few decades, especially since the dawn of the United Nations (UN), when rules and norms regulating activities carried on outside the legal boundaries of nations were developed. Numerous international agreements – bilateral, regional or multilateral – have been concluded and international customary rules, as evidence of a

47 The 1987 Montreal Protocol introduced a series of effective steps to phase out the global production and consumption of ozone-depleting substances in the 1980s. The Protocol and successor agreements are not only regarded as highly successful examples of international environmental regulatory cooperation, there are also lessons to be learned from the ozone layer experience in dealing with climate change. The Montreal Protocol has made a substantial commitment to climate goals, and there are substantial proposals on the way to increase this. Having phased out 97% of almost 100 ozone-depleting substances (ODSs) it placed the ozone layer on a path to recovery. "Because many ODSs are also potent greenhouse gases (GHGs), their phase-out under the Montreal Protocol has provided an often overlooked bonus for climate mitigation: by the end of the decade, the Montreal Protocol will have done more to mitigate climate change than the initial Kyoto Protocol reduction target, reducing emissions in terms of carbon dioxide ($CO_2$), equivalent to 135 billion tonnes between 1990 and 2010 and delayed climate impacts – including abrupt and irreversible impacts – by about 12 years". See http://www.igsd.org/montreal/index.php (also for further references, last accessed 25 November 2012.

48 This Section is largely based on Ruppel (2013).

general practice accepted as law, have been established. International agreements are binding upon states if the consent to be a party to them is expressed by a signature followed by ratification, or by accession, where the state is not a signatory to a treaty, or by declaration of succession to a treaty concluded before such a state existed. The sources of international law in general are listed in Article 38 of the Statute of the International Court of Justice (ICJ), the principal judicial organ of the United Nations. However, considering that Article 38 of the Statute of the ICJ was first drafted in 1920, these provisions no longer reflect all the sources of today's international law. New developments in respect of sources of law have to be considered in addition to those recognised in Article 38.

The 1992 United Nations Conference on Environment and Development (UNCED) in Rio de Janeiro, Brazil, reaffirmed the Declaration of the United Nations Conference on the Human Environment, adopted in Stockholm, Sweden, in 1972, seeking to build upon it with the goal of establishing a new and equitable global order through the creation of new levels of cooperation among states, key sectors of societies and people, working towards international agreements which respect the interests of all and protect the integrity of the global environmental and developmental system, recognising the integral and interdependent nature of the earth. It proclaims first and foremost that human beings are at the centre of concerns over sustainable development. They are entitled to a healthy and productive life in harmony with nature (Principle 1). Moreover, states have, in accordance with the Charter of the United Nations and the principles of international law, the sovereign right to exploit their own resources pursuant to their own environmental and developmental policies, and the responsibility to ensure that activities within their jurisdiction or control do not cause damage to the environment of other states or of areas beyond the limits of national jurisdiction (Principle 2). Thirdly, the right to development must be fulfilled so as to equitably meet developmental and environmental needs of present and future generations (Principle 3).

The UNFCCC and the Kyoto Protocol are treaties in terms of international law and Article 2.1(a) of the Vienna Convention on the Law of Treaties. International oversight and implementation of the climate regimes are only possible through an array of institutions under the UNFCCC and Kyoto regimes.[49] The COP is the supreme body of UNFCCC, which regularly re-

49 Depledge & Yamin (2009).

views the implementation of the Convention and any related legal instruments that the COP may adopt to promote the effective implementation of the Convention.

The mandate of the COP to amend the UNFCCC and the Kyoto Protocol, or adopt a new legal instrument that either supplements or replaces the Kyoto Protocol, is broadly limited by the UNFCCC's objective and guiding principles. The UNFCCC, however, only provides a general framework to combat climate change. Parties have a responsibility to protect the climate system in accordance with their common but differentiated responsibilities and respective capabilities.[50]

The UNFCCC allows for the introduction of protocols to the Convention. The first of these is the Kyoto Protocol. This agreement came into force on 16 February 2005. A number of global initiatives are being implemented to assist in the operationalisation of the UNFCCC. For example, the Global Environment Facility (GEF) serves as an operating entity of the UNFCCC financial mechanism and has been supporting the national capacity self-assessment process at national level for some time. This is aimed at providing countries with an opportunity to articulate their own capacity needs in implementing the UNFCCC, the other two Rio Conventions and other non-Rio Conventions (e.g. chemicals). The ultimate objective of the UNFCCC is –[51]

> to achieve, in accordance with the relevant provisions of the Convention, stabilization of greenhouse gas concentrations in the atmosphere at a level that would prevent dangerous anthropogenic interference with the climate system. Such a level should be achieved within a time-frame sufficient to allow ecosystems to adapt naturally to climate change, to ensure that food production is not threatened and to enable economic development to proceed in a sustainable manner.

The Convention is a framework document, identifying two major areas of action required to address climate change, namely mitigation[52] and adaptation.[53] Moreover, the Convention as a legal instrument identifies a wide range of measures (see, e.g., the diversity of measures in Article 4.1) to address climate change through other activities such as scientific and technical cooperation, technology transfer, finance etc. The UNFCCC allows

---

50 For more details see AMCEN (2011).
51 Article 2 UNFCCC; UNFCCC (2011).
52 UNFCCC (2009).
53 UNFCCC (2010).

any state to become a party, and as at 2011 has 194 signatories, making it a global instrument. Within this framework of global participation, actual obligations of parties differ substantially between industrialised and developing countries. The UNFCCC enshrines a number of key principles (Article 3) including the principles of *equity* and *common but differentiated responsibilities and respective capabilities*. Today's accumulated greenhouse gas emissions originate mainly from over 150 years of carbon-based industrial activity in developed states. Therefore UNFCCC recognises that all countries have a common responsibility to tackle climate change, but places a heavier burden on industrialised states to fulfil their historic responsibility of addressing climate change.[54]

These principles are reflected in the obligations established for developed and developing countries in the Convention, including those relating to mitigation, adaptation, technology transfer, finance as well as communication of information relating to the Convention. The Convention goes further to make provision for countries in special situations, including particularly vulnerable countries, least-developed countries and countries undergoing transition to a market economy. Article 4(4) UNFCCC, for instance, states:

> The developed country parties … shall assist the developing country parties that are particularly vulnerable to the adverse effects of climate change in meeting costs of adaptation to those adverse effects.

The Kyoto Protocol came into force in 2005 and shares the objectives and the institutions of the UNFCCC. The major distinction between the two is that while the UNFCCC only encourages industrialised countries to stabilise greenhouse gas emissions, the Kyoto Protocol obliges them to do so. Just like the UNFCCC, the Kyoto Protocol imposes a heavier burden on developed nations under the principle of common but differentiated responsibilities. This group of countries must first and foremost take domestic action to address climate change, but the Kyoto Protocol allows them a certain degree of flexibility in satisfying their emissions commitments.

Under the Kyoto Protocol, actual emissions have to be monitored – each party must keep a national register to show measures carried out under the Kyoto Protocol instruments. The secretariat keeps an independent transaction log to verify that operations are consistent with the rules of the Kyoto Protocol. The most important aspect of the Kyoto Protocol is arguably the

54 Boisson de Chazourne (2008).

creation of an aggregate target for the developed countries (Article 3) as well as legally binding and quantified individual targets set out in Annex B. It should also be noted that there are significant commitments for reporting, review, independent assessment and compliance (Articles 5, 7, 8 and 18).

Under the adaptation objective, the Kyoto Protocol, like the UNFCCC, is designed to support countries in adapting to the inevitable effects of climate change and to facilitate the development of techniques that can help increase resilience to climate change impacts. An Adaptation Fund was set up to help with concrete adaptation projects in developing countries. The Adaptation Fund is a solidarity fund in which a proportion of the revenue of CDM projects in developing countries is contributed to a fund to assist adaptation projects in other developing countries.

In the course of the United Nations Climate Change Conference held in Cancun, Mexico in 2010, a set of agreements were reached, building on the Bali Road Map[55] and the Copenhagen Accord,[56] which clearly reflect that the parties to the UNFCCC and the Kyoto Protocol had taken up the issue of climate justice. Three decisions have resulted from the Cancun Conference: one decision by the COP to the UNFCCC[57] and two decisions by the COP serving as the meeting of the Parties to the Kyoto Protocol.[58] The reduction of greenhouse gas emissions and the support for developing nations to deal with climate change are at the core of the Cancun agreements. In order to advance action regarding the aim of the reduction of greenhouse gas emissions in a mutually accountable way, national plans are formally captured at international level under the banner of the United Nations Framework Convention on Climate Change. Support for developing nations is provided for in the Cancun agreements and includes financial, technology

---

55 The Bali Road Map emerged from the 2007 Bali Climate Change Conference and includes the Bali Action Plan (Decision 1/CP.13), which launched a "comprehensive process to enable the full, effective and sustained implementation of the Convention through long-term cooperative action" along with a number of other decisions and resolutions.

56 Agreed upon by the UNFCCC Conference of the Parties, in Copenhagen on 18 December 2009 by way of Decision 2/CP.15.

57 Decision 1/CP.16 *The Cancun Agreements: Outcome of the work of the Ad Hoc Working Group on Long-term Cooperative Action under the Convention.*

58 Decision 1/CMP.6 *The Cancun Agreements: Outcome of the work of the Ad Hoc Working Group on Further Commitments for Annex I Parties under the Kyoto Protocol at its fifteenth session*; and Decision 2/CMP.6 *The Cancun Agreements: Land use, land-use change and forestry.*

and capacity-building support, which is to be realised through various mechanisms: nationally appropriate mitigation actions (NAMA); reducing emissions from deforestation and forest degradation (REDD+); the Clean Development Mechanism (CDM); the Cancun Adaptation Framework (CAF); the technology mechanism; and the Green Climate Fund (GCF).

At the COP18 to the UNFCCC and the MOP8 to the Kyoto Protocol held in Doha, Qatar in 2012, a second commitment period under the Kyoto Protocol has been launched, with 2020 as the end date. Unfortunately, several countries that had previously participated in the Kyoto Protocol have not joined the second commitment period, such as Russia, Canada, New Zealand and Japan. Although it had been agreed to work towards a universal climate change agreement covering all countries from 2020 it will still be seen whether such agreement is to be adopted by 2015.

### *II. The Intergovernmental Panel on Climate Change*

The IPCC was established by the UNEP and the World Meteorological Organisation (WMO) in 1988. The ultimate role of the IPCC is –[59]

> to assess on a comprehensive, objective, open and transparent basis the scientific, technical and socio-economic information relevant to understanding the scientific basis of risk of human-induced climate change, its potential impacts and options for adaptation and mitigation. Review by experts and governments is an essential part of the IPCC process. The Panel does not conduct new research, monitor climate-related data or recommend policies. It is open to all member countries of WMO and UNEP.

In the UNFCCC explicit reference is made to the IPCC under Article 21:

> [T]he head of the interim secretariat referred to in paragraph 1 above will cooperate closely with the Intergovernmental Panel on Climate Change to ensure that the Panel can respond to the need for objective scientific and technical advice.

The IPCC was subsequently and repeatedly included in the Kyoto Protocol to the Convention where the methodological work of the Intergovernmental Panel on Climate Change should be taken into account in formulating guidelines on verification of emission reductions.[60]

---

59 IPCC (2001).
60 See Article 3(4) of the Kyoto Protocol.

The IPCC consists of three Working Groups: The IPCC Working Group I (WG I) assesses the physical scientific aspects of the climate system and climate change. The main topics assessed by WG I include: changes in greenhouse gases and aerosols in the atmosphere; observed changes in air, land and ocean temperatures, rainfall, glaciers and ice sheets, oceans and sea level; historical and paleo-climatic perspectives on climate change; biogeochemistry, carbon cycle, gases and aerosols; satellite and other data; climate models; climate projections, causes and attribution of climate change.[61] The WG I Technical Support Unit, which manages the organisational and administrative activities of the Working Group, is hosted by the University of Berne, Switzerland, and funded by the government of Switzerland.[62]

The IPCC Working Group II (WG II) assesses the vulnerability of socio-economic and natural systems to climate change, negative and positive consequences of climate change, and options for adapting to it. It also considers the relationship between vulnerability, adaptation and sustainable development. Information is evaluated by sector (water resources; ecosystems; food and forests; coastal systems; industry; human health) and region (Africa; Asia; Australia and New Zealand; Europe; Latin America; North America; Polar Regions; Small Islands).[63] In its reports, Working Group II elaborates on the scientific, technical, environmental, economic and social aspects of the vulnerability (sensitivity and adaptability) to climate change of, and the negative and positive consequences for, ecological systems, socio-economic sectors and human health, with an emphasis on regional, sectoral and cross-sectoral issues. The WG II Technical Support Unit is housed at the Carnegie Institution for Science in Stanford, California, USA.[64]

The IPCC Working Group III (WG III) assesses options for mitigating climate change through limiting or preventing greenhouse gas emissions and enhancing activities that remove them from the atmosphere. The main economic sectors are taken into account, both in a short-term and in a long-term perspective. The sectors include energy, transport, buildings, industry, agriculture, forestry, and waste management. WG III analyses the costs and benefits of the different approaches to mitigation, considering also the avail-

61 See IPCC on Working groups / Task Force at http://www.ipcc.ch/working_groups/working_groups.shtml, last accessed 17 February 2013.

62 See https://www.ipcc-wg1.unibe.ch/, last accessed 17 February 2013.

63 See http://www.ipcc.ch/working_groups/working_groups.shtml, last accessed 17 February 2013.

64 See http://ipcc-wg2.gov/index.html, last accessed 17 February 2013.

able instruments and policy measures. The approach is more and more solution oriented.[65] The IPCC WG III Technical Support Unit is housed at the Potsdam Institute for Climate Impact Research in Potsdam, Germany.[66]

The above three working groups were intended to:[67]

> draw on slightly different scientific constituencies, since impact and responses would require factoring in research outside the physical sciences and would touch on political issues. Working Group I would be dominated by climate scientists, while Working Groups II and III would have a wider participation, including, as time went on, by economists and other social scientists.

The historical –[68]

> roots of IPCC's strength reached very deep. Most people were scarcely aware that IPCC, and virtually every other international initiative ..., relied on a key historical development: The worldwide advance of democracy. It is too easy to overlook the obvious fact that international organizations govern themselves in a republican fashion, with vigorous free debate among all members and votes in councils of elite leaders.

Often, as in IPCC, decisions among the dozens or hundreds of elite leaders are made by a negotiated consensus in a spirit of equality, of mutual accommodation, and of commitment to the community process – all of which are seldom celebrated, but essential, components of the republican political culture.[69] It has been said that it is –[70]

> an important historical fact that such international regimes have been created chiefly by governments that felt comfortable with such mechanisms at home, that is, democratic governments. Nations like Nazi Germany, Communist China, and the former SU did little to create international organizations (aside from front groups under their own thumb), and often participated in them awkwardly. Happily, in the second half of the twentieth century, nations under democratic governance became globally predominant.

That encouraged the proliferation of international institutions that were democratic, or at any rate elite-based republican, exerting an ever stronger influence in world affairs.[71] "The democratization of international relation-

65 See http://www.ipcc.ch/working_groups/working_groups.shtml, last accessed 17 February 2013.

66 See http://www.ipcc-wg3.de/, last accessed 17 February 2013.

67 Mathiason & Bhandari (2010).

68 Weart (2012).

69 Weart (1998:61).

70 Weart (2012).

71 Weart (1998:262–267).

ships was the foundation upon which IPCC took its stand."[72] In 2007, the IPCC and Albert Arnold (Al) Gore Jr. were awarded the Nobel Peace Prize "for their efforts to build up and disseminate greater knowledge about man-made climate change and to lay the foundations for the measures that are needed to counteract such change".[73]

This Prize was most probably not awarded to the IPCC without good reason.[74] Despite criticism it should not be forgotten that the IPCC is a very valuable institution that tries to help in an unprecedented way to resolve socio-political conflicts by gathering scientific knowledge and presenting it in a comprehensible manner. "The evidence shows the scientific consensus arrived at by the IPCC is a solid one, given the composition of the panel, and an innovative means of connecting science with politics."[75]

The 4th IPCC Assessment Report (AR4) – against all contrary opinions – can be considered a reliable study on the state of climate science and uncertainties in the year 2007. Although two minor mistakes had been detected in the report of several thousand pages, the rest remains valid.[76] The 5th IPCC Assessment Report (AR5) is expected to be published in 2014.[77] For AR5 the IPCC has made it a priority to engage developing countries more fully:[78]

> AR5 will be able to provide much greater regional detail than available literature has allowed in the past. We all have to make a major effort to do full justice to expectations in different parts of the world, and for this reason … we must take care of this aspect as diligently as possible. We would need to be equally diligent in going the extra mile in assessing literature in local languages where for scientific reasons we would be able to enrich the AR5 with comprehensive knowledge and information.

72 Weart (2012).

73 See http://www.nobelprize.org/nobel_prizes/peace/laureates/2007/, last accessed 04 March 2013.

74 Kowarsch (2010).

75 Mathiason & Bhandari (2010).

76 Because of doubts regarding the IPCC results US Congress has mandated a large group of scientists and representatives of the private industry in 2008 to verify the IPCC results. The outcome can be accessed at http://online.wsj.com/article/SB10001424052748704691304575254691763608402.html?mod=WSJ_hps_SECONDTopStories, last accessed 17 February 2013.

77 See http://www.ipcc.ch/activities/activities.shtml, last accessed 14 February 2012.

78 Pachauri (2009).

The IPCC gives valuable advice to national governments and international organisations.[79] By effectively and objectively assessing scientific knowledge and prevailing uncertainty, the IPCC provides the world with the best possible and much-needed evidence of climate change related impacts. Scientific authority also depends on reliable indicators.[80] In this context the IPCC plays – no doubt – a decisive role in the policy reform and political decision-making process:[81]

> Because of its scientific and intergovernmental nature, the IPCC embodies a unique opportunity to provide rigorous and balanced scientific information to decision makers. By endorsing the IPCC reports, governments acknowledge the authority of their scientific content. The work of the organization is therefore policy-relevant and yet policy-neutral, never policy-prescriptive.

The IPCC thus bridges the two fields, by getting the facts right so the policies may be effective. In effect, "if scientists cannot agree, political leaders and other stakeholders are unlikely to agree either."[82]

### *III. The UN Security Council and the Responsibility to Protect*

Only recently UN Secretary-General Ban Ki-moon made reference "to the gathering threat of climate change" at the Sorensen Distinguished Lecture on the United Nations at the Council on Foreign Relations. He said:[83]

> [S]cientists have long sounded the alarm. Top-ranking military commanders and security experts have now joined the chorus. Yet the political class seems far behind .... Too many leaders seem content to keep climate change at arm's length, and in its policy silo. Too few grasp the need to bring the threat to the centre of global security.

Framing climate change more and more –[84]

> as a security issue could serve to enhance and broaden the policy response at various governance levels by facilitating policy makers and their publics recognizing the common origins of what may otherwise appear as unconnected

79 InterAcademy Council (2013).

80 Davis et al. (2012).

81 See http://www.ipcc.ch/organization/organization.shtml#.URelrmhpvos, last accessed 17 February 2013.

82 Mathiason & Bhandari (2010:58).

83 Ki-moon (2013).

84 Scott (2012).

> phenomena. Debate about climate change is often couched in terms of a hypothetical future: by how much the temperature will rise, by how much countries should reduce their emissions, and the nightmare scenarios that may come into play if they fail to do so.

This focus on what may appear a hypothetical future renders climate change a particularly daunting and difficult policy arena for governments because, as NATO Secretary General Anders Fogh Rasmussen explained:[85]

> The science is not yet perfect. The effects are just starting to be visible, but it's difficult to pin down what's actually changing because of climate change. The timelines are not clear either. And as a politician, I know exactly what that means. When we have to choose between spending money now on schools or health care, or diverting funds to try to prevent something that will likely only hurt long after they have left office, the choice for most leaders is pretty clear. And, let me say, not hard to understand.

In 2011, the United Nations Security Council expressed concern that the possible adverse effects of climate change could, in the long run, aggravate certain existing threats to international peace and security and that the loss of territory in some states could have possible security implications.[86] In a statement read out by the then Council President, Peter Wittig of Germany, following a day-long debate on "maintenance of international peace and security: the impact of climate change", he noted that "conflict analysis and contextual information" on, among others, the "possible security implications of climate change" was important when climate issues drove conflict, challenged implementation of Council mandates or endangered peace processes.[87]

UN Secretary-General Ban Ki-moon, who opened the aforementioned 2011 Council debate, pointed to the devastating impact of extreme weather and rising seas on lives, infrastructure and budgets — an "unholy brew" that could create dangerous security vacuums. "We must make no mistake. … The facts are clear: climate change is real and accelerating in a dangerous manner," he said, declaring that it "not only exacerbates threats to international peace and security; it is a threat to international peace and security". Events in Pakistan, the Pacific islands, Western Europe, China and the Horn of Africa, among other areas, illustrated the urgency of the situation, he said. Worldwide, hundreds of millions of people were in danger of food and water

85 Rasmussen (2009).
86 United Nations Security Council (2011).
87 (ibid.).

shortages. Environmental refugees were "reshaping the human geography" of the planet.[88]

Although the aforementioned statements clearly frame climate change as a potential source of conflict, a potential threat to national and international peace and human security, the future role of the UN Security Council with regard to climate change remains to be determined. The Council would arguably be acting within its legal powers if, for example, it passed resolutions requiring governments at all levels "to prioritize adaptation strategies in their planning and national governments to contribute military or other resources to a global disaster mitigation unit".[89] Yet in 2011, as in 2007, the Security Council did not take a decision on climate change. This time, however, it did agree on a presidential statement, a non-legally binding document adopted by consensus, expressing concern that possible adverse effects of climate change may, in the long run, aggravate certain existing threats to international peace and security.[90]

At present, the UN Security Council has only 15 members – five of which are permanent and ten of which are members for two-year terms. Decisions on all but procedural matters are taken by an affirmative vote of nine members, including the concurring votes of the five permanent members.[91] A cornerstone of the United Nations Charter paradigm is the notion of collective security which is perhaps the first and most obvious manifestation of the principle of solidarity in the post World War II era.[92] In fact, it forms the political and legal foundation for the collective security system established by the UN Charter. Under Article 25 of the UN Charter, member states "agree to accept and carry out the decisions of the Security Council".[93] Article 39 stipulates that the Security Council can identify a "threat to the peace, breach of the peace, or act of aggression" and "make recommendations, or decide what measures shall be taken in accordance with Articles 41 and 42, to maintain or restore international peace and security".[94] Article 41 provides for the Council to decide on appropriate measures not involving the use of

88 (ibid.).

89 Scott (2012).

90 Statement by the President of the Security Council (20-07-2011) UN Doc S/PRST/2011/15.

91 UN Charter Article 27.3. Although not explicitly stated in the Charter, it has become accepted that this vote may include abstentions by permanent members.

92 Koroma (2012).

93 Article 25.

94 Article 39.

armed force,[95] and Article 42 provides that if the Security Council considers that such measures "would be inadequate or have proved to be inadequate, it may take such action by air, sea, or land forces as may be necessary to maintain or restore international peace and security".[96] The Security Council can thus enforce its decisions made in response to a perceived "threat to the peace, breach of the peace, or act of aggression" by use of force if it deems it necessary to do so. It is generally accepted among the international law community that it is at the Council's political discretion to define what constitutes a threat to the peace for the purposes of Chapter VII of the UN Charter.[97]

A still controversial manifestation of the notion of solidarity in international law is the emerging doctrine of the responsibility to protect. This concept was developed by the International Commission on Intervention and State Sovereignty in September 2000, after the UN Secretary General Kofi Annan emphasised the grave failure of the international community to handle gross and systematic violations of human rights such as those perpetrated in Rwanda and other areas.[98] The aforementioned concept has gained growing attention in the context of the notion of global solidarity and collective security as it aims to address legal and political dilemmas for intervention to stop or pre-empt human suffering and crimes against humanity.[99]

Under Article 52 of the UN Charter, regional organisations may undertake actions aimed at the maintenance of international peace and security. Article 53 (I) of the UN Charter specifically provides that such regional organisations may undertake enforcement measures, provided that they have the authorisation of the UN Security Council. Most obviously the crux of the responsibility to protect concept is the dilemma of state sovereignty and intervention for humanity. In light of this, current discussions focus on the duty of the international community and the territorial state in cases of natural disasters, raising the question whether the doctrine of the responsibility to protect can actually be extended to the international law relating to disaster relief and in particular to cases of grave circumstances such as severe human suffering during times of natural disasters. Unfortunately, so far for inter-

95 Article 41.
96 Article 42.
97 See for example Wood (2006).
98 Report of the Secretary-General on the Work of the Organization, document A/54/1, at 48.
99 Koroma (2012).

national law and politics it still seems to make a big difference whether human suffering is the result of a natural disaster or of an (international) armed conflict.[100] However, when responding to the question whether the doctrine of the responsibility to protect should in future be extended to the international law relating to disaster relief one could argue with Achim Steiner as follows:[101]

> There is no reason why the international community cannot avoid escalating conflicts, tensions and insecurity related to a changing climate if a deliberate, focused and collective response can be catalyzed that tackles the root causes, scale, potential volatility and velocity of the challenges emerging. In bringing forward a response that enhances global security and cooperation on the climate challenge, the world can perhaps also better manage risk from numerous other challenges and in doing so diminish tensions between nations and lay the foundations and possibilities of a more sustainable and equitable peace.

It becomes apparent from the above that climate change is moving from mere politicisation towards a state of securitisation.[102] Once an issue is successfully securitised it moves out of the sphere of normal politics to be dealt with as an emergency issue without the normal democratic processes being brought to bear, and the securitising actor can, through this process, infuse the concept of 'security' with any meaning desired.[103] Full securitisation would seem to be represented by the issue moving outside of the normal multilateral treaty framework used to manage political issues of mutual concern to the body with "primary responsibility for the maintenance of international peace and security": the United Nations Security Council.[104]

Most obviously, the nature and "impacts of climate change challenge traditional notions in international law, most notably those relating to the principle of territorial sovereignty, with its presumptions of defined territory and fixed maritime boundaries".[105] "Sovereignty in the relations between States signifies independence. Independence in regard to a portion of the globe is the right to exercise therein, to the exclusion of any other State, the functions

---

100 Thielbörger & Liburd (2012).
101 Steiner (2011).
102 See with further references Scott (2012:221).
103 See Taureck (2006:55).
104 Scott (2012:221).
105 Schrijver (2011:1285).

of a State."[106] The world is divided into clearly demarcated territories. Each territory has one government within the territory, with full jurisdiction over all persons and resources within its domain.[107] In the context of climate change it seems appropriate, however, to explore whether the law of state responsibility offers a useful paradigm to address the problem.[108] Unfortunately, however, national governments and statesmen more often than not regard themselves as –[109]

> primarily responsible not vis-a-vis an existing global order, which they all too often violate, but vis-a-vis a possible future order, which they lack the will and vision to help bring about. This is the ultimate crime against peace and justice.

The UNFCCC and the subsequent Kyoto Protocol are an articulation of how states balance their sovereign right to follow their own development agenda with their overall responsibilities under international law, including those measures aimed at avoiding harm to areas beyond the limits of national jurisdiction. This means that the global nature of climate change demands that states scale back some of their sovereignty by engaging in international cooperation and negotiation in the interest of the "common concern of humankind".[110] Efforts to curb climate change have given rise – sometimes in conjunction with developments in other environmental regimes – to the evolution of new principles and concepts of international law, including the principle of common but differentiated responsibilities, the notion of common concern of humankind, protection of vulnerable countries and others.[111] With regard to the application of the responsibility to protect doctrine to climate change it is argued here that existing relevant international obligations such as the responsibility to avoid trans-boundary harm must be seen in a broader context in order to widen the international responsibility to protect people and ecosystems at the same time.

106 Permanent Court of Arbitration, The Island of Palmas Case (or Miangas) *Unites States of America v The Netherlands* Award of the Tribunal 04 April 1928, XI UNRIAA 838.

107 Pogge (1987:429).

108 For an interesting exploration see Voigt (2008).

109 Pogge (1987:436).

110 See the Preamble to the UNFCCC.

111 Schrijver (2011:1278).

### IV. Climate Change and International Human Rights Law

As early as 1984, Karel Vasak in his inaugural lecture at the International Human Rights Institute in Strasbourg proposed the concept of solidarity or third generation rights, including the right to development, the right to peace and the right to a healthy environment.[112] Such rights –[113]

> are new in that they may both be invoked against the State and demanded of it; but above all (and herein lies their essential characteristic) they can be realized only through the concerted efforts of all the actors on the social scene: the individual, the State, public and private bodies and the international community.

The efforts that have been made so far to place rights at the centre of any future climate change dispensation have only recently started to become more human rights focused. One reason for the past silence of human rights regarding climate change is the fact that most international human rights instruments were drafted before the emergence of climate change as a common concern. However, silence is increasingly turning into salience. When looking at the most severe impacts of climate change such as drought, floods, migration and famines it becomes very clear that climate change and its effects affect large numbers of people and have an impact on a broad range of human rights; the right to life in the first place, but also the rights to health, adequate food and water, property and adequate housing, self-determination, to name only the most common and pressing ones.

When it comes to the question of the state of fulfilment of human rights in the world, statistics are frequently consulted. Only some of the respective figures will be given as examples. This seems appropriate because the negative effects of climate change will most affect those people who already appear in one or more of the following figures. In developing regions, 24% of people live on less than US$1.25 a day.[114] Globally almost 870 million people (or one in eight) are chronically undernourished, of which 852 million live in developing countries.[115] The global under-five mortality rate is 45.2 per 1000 live births,[116] 63 in developing regions.[117] One in nine people, or 780 million, lack access to an improved water source, 2.5 billion lack im-

112 Koroma (2012:108).
113 Vasak (1984:839).
114 United Nations (2012).
115 FAO et al. (2012).
116 See http://hdr.undp.org/en/data/map/, last accessed 14 February 2013.
117 United Nations (2012).

proved sanitation, and 3.4 million people die each year from a water related disease.[118] Over a billion people lack adequate housing[119] and about 1.5 billion have no access to electricity.[120] Approximately 775 million adults are illiterate[121] and around 215 million children are child labourers.[122]

There are various reasons why a human rights based approach to climate change is gaining momentum with a high relevance for the future climate change debate. The most important one is probably the cross-fertilisation of human rights and climate change effects and the related mitigation and adaptation measures. With the threats climate change poses to human and environmental security, existing legal structures are likely to come under pressure.[123] "[H]uman rights obligations may provide a legal baseline for how climate change is tackled and what must be protected from its impacts."[124] Human rights may serve as powerful tools for ensuring greater capacity to adapt to climate change. In order to design and implement a legal climate change regime that includes the policy value and the legal force of human rights it is required to introduce likely human rights impacts and outcomes of climate change.[125] The experiences gained in the field of human rights law may furthermore be useful sources of information in the processes of climate change related policy and legal drafting. Perhaps jurisprudence particularly related to the effects of climate change has not yet been established by international human rights tribunals. Jurisprudence by international human rights tribunals to address the impact of environmental harm[126] on human rights, however, may well be extended to apply also to the negative effects of climate change as global environmental harm. Furthermore, climate change impacts on human rights should be considered when adaptation and mitigation measures are being developed and implemented. Tackling the negative effects of climate change may have a positive influence on the fulfilment of human rights. The less the negative effects of climate change, the better the chances to fully enjoy all human rights and fundamental free-

118 UNICEF et al. (2012).
119 OHCHR & UN-HABITAT (2009).
120 AGECC (2010).
121 UIS (2012).
122 See http://www.ilo.org/global/topics/child-labour/lang--en/index.htm, last accessed 14 February 2013.
123 Pedersen (2012:28).
124 McInerney-Lankford (2009).
125 ICHRP (2008).
126 Knox (2009).

doms. Moreover, international human rights law places certain duties on states (in very general terms, the duty to refrain from violating human rights itself, but also to protect its citizens from human rights violations) to address the effects of climate change on human rights, irrespective of their relative contributions of greenhouse gas emissions to global warming.

In the context of climate change, three basic obligations of states can be identified, namely addressing the causes of climate change, i.e. mitigating climate change; addressing the effects of climate change, i.e. adapting to the effects of climate change by reducing risks created by climate change and vulnerabilities caused by it; and addressing the consequences of climate change, for example by protecting individuals displaced by the effects of climate change.[127]

The duty to cooperate[128] in the international protection of human rights by means of diplomacy, by institutional cooperation on the UN or regional level, or by imposing unilateral or multilateral sanctions to induce a state to comply with human rights obligations is a state obligation that could also apply to climate change related matters. To this end, the United Nations Human Rights Council adopted Resolution 19/33 in 2012, which –[129]

> [u]rges States to take necessary measures to enhance bilateral, regional and international cooperation aimed at addressing the adverse impact of consecutive and compounded global crises, such as financial and economic crises, food crises, climate change and natural disasters, on the full enjoyment of human rights.

Both the International Covenant on Civil and Political Rights (ICCPR) and the International Covenant on Economic, Social and Cultural Rights (ICESCR), which together with the Universal Declaration of Human Rights (UDHR) form part of the International Bill of Rights, call on state parties to take steps (legislative or other measures) to give effect to the rights contained therein. Both Covenants recognise the right of peoples to self-determination; both have provisions which prohibit all forms of discrimination in the exercise of human rights; and both have the force of law in the countries which have ratified them. Most of the rights and freedoms recognised in the ICCPR

---

127 Kälin & Schrepfer (2012:17).

128 For a detailed analysis of this concept see Delbrück (2012).

129 Section 15 of the Resolution on the enhancement of international cooperation in the field of human rights see A/HRC/19/L.13/Rev. 1, 20 March 2012, http://daccess-dds-ny.un.org/doc/RESOLUTION/LTD/G12/124/35/PDF/G1212435.pdf?OpenElement, last accessed 04 March 2013.

are also entrenched in national constitutions' Bill of Rights. This may include, among others, the right to dignity, the right to life, the right to health, the right to water, the right to legal representation, the guarantee against torture and other cruel or inhumane treatment or punishment, and the right to protection against discrimination on any grounds. States have obligations under international human rights law to address disadvantage and threats to human rights and to ensure that policies aimed at limiting the effects of climate change are implemented effectively and in ways that do not overburden or discriminate against specific vulnerable groups, e.g. women, children and indigenous people.[130] In 2008, the UN General Assembly adopted, by consensus, the Optional Protocol to the ICESCR, which will come into force on 5 May 2013[131] and which provides a mechanism through which persons can petition the UN Committee on Economic, Social and Cultural Rights about violations of their rights.

One starting signal for addressing the linkages between climate change and human rights on the international level has been the United Nations Human Rights Council's first resolution on human rights and climate change in 2008.[132] In 2009, a number of countries called on the Office of the United Nations High Commissioner for Human Rights (OHCHR) to conduct a detailed analytical study of the human rights dimension of climate change, taking into account the views of states and other stakeholders. This study[133] was submitted to the tenth session of the Council held in 2009. In the same year, the Council adopted resolution 10/4 on human rights and climate change, which noted the effects of climate change on the enjoyment of human rights, and reaffirmed the potential of human rights obligations and commitments to inform and strengthen international and national policy making. The Council stated that climate change and human rights are governed by international regimes that have evolved separately, with different premises underlying the legal frameworks of multilateral environmental

---

130 Ruppel (2010).

131 Three months after being ratified by 10 parties. As of 12 February 2013 the Protocol had 42 signatories and 10 parties. See http://treaties.un.org/Pages/ViewDetails.aspx?src=TREATY&mtdsg_no=IV-3-a&chapter=4&lang=en, last accessed 12 February 2013.

132 UN Doc A/HRC/7/23, 28 March 2008, available at http://ap.ohchr.org/documents/E/HRC/resolutions/A_HRC_RES_7_23.pdf, last accessed 13 February 2013.

133 UN Doc A/HRC/10/61, 15 January 2009, available at http://daccess-dds-ny.un.org/doc/UNDOC/GEN/G09/103/44/PDF/G0910344.pdf?OpenElement, last accessed 12 February 2013.

agreements (like the UNFCCC) and human rights treaties. In 2012, the Human Rights Council created a new mandate of an independent expert on the issue of human rights obligations relating to the enjoyment of a safe, clean, healthy and sustainable environment.[134] The new independent expert will among other things serve to identify human rights challenges related to climate change.

It is not only within international human rights law that climate change related issues are moving into the centre of the debate. Also within the international climate change negotiations human rights impacts have gradually become a more relevant aspect.[135]

In fact, –[136]

> climate change prompts significant questions about justice and distribution. There is an acute need for intelligent collective action focusing on the human suffering that climate change will cause in future. On the one hand, as a matter of law, the human rights of individuals need to be viewed in terms of state obligations: it is principally the state that is responsible for human rights fulfilment. On the other hand the assignation of such responsibility to only the state seems inadequate, especially in the context of climate change and human security.

This is also reflected by more recent outcomes of COP to the UNFCCC. One remarkable statement in this regard is the emphasis made by Cancun Decision 1/CP.16[137] on a human rights oriented approach to deal with all issues relating to climate change, by "[r]ecognising that climate change represents an urgent and potentially irreversible threat to human societies and the planet, and thus requires to be urgently addressed by all Parties…" and:

> [n]oting resolution 10/4 of the United Nations Human Rights Council on human rights and climate change, which recognizes that the adverse effects of climate change have a range of direct and indirect implications for the effective enjoyment of human rights and that the effects of climate change will be felt most acutely by those segments of the population that are already vulnerable owing to geography, gender, age, indigenous or minority status, or disability ….

134 UN Doc A/HRC/RES/19/10, 19 April 2012, available at http://daccess-dds-ny.un.org/doc/RESOLUTION/GEN/G12/131/59/PDF/G1213159.pdf?OpenElement, last accessed 12 February 2013.

135 Scholtz (2010).

136 Ruppel & van Wyk (2011).

137 Decision 1/CP.16 http://unfccc.int/resource/docs/2010/cop16/eng/07a01.pdf#page=2, last accessed 12 February 2013.

Moreover, the Conference of the Parties:

> [e]mphasises that Parties should, in all climate change related actions, fully respect human rights.

The inclusions of human rights wording and concepts in the Cancun Agreements represents a unprecedented recognition of the fundamental link between human rights and climate change, and the first tangible results of years of patient analysis, advocacy and alliance building by communities vulnerable to climate change. Rights have become a relevant part of this discourse.[138]

With all due respect for the importance of human rights law for the climate change related problems with which mankind is confronted, one should, however, not turn a blind eye to some of the challenges of international human rights law that might contribute to the disadvantage of those living in the regions most vulnerable to climate change, and particularly those segments of the population who are most vulnerable to the negative effects of climate change, namely women, children and indigenous people. Such challenges include insufficient enforcement mechanisms, the difficulty to establish extraterritorial responsibility and local accountability, the possibility of derogation from many human rights in times of emergency that may be declared in case of catastrophic events such as floods and droughts, or conflicting human rights, e.g. the human right to property or peaceful enjoyment of possessions to prevent or reduce action on climate change.[139]

Several international human rights mechanisms are being used to drive action on climate change.[140] Besides the Human Rights Council's Special Rapporteurs and Special Representatives of the Secretary-General, who conduct country missions, comment on country situations and receive human rights complaints, among other things, the Universal Periodic Review operating since 2008 under the umbrella of the Human Rights Council has become a useful mechanism for states particularly vulnerable to climate change to highlight the threats of global warming to people's rights. Within the process of this peer review, the degree to which a UN member state is complying with international human rights law and domestic laws and commitments is being reviewed every four-and-a-half years by other UN member states. In the period from 2008 to 2011, 31 states have raised climate

138 Cameron & Limon (2012:204).

139 For further information see ICHRP (2008:5).

140 For a detailed analysis see Cameron & Limon (2012).

change related concerns in the national reports and thereby at least placed some moral pressure on high-emitting developed states.[141] Reports by human rights treaty bodies will have "persuasive force insofar as the organs retain their independence, deliver reasoned and consistent opinions using accepted methods of treaty interpretation, and establish a pattern of compliance by State Parties."[142]

### *V. Climate Refugee Law*

In terms of international legal instruments, it must be stated that the issue of climate induced migration is only fragmentarily regulated. There is no single international agreement applicable and neither existing climate change law nor refugee law adequately provides for a consolidated legal framework. Voices asking for a stand-alone international legal regime addressing climate change induced migration are becoming louder.[143] The following two legal regimes and their scope of application show the difficulties for the international and African context.

The movement of persons across international borders due to climate change related events prompts several questions and challenges to international law. The Geneva Refugee Convention of 1951 defines a refugee as a person with a

> well-founded fear of being prosecuted for reasons of race, religion, nationality, membership of a particular social group or political opinion, is outside the country of his/her nationality and is unable or, owing to such fear, is unwilling to avail himself/herself of the protection of that country; or who, not having a nationality and being outside the country of his/her former habitual residence as a result of such events, is unable or, owing to such fear, unwilling to return to it.[144]

Unfortunately this definition provides numerous complications in attempting to classify climate refugees as refugees under international refugee law. The scope of application of the Geneva Refugee Convention for climate refugees is questionable per se; in any case, it would only be applicable to

141 Cameron & Limon (2012:214).
142 Shelton (2012:574).
143 See Hodgkinson & Young (2012).
144 Article 1.A.(2). 1951 Convention Relating to the Status of Refugees.

those migrants who have crossed borders, as it does not provide for internal displacement.

The legal distinction between those moving voluntarily (rather referred to as migrants) and those being forcibly displaced across borders (rather referred to as displaced persons) with the respective legal consequences does not adequately capture the reality of migration as an adaptation strategy, which cannot clearly be allocated under one of the two categories.[145] Once a person has migrated across an international border because of climate change related events and does not qualify as refugee, the only set of legal norms that applies is international human rights law. A right to stay on foreign territory can only be "derived from the human rights prohibition of inhuman treatment – of forcible return of people to a country where they would be exposed to serious risks to life and health",[146] and international law is lacking a set of status rights, particularly for those migrating as a measure of adaptation to climate change.

New strategies and legal frameworks will have to be developed and negotiated to adequately address climate change related cross-border movement of persons. These should particularly encompass the following aspects:[147]

1. Preventing displacement through disaster risk and vulnerability reduction and other adaptation measures;
2. Managing migration as adaptation measures;
3. Providing temporary protection status for persons displaced to other countries and permanent admission in cases where return turns out to be impermissible, impossible or cannot be reasonably be expected over time; and
4. Organizing resettlement/relocation for populations of low-lying small island states and other states losing substantial amounts of their territory.

The issue of internal displacement has been taken up by the African Union by adopting the African Union Convention for the Protection and Assistance of Internally Displaced Persons in Africa in Kampala in 2009. As of 17 January 2013, the Kampala Convention had 36 signatories, 16 countries[148] had ratified it and it has entered into force on 6 December 2012. It is the first regional legal instrument in the world containing legal obligations

145 Kälin & Schrepfer (2012:42).

146 (ibid.).

147 (ibid.:58).

148 Benin, Burkina Faso, Central African Republic, Chad, Gabon, Gambia, Guinea-Bissau, Lesotho, Mali, Nigeria, Niger, Sierra Leone, Swaziland, Togo, Uganda and Zambia.

for states with regard to the protection and assistance of Internally Displaced Persons (IDPs). The Kampala Convention defines IDPs as:[149]

> persons or groups of persons who have been forced or obliged to flee or to leave their homes or places of habitual residence, in particular as a result of or in order to avoid the effects of armed conflict, situations of generalized violence, violations of human rights or natural or human-made disasters, and who have not crossed an internationally recognized State border.

The Convention explicitly recognises its relevance for climate change induced displacement, as it is states in Article 5 that "States Parties shall take measures to protect and assist persons who have been internally displaced due to natural or human made disasters, including climate change." However, the Kampala Convention applies to all situations of internal displacement regardless of its causes (Article 15).

### *VI. Climate Change, the Oceans and the Law of the Sea*

The intersection of climate change with the law of the sea cannot be denied. Where the impacts of climate change manifest themselves within the oceans arena sovereignty questions arise and have the potential to manifest themselves in areas beyond national jurisdiction. The oceans cover more than 70% of the earth's surface and play a pivotal role in the climate change debate. On the one hand, the oceans must be seen as victims of climate change. Changes in ocean temperature and heat content, changes in ocean salinity, changes in sea level and biogeochemical changes (ocean acidification in particular) all have severe consequences, not only for marine ecosystems.[150] The last Assessment Report of the IPCC (AR4) projected sea level rise to range from 0.18 to 0.59 m (depending on the scenario) at the end of the 21st century (2090–2099).[151] Primary contributors to global average sea level change are the expansion or contraction of the ocean due to changes in temperature and the transfer of water, particularly from glaciers and ice sheets.

On the other hand, the oceans are also a part of the solution, playing a significant role in effectuating climate change impacts. The oceans are the largest sinks of $CO_2$ as well as the largest heat sinks. The oceans, by inter-

149 Article 1(k) of the Kampala Convention.
150 See Craig (2012:54).
151 IPCC (2007:13).

acting with the atmosphere, create heat circulation and wind and weather patterns, which determine the impacts of climate change on all terrestrial life.[152] The oceans absorb one quarter of human emissions of carbon dioxide annually,[153] acting to slow the rate of climate change.[154]

The law of the sea is faced with considerable challenges regarding the impacts of climate change on the oceans.[155] Fields of international law that come to mind with regard to the effects of climate change on the oceans are international fisheries law and the broader field of marine environmental law. Furthermore, sea level rise and the opening of previously ice-covered ocean areas present navigational rules, the law pertaining to the protection of sensitive polar marine environments, but in particular international law relating to entitlement to maritime zones with a number of challenges.

Besides a large set of international treaties governing various aspects of marine pollution[156] and biodiversity protection,[157] the 1982 United Nations Convention on the Law of the Sea (UNCLOS III) is the main international legal instrument in terms of marine governance.[158] With 165 parties,[159] the Convention is a broadly applicable set of rules defining the rights and responsibilities of nations in their use of the world's oceans and establishing guidelines for the environment and management of marine natural resources. However, it seems that UNCLOS does not provide sufficient rules to resolve the problems related to the effects that climate change has on the oceans.

UNCLOS III provides that states are entitled to four types of maritime zones: the territorial sea (which may not exceed 12 miles in breadth and over which the coastal state is sovereign); the contiguous zone (up to 24 miles in

152 Craig (2012:53).

153 Le Quéré et al. (2010).

154 Freestone (2009:383).

155 For an in-depth discussion see Rayfuse (2012).

156 Such as the 1954 International Convention for the Prevention of Pollution of the Sea by Oil; the 1972 London Convention on the Prevention of Marine Pollution by Dumping of Wastes and other Matter; or the 1973 International Convention for the Prevention of Pollution from Ships (MARPOL).

157 Such as the 1992 United Nations Convention on Biological Diversity.

158 Craig (2012:71).

159 As of 31 January 2013. See http://www.un.org/Depts/los/reference_files/chronological_lists_of_ratifications.htm, last accessed 20 February 2013. The United States have not acceded to the convention. It is argued that accession "would expose the United Sates to international lawsuits (including suits based on U.S. contributions to global climate change) that would harm its environmental, economic and military interests". See Groves (2012).

breadth, in which the state may exercise jurisdiction over customs, immigration and pollution); the exclusive economic zone (up to 200 miles, in which the state has exclusive rights to explore and exploit natural resources, establish artificial structures, conduct scientific research, and protect the marine environment); and the continental shelf (not exceeding 350 miles, in which the state possesses sovereign rights for the purpose of exploring and exploiting the natural resources). Besides processes such as explosions or eruptions, climate change related changes of the oceans with sea level rise leading the way are further causes for shifts in coastal geography, which in turn directly impact maritime entitlements. It is presumably attributable to the lack of sufficient knowledge of climate change at the time when UNCLOS was concluded in 1982, that the convention remains silent on whether baselines for maritime zones are ambulatory (i.e. whether maritime zones shift with the coastline) or fixed.[160]

The threats of climate change and sea level rise present international law with massive legal challenges. Sea level rise rendering small islands uninhabitable is an extreme scenario, which is certainly not applicable to all small island nations. It, however, puts to the fore the effects of climate change on socio-economic conditions and bio-physical resources and many of the challenges with which the law of the sea (and many other fields including refugee law, human rights law, etc.)[161] is confronted in the era of climate change. The options for small island states, which potentially lose statehood and maritime claims due to sea level rise, are increasingly being explored, on paper and in practice. One option to maintaining maritime zones and statehood, which has been suggested, realised and controversially discussed, not only from a legal point of view, is the construction of artificial islands.[162] However, a solution to the legal problems of the consequences of climate change induced sea level rise at international level is not yet in sight.

Lastly, new technology permits companies to exploit oil and gas reserves in the newly accessible continental shelf. Improvements in deep seabed mining technology make it feasible to extract rare earth and other minerals from the ocean floor outside of any nation's jurisdiction. Newly available oil and gas exploration, shipping, tourism and fishing in the Arctic as a result of global warming has a variety of security implications in newly accessible

160 For a detailed discussion see Lisztwan (2011).

161 It has for example been estimated that a one-meter rise in sea levels will affect 145 million people. See Anthoff et al. (2006); Barnett & Webber (2010).

162 See Gagain (2012).

Arctic sea routes as well as in other potentially contested sea lanes, i.e. in the South China Sea and in the Antarctic.

The high seas, one of the four global commons,[163] have to be protected from environmental threats caused by deep-sea mining, overfishing, ocean warming, acidification and pollution. The protection of the high seas in terms of security threats, however, also plays an important role in the international trade arena. The United Nations International Maritime Organization estimates that over 90% of world trade are carried by sea.[164] The global network of merchant ships thus provides one of the most important modes of transportation.[165]

Piracy may have serious implications for the continued economic development of many regions and is becoming a major challenge for international law. International law addresses the issue of piracy particularly in Articles 100–107 and 110 of the UNCLOS. Article 101, UNCLOS provides that:

> piracy consists of any of the following acts:
>
> a) any illegal acts of violence or detention, or any act of depredation, committed for private ends by the crew or the passengers of a private ship or a private aircraft, and directed:
>    (i) on the high seas, against another ship or aircraft, or against persons or property on board such ship or aircraft;
>    (ii) against a ship, aircraft, persons or property in a place outside the jurisdiction of any State;
>
> b) any act of voluntary participation in the operation of a ship or of an aircraft with knowledge of facts making it a pirate ship or aircraft;
>
> c) any act of inciting or of intentionally facilitating an act described in subparagraph (a) or (b).

The welfare of seafarers and the security of navigation and commerce are at risk due to acts of piracy, which may result in the loss of life, physical harm or hostage-taking of seafarers, significant disruptions to commerce and navigation, financial losses to ship-owners, increased insurance premiums and security costs, increased costs to consumers and producers, and damage to

163 The other four being the atmosphere, Antarctica, and outer space.

164 IMO (2011).

165 See Kaluza et al. (2010).

the marine environment. The Division for Ocean Affairs and the Law of the Sea has reported as follows:[166]

> In the first six months of 2012, 206 attacks were reported worldwide, compared with 316 attacks during the same period in 2011. The total number of acts or attempted acts of piracy and armed robbery at sea worldwide, as reported to IMO in 2011, was 544, compared with 489 in 2010.
> At the regional level, in 2011 IMO received 223 incident reports for East Africa; 63 for the Indian Ocean; 28 for the Arabian Sea; 113 for the South China Sea; 22 for the Straits of Malacca and Singapore; 29 for South America and the Caribbean; and 61 for West Africa.

Especially developing countries are increasingly building up their marine military forces to address current threats such as depletion of natural resources and hazards of maritime transport routes by piracy.[167] As continuous economic growth can only be achieved if a safe passage of goods, raw materials and energy is warranted, defence budgets are being increased.

China, for example, who transports 95% of its imports and exports via the oceans, has increased its budget for armament by 216% from 2000 to 2009, with upgrading the submarine fleet as a focus area. India, in its 2007 Maritime Military Strategy, recognises a direct link between national economic development and open sea routes.[168] Brazil's National Strategy of Defence provides that:[169]

> "Sea denial", "sea control" and "power projection" should focus, without defining any hierarchy for the objectives, and according to the circumstances, on the following:
>
> a. Proactive defence of the oil platforms;
> b. Proactive defence of naval and port facilities, archipelagos and oceanic islands located within the Brazilian jurisdictional waters;
> c. Promptness to respond to any threat against sea-lanes of trade, by States, or by non-conventional or criminal forces;

166 See United Nations General Assembly Oceans and the Law of the Sea Report of the Secretary General, 31 August 2012, A/67/79/Add.1, available at http://daccess-dds-ny.un.org/doc/UNDOC/GEN/N12/478/41/PDF/N1247841.pdf?OpenElement, last accessed 01 February 2013.

167 For this and the following observations on maritime armament see Grebe & Schwarz (2011).

168 (ibid.).

169 Available at http://www.defesa.gov.br/projetosweb/estrategia/arquivos/estrategia_defesa_nacional_ingles.pdf, last accessed 29 January 2013.

d. Capacity to join international peacekeeping operations outside of the territory and the Brazilian jurisdictional waters, under the aegis of the United Nations or other multilateral organizations in the region.

South Africa, as one example of a developing nation on the African continent, and considered to be the most powerful nation on the continent in military terms,[170] also considers its navy to be an important tool to secure free and safe passage for trade vessels and thereby to contribute towards regional stability. Approximately 98% of South Africa's international trade moves by sea and the prosperity of the region is highly dependent on the stability and unhindered flow of trade into and out of the region.[171]

In this context is noteworthy that Africa is now taking legal action "to liberate African coastal waters from age-old foreign dominance, and take a significant step towards a more unified continent"[172] and thus taking another significant step away from the remains of colonialism. The African Union has come up with an African Maritime Transport Charter (which still has to come into force)[173] and is about to conclude plans to establish an African Cabotage Regime, which will only allow African vessels to move cargo along the coast of the continent and prevent non-African mother vessels in African waters from using smaller vessels to move products back and forth in African waters. The aim is to support the African shipping industry by only allowing African owned vessels to trade along Africa's coast.

## *VII. Climate Change and World Trade Law*

The international trade regime under the World Trade Organisation (WTO) is also strongly related to the international climate change regime. In fact, both regimes recognise that climate change may provide opportunities as well as challenges for international development. The WTO is a remarkable example of institutional evolution and its dispute settlement system is as effective as it is impartial. However, similar to the international climate change negotiations, the so-called Doha Development Round of multilateral

170 Flemes & Costa Vaz (2011:16).

171 According to the website of the South African Navy at http://www.navy.mil.za/aboutus/role/page2.htm#01, last accessed 29 January 2013.

172 Ezeanya (2013).

173 Available at http://www.au.int/en/content/revised-african-maritime-transport-charter, last accessed 28 January 2013.

trade negotiations have been complex and without success so far. Both negotiation processes seem to be lacking the necessary consensus of the parties involved. The only difference between the two negotiation processes lies in the fact that "the climate doesn't have time for a Doha-like approach".[174] Unfortunately, after more than 10 years of repeated negotiation failures, the Doha Development Round is unlikely to be concluded in the near future. Some even contend that the "WTO risks its future by keeping Doha alive".[175]

With regard to the persistence of global poverty and socio-economic inequalities, international trade rules often allow affluent countries to continue to protect their markets – with tariffs, quotas, anti-dumping duties, export credits and huge subsidies to domestic producers – at the expense of potential agricultural and textile exports from developing countries, for example.[176] International trade should therefore be considered as a means to an end, but not as the end in itself. An effective international trade regime must first and foremost be friendly to the environment, poverty reduction and sustainable development.[177] The increasing awareness about the negative effects of climate change and the continuing communication among international institutions as well as the public dialogue necessarily lead to the rethinking and eventually to the adjustment of traditional frameworks. These also lead to fruitful discussions, for example, on new trade and climate change related measures, such as carbon labelling or similar standards or regulations on the imposition of border carbon adjustments, which impose border taxes on the embodied carbon of imported goods, set at the level of equivalent domestic taxes.[178]

In the light of the fact that the global village, with international trade as a foundation, has become a reality, it is commendable that the 'trade versus environment' debate has shifted towards the concept of mutual supportiveness between trade and environment or trade and climate change respectively, even though it might – at first glance – appear to be a forced marriage.[179]

174 Houser (2010).
175 See Miles (2011).
176 Pogge (2010:534).
177 Ruppel & Ruppel-Schlichting (2012:46).
178 Ruppel & Ruppel-Schlichting (2012).
179 (ibid.).

Again, world trade law "can both constrain and enable climate action".[180] World trade law has the potential to promote community goals, namely the enhancement of economic development.[181] A closer look at world trade law, however, –[182]

> sadly shows that accordingly solidarity is poorly implemented. The flaw is not in WTO law itself: WTO law allows developed countries to act in favour of developing countries. But developed countries can choose not to implement relevant exceptions and too often implement them poorly.

Moreover, both the policy-making and academic communities have been focusing on the role of the WTO.[183] There has been much discussion about the ways in which the WTO exerts a negative influence on climate law and policy. This includes its potential 'chilling' effect on the climate treaties, referring to the fact that parties to the climate regime have refrained from adopting multilateral trade measures – for instance, against non-compliers or non-parties.[184] While WTO law may thus seem to constrain climate ambitions, attention has increasingly shifted to ways that the organisation might contribute to climate change mitigation. One of these options is pursuing the reduction of fossil fuel subsidies,[185] as called for by the G20 in 2010.[186]

With the aim to achieve a global agreement to tackle aviation emissions, the European Union (EU) has since the beginning of 2012 included emissions from international aviation into the EU Emission Trading System (EU ETS), which applies to EU and non-EU airlines alike.[187] The recent independent action by the EU on international aviation emissions[188] has given rise to a boiling international dispute whereby the EU has been accused of

---

180 Moncel & van Asselt (2012:169).

181 Wolfrum (2006:1097).

182 Hestermeyer (2012:57).

183 See for example Doelle (2004); Hufbauer et al. (2009); Epps & Green (2010); Zelli & van Asselt (2010:79).

184 See Eckersley (2004:24).

185 Green (2006:381); Bigdeli (2008:78).

186 Paragraph 24 of the Pittsburgh Summit Declaration, available at http://www.g20.utoronto.ca/2009/2009communique0925.html, last accessed 17 February 2013.

187 In November 2012, however, the European Commission has proposed deferring the application of the scheme to flights into and out of Europe until after the ICAO General Assembly in autumn 2013 as a gesture of goodwill in support of an international solution.

188 Kulovesi (2012).

using unilateral trade measures and exercising extraterritorial jurisdiction in violation of international law,[189] and failing to adequately reflect the principle of common but differentiated responsibilities and respective capabilities in the design of its aviation scheme.[190]

Similar opposition is to be expected if the EU applies measures to emissions from international shipping. These are estimated to be responsible for 2.7% of the global $CO_2$ emissions in 2007.[191] Since the International Maritime Organisation (IMO) is struggling to agree upon global action on measures such as a levy on $CO_2$ emissions or a cap-and-trade scheme for curbing emissions from shipping, the European Commission is considering to including maritime transport emissions in the EU's greenhouse gas reduction commitment.[192] It becomes clear that powerful states can turn to unilateralism when they decide that they may achieve their foreign policy goals by unilateral action rather than by cooperation.[193] This in turn reflects that the international system is still characterised "by gross inequalities in power".[194]

While the question of response measures remains sensitive in UNFCCC negotiations, the forum could provide for a multilateral dialogue to examine the implications of unilateral climate action designed to promote the ultimate objective of the UNFCCC. In some cases, the WTO dispute settlement mechanism could also enter the scene if the measure in question falls under WTO Agreements:[195]

> In all cases, however, the focus should shift from the relatively simplistic choice between multilateral action, unilateral action or no action[196] towards exploring ways in which interaction between a plural mix of legal regimes and jurisdictions in a global context can best serve the ultimate objective of the UNFCCC to avoid dangerous anthropogenic climate change.

---

189 For an overview of legal arguments in this regard, see Kulovesi (2011:535).

190 Scott & Rajamani (2012:469).

191 See http://www.imo.org/MediaCentre/resources/Pages/Greenhouse%20gas%20emissions.aspx, last accessed 05 February 2013.

192 See http://ec.europa.eu/clima/policies/transport/shipping/index_en.htm, last accessed 05 February 2013.

193 Delbrück (2012:15).

194 Schreuer (2001:177).

195 Kulovesi (2012).

196 Similarly see Morgera (2012).

Thus, more international cooperation in economic areas is necessary in order to ensure more coherence and global welfare.[197] As stated by Delbrück, –[198]

> [I]t is not surprising that given the broad scope of subjects covered by international economic law in general and the law of the WTO in particular – cooperation in these fields show the variety of modes and mechanisms to implement obligations to cooperate.

After all, while world trade has – no doubt – contributed significantly to greenhouse gas emissions, it also offers a variety of options in terms of new technologies and services, which will be crucial in mitigating further climate change.

Lastly, climate induced migration on the scale that is expected is not unlikely to have serious repercussions socially, economically and politically. In this sense, it is worth examining the implications such displacement may have for international trade.[199] Some authors have started to approach international trade from an anti-capitalist perspective, linking trade to migration by arguing that the multilateral economic system is a capitalist one, whereby strong capitalist interests are protected by regulatory regimes such as the World Trade Organisation (WTO) to continue exploiting the ecosystem in an unsustainable way in pursuit of profit. The environmental damage, in turn, leads to the displacement of people who are forced to migrate by the lack of resources and the basics for survival.[200]

## D. *The Future We Want?*

From the aforementioned it becomes clear that the existing regimes and intersections of law may not yet suffice to assure the best possible outcomes for future generations. This, among other things, was addressed at the 2012 Rio+20 Conference on Sustainable Development, which was the biggest UN conference ever. The conference should have been a major step forward in achieving a sustainable future – the future we want. This, however, did not happen due to a number of reasons, so that "the future we want" still needs further political attention and action.

197 Tietje (2001).
198 Delbrück (2012:9).
199 Leal-Arcas (2012).
200 (ibid.) with further references; Westra (2009); Stokke (2005).

The 2012 Club of Rome Report entitled “2052 – A Global Forecast for the Next Forty Years”,[201] addresses several global goals as essential for the transition towards a sustainable, equitable and ‘happier’ world. Some of these global goals are also most relevant to the challenges of the Anthropocene: The report argues that societal values are essential for a sustainable and equitable society and that they must be fully reflected in all economic decisions. It further contends that a more equitable distribution of income both within and between countries is required. Moreover it holds that the ecology must be seen as a binding constraint for all forms of human activity and should therefore be managed in a manner which reflects its biophysical and economic value. Never should the world be in overshoot. Appropriate governance systems at a local, national and global level must be established to manage the transition into an equitable and sustainable global world.[202]

In the light of the aforementioned the following sections will reflect in more detail on the way forward and make some recommendations for the future we (may or may not) want.

## *I. Economic Development, Regional Integration and the Reduction of Poverty*

The furtherance of economic development, regional integration, and the reduction of poverty go hand in hand.[203] This interrelationship has become closer over the past few years due to increasing discussions in the world community on the issue, especially in the context of climate change. Yet, many regional integration processes around the world still face obstacles and challenges.[204] The fear of losing state autonomy, the fear of losing national identity, socio-economic disparity among members, historical disagreement, lack of vision, and unwillingness to share resources are some of the obstacles that present themselves with regard to regional integration.[205] Regional integration provides an –[206]

201 Club of Rome (2012).
202 (ibid.).
203 This section is largely based on Ruppel (2012).
204 See Ruppel (2009b).
205 Ruppel (2009a).
206 Ruppel & Ruppel-Schlichting (2012:41).

opportunity to enhance political stability by establishing regional organisations, which play an increasing role (not only in the facilitation of trade but also) in defusing conflicts within and between countries and in promoting human rights. In terms of climate change related matters, such organisations are of the utmost relevance, especially when it comes to climate change related disaster management and environmentally induced migration. In this context, regional integration may serve as a tool to maintain political stability by building trust, enhancing understanding between groups and deepening interdependence.

The triumph of market mechanisms has accelerated the process of globalisation. After the collapse of the competition between market-driven and state-commanded economies, developing countries seem to have only one option to follow for modernisation and development. Liberal democracy does not seem to have any serious competitors. Given this monolithic economic and political framework, it is not an easy task to determine where sustainable economic development actually fits in.[207] The same applies to the question regarding the relation between market, development and well-being, and the influence economic development can play on the alleviation of poverty in view of the fact that economic development is not always concomitant with greater welfare of the average individual, as the growth of the gross national product (GNP) is not a sufficient indicator with which to measure the level of security and the quality of life of people.

After all, it is a sad reality that about half of all human beings still live in severe poverty and about a quarter live in extreme or life-threatening poverty.[208] One major reason why poverty is still so prominent today is that "affluent societies are not merely helping too little, but also harming too much."[209] The principle of common but differentiated responsibilities, one of the cornerstones of the international climate change regime, explicitly referred to in the UNFCCC and the Kyoto Protocol, is meant to address this disparity. The trans-boundary nature of climate change action and impacts have environmental and developmental repercussions for all countries. The differentiation of responsibilities, however, should support even greater efforts in future,[210] especially in view of the on-going "disparity between the human and the economic magnitude of world poverty" and "the enormous extent of economic inequality in the world today".[211] About 60% of the

207 Pillay (2009).
208 Pogge (2011:20).
209 Pogge (2004:1759).
210 Garibaldi et al. (2012).
211 Pogge (2010:528).

world's population holds less than 2% of global wealth, in contrast to the top 1% of the world's population, who hold 40% of global wealth.[212] "Because of these enormous inequalities, we are now at the point where the world is easily rich enough in aggregate to abolish all poverty. We are simply choosing to prioritize other ends instead."[213] Sustainable economic development therefore depends on equity:

> In the analysis of the causes of and solutions to climate change, the quality of the equity commons and the governance rules that protect and enhance it are key elements in crafting a viable international agreement on future emissions allocation and burden-sharing of emissions mitigation and climate adaptation costs. More broadly, equity – together with so many of the public goods that provide the foundation for sustainable development – is vulnerable. Deliberate policies in favour of increasing equity over time not only improve social welfare, but also act to shore up the foundations for the equity commons of the future, by establishing and strengthening rules for its governance.[214]

### *II. Cooperative Global Climate Governance*

Although the problem of climate change is rather clear, political solutions are often far and unfair. The international community seems unable to come up with agreements that both remedy the substantive causes of climate change and the damage caused by it. An agreement that is optimal for the world and its future generations may not be optimal for some national economies, which would probably have to bear a large burden for significant domestic emissions reductions and which are not among the nation's most gravely affected and threatened by climate change. The key remaining question is how responsibility for global climate protection can be shared more equitably in future. In order for that to happen more effective and equitable legal and policy responses need to be implemented.

We live in an increasingly interconnected and interdependent world. It is a world bound together, not just by state interests, but also – and especially in the context of climate change – by an interest in more global coopera-

212 Davies et al. (2006).
213 Pogge (2010:528).
214 Stanton (2012).

tion.[215] It should thus be "in the interests of all States … to uphold the rule of law in the world."[216]

Yet, it would be irrational to accept more powerful organs of world government, without a certain decrease of national government's power.[217] According to a minimal definition of cooperation the term could mean that states are to enter into contact with each other.[218] It could further be argued that under general international law states are under an obligation to cooperate,[219] an effort for instance "to accomplish an object by joint action, where the activity of a single state cannot achieve the same result".[220] Areas where international cooperation is essential include the international protection of human rights, the duty to cooperate in international economic law and related areas, and the duty to cooperate in international dispute settlement.[221]

On the one hand international duties to cooperate are based on treaties made by the sovereign states, which leaves it in their discretion whether they adhere to that treaty or not.[222] On the other hand one can also argue that cooperation by states actually is "the most important manifestation of sovereignty",[223] rather than – "as was assumed in earlier times – an obstacle to international cooperation".[224] However, it must "be admitted that the hard law obligations to cooperate share the fate of other binding rules of international law, i.e. that some States still prefer not to comply" with them.[225]

In the analysis of the causes of and solutions to climate change more "[d]eliberate policies in favour of increasing equity over time [would] not only improve social welfare, but would also act to shore up the foundations for the equity commons of the future, by establishing and strengthening rules for its governance".[226] Yet, in the development of international law it is so far "precipitate to consider solidarity as a legally binding principle for all in

215 Koh (2012).
216 (ibid.:1237).
217 Pogge (1987:430).
218 Delbrück (2012:4).
219 (ibid.:5).
220 Wolfrum (1995:1242).
221 See Delbrück (2001).
222 Delbrück (2012:13).
223 See Schreuer (2001:179).
224 Delbrück (2012:14).
225 (ibid.).
226 Stanton (2012:407).

international law. All too often its content is too uncertain for it to work as an applicable legal norm."[227]

According to the UN General Assembly's definition of solidarity in the UN Millennium Declaration "[g]lobal challenges must be managed in a way that distributes the costs and burdens fairly in accordance with basic principles of equity and social justice. Those who suffer or who benefit least deserve help from those who benefit most."[228] Common but differentiated responsibilities as stipulated in Principle 7 of the Rio Declaration states: "States shall cooperate in a spirit of global partnership to conserve, protect and restore the health and integrity of the earth's ecosystem. In view of the different contributions to global environmental degradation, States have common but differentiated responsibilities."[229]

International law as a value-based order should go beyond mere coexistence and involving the commonly shared interests of the international community.[230] Solidarity has long been invoked as a strong moral claim but it is more and more considered to be a "value reflected in international law".[231] Solidarity involves three different, not necessarily cumulative aspects:[232] "The achievement of common objectives through common action of States, the achievement of common objectives through differentiated obligations of States and actions to benefit particular States".[233]

Yet, from the above it becomes clear that several independent international legal regimes exist, which are relevant in one way or another in the context of climate change. There are intersections between these regimes although they are fragmented. On the one hand such fragmentation and regulatory diversity may well be beneficial if the intersections of law are orchestrated in an innovative manner. On the other hand it is argued here that the law (at least as it exists today) is not enough to effectively address the challenges that accompany climate change. While there are some regimes dedicated exclusively to climate change (such as the UNFCCC), others impact deeply on climate change, yet have a primary focus dealing with quite

227 Hestermeyer (2012:48).
228 UNGA Res. 55/2 para. 6 (adopted without vote).
229 United Nations Conference on Environment and Development, Rio Declaration on Environment and Development, 14 July 1992, 31 ILM 874.
230 Wolfrum (1984).
231 Hestermeyer (2012:63).
232 Seibert-Fohr (2012).
233 Wolfrum (2010).

different subjects (human rights, world trade, the oceans framework etc.). Dealing with climate change involves creating a coherent and orchestrated international regime, a set of arrangements among states and other stakeholders designed to solve a global problem that cannot be solved by individual nation-states. While the existing international regimes rest largely on intergovernmental agreement, one dealing with climate change will have to go far beyond the capacity of governments and will need support from non-state actors as well, creating a multi-stakeholder regime.[234] For local and national action to be effective, such a global regime should aim at cooperation and solidarity, and be supportive and well designed.

Human activities seem to be moving several of the Earth's sub-systems outside the range of natural variability typical for the previous 500,000 years.[235] Human societies therefore must now change course and steer away from critical tipping points in the Earth system that might lead to rapid and irreversible change.[236] According to Biermann et al. –[237]

> [t]his requires fundamental reorientation and restructuring of national and international institutions toward more effective Earth system governance and planetary stewardship .... The world saw a major transformative shift in governance after 1945 that led to the establishment of the UN and numerous other international organizations, along with far-reaching new international legal norms on human rights and economic cooperation. We need similar changes today, a 'constitutional moment' in world politics and global governance.

At the same time international law and global governance will require more empowerment of international judicial institutions that learn to integrate inter-disciplinary tools to accommodate the inter-linkages between legal and institutional reforms and climate change policy.

## *E. Conclusion*

To conclude with a statement made by UN Secretary-General Ban Ki-moon in February 2013: "We live in an age of monumental transition – economic, demographic, political. Global interdependence is deepening. Transnational

234 Mathiason & Bhandari (2010).
235 Steffen et al. (2004); Schellnhuber et al. (2004).
236 Rockström et al. (2009).
237 Biermann et al. (2012).

threats are growing. This means we must make better use of the United Nations machinery."[238] So far, however, –[239]

> the climate change regime complex is a loosely coupled system of institutions; it has no clear hierarchy or core, yet many of its elements are linked in complementary ways. It occupies neither extreme. Instead, it is a regime complex whose elements are loosely linked to one another, between the poles of integration and fragmentation.

In the threatening context of climate change this can be interpreted as a failure of the system: More coherent, cooperative, collective action is needed to address climate change. The piecemeal, fragmentary approach to both understanding and addressing the issue of climate change is unsatisfactory. Humanity has the opportunities, tools, science, technology and insight to deal with climate change and to move into a better world. Whether we manage to do so will depend on improved mechanisms of international law and governance. The failure to bring international relations under the rule of law through the absence of more effective central mechanisms of adjudication and/or enforcement explains the pervasive ambiguity of international law.[240] In fact, what is missing is more world government, a strengthening of the central organs of the United Nations, for example, that would make it more likely that international law will be applied and enforced.[241]

Legitimate voices[242] have been aired regarding the need of a specialised international judicial body to hear and determine trans-boundary environmental matters and to provide greater coherence to the fragmented global climate governance regime. Such a judicial body could provide interpretive guidance and judicial support, which in turn would – no doubt – also be of benefit when combating climate change. It could thus contribute to coordination of the intersections of law, to legal harmonisation and to a complementation of existing fragmented climate relevant regimes.

Such a judicial body would also be in line with Article 14 of the UNFCCC (dispute resolution) and particularly Article 33(1) of the UN Charter:

> The parties to any dispute, the continuance of which is likely to endanger the maintenance of international peace and security, shall, first of all, seek a solution by negotiation, enquiry, mediation, conciliation, arbitration, judicial settlement,

238 Ki-moon (2013).
239 Keohane & Victor (2011).
240 Pogge (1987:426).
241 (ibid.:427).
242 Hockman (2010:215).

> resort to regional agencies or arrangements, or other peaceful means of their own choice.

It could resolve conflicting international law obligations and overlapping mandates of the global climate governance structures; create a model for compliance and enforcement to encourage national protection standards; and promote greater accountability and access to justice.

Lastly, existing intersections of law and more cooperative global climate governance can "develop an unforeseen dynamism, in particular if … endowed with institutions of a norm-setting and also of a judicial character".[243] However, the law only enfolds "effective force from the underlying political consensus. Without such consensus, legal devices, no matter how scrupulously they have been thought out, may be swept away by the ground forces active in international society".[244]

The threats of the very existence of humanity are obvious: In this respect, in no area of law should the common interests of mankind be clearer than when addressing climate change and the challenges in the Anthropocene. With this in mind one should reasonably think that it is possible to identify and agree upon the necessary reforms in response to the changing climate and for the survival of mankind.

## *References*

AGECC/United Nations Secretary-General's Advisory Group on Energy And Climate Change, 2012, *Energy for a Sustainable Future*, New York, AGECC, available at, http://www.un.org/wcm/webdav/site/climatechange/shared/Documents/AGECC %20summary%20report%5B1%5D.pdf, last accessed 30 May 2013.

Ahmed, Syud A., Noah S. Diffenbaugh & Thomas W. Hertel, 2009, Climate Volatility Deepens Poverty Vulnerability in Developing Countries, *Environmental Research Letters* 4 (3), 1–8.

AMCEN/African Ministerial Conference on the Environment, 2011, *Addressing Climate Change Challenges in Africa: A Practical Guide Towards Sustainable Development*, available at www.unep.org/roa/amcen/docs/publications/guidebook_Climate-Change.pdf, last accessed 5 August 2012.

243 Tomuschat (2012:1285).
244 (ibid.).

Anthoff, David, Robert Nicholls, Richard Tol & Athanasios Vafeidis, 2006, Global and Regional Exposure to Large Rises in Sea-Level: a Sensitivity Analysis, Tyndall Centre Working Paper No. 90, available at http://www.tyndall.ac.uk/sites/default/files/wp96_0.pdf, last accessed 30 May 2013.

Barnett, John & Neil Adger, 2007, Climate Change, Human Security and Violent Conflict, *Political Geography* 26, 639–655.

Barnett, John & Michael Webber, 2010a, Accommodating Migration to Promote Adaptation to Climate Change, Background Paper to the 2010 World Development Report, World Bank Policy Research Working Paper 5270, available at https://openknowledge.worldbank.org/bitstream/handle/10986/3757/WPS5270.txt?sequence=2, last accessed 30 May 2013.

Barnett, John & Michael Webber, 2010b, Migration as Adaptation: Opportunities and Limits, in: McAdam, Jane (Ed.) *Climate Change and Displacement: Multidisciplinary Perspectives*, Oxford, Hart Publishing, 37–56.

Benedict XVI, Supreme Pontiff of the Catholic Church, 2011, The Listening Heart – Reflections on the Foundations of Law, Address to the German Bundestag, Apostolic Journey to Germany 22–25 September 2011, visit to the Bundestag, 22 September 2011, available at http://www.vatican.va/holy_father/benedict_xvi/speeches/2011/september/documents/hf_ben-xvi_spe_20110922_reichstag-berlin_en.html, last accessed 05 June 2013.

Biermann, Frank, Kenneth Abbott, Steinar Andresen, Karin Bäckstrand, Steven Bernstein, Michele M. Betsill, Harriet Bulkeley, Benjamin Cashore, Jennifer Clapp, Carl Folke, Aarti Gupta, Joyeeta Gupta, Peter M. Haas, Andrew Jordan, Norichika Kanie, Tatiana Kluvánková-Oravská, Louis Lebel, Diana Liverman, James Meadowcroft, Ronald B. Mitchell, Peter Newell, Sebastian Oberthür, Lennart Olsson, Phillip Pattberg, Roberto Sánchez-Rodríguez, Heike Schroeder, Arild Underdal, Susana Camargo Vieira, Coleen Vogel, Oran R. Young, Andrea Brock & Ruben Zondervan, 2012, Navigating the Anthropocene: Improving Earth System Governance, *Science* 335 (6074), 1306–1307.

Bigdeli, Sadeq Z., 2008, Will the 'Friends of Climate' Emerge in the WTO? The Prospects of Applying the 'Fisheries Subsidies' Model to Energy Subsidies, *Carbon and Climate Law Review* 2 (1), 78–88.

Black, Richard, W. Neil Adger, Nigel W. Arnell, Stefan Dercon, Andrew Geddes & David Thomas, 2011, The Effect of Environmental Change on Human Migration, *Global Environmental Change* 21, S3–S11.

Boisson de Chazourne, Laurence, 2008, United Nations Framework Convention on Climate Change, United Nations Audio Visual Library of International Law, available at http://untreaty.un.org/cod/avl/pdf/ha/ccc/ccc_e.pdf, last accessed 30 May 2013.

Brunnée, Jutta, Silke Goldberg, Richard Lord QC & Lavanya Rajamani, 2012, Overview of Legal Issues relevant to Climate Change, in: Lord, Richard, Silke Goldberg, Lavanya Rajamani & Jutta Brunnée, *Climate Change Liability*, Cambridge, Cambridge University Press.

Cameron, Edward & Marc Limon, 2012, Restoring the Climate by Realizing Rights: The Role of the International Human Rights System, *Review of European Community and International Environmental Law* 21 (3), 204–219.

Club of Rome, 2012, *2052 – A Global Forecast for the Next Forty Years*, White River Jct., VT, Chelsea Green, some of the findings available at http://www.clubofrome.org/?p=703, last accessed 28 May 2013.

Costello, Anthony, Mustafa Abbas, Adriana Allan, Sarah Ball, Sarah Bell, Richard Bellamy, Sharon Friel, Nora Groce, Anne Johnson, Maria Kett, Maria Lee, Caren Levy, Mark Maslin, David McCoy, Bill McGuire, Hugh Montgomery, David Napier, Christina Pagel, Jiresh Patel, Jose de Oliveira, Nanneke Redclift, Hannah Rees, Daniel Rogger, Joanne Scott, Judith Stephenson, John Twigg, Jonathan Wolff & Craig Patterson, 2009, Managing the Health Effects of Climate Change, *Lancet* 373 (9676), 1693–1733.

Craig, Robin K., 2012, *Comparative Ocean Governance – Place-Based Protections in an Era of Climate Change*, Chaltenham, Edward Elgar Publishing.

Crutzen, Paul J., & Eugene F. Stoermer, 2000, The "Anthropocene", Global Change Newsletter 41, 12–13.

Cuevas, Luis E., Isabelle Jeanne, Anna Molesworth, Michael Bell, Emily C. Savory, Stephen J. Connor & Madeleine C. Thomson, 2007, Risk Mapping and Early Warning Systems for the Control of Meningitis in Africa, *Vaccine* 25, A12–A17.

Davies, James B., Susanna Sandstrom, Anthony Shorrocks & Edward N. Wolff, 2006, The World Distribution of Household Wealth, available at http://www.iariw.org/papers/2006/davies.pdf, last accessed 30 May 2013.

Davis, Kevin E., Benedict Kingsbury, Sally Engle Merry, 2012, Indicators as a Technology of Global Governance, *Law & Society Review* 46 (1), 71–104.

De Magny, Guillaume C., Jean F. Guégan, Michael Petit & Bernard Cazelles, 2007, Regional-scale Climate-variability Synchrony of Cholera Epidemics in West Africa, *BMC Infectous Diseases* 7, 20.

Delbrück, Jost (Ed.), 2001, *International Law of Cooperation and State Sovereignty, Proceedings of an International Symposium of the Kiel Walther Schücking Institute of International Law*, Berlin, Duncker & Humblot.

Delbrück, Jost, 2012, "The International Obligation to Cooperate – An Empty Shell or a Hard Law Principle of International Law? – A Critical Look at a Much Debated Paradigm of Modern International Law, in: Hestermeyer, Holger P., Doris König, Nele Matz-Lück, Volker Röben, Anja Seibert-Fohr, Peter-Tobias Stoll & Silja Vöneki (Eds), *Coexistence, Cooperation and Solidarity: Liber Amicorum Rüdiger Wolfrum*, Leiden, Martinus Nijhoff Publishers, 3–16.

Depledge, Joanna & Farhana Yamin, 2009, The Global Climate Regime: A Defence, in: Helm, Dieter & Cameron Hepburn (Eds) *The Economics and Politics of Climate Change*, Oxford, Oxford University Press, 433–453.

Doelle, Meinhard, 2004, Climate Change and the WTO: Opportunities to Motivate State Action on Climate Change through the World Trade Organization, *Review of European Community and International Environmental Law (RECIEL)* 13 (1), 85–103.

Eckersley, Robin, 2004, The Big Chill: The WTO and Multilateral Environmental Agreements, *Global Environmental Politics* 4 (2), 24–50.

Epps, Tracey & Andrew Green, 2010, *Reconciling Trade and Climate: How the WTO can Help Address Climate Change*, Cheltenham, Edward Elgar Publishing.

Evans, Gareth, 2004, Rethinking Collective Action: The Responsibility to Protect and a Duty to Prevent – The Policy Challenge, *American Society of International Law Proceedings* 98, 78–89.

Ezeanya Chika, 2013, The Need for a Sea Change in Africa's Coastal Laws, *Think Africa Press*, 22 January 2013, available at http://thinkafricapress.com/economy/search-independence-african-coastal-waters, last accessed 28 January 2013.

FAO/Food and Agriculture Organization of the United Nations, World Food Programme & International Fund for Agricultural Development, 2012, *The State of Food Insecurity in the World 2012*, Rome, FAO, available at http://www.fao.org/publications/sofi/en/, last accessed 30 May 2013.

Flemes, Daniel & Alcides Costa Vaz, 2011, Security Policies of India, Brazil and South Africa – Regional Security Contexts as Constraints for a Common Agenda, GIGA Working Papers No. 160, available at http://www.giga-hamburg.de/dl/download.php?d=/content/publikationen/pdf/wp160_flemes-vaz.pdf, last accessed 29 January 2013.

Foresight, 2011, *Migration and Global Environmental Change, Final Project Report*, London, The Government Office for Science.

Freestone, David, 2009, Climate Change and the Oceans, *Carbon and Climate Law Review* 4, 383–386.

Gagain, Michael, 2012, Climate Change, Sea Level Rise, and Artificial Islands: Saving the Maledives' Statehood and Maritime Claims Through the 'Constitution of the Oceans', *Colorado Journal of International Environmental Law and Policy* 23 (1), 77–120.

Gardiner, Stephen M., 2004, Ethics and Global Climate Change, *Ethics* 114, 555–600.

Garibaldi José Alberto, Mónica Araya & Guy Edwards, 2012, Shaping the Durban Platform: Latin America and the Caribbean in a future High Ambition Deal, Policy Brief, available at http://www.intercambioclimatico.com/wp-content/uploads/Shaping-Durban-Platform_Final-April_2012.pdf, last accessed 17 February 2013.

Grebe, Jan & Christoph Schwarz, 2011, Die maritime Aufrüstung der Schwellenländer. Strategische und friedenspolitische Implikationen, in: Johannsen, Margaret, Bruno Schoch, Corina Hauswedell Tobias Debiel & Christiane Fröhlich (Eds), *Friedensgutachten 2011*, Berlin, LIT Verlag, 316–327.

Green, Andrew, 2006, Trade Rules and Climate Change Subsidies *World Trade Review* 5 (3), 377–414.

Groves, Steve, 2012, Accession to the U.N. Convention on the Law of the Sea Would Expose the U.S. to Baseless Climate Change Lawsuits, Backgrounder No. 2660, available at http://report.heritage.org/bg2660, last accessed 08 February 2013.

Hestermeyer, Holger P. 2012, Reality or Aspiration? – Solidarity in International Environmental and World Trade Law, in: Hestermeyer, Holger P., Doris König, Nele Matz-Lück, Volker Röben, Anja Seibert-Fohr, Peter-Tobias Stoll & Silja Vöneki (Eds), *Coexistence, Cooperation and Solidarity: Liber Amicorum Rüdiger Wolfrum*, Leiden, Martinus Nijhoff Publishers, 45–63.

Hockman, Stephen, 2010, The Case for an International Court for the Environment, *Journal of Court Innovation* 3 (1), 215–230.

Hodgkinson David & Lucy Young, 2012, 'In the Face of Looming Catastrophe': A Convention for Climate Change Displaced Persons, available at http://www.ccdpconvention.com/documents/Climate%20change%20displacement%20treaty%20proposal.pdf, last accessed 30 May 2013.

Hope Sr., Kempe R., 2009, Climate Change and Poverty in Africa, *International Journal of Sustainable Development & World Ecology* 16 (6), 451–461.

Houser, Trevor, 2010, Copenhagen, the Accord, and the Way Forward, Policy Brief 10 (5) Peterson Institute for International Economics, available at http://www.piie.com/publications/interstitial.cfm?ResearchID=1508, last accessed 17 February 2013.

Hufbauer, Gary C., Steve Charnovitz & Jisun Kim, 2009, *Global Warming and the World Trading System*, Washington, Peterson Institute for International Economics.

ICHRP/International Council on Human Rights Policy, 2008, *Climate Change and Human Rights: A Rough Guide*, Versoix, ICHRP, available at http://www.ichrp.org/files/reports/45/136_report.pdf, last accessed 30 May 2013.

IDMC/Internal Displacement Monitoring Centre, 2011, *Displacement Due to Natural Hazard-induced Disasters, Global Estimates for 2009-2010*, Geneva, IDMC.

IMO/International Maritime Organization, 2011, IMO and the Environment, available at http://www.imo.org/ourwork/environment/documents/imo%20and%20the%20environment%202011.pdf, last accessed 01 February 2013.

InterAcademy Council, 2013, IAC in Brief, available at http://www.interacademycouncil.net/23450/27799.aspx, last accessed 17 February 2013.

IPCC/Intergovernmental Panel on Climate Change, 2001, Tenth Anniversary Brochure, available at http://www.ipcc.ch/pdf/10th-anniversary/anniversary-brochure.pdf, last accessed 17 February 2013.

IPCC/Intergovernmental Panel on Climate Change, 2007, Summary for Policymakers, in: Susan Solomon, Dahe Qin, Martin Manning, Zhenlin Chen, Melinda Marquis, Kristen B. Averyt, Melinda Tignor & Henry L. Miller (Eds), *Climate Change 2007: The Physical Science Basis. Contribution of Working Group I to the Fourth Assessment Report of the Intergovernmental Panel on Climate Change*, Cambridge, Cambridge University Press, 1–18.

IPCC/Intergovernmental Panel on Climate Change, 2012, *Managing the Risks of Extreme Events and Disasters to Advance Climate Change Adaptation. A Special Report of Working Groups I and II of the Intergovernmental Panel on Climate Change*, Field, Christopher B., Vicente Barros, Thomas F. Stocker, Dahe Qin, David J. Dokken, Kristie L. Ebi, Michael D. Mastrandrea, Katharine J. Mach, Gian-Kasper Plattner, Simon K. Allen, Melinda Tignor & Pauline M. Midgley (Eds), Cambridge, Cambridge University Press.

Jansen, Cassie C. & Nigel W. Beebe, 2010, The Dengue Vector Aedes Aegypti: What Comes Next, *Microbes Infect* 12, 272–279.

Kälin, Walter & Nina Schrepfer, 2012, Protecting People Crossing Borders in the Context of Climate Change. Normative Gaps and Possible Approaches, Legal and Protection Policy Research Series, available at http://www.unhcr.org/protect, last accessed 19 February 2013.

Kaluza Pablo, Andrea Kölzsch, Michael T. Gastner & Bernd Blasius, 2010, The Complex Network of Global Cargo Ship Movements, *Journal of the Royal Society*, available at http://rsif.royalsocietypublishing.org/content/early/2010/01/19/rsif.2009.0495.full, last accessed 30 May 2013.

Keohane, Robert O. & Dawid G. Victor, 2011, The Regime Complex for Climate Change, *Perspectives on Politics* 9 (1), 7–23.

Ki-moon, Ban, 2013, Remarks at the Sorensen Distinguished Lecture on the United Nations at the Council on Foreign Relations, New York, 11 February 2013, available at http://www.un.org/sg/statements/index.asp?nid=6595, last accessed 15 February 2013.

Kingsbury, Benedict, Nico Krisch & Richard B. Stewart, 2005, The Emergence of Global Administrative Law, *Law and Contemporary Problems* 68 (3–4), 15–61.

Knox, John H., 2009, Climate Change and Human Rights Law, *Virginia Journal of International Law* 50 (1), 1–59.

Koh, Tommy, 2012, Is there a Role for Law in a World Ruled by Power?, in: Hestermeyer, Holger P., Doris König, Nele Matz-Lück, Volker Röben, Anja Seibert-Fohr, Peter-Tobias Stoll & Silja Vöneki (Eds), *Coexistence, Cooperation and Solidarity: Liber Amicorum Rüdiger Wolfrum*, Leiden, Martinus Nijhoff Publishers, 1235–1239.

Koroma, Abdul D., 2012, Solidarity: Evidence of an Emerging International Legal Principle, in: Hestermeyer, Holger P. Doris König, Nele Matz-Lück, Volker Röben, Anja Seibert-Fohr, Peter-Tobias Stoll & Silja Vöneki (Eds), *Coexistence, Cooperation and Solidarity: Liber Amicorum Rüdiger Wolfrum*, Leiden, Martinus Nijhoff Publishers, 103–129.

Kotir, Julius H., 2010, Climate Change and Variability in Sub-Saharan Africa: A Review of Current and Future Trends and Impacts on Agriculture and Food Security, *Environment, Development and Sustainability* 13 (3), 587–605.

Kowarsch, Martin, 2010, Wie neutral ist die Klimawissenschft? Über die Aktuellen Vorwürfe an den IPCC und die Klimaforschung, available at http://dialog.hfph.mwn.de/Members/m_kowarsch/wie-neutral-ist-die-klimawissenschaft/, last accessed 17 February 2013.

Kulovesi, Kati, 2011, Make Your Own Special Song Even If Nobody Else Sings Along: International Aviation Emissions and the EU Emissions Trading Scheme, *Climate Law* 2 (4), available at http://papers.ssrn.com/sol3/papers.cfm?abstract_id=1952555, last accessed 30 May 2013.

Kulovesi, Kati, 2012, Addressing Sectoral Emission outside the United Nations Framework Convention on Climate Change: What Roles for Multilateralism, Minilateralism and Unilateralism, *Review of European Community and International Environmental Law* 21 (3), 220–230.

Le Quéré, Corinne, Taro Takahashi, Erik T. Buitenhuis, Christian Rödenbeck & Stewart C. Sutherland, 2010, Impact of Climate Change and Variability on the Global Oceanic Sink of $CO_2$, *Global Biogeochemical Cycles* 24 (4), 1–10, available at http://www.imedea.uib-csic.es/master/cambioglobal/Modulo_1_02/Ocean%20Sinks/biblio/LeQuere_GBC_2010_change_CO2.pdf, last accessed 30 May 2013.

Leal-Arcas, Rafael, 2012, On Climate Migration and International Trade, *Vienna Journal of International Constitutional Law* 6 (3 and 4), 410–440.

Lecocq, Franck & Zmorak Shalizi, 2007, How Might Climate Change Affect Economic Growth in Developing Countries? A Review of the Growth Literature with a Climate Lens, The World Bank Development Research Group Sustainable Rural and Urban Development Team Policy, Research Working Paper 4315.

Lisztwan, Julia, 2011, Stability of Marine Boundary Agreements, *The Yale Journal of International Law* 37 (1), 153–200.

Macer, Darryl R., Moss, Jeremy, Michael McMann, Jessica Rae, Andrea Zipprich, Aori Nyambati, Diana Ngo, Ming M. Cheng, N Manohar & Gregor Wolbring, 2011, *Energy Equity and Environmental Security, Ethics and Climate Change in Asia and the Pacific*, Bangkok, UNESCO, available at http://unesdoc.unesco.org/images/0021/002182/218271E.pdf, last accessed 30 May 2013.

Mathiason, John & Medani P. Bhandari, 2010, Getting the Facts Right: The Intergovernmental Panel on Climate Change and the New Climate Regime *Journal of International Organization Studies* 1 (1), 57–71.

McInerney-Lankford, Siobhan, 2009, Climate Change and Human Rights: An Introduction to Legal Issues, *Harvard Environmental Law Review* 33, 431–438.

McKinnon, Catriena C., 2009, Runaway Climate Change: A Justice-based Case for Precautions, *Journal of Social Philosophy* 40 (2), 187–203.

Miles, Tom, 2011, WTO Risks Future by Keeping Doha Alive, *Taipei Times,* 31 December 2011, available at http://www.taipeitimes.com/News/editorials/archives/2011/12/31/2003522031, last accessed 17 February 2013.

Moncel, Remi & Harro van Asselt, 2012, All Hands on Deck! Mobilizing Climate Change Action beyond the UNFCCC, *Review of European Community and International Environmental Law (RECIEL)* 21 (3), 163–176.

Morgera, Elisa, 2012, Bilateralism at the Service of Community Interests? Non-judicial Enforcement of Global Public Goods in the Context of Global Environmental Law" *European Journal of International Law* 23 (3), 743–767.

Nordås, Ragnhild & Nils P. Gleditsch, 2007, Climate Change and Conflict, *Political Geography* 26 (6), 627–638.

OHCHR/Office of the United Nations High Commissioner for Human Rights & UN-HABITAT, 2009, The Right to Adequate Housing, Fact Sheet No. 21, available at http://www.ohchr.org/Documents/Publications/FS21_rev_1_Housing_en.pdf, last accessed 14 February 2013.

Pachauri, Rajendra K., 2009, Speech at the Opening Ceremony of the Scoping Meeting for the Fifth Assessment Report (AR5) of the IPCC, 13 July 2009, available at http://www.ipcc.ch/graphics/speeches/rkp-speech-venice-07-2009.pdf, last accessed 20 August 2010.

Pedersen, Ole W., 2012, The Janus-Head of Human Rights and Climate Change: Adaptation and Mitigation, available at http://www.nccr-climate.unibe.ch/conferences/climate_economics_law/papers/Pedersen_OleW.pdf, last accessed 31 December 2012.

Pillay, Ariranga G., 2009, The Optional Protocol to the International Covenant on Economic, Social and Cultural Rights, in: Roschmann, Christian & Peter Wendoh (Eds) *Conference Report: Economic Rights as Human Rights*, Nairobi, Konrad-Adenauer-Foundation, available at http://www.kas.de/wf/doc/kas_17782-1522-2-30.pdf?091013125641, last accessed 30 May 2013.

Pogge, Thomas W., 1987, International Relations as a Modus Vivendi, *American Proceedings of International Law* 81, 429–456.

Pogge, Thomas W., 2004, The Incoherence Between Rawls's Theories of Justice, *Fordham Law Review* 72, 1739–1759.

Pogge, Thomas W., 2010, Keynote Address: Poverty, Climate Change, and Overpopulation, *Georgia Journal of International and Comparative Law* 38, 525–542.

Pogge, Thomas W., 2011, Are We Violating the Human Rights of the World's Poor? *Yale Human Rights and Development Law Journal* 14 (2), 1–33.

Raleigh, Clionadh, 2010, Political Marginalization, Climate Change, and Conflict in African Sahel States *International Studies Review* 12, 69–86.

Raleigh, Clionadh & Hendrik Urdal, 2007, Climate Change, Environmental Degradation and Armed Conflict, *Political Geography* 26 (6), 674–694.

Rasmussen, Anders Fogh, Speech on Emerging Security Risks, Lloyd's of London, 1 October 2009, available at http://www.nato.int/cps/en/natolive/opinions_57785.htm, last accessed 17 February 2013.

Rayfuse, Rosemary, 2012, Climate Change and the Law of the Sea, in: Rayfuse Rosemary & Shirley V. Scott (Eds), *International Law in the Era of Climate Change*, Cheltenham, Edward Elgar, 147–174.

Rayfuse, Rosemary & Shirley V. Scott, 2012, Mapping the Impact of Climate Change on International Law, in: Rayfuse, Rosemary & Shirley V. Scott (Eds) *International Law in the Era of Climate Change*, Cheltenham, Edward Elgar, 3–25.

Rockström, Johan, Will Steffen, Kevin Noone, Åsa Persson, F. Stuart Chapin, Eric F. Lambin, Timothy M. Lenton, Matren Scheffer, Carl Folke, Hans J. Schellnhuber, Björn Nykvist, Cynthia A. de Wit, Terry Hughes, Sander van der Leeuw, Henning Rodhe, Sverker Sörlin, Peter K. Snyder, Robert Costanza, Uno Svedin, Malin Falkenmark, Louise Karlberg, Robert W. Corell, Victoria J. Fabry, James Hansen, Brian Walker, Diana Liverman, Katherine Richardson, Paul Crutzen & Jonathan A. Foley, 2009, A Safe Operating Space for Humanity, *Nature* 472–475.

Ruppel, Oliver C., 2009a, Regional Economic Communities and Human Rights in East and Southern Africa, in: Bösl, Anton & Joseph Diescho (Eds), *Human Rights in Africa*, Windhoek, Macmillan Education Namibia, 275–318.

Ruppel, Oliver C., 2009b, The SADC Tribunal, Regional Integration and Human Rights: Major Challenges, Legal Dimensions and Some Comparative Aspects from the European Legal Order, *Recht in Afrika* 2, 203–239.

Ruppel, Oliver C., 2010, Climate Change and Human Vulnerability: Law and Policy, Unpublished paper presented at an expert workshop on *Identifying Priorities for Climate Change,* organised by the Human Rights and Documentation Centre (HRDC) of the University of Namibia (UNAM) in cooperation with the Tyndall Centre for Climate Change (University of Oxford) held at the Safari Court Conference Centre, Windhoek 09 March 2010.

Ruppel, Oliver C., 2012, The WTO Chairs Programme and the Role of World Trade and Economic Development for Regional Integration, the Reduction of Poverty and the Promotion of Economic, Social and Cultural Rights in Africa, *Recht in Afrika* 1, 143–156.

Ruppel, Oliver C., 2013, International Legal Climate Change Regimes and Climate Finance from a Southern African Perspective, in: Ruppel, Oliver C. & Katharina Ruppel-Schlichting (Eds), *Environmental Law and Policy in Namibia. Towards Making Africa the Tree of Life*, Windhoek/Essen, Orumbode Press/Welwitschia Verlag Dr. A. Eckl, 287–308.

Ruppel, Oliver C. & Katharina Ruppel-Schlichting, 2012, Climate Change and Human Security: Relevant for Regional Integration in SADC?, in: Hartzenberg, Trudi, Gerhard Erasmus & Andre Du Pisani (Eds), *Monitoring Regional Integration in Southern Africa – 2011 Yearbook for Regional Integration*, Stellenbosch, tralac, 32–71.

Ruppel, Oliver C. & Sanita van Wyk, 2011, The Effects of Climate Change on Human Security and Human Rights in Africa, *China Monitor*, 10–16.

Scheffran, Jürgen & Antonella Battaglini, 2011, Climate and Conflicts: The Security Risks of Global Warming, *Regional Environmental Change* 11, S27–S39.

Schellnhuber, Hans J., Paul J. Crutzen, William C. Clark, Martin Claussen & Hermann Held (Eds), 2004, *Earth System Analysis for Sustainability*, Cambridge, MIT Press.

Scholtz, Werner, 2010, The Promotion of Regional Environmental Security and Africa's Common Position on Climate Change, *African Human Rights Law Journal* 10, 1–25.

Schreuer, Christoph, 2001, State Sovereignty and the Duty of States to Cooperate – Two Incompatible Notions?, in: Delbrück, Jost (Ed.) *International Law of Cooperation and State Sovereignty, Proceedings of an International Symposium of the Kiel Walther Schücking Institute of International Law*, Berlin, Dunker & Humblot, 162–180.

Schrijver Nico, 2011, The Impact of Climate Change: Challenges for International Law, in: Fastenrath, Ulrich, Rudolf Geiger, Daniel E. Kahn, Andreas Paulus, Sabine von Schorlemer & Christoph Vedder (Eds), *From Bilateralism to Community Interest – Essays in Honour of Judge Bruno Simma*, Oxford, Oxford University Press, 1278–1297.

Scott, Shirley V., 2012, The Securitization of Climate Change in World Politics: How Close have We Come and would Full Securitization Enhance the Efficacy of Global Climate Change Policy?, *Review of European Community and International Environmental Law* 21 (3), 220–230.

Scott, Joanne & Lavanya Rajamani, 2012, EU Climate Change Unilateralism, *European Journal of International Law* 23 (2), 469–494.

Seibert-Fohr, Anja, 2012, The International Covenant on Civil and Political Rights: Moving from Coexistence to Cooperation and Solidarity, in: Hestermeyer, Holger P., Doris König, Nele Matz-Lück, Volker Röben, Anja Seibert-Fohr, Peter-Tobias Stoll & Silja Vöneki (Eds), *Coexistence, Cooperation and Solidarity: Liber Amicorum Rüdiger Wolfrum*, Leiden, Martinus Nijhoff Publishers, 517–552.

Shelton, Dinah, 2012, The Legal Status of Normative Pronouncements of Human Rights Treaty Bodies, in: Holger P., Doris König, Nele Matz-Lück, Volker Röben, Anja Seibert-Fohr, Peter-Tobias Stoll & Silja Vöneki (Eds), *Coexistence, Cooperation and Solidarity: Liber Amicorum Rüdiger Wolfrum*, Leiden, Martinus Nijhoff Publishers, 553–575.

Spier, Jaap, 2012, *Shaping the Law for Global Crises*, Den Haag, Eleven International Publishing.

Stanton, Elizabeth A., 2012, The Tragedy of Maldistribution: Climate, Sustainability, and Equity, *Sustainability* 4, 394–411.

Steffen, Will, Angelina Sanderson, Peter D. Tyson, Jill Jäger, Pamela A. Matson, Berrien Moore, Frank Oldfield, Katherine Richardson, Hans J. Schellnhuber, B.L. Turner & Robert J. Wasson, 2004, *Global Change and the Earth System*, Berlin, Springer, Executive Summary available at http://www.igbp.net/download/18.1b8ae20512db692f2a680007761/, last accessed 30 May 2013.

Steiner, Achim, Address at the UN Security Council Debate on the Impact of Climate Change on Maintaining International Peace and Security, 20 July 2011, available at http://www.unep.org/Documents.Multilingual/Default.Print.asp?DocumentID=2646&ArticleID=8817&l=en, last accessed 17 February 2013.

Stokke, Olav Schram, Jon Hovi & Geir Ulfstein (Eds), 2005, *Implementing the Climate Regime: International Compliance*, Oxford, Earthscan.

Tamiotti, Ludivine, 2009, *Trade and Climate Change: A Report by the United Nations Environment Programme and the World Trade Organization*, Geneva, UNEP & WTO, available at http://www.wto.org/english/res_e/booksp_e/trade_climate_change_e.pdf, last accessed 30 May 2013.

Taureck, Rita, 2006, Securitization Theory and Securitization Studies, *Journal of International Relations and Development* 9 (1), 53–61.

*The Economist*, The Anthropocene: A man-made world – Science is Recognising Humans as a Geological Force to be Reckoned with, *The Economist*, 26 May 2011, available at http://www.economist.com/node/18741749, last accessed 16 February 2013.

Theisen, Ole M., 2008, Blood and Soil? Resource Scarcity and Internal Armed Conflict Revisited, *Journal of Peace Research* 45 (6), 801–818.

Thielbörger, Pierre & Medgy Liburd, 2012, Haiti Two Years After the Earthquake – A Case for the Responsibility to Protect?, *BOFAXE* 399E.

Tietje, Christian, The Duty to Cooperate in International Economic Law and Related Areas, in: Delbrück, Jost (Ed.) *International Law of Cooperation and State Sovereignty, Proceedings of an International Symposium of the Kiel Walther Schücking Institute of International Law*, Berlin, Duncker & Humblot, 45–65.

Tomuschat, Christian, 2012, Risk and Security in International Law, in: Hestermeyer, Holger P., Doris König, Nele Matz-Lück, Volker Röben, Anja Seibert-Fohr, Peter-Tobias Stoll & Silja Vöneki (Eds), *Coexistence, Cooperation and Solidarity: Liber Amicorum Rüdiger Wolfrum*, Leiden, Martinus Nijhoff Publishers, 1283–1308.

Tschakert, Petra & Raymond Tutu, 2010, Solastalgia: Environmentally Induced Distress and Migration Among Africa's Poor Due to Climate Change, in: Afifi, Tamir & Jager Jäger (Eds), *Environment, Migration and Social Vulnerability*, Berlin, Springer, 57–72.

UIS UNESCO Institute for Statistics, 2012, Adult and Youth Literacy, UIS Factsheet No. 20, available at http://www.uis.unesco.org/FactSheets/Documents/fs20-literacy-day-2012-en-v 3.pdf, last accessed 14 February 2013.

UN/United Nations, 2012, *The Millennium Development Goals Report 2012*, New York, United Nations.

UNDP/United Nations Development Programme, 1994, *1994: Human Development Report*, New York, UNDP.

UNFCCC/United Nations Framework Convention on Climate Change, 2009, The Need for Mitigation, available at http://unfccc.int/files/press/backgrounders/application/pdf/press_factsh_mitigation.pdf, last accessed 17 February 2013.

UNFCCC/United Nations Framework Convention on Climate Change, 2010, The Need for Adaptation, available at http://unfccc.int/files/press/application/pdf/adaptation_fact_sheet.pdf, last accessed 17 February 2013.

UNFCCC/United Nations Framework Convention on Climate Change, 2011, An Introduction to the United Nations Framework Convention on Climate Change (UNFCCC) and the Kyoto Protocol, available at http://unfccc.int/files/press/backgrounders/application/pdf/unfccc_and_kyoto_protocol.pdf, last accessed 17 February 2013.

UNICEF/United Nations Children's Fund & WHO/World Health Organization, 2012, *Joint Monitoring Programme (JMP) for Water Supply and Sanitation. Progress on Sanitation and Drinking-Water*, Geneva, WHO.

United Nations General Assembly, 2012, Oceans and the Law of the Sea Report of the Secretary General, 31 August 2012, A/67/79/Add.1, available at http://daccess-dds-ny.un.org/doc/UNDOC/GEN/N12/478/41/PDF/N1247841.pdf?OpenElement, last accessed 01 February 2013.

United Nations Security Council, 2011, Press Statement 20 July 2011, SC/10332, available at http://www.un.org/News/Press/docs/2011/sc10332.doc.htm, last accessed 04 February 2013.

US National Intelligence Council, 2010, Effects of Future Climate Change on Cross-border Migration in North Africa and India, *Population and Development Review* 36, 408–412.

Vasak, Karel, 1984, Pour une troisième generation des droits de l'homme, in: Swinarski, Christophe (Ed.), *Studies and Essays on International Humanitarian Law and Red Cross Principles*, The Hague, Martinus Nijhoff Publishers, 837–850.

Voigt, Christina, 2008, State Responsibility for Climate Change Damages, *Nordic Journal for International Law* 77, 1–22.

Warner, Koko, Mohammed Hamza, Anthony Oliver Smith, Fabrice Renaud & Alex Julca, 2010, Climate Change, Environmental Degradation and Migration, *Natural Hazards* 55, 689–715.

Weart, Spencer R., 1998, *Never at War: Why Democracies Will Not Fight One Another*, Yale, Yale University Press.

Weart, Spencer R., 2012, The Evolution of International Cooperation in Climate Science, *Journal of International Organization Studies* 3 (1), 41–59.

Westra, Laura, 2009, Environmental Justice and the Rights of Ecological Refugees, Oxford, Earthscan.

White, Gregory, 2011, *Climate Change and Migration: Security and Borders in a Warming World*, Oxford, Oxford University Press.

Wolfrum, Rüdiger, 1984, *Die Internationalisierung staatsfreier Räume*, London, Springer.

Wolfrum, Rüdiger, 1995, International Law of Cooperation, in: Bernhardt, Rudolph (Ed.) *Encyclopedia of Public International Law (EPIL)* Vol. II, 1242–1247.

Wolfrum, Rüdiger, 2009, Solidarity Amongst States: An Emerging Structural Principle of International Law, Indian Journal of International Law 49 (1), 8–20.

Wolfrum, Rüdiger, 2010, Solidarity amongst States: An Emerging Structural Principle of International Law, in: Piovesan, Flavia & Ines Virginia Prado Soares (Eds), *Direito ao Desinvolvimento*, Belo Horizonte, Editora Fórum 2010, 57–72.

Wood, Michael, The UN Security Council and International Law, Hersch Lauterpacht Memorial Lecture, Cambridge, 08 November 2006, available at http://www.lcil.cam.ac.uk/Media/lectures/pdf/2006_hersch_lecture_2.pdf, last accessed 17 February 2013.

World Bank, 2010, *World Development Report 2010: Development and Climate Change*, Washington, World Bank.

World Bank, 2012, New Report Examines Risks of 4 Degree Hotter World by End of Century, Press Release, 18 November 2012, available at http://www.worldbank.org/en/news/2012/11/18/new-report-examines-risks-of-degree-hotter-world-by-end-of-century, last accessed 28 May 2013.

Zelli, Fariborz & Harro van Asselt, 2010, The Overlap Between the UN Climate Regime and the World Trade Organization: Lessons for Post-2012 Climate Governance, in: Biermann, Frank, Philippe Pattberg & Fariborz Zelli (Eds), *Global Climate Governance Beyond 2012: Architecture, Agency and Adaptation*, Cambridge, Cambridge University Press, 79–96.

2

# International Climate Change Policy: Where do we Stand?

*Nadia von Bassewitz*

*Abstract*

The article begins with a definition of what is meant by *climate change*, from the perspective of an Organisation for Economic Co-operation and Development (OECD) publication. The second section identifies the major legal instruments of international climate change policy adopted before 2012, followed by an analysis of the post-2012 international legal framework in the third section. The next three sections investigate what policy instruments and methods individual countries use to tackle global warming. Two industrialised societies are looked at in detail, namely the European Union, which is a Kyoto Protocol Annex I party, and the United States, the second-largest emitter of greenhouse gases (GHGs), but not a party to the Kyoto Protocol. The investigation then directs its attention to a member of the BASIC group (Brazil, South Africa, India and China) of newly industrialised countries, namely China, as a Kyoto Non-Annex I party and the highest current emitter of GHGs. In the last section, the discussion turns to the developing world, with a special focus on least-developed countries.

## *A. Introduction*

Very few scientists today doubt that climate change is taking place and that human activities are contributing to this trend.[1] But what is actually meant by the term *climate change*? The 2006 *Adaptation to Climate Change: Key*

1 According to a web-based poll undertaken by the Institute for the Study of Earth, Oceans and Space of the University of New Hampshire in 2009, 82% of earth scientists and 97% of climate scientists are of the opinion that global warming is real and that human activities are a major contributing factor to the warming.

*Terms* paper by the Organisation for Economic Co-operation and Development identifies four definitions of it:[2]

Climate Change [refers] –

- … to any change in climate over time, whether due to natural variability or as a result of human activity. (IPCC TAR, 2001 a)
- … to a statistically significant variation in either the mean state of the climate or in its variability, persisting for an extended period (typically decades or longer). Climate change may be due to natural processes or external forcing, or to persistent anthropogenic changes in the composition of the atmosphere or in land-use. (IPCC TAR, 2001 b)
- [to a] change of climate which is attributed directly or indirectly to human activity that alters the composition of the global atmosphere and which is[,] in addition to natural climate variability[,] observed over comparable time periods. (UNFCCC, Article 1)
- [t]he climate of a place or region [changing] if over an extended period (typically decades or longer) there is a statistically significant change in measurements of either the mean state or variability of the climate for that place or region. (Changes in climate may be due to natural processes or to persistent anthropogenic changes in atmosphere or in land use. Note that the definition of climate change used in the United Nations Framework Convention on Climate Change is more restricted, as it includes only those changes which are attributable directly or indirectly to human activity). (UN/ISDR, 2004)

## B. *Major Legal Documents of International Climate Change Policy Adopted Before 2012*

From its inception, international climate change policy and legislation was firmly rooted in the United Nations (UN) system. During the mid- to late 1980s, for the first time ever, research was able to demonstrate that man-made global warming was indeed happening. Against this background, the United Nations Environment Programme (UNEP) and the World Meteorological Organization (WMO) set up an Intergovernmental Panel on Climate Change (IPCC) in 1988 to gather the scientific evidence for (or against) human-induced global warming.

The IPCC's first Assessment Report, which appeared in 1990, presented scientific evidence that global warming was a reality, triggering worldwide concern:

2 OECD (2006:12).

The Report identified as main factors affecting climate (i) atmospheric gases, so-called greenhouse gases, some of which occur naturally (e.g. water vapour, ozone, carbon dioxide ($CO_2$) methane ($CH_4$), nitrous oxide ($N_2O$) and some of which are man-made (e.g. chlorofluorocarbons, CFC) as well as (ii) aerosols which are tiny particles within the atmosphere.[3]
The Report presented evidence that, for a thousand years prior to the industrial revolution, the concentration of GHG was relatively constant and that however since the beginning of the industrialisation in the 18th century the concentration of several GHG, in particular $CO_2$, methane, $N_2O$ and CFCs, have been increasing markedly primarily due to man's activities.[4]
As a result of the enhanced greenhouse effect, the Earth's surface and atmosphere are warming up. The IPCC Report warned that in a Business-as-Usual scenario (i.e. few or no steps are taken to limit GHG emissions) future human-made emissions will result in a likely increase in global mean temperature of about 1°C above the present value (about 2°C above that in the pre-industrial period) by 2025 and 3°C above today's (about 4°C above pre-industrial) before the end of the 21st century.[5]

### *I. United Nations Framework Convention on Climate Change, 1992*

Based on the findings of the 1990 IPCC Assessment Report, the UN Framework Convention on Climate Change (UNFCCC) was signed during the Rio Earth Summit in June 1992, and entered into force on 21 March 1994. The major accomplishment of the UNFCCC was that it recognised, for the first time, that there was indeed a man-made problem of climate change at a moment when there was still considerable doubt regarding the causes of global warming, its extent and impact.[6]

The ultimate objective of the UNFCCC is to stabilise greenhouse gas (GHG) concentrations "at a level that would prevent dangerous anthropogenic interference with the climate".[7] The stabilisation level is not quantified in the UNFCCC. The latest climate analysis has identified a stabilisation range of 450 to 500 parts per million (ppm) of carbon dioxide in the atmosphere. According to the UNFCCC, this level should be reached within a time frame which allows ecosystems to adapt naturally to global warming,

3 IPCC (1990:13–14).
4 (ibid.:15–16).
5 (ibid.:22).
6 Bothe (2003:240).
7 UNFCCC (2012a:1).

while making sure that food production is not at risk and that development occurs in a sustainable manner.

The UNFCCC is a framework document which introduces two main policy approaches which are intrinsically linked in order to address global warming, namely mitigation and adaptation. While *mitigation* tackles the very cause of climate change, *adaptation* deals with the unavoidable effects of global warming. The less mitigation takes place, the more adaptation is required, and vice versa:

- *Climate mitigation* refers to any action taken to eliminate or decrease the long-term impact of global warming on human life or property.[8] According to the UNFCCC Glossary, mitigation is understood as "human intervention to decrease the sources of GHG or enhance their reabsorption".[9]
- *Climate adaptation* involves initiatives or measures to reduce the vulnerability of individuals, groups and natural systems to the negative effects of climate change.[10] According to the UNFCCC Glossary, the term involves the adaption of natural or human systems in response to climatic stimuli or their impact, which moderates damages or exploits beneficial opportunities.[11]

Any country can become a party to the UNFCCC,[12] thus making it a global instrument. Within this framework of global participation, however, states parties' obligations vary substantially between developed and developing countries. The UNFCCC notes that –

- the largest share of accumulated GHG emissions has originated in developed countries, as opposed to developing countries, and
- per capita GHG emissions in developed countries are much higher than in developing countries.

As a consequence, the UNFCCC introduces the principle of "common but differentiated responsibilities and respective capabilities" or *CBDR principle* for its member states.[13] The CBDR principle is based on the general

8 UNFCCC (2009a).
9 UNFCCC (2012b).
10 UNFCCC (2009c).
11 UNFCCC (2012b).
12 Article 22.1, UNFCCC.
13 Article 3.1, UNFCCC.

principles of equity of international law and includes two elements: while all states together have a common responsibility for the protection of the environment, there are nonetheless differences between the states in terms of their historical contribution to global warming and their ability to fight it, which is why states need to bear different responsibilities.[14] In view of this, the UNFCCC foresees asymmetrical obligations and places the heaviest burden on the wealthier industrialised states.[15] At the same time, the UNFCCC recognises that developing nations, especially least-developed countries (LDCs) and small island developing states (SIDSs) are more vulnerable to climate change, in part because of their greater exposure to climate trends and in part because of their low adaptation skills.[16]

In line with the CBDR principle, the UNFCCC divides states parties as follows:

- *Annex I parties* include the 41 industrialised countries, covering members of the Organisation for Economic Co-operation and Development (OECD) in 1990 and the former Soviet Bloc (economies in transition, or EITs),[17] and
- *Non-Annex I parties*, which are mostly developing countries.

However, it is important to understand that the CBDR principle is not deemed absolute under the UNFCCC and that the UNFCCC provides, to a certain degree, for a transition from the Non-Annex I group to the Annex I group.

Based on the CBDR principle, the UNFCCC imposes voluntary mitigation targets for Annex I parties, according to which they were supposed to reduce their GHG emissions to 1990 levels by the year 2000. As economic development is vital for the world's poorest countries, the UNFCCC accepts that GHG emissions originating in those countries would grow in the near future, as a result of which Non-Annex I parties were not subject to mitigation targets. The UNFCC aims at helping the developing countries limit emissions in ways that will not restrict their development.

14 CISDL (2002:1–2).
15 Boisson de Chazourne (2008:2).
16 Preamble, UNFCCC.
17 There is a sub-category of so-called Annex II parties, which consist of the OECD members of Annex I but not the EITs. Only these Annex II states parties are obliged to make funding available for Non-Annex I parties.

The UNFCCC obliges OECD members (the Annex II states parties) to support developing countries in the elaboration of national adaptation plans.[18] Moreover, industrialised members agreed to share adaptive know-how and technology to offer urgently needed capacity-building for Non-Annex 1 states parties.[19]

The UNFCCC's institutions and procedures are drawn from the UN system, with a Conference of the Parties (COP) being the highest decision-making organ.[20] Decisions are adopted by way of negotiation within the COP, which convenes once a year to review the UNFCCC's implementation.[21] Procedures are governed by the rules included in the UNFCCC itself, as well as the Draft Rules on Procedure – even though the latter have never been formally adopted. This is why most of the decisions of the COP can only be taken by consensus.[22]

### *II. The Kyoto Protocol, 1997*

The publication of the Second IPCC Assessment Report in 1995[23] showed that the measures taken up to that point under the UNFCCC to fight global warming were insufficient. As a consequence, on 11 December 1997, the UNFCCC parties signed the Kyoto Protocol to the UNFCCC,[24] which only came into operation on 16 February 2005. At the time, more than 55 countries had ratified Kyoto, accounting for more than 55% of global carbon dioxide emission in 1990. This included all OECD countries, but with the important omission of the United States of America (USA/US): the biggest emitter in 1990, the US signed but has never ratified the Protocol. As at the time of writing, 190 countries and the European Union (EU) have ratified Kyoto, while Canada withdrew in December 2011.

18 Article 4.3, UNFCCC.
19 Boisson de Chazourne (2008:4).
20 Articles 7–10, UNFCCC.
21 UNFCCC (2012c).
22 Depledge & Yamin (2009:438).
23 IPCC (1995).
24 Hereinafter also *Kyoto* or *the Protocol.*

In brief, the Protocol operationalises the UNFCCC:[25]

- It shares the objectives, instruments and the institutions of the UNFCCC. Even more importantly, Kyoto replicates the CBDR doctrine formulated in the UNFCCC. However, as opposed to the UNFCCC, Kyoto excludes a transition from the Non-Annex I category to the Annex I category, hence introducing a so-called firewall between these two groups of countries.
- The major distinction between the UNFCCC and Kyoto is that, for the first time ever, a UN instrument imposes legally binding mitigation targets as opposed to the non-binding goals under the UNFCCC. In line with the CBDR principle, only Annex I parties take on binding mitigation objectives, while the Non-Annex I parties are expected to carry out voluntary mitigation actions.

*1. Mitigation*

Kyoto introduces binding quantified emission reduction targets for the industrialised countries. Under the Protocol, 41 industrialised countries – including EITs and the EU – are obliged to reduce their GHG emissions by 5.2% compared with 1990 levels over the first commitment period from 2008 to 2012.[26] The individual national targets include, from the 1990 base year, an 8% decrease for the EU,[27] 6% each for Canada and Japan, no decrease for Russia, and an 8% increase for Australia.[28]

The Kyoto Protocol allows for more flexibility as to how to meet binding GHG emission reduction targets by designing three innovative instruments. Under the Clean Development Mechanism (CDM), countries with Kyoto targets may implement an emission reduction project in developing countries, based on which they obtain certified emission reduction (CER) units, which count towards fulfilling their Kyoto obligations.[29] A CDM project is obliged to confer measurable and verifiable emission reductions that are

25 UNFCCC (2012d).

26 Article 3.1, Kyoto Protocol.

27 This value includes reduction targets of 21% for Germany, 12.5% for the United Kingdom, and 0% for France, while Spain may increase its emissions by 15%.

28 Cf. http://unfccc.int/kyoto_protocol/items/3145.php, last accessed 8 October 2012.

29 Article 12, Kyoto Protocol.

additional to what would otherwise have occurred without the CDM.[30] Join implementation is a mechanism similar to the CDM, but the emission reduction project has to be implemented in industrialised countries.[31] CDM and joint implementation are the first global investment tools of their kind, stimulating foreign investment and knowledge transfer in the host country, while offering industrialised countries flexible and cost-effective ways of meeting a part of their Kyoto obligations.[32] Emissions trading is based on the idea that the mitigation targets under Kyoto are formulated as levels of permitted GHG emissions over the 2008–2012 commitment period. As laid down in Article 17 of Kyoto, emissions trading permits countries with CER units to spare to trade such units with other countries that have exceeded their CER allowance.[33]

Under Kyoto, states parties are obliged to monitor their GHG emissions and to keep a national register of trades carried out under Kyoto.[34] The UNFCCC Secretariat keeps an independent transaction log to verify that operations are consistent with the Kyoto Protocol. Furthermore, a compliance mechanism has been established to verify the implementation of the Protocol by its members.

### 2. *Adaptation*

Under the adaptation objective, the Kyoto Protocol, like the UNFCCC, is designed to support developing countries, especially LDCs and SIDSs, in adapting to the inevitable impacts of climate change and in facilitating the development of know-how and technologies that could help increase resilience to climate change impacts.[35] A range of funds have been created through the UNFCCC which are managed by the Global Environment Facility (see the discussion under Section D below).

30 Hepburn (2009:412).

31 Article 6, Kyoto Protocol.

32 UNFCCC (2012g).

33 UNFCCC (2012f.).

34 UNFCCC (2012e).

35 The IPCC defines *resilience* as the ability of a social or ecological system to absorb disturbances while retaining the same basic structure and ways of functioning, the capacity for self-organisation, and the capacity to adapt to stress and change.

## C. The Post-2012 International Legal Framework – Which Way Forward?

### I. Assessment of the Kyoto Protocol

When evaluating Kyoto toward the end of its first commitment period in 2012, we have to recall the logic behind the agreement. Kyoto was always deemed an initial step towards a low-carbon future,[36] introducing humble reduction targets of 5% for industrialised countries over a short period of five years only, until 2012. After 2012, Kyoto was to be followed by a chain of other agreements to impose ever wider and deeper reductions for Annex I parties. Developing countries were expected to follow suit in time, so that at last, all countries would have binding GHG emission reduction goals.

However, the results of Kyoto are mixed, to say the least. Economists agree that the Protocol imposes relatively high costs and generates negligible benefits, while failing to provide a real solution.[37] Additionally, most climate researchers warn that the Protocol has failed to decrease GHG emissions to the extent necessary.[38]

Progress towards the mitigation targets under Kyoto has also not been satisfying. According to the Netherlands Environmental Assessment Agency (NEAA), industrialised countries will, as a group, probably meet the GHG emission reduction goals imposed under Kyoto.[39] When extrapolating the trend of the years 2000–2007 to the period 2008–2012, NEAA forecasts an average emission reduction of almost 16% for this group of countries in the first Kyoto commitment period.[40] However, the NEAA also indicates that the expected decrease of 16% is mainly attributable to large GHG emission reductions of about 40% in Germany and the EITs after the fall of the Berlin Wall.[41] The World Bank reports that there are significant differences in performance across individual countries.[42] If one looks at the individual state level, the compliance gap for many of them is quite noteworthy.[43]

36 UNFCCC (2012e).
37 Olmstead & Stavins (2006:1).
38 Helm (2009b:16).
39 NEAA (2012).
40 (ibid.).
41 (ibid.).
42 World Bank (2008:6).
43 Barrett (2009:62).

Various factors have contributed to this underachievement. Some are linked to the Kyoto Protocol itself, others go beyond the scope of Kyoto, as follows:

- An important deficiency of the Kyoto regime itself is the lack of broad participation, i.e. the number of countries willing to take real action via obligatory mitigation objectives has always been quite small.[44] The world's largest GHG emitter at the time, the US, has never ratified Kyoto. Canada, an Annex 1 state party, left Kyoto in December 2011. Moreover, the largest increase in GHG emissions today originates from six newly industrialised countries (NICs), i.e. including the BASIC group made up of Brazil, China, India and South Africa, as well as Indonesia and Mexico. These six countries ratified Kyoto as Non-Annex I parties. As a consequence, nowadays, Kyoto addresses only 30% of GHG emissions in the world.[45]
- Another great – if not the greatest – weakness of Kyoto is its inflexible partition of countries into two groups in line with the CBDR dogma, building the so-called firewall between Annex I and Non-Annex I members, which has reinforced the existing ideological North–South divide.[46] Kyoto has no graduation process by which to verify whether some of the NICs such as the BASIC group are ready to join the Annex 1 group.[47] This split between richer and poorer nations under Kyoto is clearly outdated and inaccurate, with 50 Non-Annex I countries now having a larger per capita income than some of the Annex I countries.[48] But, more importantly, the partition means that today's biggest GHG emitter, China, remains unconstrained in its emissions output, implying that half of all worldwide emissions will in the near future be generated in a country without binding mitigation targets.[49]
- Some observers indicate that the very methodology of Kyoto is flawed in that it takes a geographical approach to GHG emission responsibilities, i.e. the so-called production basis methodology, as opposed to the consumption methodology.[50] Emissions are attributed to states on the basis

44 (ibid.:61).
45 Gao (2007:7).
46 Gosh & Woods (2009:454).
47 Depledge & Yamin (2009:443).
48 Olmstead & Stavins (2010:2).
49 (ibid.).
50 Helm (2009b:20).

of territory.[51] All emissions produced within a state account for its emissions total. Hence, Kyoto places the burden of GHG emission reduction on those states which produce emission-intensive goods, rather than those which import and finally consume such goods. The weakness of this methodology is that industrialised countries, which are subject to binding emission reduction targets, can relocate carbon-dioxide-intensive production abroad to developing countries (with no such goals) in order to meet their Kyoto targets.[52]

- Another important shortcoming of Kyoto relates to the lack of compliance incentives and enforcement mechanisms to deter non-participation and non-compliance.[53] The UNFCCC, like Kyoto, includes rules for monitoring compliance, in particular for the GHG emission reduction targets of Annex I countries. But monitoring is still inadequate, both in terms of linking it to effective implementation and including issues of importance for developing countries.[54]

In defence of Kyoto, however, it must be said that the problem of global warming does not easily lend itself to a binding international agreement. To name but a few issues, the allocation of responsibility for the existing level of GHGs is complex; the measurement of emissions is at best weak; GHG emissions per capita are low in those nations most rapidly increasing their overall emissions; and the impact of global warming varies between countries.[55] Additionally, the complexity of global warming is increasing all the time and is having a severe impact on international negotiations,[56] which is why each negotiation round becomes more demanding.

51 Pan et al. (2009:145).

52 A country such as the United Kingdom (UK) could produce low-GHG-intensive goods, i.e. services instead of manufacture, and import high-GHG-intensive goods such as aluminium and steel from abroad. By transferring energy-intensive production to China, India or other developing countries, the UK could meet its Kyoto obligations without making any noticeable difference to climate change.

53 Barrett (2009:63); Aldy & Stavins (2009:8).

54 Gosh & Woods (2009:463).

55 Helm (2009b:19).

56 Depledge & Yamin (2009:446).

### *II. COP12, Montreal, Canada, 2005*

Recognising that Kyoto on its own was insufficient to fight global warming, the COP in Montreal adopted a set of decisions which laid the foundation for an innovative dual-track climate negotiation process:[57]

The first track, the so-called Kyoto track, is about negotiating obligatory emission reduction targets for Annex I parties for a second commitment period (CP2) beyond 2012. This negotiation path is only open for Annex I parties. The negotiation track is supervised by the Ad Hoc Working Group on Further Commitments for Annex I Parties under the Kyoto Protocol (AWG–KP).

The second track, the so-called UNFCCC track, involves the negotiation of emission reduction goals for industrialised countries that have not ratified Kyoto, first and foremost the US. This track also covers negotiations for nationally appropriate mitigation actions (NAMAs) to be undertaken by developing countries. This track is open to all UNFCCC states parties. The negotiation path is overseen by the Ad Hoc Working Group for Long-term Cooperative Action under the UNFCCC (AWG–LCA), which was set up at COP13 in Bali in 2007.

This two-track negotiation and decision-making process aims at broadening the participation and improving the effectiveness of the international climate regime.

### *III. COP13, Nusa Dua, Bali, Indonesia, 2007*

In 2007 the IPCC's Fourth Assessment Report substantially increased the pressure on the international community to urgently address global warming:

> The IPCC Report indicates that the warming of the climate system is unequivocal, as is now evident from observations of increases in global average air and ocean temperatures, widespread melting of snow and ice and rising global average sea level.[58]
> The Report confirms the findings of earlier Assessment Report that atmospheric concentrations of $CO_2$, methane ($CH_4$) and nitrous oxide ($N_2O$) have increased markedly as a result of human activities since 1750 and now far exceed pre-industrial values determined from ice cores spanning many thousands of

57 UNFCCC (2005).
58 IPCC (2007:2).

years.[59] The Report notes that most of the observed increase in global average temperatures since the mid-20th century is very likely due to the observed increase in anthropogenic GHG concentrations.[60]
The Report projects an increase of global average temperature of about (i) 0.2°C per decade for the next two decades for all four GHG emissions scenarios and (ii) 1.8°C to 4°C by the year 2100 depending on the GHG emissions scenario.[61]

### 1. *Outcome*

COP13 in Bali in 2007 decided to uphold the dual negotiations path under both the UNFCCC and Kyoto, with the expectation that the two tracks should be unified in Copenhagen in 2009.

The COP adopted the Bali Road Map, which is an overarching term to include all the decisions made in Bali, identifying the challenges under the two negotiation streams.[62] The main objective of the Road Map was to achieve a legally binding, inclusive climate agreement in Denmark 2009, which was to replace Kyoto after 2012 and would ideally include all major GHG emitters.

The Bali Road Map includes the Bali Action Plan (BAP) which lays down the mandate of the AWG–LCA to supervise the UNFCCC negotiation track.[63] The BAP is built on five pillars, i.e. a shared long-term vision and enhanced action on mitigation, adaptation, technology and funding, which has determined the agenda of any COP to come. Some of the BAP's highlights are as follows:

- The BAP calls for a "shared vision of long-term action" on global warming recognises the need for an overall long-term mitigation objective beyond 2012.[64]

59 (ibid.:5).
60 (ibid.).
61 (ibid.:7-8).
62 CCES (2007c:2).
63 UNFCCC (2012i).
64 Initial proposals for the BAP supported by the EU foresaw that developed countries would have to reduce GHG emissions by 25–40% below 1990 levels by 2020. Due to strong opposition from the US, but also Canada and Japan, the final decision only asks for "deep cuts in global emission". See also TWN (2007).

- As to the mitigation pillar, the BAP urges that national mitigation commitments/actions by all states have to be "measureable, reportable and verifiable" (MRV).[65] For the first time, developing countries pronounced their willingness to consider taking national appropriate mitigation action, hence softening the rigid CBDR viewpoint they had held before.[66]
- Vis-à-vis the adaptation pillar, the BAP recognises the need for enhanced international cooperation to support urgent implementation of adaptation measures, taking into account the immediate needs of vulnerable developing countries.[67]

Furthermore, the COP13 decided to operationalise the Adaptation Fund under Kyoto in support of LDCs and SIDSs.[68] The states parties established a 16-member Adaptation Fund Board to manage the fund on behalf of the COP with the Global Environment Fund operating as secretariat. For the first time, the COP managed to put deforestation on the international climate agenda, which accounts for 20% of all GHG emissions. The states parties agreed to study the issue, especially on how to measure GHG emissions from deforestation with a view to launch a UN Collaborative Programme on Reducing Emissions from Deforestation and Forest Degradation in Developing Countries (UN-REDD) initiative.

### 2. *Assessment*

When the COP13 in Bali was evaluated, its outcomes were considered a leap forward in many respects. The BAP was notable in being the first international climate decision after Kyoto which the US joined, despite vigorous earlier opposition. This gave hope that the US could be re-engaged in the post-2012 climate negotiations.

Yet, many vital issues remained open in Bali:[69]

- The AWG–LCA of the UNFCCC track did not succeed in determining what kind of procedure – i.e. formal negotiations, informal dialogues, or

65 BAP, clause 1.(b)(i) and (ii.).
66 UN-NGLS (2008:2).
67 BAP, clause 1.(c)(i.).
68 CCES (2007c:6).
69 UN-NGLS (2008:3).

both – would be adopted in the working group. The objectives to be achieved under the BAP are not quantified, especially with respect to –
  - the long-term mitigation objective under the UNFCCC, and
  - the funding by industrial countries of the adaptation and mitigation efforts by developing countries.
- Even more significantly, COP13 was unable to reach an understanding on a question which has dominated the international climate negotiations since then, i.e. the terms of a post-Kyoto climate agreement after the expiry of the first commitment period in 2012.

### *IV. COP15, Copenhagen, Denmark, 2009*

The COP15 in Denmark in 2009 was meant to finalise two years of intense negotiations which had been launched with the 2007 Bali Road Map and, more importantly, consolidate the two negotiation tracks under the UNFCCC and Kyoto.

#### *1. Expectations*

Expectations before the start of the COP15 were running high. It was hoped that the Copenhagen meeting could streamline the two negotiation tracks and produce a new, legally binding global climate agreement to replace the Kyoto Protocol after 2012: one that would involve all the major GHG emitters – or at least a strong commitment by all states parties to work toward this goal.[70]

Furthermore, under the UNFCCC track, COP15 was expected to recognise the imperative of limiting global warming to 2°C.[71] The meeting was also meant to decide on new instruments to generate sustainable funds to bolster mitigation and adaptation efforts by developing countries.[72] In terms of mitigation, it was hoped that the industrialised states parties could agree to a quantified long-term mitigation objective such as a 50% reduction by 2050 ('50 by 50') goal formulated by the G8.[73] The largest developing states,

70 Stavins & Stowe (2010:2).
71 UNFCCC (2009a:2).
72 WRI (2002:4).
73 The world's eight wealthiest countries; Stavins & Stowe (2010:2).

such as Brazil, China, India and Mexico, were to take on a wider range of quantifiable, policy-based mitigation commitments, e.g. sectoral GHG emission reduction targets and intensity goals.[74] The agreement was also intended to establish the terms for reporting on and verifying states parties' mitigation actions.

### 2. *Outcome*

With the negotiations on the verge of breaking down and after an all-night final session, all the COP15 managed to deliver was the informal Copenhagen Accord,[75] a three-page document with two empty annexes addressing the following:

- Under the UNFCCC negotiation track, the Accord recognises the scientific view as outlined in the IPCC's Fourth Assessment Report that global warming has to be limited to 2°C in order to prevent the most damaging impacts.[76] Copenhagen introduced a new instrument in the form of "national pledges", which every UNFCCC member had to submit by the end of January 2010.[77] The Accord is labelled a *portfolio of national commitments*, under which each state commits and enrols to observe its domestic GHG mitigation targets, whether those are in the form of laws, regulations or multi-year action plans.[78] The Accord also highlights the need for scaled-up and predictable funding to developing countries. It called for US$30 billion over a three-year period from 2010 to 2012, to be split equally between mitigation and adaption, and identified a "goal" of US$100 billion per year as from 2020.[79]
- Copenhagen endorsed the maintenance of the two negotiating tracks by prolonging the Kyoto as well as the AWG–LCA. While some states parties had hoped these negotiating forums could be used to operationalise the new Accord, no formal link between the working groups and the Accord was established.

---

74 CCES (2009b:2).
75 Hereinafter *Copenhagen* or *Accord.*
76 Egenhofer & Giorgiev (2009:3).
77 (ibid.).
78 Stavins & Stowe (2010:2).
79 Egenhofer & Giorgiev (2009:2).

*3. Assessment*

The gathering in Copenhagen drew a level of political attention well beyond expectations in that an unprecedented number of heads of state and government – almost 120 – decided to meet to provide the final input for a new global climate regime. Did this strategy pay off?

Initially, most observers showed profound frustration, because for them the three-page Copenhagen Accord represented all that was wrong with international climate negotiations:[80]

- It was felt that the COP had covered no ground whatsoever towards a binding post-2012 climate agreement, as neither of the two working groups, i.e. the AWG–KP and AWG–LCA, had been able to reach a formal decision. Instead, the Copenhagen Accord is a non-binding informal decision, which was not even supported by all UNFCCC members. Due to opposition by a handful of states at the 11th hour, the COP only "took note" of the decision.[81]
- The Accord does not impose a long-term binding mitigation goal on all industrialised countries; nor does it foresee mitigation commitments for the NICs. Instead, the Accord introduces an "open enrolment" framework under which states can register their voluntary domestic mitigation pledges.[82] From the onset, experts have been wary about the quality of the voluntary mitigation pledges. Indeed, today, experts are unanimously of the view that the pledges do not suffice to keep global warming at 2°C below pre-industrial level; instead, the pledges would reflect a target of approximately 3°C.[83] Highlighting the gap between the pledged and the necessary GHG emission reductions (the so-called ambition gap), experts since have called for increased reduction targets. Moreover, most of the pledges that are deemed insufficient are listed with conditions.[84]
- What is more, like no other COP before it, Copenhagen revealed the dividing lines on the terms of a post-2012 climate agreement. These were as follows:

80 Falkner et al. (2010:252).
81 Anderson (2009:2).
82 Cao (2010:3).
83 UBA (2010:5).
84 Diringer (2010a:1).

- While the developing world insists that the CBDR principle be continued as formulated in Kyoto, i.e. excluding mitigation targets – be they voluntary or otherwise – for developing countries,[85] industrial countries argue that the CBDR principle needs to be revised, given that emissions levels are increasing quickly in some developing countries.[86]
- Annex I parties are unwilling to go any further without noteworthy commitments from the US and the large developing countries.[87] They want Kyoto replaced by a new binding agreement, ideally covering all large GHG emitters, including the US and China. For its part, the US indicated that any new climate agreement should be very different from the obligatory top-down Kyoto model and called for a voluntary bottom-up model instead, including all large GHG emitters.[88] On the other hand, the Group of 77/China insist on the extension of Kyoto in its present form, together with a separate agreement under the AWG–LCA.[89]

However, observers soon realised that, with the Copenhagen Accord, a total breakdown of the climate negotiations had been prevented.[90] By the end of 2010, the Accord had become the first-ever vehicle to include explicit, albeit not unconditional, mitigation pledges from all the world's major economies, including China, India and other large developing states.[91] Experts now believe that the Accord is a compromise of what was realistically possible, given the political impasse.[92] Other observers go as far as implying that Denmark could have seen the emergence of a new climate architecture – moving away from the top-down model of Kyoto with its internationally agreed obligatory emission limits and designated instruments, towards a bottom-up model relying on voluntary national pledges and using flexible instruments.[93]

85 Sterk et al. (2011a:4).
86 (ibid.).
87 Sterk et al. (2011b:5).
88 (ibid.:6).
89 Sterk et al. (2012:5).
90 Falkner et al. (2011:252).
91 (ibid.:253).
92 (ibid.).
93 Egenhofer & Giorgiev (2009:3).

#### 4. *Beyond Copenhagen*

Undoubtedly, the weak outcome in Denmark raised important questions about the future of the international climate negotiations.

The first question was how to incorporate Copenhagen, which is an informal political decision reached outside the UNFCCC process, into the UN legal framework.[94] After Copenhagen, three possible scenarios were discussed:

- Using Copenhagen as an alternative negotiation path – a route favoured by the US
- Ignoring Copenhagen and moving on with the UNFCCC/Kyoto tracks only, and
- Integrating successful elements of Copenhagen into the UNFCCC/Kyoto tracks, which was the EU view.[95] For most observers, this third option appeared to be the most realistic for COP18 in Cancun, Mexico, in 2010.

An even more important question concerned the form of a post-2012 climate treaty and how best to reach it. In view of the weak COP outcome in Copenhagen, quite a few observers argued that the EU's "global climate deal" strategy[96] was obsolete, and that a new approach to formulating a climate treaty was necessary.[97]

Indeed, Denmark had exposed major hurdles on the way to a new legally binding global climate agreement:[98]

- Of the major GHG emitters that account for two thirds of carbon dioxide emissions (China, the EU, India, Russia and the US), only the EU unequivocally supports the idea of a new global legally binding treaty
- The EU was, however, unable to exercise leadership at Copenhagen, and
- Instead, the negotiations were overshadowed by the political impasse between the US and China: Washington made its ratification of a new

94 Diringer (2010b:2).

95 (ibid.).

96 A *global climate deal strategy* entails deriving a package deal with legally binding quantified targets on all the key issues – mitigation, adaptation and funding – and is universal in its application; see Falkner et al. (2010:256). This strategy was first used when the 1985 Vienna Convention on Ozone Layer Protection was negotiated, and it dominated international climate-related policy until Kyoto.

97 Falkner et al. (2010:256).

98 Bodansky et al. (2010:3).

binding treaty contingent on obligatory GHG emission reduction goals for key developing countries, such as those of the BASIC group.

On the other hand, China remained opposed to any kind of binding objectives unless the US took the initiative in limiting its GHG emissions.[99] Experts have, before and after Copenhagen, investigated various other options for a post-2012 climate treaty. There is unfortunately only room to discuss two pertinent aspects here:[100]

- ***Top-down v Bottom-up:*** The EU and developing countries, particularly LDCs and SIDSs, insist that a top-down model à la Kyoto, with internationally agreed obligatory emission reduction goals, is the only way to meet the long-term 2°C objective under the UNFCCC. In contrast, the US and others are of the view that a bottom-up approach with domestic voluntary pledges is likely to be more effective, as countries will submit only what they can actually realise. A growing number of observers postulate that what is really needed is something in-between: an international legal instrument that is flexible enough to guarantee wide participation, and binding enough so that states parties can be reasonably confident that others will fulfil their obligations.[101] Although dispensing with the idea of a legally binding climate deal, this option maintains the need for a strong international climate framework by embedding national pledges in a wider international regime.[102]
- ***All-inclusive treaty v Fragmented agreements:*** Instead of waiting for a comprehensive post-2012 climate deal, which includes all pillars of the Bali Road Map, many observers today suggest a fragmentation of the climate negotiations.[103] They want to disaggregate the key issues into components that can be developed in a more flexible way through parallel agreements using various sets of instruments, institutions and methods which are only integrated and linked over time.[104] They favour negotiating –

99 Barriaux (2010).
100 For an overview, see Kuik et al. (2008).
101 Diringer (2010b:1).
102 Falkner et al. (2010:259).
103 Falkner (2011:258); Bodansky & Day O'Connor (2011:10).
104 Bodansky et al. (2010:10).

- issue-specific agreements
- agreements which target specific industries or specific policies, or
- agreements involving only a few like-minded countries.[105]

Specifically, considering the logjam between the US and China, there is a growing number of experts who recommend that a 'coalition of the willing' including the EU, Japan and Russia and progressive developing countries such as Indonesia, Korea and Mexico should continue with Kyoto beyond 2012.[106]

The last major question that needed to be looked at after Denmark involved the obvious weaknesses of the UN negotiation framework. Observers agreed that there was an urgent need for more effective decision-making rules, which simultaneously guaranteed participation and inclusiveness.[107] After Copenhagen, various options were discussed. Under the first main option, negotiations would continue under the UN umbrella, but the decision-making procedure would be overhauled by –

- introducing majority rule, or
- keeping voting rules as they were, but using more exclusive negotiation groups.[108]

Other experts looked beyond the UN as a negotiation platform and suggested using other forums.[109] Alternatives included specialist institutions, i.e. the International Maritime Organization, the International Civil Aviation Organization or other broader international institutions, like the G8, the G20,[110] or the Major Economies Forum on Energy and Climate.[111]

105 Kuik et al. (2008:327).
106 Falkner et al. (2010:259).
107 UBA (2010:28).
108 Falkner et al. (2010:258).
109 Bodansky & Day O'Connor (2011:3–10).
110 Group of 20 major economies.
111 Bodansky et al. (2010:19).

## V. COP16, Cancun, Mexico, 2010

### 1. Expectations

In light of the ongoing discussion about the right approach forward, and with the gridlock between the US and China unresolved, negotiators began organising the next COP in Mexico. Expectations were low. In particular, the US had nothing to bring to the table, given that any new climate deal would not reach a quorum in the US Senate. For this reason, as opposed to a year earlier, COP16 was not thought capable of producing a new legally binding climate deal but merely a set of informal decisions to move forward on key elements of climate-related policy.[112] It was expected that states parties would at least declare their intention to work towards a new binding climate deal, while leaving on the table all options regarding a specific legal framework, including new obligations under Kyoto.

Observers hoped that the COP in Mexico would consolidate the various discussions and negotiation streams which existed: the two formal negotiation streams, supervised by the Kyoto and AWG–LCA, and the more informal Copenhagen Accord.[113] The key issue to resolve in this regard was how to integrate the non-binding Copenhagen Accord into the wider UNFCCC framework. Under the UNFCCC track, COP16 was expected to at least decide on the basic parameters of new or improved mechanisms in those areas where negotiations had reached a certain level, e.g. the REDD+ initiative by UN-REDD and others in respect of deforestation, or initiatives regarding finance or technology.[114]

### 2. Outcome

The Cancun Agreements did extremely well to integrate the key elements of the Copenhagen Accord into the UNFCCC. For example, –

- Cancun includes, for the first time in an official UN document, the objective to limit the temperature increase below 2°C. Unlike the year be-

112 CCES (2010b:1).
113 Stavins (2010a:3).
114 (ibid.).

fore, in Cancun the states parties formally agreed to the goal instead of only taking note of it.

- by incorporating the mitigation targets and actions pledged under Copenhagen into the UNFCCC, the new agreements set GHG emission reduction goals for some 80 countries.[115] As a consequence, for the first time, all the large GHG emitters – including Brazil, China, the EU, India and the US – have signed up under the UNFCCC for targets and actions to reduce GHG emissions by 2020.[116]
- under the UNFCCC track, the COP inaugurated the Cancun Adaptation Framework in order to improve adaptation efforts and instituted an Adaptation Committee to provide technical support to LDCs on adaptation-related matters.[117] While the emphasis of Kyoto was on mitigation, Cancun put adaptation firmly on the table in line with the IPCC's Fourth Assessment Report. The funding goals set in the Copenhagen Accord were reiterated.[118] A Green Climate Fund was established and designated as an operating entity of the financial mechanisms of the UNFCCC, and will be operated under the guidance of the COP with the World Bank as its interim trustee.[119]
- on the Kyoto track, while the decision about extending Kyoto beyond 2012 was once again referred to the next COP, progress was made on other issues,[120] such as the agreement to use 1990 as a base year and to continue emissions trading and other market-based instruments inaugurated by Kyoto.

### 3. *Assessment*

Despite very gloomy predictions ahead of the Cancun COP, the participants achieved unprecedented consensus on a range of issues going forward.

Observers suggest that much of the progress reached could be linked to a somewhat changed negotiation approach. Importantly, Cancun knocked a

115 Stavins (2010b:1).
116 CCES (2010c:1).
117 (ibid.:3).
118 Copenhagen foresaw the joint commitment by developed countries to provide US $30 billion in start-up finance for developing countries in 2010–2012, and their willingness to try to mobilise US$100 billion a year as from 2020.
119 UNFCCC (2012i).
120 CCES (2010c:6).

hole in the firewall between Annex I and Non-Annex I parties – a key step in overcoming the rich–poor gulf which has hobbled climate negotiations for many years.[121] The Cancun Agreements formulate two principles on which all countries are –[122]

- obliged to recognise their historic GHG emissions, i.e. the industrialised world, and
- liable for their future GHG emissions, i.e. the industrialised and the larger developing countries.

Moreover, the states parties seem to have recognised, at least implicitly, that moving forward in incremental steps is going to be more effective than holding out for an all-inclusive global climate deal.[123] Similarly, after the hostile recriminations between the US and China which deadlocked the COP in Copenhagen, these countries adopted a more productive tone in Mexico, with India as a key broker.[124]

## *VI. COP17, Durban, South Africa, 2011*

### *1. Expectations*

Despite the progress achieved in Mexico, it was still far from clear where the negotiations would be headed after Cancun.[125] Indeed, as outlined before, COP16 was once more unable to solve the major question of what to do with the Kyoto Protocol beyond 2012, once the first commitment period ran out. The issue of extending Kyoto for Annex 1 countries had been discussed during COP13 in Bali in 2007, but it has been put off ever since. However, with little more than a year to go before the end of Kyoto, observers pointed out that the question could not be delayed further, given that the necessary ratification would itself take at least a year.[126] This made Durban the last genuine opportunity to extend Kyoto into a second commitment period (CP2) and thereby prevent a so-called commitment gap.

121 Stavins (2011a:1).
122 Stavins (2010b:3).
123 (ibid.:2).
124 CCES (2010c:1).
125 (ibid.:2).
126 Stavins (2011b:3).

In the run-up to the COP17, experts discussed various possible negotiation options for Durban, ranging from a worst-case scenario to a very ambitious one:

- At worst, states parties would be unable to agree on prolonging Kyoto beyond 2012, not even informally. Many experts thought this was the most likely option going into the negotiations in light of the unresolved gridlock, described earlier. Even those Kyoto states parties that had supported its formal extension were no longer inclined to do so without reciprocal commitments by other nations.[127] For example, the EU, which was generally willing to sign up to CP2 of Kyoto, would only do so if negotiations for a new binding climate deal including China and the US were to be launched.[128] Then again, the US would only agree to negotiate such a new deal if it included GHG emission reduction goals of the "same legal force" for all of the major emitters – including China and India – although these obligations could be differentiated.[129] China, on the other hand, seemingly kept on opposing any binding targets, no matter how differentiated they were.
- In the very ambitious scenario, industrialised countries would agree in Durban to a formal extension of Kyoto. For the reasons explained above, most observers viewed this option as highly improbable. Moreover, with Canada, Japan and Russia's declaration to oppose a CP2 of Kyoto – a position which these countries declared just before Durban – a prolonged Kyoto would cover even less GHG emission reduction than the original Kyoto.[130]
- Viewed by most observers as the best-case scenario, states parties would reach an informal agreement launching a CP2 of Kyoto in which GHG emission reduction goals would be political rather than legally binding obligations.[131] Due to the political nature of the new targets, their endorsement would not require a formal amendment of Kyoto, but only a simple decision.

127 Bodansky (2011:5).
128 Stavins (2011b:3).
129 Bodansky (2011:2).
130 (ibid.).
131 (ibid.).

On top of the issue of an extension of Kyoto beyond 2012, it was hoped that COP17 in Durban COP would advance the UNFCCC negotiation stream and include –

- the review of the long-term objective of keeping global warming at 2°C, and
- the elaboration of incremental steps to implement Cancun – involving the overall level of the mitigation pledges, improved MRV, REDD+, adaptation issues, technology and finance.

## *2. Outcome*

In spite of the doom and gloom at COP17's opening, the states parties adopted the Durban Agreements on a range of issues that may lead to a historic breakthrough in international climate negotiations. What was sure was that COP17 kept the discussion of global climate efforts from breaking down and instead moved it in the right direction.[132]

Thirty hours after the scheduled end of the Durban COP, on the Kyoto track, the states parties agreed to a prolongation of Kyoto beginning in 2013. However, the details of the new reduction targets including the length of the new commitment period would be established at COP18. The BASIC group (and other NICs) remained Non-Annex I parties without binding targets.[133] States parties to CP2 of Kyoto would have to submit their quantified reduction targets by May 2012. However, Canada, Japan and Russia indicated that they would not participate in CP2 of Kyoto.[134]

The extension of Kyoto beyond 2012 was combined with the launch of new road map for the negotiation of a post-2020 climate agreement by way of "a protocol, another legal instrument or an outcome with legal force".[135] Negotiations are supervised by a newly formed Ad Hoc Working Group on the Durban Platform for Enhanced Action (ADP), and are to finish no later than 2015.[136] The new climate deal is to come into effect only from

132 Diringer (2011b:1).
133 Boyle (2011:4).
134 (ibid.).
135 CCES (2011:1).
136 Sterk et al. (2011b:8).

2020 onwards. Negotiations within the AWG–LCA stream were to continue for at least another year, until the COP in Doha.[137]

On the UNFCCC track, the COP took various significant steps to further the implementation of the Cancun Agreements:

- A major outcome was the operationalisation of the Green Climate Fund to serve as a key vehicle for climate funding.[138] For years, funding has been an issue of ongoing conflict between developing countries and their more industrialised counterparts. Though the UNFCCC includes funding mechanisms, it has a very weak role, and sums pledged to the three financing tools – the Least Developed Countries Fund, the Special Climate Change Fund, and the Adaptation Fund – are notoriously low.[139] However, while states parties were able to decide on the institutional structure of the Green Climate Fund, the decision does not indicate where the money for the Fund will actually come from beyond 2012.[140]
- Importantly, in order to further improve monitoring under the UNFCCC, Durban's COP17 introduced –[141]
  - a voluntary international assessment and review mechanism for developed countries, and
  - a non-binding international consultation instrument for developing countries.
- Finally, states parties agreed on various other operational actions including, for example, –[142]
  - the upcoming inauguration of the new technology instrument in 2012, and
  - funding and technical aspects of the REDD+ initiative.

### 3. *Assessment*

Observers believe that Durban kept the international climate negotiations intact and moving in the right direction, thus increasing the likelihood of

137 Boyle (2011:1).
138 (ibid.:7).
139 Sterk et al. (2011b:23).
140 (ibid.:25).
141 CCES (2011:3).
142 Boyle (2009:10–12).

meaningful long-term climate action.[143] South Africa reopened the door to legally binding GHG reduction targets for all major emitters – a door which seemed to have been closed after Denmark.[144] For China, India and the US to even consider an inclusive and robust legal agreement beyond 2012 was certainly an important move.[145] Durban was a step forward in more than one respect:

- The key importance of the innovative ADP was not so much that a further negotiation stream had been launched: it would have been sufficient to prolong the UNFCCC track beyond 2012. It was more that, for the first time, Durban offered states parties the option of a more symmetrical future climate agreement.[146] Thus, while the UNFCCC track is obliged to adhere to the CBDR principle, the ADP does not necessarily have to do so.[147] This is because the mandate to launch the Durban track does not include a reference to the CBDR principle of the UNFCCC or Kyoto, in spite of the insistence by Non-Annex I parties to do so.[148]
- That Kyoto will live to see yet another day has been hailed as a major realpolitik victory for progressive countries, including the EU.[149] Despite tough opposition, the EU managed to form a "green coalition" with the most vulnerable nations, LCDs and two of the BASICs (Brazil and South Africa) in order to obtain a road map for a new universal climate treaty.[150] In doing so, the EU's interventions were vital in avoiding another Copenhagen outcome, which, at times, was a very real possibility in Durban.[151] According to observers, had there been no agreement on a road map, the EU would not have agreed to commit to a CP2 of Kyoto; this, in turn, would have led to developing countries blocking any headway to be made along the AWG–LCA stream, with potentially devastating consequences for the overall international climate negotiations.[152]

143 Diringer (2011b:1); Stavins (2011c:1).
144 Climatico (2011:1).
145 Boyle (2011:1).
146 Diringer (2011:1).
147 Stavins (2011c:2).
148 Rajamani (2011:1).
149 Sterk et al. (2011b:4).
150 (ibid.:31).
151 (ibid.).
152 (ibid.).

- What is also significant is the fact that old alliances were crumbling. In the past, developing countries used to negotiate as one G77 bloc, but were often hijacked by the larger developing countries. In Durban, for example, the LDCs and SIDSs formed a coalition with the EU.[153] Similarly, the BASIC group fractured in Durban: first South Africa and then, later, Brazil insisted on a road map for a new universal treaty, while China indicated its willingness to ponder adopting binding goals after 2020.[154]

Yet again, observers raised various concerns. For one, great ambiguity remains over the legal nature of the post-2020 climate deal, particularly the binding force of the targets. The agreement to launch the ADP, the new negotiation platform, undertakes "to develop a protocol, another legal instrument or an agreed outcome with legal force under the UNFCCC."

However, apart from the question of what is meant by an "agreed outcome with legal force" – which is a novel term in international law – even a legally binding agreement need not contain legally binding commitments.[155] The Durban forum may well have provided for a legally binding agreement by referring to an "agreed outcome with legal force", but nowhere does it state that this new agreement is to include legally binding mitigation targets. It can be expected that the US and others will continue to fight tooth and nail against having quantified binding goals in a 2020 agreement.[156] Observers also warned that the new emission reduction targets under a CP2 of Kyoto would be much weaker than the ones under the first commitment period (CP1) of Kyoto because, for many countries, they would consist of voluntary national pledges which are not linked to an overall emission reduction objective.[157] The large developing countries are still without binding targets, Russia and Japan will not participate in CP2 of Kyoto, and Canada withdrew from Kyoto for good just after COP17 in Durban in 2011.

153 (ibid.:33).
154 (ibid.).
155 Bodansky (2012:1).
156 Sterk et al. (2011b:33).
157 Boyle (2011:2).

## VII. COP18, Doha, Qatar, 2012

### 1. Expectations

In terms of public and media interest, Doha was expected to be a lower-key meeting than its predecessors in Cancun, Mexico, in 2010, and in Durban, South Africa, in 2011.[158] After three years of rather 'big moment' COPs, i.e. in Copenhagen, Denmark, in 2010, and those in Mexico and South Africa, the COP in Doha, Qatar, in 2012 was more about giving operational momentum to the agreements reached in Durban.[159] The progress required in Doha was deemed quite modest, so it was important that expectations were not too high in order to avoid it being labelled "another failure".[160] Although no new initiatives or processes were expected to be launched, much remained at stake. COP18 in Doha came at a particularly important time, namely the end of CP1 of the Kyoto Protocol on 31 December 2012. Hence, Doha would be vital in ensuring a seamless continuation of the Protocol as of 1 January 2013 and, thus, the maintenance of the only international legally binding instrument that existed in respect of tackling global warming.[161] Securing final agreement for the immediate launch of a CP2 of the Kyoto Protocol from 1 January 2013 would be a defining feature of the meeting.[162] Critical issues included the following:

- *CP2 timeframe of five or eight years:* Parties were divided as to whether CP2 should end in 2017 or 2020.[163]
- *The level of ambition:* Despite clear scientific evidence for the kind of reductions required from developed countries, i.e. a 25–40% cut in 1990 emissions by 2020, the level of ambition for a CP2 remained uncertain. Before COP18, the only numbers on the table were those pledged in the aftermath of the Copenhagen summit.[164]
- *Market-based instruments:* Developing countries argued that industrialised countries which were not part of a CP2 of Kyoto should not be

158 The Climate Group (2012a:1).
159 WRI (2012a).
160 House of Commons (2012:6).
161 IISD (2012:3).
162 The Climate Group (2012a:1).
163 FIELD (2012:7).
164 The Climate Group (2012a:4).

entitled to use the CDM or other market-based tools instituted in the CP1 of Kyoto,[165] and

- *Hot air targets:* Many countries – both developed and developing – requested avoiding the granting of hot air targets under a CP2 by limiting the number of surplus emissions units that could be carried over from a CP1 to a possible CP2.[166]

Furthermore, COP18 was expected to conclude the work on the UNFCCC track inaugurated in Bali in 2007 and supervised by the AWG–LCA since then.[167] The three key issues under negotiation in Doha, therefore, included mitigation (for both developed and developing countries), funding, and MRV – although, in the run-up to Doha, states parties were sharply divided over what issues needed finalisation at the COP18.[168] As regards funding, the vital question was whether developed countries would make new commitments for the 2013–2020 period to bridge the gap between the 'start finance' period, which was to end in 2012, and the US$100 billion commitment made in Copenhagen, which begins in 2020.[169] The task in Doha was either to formally conclude work on each of these areas or move them to the two UNFCCC subsidiary bodies or to the ADP itself.

Finally, under the ADP, much needed to be concretised, e.g. starting to define the details regulating the negotiations expected to set up the new "protocol, another legal instrument or a legal outcome under the [UNFCCC]" applicable to all states parties by 2015.[170] As a negotiation track, the objective in Doha was to consolidate the understanding among countries on the negotiation issues for the next three years under both so-called workstreams:[171]

- Under Workstream 1, which focused on the post-2020 treaty, countries were to discuss how the UNFCCC's existing framework would apply to the new 2020 treaty; for example, developing countries were insisting on upholding the CBDR rule and the so-called firewall between Annex I and Non-Annex I parties, and

165 The Climate Institute (2012:16).
166 The Climate Group (2012a:5).
167 ICCG (2012:2).
168 The Climate Group (2012a:2).
169 The Climate Institute (2012:6).
170 ICCG (2012:2).
171 The Climate Group (2012a:9).

- Under Workstream 2, countries were to look at how to improve the level of existing pledges for the 2013–2020 period.

### 2. Outcome

The Doha Climate Gateway successfully managed to end two long-standing negotiation streams, i.e. Kyoto and the UNFCCC track, and to progress to a unified track – the ADP – with the objective of an inclusive legal climate agreement by 2015.

First and foremost, the COP agreed on the revision of the Kyoto Protocol to formally establish the CP2.[172] Having been launched in 2005, the AWG–KP thus terminated its work in Doha. States parties also decided that CP2 would run for eight years, i.e. from 2013 through 2020.[173]

In addition, the extended Kyoto features an ambition trigger, requiring that states parties verify and increase their emission reduction targets by 2014 in line with the IPCC's Fourth Assessment Report.[174] In the face of strong Russian objections, as the main benefactor of hot air targets under KP1, states parties decided to restrict the use of the surplus emission allowance gained under KP1.[175] For Non-CP2 parties such as Canada, Japan, New Zealand and Russia, COP18 agreed to restrict their eligibility to market-based instruments, including the CDM and emissions trading.[176] Furthermore, the states parties agreed to terminate the Convention track under the AWG–LCA. As expected, the COP simply took note of the pledges already listed under Copenhagen, while launching a one-year work programme to verify those pledges. Efforts to improve the accounting rules for the MRV were unsuccessful.

A major outcome of the UNFCCC track was the agreement to look towards establishing some kind of 'loss and damage' mechanism in favour of the most vulnerable countries at COP19 in 2013.[177] On the other hand, industrial countries refused any new funding commitment, and only agreed to maintain through 2015 the average finance levels provided during 2010–

172 Marcu (2012:1).
173 CCES (2012:2).
174 WRI (2012b).
175 CCES (2012:2).
176 WRI (2012b).
177 The Climate Group (2012b:20).

2012 – roughly US$10 billion a year.[178] At least a few European countries, i.e. Denmark, France, Germany, Sweden, and the UK, pledged to somewhat augment their funding post-2012.

Under the ADP track, China, India and other developing countries again tried to introduce the CBDR principle explicitly into the ADP framing, which the US and others had objected to include in the mandate.[179] The COP decided to establish a one-year work programme to think through the application of the UNFCCC principles.[180] The COP also stated that the ADP should consider "elements for a draft text" for the new agreement no later than COP20, "with a view to making available a negotiating text before May 2015".[181] For the rest, the ADP decision was mainly of a procedural nature.

### 3. *Assessment*

> This round of international climate change talks was a modest step forward. We always knew they would be very tough after the breakthrough at the same conference in Durban last year.[182]
> The success in agreeing a second Kyoto commitment period, although important politically, is also something of a Pyrrhic victory for its supporters.[183]

Despite the devastation wrought by Hurricane Sandy and President Barack Obama's re-election, the US was once again less than helpful in moving ahead, declining any proposal to increase their emission reduction targets or to commit to new funding.[184] The EU is similarly to blame for a lack of progress on this score, since the traditional frontrunner arrived at Doha with a reduction objective it had basically already met, with no joint funding commitment, and with divergent positions among the EU member states on various issues.[185] With the industrialised world unwilling to increase their targets or to improve on funding, there was no incentive for the likes of China or India to better their voluntary emission reduction goals.[186]

178 WRI (2012b).
179 CCES (2012:3).
180 WRI (2012b).
181 CCES (2012:4).
182 Davey (2012); Sterk et al. (2012a:1).
183 The Climate Group (2012c).
184 Sterk et al. (2012b:39).
185 The Climate Group (2012b:2).
186 Sterk et al. (2012b:38).

- Observers warn that the outcome of COP18 was even more modest than would have been necessary. CP2 will be a far cry from CP1 in terms to emission reductions: under the original Kyoto, all industrialised countries (39 at the time) – representing more than 55% of all global emissions – committed to reducing those emissions. With Canada having withdrawn from the treaty entirely, and Japan and Russia declining to sign up to CP2, this left the EU27, plus Australia (subject to conditions), Belarus, Iceland, Kazakhstan, New Zealand (possibly), Norway, and the Ukraine as members[187] – representing only around 15% of global emissions. Also noteworthy is that, at the time of writing, the 2020 target of the largest party (the EU) had almost already been met.[188] Moreover, the overall emission reduction to be achieved under CP2 will be approximately 18% by 2020 from 1990 levels and, hence, significantly less than the 25–40% range recommended by the IPCC.[189] The net result is that Doha left the world firmly on track to 4°C or more of warming by 2100.[190]
- In terms of the UNFCCC track, mitigation ambitions remained low. As expected, funding proved to be the most difficult issue to resolve in Doha. There was no joint commitment by Annex I parties in terms of mid-term funding from 2013 to 2020. The relevant decision simply "urges", "invites" and "encourages". Annex I parties to increase their funding, but when their "financial circumstance permit".[191]

Yet again, Doha also produced some positive results, as follows:

- Its main objective was to streamline the complex, multi-track negotiating process. The achievement in reducing the overall negotiations down to one unique track from 2013 onwards should not be underestimated:[192] it allows states parties to concentrate on the discussions at hand, and frees up time and resources for states parties to the UNFCCC.[193]
- As for the level-of-ambition discussion, there remains hope. Workstream 2 under the ADP, which was instituted to increase pre-2020 ambitions of the states parties, and the review of the CP2 states parties' targets

187 The Climate Group (2012c).
188 The Climate Group (2012b:2).
189 Sterk et al. (2012b:6).
190 WRI (2012b).
191 The Climate Group (2012b:3).
192 The Climate Group (2012c); CCES (2012:1).
193 Marcu (2012:1).

envisaged between 2013 and 2015, provide the tools for jump-starting mitigation efforts.[194]

- The idea of setting up a kind of 'loss and damage' instrument at COP19 in 2013 is a major achievement for LDCs, particularly for those most vulnerable to the long-term impacts of global warming.[195] Where mitigation and adaptation fail, people may suffer damages to their assets and health due to global warming. Yet, industrialised countries, notably the US, remain extremely wary of such a legal mechanism, fearing that, as traditionally high emitters, they may be held liable for damages of potentially unlimited economic value whose attribution to global warming may still be unclear.[196]

However, following Doha, the Wuppertal Institute for Climate, Environment and Energy pointed out the following:[197]

> The decisions required in 2015 will be momentous: to raise collective global ambition for 2020–30 to meet the 2°C pathway; to agree a new, legally binding framework; to identify the sources of finance that can meet the goal of providing US$100 billion in climate assistance to the poorest countries by 2020; and to agree a new international collaboration on the development, demonstration and deployment of low-carbon technologies.

With just three years to go, there is no time to lose. The next three sections therefore investigate what policy instruments and methods individual countries use to tackle global warming. Two industrialised societies are looked at in detail, namely the EU, which is a CP1 and CP2 Kyoto Annex I party; and the US, the second largest GHG emitter, but not a party to Kyoto. The investigation then directs its attention to a member of the BASIC group of NICs, namely China, as a Non-Annex I party and the highest current emitter of GHGs. Based on Kuik et al.'s synthesis article of 2008,[198] the investigation aims to outline the main climate policy instruments used by these three countries. Kruik et al. examine various policy approaches for a new climate treaty along the lines of what they regard as five key policy dilemmas:[199]

194 The Climate Group (2012b:1).
195 (ibid.:3).
196 Morgan (2012).
197 Sterk et al. (2012b:43).
198 Kuik et al. (2008).
199 (ibid.).

- *The Carrot v The Stick*, i.e. taxes, permits and standards v financial incentives such as subsidies and tax credits
- *The 'Front Door' v The 'Back Door'*, i.e. direct climate policy (the front door) v indirect climate policy as a side benefit of other policies such as those on energy, air quality, technology and security (the back door)
- *Market-based Instrument v Direct Regulation*
- *Multilateral Parties v Small (Unilateral) Parties; and*
- *Mitigation v Adaptation.*

The discussion then turns to the developing world, with a special focus on LDCs.

## *D. Climate Change and Industrialised Countries*

### *I. The European Union*

The EU, representing 27 member states, with a combined output of 4,050 Mt of carbon dioxide in 2010, is the world's third largest GHG emitter after China and the US, accounting for 12.3% of global carbon dioxide emissions, and having shown an increase of 3% from 2009 (and a decrease of 5% from 1990).[200] The EU27's emission intensity of 0.26 kg carbon dioxide per unit of GDP[201] in 2010 is about a third lower than the US's (0.41 kg) and China's (0.77 kg), and has decreased by 37.2% from 1990.[202] Per capita carbon dioxide emissions in the EU27 totalled 7.29 t of carbon dioxide in 2010 (compared with 17.3 t for the US and 5.4 t for China), which signifies a reduction of 17.3% for the EU27 from 1990.[203] The EU has long been at the vanguard of international efforts to address global warming, and has been a state party to the UNFCCC since 1993 and the Kyoto Protocol since 2002. Unlike the US, the EU follows mainly a 'stick' climate policy approach, i.e. a permit and tax system, combined with direct regulations such as fuel efficiency standards. This top-down approach is complemented by financial incentives to stimulate innovation and market-based instruments. The European Climate Change Programme (hereinafter ECCP I) of 2000, replaced in

200 NEAA (2011:11, 14).

201 In 2005 US Dollars, using purchasing power parities.

202 IAE (2011:96, 98). Note that the numbers relate to carbon dioxide emissions from fossil fuel combustion, which account for over 90% of all such emissions.

203 (ibid.:99, 101).

2005 by the ECCPII, serves as the main EU climate policy document, which is an example for a 'front door' approach.

### *1. Mitigation Policy*

#### a) Binding Mitigation Targets?

Under Kyoto, the 15 older EU member states (EU-15) took on the obligation of reducing their collective GHG emissions by 8% on average in the first commitment period until 2012, compared with its 1990 base-year emissions.[204] While the EU's Fifth National Communication under the UNFCCC in 2009 projected that the EU-15 would overshoot their Kyoto target by 5.8% in 2010,[205] individual members had varied results in achieving their targets. Though some member states, like France, Greece and the United Kingdom, have reduced domestic GHG emissions beyond their Kyoto targets, others like Austria, Denmark, Italy, Portugal and Spain are lagging in compliance.[206]

Regarding the 2013–2020 follow-up, in March 2007 the EU took the unilateral decision to reduce its GHG emissions by at least 20% by 2020 compared with its 1990 levels. In order to implement this obligation, in June 2009 the EU adopted a Climate and Energy Package,[207] in which it reiterated the overall target of a 20% GHG emission reduction, with an additional commitment to push this up to 30% if a satisfactory international agreement involving all big GHG emitters is reached. Apart from these new mitigation targets, the package obliged EU members to generate 20% of their energy from renewables by 2020. In a communication of May 2010[208] and a Staff Work Document of February 2012,[209] the European Commission (hereinafter *Commission*) investigated various options for moving towards a 30% GHG emission reduction, underlining that the 20% target was well within reach, given a much quicker GHG emission decrease than originally anticipated due to high oil prices as well as the ongoing global financial crisis.

204 EC (2011c).
205 EC (2009a:vi).
206 Barrett (2009:62).
207 EC (2007).
208 EC (2010a).
209 EC (2012a).

Yet, due to opposition from some EU members, the EU was unable to push its emission reduction target to 30% at COP18 in Doha, while observers indicated that the 20% target had already been met.

With regard to the long-term ambition of keeping the global temperature increase to below 2°C, the Climate and Energy Package contains an EU objective of reducing domestic GHG emissions by 80–95% below 1990 levels by 2050, which was reconfirmed by the European Council (hereinafter *Council*) in February 2011. In March 2011, the Commission adopted a *Roadmap for Moving to a Competitive Low Carbon Economy in 2050*[210] outlining scenarios on how to achieve this target.

### b) Policy Instruments

In order to achieve the binding mitigation goals, the EU uses various regulatory instruments, central to which is the trade in GHG emission permits, i.e. the EU Emissions Trading System (EU ETS) launched in 2005, which now in its third phase, running till 2020.[211]

The 2005 Directive introduced a mandatory 'cap and trade' regime. According to this regime, a GHG emission limit for the whole ETS is fixed, and for each year a certain quantity (a 'cap') of GHG emission allowances is granted to EU members, who in turn distribute ('trade') these via public sale to the 10,800 installations throughout the bloc participating in the scheme.[212] If an installation emits more than the allowances it has obtained, it has to buy unused allowances from other installations under the ETS.[213] In 2009, a revised ETS directive was adopted[214] to further improve the EU scheme for a third phase running from 2013 to 2020. Moreover, the Commission wants to promote the creation of a robust OECD-wide carbon market by 2015, to be further extended to the larger developing countries by 2020.

210 EC (2011b); hereinafter *Roadmap for 2050*.

211 Directive 2003/87/EC of the European Parliament and the Council of 13 October 2003 Establishing a Scheme for Greenhouse Gas Emission Allowance Trading within the Community, *Official Journal L 275*, 25 October 2003, 32–46.

212 EC (2009a:84–86).

213 (ibid.).

214 Directive 2009/29/EC of the European Parliament and the Council of 23 April 2009 Amending Directive 2003/87/EC to Improve and Extend the Greenhouse Gas Emission Allowance Trading Scheme of the Community, *Official Journal 140*, 5 June 2009, 63–87.

Up until 2012, the EU ETS only involved carbon dioxide emissions in the power- and heat-generating sector as well as a few other emissions-intensive industries (oil, iron and steel) across the 27 EU members plus Iceland, Liechtenstein and Norway. While overall ETS guidelines are set at EU level, allocation rules and GHG emissions caps are determined at national level. The firms in question obtain allowances to emit a certain tonnage of GHG each year,[215] but the overall emissions level is reduced over time. It is currently some 6.5% below the 2005 level, and by 2020 will be 21% lower. In 2008, 3 billion t of carbon dioxide were traded at a market value of US$92 million.[216]

While the EU ETS is generally acknowledged for its pioneering role, observers outline a number of flaws: at the beginning, it was unclear whether it was indeed more effective than carbon taxes in terms of reducing GHG emissions.[217] In Phase II, the ETS only covered 40% of all EU GHG emissions, notably excluding the transport sector, which was an important omission.[218] Other weaknesses included the lack of harmonised rules on allocation within the EU, the lack of strict enforcement tools, and the missing linkage between the EU ETS and other ETSs in third countries.[219] In the revised EU ETS Directive of 2009, coverage of the ETS was broadened for Phase III in order to include carbon dioxide emissions from the chemical industry and the aviation sector as well as certain other GHGs such as nitrous oxide. Moreover, caps are to be set at EU level and allocation rules harmonised across the EU. In spite of these structural reforms the ETS has been facing its most important challenge to date over the past months: the growing surplus of allowances due the economic crisis risks undermining the orderly functioning of the regime.

As to the other main regulatory instrument usually discussed by observers, namely carbon taxes, notably, quite a few EU members have introduced some form of eco-taxation;[220] however, there is no EU-wide carbon tax as

---

215 For an overview, see Massai (2007:18).

216 CCES (2009a:3).

217 Helm (2009a:229).

218 Farnsworth (2007:29).

219 Massai (2007:21).

220 Finland and The Netherlands (1990), Norway and Sweden (1991), the UK (1993), Germany (1999), Denmark (2002) and Ireland (2010) have a carbon dioxide tax in place; Austria, Belgium and Slovenia have some kind of carbon elements in their tax regime; in 2010 Spain was investigating the options for introducing carbon taxations; see Wilkinson (2012).

yet. However, proposals for a carbon taxation dispensation have been floated since 1999, prompting the Commission in 2010 to propose the amendment of the Energy Taxation Directive of 2003 which lays down common rules for the taxation of energy products.[221] The proposal aimed to introduce a carbon tax mainly for those sectors not yet included in the ETS.[222] However, the proposal has yet to reach a majority in the Council due to considerable lobbying by industry.

These regulatory instruments are accompanied by incentive-based mechanisms which promote the transition to a low-carbon society. The Energy Review of 2008, which stress that the EU's climate goals for 2020 will necessitate an overhaul of EU energy arrangements, looks at the challenges facing the bloc between 2020 and 2050 and formulates an EU Strategic Energy Plan.[223] This Plan tries to speed up the development of innovative, inexpensive, low-carbon technologies, and is built on a wide-ranging research and development scheme.[224] The initiative is complemented by the European Energy Programme for Recovery, which allocated €1 billion for carbon capture and storage installations and €50 million for offshore wind installations.[225] In 2011, the Commission proposed its *Roadmap for 2050*, which is based on the view that innovative ideas are needed to scale up investments in energy, transport, industry and information technologies and that more focus is necessary to combat energy inefficiency.[226] Together with the Energy Efficiency Plan of 2011 and the White Paper on Transport of 2011, the *Roadmap for 2050* is a key deliverable to achieve the EU's long-term objective of reducing GHG emissions up to 95% by 2050.[227]

### 2. *Funding for Developing Countries*

As an Annex I party under the UNFCCC, the EU is obliged to assist developing countries to tackle global warming, both in respect of reducing GHG emissions and in adapting to the unavoidable impacts of climate change.

221 EC (2009a:109).
222 FSB (2010:2).
223 EC (2012b).
224 EC (2010b).
225 EC (2012c).
226 EC (2011a:3).
227 (ibid.:4).

Before COP15 in Denmark in 2009, the Commission adopted a communication *Stepping Up International Climate Finance: A European Blueprint* for the Copenhagen deal, recognising that supporting developing countries was vital to reaching an ambitious outcome at COP15. The blueprint identified that, by 2020, developing countries would incur yearly costs of €100 billion to finance their mitigation and adaption activities, and proposed that industrialised nations and larger developing countries grant them funding to the tune of some €22–50 billion a year, with the remaining €50 billion coming from national sources and expanded international carbon dioxide markets.[228]

At COP15 in Denmark in 2009 and COP16 in Mexico in 2010, the EU and other industrialised countries pledged to jointly grant nearly US$30 billion from 2010 to 2012 to kick-start the scheme, and offered to mobilise US $100 billion a year by 2020. Despite budgetary constraints, the EU did, in fact, manage to award €2.34 billion in 2010 as well as in 2011, bringing its contribution to €4.68 billion, or 65% of the overall pledge for 2010–2012, most of which was deployed through existing instruments.[229]

With the assistance of the EU, COP17 in Durban in 2011 launched the Green Climate Fund, the new funding instrument intended to serve as the key long-term financing vehicle. However, states parties were unable to reach consensus on where the money for the Fund would come from, in either the medium (beyond the fast-start resources) or long term.[230] In recent years, three funding mechanisms have been set up by the Commission, i.e. the Global Climate Change Alliance, the Global Energy Efficiency and Renewable Energy Fund, and the Climate Change Windows, pooling more than €1.5 billion in grants from the EU budget and EU members' national budgets. These tools are estimated to leverage around €14 billion in climate finance by 2013.[231]

### *II. The United States of America*

The US, with its output of 5,250 Mt of carbon dioxide in 2010 alone, is the world's second largest GHG emitter after China and ahead of the EU in third

228 EC (2009a:244).
229 EC (2012d).
230 Boyle (2011:2).
231 EC (2012d).

position. The US accounts for 15.9% of global carbon dioxide emissions, showing an increase of 4% from 2009 (and an increase of 5% from 1990).[232]

The US's emissions intensity of 0.41 kg carbon dioxide per unit of GDP is almost half as low as China's (0.77 kg) and about a 40% higher than the EU27's (0.26 kg), and has decreased by 30.6% from its 1990 levels.[233] Per capita emissions in the US totalled 17.3 t of carbon dioxide in 2010, which is more than double the value of the EU27's 7.29 t and more than three times the value of China's 5.4 t; for the US, this signifies a reduction of 11.0% since 1990.[234]

While the US became a state party to the UNFCCC in 1992, it declined to ratify Kyoto in 2003, making it the only major industrialised country – and the world's largest GHG emitter at the time – to do so. Unlike the EU, against the background of voluntary mitigation goals, the US pursues a 'carrot' climate policy approach with various incentive-driven instruments rewarding energy innovation, such as its Climate Change Technology Program, while favouring market solutions. This bottom-up approach is – albeit only recently – complemented by direct regulations, i.e. energy standards and mandates. Unlike the EU, the US lacks an explicit climate policy and strives to attain reduction of GHG emissions by indirect means such as air quality and energy efficiency policies.

### *1. Mitigation Policy*

#### a) Binding Mitigation Targets?

Since the US has never ratified Kyoto, it has, to date, never signed up to binding mitigation targets at international level. Following COP15 in Denmark in 2009, the US signed up to the Copenhagen Accord and took on the voluntary goal of reducing its GHG emissions by 17% below 2005 levels. With regard to the long-term vision of keeping global warming at 2°C, at COP15 the US Government stated that it sought to voluntarily reduce its GHG emissions by 85% by 2050, compared with its 2005 levels.

---

232 NEAA (2011:11,14).
233 IAE (2011:96, 98).
234 (ibid.:99, 101).

Neither has the US, at federal level, adopted an explicit climate policy ('front door' approach). In order to reach the voluntary goal agreed at COP15, the Obama Administration made an all-inclusive obligatory Climate Bill one of its legislative priorities. The so-called American Clean Energy and Security (ACES) Bill was passed by the House of Representatives in June 2009,[235] but was defeated in the Senate in June 2010.[236] The ACES Bill would have established a nationwide cap-and-trade scheme covering 85% of US GHG emissions, with the long-term goal of achieving an 80% reduction in GHG emissions relative to 2005 levels by 2050.[237] Emissions limits would have been placed on power generation, oil refining, natural gas supply, and other energy-intensive industries, such as iron and steel, cement and paper, covering approximately 85% of US GHG emissions by 2016.[238]

Since the failure of the ACES Bill, the Clean Air Act (CAA) has become vital for developing a federal climate policy ('back door' approach).[239] The CAA formulates the broad authority of the Environmental Protection Agency (EPA) to develop regulations to mitigate harm from air pollution. In 2011, the EPA succeeded in implementing regulations imposing mandatory fuel standards for vehicles, thus introducing, for the first time, obligatory emission reduction targets at national level via the 'back door' of the CAA.[240]

For lack of binding mitigation targets at federal level, individual states have, over the years, introduced obligatory emission objectives. Since November 2009, 23 of the 50 states had adopted a state GHG emission reduction target, although these vary in stringency, timing, and enforceability.[241] As for carbon dioxide trading systems, the most important is the Regional Greenhouse Gas Initiative involving 10 north-east US states, which was launched in 2009 as the first-ever obligatory cap-and-trade programme.[242] Emissions from large power generators in 10 north-east and mid-Atlantic states are capped at 2009 levels, and the cap will be reduced by 2.5% in each of the four years from 2015 through 2018, for a total reduction of 10% below 2009 levels.[243]

235 CCES (2009c).
236 Zusman et al. (2012:3).
237 (ibid.).
238 (ibid.).
239 Burtraw (2011:1).
240 US (2010:44).
241 (ibid.:62).
242 Burtraw (2011:1).
243 US (2010:61).

### b) Policy Instruments

The US mitigation policy at federal level is essentially incentive-driven and aims to achieve a reduction of GHG emissions by way of investing in low-carbon technology and renewable energy. The main policy instrument is the American Recovery and Reinvestment Act of 2009, through which the US Government offers subsidies and tax incentives of more than US$90 billion for investment in sustainable energy technologies.[244]

Key aspects include the following:[245]

- Appropriating funding for numerous grant programs and tax incentives for clean energy technologies;
- A 30% tax credit for residential energy investments, as well as mandates for improved energy standards for heating facilities;
- Increasing the investments allocated to new clean renewable energy bonds and qualified energy conservation bonds;
- Investing in critical energy infrastructure by providing loan guarantees for new or upgraded electric power transmission projects, and by providing funding for the Smart Grid and new Smart Grid technologies;
- Asserting an energy efficiency leadership role for the federal government, investing in the "green" conversion of federal facilities, and purchasing vehicles for government use with higher fuel economy, including hybrid and electric vehicles.

Another very important incentive scheme is the Energy Improvement and Extension Act adopted in 2008, which offers a set of incentives for renewable energy production, clean coal and carbon sequestration, as well as energy-efficient transportation.[246]

Recently, this incentive-based policy has been complemented by regulatory instruments which are designed and implemented by EPA, based on the CAA:[247]

- The first involves new vehicle fuel economy standards regulations that took effect in January 2011, affecting all vehicles beginning with the 2012 year model
- The second instrument introduces permits for the construction of and major alterations to new sources of GHG emission. Since January 2011,

---

244 (ibid.:41).

245 For more information, see (ibid.:40).

246 (ibid.:43).

247 Burtraw (2011:2).

this instrument has applied to about 900 construction projects per year at sites that emit large quantities of GHGs, and

- The third and most important tool concerns performance standards that apply to the operation of new GHG-emitting sources in various categories.

*2. Funding for Developing Countries*

As with the EU, the US is committed to helping developing countries in their mitigation and adaptation efforts. Since 1991, the US Agency for International Development (USAID) has included climate change funding mechanisms in its development funding, spending approximately US$2.6 billion on climate-related development programmes.[248] However, in its evaluation of the Fourth US National Communication in 2009, the UNFCCC noted that US resources, which expressly target developing countries, in particular LCDs and SIDSs were modest.[249] This is mainly due to the US's 'back door' approach according to which climate goals are embedded in a wider development agenda.

Against this background, the Obama Administration passed the Consolidated Appropriations Act of 2010, which nearly tripled climate-related foreign assistance to over US$1 billion in 2010, including a first-ever US contribution of US$50 million to the Least Developed Country Fund and Special Climate Change Fund; a contribution of US$375 million to the World Bank-managed Climate Investment Funds; and substantially increased funding for the USAID climate programmes.[250] At COP16 in Mexico in 2010, the US pledged to contribute US$1 billion between 2010 and 2012 in aid to reduce GHG emissions from deforestation, land degradation, and other activities.[251] Furthermore, in 2004, the US's Millennium Challenge Account (MCA) was launched. To date, agreements with 20 countries totalling nearly US$7.2 billion have been signed under the MCA.[252]

248 US (2010:77).
249 UNFCCC (2009b:30).
250 US (2010:98).
251 (ibid.:99).
252 (ibid.:102).

### *E. Newly Industrialised Countries: China*

China, with its output of 8,950 Mt of carbon dioxide in 2010 alone, is the world's largest GHG emitter, accounting for 27.1% of global carbon dioxide emissions, showing an increase of 10% from 2009 (and of 257% from 1990).[253] Coal constitutes 70% of China's primary energy – more than twice the international average.[254] China's emissions intensity of 0.77 kg carbon dioxide per unit of GDP in 2010 is almost double the US's value of 0.41 kg and almost three times higher than the EU's value of 0.26 kg, but has decreased by 52.4% since 1990.[255] Per capita carbon dioxide emissions in China totalled 5.4 t of carbon dioxide in 2010, which is still below the value of the EU's 7.29 t and less than one third of the US's value of 17.3 t; for China, however, this signifies an increase of 174.3% since 1990.[256] Even more importantly, the NEAA forecasts that, by 2017, China will have overtaken the US as the highest per capita GHG emitter.[257]

China employs mainly a 'stick' climate policy approach through a permit and tax system using various direct regulations, supplemented by some market-based instruments. Like many other developing countries, China tries to achieve its climate goals indirectly, as side effects of a general development policy (the 'back door' approach).[258]

#### *I. Binding Mitigation Targets?*

China has participated actively in the international climate negotiations since the beginning and has ratified the UNFCCC as well as the Kyoto Protocol. However, it is important to remember that China, as a Non-Annex 1 (developing) country, did not have to take on quantified binding mitigation targets under Kyoto.[259]

---

253 NEAA (2011:11, 14).
254 CELP (2012:1).
255 IAE (2011:96, 98). Note that the numbers relate to carbon dioxide emissions from fossil fuel combustion, which accounts for over 90% of all carbon dioxide emissions.
256 NEAA (2011:14).
257 (ibid.:12).
258 Lewis (2007:1).
259 China, an active participant in the CDM, is by far the largest source of CDM credits, accounting for over 40% of those generated to date.

For years, China, together with India, followed a rather inflexible policy at the climate negotiations, rejecting each attempt to commit it to setting (binding) GHG emission reduction goals, emphasising the historical liability of the industrialised nations and its own development needs.[260] While this line was reiterated once more at the G8 Summit in Germany in 2007, where President Hu Jintao argued that climate-related policy was essentially development policy, after the COP13 in Bali in 2007 there were indications that China was willing to take on a more proactive role in climate negotiations.[261]

In November 2009, China stated its intention to reduce the intensity of carbon dioxide emissions per unit of GDP by 40–45% by 2020, compared with 2005 levels.[262] According to China, this is a "domestic voluntary action" which will be included as a compulsory indicator in its medium- and long-term planning for economic development.[263] In January 2010, China followed up on its statement and voluntarily pledged to reduce its GHG emissions intensity by up to 45% by 2020, a target which was reiterated at COP16 in Mexico in December 2010.

In Mexico 2010, China – at least officially – continued to put forward its view that any legally binding climate change mitigation objectives were unacceptable unless the US accepted them as well.[264] However, at COP17 in Durban in 2011, for the first time, China stated its willingness to participate in a legally binding international climate treaty, depending on the outcome of negotiations.[265] Accordingly, China would be willing to take on legally binding commitments matched with its economic development and in line with the CBDR principle under the UNFCCC,[266] but the country set out five specific preconditions:[267]

- Parties must extend Kyoto through a second commitment period;
- Developed countries must meet their funding obligations under the Green Climate Fund;
- The consensus reached in Durban on funding, technology, REDD+, adaptation, and MRV measures must be institutionalised;

260 Gupta (2007:167).
261 Oberheitmann & Sternfeld (2009:141).
262 Xinhuanet (2009).
263 (ibid.).
264 Gupta (2007:177).
265 Hsu (2011:2).
266 Xinhuanet (2011).
267 Hsu (2011:2).

- The obligation to review the adequacy of long-term goals scheduled to take place between 2013 and 2015 must be completed, and
- A framework for a post-2020 agreement must be defined that upholds the CBDR, respective capacities, and environmental integrity.

## *II. Policy Instruments*

With growing political attention focused on the impacts of global warming, China's first National Report on Climate Change was issued in late 2006. In June 2007, China adopted a National Climate Change Programme, outlining a list of key measures until 2010.[268] With this step, China became the first developing country to have an overarching climate strategy.[269] In 2008, the State Council also issued a White Paper on Climate Change.

China's climate-related policy focuses on two main aspects:

- Lowering the energy intensity,[270] while acknowledging that coal will be the primary energy source for many more years, and
- Improving the use of green forms of energy, including nuclear energy and renewables, but also carbon capture and storage.[271]

China's 11th Five-Year Plan, covering 2006–2011, and the Medium- and Long-term Development Plan for Renewable Energy in China (DPRE) of 2007 introduced binding goals for energy intensity and the use of renewables, and describe various means to achieve these objectives.

In March 2011, the Twelfth Five-Year Plan, covering 2012–2016 (FYP12), was revealed. This formulates new targets for 2015, and outlines key measures to achieve them:

- FYP12 sets two national reduction targets: one target for reducing overall energy intensity by 16% below 2010 levels by 2015, and the other specifically for lessening carbon dioxide intensity per unit of GDP by 17%, compared with 2010, by 2015.[272] These goals are deemed in line with

268 CCES (2007c).

269 CELP (2012:1).

270 *Energy intensity* is generally defined as the amount of energy used in producing a given level of output or activity. It is measured by the quantity of energy required to perform a particular activity (service), expressed as energy per unit of output or activity measure of service. See USDA (2012).

271 Oberheitmann et al. (2008:143).

272 Seligsohn & Hsu (2011a:1).

the voluntary pledges China submitted in Denmark in 2009, and reaffirmed in Mexico in 2010.[273] In order to meet these goals, the intention is to broaden the 'Top 1,000 Enterprises' scheme to include 10,000 large energy-intensive enterprises, thus covering the entire national industrial complex.[274] Even more importantly, China announced it would undertake measures to gradually introduce provincial and regional voluntary carbon-trading schemes until 2015, by drawing on the experiences of international carbon-trading markets.[275]
- As to renewables, the FYP12 provides for increasing the share of non-fossil energy to 11.4% of overall primary energy consumption by 2015 through optimising the energy mix and the development of clean energy.[276] Measures to achieve this goal involve the installation of an additional 70 GW of wind power and an additional 30 GW of solar power by 2015. As regards nuclear power, the object is to install an additional 40 GW by 2015, which would mean China could have the world's largest installed capacity of nuclear energy by 2020.[277] However, the Fukushima nuclear disaster in Japan in March 2011 may somewhat limit China's appetite for nuclear expansion.[278] Other measures include the development and accelerated use of clean coal technology and unconventional gas-oil resources such as coal-bed gas and shale gas.

In order to implement the targets set out in FYP12, in July 2011 the Chinese Government issued a work plan under which the two national reduction targets were broken down by assigning obligatory overall energy and specific carbon dioxide reduction targets to the 28 Chinese provinces, with reductions ranging from 10% to 18%.[279] It is important to note that the work plan underscores China's intention to launch carbon-trading pilot projects, although only on a voluntary basis.[280] In July 2011, the Chinese Government announced that it would launch pilot project trading schemes in six provinces by 2013.

---

273 Wyns (2012:1).
274 Seligsohn (2011).
275 Xinhuanet (2011:VIII).
276 Seligsohn & Hsu (2011a:3).
277 (ibid.).
278 Seligsohn & Hsu (2011b:2).
279 Finamore (2011:1–2).
280 (ibid.:3).

Besides piloting carbon-trading schemes, the Chinese Government is also reportedly considering implementing a nationwide carbon tax, according to Chinese media in January 2012.[281] While plans have yet to be finalised, the proposed tax could become operational before 2015 and could start with a rather low amount of US$1.50 per t of carbon dioxide emitted by large industrial installations, with the tax amount to be increased quickly thereafter.[282] Discussions about introducing a carbon tax have been around since the release of the FYP12 while a national natural resource tax was introduced on crude oil and gas late in 2011. In addition to these regulatory instruments, China also uses public finance incentives, tax breaks and financial support schemes to reach its mitigation objectives.

### *F. Least-developed Countries*

Although global warming potentially threatens all countries, observers and experts concur that developing states, in particular the LDCs and the SIDSs, will be most affected.[283] Poor nations are more affected than the rich ones because of their greater vulnerability to climate shocks, and their lower adaptive capacities.[284] The biggest vulnerability is that rising temperatures affect developing countries' most important sources of national income – farming and tourism.[285] The World Bank forecasts that a 2°C warming above pre-industrial temperatures, which is the minimum to expect, could result in a permanent loss in GDP of 4–5% in Africa and South Asia, as opposed to minimal losses in high-income countries, and a 1% loss in world average GDP.[286] World farming production could fall by 16% by 2080, but by as much as 21% in developing countries.[287] Increasing temperature will have severe effects on health, particularly in developing countries, where people will be exposed to life-threatening illnesses such as malaria even more than they are at present.[288] Lastly, whereas for other regions the impacts of global

281 Gera (2012:1).
282 (ibid.).
283 Collier et al. (2009:127).
284 World Bank (2009:40).
285 *Economist* (2009:1).
286 World Bank (2009:5).
287 *Economist* (2009:2).
288 World Bank (2009b:42).

warming will only unfold in the future, in developing countries, especially those in Africa, many of the negative effects are already visible.[289]

Compounding this vulnerability to global warming is the fact that developing countries lack the institutional capacity, financial resources and technical expertise necessary to address the ever-increasing impacts of climate change. Recognising this, based on the CBDR principle, the UNFCCC obliges industrialised nations to support the developing world with finance and access to innovative technology in order to increase such vulnerable countries' adaptive capacities. It is important to keep in mind that developing countries' higher vulnerability to global warming is ironically accompanied by a much lower responsibility for GHG emissions.[290] The present carbon footprint of developing nations is extremely low.[291] The carbon dioxide per capita of a low- or middle-income country is 1.3 t to 4.5 t of carbon dioxide, respectively, compared with 15.3 t for high-income countries. [292]

### *I. Adaptation*

Whereas the key concern for industrialised states is mitigating their GHG emissions, for developing nations, due to their higher vulnerability, the overriding concern is adaptation to the inevitable impacts of global warming. The UNFCCC urges all states parties to formulate and implement national adaptation measures as well as to cooperate internationally on adaptation issues. Article 4.9 of the UNFCCC recognises the specific needs of LDCs, in that they do not have the necessary capacities to deal with adaptation to global warming.

In 2001, at COP7 in Morocco, in order to implement Article 4.9, states parties established a working plan for LDCs in particular, including what were termed *National Adaptation Programmes of Action* (NAPAs), with the objective of communicating LDCs' urgent and immediate adaptation needs.[293] The main content of NAPAs constitutes a list of ranked adaptation measures aimed at facilitating the development of projects to implement the

289 For an overview of regional impacts of climate change, see World Bank (2009).
290 Collier et al. (2009:125).
291 World Bank (2009b:39).
292 (ibid.:44).
293 UNFCCC (2012k).

NAPAs.[294] Once a NAPA has been submitted to the UNFCCC Secretariat, the country concerned becomes eligible to apply for financial support for NAPA projects under the LDC Fund.[295] By January 2012, the UNFCCC Secretariat had received 47 NAPAs.

At COP12 in Nairobi, Kenya, in 2005, the UNFCCC Secretariat launched the Nairobi Work Programme on Impacts, Vulnerability and Adaptation. The Nairobi Work Programme is an international framework that initially operated from 2005 to 2010. It was extended in Mexico at COP16 in 2010 to assist developing countries in particular to better understand adaptation and to make informed decisions on practical adaptation options.[296] In Bali 2007, the Bali Action Plan identified adaptation as one of the cornerstones of the sustained implementation of the UNFCCC.[297] Since Bali, the Working Group on Long-Term Cooperative Action has mainly dealt with adaptation issues.

Also in Mexico in 2010, states parties established the Cancun Adaptation Framework (CAF). Its objective was to intensify cooperation on adaptation issues under the UNFCCC while confirming that adaption is obliged to enjoy the same priority as mitigation.[298] The CAF is meant to enable LDCs to formulate and implement national adaptation plans in order to identify their intermediate and long-term adaption needs, and to develop strategies to tackle those needs, building on their experience with the NAPAs.[299] However, as opposed to NAPAs, national adaptation plans can also be submitted by developing countries other than LDCs. As part of the CAF, the COP established a programme to investigate mechanisms and systems such as climate risk insurance to address potential damage caused by global warming.[300] In order to further the implementation of the CAF, at COP18 in Durban 2011, states parties agreed on the procedure, the work modalities, and the institutional structure of a new Adaptation Committee, as well as on guidelines for the aforementioned national adaptation plans.

294 (ibid.).
295 UNFCCC (2012l).
296 UNFCCC (2012m:1).
297 UNFCCC (2012l).
298 UNFCCC (2012n).
299 UNFCCC (2012o).
300 UNFCCC (2012n).

### *II. Funding*

Funding is vital in order for most developing countries, in particular for LDCs and SIDSs, to implement adaptation and mitigation initiatives.

#### *1. Adaptation Funding*

Estimates for funding for adaptation initiatives vary considerably. In 2010, the World Bank projected that it would cost US$70–100 billion each year (at 2005 prices) to adapt to global warming.[301] The UNFCCC predicted that adaptation would require supplementary investments of US$60–182 billion a year. Of these funds, developing countries would need US$28–67 billion annually.[302] There has been a significant increase in adaptation funding, from US$587 million (8% of total climate funding) in 2010, to US$957 million (21% of total funding) by 2011.[303]

There are presently five multilateral funds that support adaptation in developing countries, including the following three:

- The Global Environmental Facility, the independently run funding institution established in 1991, with the World Bank serving as its Trustee, operates two climate funds, both established in 2001:
  - The Least-developed Countries Fund is tasked with funding the preparation and implementation of NAPAs. As of December 2011, this Fund had approved some US$217 million for short-term NAPA projects and mobilised millions more in co-financing arrangements. A total of 52 projects had been funded as at that date.[304]
  - The Special Climate Change Fund was established to fund long-term adaptation measures and access to innovative technology. This Fund has approved some US$150 million for 39 projects, of which US$80 million has been disbursed.[305] The Fund has also leveraged about US

301 World Bank (2009).
302 UNFCCC (2007:38).
303 Nakhooda et al. (2011a:2).
304 GEF (2012a).
305 Nakhooda et al. (2011a:3).

$1 billion in co-financing deals.[306] The demand for support from this Fund is high and exceeds current resources.[307]

- The Adaptation Fund under the Kyoto Protocol began operating in 2008 and is managed by the Adaptation Fund Board, with the Global Environmental Facility as its Secretariat and the World Bank as its Trustee. The Fund was established to support adaptation projects in developing countries and is fed by a 2% levy on sales under the CDM and through contributions by governments, businesses and individuals.[308] Since 2010, the Fund has approved financial support packages of over US$100 million for 17 adaptation projects,[309] of which US$22 million has been disbursed.[310] Projections indicated that demand for funding would be US $341 million in 2012, while the Fund only had US$146 million available at the time of writing.[311]

Apart from multilateral funding vehicles, climate funding in general – including those for adaptation – are increasingly received through bilateral instruments or national trust funds.[312]

### 2. *Mitigation Funding*

Estimates of the costs of climate change mitigation initiatives vary considerably. The UNFCCC projected in 2007 that US$176 billion would be required by 2030 to fund mitigation activities. Between 2004 and 2011, US $2.97 billion was approved for mitigation initiatives, of which US$1.17 billion has been disbursed.[313] Today, mitigation represents about 66% of total climate funding, much of which has been directed at India and China, where emissions are growing rapidly.[314]

The CDM, which was established under the Kyoto Protocol and has been operational since 2008, provides an incentive-based mechanism linking mit-

306 GEF (2012a).
307 (ibid.).
308 AF (2012a).
309 AF (2012b).
310 Nakhooda et al. (2011a:3).
311 (ibid.).
312 For more information, see Nakhooda et al. (2011b).
313 Nakhooda et al. (2011c:2).
314 (ibid.).

igation to financing for sustainable development.[315] The CDM has already registered more than 3,497 projects in 72 countries and has issued CERs of almost 750 Mt. Transactions involving CERs were valued at approximately US$20 billion in 2010.[316] However, a major difficulty of this instrument today is that the price of one CER has fallen below €1 due to a vast oversupply on the carbon market. Unfortunately, CDM ventures are mainly located in a few of the larger developing countries of the BASIC group, especially in China (46%), India (21%) and Brazil (6%), while LDCs have generally been bypassed: Africa, for example, represents less than 2% of CDM projects.[317] The scope of CDM, which is currently very restrictive, has to be broadened not only to include the power-generating industry (hydropower), but also to exclude deforestation, thus opening up new opportunities – especially for Africa.[318]

A number of additional multilateral funds to support mitigation efforts in developing countries have emerged since 2008, including the Clean Technology Fund and the Scaling Up Renewable Energy Program, both of which operate under the World Bank's Climate Investment Funds.[319]

### 3. *Funding Gap*

However, on the whole, the funding that is presently offered under the UNFCCC and the Kyoto Protocol is minimal compared with the scale of the adaptation and mitigation costs identified. In 2007, the UNFCCC estimated that the annual funds needed in 2030 are likely to be around US$28–67 billion for adaptation, and US$176 billion for mitigation. At present, there is a vast funding gap that needs to be closed. For instance, funding for adaptation and mitigation efforts in developing nations – as per the commitments made by wealthier nations – amounts to less than 5% of what may be needed annually by 2030.

In light of this funding gap, at COP15 in Copenhagen in 2009, industrialised countries pledged to grant new and additional resources of US$30 billion, to be operational almost immediately, from 2010 to 2012, and to

315 UNFCCC (2012o).
316 CDM Executive Board (2011:4–5).
317 Nakhooda et al. (2011d:1).
318 Collier et al. (2009:138).
319 Nakhooda et al. (2011c:2).

mobilise long-term funding together with developing countries of US$100 billion a year by 2020. The informal pledge was incorporated into the final decisions of COP16 in Mexico in 2010, and the states parties reaffirmed that funding for climate change adaptation would be prioritised for the most vulnerable developing countries, namely LDCs, SIDSs and Africa. While COP17 in Durban in 2011 saw the launch of the Green Climate Fund, the question of medium- and long-term funding remains unresolved and needs urgent attention. As outlined before, states parties in Qatar at COP18 in 2012 were unable to make any headway whatsoever, especially with a view to committing to the mid-term funding between 2013 and 2020.

## G. Concluding Remarks

Regardless of having created a new window of opportunity, the success of the Doha Gateway is far from certain, given the existing differences between the major GHG emitters – the BASIC group, the EU, and the US. Hence, some observers argue that the issue is not the format of the international negotiations process – the top-down Kyoto approach or the bottom-up Copenhagen model – but the lack of national political will in some countries to tackle global warming and climate change.[320] Canada, Russia and the US are bringing nothing constructive to the table at the moment, and the only intention of Saudi Arabia and other oil-exporting countries is to block any further progress on mitigation measures that impact on their sales.[321]

In addition, what unfortunately has become clear over time is that the voluntary climate change mitigation targets and NAMAs pledged by states parties and integrated into the Cancun Agreements are not enough to limit the temperature increase to 1.5°C or 2°C over the course of the 21st Century and urgently need to be intensified. According to most climate scientists, global GHG emissions need to peak by 2015 at the very latest in order for there to be a meaningful chance of limiting the global temperature rise to 2°C. Instead, GHG emissions actually rose by 6% in 2010.[322]

320 Sterk et al. (2011b:35).
321 (ibid.).
322 Bodanksy (2012:1).

## References

AF/Adaptation Fund, 2012a, *About the Adaptation Fund*, available at http://www.adaptation-fund.org/about, last accessed 5 July 2012.

AF/Adaptation Fund, 2012b, *Adaptation Fund*, available at http://www.adaptation-fund.org/sites/default/files/ADAPTATION%20FUND.7.singles.pdf, last accessed 5 July 2012.

Anderson, Frederick R., 2009, What Can We Make of the Copenhagen Accord?, *Bloomberg Bureau of National Affairs' World Climate Change Report*, 21 December, available at http://www.mckennalong.com/media/resource/1116_ANDERSON_s%20NOTEBOOK%20-%2012.21%20-%20What%20Can%20We%20Make%20of%20the%20Copenhagen%20Accord.pdf, last accessed 7 February 2011.

Barrett, Scott, 2009, Climate Treaties and the Imperative of Enforcement, in: Helm, Dieter and Cameron Hepburn (Eds), *The Economics and Politics of Climate Change*, Oxford, Oxford University Press, 58–80.

Barriaux, Marianne, 2010, China Calls on USA to Take the Lead in Climate Talks, *Agence France Presse*, 23 November, available at http://news.asiaone.com/News/AsiaOne%2BNews/World/Story/A1Story20101123-248728.html, last accessed 5 July 2012.

Bodansky, Daniel, 2009, *Legal Form of a New Climate Change Agreement: Avenues and Options*, April, available at http://www.c2es.org/docUploads/legal-form-of-new-climate-agreement-paper.pdf, last accessed 3 May 2013.

Bodansky, Daniel, 2011, *W(h)ither the Kyoto Protocol: Durban and Beyond*, available at http://belfercenter.ksg.harvard.edu/files/Bodansky_Viewpoint-Final.pdf, last accessed 5 July 2012.

Bodansky, Daniel, 2012, Evaluating Durban, available at http://www.c2es.org/blog/bodanskyd/evaluating-durban, last accessed 5 July 2012.

Bodansky, Daniel & Sandra Day O'Connor, 2011, Multilateral Climate Efforts beyond the UNFCCC, Centre for Climate and Energy Solutions, available at http://papers.ssrn.com/sol3/papers.cfm?abstract_id=1963928, last accessed 5 July 2012.

Bodansky, Daniel, Sandra Day O' Connor & Elliot Diringer, 2010, The Evolution of Multilateral Regimes: Implication for Climate Change, available at http://www.c2es.org/docUploads/evolution-multilateral-regimes-implications-climate-change.pdf, last accessed 5 July 2012.

Boisson de Chazourne, Laurence, 2008, United Nations Framework Convention on Climate Change, available at http://www.un.org/law/avl, last accessed 27 November 2010.

Bothe, Michael, 2003, The United Nations Framework Convention on Climate Change – An Unprecedented Multilevel Regulatory Challenge, *Zeitschrift für ausländisches öffentliches Recht und Völkerrecht* 63, 239–254, available at http://www.zaoerv.de/63_2003/63_2003_2_a_239_254.pdf, last accessed 3 May 2013.

Boyle, Jessica, 2011, Assessing the Outcome of COP17 – In Pursuit of a Binding Climate Agreement: Negotiators Expand the Mitigation Tent but Reinforce the Ambition Gap, Manitoba, International Institute for Sustainable Development, December, available at http://www.iisd.org/pdf/2011/assessing_the_outcomes_cop17.pdf, last accessed 5 July 2012.

Burtraw, Dallas, 2011, US Climate Policy Efforts, Policy Brief No. 255, Washington, DC, Center for European Policy Studies, September, available at http://aei.pitt.edu/32485/1/PB_No_255_Burtraw_on_US_climate_change_policy_efforts.pdf, last accessed 5 July 2012.

Cao, Jing, 2010, Beyond Copenhagen: Reconciling International Fairness, Economic Development and Climate Protection, *Discussion Paper 2010–44*, Cambridge, MA, Harvard Project on International Climate Agreements, Belfer Center for Science and International Affairs, Harvard Kennedy School, October, available at http://belfercenter.ksg.harvard.edu/files/CaoHPICADP44.pdf, last accessed 5 July 2012.

CCES/Center for Climate and Energy Solutions, 2007a, *Key Country Policies Brief: China*, June, available at http://www.c2es.org/docUploads/ChinaNationalClimateChangeProgramme%20June%2007.pdf, last accessed 5 July 2012.

CCES/Center for Climate and Energy Solutions, 2007b, *Key Country Policies Brief: Fact Sheet China*, available at http://www.c2es.org/publications/climate-change-mitigation-measures-china, last accessed 5 July 2012.

CCES, Center for Climate and Energy Solutions, 2007c, *International Negotiations Policy Brief: A Summary of the 13th COP in Bali*, December, available at http://www.c2es.org/docUploads/Pew%20Center_COP%2013%20Summary.pdf, last accessed 5 July 2012.

CCES/Center for Climate and Energy Solutions, 2009a, *International Country Brief: Fact Sheet European Union*, available at http://www.c2es.org/docUploads/eu-fact-sheet-12-05-09.pdf, last accessed 5 July 2012.

CCES/Center for Climate and Energy Solutions, 2009b, *International Negotiations Policy Brief: A Copenhagen Agreement*, November, available at http://www.c2es.org/docUploads/copenhagen-agreement-nov 2009_0.pdf, last accessed 5 July 2012.

CCES/Center for Climate and Energy Solutions, 2009c, *International Negotiations Policy Brief: A Summary of the Waxman-Markey Bill*, available at http://www.c2es.org/federal/analysis/congress/111/acesa/waxman-markey-short-summary, last accessed 5 July 2012.

CCES/Center for Climate and Energy Solutions, 2010a, *International Negotiations Policy Brief: A Summary on Mitigation Targets and Actions Pledged under Copenhagen*, May, available at http://www.c2es.org/docUploads/targets-and-actions-copenhagen-accord-05-24-2010.pdf, last accessed 5 July 2012.

CCES/Center for Climate and Energy Solutions, 2010b, *International Negotiations Policy Brief: A Pre-view on the 16th COP in Cancun*, December, available at http://www.c2es.org/docUploads/making-progress-cancun.pdf, last accessed 5 July 2012.

CCES/Center for Climate and Energy Solutions, 2010c, *International Negotiations Policy Brief: A Summary of the 16th COP in Cancun*, December, available at http://www.c2es.org/docUploads/cancun-climate-conference-cop16-summary.pdf, last accessed 5 July 2012.

CCES/Center for Climate and Energy Solutions, 2011, *International Negotiations Policy Brief: A Summary of the 17th COP in Durban, December*, available at http://www.c2es.org/docUploads/COP17_Summary.pdf, last accessed 5 July 2012.

CCES/Centre for Climate and Energy Solutions, 2012, *Science FAQ*, available at http://www.c2es.org/science-impacts/basics/faqs/climate-science, last accessed 5 July 2012.

CCES/Centre for Climate and Energy Solutions, 2012, *Outcomes of the UN Climate Change Conference in Doha – Qatar, Eighteenth Session of the Conference of the Parties to the United Nations Framework Convention on Climate Change (COP 18) and Eighth Session of the Meeting of the Parties to the Kyoto Protocol (CMP 8)*, 27 November–7 December, available at http://www.c2es.org/docUploads/c2es-cop-18-summary.pdf, last accessed 21 January 2013.

CDM Executive Board, 2011, *Annual Report of the Executive Board of the Clean Development Mechanism to the Conference of Parties*, November, available at http://unfccc.int/resource/docs/2011/cmp7/eng/03p01.pdf, last accessed 5 July 2012.

CELP/Center for Environmental Law and Policy, 2012, *Climate Policy and Emissions Data Sheet: China*, available at http://envirocenter.yale.edu/uploads/pdf/China_Climate_Policy_Data_Sheet.pdf, last accessed 5 July 2012.

CISDL /Centre for International Sustainable Development Law, 2002, A CISDL Legal Brief: *The Principle of Common But Differentiated Responsibilities: Origins and Scope.* For the World Summit on Sustainable Development 2002. Johannesburg, 26 August, available at http://cisdl.org/public/docs/news/brief_common.pdf, last accessed on 20 May 2013.

Climatico, 2011, COP 17 De-briefing: Enhancement, Decisions and the Durban Package, December, available at http://www.climaticoanalysis.org/wp-content/uploads/downloads/2012/01/COP-17-De-briefing.pdf, last accessed 5 July 2012.

Collier, Paul, Gordon Conway & Tom Venables, 2009, Climate Change in Africa, in: Helm, Dieter & Cameron Hepburn (Eds), *The Economics and Politics of Climate Change*, Oxford, Oxford University Press, 125–141.

Davey, Edward, 2012, COP18: Statement from UK Energy and Climate Change Secretary, Oral Statement, 9 December 2012, available at https://www.gov.uk/government/speeches/cop18-statement-from-uk-energy-and-climate-change-secretary, last accessed 21 January 2013.

Depledge, Joanna & Farhana Yamin 2009, The Global Climate Regime: A Defence, in: Helm, Dieter & Cameron Hepburn (Eds), *The Economics and Politics of Climate Change*, Oxford, Oxford University Press, 433–453.

Diringer, Elliot, 2010a, *Copenhagen Accord: Act II*, March, available at http://www.c2es.org/blog/diringere/building-on-copenhagen, last accessed 5 July 2012.

Diringer, Elliot, 2010b, From Copenhagen to Cancun: Interview, March, available at http://www.c2es.org/docUploads/argus-media-interview-diringer-01-10-2011.pdf, last accessed 5 July 2012.

Diringer, Elliot, 2011a, The Question of Binding, May, available at http://www.c2es.org/blog/diringere/durban-question-of-binding, last accessed 5 July 2012.

Diringer, Elliot, 2011b, Durban – How Big a Deal?, December, available at http://theenergycollective.com/seidel/72227/durban-how-big-deal, last accessed 5 July 2012.

*Economist*, 2009, Developing Countries and Global Warming: A Bad Climate for Development, 17 September, available at http://www.economist.com/node/14447171, last accessed 5 July 2012.

Egenhofer, Christian & Anton Georgiev, 2009, *The Copenhagen Accord: A First Stab at Deciphering the Implications for the EU*, Brussels, Centre for European Policy Studies, 25 December, available at http://www.ceps.be/book/copenhagen-accord-first-stab-deciphering-implications-eu, last accessed 5 July 2012.

EC/European Commission, 2000, *Communication from the Commission to the Council and the European Parliament: On EU Policies and Measures to Reduce GHG Emissions – Towards a European Climate Change Programme*, COM (2000:88), 8 March, available at http://eur-lex.europa.eu/LexUriServ/LexUriServ.do?uri=COM:2000:0088:FIN:EN:PDF, last accessed 5 July 2012.

EC/European Commission, 2005, *Communication from the Commission to the Council and the European Parliament: Winning the Battle against Climate Change*, COM (2005)35, 9 February, available at http://eur-lex.europa.eu/LexUriServ/site/en/com/2005/com2005_0035en01.pdf, last accessed 9 February 2011.

EC/European Commission, 2007, *Communication from the Commission to the European Council and the European Parliament: An Energy Policy for Europe*, COM(2007) 1, 10 January, available at http://ec.europa.eu/energy/energy_policy/doc/01_energy_policy_for_europe_en.pdf, last accessed 9 February 2011.

EC/European Commission, 2008, *Commission Staff Working Group: The Support of Electricity from Renewable Energy Sources*, at http://ec.europa.eu/energy/climate_actions/doc/2008_res_working_document_en.pdf, last accessed 14 June 2012.

EC/European Commission, 2009a, *Fifth National Communication from the European Community under the UNFCCC*, available at http://unfccc.int/resource/docs/natc/ec_nc5.pdf, last accessed 5 July 2012.

EC/European Commission, 2009b, *White Paper: Adapting to Climate Change – Towards a European Framework for Action*, COM (2009) 147/4, April, available at http://eur-lex.europa.eu/LexUriServ/LexUriServ.do?uri=COM:2009:0147:FIN:EN:PDF, last accessed 5 July 2012.

EC/European Commission, 2010a, *Communication from the Commission to the European Parliament, the Council, the European Economic and Social Committee and the Committee of the Regions: Analysis of Options to Move beyond 20% GHG Emission Reductions*, COM (2010) 265 final, 26 May, available at http://eur-lex.europa.eu/LexUriServ/LexUriServ.do?uri=COM:2010:0265:FIN:EN:PDF, last accessed 5 July 2012.

EC/European Commission, 2010b, *The European Strategic Energy Technology Plan*, available at http://ec.europa.eu/energy/publications/doc/2010_setplan_brochure.pdf, last accessed 5 July 2012.

EC/European Commission, 2011a, *Communication from the Commission to the European Parliament, the Council, the European Economic and Social Committee and the Committee of the Regions: Roadmap for Moving to a Competitive Low Carbon Economy in 2050*, COM (2011) 112 final, 3 March, available at http://eur-lex.europa.eu/LexUriServ/LexUriServ.do?uri=COM:2011:0112:FIN:EN:PDF, last accessed 5 July 2012.

EC/European Commission, 2011b, *White Paper: Roadmap to a Single European Transport Area*, COM (2011) 144 final, 28 March, available at http://ec.europa.eu/transport/strategies/doc/2011_white_paper/white_paper_com(2011)_144_en.pdf, last accessed 5 July 2012.

EC/European Commission, 2011c, *Climate Change: A Fact Sheet*, April, available at http://ec.europa.eu/clima/publications/docs/factsheet-climate-change_en.pdf, last accessed 5 July 2012.

EC/European Commission, 2012a, *Commission Staff Working Paper: Analysis of Options beyond 20% GHG Emission Reductions*, SWD (2012) 5 final, 1 February, available at http://ec.europa.eu/clima/policies/package/docs/swd_2012_5_en.pdf, last accessed 5 July 2012.

EC/European Commission, 2012b, *Brief on the European Strategic Energy Technology Plan*, available at http://europa.eu/legislation_summaries/energy/european_energy_policy/en0019_en.htm, last accessed 5 July 2012.

EC/European Commission, 2012c, *Brief on European Energy Programme for Recovery*, available at http://europa.eu/legislation_summaries/energy/european_energy_policy/en0012_en.htm, last accessed 5 July 2012.

EC/European Commission, 2012d, *Brief on International Climate Finance*, available at http://ec.europa.eu/clima/policies/finance/international/index_en.htm, last accessed 5 July 2012.

Falkner, Robert, Hannes Stephan & John Vogler, 2010, International Climate Policy after Copenhagen: Towards a 'Building Blocks' Approach, Working Paper No. 25, London, Centre for Climate Change Economics and Policy, July, available at http://www.cccep.ac.uk/Publications/Working-papers/Papers/20-29/WP25_climate-policy-copenhagen.pdf, last accessed 5 July 2012.

Farnsworth, Norman, 2007, The EU Emissions Trading Directive: Time For Revision?, in: Douma, Waybe T., Leonardo Massai & Massimiliano Montini (Eds), *The Kyoto Protocol and Beyond*, The Hague, T.M.C. Asser Press, Chapter 4.

FIELD/Foundation for International Environmental Law and Development, 2012, *In Preparation for Doha: Defining "Legal Instrument", The 2013–2015 Review, and the Second Commitment Period under the Kyoto Protocol*, August, available at http://www.field.org.uk/files/field_doha_negotiations_paper_august_2012.pdf, last accessed 21 January 2013.

Finamore, Barbara, 2011, Province-by-province Details of China's Five-Year Energy and Environment Priorities, available at http://www.nrdc.cn/eblog/bfinamore/2011/09/15/province-by-province-details-of-chinas-five-year-energy-and-environment-priorities/, last accessed 5 July 2012.

Gao, Feng, 2007, The International Climate Regime: Where do we stand?, in: Douma, Waybe T., Leonardo Massai & Massimiliano Montini (Eds), *The Kyoto Protocol and Beyond*, The Hague, T.M.C. Asser Press, Chapter 2.

GCCA/Global Climate Change Alliance, 2012a, *What is the GCCA?*, at http://www.gcca.eu/about-the-gcca/what-is-the-gcca, last accessed 5 July 2012.

GCCA/Global Climate Change Alliance, 2012b, *Brief on Funding*, available at http://www.gcca.eu/pages/41_2-GCCA-Beneficiaries.html, last accessed 5 July 2012.

GEF/Global Environmental Facility, 2012a, *Brief on Least Developed Countries Fund*, available at http://www.thegef.org/gef/LDCF, last accessed 5 July 2012.

GEF/Global Environmental Facility, 2012b, *Brief on Special Climate Change Fund*, available at http://www.thegef.org/gef/SCCF, last accessed 5 July 2012.

Gera, R, 2012, Is China Preparing to Implement a Carbon Tax?, 6 January, available at http://cleantechnica.com/2012/01/06/is-china-preparing-to-implement-a-carbon-tax/, last accessed 5 July 2012.

Gosh, Arunabha & Ngaire Woods, 2009, Governing Climate Change: Lessons from other Governance Regimes, in: Helm, Dieter & Cameron Hepburn (Eds), *The Economics and Politics of Climate Change*, Oxford, Oxford University Press, 454–477.

Gupta, Joyeeta, 2007, Developing Countries and the Post-Kyoto Regime, Breaking the Tragic Lock-in Waiting for Each Other's Strategy, in: Douma, Waybe T., Leonardo Massai & Massimiliano Montini (Eds), *The Kyoto Protocol and Beyond*, The Hague, T.M.C. Asser Press, Chapter 13.

Helm, Dieter, 2009a, EU Climate Change Policy – A critique, in: Helm, Dieter & Cameron Hepburn (Eds), *The Economics and Politics of Climate Change*, Oxford, Oxford University Press, 222–246.

Helm, Dieter, 2009b, Climate-change Policy: Why has so Little been Achieved?, in: Helm, Dieter & Cameron Hepburn (Eds), *The Economics and Politics of Climate Change*, Oxford, Oxford University Press, 9–35.

Hepburn, Cameron & Nicholas Stern, 2009, The Global Deal on Climate Change, in: Helm, Dieter & Cameron Hepburn (Eds), *The Economics and Politics of Climate Change*, Oxford, Oxford University Press, 36–57.

House of Commons, Energy and Climate Change Committee, 2012, *The Road to UNFCCC COP 18 and Beyond, Second Report of Session 2012–13, Report, together with Formal Minutes, Oral and Written Evidence, Ordered by the House of Commons, to be Printed 17 July*, available at http://www.publications.parliament.uk/pa/cm201213/cmselect/cmenergy/88/88.pdf, last accessed 3 May 2013.

Hsu, Angel, 2011, Propelling the Durban Climate Talks – China Announces Willingness to Consider Legally Binding Commitments Post-2020, 6 December, available at http://www.chinafaqs.org/blog-posts/propelling-durban-climate-talks-china-announces-willingness-consider-legally-binding-comm, last accessed 5 July 2011.

IEA/International Energy Agency, 2011, *$CO_2$ Emissions from Fossil Fuel Combustion: Highlights, 2011 Edition*, available at http://www.iea.org/co2highlights/CO2highlights.pdf, last accessed 5 July 2012.

ICCG/International Center for Climate Governance, 2012, COP-18 delivers the Doha Climate Gateway, *Climate Policy News*, 3–9 December 2012, available at http://www.iccgov.org/FilePagineStatiche/Files/Publications/cpn/ICCG_climate-policy-news_2012_12_3-9.pdf, last accessed 21 January 2013.

IISD/International Institute for Sustainable Development, 2012, Summary of the Doha Climate Change Conference: 26 November–8 December 2012, *Earth Negotiations Bulletin, A Reporting Service for Environment and Development Negotiations*, 12, 567, available at http://www.iisd.ca/download/pdf/enb12567e.pdf, last accessed 21 January 2013.

IPCC/Intergovernmental Panel on Climate Change, 1990, *First Assessment Report*, available at http://www.ipcc.ch/publications_and_data/publications_and_data_reports.shtml#1, last accessed 9 February 2011.

IPCC/Intergovernmental Panel on Climate Change, 1995, *Second Assessment Report*, available at http://www.ipcc.ch/pdf/climate-changes-1995/ipcc-2nd-assessment/2nd-assessment-en.pdf, last accessed 20 May 2013.

IPCC/Intergovernmental Panel on Climate Change, 2007, *Climate Change 2007: Synthesis Report. Contribution of Working Groups I, II and III to the Fourth Assessment Report of the Intergovernmental Panel on Climate Change*, Core Writing Team: Pachauri, Rajendra K. & Andy Reisinger (Eds), Geneva, IPCC, available at http://www.ipcc.ch/publications_and_data/publications_ipcc_fourth_assessment_report_synthesis_report.htm, last accessed 13 January 2013.

Kuik, Onno, Jeroen Aerts, Frans Berkhout, Frank Biermann, Jos Bruggink, Joyeeta Gupta & Richard Tol, 2008, Post-2012 Climate Policy Dilemmas: A Review of Proposals, *Climate Policy* 8, 3, available at http://www.ivm.vu.nl/en/Images/2008_Kuik%2Betal-Post-2012_climate_policy_dilemmas_A_review_of_proposals-Climate_Policy_8_tcm53-253805.pdf, last accessed 5 July 2012.

Lewis, Joanna, 2007, China's Strategic Priorities in International Climate Negotiations, available at http://www.c2es.org/newsroom/articles/chinas-strategic-priorities-international-climate-change-negotiations, last accessed 5 July 2012.

Marcu, Andrei, 2012, Doha/COP 18: Gateway to a New Climate Change Agreement, Brussels, Centre for European Policy Studies, 13 December, available at http://www.ceps.eu/book/dohacop-18-gateway-new-climate-change-agreement, last accessed 21 January 2013.

Massai, Leonardo, 2007, Legal Challenges in European Climate Policy, in: Douma, Waybe T., Leonardo Massai & Massimiliano Montini (Eds), *The Kyoto Protocol and Beyond*, The Hague, T.M.C. Asser Press, Ch. 3.

Morgan, Jennifer, 2012, *Reflections on COP 18 in Doha: Negotiators Made Only Incremental Progress*, 14 December, available at http://insights.wri.org/news/2012/12/reflections-cop-18-doha-negotiators-made-only-incremental-progress, last accessed 3 May 2012.

Nakhooda, Smita, Alice Caravani, Neil Bird & Liane Schalatek, 2011a, Climate Finance Fundamentals: Adaptation Finance, Heinrich Böll Stiftung North America and Overseas Development Institute Brief 3, November, available at http://www.odi.org.uk/resources/docs/7469.pdf, last accessed 5 July 2012.

Nakhooda, Smita, Alice Caravani, Allister Wenzel & Liane Schalatek, 2011b, The Evolving Architecture of Global Climate Finance, Heinrich Böll Stiftung North America and Overseas Development Institute Brief 2, November, available at http://www.odi.org.uk/resources/docs/7468.pdf, last accessed 5 July 2012.

Nakhooda, Smita, Alice Caravani, Neil Bird, Maria Apergi, Liane Schalatek & Allister Wenzel, 2011c, Climate Finance Fundamentals: Mitigation Finance, Heinrich Böll Stiftung North America and Overseas Development Institute Brief 4, November, available at http://www.odi.org.uk/resources/docs/7471.pdf, last accessed 5 July 2012.

Nakhooda, Smita, Alice Caravani, Neil Bird & Liane Schalatek, 2011d, Regional Briefing: Sub-Saharan Africa, Heinrich Böll Stiftung North America and Overseas Development Institute Brief 7, November, available at http://www.odi.org.uk/resources/docs/7474.pdf, last accessed 5 July 2012.

NEAA/Netherland Environmental Assessment Agency, 2011, *Long-term Trends in Global $CO_2$ Emission: Report 2011*, available at http://www.pbl.nl/en/publications/2011/long-term-trend-in-global-co2-emissions-2011-report, last accessed 5 July 2012.

NEEA/Netherlands Environmental Assessment Agency, 2012, *Frequently Asked Questions on Climate Change*, available at http://www.pbl.nl/en/dossiers/climatechange/faqs, last accessed 5 July 2012.

Oberheitmann, Andreas & Eva Sternfeld, 2009, Climate Change in China – The Development of China's Climate Policy and its Integration into a New International Post-Kyoto Climate Regime, *Journal of Current Chinese Affairs* 38, 3, available at http://hup.sub.uni-hamburg.de/giga/jcca/article/view/64/64, last accessed 5 July 2012.

OECD & IEA/Organisation for Economic Co-operation and Development and International Energy Agency, 2006, *Adaptation to Climate Change: Key Terms*, April, available at http://www.oecd.org/dataoecd/36/53/36736773.pdf, last accessed 5 July 2012.

Olmstead, Sheila M. & Robert N. Stavins, 2006, An International Policy Architecture for the post-Kyoto Era, *American Economic Review Paper and Proceedings* 96 (2), 35–38, available at http://www.aeaweb.org/annual_mtg_papers/2006/0106_1015_0604.pdf, last accessed 3 May 2013.

Pan, Jiahua, Jonathan Phillips & Ying Chen, 2009, China's Balance of Emissions: Approaches to Measurement and International Responsibility, in: Helm, Dieter & Cameron Hepburn (Eds), *The Economics and Politics of Climate Change*, Oxford, Oxford University Press, 142–166.

Parker, Larry, John Blogett & Brent Yacobucci, 2011, *US Global Climate Change Policy: Evolving Views on Cost, Competitiveness, and Comprehensiveness*, Washington, DC, Congressional Research Service, February, available at http://www.fas.org/sgp/crs/misc/RL30024.pdf, last accessed 5 July 2012.

Rajamani, Lavanya, 2011, Decoding the Durban Platform: Blog of the European Journal of International Law – Talk, December, available at http://www.ejiltalk.org/author/lrajamani/, last accessed 5 July 2012.

Seligsohn, Deborah, 2011, China Moving Forward on 12th Five Year Plan Climate and Energy Implementation: Targets, Taxes, Emissions Trading Plans in Development, 2 August 2011, available at http://www.chinafaqs.org/blog-posts/china-moving-forwa rd-12th-five-year-plan-climate-and-energy-implementation-targets-taxes-, last accessed 5 July 2012.

Seligsohn, Deborah & Angel Hsu, 2011a, How does China's 12th Five Year Plan Address Energy and Environment?, 7 March 2011, available at http://www.wri.org/stories/20 11/03/how-does-chinas-12th-five-year-plan-address-energy-and-environment, last accessed 5 July 2012.

Seligsohn, Deborah & Angel Hsu, 2011b, Looking to Durban: China's Climate Change Policy Progress since Cancun, 23 November 2011, available at http://insights.wri.or g/node/188, last accessed 5 July 2012.

Stavins, Robert N., 2009, Addressing Climate Change with a Comprehensive US Cap and Trade System, in: Helm, Dieter & Cameron Hepburn (Eds), *The Economics and Politics of Climate Change*, Oxford, Oxford University Press, 197–221.

Stavins, Robert N., 2010a, Defining Success for Climate Negotiations in Cancun, November, available at http://www.robertstavinsblog.org/2010/11/19/defining-succe ss-for-climate-negotiations-in-cancun/, last accessed 5 July 2012.

Stavins, Robert N., 2010b, What Happened (and Why): An Assessment of the Cancun Agreements, December, available at http://www.robertstavinsblog.org/2010/12/13/s uccessful-outcome-of-climate-negotiations-in-cancun/, last accessed 5 July 2012.

Stavins, Robert N., 2011a, Why Cancun trumped Copenhagen: Warmer Relations on Rising Temperatures?, January, available at http://www.robertstavinsblog.org/2011/ 01/01/why-cancun-trumped-copenhagen/, last accessed 5 July 2012.

Stavins, Robert N., 2011b, Can the Durban Climate Negotiations Succeed?, November, available at http://www.robertstavinsblog.org/2011/11/28/can-the-durban-climate-ne gotiations-succeed/, last accessed 5 July 2012.

Stavins, Robert N., 2011c, Assessing the Climate Talks: Did Durban succeed?, December, available at http://www.robertstavinsblog.org/2011/12/12/assessing-the-climate -talks-did-durban-succeed/, last accessed 5 July 2012.

Stavins, Robert N. & Robert C. Stowe, 2010, What hath Copenhagen Wrought? A Preliminary Assessment, *Environment* 52, 3, May/June, available at http://www.hks.har vard.edu/fs/rstavins/Selected_Articles/Stavins_&_Stowe_Environment.pdf, last accessed 5 July 2012.

Sterk, Wolfgang, Christof Arens, Urda Eichhorst, Florian Mersmann & Hanna Wang-Helmreich, 2011a, Processed, Refried – Little Substance Added: Cancun Climate Conference Keeps United Nations Process Alive but Raised More Questions than it Answers, Wuppertal, Wuppertal Institute for Climate, Environment and Energy, March, available at http://wupperinst.org/uploads/tx_wupperinst/COP16-report.pdf, last accessed 5 July 2012.

Sterk, Wolfgang, Christof Arens, Urda Eichhorst, Hanna Wang-Helmreich & Timon Wehnert, 2011b, On the Road Again: Progressive Countries Score a Realpolitik Victory in Durban While the Real Climate Continues to Heat Up, Wuppertal, Wuppertal Institute for Climate, Environment and Energy, December, available at http://www. wupperinst.org/uploads/tx_wibeitrag/COP17-report.pdf, last accessed 5 July 2012.

Sterk, Wolfgang, Christof Arens, Nicolas Kreibich, Florian Mersmann & Timon Wehnert, 2012a, Same Procedure as Last Year? Same Procedure as Every Year: The Doha Climate Conference Once Again Saves the UN Climate Process While Real Climate Action Is Once Again Shelved for Later, A First Assessment of the Climate Conference in Doha, Wuppertal, Wuppertal Institute for Climate, Environment and Energy, 13 December, available at http://wupperinst.org/uploads/tx_wupperinst/doha-results.pdf, last accessed 21 January 2013.

Sterk, Wolfgang, Christof Arens, Nicolas Kreibich, Florian Mersmann & Timon Wehnert, 2012b, Sands Are Running Out for Climate Protection: The Doha Climate Conference Once Again Saves the UN Climate Process While Real Climate Action Is Shelved for Later, Wuppertal, Wuppertal Institute for Climate, Environment and Energy, December, available at http://wupperinst.org/uploads/tx_wupperinst/doha-report.pdf, last accessed 21 January 2013.

The Climate Group, 2012a, Doha: COP18 Briefing, Briefing Note, 19 November 2012, available at http://thecleanrevolution.org/_assets/files/COP18-Pre-COP-Briefing.pdf, last accessed 21 January 2013.

The Climate Group, 2012b, Doha: Post-COP18 Briefing, Briefing Note, 14 December 2012, available at http://thecleanrevolution.org/_assets/files/COP18-Post-COP-Briefing-v 2--14Dec12.pdf, last accessed 21 January 2013.

The Climate Group, 2012c, COP 18 Wrap-up: Weak Doha Outcome Underlines Importance of Clean Revolution: News and Blogs, 10 December 2012, available at http://www.theclimategroup.org/what-we-do/news-and-blogs/cop-18-wrap-up-weak-doha-outcome-underlines-importance-of-clean-revolution-leadership/, last accessed 21 January 2013.

The Climate Institute, 2012, Doha Climate Summit: Time to Focus Global Climate Talks, Policy Brief, November 2012, available at http://www.climateinstitute.org.au/verve/_resources/TCI_DohaClimateSummit_PolicyBrief_November2012.pdf, last accessed 21 January 2013.

UBA/Umweltbundesamt, 2010, Copenhagen and Beyond: Reshuffling the Cards, Discussion Paper 6/2010, available at http://www.umweltbundesamt.de/uba-info-medien/4003.html, last accessed 27 November 2010.

UNDP/United Nations Development Programme, 2007, *Human Development Report 2007/2008: Fighting Climate Change: Human Solidarity in a Divided World*, New York, UNDP, available at http://hdr.undp.org/en/reports/global/hdr2007-2008/ last accessed 27 November 2010.

UNFCCC/United Nations Framework Convention on Climate Change, 2005, *The United Nations Climate Change Conference Agrees on Future Critical Steps to Tackle Climate Change*, 10 December, available at http://www.unfccc.int, last accessed 27 November 2010.

UNFCCC/United Nations Framework Convention on Climate Change, 2007, *Climate Change: Impacts, Vulnerability and Adaptation in Developing Countries*, available at http://www.unfcc.int, last accessed 27 November 2010.

UNFCCC/United Nations Framework Convention on Climate Change, 2009a, *The Need for Mitigation*, at http://www.unfccc.int, last accessed 27 November 2010.

UNFCCC/United Nations Framework Convention on Climate Change, 2009b, *Report of the In-depth Review of the 4th National Communication of the United States of America, 2009*, FCCC/IDR.4/US, 17 February, available at http://www.unfcc.int, last accessed 27 November 2010.

UNFCCC/United Nations Framework Convention on Climate Change, 2009c, *The Need for Adaptation*, available at http://www.unfccc.int, last accessed 27 November 2010.

UNFCCC/United Nations Framework Convention on Climate Change, 2012a, *Glossary on Climate Change Acronyms*, available at http://unfccc.int/essential_background/glossary/items/3666.php, last accessed 5 July 2012.

UNFCCC/United Nations Framework Convention on Climate Change, 2012b, *Brief on the Conference of Parties*, available at http://unfccc.int/bodies/body/6383.phplast accessed 5 July 2012.

UNFCCC/United Nations Framework Convention on Climate Change, 2012c, *Introduction to the Kyoto Protocol*, available at http://unfccc.int/essential_background/kyoto_protocol/items/6034.php, last accessed 5 July 2012.

UNFCCC/United Nations Framework Convention on Climate Change, 2012d, *Brief on the Kyoto Protocol*, available at http://unfccc.int/kyoto_protocol/items/2830.php, last accessed 5 July 2012.

UNFCCC/United Nations Framework Convention on Climate Change, 2012e, *Brief on Emissions Trading*, available at http://unfccc.int/kyoto_protocol/mechanisms/emissions_trading/items/2731.php, last accessed 5 July 2012.

UNFCCC/United Nations Framework Convention on Climate Change, 2012f, *The Clean Development Mechanism (CDM)*, available at http://unfccc.int/kyoto_protocol/mechanisms/clean_development_mechanism/items/2718.php, last accessed 5 July 2012).

UNFCCC/United Nations Framework Convention on Climate Change, 2012g, *Brief on Joint Implementation*, available at http://unfccc.int/kyoto_protocol/mechanisms/joint_implementation/items/1674.php, last accessed 5 July 2012.

UNFCCC/United Nations Framework Convention on Climate Change, 2012h, *Brief on the Bali Road Map*, available at http://unfccc.int/essential_background/bali_road_map/items/6072.php, last accessed 5 July 2012.

UNFCCC/United Nations Framework Convention on Climate Change, 2012i, *Brief on the Cancun Agreements*, available at http://unfccc.int/essential_background/cancun_agreements/items/6132.php, last accessed 5 July 2012.

UNFCCC/United Nations Framework Convention on Climate Change, 2012k, *Brief on Adaptation*, available at http://unfccc.int/adaptation/items/4159.php, last accessed 5 July 2012.

UNFCCC/United Nations Framework Convention on Climate Change, 2012l, *Brief on the Nairobi Work Programme*, available at http://unfccc.int/adaptation/nairobi_work_programme/items/3633.php, last accessed 5 July.

UNFCCC/United Nations Framework Convention on Climate Change, 2012m, *Brief on the Cancun Adaptation Framework*, available at http://unfccc.int/adaptation/cancun_adaptation_framework/items/5852.php, last accessed 5 July 2012.

UNFCCC/United Nations Framework Convention on Climate Change, 2012n, *Brief on National Adaptation Plans*, available at http://unfccc.int/adaptation/cancun_adaptation_framework/national_adaptation_plans/items/6057.php, last accessed 5 July 2012.

UNFCCC/United Nations Framework Convention on Climate Change, 2012o, *What is the CDM?*, available at http://cdm.unfccc.int/about/index.html, last accessed 5 July 2012.

UN-NGLS/United Nations Non-governmental Liaison Service, 2008, UN Climate Change Talks in Bali – A New Climate Deal?, *NGLS Roundup*, 131, January, available at http://www.un-ngls.org, last accessed 12 February 2011.

US/United States, 2006, *Fourth National Communication under the UNFCCC (as required by Article 12 UNFCCC) US Fourth Climate Action Report*, available at http://www.unfcc.int, last accessed 27 November 2010.

US/United States, 2010, *Fifth US Climate Action Report: Fifth National Communication of the United States of America under the UNFCCC*, June, available at http://unfccc.int/resource/docs/natc/usa_nc5.pdf, last accessed 5 July 2012.

USDA/United States Department of Energy, 2012, *Energy Intensity Indicators: Terminology and Definitions*, 17 September, available at http://www1.eere.energy.gov/analysis/m/eii_trend_definitions.html, last accessed 3 May 2013.

Wilkinson, Peter, 2012, Carbon Taxes Made Simple #4: Carbon Taxes Around the World, 12 May, available at http://sustainabilitypr.net.au/2012/05/carbon-tax-made-simple-4-carbon-taxes-around-the-world/, last accessed 5 July 2012.

World Bank, 2008, *Development and Climate Change: A Strategic Framework for the World Bank Group, Technical Report*, available at http://www.beta.worldbank.org/overview/strategic-framework-development-and-climate-change, last accessed 27 November 2010.

World Bank, 2009b, *World Development Report 2010: Understanding the Links between Climate Change and Development*, available at http://www.worldbank.org/wdr2010, last accessed 27 November 2010.

WRI/World Resource Institute, 2002, *Foundation for a Low Carbon Future: Essential Elements of a Copenhagen Agreement*, available at http://www.wri.org, last accessed 27 November 2010.

WRI/World Research Institute, 2012a, Issues to Watch at the Doha Climate Negotiations (COP 18), *WRI Insights*, 19 November 2012, available at http://insights.wri.org/news/2012/11/issues-watch-doha-climate-negotiations-cop-18, last accessed 21 January 2013.

WRI/World Research Institute, 2012b, Reflections on COP 18 in Doha: Negotiators Made Only Incremental Progress, WRI Insights, 14 December 2012, available at http://insights.wri.org/news/2012/12/reflections-cop-18-doha-negotiators-made-only-incremental-progress, last accessed 21 January 2013.

Wyns, Tomas, 2012, The Year of the Dragon: Climate Action in China in 2012 and Beyond, 5 February 2012, available at http://europeanclimatepolicy.eu/?p=59, last accessed 5 July 2012.

*Xinhuanet*, 2009, China Announces Targets on Carbon Emission Cuts, 26 November 2009, available at http://news.xinhuanet.com/english/2009-11/26/content_12544181.htm, last accessed 5 July 2012.

*Xinhuanet*, 2011, China's Policies and Actions for Addressing Climate Change, November, available at http://news.xinhuanet.com/english2010/china/2011-11/22/c_131262368.htm, last accessed 5 July 2012.

Zusman, Eric, Koji Fukuda, Makoda Yoshini & Jun Ichihara, 2012, Why the United States Lacks a Federal Climate Policy: Collective Action Problems, Tea Parties and Blue Dogs, Institute for Global Environmental Strategies Working Paper CC-2011-5, May 2012, available at http://www.iges.or.jp/en/cp/pdf/activity20120106/Zusman.pdf, last accessed 5 July 2012.

3

# Climate Change Lawfare*

*Siri Gloppen & Asuncion Lera St. Clair*

*Abstract*

Climate change poses unprecedented challenges. No one knows how suitable a planet three or four degrees warmer on average will be for human life. What we do know is that climate change impacts are already changing socio-ecological systems and will lead to profound changes, as well as conflicts over resources. This article looks at how issues concerning the injustice and human rights violations caused by climate change are transformed and manifested in legal conflicts. The role of law in social transformation has been growing and evolving over the last few decades. Law is at the centre of efforts by national, local and international actors – state and non-state – to transform and develop societies. Drawing from literature on social lawfare – which refers to the diverse strategies in which rights and legal institutions are adopted intentionally and strategically with the aim of helping to deliver, or at least catalyse, social transformation – we have coined the term *climate change lawfare*. This term theorises on how emerging rights-related issues around climate change manifest themselves in legal strategies. Climate change lawfare aims to capture the diverse strategies in which rights and legal institutions figure prominently, are adopted intentionally, and are used strategically with the aim of helping to deliver, or at least catalyse, social transformation in relation to climate change. This includes both legal reform strategies and diverse forms of legal activism from 'below'. This article develops the concept of *climate change lawfare* and constructs a typology by systematising emerging material on climate-related legal conflicts. This may, in turn, provide a better starting point for systematic investigations into

* The authors wish to thank Catalina Vallejo of the Chr. Michelsen Institute for valuable research assistance. An earlier version of this text appeared in *Social Research*, 2012, 79(4), 899–930.

the conditions that place rights and courts at the centre of such disputes, and of the effects and impacts of various legal strategies.

## *A. Introduction: Climate Change Goes to Court*

The aim of this article is to investigate and map climate change discourses in legal spaces. We look at how issues concerning the injustice and human rights violations caused by climate change and by mitigation and adaptation policies and strategies are transformed and manifested in legal conflicts. The article investigates ways in which climate change impacts are legalised, judicialised and debated in relation to concrete conflicts about natural resources and environmental harm, which we refer to as *conflicts over climate justice and sustainability*. Today, the law is in many ways 'the new politics', in the sense that the legal field is expanding in social and political significance, not least in the contexts where other governance structures are weak.[1]

The article starts with an overview of the likely impacts of climate change and the governance problems involved in addressing such impacts. Noting the use of legal strategies by climate justice activists, and drawing on earlier work on the role of law in socio-economic justice, we develop the concept of *climate change lawfare*.[2] This concept refers to various uses of law, including attempts at improving climate governance through legislation and other forms of regulation, as well as strategies that use existing legal norms and structures. We acknowledge that there is significant rights-based mobilisation outside legal institutions ('rights talk'), and that examining this

1 The increasing importance of law in politics – often referred to as the *judicialisation*, *legalisation* or *juridification* of politics – or, more polemically, as the *juristocracy* – is noted by a number of scholars in relation to a wide range of fields and geographical areas; see e.g. Comaroff & Comaroff (2006, 2009, 2011); Couso et al. (2010); Ferejohn (2002); Gauri & Brinks (2008); Hirschl (2004, 2006); Shapiro & Stone Sweet (2002); Sieder et al. (2005); Tate & Vallinder (1995); Yamin & Gloppen (2011). Much of the literature is critical of this development, seeing it as an undemocratic takeover of political decision-making by unelected judges and bureaucrats, and fearing that "the haves always come out ahead in court" – to paraphrase Galanter (1974). Others hold that legal processes also open up space for democratic deliberation and may enable marginalised voices to be heard; thus, they potentially provide an institutional avenue for poor and stigmatised groups; see e.g. Gargarella et al. (2006).

2 Gloppen et al. (2011).

mobilisation and its drivers is important in its own right.[3] However, in this article, we mainly focus on legal mobilisation, and from there we analyse and theorise on climate change lawfare in courts and quasi-judicial bodies.

Our goal in this article is to offer a provisional synthesis of, and a theoretical lens for investigating, the legalisation of climate change impacts using the concept of *lawfare*, rather than provide a comprehensive analysis of climate-change-related legislation and litigation. By summarising and categorising a variety of cases, we develop a typology that offers a better understanding of the increasing legalisation of climate change politics, and what form climate change issues take when formulated as legal claims. This can serve as a starting point for systematic investigations into the conditions and driving forces that place rights and courts at the centre of climate change conflicts, and the effects and impacts – material, symbolic and political – of various legal strategies. The broader ambition is to understand the transformative potential of using the law to address problems of sustainability and social justice in the context of climate change. As such, this article attempts to illustrate what has been done and lays the foundation for further work on how the law can contribute towards responding to the challenges posed by climate change.

## *B. Climate Change Impacts, Governance, and Justice Challenges*

Climate change has impacted both human and natural systems, and will continue to do so substantially in the next few decades. Recent scholarship shows that large parts of Canada, Eurasia and North Africa have a high likelihood of passing the 2°C threshold by 2030, and that the whole planet is likely to do so by 2050.[4] Others consider that we may have to adapt to temperature increases of 4°C or more in the course of the 21st Century.[5] The literature on the likelihood and type of impacts we can expect by 2050, related to overall increases in temperature, was thoroughly assessed in the Intergovernmental Panel on Climate Change (IPCC) Fourth Assessment Report.[6] The Report shows that climate change will impact fundamental natural resources, such as water, in a very substantial way. There is a high

3 Dugard et al. (2012).
4 Joshi et al. (2011).
5 New et al. (2011); Stafford Smith et al. (2011).
6 IPCC (2007).

degree of certainty that increased water flows in high latitudes will lead to floods. At the same time, there is a high degree of confidence that less water will be available in dry areas in mid-latitudes, and that there are likely to be more drought-affected areas. Also, water stored in glaciers and snow will decrease and change river-flows downstream – one of the clearest challenges posed by climate change in regions such as Asia. Thus, climate change is likely to lead to substantial water scarcity and water-related damage. Shortages of water will also lead to energy-related scarcity, and compromise the production of hydroelectric power.[7]

Higher-than-average temperatures – along with a higher concentration of carbon dioxide in the atmosphere – affect the onset and end of seasons, change disease vectors, and influence the biological processes that govern ecosystems in most regions of the world. Increases in wildfires, changes in insects' life cycles, changes in the onset and end of seasons, and changes in the structure of ecosystems, which in turn lead to less biodiversity, are all expected results of small overall increases in temperature of about 1.5–2.5°.[8] Along with other global shifts, such as changes in land use, increased urbanisation and deforestation, these changes are likely to have substantial impacts on the availability of agricultural land. Climate change also affects the oceans and sea-level rise. Ocean acidification, destruction of coral reefs, and salination of coastal areas affects fisheries and ecosystems located close to seas and oceans. There is also an increasing amount of evidence showing strong correlations between higher temperatures and mutations (changes) in diseases such as malaria, dengue fever, and cardiovascular problems related to heat. Furthermore, "…climate change affects the fundamental requirements for health, clean air, safe drinking water, sufficient food and secure shelter."[9]

A report released by the IPCC in 2012 shows evidence of increased heat waves and other extreme events such as storms and droughts.[10] The report explores the interactions between human, environmental and climatic factors, and argues that the capacity to respond to such extreme events is determined not only by the magnitude of the natural event, but also – and often perhaps more importantly – by the social and human conditions of the re-

7 Gleick (2010).
8 IPCC (2007).
9 WHO (2012).
10 IPCC (2012).

gions affected.[11] Overcrowded cities, many of which host millions of poor and marginalised people in slums, are much more vulnerable to relatively small weather events because of the poor construction of dwellings and their location, which is often in a high-risk area.[12] Some 90% of all the deaths that have occurred since the 1990s because of extreme weather events occurred in developing countries.

These impacts merge with the following existing factors: socio-economic vulnerabilities, inequalities regarding access to resources and services, lack of power of minorities and marginalised groups, and negative impacts of natural climate variability on societies. The human costs of climate change are already high. A United Nations report quantifies climate-related deaths to around 300 million people per year.[13] If emissions are not drastically reduced, the human costs will be very high in the next decades in terms of water and food security, people displaced from their homes and communities, and protection of basic needs, such as employment, housing and health. Regardless of the obvious scientific uncertainty, if greenhouse gases (GHGs) are not substantially reduced in a short period of time, impacts in the near future are likely to be dramatic and perhaps irreversible.

### *I. Challenges to Governance*

As resources become scarcer, conflicts are likely to increase, but alternative modes of cooperation and alliances may also emerge. A key factor in the design and successful functioning of any solutions to climate change – both to create incentives for mitigation and to regulate adaptation – is the availability of suitable governance structures. Yet political institutions are generally not well equipped to regulate issues that are transboundary, or are fraught with unknowns, or that require long-term thinking. And we are far from overcoming elected politicians' institutional incentives for inaction. Furthermore, we lack conceptual and theoretical tools for thinking about politics in relation to climate change.[14] Until now, global governance structures have not reached the needed international agreements on mitigation policies, and are deadlocked over the impossibility of reconciling the inter-

11 (ibid.).
12 (ibid.).
13 GHF (2009).
14 Gardiner (2010, cited in O'Brien et al. 2010).

ests of advanced economies with those of less-developed countries and emergent economies. At the United Nations Framework Convention on Climate Change (UNFCCC) Conference of the Parties held in Durban, South Africa, in late 2011, states parties were only able to agree on postponing the decision on what to do after the Kyoto Protocol expired.

Governance for adaptation is part of the work done by local-level politics in the vast majority of advanced economies. It merges with existing policies to regulate natural resource use, the construction of protective mechanisms, taxation measures, etc. For developing countries, the governance for adaptation is muddled with existing development and poverty reduction policies and strategies. Despite some progress having been made at the local level, e.g. by municipalities, in general it is clear that governance measures to reduce emissions and prevent dangerous climate change has yet to emerge.

Within the literature addressing the problems related to climate negotiations, there is considerable focus on the role of market tools to regulate and create incentives for decreased GHG pollution.[15] The focus is on carbon only, however, and on markets as key tools for change. Carbon markets, Clean Development Mechanisms, the Special Climate Change Fund, and emerging National Adaptation Plans for Action are all primarily market-driven policy recommendations. They do not address the underlying causes of climate change, which are rooted in conceptions of development and progress that have made consumption the overarching measure of a well-functioning economy and of people's subjective perception of well-being. Nor do these market solutions provide the new institutions needed to fill the major governance gaps required to govern unavoidable climate change impacts. In the case of developing countries, the management of these policy mechanisms is in the hands of global development institutions, which are mainstreaming matters related to climate into their ongoing programmes and, thus, perpetuating existing neoliberal, market-based solutions to both issues.[16]

Thus, governance problems raised by climate change impacts relate not only to transnational relations and claims between countries, but also to internal conflicts over public spending, allocation of resources, costs and responsibilities, and prioritisation of some issues over others. In general terms, one could say that a changing climate adds an extra layer of complexity and

15 Aldy & Stavins (2009); Barrett (2006); Victor (2011).
16 Gasper et al. (2013).

friction to existing transnational relations. Nationally, welfare provisions have to compete with the requirements for coping with climate change impacts, and shifting resources to a transition towards more sustainable economic activities and lifestyles.

Firstly, climate change requires the reduction of GHGs, a process usually referred to as *mitigation*. The mitigation of GHGs requires carbon to be taxed and investments to be made into renewable energy and public infrastructure, such as public transport. *Adaptation* – the process of adjustment to actual or expected climate change and its effects in order to moderate harm or exploit beneficial opportunities – will place further demands on public funds. For example, building protection against sea-level rise, landslides or floods involves costly public works. Equally importantly, adaptation may entail the moving of populations from one area to another, and/or the special protection of poor social sectors. Furthermore, climate change impacts themselves are costly, and are likely to put pressure on human and economic public resources. For example, public funds will partly have to cover damage to public roads and other infrastructure by storms or other severe weather events, or increases in public health costs in cases where new diseases are brought about by increased temperatures. Thus, climate change calls for a serious rethinking of priorities in social policy.[17] In cases where basic needs are still not being met for large numbers of people, climate change challenges and impacts further complicate and pressurise the unequal distribution of and access to resources and services. The complexity of the challenges, and the shortcomings of political bodies and the market to come up with solutions, has brought attention to rights and the possibilities for addressing them by way of law and legal arenas.

### *II. Climate Justice and Human Rights*

Firstly, climate change impacts challenge existing frameworks for rights protection because they most strongly affect sectors of the population that are already vulnerable in ways that compromise their constitutional rights. Socio-economic rights, in particular, are very likely to suffer unless appropriate governance structures and protection systems are put in place. Fundamental rights taken for granted by most countries may also be compro-

17 Gough (2011).

mised; these include property rights, rights to national territorial sovereignty, and the right not to be forcibly displaced. Secondly, climate change is a matter of justice because the regions of the world that are likely to be the most severely affected are those that have contributed the least to increased GHGs and, thus, are the least responsible for the key drivers of climate change, i.e. industrialisation and modernisation. The poorest sectors of all societies are, de facto, paying the price of unfettered consumerism in other countries or by other social classes within their own countries. This double injustice is aggravated by the lack of careful understanding of the context in which market-driven solutions to climate change are being implemented. Neither mitigation polices nor adaptation strategies create win-win solutions. They involve choices regarding the distribution of harms and benefits, and choices between the short- and long-term needs of both humans and the natural environment. For example, mitigation strategies associated with reforestation may negatively affect vulnerable people, who may lose their farmland. And, like other markets, markets created to value the true costs of carbon in the atmosphere and to treat it as a pollutant are vulnerable to problems of externalisation.[18] In addition, existing adaptation plans suffer from the same problems that development aid has suffered from since its inception after World War II, i.e. development planning involves difficult dilemmas and leads to winners and losers; economic resources, although needed, may not protect people, for example, when corruption is rampant.[19]

Thus, mitigating for and adapting to climate change may lead to a more unequal if perhaps more sustainable world, or it may lead to the emergence of authoritarian regimes because of the urgency for change and the lack of democratic governance tools to promote such change. It is a gross simplification to presume that all measures to cope with or prevent climate change will be good for poor people, or will lead to a more equitable and fair world. The opposite is also a possibility; many activists proposing climate justice argue that the opposite is, in fact, more likely. These activists demand a move beyond a scientific framing of climate change and towards a social and human understanding of the problems involved. From this perspective, both climate change impacts and many mitigation and adaptation strategies vio-

18 Bond (2010). *Externalities* in economics generally refer to (positive or negative) effects on third parties who are not involved in an activity or transaction. Here, *externalisation* means that polluting actors are able to avoid (the full cost of) their responsibilities.

19 Petherik (2012).

late widely recognised rights; these include the right to food, health, housing, not being forcibly displaced and – even – life.

Different kinds of rights talk form part of various climate-change-related mobilisation efforts and political strategies; legal mobilisation, more narrowly conceived, is emerging as an important institutional space for contestation over climate chance governance. It is important, therefore, to better understand the evolving role of laws and legal institutions as the default regulators of climate change. Legal mobilisation is intertwined with the emergence and increased relevance of social movements for climate justice, and figures centrally among the strategies adopted by climate justice activists. This reflects a general trend towards legal mobilisation for social justice, including for access and entitlement to natural resources such as water, or to services such as electricity, housing and health.[20] Clearly, legislation is a product of political bodies; but our focus is on the law as a tool and as a space for contestation for other social actors. In the following discussion, we focus on the law and its institutions such as courts as tools – both for preventing harm to vulnerable groups and for transformative change.

## *C. Climate Change Lawfare*

The concept of *climate change lawfare* builds on the concept of *social lawfare.*[21] The role of the law in social transformation has been growing and evolving over the last few decades. An array of diverse factors – operating very differently in different contexts – has combined to increase the importance of rights, courts, and various legal and quasi-legal institutions as sites of political struggle. These include systematic weakness in political systems, with (more or less) democratic institutions marked by elite capture and lack of responsiveness. This has resulted in a consequent unwillingness or in-

20 Gargarella et al. (2006); Gauri & Brinks (2008); Gloppen et al. (2010); Yamin & Gloppen (2011).

21 The notion of *social lawfare* was developed by a group of scholars (including the authors) as part of an effort to create the conceptual foundation for a new collaborative Global Centre for Law and Social Transformation. The main focus of the Centre is to better understand the effects and impacts – desired and undesired – of social lawfare strategies. The Centre is coordinated from the Chr. Michelsen Institute (CMI) in Bergen, Norway. For a semiotic analysis of the concept of *lawfare*, see Tiefenbrun (2011:29).

ability to tackle pressing social problems, from severe poverty and inequality to environmental challenges. Nonetheless, alongside a sometimes-deteriorating political opportunity structure, the legal opportunity structure has, in many cases, improved. Many countries have adopted new rights-rich constitutions, many policy areas see denser national and international regulation, many judiciaries have been reformed, and many places are experiencing a stronger rights consciousness. In some places, of course, the law remains very distant from these debates, but where the debates are present, actors within civil society and the state – nationally and internationally – have turned to legal strategies and arenas to fight battles that, traditionally, had been resolved in the political domain. This battling of legal perspectives and use of the law is what is meant by the concept *lawfare*.[22] Included in this concept is the notion that "…the weak may use the law strategically to thwart the will of the powerful".[23]

*Social lawfare* refers to the diverse strategies in which rights and legal institutions figure prominently, are adopted intentionally, and are used strategically with the aim of helping deliver, or at least catalyse, social transformation and human development. Visions of social transformation and human development differ, as do views on means for getting there. The concept of *social lawfare* also includes legal strategies for maintaining the status quo in response to pressures for transformation sought by others, and furthering aims that proponents of liberal democracy or human rights would deem reactionary.[24]

*Social lawfare*, understood as the strategic use of law to bring about or resist social transformation, occurs in two main forms. One set of social lawfare strategies seeks social change by way of changing the law. While legal change normally involves legislators and politicians, it may be motivated or initiated from outside by international actors and institutions, as

22 Gloppen et al. (2011).

23 Scobbie (2006). This use of *lawfare* – where law is potentially a tool for progressive change that may also be used by poor and marginalised people to advance their causes – differs, for example, from the way the term is employed by Comaroff and Comaroff (2006), who use it to describe authorities' use of "the violence inherent in the law" for purposes of dominance and discipline. *Lawfare* is also used by The Lawfare Project to describe "*negative* manipulation of international and national human rights laws" [emphasis in original], with reference to the attempts by non-governmental organisations (NGOs) to use international law to delegitimise Israel; see www.thelawfareproject.org, last accessed 3 May 2013.

24 Gloppen et al. (2011).

well as by rights-based mobilisation for legal change 'from below', i.e. by ordinary citizens and civil society activists, which is the focus of this article.

The other main form of social lawfare seeks social change by mobilising within existing legal structures. This includes litigation before courts and other complaint mechanisms and monitoring bodies, and the judgements arising from such efforts. It also includes strategies which involve the use of rights talk to mobilise public opinion and pressure for compliance and rights realisation.

In both cases, lawfare may, in the first instance, change the operation of what we may call *intermediary mechanisms* of social transformation. It may lead to the establishment of new institutional and organisational structures, or changes to existing structures. It may change the set of actors – and, thus, the type of knowledge and experience – influencing processes and decisions, and the power relations between and among those actors. Lawfare may also lead to changes in discourses and ideas. This, in itself, could constitute important transformation, but might also lead to more lasting and tangible changes in the various dimensions which those engaging in these strategies would generally be aiming to bring about, i.e. lawfare may change the societal goals and values in relevant areas, the processes of decision-making, the conditions of sustainability, and policy outcomes which have a material effect on the ground. Furthermore, in addition to changes that would be intended and wanted by those driving the lawfare, it may also have unintended and unwanted effects.[25]

25 Unintended negative consequences may take different forms. For example, public interest litigation seeking to reduce urban pollution by moving large industrial emitters out of city centres may take away the livelihood of poor urban dwellers who cannot afford the commute. Litigation for medication and health services may skew resources towards high-cost interventions and potentially away from preventive care and basic services that are essential to the health of poor people (Ferraz 2011). Successful litigation for sexual and reproductive rights (e.g. abortion or same-sex marriage) may produce a political backlash. Efforts to hold political leaders accountable for human rights abuses (e.g. convicting the former Liberian President, Charles Taylor, in the Special Court for trying war crimes in Sierra Leone; indicting the incumbent Sudanese President, Omar al-Bashir, at the International Criminal Court) could make dictators and warlords cling to power at all costs, making negotiated deals impossible; and 'shaming' campaigns to free prisoners of conscience could prompt repressive regimes to kill dissidents instead of imprisoning them. Individual titling of land – recommended by Peruvian economist Hernando de Soto as a strategy for development, based on the reasoning that it enabled use of property as collateral for credit – has in some cases been found to exacerbate poverty by facilitating the per-

***Table 1: Lawfare Strategies and Effects***[26]

| **Social, institutional and political causes lead to …** | **… two strategies …** | **… that work through changes in three intermediate mechanisms, namely –** | **… to produce ultimate changes in three areas, namely –** |
|---|---|---|---|
| | 1. rights-based mobilisation for legal reform | a. institutional/ organisational | i. societal goals and values |
| | 2. legal mobilisation within the existing framework, including litigation | b. actors and power relations | ii. processes of decision-making |
| | | c. discourses and ideas | iii. policy outcomes, material changes and sustainability |

As climate change enters the domain of contestation for limited resources and conflicts of welfare versus environmental rights, and as the boundaries of environment and the planet lead to a rethinking of the suitability and gaps in existing law and global norms, we see a new form of lawfare emerging, analogous to the discussion outlined above. Drawing from this concept of *social lawfare*, we coined the term *climate change lawfare* to theorise on emerging rights-related issues around climate change that manifest themselves in legal strategies. Like *social lawfare*, the notion *climate change lawfare* aims to capture the diverse strategies in which rights and legal institutions figure prominently, are adopted intentionally, and are used strategically with the aim of helping to deliver, or at least catalyse, social transformation and human development – with the additional dimension of these strategies being related to climate change. In using an analogous distinction to the one used to frame the concept of *social lawfare*, it is useful to distinguish between two distinct climate change lawfare strategies. The first seeks transformation though mobilisation aimed at changing the law, where sustainability and rights protection in the context of climate change are sought though legal reform. The second seeks transformation through mobilisation within existing legal structures, such as courts and various complaint mechanisms and treaty bodies, or through rights-based civil society activism aimed at compliance though shaming and public opinion. Our main focus in this article is on the latter, but we also illustrate what climate change lawfare though the former, i.e. legal reform, may entail.

manent sale of property by poor people in situations of need, leaving them without a livelihood (Davis 2006). For a discussion of how the use of law intending to protect vulnerable groups may end up turning against them, see also Comaroff and Comaroff (2006, 2009).

26 Gloppen et al. (2011).

### *I. Climate Change Lawfare by Way of 'Engineering': Altering Legal Regimes*

Reform of the regulatory framework, or the rules of the game, in response to climate change may be initiated by national or international governmental actors or may result from lawfare by national or transnational non-governmental organisations (NGOs) and corporate (and other) actors. Such lawfare may aim at different levels, seeking to change –

- the international or regional treaty system
- national constitutions, or
- statutory law and/or administrative regulatory regimes, i.e. regimes affecting regulation and policy affecting the space for climate change mitigation and adaptation in a range of ways.

We are particularly interested in the use of rights in regulating socio-environmental conflicts in the context of climate change.

Arguably, the Andes are showing the most radical and interesting cases of climate change lawfare aiming to change the rules of the game in order to transform the political playing field in a way that simultaneously gives prominence to ecological sustainability and human rights. In Bolivia, temperatures have been rising steadily for 60 years, and an expected 3.5–4°C increase over the next 100 years would turn much of the country into a desert. Bolivians are already struggling to cope with melting glaciers and more frequent extreme weather events such as floods, droughts, frosts and mudslides. A much smaller ice cap will cause a farming crisis and serious water shortages.[27] The election in 2005 of Evo Morales, Latin America's first indigenous president, marked a turn to what can be seen as a form of climate change lawfare. Much of the legal system was restructured, starting with the adoption of a new constitution in 2009, influenced by indigenous Andean worldviews and cosmology, which place the environment and the earth deity at the centre and consider humans equal to all other living entities. This indigenous ontology also has an associated alternative paradigm of development and well-being, called *Buen Vivir* ("living well"). Buen Vivir reframes the conditions for a good life – not in terms of consumerism, but in terms of a balanced relation to one's environment.[28] The source of this paradigm is

27 Vidal (2011).
28 Heinrich Böll Foundation (2011).

the indigenous ontology where being human and having rights as a human being cannot be conceived in isolation from the environment. This eventually translated into an innovative set of laws, namely the Law of Mother Earth, passed in 2012.

The Law of Mother Earth, which declares that humans and all elements of nature have equal rights,[29] was pushed for and drafted in collaboration with indigenous and *campesino*[30] organisations. Initiated by the World People's Conference on Climate Change and the Rights of Mother Earth in Cochabamba in April 2010, the Law redefines the country's rich mineral deposits as 'blessings', and establishes new rights for nature. These include –

- the right of nature to life and to exist
- the right to continue vital cycles and processes free from human alteration
- the right to pure water and clean air
- the right to balance
- the right not to be polluted
- the right not to have cellular structure modified or genetically altered, and
- the right to not be affected by infrastructure and development projects that impact ecosystems and local inhabitant communities.

It is very important to note, that this characterisation of the components of Mother Earth defines *ecosystems* in ways that explicitly include the social, cultural and economic dimensions of human communities. This reflects the ontological view underpinning the law, which sees no dualism between nature and society. In addition, Mother Earth is established as a juridical person, as a collective subject of public interest, and legal action can be brought to defend her rights. The Law also proclaims the creation of an ombudsman for Mother Earth (*Defensoría de la Madre Tierra*) as a counterpart to the human rights ombudsman.[31]

However, the first country to recognise the legally enforceable rights of nature or of ecosystems in its constitution was not Bolivia, but Ecuador. A new Ecuadorian Constitution was adopted by a constitutional referendum in September 2008. Again, powerful indigenous groups were instrumental in the drafting of this supreme law of Ecuador. They pressured for constitutional change to give nature the right to exist, persist, maintain, and regen-

29 Law 071 of the Plurinational State.

30 Peasant.

31 Ley (Corta) de Derechos de Madre Tierra, December 2010, Article 10.

erate its vital cycles, structure, functions and processes in evolution. As in Bolivia, indigenous groups in Ecuador have ontological frameworks that do not make the (Western) distinction between humans and nature.

in the fight for sustainability and climate justice, These examples can be seen as attempts to use and extend the anthropocentric idea of human rights and the force and protection they provide. This philosophy has, in large part, been driven by indigenous organisations that derive power and direction from their ontological worldviews, in conjunction with the increased political recognition of indigenous groups as social actors. However, for cultures that do not have such ontological beliefs, the granting of rights to Mother Earth may be regarded not only with scepticism, but also as an appropriation of the planet, which belongs to others as well. In addition, for those who consider the earth to be a deity, the granting of rights that are anthropocentric in nature to the earth deity can be seen as arrogant. But regardless of these ontological and religious distinctions, the cases of Bolivia and Ecuador represent, in fact, radical forms of climate change lawfare which seek social change in pursuit of sustainability and climate justice though changing constitutional structures. Furthermore, they illustrate that legal reform can result from pressure 'from below', i.e. from civil society and, in these cases, from indigenous organisations and previously marginalised groups. It should also be noted that both Bolivia and Ecuador engage closely with the climate justice movement and are perceived by actors in this increasingly visible global movement as examples of the feasibility of a transition to sustainability that merges socio-economic and environmental claims for justice.[32]

What do we know about the effects of these development models and efforts? If one goes back to the categories outlined in Table 1 and look first at the changes in intermediary mechanisms, one can see how these processes have involved new actors in constitution- and lawmaking, changing the processes and sites through which reforms are drafted and debated, and engaging radically new discourses and ideas around rights, sustainability and justice, including the incorporation of ontologies and cosmologies that had previously been marginalised as being non-scientific. However, so far, the new constitutional rights in Ecuador have not led to new laws to implement them. Nor have they "stopped oil companies from destroying some of the most biologically rich areas of the Amazon".[33] In Bolivia, the Law of Mother

32 Martinez-Alier et al. (2011); Bassey (2012); Bond (2010).
33 Vidal (2011).

Earth was expected to lead to radical conservation and social measures to reduce pollution and control industry, but this has not yet materialised. Nevertheless, the ways in which indigenous peoples, their organisations and cosmology have been engaged in these processes of legal reform alone constitute an instance of climate change transformation. Furthermore, there seem to be transformative outcomes materialising from these processes in the form of changes in societal goals and values. These are embedded in development models such as Buen Vivir. It remains to be seen to what extent these lawfare efforts will effectively change decision-making processes and policy outcomes in the long term, whether they will change conditions for sustainability and climate justice, and whether they will have material effect on the ground.

## *II. Climate Change Lawfare within Existing Legal Regimes*

The second category of climate change lawfare strategies engages existing legal frameworks and bodies in struggles around sustainability and climate justice. This category covers an untidy universe of actions related in various ways to legal structures and bodies at international, national and local levels. The rest of this article seeks to put some order into this universe by developing a typology of these climate change lawfare strategies in order to provide a better basis for subsequent analysis of the phenomenon. We start by outlining the universe of cases and issues at stake, as well as conditions of justiciability. This is not an exhaustive description, but rather a set of cases drawn from jurisdictions on various continents to illustrate the diversity of issues, legal frameworks and bodies involved. A first distinction is made between legal institutions and lawfare strategies that engage international laws and bodies, and climate change lawfare that takes place at the national level, involving domestic laws.

### *1. International Climate Change Lawfare*

At the international level, climate change issues have been argued before the United Nations Educational, Scientific and Cultural Organisation (UNESCO) World Heritage Committee, the Kyoto Committee, and the Inter-American Human Rights Commission (IAHRC). While these institutions are not courts, and the cases do not represent litigation in the strict sense, they represent formal compliance mechanisms and engage legal norms of various kinds. Most importantly, the cases highlight the emergent links between climate change, rights, and the strategic and deliberate use of the law.

Since 2005, several NGOs have submitted petitions to the UNESCO World Heritage Committee demanding that a number of sites be added to the World Heritage Danger List as a result of glacial degradation caused by climate change. These include the Sagarmatha National Park (Everest), the coral reefs off the coast of Belize, the glaciers in Peru, Australia's Blue Mountains and Great Barrier Reef, and the Waterton-Glacier International Peace Park in Canada. The NGOs argued, for example, that melting glaciers could potentially destroy the natural and cultural value of these sites, and place thousands of lives at risk. From the perspective of this article, these efforts are interesting as an example of legal strategies by civil society organisations to create and force commitments to sustainability by making use of established legal mechanisms for protection – in this case, the protection of our natural and cultural heritage – which appeal directly against the existing and future harms brought about by climate change. Although the petitions were rejected in the legal sense, they have led to substantial work within UNESCO and the World Heritage Committee to better integrate scientific evidence of climate change impacts and to produce policy papers and strategic documents highlighting the dangers posed by climate change on World Heritage Sites.

Another example of international climate change lawfare is the legal initiative lodged in October 2006 by Friends of the Earth (FoE) Canada, FoE International, and the Climate Justice Programme to require Canada to comply with the Kyoto Protocol.[34] Claiming that Canada violates the Kyoto Protocol and the UNFCCC, they required action under the Canadian Environmental Protection Act to control GHG emissions. They referred to a report by the Canadian Commissioner of the Environment and Sustainable Development, showing that the gap between Canada's GHG emissions and its Kyoto commitments is growing: the 2004 emissions were 26.6% above the 1990 levels, resulting in a gap of 34.6% from Canada's Kyoto target of a 6% reduction by 2008–2012.[35] Per capita, Canadians are among the highest emitters in the world, with the production and consumption of fossil fuels accounting for 80% of these emissions domestically.[36] This case is an in-

34 FoE (2006).
35 (ibid.).
36 (ibid.).

teresting example of lawfare seeking to enforce compliance with international agreements into which states have voluntarily entered, but where the enforcement mechanisms are weak. Although the petition did not have legal consequences, and while we do not yet see a human-rights-based case emerging from these types of claims, they are powerful in raising awareness of the lack of compliance with international regulations. By naming and shaming they also raise moral awareness of the lack of political leadership.

One of the most striking international climate change lawfare efforts, and one of the first cases directly linking climate change with violations of human rights, is the petition brought before the IAHRC by the Inuit Circumpolar Conference (ICC 2005) against the United States of America (US). On behalf of all Inuit of the Arctic regions of the US and Canada, the ICC – assisted by the Center for International Environmental Law and Earthjustice – sought relief from human rights violations resulting from global warming caused by acts of omission and commission by the US. The case draws from the key conclusions presented by the Arctic Climate Impact Assessment, which documents and projects climate change impacts in the Arctic region.[37] The case pointed to the US as being responsible for 25% of the world's carbon dioxide emissions, and to that country's lack of participation in international efforts to combat climate change through the Kyoto Protocol. Although the IAHRC decisions are not enforceable, it was hoped that a declaration recognising that human-induced climate change has infringed on the human rights of the Inuit would contribute towards creating a new foundation under international law for linking environmental degradation to human rights claims. It was hoped that a ruling establishing the liability of the US for its contributions to climate change might push the country towards international collaboration on climate change issues and raise awareness about the human rights consequences of climate change. The petition pointed to the obligation of the US towards its neighbours, both as a member of the Organisation of American States and as a signatory to the American Declaration of the Rights and Duties of Man, which includes the protection of the rights to life, work, residence and movement, inviolability of the home, preservation of health and well-being, and the benefits of culture.

37 ACIA (2006).

The claim was not for monetary reparation, but was –[38]

> … about encouraging the United States of America to join the world community to agree to deep cuts in greenhouse gas emissions needed to protect the Arctic environment and Inuit culture and, ultimately, the world. We [the Inuit] submit this petition not in a spirit of confrontation, that is not the Inuit way, but as a means of inviting and promoting dialogue … within the context of the Climate Change Convention … I [ICC Chair, Ms Sheila Watt-Cloutier] invite governments and non-governmental organizations worldwide to support our petition and to never forget that, ultimately, climate change is a matter of human rights.

The IAHRC dismissed the petition in November 2006 on the basis that it failed to establish whether the alleged facts would tend to characterise a violation of the rights protected by the American Declaration. The role that scientific uncertainty and limited interpretations of human rights may have had in this ruling is an issue for further analysis, but it is noted here that this petition represents a very significant climate change lawfare effort. Despite its dismissal, the petition succeeded in bringing attention to the relationship between climate change and human rights, the nature of the problem, and the weakness of existing governance mechanisms to control the negative impacts of climate change.

### *2. National Climate Change Lawfare*

Legal mobilisation on climate change issues can also be found at the national level, with cases brought before both courts and quasi-judicial bodies. While the international cases discussed above were unsuccessful in legal terms, a number of domestic cases have been won in court. While a proper analysis of this body of litigation and the remedies provided is beyond the scope of this article, Table 2 shows the emergence of legal mobilisation on climate change issues in a number of countries on all continents.

38 ICC (2005).

***Table 2: Climate-change-related Litigation in Domestic Courts***

| Country | Parties | Matter | Case Decision |
|---|---|---|---|
| **Argentina** | *Argentine Citizens v State* (2003) | **Government inaction on climate change adaptation**<br>Argentina's Acción Informativa mechanism and Article 6 of the UNFCCC was used to demand access to information on climate change actions (after severe flooding) to force the government to admit failure to adapt | Case revealed that plans to prevent flooding were developed but not implemented |
| **Australia** | *Australian Conservation Foundation v Minister for Planning* (2004) | **Government inaction**<br>Greenhouse gas (GHG) emissions have to be considered before approving coal mine expansion | Favourable to plaintiffs |
| | *Peter Gray v Minister for Planning* (2006) | **Intergenerational equity**<br>Assessment of greenhouse gas impact and ecologically sustainable development required for new coal mines | Favourable to plaintiffs |
| | *Taralga Landscape Guardians Inc. v Minister for Planning* (2007) and similar cases | **Challenging of wind farms**<br>Growing area of litigation; wind farm developments are being challenged by local communities concerned over the amenity, landscape and health effects | Mixed<br>Some proposals were refused, others were approved or modified |
| **Czech Republic** | *Micronesia Request to Czech Environment Ministry* (2009) | **Request for a transboundary environmental impact assessment (TEIA)**<br>Micronesia requested the Ministry to conduct a TEIA in respect of a plan to modernise the Prunerov II power plant, which would emit 0.021% of global carbon dioxide for 25 years; there were reasonable grounds to believe that it would affect the territory of Micronesia | The TEIA found that the plans did not entail the "best available techniques"; plans were delayed and modified, but approved |
| **Germany** | *Germanwatch & BUND (Friends of the Earth Germany) v Ministry of Economics and Labour* (2004) | **Support of export credit agencies (Hermes) to fossil fuel projects**<br>Contention that financial support for projects overseas contributing to climate change (coal power plants, mining) violates the Kyoto Protocol | Favourable to plaintiffs |

| Country | Parties | Matter | Case Decision |
|---|---|---|---|
| | *Germanwatch v Volkswagen* (Complaint to Ministry of Economics and Technology) (2007) | **Violation of Organisation for Economic Co-operation and Development (OECD) Guidelines for Multinational Enterprises**<br>The Volkswagen Corporation (a member of the resource-intensive transport sector) is obliged to aim at a business strategy to avoid climate change and to implement climate protection goals | Complaint rejected; considered beyond the scope of the Guidelines |
| **New Zealand** | *Environmental Defence Society Inc. v Auckland Regional Council and Contact Energy Limited* (2002) | **Offsetting conditions on carbon dioxide emissions**<br>Permission to construct and operate a gas-fired combined-cycle power station should be conditioned on the energy company's offsetting of carbon dioxide emissions (planting of trees to act as carbon sinks) | Partially favourable to plaintiffs; accepted scientific consensus on anthropogenic climate change |
| **Nigeria** | *Individuals and Communities against Shell, Chevron, Agip and Total* (2005) and other similar cases | **Gas flaring violates human rights**<br>All major multinational oil companies in Nigeria flare gas; this damaging and wasteful practice of carbon dioxide and methane emission contributes to global warming, and the harmful environmental impact in communities violates their constitutional rights to life and dignity | Successful case against Shell in 2005 overturned in 2006 (appeal process – also covers actions of other companies) |
| **United Kingdom** | *World Development Movement et al. v United Kingdom (UK) Treasury* (2009) | **Government inaction on investments contributing to climate change**<br>UK Treasury allows public money to be invested (through the Royal Bank of Scotland) in energy companies and projects contributing towards climate change | Favourable to defendants |
| **United States of America** | *Inupiat village of Kivalina Alaska v ExxonMobil Corp. et al.* (9 oil companies) (2008) | **Against industry: Federal common law claim of nuisance**<br>The oil company's excessive carbon dioxide emissions contribute to global warming; ice protecting the coast has diminished; erosion and destruction will require Kivalina's residents to be relocated | Favourable to defendants; claim barred by 'political question doctrine' and lack of jurisdiction |
| | *Massachusetts v Environmental Protection Agency* (EPA) (2007) | **Regulation of carbon dioxide emission standards under the Clean Air Act**<br>States, local governments and private organisations alleged that the EPA had abdicated its responsibility to regulate GHG emissions | Favourable to plaintiffs; Supreme Court ruled that global warming existed, was caused by humans, was a threat, and had caused the loss of shoreline |

| Country | Parties | Matter | Case Decision |
|---|---|---|---|
| | *Center for Biological Diversity* (CBD) petition to *National Marine Fisheries Service* (NMFS) (2005) | **Petition under Endangered Species Act**<br>The CBD petitioned the NMFS to list three species of coral; the listing required a recovery plan, including consideration of sources of GHG emissions and the impact of global warming | Favourable to petitioners; listing to be warranted for two of the coral species |
| | *Friends of the Earth et al. v Export Credit Agencies* (Mosbacher) (2007) | **Export credit agencies' obligations under the National Environment Policy Act**<br>Such agencies have to consider the impact of GHGs emitted by projects they support | Favourable to defendants |
| | *Center for Biological Diversity et al. v National Highway Traffic Safety Administration* (2007) | **Regulation of fuel-economy standards under** the **Energy Policy and Conservation Act**<br>Poor environmental review of gas-mileage standards; exempting sport utility vehicles and pick-up trucks violated the law | Favourable to plaintiffs |
| | *Comer v Murphy Oil* (2007) | **Against industry: Large carbon-dioxide emitters**<br>Victims of Hurricane Katrina sought compensation for loss of private property and use of public property | Favourable to defendants; Supreme Court declined case in 2011 |
| | *Alec Loorz et al. v US Government* (2011) | **Government inaction violating the Public Trust Doctrine**<br>The state had a duty to protect and preserve the atmosphere, including establishing and enforcing GHG emissions limitations to mitigate climate change caused by humans | Favourable to defendants; failed to establish emergent factors requiring immediate attention |

The claims vary enormously in the issues they raise, the way the issues are framed, and how they are treated. The sample cases presented in Table 2 illustrate the range of claims that have been made. They show a landscape of actors and reasons, including –

- citizens suing industry and states for nuisance and carbon dioxide emissions
- governments and/or NGOs suing export credit agencies for funding the fossil fuel industry
- governments suing the power industry for a proposed plant's carbon dioxide emissions
- industry suing governments for lack of scientific evidence on carbon dioxide relations to climate change, and
- communities suing oil companies because their emissions cause global warming.

The largest number of cases is brought in the US, but cases are increasingly emerging in other countries. Based on these cases, Tables 3 and 4 develop an analytical framework aimed at better understanding various climate change lawfare efforts. A first important distinction is made between the lawfare that primarily aims at a judicial decision – in court, or in a court-like environment – and cases primarily aimed at raising awareness, earlier referred to as *rights talk.* These are not mutually exclusive categories. Out-of-court mobilisation may accompany a legal case, and some cases lodged before courts, particularly before other bodies with little or no enforcement powers, primarily aim at shaming as well as strengthening the focus and attention of a broader mobilisation process. As discussed above, lost cases may lead to important gains in a broader perspective; conversely, a court victory may achieve little unless followed up by other efforts.

A second important distinction is between *direct* and *indirect* climate change lawfare. As is shown in Table 2, some cases directly address climate change issues in the form of responsibility for global warming (most commonly, carbon dioxide emissions) and seek to establish accountability for the effects of climate change (e.g. on sea-level rise). Other cases focus on the responsibility for environmental harms associated with climate change, or with mitigation or adaptation efforts (rather than on climate change itself), and of the effect of these. Table 3 draws a typology of climate lawfare cases along these lines.

***Table 3: Climate Change Lawfare Typology***

| Type | Claim | In Court (or quasi-judicial Body) | Out of Court Mobilisation |
|---|---|---|---|
| Direct | Responsibility for climate change | | |
| | Accountability for climate-change-related damages to livelihoods/health/cultural rights… | | |
| Indirect | Responsibility for climate-change-related environmental degradation | | |
| | Accountability for resulting damages | | |

***Table 4: Direct, Court-centred Climate Change Lawfare: Aims, Claims, Issues and Remedies***

| Aims | Climate change Claims | Counter-claims | Core issues | Remedies |
|---|---|---|---|---|
| Establish **responsibility** for climate change/global warming → mitigation | Regulatory failure (weak mitigation measures fail precautionary principle) | Regulatory failure (too harsh, not scientifically supported) | • Precautionary principle, extent, onus of proof<br>• Scientific knowledge, validity, relevance<br>• Responsibility | Declaratory<br>Mandatory<br>Structural |
| | Compliance failure | | | |
| **Accountability** for damages (human rights violations) from climate change → and from climate policy | Adaptation measures, compensation | | • Causal links<br>• **Attribution** of responsibility | |
| | | Remedies for negative (human rights) effects of mitigation and adaptation measures | • Trade-offs of social/ environmental rights v climate | |

In the following discussion, we focus on the direct, court-centred climate lawfare (the dark area in Table 3). While the cases are diverse in nature, Table 4 shows how the aim is either to establish *responsibility* for climate change in order to strengthen mitigation efforts, or to establish *accountability* for damages and human rights violations resulting from climate change. The latter category includes violations arising from mitigation or adaptation efforts that have negative consequences for some groups, or for nature (the dark cell). In Table 2, this type is represented by windmill cases from Australia, where individuals and communities have sought – and to varying degrees have succeeded – to stop the construction of windmills due to the harm caused to individuals, communities or nature. An issue here is how significant the climate mitigation effects have to be in order to outweigh the inconvenience caused. In a similar case in New Zealand, the court ruled that even very small mitigation effects (arising from windmills) should be taken into account. Similar, more or less legalised conflicts are found, for example, around hydropower stations and reforestation projects.

Cases aiming to establish responsibility for climate change have, to a large extent, focused on carbon dioxide emissions. In Table 2, these are represented by two main types of cases: the first focuses on governmental failure to set emission standards, particularly related to the establishment of a new high-emission industry such as fossil fuel plants, or in relation to the granting of export credits to fund high-emission projects. The second focuses on failures by industry or government to comply with existing commitments; this type of case includes one from Canada related to the Kyoto Protocol. Core issues in both types of cases are –

- the extent to which sufficiently strong scientific knowledge exists to support the regulation in question
- the responsibility laid on individual emitters, and
- in cases of uncertainty, how the precautionary principle should weigh in, i.e. who should bear the onus of proof?

Similarly, in cases related to accountability for human rights violations and other damages caused, the core issues relate to the scientific bases for establishing the required causal links, i.e. how clear is it that the concrete damages in question – e.g. for the Inuit who risk losing their village after the decrease of protective ice – are a result of climate change and not a result of normal weather variability? Furthermore, even if accountability can be sufficiently established, how sound is the basis for attributing responsibility to particular companies or governments (large carbon dioxide emitters)? In the 'grey cell' in Table 4, cases of damages arising from mitigation or adaptation measures, what should the trade-offs be between concern for climate change and other rights?

As Table 2 shows, much of the climate lawfare to date has failed in legal terms – although to varying degrees it has still contributed towards advancing the cause out of court. Nonetheless, some cases have succeeded. Judges in the US and elsewhere have confirmed the scientific consensus on anthropogenic climate change, and have found sufficient proof of causal links to order regulatory measures and attribute responsibility. In most of the cases represented here, the remedies provided by the court have been quite simple declaratory ('fix this') or mandatory ('do that') orders, directed at regulatory authorities or industry. However, based on legal developments in other areas, we could expect courts to adopt more structural approaches, such as that taken by the Argentine Supreme Court in the Mantaza-Riachuelo case.[39] The problem presented in this case related to the century-long pollution of the Mantaza-Riachuelo River Basin in Buenos Aires, threatening the life and health of hundreds of thousands of poor people living on the banks of the river. Solutions were complicated by the fact that a dozen municipalities shared jurisdiction, and there were many hundreds of polluters involved. The court ordered, and continues to supervise, a process of devising and implementing a coordinated plan involving all major stakeholders. While end results have been sluggish, the judgment has resulted in increased awareness and understanding, as well as institutional and organisational reform that, over time, may bring more sustainable solutions.[40] Climate change cases present similar problems related to jurisdiction and coordination. It will be interesting to see whether more sophisticated structural remedies are developed.

## D. Concluding Remarks

Our aim in this article was to offer a theoretical perspective that would allow an investigation of the legalisation of climate change conflicts. Using the concept of *climate change lawfare*, we have shown how, in the context of impotent governance structures, the law may develop into a powerful arena for transformative change. Among activists and decision-makers, and to some extent the public, rights talk and legal challenges are already changing understandings of climate change problems, responsibilities and accountability, and are transforming legal structures – old and new. While few of the cases have thus far been won in court, they have raised awareness and, in some cases, achieved at least some of their aims out of court. Moreover, as an increasing number of cases serve to familiarise the judicial community with such issues, and as new principles and remedies are developed, the law and the courts are likely to become a major arena of contestation over climate justice.

39 Staveland-Sæter (2011).
40 (ibid.).

This article does not aim to provide a comprehensive analysis of climate-change-related legislation and litigation; yet it is hoped that the summary of cases presented, and the climate change lawfare typology developed herein, may offer a starting point for systematic investigations into the conditions and driving forces that place rights and courts at the centre of climate change conflicts, as well as of the effects and impacts – material, symbolic and political – of various legal strategies. The broader ambition is, ultimately, to understand the transformative potential of the law to address problems of sustainability and social justice in the context of climate change. As such, this article illustrates and lays the foundation for further work on the contributions the law can make in responding to the challenges posed by climate change.

## *References*

Aldy, Joseph F. & Robert N. Stavins (Eds), 2009, *Post-Kyoto International Climate Policy: Summary for Policymakers*, Cambridge, Cambridge University Press.

ACIA/Arctic Climate Impact Assessment, 2006, *Arctic Climate Impact Assessment – Scientific Report*, Cambridge, Cambridge University Press.

Barrett, Scott, 2006, *Environment and Statecraft: The Strategy of Environmental Treaty-making*, Oxford, Oxford University Press.

Bassey, Nnimmo, 2012, *To Cook a Continent: Destructive Extraction and Climate Crisis in Africa*, Oxford, Pambazuka Press.

Bond, Patrick, 2010, *The Politics of Climate Justice: Paralysis Above, Movement Below*, Durban, University of KwaZulu-Natal Press.

Comaroff, John L., 2001, Colonialism, Culture, and the Law: A Foreword, *Law & Social Inquiry* 26 (2), 305–314.

Comaroff, Jean & John L. Comaroff (Eds), 2006, *Law and Disorder in the Postcolony*, Chicago, University of Chicago Press.

Comaroff, John L. & Jean Comaroff, 2009, *Ethnicity, Inc.*, Chicago, University of Chicago Press.

Comaroff, John L. & Jean Comaroff, 2011, *Theory from the South: Or, How Euro-America is Evolving Toward Africa,* Boulder, CO, Paradigm Publishers.

Couso, Javier, Alexandra Huneeus & Rachel Sieder (Eds), 2010, *Cultures of Legality: Judicialization and Political Activism in Latin America*, New York, Cambridge University Press.

Davis, Mike, 2006, *Planet of Slums*, New York, Verso.

Doney, Scott C., Victoria J. Fabry, Richard A. Feely & Joan A. Kleypas, 2009, Ocean Acidification: The Other $CO_2$ Problem, *Annual Review of Marine Science* 1, 169–192.

Dugard, Jackie, Jennifer MacLeod & Anna Alcaro, 2012, A Rights-based Examination of Residents' Engagement with Acute Environmental Harm across four Sites on South Africa's Witwatersrand basin, *Social Research* 79 (4), 931-956.

Ferejohn, John, 2002, Judicializing Politics, Politicizing Law, *Law and Contemporary Problems* 65, 41–68.

Ferraz, Octavio L.M., 2011, Brazil: Health Inequalities, Rights, and Courts: The Social Impact of the 'Judicialization of Health', in: Yamin, Alicia E. & Siri Gloppen (Eds), *Litigating Health Rights: Can Courts Bring More Justice to Health?*, Oxford, Harvard Law School Human Rights Program Series with Harvard University Press, 26–102.

Friends of the Earth, 2006, *Submission of Information in Respect of Canada*, available at https://docs.google.com/viewer?a=v&q=cache:x5Qmu9QoX8oJ:www.reseauactionclimat.ca/archive/e/cop-12/unfccc-compliance-letter-foe.pdf+Friends+of+the+Earth+Canada+and+Friends+of+the+Earth+International+and+the+Climate+Justice+Programme+to+require+Canada+to+comply+with+the+Kyoto+Protocol+200&hl=en&gl=no&pid=bl&srcid=ADGEESg-JU_v 1LVe4hYfJiMY4ueXEuM1wU5WK36 8l0LXNrPVThwxVujGnO4YUTK79oFzCvnVH90P4xI19NQFuaSwekOWUAM5udHD_T0aFjGPET8hjETZgsQRFERxtdGFhgtVw9_eVJPY&sig=AHIEtbSfxp1mZDW--Ip_ouJUhaEtANGGMA, last accessed 30 March 2012.

Galanter, Marc, 1974, Why the Haves Come Out Ahead: Speculations on the Limits of Legal Change, *Law and Society Review* 9 (1), 95–160.

Gargarella, Roberto, Pilar Domingo & Theunis Roux, 2006, *Courts and Social Transformation in New Democracies*, London, Ashgate.

Gasper, Des, Ana Victoria Portocarrero & Asuncion Lera St Clair, 2013, Climate Change and Development Framings: A Comparative Analysis of the Human Development Report 2007/8 and the World Development Report 2010, *Journal of Global Environmental Change,* available at http://dx.doi.org/10.1016/j.gloenvcha.2012.10.014, last accessed 2 May 2013.

Gauri, Varun & Daniel M. Brinks, 2008, *Courting Social Justice: Judicial Enforcement of Social and Economic Rights in the Developing World*, Cambridge/New York, Cambridge University Press.

Gauri, Varun & Siri Gloppen, 2012, Human Rights-Based Approaches to Development: Concepts, Evidence, and Policy, *World Bank Policy Research Working Paper No. 5938*, available at http://ssrn.com/abstract=1982447, last accessed 12 October 2012.

Gleick, Peter H., 2010, *Bottled and Sold: The Story Behind Our Obsession with Bottled Water*, Washington, DC, Island Press.

GHF/Global Humanitarian Forum, 2009, *Human Impact Report – Climate Change: The Anatomy of a Silent Crisis*, Geneva, GHF, available at http://www.ghf-ge.org/human-impact-report.php, last accessed 12 October 2012.

Gloppen, Siri, Bruce M. Wilson, Roberto Gargarella, Elin Skaar & Morten Kinander, 2010, *Courts and Power in Latin America and Africa*, London, Palgrave Macmillan.

Gloppen, Siri, Daniel M. Brinks & Malcolm Langford, 2011, *Collaborative Global Centre for Law and Social Transformation: Concept Note*, November, Bergen, Chr. Michelsen Institute.

Gough, Ian, 2011, Climate Change, Double Injustice and Social Policy: A Case Study of the United Kingdom, *United Nations Research Institute for Social Development (UNRISD) Occasional Papers on Green Economy and Sustainable Development*, available at http://personal.lse.ac.uk/goughi/UNRISD%20Gough.pdf, last accessed 12 October 2012.

Heinrich Böll Foundation (Eds), 2011, *Buen Vivir*: Latin America's New Concepts for the Good Life and the Rights of Nature, *Publication Series Ecology 17*, Berlin, Heinrich Böll Foundation.

Hirschl, Ran, 2004, *Towards Juristocracy: The Origins and Consequences of the New Constitutionalism*, Cambridge, MA, Harvard University Press.

Hirschl, Ran, 2006, The New Constitutionalism and the Judicialization of Pure Politics Worldwide, *Fordham Law Review* 75 (2), 721–754.

IPCC/Intergovernmental Panel on Climate Change, 2007, *IPCC Fourth Assessment Report, Climate Change 2007,* available at http://www.ipcc.ch/publications_and_data/publications_and_data_reports.shtml#.UO1mxG_Ad8E, last accessed 12 October 2012.

IPCC/Intergovernmental Panel on Climate Change, 2012, *Special Report: Managing the Risks of Extreme Events and Disasters to Advance Climate Change Adaptation*, available at http://www.ipcc.ch/publications_and_data/publications_and_data_reports.shtml#.UO1mxG_Ad8E, last accessed 12 October 2012.

ICC/Inuit Circumpolar Conference, 2005, *Inuit Petition to the Inter-American Commission on Human Rights to Oppose Climate Change Caused by the United States of America*, available at http://inuitcircumpolar.com/index.php?ID=316&Lang=En, last accessed 30 March 2012.

ICC/Inuit Circumpolar Conference, 2005, *Petition to the Inter-American Commission on Human Rights seeking Relief from Violations Resulting from Global Warming caused by Acts and Omissions of the United States*, 7 December, available at http://inuitcircumpolar.com/files/uploads/icc-files/FINALPetitionICC.pdf, last accessed 12 October 2012.

Joshi, Manoj, Ed Hawkins, Rowan Sutton, Jason Lowe & David Frame, 2011, Projections of When Temperature Change will Exceed 2°C Above Pre-industrial Levels, *Nature Climate Change* 1, 407–412.

Martinez-Alier, Joan, 2011, The EROI of Agriculture and its Use by the Via Campesina, *The Journal of Peasant Studies*, 38(1), 145–160.

New, Mark, Diana Liverman, Heike Schroder & Kevin Anderson, 2011, Four Degrees and Beyond: The Potential for a Global Temperature Increase of Four Degrees and its Implications, *Philosophical Transactions of the Royal Society A* 369 (1934), 6–19.

O'Brien, Karen L., Asuncion Lera St. Clair & Berit Kristoffersen (Eds), 2010, *Climate Change, Ethics and Human Security*, New York, Cambridge University Press.

Peel, Jacqueline, 2010, *The University of Melbourne, Inuit Petition to the Inter-American Commission on Human Rights*, available at http://blogs.unimelb.edu.au/peel_climatechange/2010/05/24/inuit-petition-to-the-inter-american-commission-on-human-rights/, last accessed 12 October 2012.

Petherik, Anna, 2012, A Note of Caution, *Nature Climate Change* 2, 144–145.

Scobbie, Iain, 2006, Jurisphilia/Jurisphobia: US Approaches to International Law Theory, in: American Society of International Law (Eds), *Proceedings of the Annual Meeting (American Society of International Law)* 100, 166–169.

Shapiro, Martin & Alec Stone Sweet, 2002, *On Law, Politics, and Judicialization*, New York, Oxford University Press.

Sieder, Rachel, Line Schjolden & Alan Angell (Eds), 2005, *The Judicialization of Politics in Latin America*, New York, Palgrave Macmillan.

Stafford Smith, Mark, Lisa Horrocks, Alex Harvey & Clive Hamilton, 2011, Rethinking Adaptation for a 4°C World, *Philosophical Transactions of the Royal Society A* 369 (1934), 196–216.

Staveland-Sæter, Kristi Innvær, 2011, Litigating the Right to a Healthy Environment: Assessing the Policy Impact of 'The Mendoza Case', *Chr. Michelsen Institute (CMI) Report*, 6, 62.

Tate, C. Neal & Torbjorn Vallinder (Eds), 1995, *The Global Expansion of Judicial Power*, New York, New York University Press.

Tiefenbrun, Susan W., 2011, Semiotic Definition of 'Lawfare', *Case Western Reserve Journal of International Law*, 43.

Victor, David G., 2011, *Global Warming Gridlock: Creating More Effective Strategies for Protecting the Planet,* Cambridge, Cambridge University Press.

Vidal, John, 2011, Bolivia Enshrines Natural World's Rights with Equal Status for Mother Earth, *The Guardian*, 10 April, available at www.guardian.co.uk/environment/2011/apr/10/bolivia-enshrines-natural-worlds-rights, last accessed 12 October 2012.

WHO/World Health Organization, 2012, Climate Change and Health, Fact Sheet No. 266, available at http://www.who.int/mediacentre/factsheets/fs266/en/, last accessed 12 October 2012.

Yamin, Alicia E. & Siri Gloppen (Eds), 2011, *Litigating Health Rights: Can Courts Bring More Justice to Health?*, Cambridge, MA, Harvard Law School Human Rights Program Series with Harvard University Press.

# 4
# Greening Permanent Sovereignty through the Common Concern in the Climate Change Regime: Awake Custodial Sovereignty!

*Werner Scholtz*

*Abstract*

The preamble of the United Nations Framework Convention on Climate Change (UNFCCC), on the one hand, designates climate change and its consequences as the common concern of humankind and, on the other hand, affirms that states have the sovereign right to exploit their own resources. An important consequence of the common concern is that it *globalises* certain natural resources, which may be in conflict with the sovereign right of states concerning their natural resources. The UNFCCC is silent on the manner in which this potential conflict should be dealt with. It is accordingly the primary objective of this essay to reconcile the aforementioned notions pursuant to the needs of the international community in the era of climate change. Thus, an analysis of the legal content and consequences of permanent sovereignty and the notion of the common concern provide an understanding of how the common concern moulds the sovereign rights of states over their natural resources in the current phase of globalisation. The author proposes that common concern results in the development of custodial obligations for states, which lead to the emergence of custodial sovereignty.

## *A. Introduction*

The preamble of the United Nations Framework Convention on Climate Change (UNFCCC) affirms that the "change in the Earth's climate and its adverse effects" are the common concern of humankind CCH.[1] According to Boyle et al., the phrase *common concern* indicates a legal status which is particularly different from permanent sovereignty, and its main consequence

1 See also UN GAR 43/53 of 6 December 1988.

is that it gives the international community a "legitimate interest in resources of global significance and a common responsibility to assist in their sustainable development".[2] However, the preamble of the UNFCCC affirms that states have the sovereign right to exploit their own resources. This reflects the notion of permanent sovereignty in international law, which entails the right of states freely to dispose of their natural resources.[3] The notion of common concern globalises certain natural resources and accordingly may be in conflict with the notion of permanent sovereignty, since the right of states over their natural resources must be exercised within the confines of the aforementioned global responsibilities. The UNFCCC, however, does not provide any clarity concerning the relationship between common concern and permanent sovereignty.

How should common concern and permanent sovereignty in the international climate change regime be reconciled? It is the primary aim of this brief essay to address this question.[4] It is this author's point of departure that the inclusion of the notion of the "common concern of humankind" in the UNFCCC invites reconciliation between permanent sovereignty and the global needs of the international community in relation to climate change. The author accordingly briefly reflects on the notion of permanent sovereignty. Furthermore, the author discusses the potential legal consequences of the common concern and subsequently determines how the common concern moulds permanent sovereignty in accordance with the needs of the international community in the current phase of globalisation. Accordingly it is the view of the author that the emergence of common concern in the context of the paramount importance of sustainable development in international environmental law further develops the content of the duties component of the right of states freely to dispose of their natural resources. This affirms that permanent sovereignty does not merely entail rights, but also global obligations for states, and accordingly imposes constraints on the exercise of permanent sovereignty. The author argues that common concern necessitates a custodial element. This reconfiguration of permanent sovereignty ensures the *greening* thereof in order to accommodate the pursuit of sustainable development through the exercise of sovereignty over natural resources. This

2 Birnie et al. (2009:128).

3 For an extensive analysis see Schrijver (1997) and Hossain & Chowdhury (1984).

4 I have dealt with this issue in general terms in a previous publication which constitutes the basis for my arguments in this publication. I shall therefore not repeat my previous arguments in detail. See Scholtz (2008:323).

marks a further development, since the needs of the international community require global cooperation in relation to climate change, and permanent sovereignty needs to respond accordingly. The article concludes with final remarks.

## B. The Marriage between Permanent Sovereignty and Common Concern

### I. Permanent Sovereignty is not Permanent

The so-called economic side of political sovereignty[5] has been included in several Multilateral Environmental Agreements, soft law documents and international declarations.[6] However, the origins of this notion can be traced back to the New International Economic Order,[7] where it was used by developing states as an important mechanism to overcome economic disparities and to curtail colonialist interference in the economic affairs of newly independent states.Article 2 of the Charter of Economic Rights and Duties of States encapsulates the core content of permanent sovereignty as it reads that "Every State has and shall freely exercise full permanent sovereignty, including possession, use and disposal, over all its wealth, natural resources and economic activities".

The principle of sovereignty over natural resources has, however, evolved since its genesis during the post-war area. The rights-based focus of permanent sovereignty gradually changed to make way for the recognition that duties emanate from permanent sovereignty.[8] The development of interna-

5 Perrez (2000:97). See, however, Brehme (1967:8 note 9).

6 Examples include: Articles 3 and 15 of the Convention on Biological Diversity (1997); common Article 1 of the 1966 International Covenant on Civil and Political Rights and the 1966 International Covenant on Economic, Social and Cultural Rights; Principle 2 of the Rio Declaration; Para. 1.1 of the ILA New Delhi Declaration of Principles of International Law Relating to Sustainable Development (2002).

7 Para. 4(e) of the Declaration on the Establishment of a New International Economic Order UN GAR 3201-S.VI of 1 May 1974 and Article 2 of the Charter of Economic Rights and Duties of States, UN GAR 3281-XXIX of 12 December 1974. See for a discussion of the New International Economic Order: Bedjaoui (1979); Makarczyk (1988) and Hossain (1980).

8 Resolution 1803, for example, requires that permanent sovereignty must be exercised in the interest of the national development and well-being of the people. UN GAR 1803 (XVII) 14 December 1962.

tional environmental law and the prominence of sustainable development[9] have had a profound impact on the interpretation of permanent sovereignty.[10] For example, good neighbourliness imposes restrictions on the manner in which states may exercise their sovereign rights over their natural resources.[11] Furthermore, the interdependence[12] of states has resulted in the increasing emergence of international legal regimes for the cooperative management of natural resources pursuant to sustainable development.[13]

## *II. Common Concern: Chrysalis of Change*

The need for concerted global action based on the common concern of humankind has become an important aspect of the management of resources[14] and accordingly has a further influence on the content of permanent sovereignty.[15] The designation of causes and/or responses as a common concern results in various interesting consequences.[16] Firstly, the common concern affirms the importance of fair and equitable burden-sharing.[17] Thus, legal measures concerning the common concern are characterised by differential treatment provisions.[18] The climate change regime provides perhaps one of the best examples of a differential treatment regime that relates

9 Sustainable development is viewed as the single most important concept in international environmental law in the "sense that the whole international environmental law has to be developed further under an overall sustainable development umbrella". Beyerlin (1996:112).

10 This is also recognised in the preamble of the United Nations Framework Convention on Climate Change (UNFCCC), which affirms that permanent sovereignty must be exercised pursuant to environmental and development policies and should not cause damage to other States.

11 The Stockholm Declaration of the United Nations Conference on the Human Environment of 1972, Article 21 (hereinafter The Stockholm Declaration).

12 See, however, Greig (2002).

13 Schachter (1977).

14 See Perrez (2000).

15 Scholtz (2008:323).

16 See Biermann (1996:431). See also Timoshenko (1995).

17 In accordance with The Hague Recommendations on International Environmental Law "costs should be shared equitably among states, taking into account historic responsibilities and present technical and financial capabilities". Para. 3 of The Hague Recommendations on International Environmental Law of 16 Augustus 1991, in: Bilderbeek (1992:194-202).

18 For an analysis of differential treatment, see Rajamani (2006).

to the common concern, since the common but differentiated responsibilities and respective capabilities principle is a core principle of the international climate change regime.[19]

It must also be borne in mind that *humankind* includes all members of the human species as a whole (present and future generations) and in this manner affirms intergenerational and intragenerational equity. Thus, the existence of the common concern seems to imply that permanent sovereignty should be exercised for the benefit of humankind, which consists of current and future generations.[20] This implies a departure from a state-centred exercise of sovereignty pursuant to a narrow national self-interest towards a more universalist approach, which takes cognisance of the common interest of the international community in relation to common concerns. Common concern opens a gateway for the importation of cosmopolitan[21] ideals in which the pursuit of global well-being plays an important role. It furthermore serves as an affirmation that other participants[22] in the international arena have an important role to play in relation to the common concern of these actors. Thus, CCH may serve as a catalyst for the further development of traditional legal subjectivity in international law[23] and give rise to legal obligations and

19 Article 3(1) of the UNFCCC reads that "The Parties should protect the climate system for the benefit of present and future generations of humankind, on the basis of equity and in accordance with their common but differentiated responsibilities and respective capabilities. Accordingly, the developed country Parties should take the lead in combating climate change and the adverse effects thereof". The Kyoto protocol of 1997 under the UNFCCC reflects the Common but Differentiated Responsibilities and Respective Capabilities principle. In terms of Art. 3, Annex I parties (developed countries) are obliged to reduce their greenhouses gas emissions to at least five per cent below 1990 levels by 2008–2012, while developing countries are not under such an obligation. Furthermore, Article 10 structures certain obligations of the parties according to CBDR. Article 10(c), for instance, instructs developed countries to "take all practicable steps to promote, facilitate and finance the transfer of, or access to, environmentally sound technologies, know-how, practices and processes … in particular to developing countries.".

20 See in this regard, Trindade (2010:327–352).

21 See Pierik & Werner (2010).

22 Eminent scholars, such as Higgins have criticised the usage of the term *subjects* of international law and rather prefer other terms, such as *participants*. See Higgins (1995:39). See further Schreuer (1993:447).

23 According to Brunnée, *common interest* is a generic term. In some instances *common interest* may result in an international law rule that entails certain duties. In these instances "[w]e are faced with the phenomenon of a common interest so compelling that it alone formulates the rule and coincides with the rule's content." This means

rights that apply not only to states, but also to non-state actors. Non-state actors have an important role to fulfil concerning the pursuit of the common concern. The emergence of non-state actors, in particular environmental non-governmental organisations (NGOs), must be viewed in the context of the critique that exists that states are ill-equipped to meet the challenges posed by global environmental degradation, such as climate change.[24] This statement does not equate the primary and indispensable role of states in the international arena with that of other participants.[25] It is, however, indicative of an increasingly important participation of other non-state actors pursuant to the common concern in international environmental law.[26] This also means that the orthodox positivist doctrine of international legal personality,[27] which recognised only states as subjects of international law, will undergo further changes[28] in order to accommodate the proliferation of non-state actors. Thus, the common concern of humankind imposes a further qualification on permanent sovereignty in the sense that its exercise should be steered by the interests of humankind and further that not merely state cooperation is required in an age of interdependence, but that other non-state actors also have an important (albeit a different or complementary) role to play in pursuit of the well-being of humankind.

The complexity in relation to the reconciliation of the common concern with the principle of permanent sovereignty is illustrative of the tension between the international law doctrine preoccupied with a state-centred sovereignty and the need for changes toward a more universalist international community responsive to the challenges of globalisation.[29]

---

that CCM is a facet of *common interest. Common interest* therefore serves as a driving force in the development of rules. Brunnée (1989).

24 Camilleri & Falk (1992:192).

25 This also does not mean that non-state actors should deserve the same legal recognition as states since "the subjects of law in any legal system are not necessarily identical in their nature or in the extent of their rights, and their nature depends upon the needs of the community". *Reparations for Injuries Suffered in the Service of the United Nations*, Advisory Opinion, ICJ Reports (1949), 175.

26 See Yamin (2001:149).

27 See Cutler (2001).

28 This refers to the increasing acceptance of the legal recognition of other entities, such as international governmental organisations. See Klabbers (2003:353).

29 For an extensive discussion, see Bederman (2008).

### *III. Reconfiguration of Permanent Sovereignty towards Custodianship*

What does the acceptance of CCH mean for permanent sovereignty? The UNFCCC recognises both concepts, without providing any answers in relation to the potential conflict between the concepts. The *concern* element does not carry with it any proprietary meaning, but relates to the causes as well as the responses[30] to global concerns. CCH may not as such have any direct proprietary meaning in relation to resources, but it nonetheless has an impact on territorial sovereignty. The common concern exists in relation to the consequences of climate change. The consequences of climate change are the result of the greenhouse gas emissions that occur in states and the way in which states regulate or omit to regulate the latter in terms of their territorial sovereignty. In the author's opinion, the atmosphere may be viewed as a global environmental resource, since it does not fit in any of the other categories, such as *shared resources*[31] or *global commons*,[32] where the *common heritage of humankind* applies.[33] The fact that CCH creates a legitimate interest[34] in relation to the actions (or omissions) of states in their own territories concerning global environmental resources may be difficult to reconcile with the right of states and peoples freely to dispose of their natural resources. Does the acceptance of the common concern mean that permanent sovereignty as a component of state sovereignty is redundant? The point of departure of the Meeting of the Group of Legal Experts to Examine the Concept of the Common Concern of Mankind in Relation to Global Environmental Issues was that *common concern* does not imply a

30 The CBD refers to the conservation of biological diversity as a common concern.

31 *Shared resources* is more relevant in situations of bilateral or regional transboundary pollution. Biermann (1995:9–10).

32 See, however, Vogler (1995:2ff.). The areas to which the common heritage of humankind is applicable are not subject to appropriation. This means that common heritage areas are owned by no one and states cannot make territorial claims to these areas. Joyner (1986). Global environmental resources may, however, be found in the territories of states.

33 Scholtz (2008:336). In my opinion, a global environmental resource is a renewable natural resource of which a part or the whole of the resource is located in the territory of a state, but which is needed and enjoyed by the whole of humankind. I have borrowed this term from Glennon (1990:34). The legal status of the atmosphere in international law is unclear. See Boyle (1991:7–13). It should also be borne in mind that the atmosphere should be distinguished from the territorial airspace of a state. The atmosphere refers to the layer of air above the territory of a state.

34 Birnie et al. (2009:128).

departure from state sovereignty, since states still possess permanent sovereignty over natural resources.[35] The author agrees with this assumption. States still have a right freely to dispose of their natural resources. Developing countries are in need of economic growth in order to alleviate poverty. Developing and developed states still have different environmental agendas. The developing world is plagued by *environmental problems of poverty* and the developed world by environmental problems deriving from the *excess of affluence*.[36] Recalling permanent sovereignty or rendering it obsolete will mean that the developing world will be unable to address its environmental problems. It may also leave the developing world vulnerable to 'eco-imperialist' motives of the more powerful developed world.[37] This means that permanent sovereignty still has an important role to play in accordance with its envisaged goal as a component to pursue development. However, in order to address its *environmental problems of poverty*, developing states must follow a path of development that is sustainable. This means that the importance of sustainable development in international environmental law must be taken into account when one interprets permanent sovereignty and its relationship with the common concern. It is therefore imperative to 'green' the economic side of sovereignty rather than to declare it obsolete. The question would accordingly rather be how the common concern changes[38] the right of states freely to dispose of their natural resources in order to respond to the challenges of global environmental degradation, such as climate change. This approach pursues the strengthening of international (environmental) law since it reconfigures permanent sovereignty pursuant to the overall objectives of sustainable development in international environmental law.

It must be borne in mind that the sovereign rights of states over natural resources have never been absolute.[39] This is recognised in the preamble of

35 Report of the Meeting of the Group of Legal Experts to Examine the Concept of the Common Concern of Mankind in Relation to Global Environmental Issues. See Horn (2004:237).

36 See Ntambirweki (1991:907).

37 Scholtz (2008:328).

38 This is in line with the idea that sovereignty as such is a dynamic concept which "can have a different meaning in different historical periods although certain essential characteristics remain". Schrijver (1999:70).

39 Permanent sovereignty has to be exercised in the interest of the people and subject to general international law. See UN GAR Resolution 1803 (XVII) 14 December of 1962. Article 21 of the Stockholm declaration imposes the principle of good neigh-

the UNFCCC, which affirms that permanent sovereignty must be exercised pursuant to environmental and development policies and should not cause damage to other states. It is accordingly the author's opinion that the common concern further moulds the interpretation of permanent sovereignty towards the duties[40] which this notion imposes on states in order to pursue sustainable development. But what does this mean for permanent sovereignty? The prominence of sustainable development and the development of international environmental law clearly have influenced the interpretation of permanent sovereignty towards an affirmation of the environmental duties of this notion. CCM, however, induces a further development in relation to permanent sovereignty since it relates to the territorial nature thereof. The fact that CCM creates a legitimate interest in the territorial acts of a state in relation to its global environmental resources cannot be easily reconciled with some of the elements[41] of sovereignty, such as territorial integrity and territorial sovereignty,[42] which allow for an exclusive claim over state territory. In general international law it is the primary objective of territorial jurisdiction to avoid conflicts of extraterritorial jurisdiction pursuant to the promotion of sovereign equality and non-intervention. The existence of CCM requires both an affirmation and qualification of permanent sovereignty. It necessitates the right of states freely to dispose of their natural resources, but also invokes the affirmation of the legitimate interest of other states in relation to global environmental resources and as such the custodianship of states over global environmental resources in their territories.

The author has previously coined the notion of custodial sovereignty in order to provide an answer to this messy question.[43] In accordance with this approach, a state is the custodian of its global environmental resources. Other states have an expectation that the relevant state will protect these resources

---

bourliness as a restriction on the manner in which States may exercise their sovereign rights over their natural resources.

40 This is in line with the thinking of Van Staden & Vollaard (2002). This is akin to the line of thought of the ICISS (2001). It must be borne in mind that this report is primarily concerned with the issue of military humanitarian intervention in cases of atrocities such as the large-scale loss of life or large-scale ethnic cleansing. The idea that sovereignty entails responsibility is, however, similar to the foundation of custodial responsibility.

41 See Steinberger, (2000:513). The updated version in the latest electronic format of the encyclopedia is still being developed.

42 Shaw (2008:489ff.).

43 Scholtz (2008:323).

for the whole of humankind. The custodial state has the right to dispose freely of its global environmental resources, but the latter right is restricted by the expectations and interests of other states. Thus, the custodial state has a duty to pursue sustainable development in its exercise of the right. In this manner the reconfiguration of permanent sovereignty acknowledges the primary right of custodial states over their natural resources. This right is, however, not absolute since it is balanced by custodial duties. This reinterpretation of permanent sovereignty respects the territorial integrity of states pursuant to sovereign equality and non-intervention, but takes cognisance of the realities of a single biosphere in a globalised world. Furthermore, other states are burdened with the duty to support the custodial state to fulfil its obligations in a cooperative manner. Two fundamental elements constitute the bedrock of the notion of custodial sovereignty. The first element concerns the common (global) responsibility of all states for the protection of global environmental resources. The second element concerns the differentiated responsibilities of states' contributions to the protection of these resources. Thus, differential treatment provisions are vital for the custodial model.[44] The common but differentiated responsibilities and capabilities principle in the UNFCCC and Kyoto Protocol therefore gives concrete expression to the elements of custodial sovereignty.[45]

## *C. Concluding Remarks*

The reconciliation of common concern and permanent sovereignty introduces the imposition of a custodial element, which 'greens' permanent sovereignty since it ensures that permanent sovereignty may be exercised pursuant to sustainable development. It accordingly takes into account developments in relation to factual realities in a global and interdependent world, as well as the development of international law and the prominence of sustainable development. The impacts of climate change are oblivious to state borders and require global cooperation. The obsession of permanent

44 Cullet (1999:551). See furthermore Rajamani (2006) and French (2000:46).

45 Article 3(1) includes this principle as one of the fundamental principles of the international climate change regime. The international climate change regime is considered to be the "clearest attempt to transform, activate and operationalize common but differentiated responsibility from a legal concept into a policy instrument". See remarks by Joyner (2002:358).

sovereignty with territorial integrity leads one to question the relevance of this notion in a globalised world confronted by global environmental degradation, in particular climate change. Sovereignty does not accord with the factual reality of a single biosphere oblivious to state borders. This question is even more acute when one recalls the historical context of the development of permanent sovereignty. The latter concept constituted an important principle in the call for a New International Economic Order, which focused on the economic development and independence of decolonised states. It is commonplace that the goals of the NIEO have not been fulfilled and that equity and economic freedom have not been achieved in international law. It therefore also ensures that permanent sovereignty does not become an outdated relic of a bygone era which focused on the pursuit of a New International Economic Order in a cold-war context. It reconfigures permanent sovereignty in accordance with the demands and needs of the international community. Thus, it provides for a reflection of the sustainable development side of sovereignty in order to ensure that sovereignty reflects "not obstacles ... but responsibility and opportunities to secure human values".[46]

## *References*

Bederman, David J., 2008, *Globalization and International Law*, New York, Palgrave MacMillan.

Bedjaoui, Mohammed, 1979, *Towards a New International Economic Order*, Paris, Holmes & Meier Publishers.

Beyerlin, Ulrich, 1996, The Concept of Sustainable Development, in: Wolfrum, Rüdiger (Ed.), *Enforcing Environmental Standards: Economic Mechanisms as Viable Means*, London, Springer, 95–121.

Biermann, Frank, 1995, *Saving the Atmosphere. International Law, Developing Countries and Air Pollution*, New York, Peter Lang Publishers Inc.

Biermann, Frank, 1996, Common Concern of Humankind: The Emergence of a New Concept of International Environmental Law, *Archiv Des Völkerrechts* 34, 426–481.

Bilderbeek, Simone (Ed.), 1992, *Biodiversity and International Law: The Effectiveness of International Environmental Law*, Amsterdam, IOS Press.

Birnie, Patricia, Alan Boyle & Catherine Redgwell, 2009, *International Law and the Environment* (Third Edition), Oxford, Oxford University Press, 2009, 128.

46 Henkin (2000:14).

Boyle, Alan E., 1991, International Law and the Protection of the Global Atmosphere: Concepts, Categories and Principles, in: Churchill, Robin (Ed.), *International Law and Global Climate Change*, London, Graham & Trotman, Martinus Nijhoff Publishers, 7–21.

Brehme, Gerhard, 1967, *Souveränität der jungen Nationalstaaten über Naturreichtümer. Die ständige Souveränität über die natürlichen Reichtümer und Hilfsquellen –Völkerrechtliche Probleme des Kampfes der jungen Nationalstaaten um ökonomische Unabhängigkeit*, Berlin, Staatsverlag der Deutschen Demokratischen Republik.

Brunnée, Jutta, 1989, Common Interest – Echoes from an Empty Shell? Some Thoughts on Common Interest and International Environmental Law, *Zeitschrift für ausländisches öffentliches Recht und Völkerrecht* 37, 791–808.

Camilleri, Joseph A. & Jim Falk, 1992, *The End of Sovereignty? The Politics of a Shrinking and Fragmenting World*, London, Edward Elgar Publishing.

Cullet, Philippe, 1999, Differential Treatment in International Law: Towards a New Paradigm of Inter-state Relations, *European Journal of International Law* 10, 549–582.

Cutler, A. Claire, 2001, Critical Reflections on the Westphalian Assumptions of International Law and Organization: A Crisis of Legitimacy, *Review of International Studies* 27 (2), 133–150.

French, Duncan, 2000, Developing States and International Environmental Law: The Importance of Differentiated Responsibilities, *International and Comparative Law Quarterly* 49, 35–60.

Glennon, Michael J., 1990, Has International Law Failed the Elephant?, *American Journal of International Law* 84, 1–43.

Greig, Don, 2002, International Community, Interdependence and All That … Rhetorical Correctness?, in: Kreijen, Gerard (Ed.), *State, Sovereignty and International Governance*, Oxford, Oxford University Press, 522–603.

Henkin, Louis, 2000, That 'S' Word: Sovereignty, and Globalization, and Human Rights, Et Cetera, *Fordham Law Review* 68 (1), 1-14

Higgins, Rosalyn, 1995, *Problems and Processes: International Law and How We Use It*, Oxford, Oxford University Press.

Horn, Laura, 2004, The Implications of the Concept of Common Concern of Mankind on a Human Right to a Healthy Environment, *Macquarie Journal of International and Comparative Environmental Law* 1, 233–269.

Hossain, Kamal (Ed.), 1980, *Legal Aspects of the New International Economic Order*, London, Frances Pinter.

Hossain, Kamal & Roy Chowdhury (Eds), 1984, *Permanent Sovereignty over Natural Resources in International Law Principle and Practice*, London, Pinter.

ICISS/International Commission on Intervention and State Sovereignty, 2001, The Responsibility to Protect, available at http://www.responsibilitytoprotect.org/files/R2P Summary.pdf, last accessed 30 July 2012.

Joyner, Christopher C., 1986, Legal Implications of the Concept of the Common Heritage of Mankind, *International and Comparative Law Quarterly* 35, 190–199.

Joyner, Christopher C., 2002, Transcript of Panel Discussion on 'Common but Differentiated Responsibility', *American Society of International Law Procedures* 96, 358ff.

Klabbers, Jan, 2003, (I Can't Get No) Recognition: Subjects Doctrine and the Emergence of Non-State Actors, in: Petman, Jarna & Jan Klabbers (Eds), *Nordic Cosmopolitanism – Essays in International Law for Martti Koskenniemi*, Leiden, Brill, 351–370.

Makarczyk, Jerzy, 1988, *Principles of a New International Economic Order: A Study of International Law in the Making*, Dordrecht, Martinus Nijhoff Publishers.

Ntambirweki, John, 1991, The Developing Countries in the Evolution of an International Environmental Law, *Hastings International and Comparative Law Review* 14, 905–928.

Perrez, Franz Xaver, 2000, *Cooperative Sovereignty: From Independence to Interdependence in the Structure of International Environmental Law*, The Hague, Kluwer Law International.

Pierik, Roland & Wouter Werner (Eds), 2010, *Cosmopolitanism in Context: Perspectives from International Law and Political Theory*, Cambridge, Cambridge University Press.

Rajamani, Lavanya, 2006, *Differential Treatment in International Environmental Law*, Oxford, Oxford University Press.

Schachter, Oscar, 1977, *Sharing the World's Resources*, Columbia, Columbia University Press.

Scholtz, Werner, 2008, Custodial Sovereignty Reconciling Sovereignty and Global Environmental Challenges Amongst the Vestiges of Colonialism, *Netherlands International Law Review* 55 (3), 323–341.

Schreuer, Christoph, 1993, The Waning of the Sovereign State: Towards a New Paradigm for International Law, *European Journal of International Law* 4, 447–471.

Schrijver, Nico, 1997, *Sovereignty over Natural Resources: Balancing Rights and Duties*, Cambridge, Cambridge University Press.

Schrijver, Nicolaas, 1999, The Changing Nature of State Sovereignty, *The British Yearbook of International Law* 70 (1), 65–98.

Shaw, Malcolm N., 2008, *International Law*, Cambridge, Cambridge University Press.

Steinberger, Helmut, 2000, Sovereignty, in: Bernhardt, Rudolf (Ed.), *Encyclopedia of Public International Law Volume IV* (Sixth Edition), Heidelberg, Max Planck Institute for Comparative Public Law, 500–521.

Timoshenko, Alexandre S., 1995, Responses to Environmental Challenges: UNEP Experience, in: Al-Nauimi, Najeeb & Richard Meese (Eds), *International Legal Issues Arising under the United Nations Decade of International Law*, Dordrecht, Martinus Nijhoff Publishers, 169–170.

Trindade, Antônio Augusto Cançado, 2010, *International Law for Humankind*, Dordrecht, Martinus Nijhoff Publishers.

van Staden, Alfred & Hans Vollaard, The Erosion of State Sovereignty: Towards a Post-territorial World?, in: Kreijen, Gerard (Ed.), *State, Sovereignty and International Governance*, Oxford, Oxford University Press, 522–603.

Vogler, John, 1995, *The Global Commons: A Regime Analysis*, Chichester, John Wiley & Sons.

Yamin, Farhana, 2001, NGOs and International Environmental Law: A Critical Evaluation of their Roles and Responsibilities, *Review of European Community and International Environmental Law* 10, 149–162.

# PART II: GLOBAL CLIMATE GOVERNANCE AND DIPLOMACY

5

# Power in Global Climate Governance

*Babette Never*

*Abstract*

This article assesses the distribution of power in international climate negotiations and beyond. Using the political science categories of *instrumental*, *structural* and *discursive* power, the article compares and contrasts the power of the central actors within the climate regime, in clean technology markets, in bilateral agreements, and at the interface of energy and climate governance. The multidimensional, relational quality of power, as well as the contrast between active and passive power, draw a differentiated picture of the behaviour of central Northern actors, i.e. the European Union (EU), Germany, Norway and the United States (US), and the emerging economies of the South, namely Brazil, China, India and South Africa. The article finds that China, in particular, is gaining power in the structural dimension, but is in a negative balance of power with the US in international climate negotiations. The EU and Germany have more 'green power' potential than they are actually using, while the least-developed countries primarily have some moral discursive power in negotiations only. Since this discursive power is not backed up by similar power potential in the other two dimensions, they currently cannot use it to their full advantage.

## *A. Global Climate Governance after Doha*

In December 2012, international climate negotiations once again came close to failure. While the outcomes of this 18th Conference of the Parties (COP18) to the United Nations Framework Convention on Climate Change (UNFCCC) were rather limited, some of the primary goals of this round of negotiations were actually attained: the closure of two negotiation tracks,[1] the agreement on a second commitment period of the Kyoto Protocol from 2013 to 2020, and some propulsion forward on a trajectory to a new agreement under the Durban Platform for Enhanced Action. Thus, COP18 counts as a 'transitional' or 'housekeeping' conference which kept the bureaucratic proceedings going, but it was devoid of any real progress.

Under the Durban Platform, a new binding agreement is supposed to come about by 2015. To what extent this new agreement will entail emission reduction targets for both industrialised countries and emerging economies depends on the interests and power relations among the central actors. These include the European Union (EU), the BASIC countries – Brazil, South Africa, India, China – and the United States of America (USA), but also other emerging economies such as Indonesia, Mexico and South Korea. In the light of this shift in relevant actor constellations, the question arises as to who actually has power in current global climate governance and what it looks like. This article offers some answers by analysing the power distribution, both within the international climate regime and beyond.

On the one hand, the decisions summarised as the Doha Climate Gateway and the proceedings of the Durban Platform kept international dialogue alive and, thus, averted a complete failure of the climate regime. On the other hand, many contested issues were simply adjourned, such as measurement and verification mechanisms, or they ended in very flexible wording. Moreover, the number of participants to the second commitment period under the Kyoto Protocol has diminished. Only the EU countries, Australia and, possibly, Belarus, Kazakhstan and Ukraine still take part. Given this perpetually slow progress, it is questionable whether such ever-increasing negotiations – in respect of the number of actors attending, their scope, and their complexity – still make sense. In addition, the decrease in global warming to a manageable level is becoming more unrealistic, at least as long as the international regime counts as the central forum for solutions.

Already, the active hubs of global climate governance are found at other levels, for example in the clean technology markets, transnational partnerships, or bilateral and national climate funds. Here, the emerging economies of the South play an increasingly important role as well. Their actions include both state and non-state actors. A complete analysis of the power distribution

1 The Ad Hoc Working Group on Long-term Cooperative Action under the Convention (AWG-LCA), and the Ad Hoc Working Group on Further Commitments for Annex I Parties under the Kyoto Protocol (AWG-KP).

in global climate governance, therefore, requires a differentiated view beyond the UNFCCC negotiations. Moreover, since climate governance is closely related to other fields such as energy or poverty reduction, the type and distribution of power may trigger spillover effects. If the power constellation changes in one field, it impacts other policy fields.

Against this background, the following assumptions form the starting points for the analysis:

- ***Power is multidimensional:*** *Instrumental* power enables an actor to directly influence or coerce others. *Structural* power means that an actor shapes the context and rules affecting others according to his/her own interests. *Discursive* power means that an actor can indirectly shape the identity, perceptions and preferences of other actors. These three faces of power[2] entail hard and soft resources that can be combined to form soft power strategies.[3]
- ***Power is relational:*** It always exists in relation to others and, thus, needs to be understood as a process in a particular context.
- ***To have power does not mean*** using ***it:*** Passive behaviour can have widespread consequences, especially regarding global public goods.

Section B that follows below analyses the power distribution within the climate regime, focusing on the international negotiations. Section C provides a complementary look beyond the regime. It compares and contrasts the power of central actors and new players in clean technology markets, the renewable energy arena, and in bilateral agreements. Section D introduces the concept of *green power* and provides an outlook on global climate governance and the relevance of green power for prospective change.

## *B. Power Distribution in the Climate Regime*

In the climate regime, it is not only industrialised countries that are in a strong structural power position anymore: developing countries with significant greenhouse gas emissions have structural power of veto because a new climate treaty without their participation would hardly be effective. This group includes the BASIC countries, particularly China and India, but also other

2 Lukes (1974).

3 Nye (2010). According to Nye (ibid.), *soft power* entails the ability to attract and coopt, rather than coerce. Soft-power resources are the assets that produce such attraction.

emerging economies such as Indonesia, Iran, Mexico and South Korea. At the 'transitional' Doha Conference, bureaucratic processing rather than political acting was at the forefront, leaving underlying power distributions largely intact. Focusing particularly on emerging economies, the following provides an analysis of the structural, instrumental and discursive power distribution that characterises such international climate negotiations.

The structural power of Brazil and Indonesia in the climate regime differs somewhat from the others because of the large areas of rainforest they possess. In the past, the rainforest nations exerted instrumental power by successfully setting a financial compensation mechanism for forest conservation on the negotiation agenda, namely the United Nations Collaborative Programme for Reducing Emissions from Deforestation and Forest Degradation in Developing Countries (UN-REDD). Costa Rica and Papua New Guinea have been particularly active in this regard.[4] In Doha, Papua New Guinea pushed for the establishment of a REDD Committee, but the issue was postponed to the June 2013 negotiation round. Also postponed was the decision about the controversial REDD verification mechanisms, with Brazil and Norway having opposing ideas about what such mechanisms should entail.

The heavyweights Brazil and Indonesia still support REDD, but both are now active in other ways in case the climate regime fails. Both countries have set up national trust funds to which donor countries have already made substantial pledges. Ecuador and Guyana have followed suit. This limits the structural power of the rainforest coalition in the international negotiations, and presents a particular disadvantage to those countries which do not have the means to pursue their interests within and outside the regime at the same time. The Democratic Republic of the Congo, with its insufficient state structures, is such a case. Therefore, it is not possible to speak of a general rise of the South in the climate negotiations.[5]

Moreover, the BASIC countries are not as uniform a group as they may seem at first. Indeed, they do not constitute a stable block of power in the climate regime per se.[6] BASIC exerted some direct instrumental power at the Copenhagen negotiations in 2009 when they managed to get their way against the EU and shape the Copenhagen Accord largely according to their own interests. Since then, however, differences in the relational quality of

4 Lederer (2012).

5 (ibid.).

6 Hallding et al. (2011).

power are becoming more apparent not only among the BASIC countries, but also between them and their respective regions. In addition, their power is limited by the structural and instrumental power of the industrialised countries and the moral discursive power of the least-developed countries (LDCs) and the Alliance of Small Island States (AOSIS). The LDCs and AOSIS will be hit hardest by the impacts of climate change, but have not caused it. They also do not accelerate climate change with their current emissions, as the BASIC countries do. Some of the members of AOSIS such as the Maldives are even threatened in their territorial existence. While this has led to a sense of responsibility and financial support by some industrialised countries, the framing of international equity has not yet turned into substantial financial commitment by all such countries. In Doha, only Denmark, the EU and Sweden announced concrete financial pledges up to 2015.

Brazil and South Africa are generally more open to binding mitigation targets than India and China. However, each of these four countries only commits to those voluntary emission reductions that they can reach with minimal extra effort according to the calculations of their own experts. Brazilian and South African experts favour a burden-sharing approach, while China and India base their calculations on a global carbon budget. The latter has been proposed in a similar form by the German Advisory Council on Global Change.[7] Within these four countries' similar approaches, there are of course differences.[8] Owing to these internal discrepancies, the BASIC countries are missing out on the possibility of strengthening their power position as a group.

In relation to India, China has more active discursive power, even though it did not use it in Doha. Right at the beginning of the Durban negotiations in 2011, the Chinese government envisioned participation in a post-2020 climate treaty as long as the principle of common but differentiated responsibility and a second commitment period in respect of the Kyoto Protocol would be honoured. At first, India did not want to participate in a new treaty at all and, thus, stood apart. The cautiously progressive steps India had taken before largely depended on the previous Minister of Environment, Jairam Ramesh;[9] their apprehension seems to have been justified with the fall-back of the Indian delegation into blocking mode after Ramesh's dismissal in July

7 Wissenschaftlicher Beirat der Bundesregierung Globale Umweltveränderungen (WBGU).
8 Winkler et al. (2011).
9 Michaelowa & Michaelowa (2011).

2011. In contrast to the other three BASIC countries, India's power derives less from active shaping and influencing than from blocking. The only exception is India's active engagement for the setting up of a centre for clean technology transfer because such transfer is clearly in India's interest. Similar to China and South Africa the domestic coal and oil industry has a veto power that should not be underestimated in its influence on governmental decision-making.[10] Sometimes representatives of these corporations even take part in the international negotiations as observers or consultants for the delegations, e.g. the South African coal-to-liquid giant, Sasol.

Nonetheless, the chance is rather low that India will be successful by continuing to insist on equity and by strengthening its instrumental and structural power position with new partners in this way. Even together with the LDCs, the establishment of a discourse on climate justice has not been successful in that it has not influenced the shaping of the climate regime in a decisive way.[11]

China and the US are in a negative balance of power. Neither will take a decisive step forward in the climate negotiations without the other doing the same. Both countries are in an extremely strong power position, which they could use in many ways to shape the prospective climate regime. However, they do not use this potential for domestic political and economic reasons. For example, the chance of passing any kind of federal climate-related policy through the US Congress are currently minimal, even though various climate governance actions are being taken at state and local levels.[12] In addition, the balance of power between the US and China in international negotiations impedes political moves forward in the short term.

Only the EU has some instrumental power – which it used more actively in Durban in 2011 than in Doha in 2012. The EU achieved their negotiation goals in Durban by gaining consensus on a road map for a new climate treaty as well as a second commitment period under the Kyoto Protocol. The most prominent display of its instrumental power was the huddle with India on the last day of negotiations, in which the EU emerged as the winner. Thus, the EU managed to make up on its loss of power at the Copenhagen negotiations in 2009 and the simultaneous loss of its leadership role, at least to some extent. Since Durban, increasing internal discord among EU member countries – particularly Poland's defensive stance – have been weakening

10 Never (2012).
11 Roberts (2011).
12 Schreurs (2012:10).

the region's general power position. In Doha, the EU achieved its goal of ensuring the second commitment period to the Kyoto Protocol was ready for ratification, but it did not opt for the unilateral move towards raising its emission reduction targets. Moreover, the EU slightly increased its attraction through soft power to developing countries, being one of the few negotiating parties that continued so-called fast-start finance beyond 2013. In Doha, the EU confirmed new financial pledges in spite of the European economic crisis. However, neither the EU nor Germany uses its full power potential because both still do not apply pressure to their traditional transatlantic allies, Canada and the US. The EU's quest for leadership in the climate regime is, therefore, overshadowed by greater diplomatic-strategic alliances.

The exit of Canada, Japan, New Zealand and Russia – and, possibly, Belarus, Kazakhstan and Ukraine – from the Kyoto Protocol has two sides to it. On the one hand, the Protocol has been saved from complete failure by this exit: it could be counted as the direct successful application of instrumental power by Protocol supporters. On the other hand, the exit is negative with respect to managing climate change owing to the high emissions and structural power of the exiting parties. At Doha in 2012, power and interest struggles turned fierce when it came to the transfer of any surplus emission rights from the first Protocol period to the second, and/or the possibility of selling such rights without participating in the second period. Finally, a compromise emerged: only those countries participating in the second period could transfer or sell their surplus rights; Poland is among these. While this reflects a slight gain in instrumental power for beneficiaries such as Poland, it is a power loss for Russia. At the time of writing, it is unclear whether Belarus, Kazakhstan and the Ukraine will exit the Protocol as well. AOSIS managed to insert a paragraph in the final Doha Decision text that sets a cap on the emissions of the latter three countries at 2008–2010 levels during the second Protocol commitment period.[13] If these three countries continue to take part, it would signify a direct instrumental power gain by AOSIS. If the three withdraw, nothing positive will have been achieved: neither for AOSIS, nor for the management of global warming. In any case, the climate regime is further weakened by the numerous withdrawals from the Protocol.

South Africa is another interesting case in terms of the relational quality of power. The country is torn between the interests of the BASIC countries, legitimating itself as a representative of sub-Saharan Africa, and the demand

13 Allan & Kruppa (2012).

to be a reliable partner for the North.[14] In relation to the other BASIC countries, South Africa's power is on a smaller scale; but in relation to the southern African region, its power is greater. The latter primarily concerns the discursive dimension: South Africa was able to foster transparency in the negotiations and to increase the participation of LDCs and civil society in the Durban negotiation round in 2011. But the capacity to produce a feel-good effect through indabas will not limit global warming. Moreover, taking this bridging function between industrialised and developing countries is becoming harder for South Africa.

There is a new framing or even an informal norm under way as the LDCs and AOSIS increase the pressure on the emerging economies to do their share as well. Indeed, the break-up of the formerly united G77 became very obvious in Doha. While the LDC and AOSIS groups remain, a new Association of Independent Latin American and Caribbean states (AILAC), comprising Chile, Colombia, Costa Rica, Guatemala, Panama and Peru, was formed in the 2012 mid-year negotiations in Bonn. AILAC calls for the mitigation of emissions by industrialised and developing countries as well as an incentive system for all countries to do so. Opposing AILAC is the new group of 'Like-minded Countries', comprising members of the Arab Group, Argentina, Bolivia, Ecuador and Venezuela, as well as India and China. They continue to advocate for international equity and the historical responsibility of the industrialised countries.[15] These alliances have started to shift the relational power between negotiating parties. However, no gain in power for any negotiating party can be expected in the immediate future that would change the general stand-off.

### *C. All Power to the Market?*

The practice of climate governance happens outside the international negotiations. At the state level, the direct, instrumental power relations can be explained by using the example of climate funds. Because of their financial resources, the donor countries Australia, Norway and Sweden, which have pledged to developing countries' national funds, but also Germany, Japan and the United Kingdom (UK), which have pledged via their own funds or

14 Atteridge (2010).
15 IISD (2012).

through emissions trading, were initially in a strong position in the bilateral and multilateral negotiations. But the structural and moral discursive resources of the receiving countries, such as Brazil, Ecuador, Indonesia and Tanzania, have a countering effect. This has led to a balancing of interests.[16]

Norway, in particular, benefits from its active leading role. It has generated a positive attraction in the sense of soft power. Norway is the driving force behind the new Energy+ Partnership, which includes Denmark, France, the Netherlands, South Korea, Switzerland and the UK. These donor countries give energy-related help to countries such as Bhutan, Ethiopia, Kenya, Liberia, the Maldives, Morocco, Nepal, Senegal and Tanzania.[17] Thus far, the latter group of countries has only benefitted marginally from the Clean Development Mechanism (CDM). In Indonesia, however, the soft power benefits are currently somewhat at risk for Norway. Already in 2010, the Norwegian government promised up to US$1 billion for the development and implementation of REDD projects in Indonesia, but progress on these projects has been slow. The Indonesian government struggles with the implementation of a forest logging moratorium, and Norway has been discovered to own a small share in a palm oil company that is involved in illegal logging.[18] If the bilateral deal is successful in the end, both Indonesia and Norway will come out more powerful.

The instrumental power of private CDM project developers and consulting firms from Brazil, China and India is increasing in the carbon markets. This is both visible in relation to European companies and as an influential voice to their own governments, which the local CDM industry is pushing towards sustaining emission trading in the international climate negotiations. Since the Kyoto Protocol is the only legally binding instrument under the UNFCCC and the CDM is a means for technology transfer, developing countries and emerging economies have always supported it.

Power struggles to date only happen along the North–South axis. In 2011, the introduction of an aviation carbon tax demonstrated the EU's instrumental power. This led not only to considerable protest from Chinese and Indian airlines, but also to a controversy with the US. US airlines even filed a complaint, which was, however, dismissed in December 2011. If the EU had continued along these lines of imposing taxes unilaterally, its role as a

16 Lederer (2012).
17 Reuters (2012).
18 Sukma Sawitri (2012).

leader would have been significantly strengthened again. However, facing this strong resistance, the EU actually revoked its decision in the run-up to Doha in order to keep the door open to international solutions. The EU expects future regulations to be imposed by the International Civil Aviation Organization or an international agreement within the climate regime.[19] If neither of these solutions comes about in the near future, the EU may revert to its unilateral move

Structural power shifts that involve private actors are particularly evident at the interface of climate and energy policy. Here, general market developments in renewable energy and energy efficiency and the technological power of individual multinational corporations are prominent. Technological power is a form of structural power which is mostly exercised by private actors. Corporations have technological power if they have privileged access to technical information and if they can steer innovation processes through their superior expertise and material resources.[20] With regard to carbon capture and storage, current technological power is in the hands of the European (e.g. BP and Statoil) and US (e.g. Exxon Mobil) oil and gas industry.[21] Here, the power of neither the EU nor the US is in decline.

If we look at the whole sector of renewable energy, the South is catching up quickly, as new public and private investments make clear (Figure 1).

***Figure 1: New Financial Investments in Renewable Energy: Industrialised v Developing (including BASIC) Countries, 2004–2010***

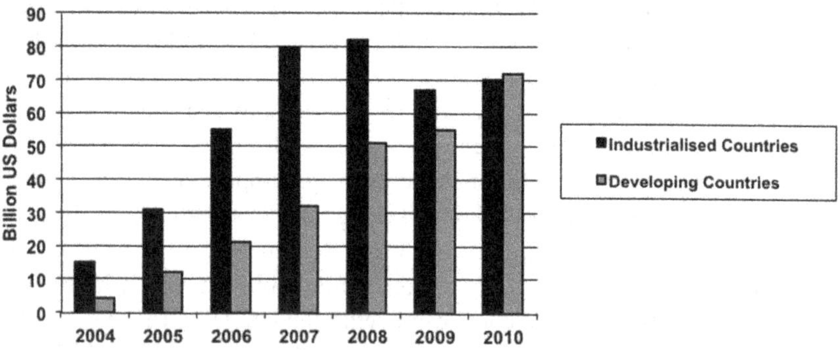


Source: Bloomberg New Energy Finance & UNEP (2011)

19 Kohn (2012).
20 Falkner (2005:105–134).
21 Tjernshaugen (2012).

The respective power of the coal and oil industries does not hinder these new investments in renewable energy because the diversification of energy sources is both an economic and climate-change-related necessity. The increase in electricity supply shortages in recent years, especially from 2008 onwards in India or South Africa, for example, makes a clear case for adding alternative energy sources to fossil fuels. The need to balance development needs with the required switch to a low-carbon path makes a quick *Energiewende*[22] to renewables following the German example unlikely in the near future.

In 2010, China has been leading global investments in clean technology with US$50 billion, followed by Germany and the US. The majority of German investments went into small scale projects such as solar roofing. Egypt, Kenya, Mexico and Pakistan also invested more than US$1 billion. India saw the strongest growth of investments (52%) in 2011, compared to the previous year. The first financial commitments under the national solar mission, which is part of the Indian domestic climate-related policy, accounted for the majority of US$10.3 billion.

The investments underline the current dynamics in the renewable energy market and the potential companies ascribe to it. Even though the switch to renewables will continue to be difficult for the BASIC countries – apart from hydropowered Brazil – current developments give reason for some hope. With the right combination of incentives, regulation and control, even China and India could surpass their self-set goals for renewable energy, energy efficiency and the carbon intensity of the economy.

Several Chinese and Indian companies belong to the Top 10 in the global solar and wind energy market (Figure 2). Together, Chinese companies have a 30% share of the global market for wind turbine producers. Moreover, China has the highest installed wind capacity in the world with 63 MW, followed by the US, Germany, Spain and India in order of MW installed.[23]

22 Best translated as "energy revolution".

23 World Wind Energy Association (2012).

***Figure 2: Shares of the Top 10 Wind Turbine Producers in the Global Market in 2010 (%)***

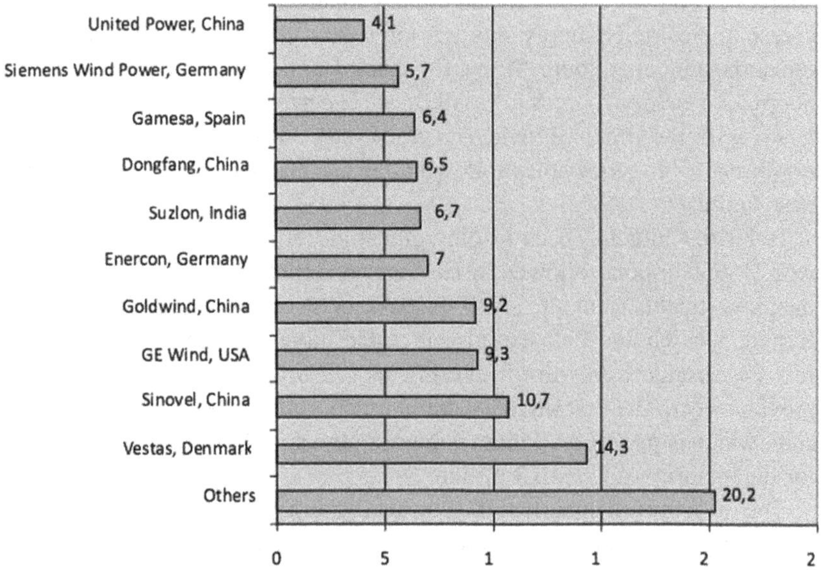


Source: REN21 (2011)

Among the Top 15 global producers of photovoltaic (PV) cells 7 are Chinese corporations. Increasingly, European and US PV producers are suffering from the Chinese competition, which already produces 55% of PV cells globally. The US, therefore, started to impose import tariffs for Chinese solar cells on a low level. Following a lawsuit by European solar firms – among them the German producer Solarworld – the European Commission initiated an anti-dumping measure against Chinese producers in September 2012. Chances for success remain questionable since many German firms are closely associated with China through supply and value chains.

Therefore, China's structural power – and to a lesser extent that of India – is increasing strongly in the wind and solar markets. Working against it is Europe's structural power, especially Germany's. Germany still has the largest market for renewable energy as well as the most installed capacity and transmission lines. The US is in a similar situation. Several European countries, including Germany, have reacted to the current surplus capacities of the solar market by cutting government subsidies. From a global climate political viewpoint, this makes sense because reasonable market and price developments are necessary to support as many countries as possible in their switch to renewable energy sources. However, this can only happen if China also cuts its subsidies and, therefore, allows for a completely free market. Up to now, Beijing has only taken a few small steps in this direction: in 2012, for example, the government cut the subsidies for solar pilot projects by only 21%, following declining component prices.[24] From now until 2015, the government also plans to reduce the solar industry's reliance on exports, and to lend strong support to developing a domestic market. Currently, 90% of Chinese PV cells are exported.[25]

In the short term, the shift of the global renewable energy market would result in a loss of structural power in climate governance for Germany in particular, and for Europe in general. In the long run, it could pay off to reduce the number of German companies to those with a substantial technological advantage. By now, the quality of Chinese PV cells is comparable to its German counterparts. If the technological power of German companies in renewable energy and energy efficiency were to be strengthened, Germany's structural power in global climate governance would be secured. To achieve this, more investments in research and development and a systematic use of green innovation potentials are required. Overall, Europe, Germany, and the US still invest a lot more in research and development than the emerging economies of the South. However, the solar industry has been neglected in this regard.

For those developing countries that are already participating in clean technology markets, the structural power and the economic dominance of Asian, European and US companies is too strong to play a significant role in the markets themselves. Even though investments are partly increasing and more renewable energy and energy efficiency projects exist, developing countries have not yet been able to influence this area of global climate governance. Owing to their lower level of development, many of these countries face other issues and have other priorities such as the extension of electricity transmission lines and the electrification of all households. Pioneers like Costa Rica, which plans to be the first carbon-neutral country in Latin America, are at a structural disadvantage because of the relational quality of power. Costa Rica is able to use its pioneering role more within the discursive dimensions in the international climate negotiations – together

24 Shen (2012).
25 Juan (2012).

with the LDCs – to apply moral pressure to the industrialised and the BASIC countries.

## D. The Future of Green Power

In a simplified version, *green power* as a complement to a *green economy* would require a combination of instrumental, structural and discursive elements with a positive orientation towards the global public good *climate*. Currently, none of the central actors possesses this green power in all of these dimensions. Instead, global climate governance at present consists of a polycentric system in which different actors are active with different means and ends. Some authors even call the global climate governance system *fragmented.*[26]

In its complete version, the concept of *green power* as developed by this author[27] not only includes the three relational power dimensions mentioned thus far herein, but also makes the connections to innovation an important factor for the transformation towards a sustainable, greener economy. The concept of *green power* additionally includes the share of global commons a country possesses (forests, biodiversity, marine life, etc.), its technological capabilities, its absorptive capacity for innovation, its integration into global green value chains, and its ability for 'smart governance'. *Smart governance* means the ability to coordinate and implement energy innovation and environmental governance in a way that goals do not contradict each other or lead to unintended negative effects in one of the issue areas. While it would go beyond the scope of this article to apply this concept to the research questions posed here, as stated earlier, none of the central actors discussed possesses active green power in all these dimensions. In general, the industrialised countries still have and could exert more green power, but Brazil, China and India – and even Costa Rica – are catching up in some areas.[28]

The differentiation between *active* and *passive* green power is particularly relevant in the current polycentric system. As we have seen in this article, a country may behave passively in international negotiations, but may actively build and exert its green power at other levels and through other channels.

26 Biermann & Pattberg (2008); Van Asselt & Zelli (2012).
27 Never (2013).
28 (ibid.).

In spite of the current polycentric system in global climate governance, the continuation of the international negotiations makes sense for three reasons:

- The participation of LDCs in global climate governance and their access to resources for the adaptation to climate change is ensured
- Governments can represent their positions and enter into those power struggles that matter to a domestic audience, particularly in a discursive, symbolic way. Simultaneously, this opens up more space for action at other levels, and
- The topic of climate change remains on the political agenda and in the public consciousness. The shift of decision to smaller actor circles or clubs like the G20 is only useful in respect of emission reductions: all other areas require a global participatory approach for climate justice reasons.

China's power is generally increasing in the structural dimension, but the country does not use its instrumental or discursive power in the negotiations to establish a global leadership position. Instead, it stands in a balance of power with the US. India is the poster child for the power of the veto: the country has not been that successful in the discursive dimension, but it is increasing its power in the structural dimension outside the climate regime. Europe has again increased its power in international negotiations, and has started positioning itself against its ally the US on other climate governance levels. In principle, Europe and Germany have the potential to become green powers but they are too hesitant to do so because of other, conflicting foreign policy objectives and domestic interest struggles – besides being restricted by China's structural power gain. Germany still possessed structural power, particularly in respect of renewable energy. Its structural power is increasingly being met by China and India.

## *References*

Allan, Andrew & Martin Kruppa, 2012, Belarus Negotiator Hints at Kyoto Exit, Says Others Could Follow, *Reuters*, 10 December, available at http://www.reuters.com/article/2012/12/10/us-climate-talks-kyoto-idUSBRE8B90ZY20121210, last accessed 10 December 2012.

Atteridge, Aaron, 2010, Multiple Identities: Behind South Africa's Approach to Climate Diplomacy, SEI Policy Brief, Stockholm, Stockholm Environment Institute.

Biermann, Frank & Philipp Pattberg, 2008, Global Environmental Governance: Taking Stock, Moving Forward, *Annual Review: Environmental Resources* 33, 277–294.

Bloomberg New Energy Finance and UNEP/United Nations Environment Programme, 2011, *Global Trends in Renewable Energy Investment 2011*, Nairobi, UNEP.

Falkner, Robert, 2005, The Business of Ozone Layer Protection, in: Levy, David & Peter Newell, *The Business of Environmental Governance*, Cambridge, MIT Press, 105–134.

Hallding, Karl, Marie Olsson, Aaron Atteridge, Antto Vihma, Marcus Carson & Mikael Román, 2011, *Together Alone: BASIC Countries and the Climate Change Conundrum*, Copenhagen, Nordic Council of Ministers.

IISD/International Institute for Sustainable Development, 2012, *Earth Negotiations Bulletin*, Winnipeg, IISD Reporting Services.

Juan, Du, 2012, China's Domestic Solar Market to Expand, *China Daily*, 25 February 2012, available at www.chinadaily.com.cn/business/ 2012-02/25/content_14691414.htm, last accessed 25 February 2012.

Kohn, Andreas, 2012, Country Positions: The EU, UN Weltklimakonferenz: Future Camp Blog, 30 November, available at http://future-camp.de/news/story/352/view/?locale=de_DE, last accessed 30 November 2012.

Lederer, Markus, 2012, Die Macht des Südens in der globalen Klimapolitik, Unpublished Paper Presented at the Conference on *Politik und Ökonomie in globaler Perspektive: Der (Wieder)Aufstieg des globalen Südens*, held at Goethe University, Frankfurt am Main, 5–7 March 2012.

Lukes, Steven, 1974, *Power: A Radical View*, London, Macmillan.

Michaelowa, Katharina & Axel Michaelowa, 2011, India in the International Climate Negotiations: From Traditional Nay-sayer to Dynamic Broker, Working Paper No. 70, Zürich, Centre for Comparative and International Studies.

Never, Babette, 2012, Who Drives Change? Comparing the Evolution of Domestic Climate Governance in India and South Africa, *Journal of Environment and Development* 21 (3), 362–387.

Never, Babette, [Forthcoming], Towards the Green Economy: Assessing Countries' Green Power, GIGA Working Paper, Hamburg, German Institute of Global and Area Studies.

Nye, Jospeh S., 2010, The Future of American Power, *Foreign Affairs* 89 (6), 2–12.

REN21/Renewable Energy Policy Network for the 21st Century, 2011, *Renewables 2011: Global Status Report*, Paris, REN21 Secretariat.

*Reuters*, Norway Pledges $300 Million to Green World's Power, available at www.reuters.com/article/2012/01/18/us-norway-carbon-idUSTRE80H1LT20120118, last accessed 18 January 2012.

Roberts, Timmons J., 2011, Multipolarity and the New World (Dis)Order: US Hegemonic Decline and the Fragmentation of the Global Climate Regime, *Global Environmental Change* 21 (3), 776–784.

Sawitri, Adisti Sukma, 2012, Indonesia's REDD Reality, *The Jakarta Post*, 3 June, available at http://www.thejakartapost.com/news/2012/06/03/indonesia-s-redd-reality.html, last accessed 12 December 2012.

Schreurs, Miranda, 2012, Breaking the Impasse in the International Climate Negotiations: The Potential of Green Technologies, *Energy Policy* 48, 5–12.

Shen, Feifei, 2012, China Reduces Solar Subsidy on Declining Costs of Components, available at www.bloomberg.com/news/2012-05-03/china-cuts-subsidies-for-solar-projects-21-on-declining-costs.html, last accessed 7 September 2012.

Tjernshaugen, Andreas, 2012, Technological Power as a Strategic Dilemma: $CO_2$ Capture and Storage in the International Oil and Gas Industry, *Global Environmental Politics* 12 (1), 8–29.

van Asselt, Harro & Fariborz Zelli, 2012, Connect the Dots: Managing the Fragmentation of Global Climate Governance, Earth System Governance Working Paper 25, Lund/Amsterdam, Earth System Governance Project.

Winkler, Harald, T. Jayaraman, Jiahua Pan, Adriano Santhiago de Oliveira, Yongsheng Zhang, Girish Sant, Jose Domingos Gonzalez Miguez, Thapelo Letete, Andrew Marquard & Stefan Raubenheimer, 2011, *Equitable Access to Sustainable Development: Contribution to the Scientific Body of Knowledge*, Beijing/Brasilia/Cape Town/Mumbai, BASIC Experts Group, available at http://www.erc.uct.ac.za/Basic_Experts_Paper.pdf, last accessed 6 May 2013.

World Wind Energy Association, 2012, World Market Recovers and Sets New Record, 7 February, available at www.wwindea.org/, last accessed 12 February 2012.

# 6
# CBDR as a Principle of Inspiring Actions rather than Justifying Inaction in the Global Climate Change Regime

*Achala C. Abeysinghe & Gilberto Arias*

*Abstract*

The principle of common but differentiated responsibilities (CBDR) recognises the existence of a common environmental goal and the need to differentiate between countries in the actions required to achieve that goal. The CBDR principle is at the centre of the current negotiations under the United Nations Framework Convention on Climate Change's Ad-hoc Working Group on the Durban Platform for Enhanced Action, which deliberates on multilateral arrangements for the post-2020 period. This contribution analyses the CBDR principle in the context of the current climate change regime, and discusses the ways in which the principle can be applied in order to encourage increased global actions to address climate change in the post-2020 regime. It argues that the CBDR principle can be operationalised for increased climate action by focusing not only on differentiated responsibilities, but also on the respective intrinsic and supported capabilities of countries in the new regime.

## *A. Introduction*

The principle of common but differentiated responsibilities (CBDR) is one of the cornerstones of the global climate change regime. The CBDR principle appears explicitly in the United Nations Framework Convention on Climate Change (UNFCCC) and informs the associated Kyoto Protocol. The principle has been applied to climate change actions by states parties through various UNFCCC decisions. The increasing impacts of climate change prove that no state is a sealed-off island with impassable boundaries. The interconnectedness and complex nature of the climate change problem requires each state to bear responsibility for all its actions and inaction. The key question to answer is how global responsibilities should be divided between and encouraged within nation states. The UNFCCC applies CBDR for this purpose.

The CBDR principle recognises the existence of a common environmental goal and the need to differentiate between countries in the actions required to achieve the said common goal. It includes two fundamental elements: the first concerns the common responsibility of states to protect the environment at a global level; the second recognises states' different obligations in the actions required to achieve the common goal. The ultimate objective of the UNFCCC is the "... stabilization of greenhouse gas concentrations in the atmosphere at a level that would prevent dangerous anthropogenic interference with the climate system."[1]

The challenge for the UNFCCC is in devising enough action under its terms and among its states parties so as to achieve this ultimate objective. As such, the interpretation and application of CBDR have been heavily debated in UNFCCC negotiations. This debate has found its place at the core of the current negotiations undertaken by the Ad Hoc Working Group on the Durban Platform for Enhanced Action, which deliberates on multilateral arrangements for the post-2020 period.

The aim of this article is to analyse the CBDR principle in the context of the current climate change regime and discuss the ways in which the principle can be applied in order to encourage increased global actions to address climate change. The article argues that the CBDR principle can provide profound and compelling reasons for countries to be responsible and act in a certain manner. The article also argues that the CBDR principle can facilitate a basic and inevitable logic to emphasise that states have to act in a certain manner, and to guide countries in selected directions for safeguarding the climate for future generations.

The article is organised into three main parts. Firstly, it analyses the theoretical basis of the CBDR principle; secondly, it describes the evolution of CBDR in the global climate change regime; and thirdly, it sets out how CBDR could be used in a future regime to encourage more common global actions, and presents conclusions.

1 Article 2, UNFCCC.

## *B. Theoretical Basis of the CDBR Principle*

The CBDR principle has two main elements. The first is *common responsibility*, which describes the shared obligations of two or more states towards the protection of a particular environmental resource.[2] Common responsibility is likely to apply where the resource is shared, under the control of no state, or under the sovereign control of a state, but subject to a common legal interest, such as biodiversity or climate.[3]

The second element is *differentiated responsibility*, which recognises the different obligations of states in the actions required to achieve the common goal – protecting the environment. In general, the differential standards are set on the basis of a range of factors. These include historical contributions to the evolution of a particular environmental problem; the ability and capacity to prevent, reduce and control the threat by taking response measures; special needs and circumstances; and states' future economic development needs. In most of the international environmental legal regimes, differentiated responsibility places weightier environmental obligations and standards on developed countries and, thus, brings a unique proposition to international law by establishing substantive equality.[4]

Although the idea of differentiated responsibilities predates the 1992 Rio Declaration, CBDR was first clearly articulated by Principle 7 of that Declaration, which reads as follows:

> States shall cooperate in a spirit of global partnership to conserve, protect and restore the health and integrity of the Earth's ecosystem. In view of the different contributions to global environmental degradation, States have common but differentiated responsibilities. The developed countries acknowledge the responsibility that they bear in the international pursuit to sustainable development in view of the pressures their societies place on the global environment and of the technologies and financial resources they command.

The mounting number of international instruments that recognise CBDR indicates that it is moving from being a 'soft' international legal principle to an increasingly robust component of international law. But without further

2 Sands (2003:286).

3 CISDL (2002).

4 *Substantive equality* emphasises the treatment which is based on different factors and takes into account the welfare of disadvantaged members of the society. The core idea of this concept is that entities that are alike should be treated alike, and those that are different should be treated differently according to their different circumstances.

operationalisation, CBDR does not by itself generate any legal obligations for states: it is principally an obligation to cooperate in developing international law.[5] CBDR has no strictly fixed content,[6] and, therefore, has to be elaborated to have legal force.

Most international legal obligations come from customary international law or treaties. For a principle to become legally binding as part of customary international law, there needs to be widespread state practice adhering to the principle, and states need to act out of legal obligation. Most commentators agree that these two requirements have not been met for CBDR, and that CBDR is therefore not in itself legally binding.[7] On the other hand, once treaties operationalise the principle, as many multilateral environmental agreements do, states parties to these agreements become legally bound by them.

Recent literature on the CBDR principle demonstrates that it has attracted much attention, especially in relation to agreements to combat transboundary environmental problems.[8] Hey notes that the CBDR principle in international environmental law entails that, while pursuing a common goal, states take on different obligations, depending on their socio-economic situation and their historical contribution to the problem at stake.[9] Rajamani notes that, by building on the acknowledgement by industrial countries that they bear the primary responsibility for having created many global environmental problems and, hence, by taking into account the contributions of states to environmental degradation in determining their levels of responsibility, the principle recognises broad distinctions between states, whether on the basis of economic development or the level of consumption.[10] However, Birnie and others suggest that the CBDR principle is not intended to be a permit for developing countries to pollute – even though obligations of developing states are conditional on the provision of technical and financial assistance from developed country parties.[11]

5 Birnie et al. (2009:133).
6 Honkonen (2009:258).
7 Sands (2003:Ch. 6); Bodansky (1993:501–502, in Stone 2004:299–300).
8 Environmental problems that span administrative boundaries and are felt regionally and globally.
9 Hey (2009).
10 Rajamani (2005:133).
11 Birnie et al. (2009).

## *C. Evolution of the CBDR Principle in the UNFCCC*

The climate change problem is a global phenomenon that has been created by the use of the global atmosphere as a free good, and – in part because of a lack of foreseeability – without consideration of the consequences for the environment, the economy or future generations. The UNFCCC aims to address this problem by stabilising greenhouse gas (GHG) concentrations in the atmosphere "at a level that would prevent dangerous anthropogenic interference with the climate system". The Convention aims to do so "… within a time frame sufficient to allow ecosystems to adapt naturally to climate change, to ensure that food production is not threatened and to enable economic development to proceed in a sustainable manner."[12]

The formulation of CBDR in the UNFCCC is noticeably different from other Rio Conventions, namely the Convention on Biological Diversity and the Convention to Combat Desertification. Unlike Principle 7 of the Rio Convention, Article 3(1) of the UNFCCC does not refer to the greater contribution of developed country parties to climate change, and emphasises respective capabilities together with the principle of CBDR.[13]

As such, Article 3.1 of the UNFCCC reads as follows:

> The Parties should protect the climate system for the benefit of present and future generations of humankind, on the basis of equity and in accordance with their common but differentiated responsibilities and respective capabilities. Accordingly, the developed country Parties should take the lead in combating climate change and the adverse effects thereof.

In addition to Article 3.1, the preamble of the UNFCCC acknowledges –

> … that the global nature of climate change calls for the widest possible cooperation by all countries and their participation in an effective and appropriate international response, in accordance with their common but differentiated responsibilities and respective capabilities and their social and economic conditions.

Accordingly, developed country parties should "take the lead in combating climate change and the adverse effects thereof". The preambular paragraphs of the UNFCCC also recognise that the largest share of historical and current global emissions of GHGs has originated in developed countries, and that per capita emissions in developing countries are still relatively low. The

12 Article 2.

13 Rajamani (2005:101).

*common responsibility* element of the CBDR is reflected in the preambular paragraphs with phrases such as "the Earth's climate and its adverse effects", which is a "common concern of humankind".[14]

Paragraph 18 of the Preamble requires "... developed countries to take immediate action ... as a first step towards comprehensive response strategies at the global, national and where agreed, regional levels." As such, under the UNFCCC, developed countries have differentiated, higher responsibilities with respect to assistance (e.g. financial assistance or technology transfer[15]), implementation, and central obligations (e.g. Annex I emission reduction commitments) towards the ultimate aims of the Convention. This higher responsibility emanates from their historical contributions to the problem and their greater respective capability for action; however, debate on the issue is complicated because of the evolving GHG emission profiles of developing countries – at present, the world's largest emitter is a developing country.[16]

Under the UNFCCC, all parties are subject to general commitments to achieve its central objective: they are required to compile an inventory of their GHG emission and submit reports, known as *national communications*, on actions they are taking to implement the Convention.[17] However, in accordance with the CBDR principle, although the core elements of the national communications for both Annex I and non-Annex I parties deal with information on emissions, the removal of GHGs, and details of the activities each party has undertaken to implement the Convention, elements from Annex I parties have to contain information on policies and measures,[18] in addition to the information on national circumstances; vulnerability assessment; financial resources and transfer of technology; and education, training and public awareness, which should be submitted by all parties.[19] Moreover,

14 UNFCCC, Preamble.

15 See *Procedures and Mechanisms Relating to Compliance under the Kyoto Protocol*, Decision 24/CP.7., UNFCCC/CP/2001/13/Add.3, available at http://unfccc.int/resource/docs/cop7/13a03.pdf, last accessed 29 October 2009.

16 UN Statistics, available at http://mdgs.un.org/unsd/mdg/SeriesDetail.aspx?srid=749&crid=, last accessed 15 April 2013.

17 Article 4(1) and Article 12.

18 Article 4(2)(a) and (b).

19 Most of the 41 Annex I parties have submitted four National Communications since 1994. Since 1996, Annex I parties have also been required to submit an annual inventory of their GHG emissions to the Secretariat. Separate reporting and review procedures have been established for Annex I GHG inventories. For submitted

Annex I parties that have ratified the Kyoto Protocol[20] are required to include supplementary information in their national communications and their annual inventories of emissions and removal of GHGs to demonstrate compliance with the Protocol's commitments. In terms of timetables, Annex I parties are required to submit information on their national inventories annually, and to submit national communications periodically according to dates set by the Conference of the Parties (COP). These measures seek to articulate that "developed countries are taking the lead in modifying longer-term trends in anthropogenic emissions", as mentioned in Article 4(2)(a) of the UNFCCC. Under the Convention, developing countries submit their national communications within four years of the initial disbursement of financial resources to assist them in preparing such communications.[21] At COP16, with guidelines adopted at COP17, developing countries agreed to report updates of national GHG inventories, including a national inventory report and information on mitigation actions, as well as needs and support received.

Moreover, the Convention supports CBDR in terms of the provision of financial assistance to developing countries. The Convention establishes differentiated "general obligations" on developed countries to assist developing countries in mitigation and adaptation through its financial mechanism, while the Kyoto Protocol obliges Annex I countries with quantified emissions reduction obligations. Additionally, both the UNFCCC and the Kyoto Protocol establish general obligations of cooperation towards technology transfer, and provide developing countries with financial assistance for mitigation and adaptation. Article 10 of the Kyoto Protocol explicitly requires taking the CBDR and specific national and regional development priorities, objectives and circumstances into account: the Article obliges parties to engage in management and other programmes to reduce GHG emissions on the basis of the CBDR principle.

---

National Communications, see UNFCCC's *National Communications Annex I* available at http://unfccc.int/national_reports/annex_i_natcom_/items/1095.php, last accessed 8 May 2013.

20 Kyoto Protocol to the United Nations Framework Convention on Climate Change (adopted 11 December 1997, entered into force 16 February 2005) 2303 UNTS 148 (Kyoto Protocol).

21 For National Communications submitted by non-Annex I states parties, see UNFCCC's *National Communications from Non-Annex I Parties*, available at http://unfccc.int/national_reports/non-annex_i_natcom/items/2716.php, last accessed 8 May 2013.

The strong application of the CBDR principle in the climate change regime has led some to argue that the CBDR is "best reflected" in the climate regime,[22] while others consider the climate regime to be the "clearest attempt to transform [CBDR] from a legal concept into a policy instrument".[23] Some see the UNFCCC as the "most dramatic stage"[24] for CBDR while some refer to the Kyoto Protocol as "CBDR in its most rigid application".[25]

As the climate change negotiations evolve, the focus on CBDR has grown stronger. Parties have sought to reframe discussions, strengthen their negotiating positions and ensure a fairer outcome by invoking different elements of CBDR. As a result, the elements of historical responsibility as well as respective capabilities have become a central part of the current climate debate. While some say that the developed countries should take greater responsibilities due to their historical contributions to the problem,[26] others argue that climate change is indeed a collective global problem that can only be combated if all countries put in every effort to resolving the problem.[27] In particular, for most countries that are vulnerable to climate change, the main argument is that all the parties need to act in order to achieve a stabilisation of GHG concentrations in the atmosphere, so that the problem will not create further damage to the countries in question.

### *D. Application of CBDR in a Future Regime*

Any metric ton of GHG emissions emitted anywhere in the world will affect the climate system equally. During 2012, extreme weather and climate events in the form of hurricanes, heat waves, droughts, fires, and flooding have been recorded all around the world. Just in the second half of 2012, millions of people have been affected across the globe – from Europe suffering from the worst cold snap in a quarter century; extreme flooding in

22 Rajamani (2005:176).
23 Joyner (2002:358).
24 Stone (2004:276, 281).
25 Weisslitz (2002:473, 483).
26 Declaration on 21 November 2012 by Ministers of the BASIC countries (Brazil, South Africa, India and China), available at http://www.indianembassy.org.cn/newsDetails.aspx?NewsId=381, last accessed 15 April 2013.
27 See IEA (2012) and other International Energy Agency (IEA) reports from 2012 on.

Australia, Brazil, China, and the Philippines; to drought in the Sahel.[28] While Hurricane Sandy caused billions of US Dollars' worth of damage in the United States (US), Australia suffers from record-breaking heatwaves.

Evidently, the most crucial prerogative under the current climate policy regime must be the urgent implementation of global climate action policies, understood as climate action programmes in general, attending to, inter alia, mitigation and adaptation policies directed to low-carbon development. The international discussion is currently converging around a 2°C (3.6°F) target (corresponding to a concentration of GHGs in the atmosphere of approximately 450 parts per million (ppm) of carbon dioxide equivalent compared with pre-industrial times, to avoid unmanageable climate risks. The Intergovernmental Panel on Climate Change scenarios suggest that, if GHGs could be reduced (in relation to 1990 levels and without Land Use, Land-use Change and Forestry – LULUCF) by 25–40% by 2020 and by 80–95% by 2050, global warming could be stabilised at the 2°C threshold. To date, several countries have announced possible GHG emission limitations and reductions, but most argue that even the most ambitious of those pledges are not sufficient to achieve the Convention's ultimate objective.[29]

Faced with the overwhelming evidence of enormous impact costs emanating from current warming, even if it is a fraction of the 2°C target the multilateral process has identified as a ceiling, climate action becomes an imperative when we note the increase in incidence in just the 2000–2009 decade, where, even if focusing only on flood and storm events, Latin America has seen a tripling of incidents compared with those that occurred between 1990 and 1999.[30] More often, extreme climate events remind us that enhanced and bold actions are required from all countries.[31] It is clear that countries need to act very quickly in taking action to cut emissions and to

28 For a timeline on extreme weather events in 2012, see http://insights.wri.org/news/2012/09/timeline-extreme-weather-events-2012, last accessed 1 February 2013.

29 See UNEP Gap Report 2012 at http://www.unep.org/pdf/2012gapreport.pdf; last accessed 15 April 2013.

30 See e.g. UNEP & ECLAC (2010); see also EM-DAT, the Centre for Research on the Epidemiology of Disasters (CRED) International Disaster Database, available at http://www.emdat.be/database, last accessed 8 May 2013.

31 As Mr Kapil Sibal, the Indian Minister of Science and Technology and Head of the Indian Delegation, mentioned in his closing statement in Bali, "It is not a question of what you will commit or what I will commit. It is a question of what we will commit together to meet that challenge!"; authors' personal notes.

close the mitigation gap before all opportunities[32] evaporate. The science is clear and continues forecasting that it is still possible to correct the global GHG emissions trajectory and avert irreversible climate change, but time is fast running out.

Many developing countries have stated that a temperature increase of 2°C on average would have devastating impacts on their countries and economies and have, therefore, called for GHG concentrations to be limited to well below 350 ppm of carbon dioxide equivalents, and a temperature increase limited to below 1.5°C. These countries have also called on Annex I countries to undertake considerably more extensive domestic emission reductions – more than 40% by 2020, and 95% by 2050 – as a contribution to this goal.

Achieving such high expectations requires greater contributions to collective efforts. In order to stabilise climate change to the level that science requires, it will be necessary not only to reduce emissions in high GHG-emitting countries radically, but also to diverge considerably from a conventional, fossil-intensive and highly GHG-emitting development trajectory in developing countries. However, the required global resolve will only materialise within an equitable framework that reflects leadership by developed countries and provides new strategies, tools and resources to incentivise the facilitation of environmentally friendly technologies and scientific know-how to poor and vulnerable developing countries, and especially climate action by all. This makes the ongoing climate negotiations even more important, since they will determine how CBDR is operationalised in the climate change regime, with a view to truly global, yet differentiated, climate action.

### *E. Countries' Common Responsibility and Common Purpose*

Common responsibility in addressing the global climate change problem is rooted in the principle of cooperation, a principle that posits that states are obliged, in the spirit of solidarity, to cooperate in preventing the climate change problem. The element of common responsibility that requires states to take action driven by a sense of common rather than national interest exerts an interest as well as a pressure on countries that were or are reluctant to take

32 See http://climatechange.worldbank.org/sites/default/files/Turn_Down_the_heat_Why_a_4_degree_centrigrade_warmer_world_must_be_avoided.pdf, last accessed 1 February 2013.

measures to protect the environment. Such interests and pressures based on the international community's common concern to combat the global climate change problem, for example, could be exerted on the US to take more responsibility and on major developing countries to take more actions to mitigate their rapidly accelerating GHG emissions.

It is now evident that climate change and its impacts present real threats to all states' development potential and development opportunities. For example, the DARA climate vulnerability monitor demonstrates with piercing clarity that the least-developed countries will suffer increasing economic losses unless all countries rapidly reduce GHG emissions. The DARA report[33] confirms many earlier scientific assessments, including those by the Intergovernmental Panel on Climate Change, but adds hard economic numbers to the earlier qualitative conclusions. According to the report, least-developed countries will lose 8% of their gross domestic product by 2030.[34]

As discussed, climate change impacts affect all countries, albeit in different forms, and require each state to plan, prioritise and implement particular adaptation and resilience strategies affecting its particular population, industry and resources. But because of the cross-cutting nature of climate change impacts, states are forced to give immediate priority to the rebuilding of destroyed infrastructure, usually as a matter of emergency, and so must postpone or subordinate investment in new infrastructure or in new resilient infrastructure for the welfare of the country, as studies have noted for Latin America.[35] Thus, this regular prioritisation of emergency repair and replacement comes at the cost of other state actions which may, *ceteris paribus*, have proven more efficient in national development in the longer term. Similar examples can be seen in other countries such as Bangladesh and Pakistan, particularly in respect to vulnerability to flooding or storm surges associated with annual climactic cycles. This is especially burdensome for smaller developing countries, in that repair and adaptation costs can amount to a far greater proportional expense than in larger economies.

Indeed, an informed, sensible argument is that, inasmuch as impact costs are the principal expense to be borne by developing countries from climate change, then these are minimised in a regime with collective, high-ambition strategies to which all countries contribute substantially, rather than in a

33 See http://daraint.org/climate-vulnerability-monitor/climate-vulnerability-monitor-2012/report/, last accessed 18 April 2013.

34 (ibid.).

35 See UNDP & ECLAC (2012:53).

collective low-ambition regime to which countries contribute nothing.[36] This control of the overriding burden of impact costs can be construed as an alternative aim of the Convention: if the impact costs could be managed, there would be no need for an international convention on stabilising emissions; the Convention would simply be on adaptation and impact costs. In this sense, beyond the equitable, beyond the transboundary elements that were discussed at the beginning of this article, the imperative for global climate action also needs to include support from a simple economic angle, applicable to each country, in that a disposition to inaction invites a subsidy to the principal cost of climate change. While it is true that these costs will not fall to all countries equally, it is entirely inequitable to subsidise those costs on others, or for future generations.

It is clear that national, domestic climate action policies which do not curb emissions imply transboundary effects: not only in connection with emissions, but also in the suboptimal mitigation efforts by any country. This also implies effects and costs which will impinge not only on the particular country's development potential, but also on the development priorities of every other country on the planet. This is perhaps not an innovative element as multilateralism in the climate action space has to deal with the domestic political temptation to a 'free ride', that is, to do as little as possible domestically and to benefit from the actions of others. Yet, as we have noted, inaction or 'free-riding' attracts domestic impact cost consequences as well.

Following from this, every nation's *common* responsibility in respect of climate action must also include a responsibility not only to its own future development paths, but also to the future development paths of all nations. This argument is clearly stronger among smaller states, as larger states may decide that their internal development priorities override any consideration – even if such policies will ultimately drive up their own internal impact costs. However, as proposed here, impact costs will outstrip the economic benefits of mitigation-averse policies, especially in the context of a globally interconnected mercantile system, where national prosperity – even by larger states – is ultimately linked to economic development in other states.

Doubtless, we will find different positions on where a given state's actions have affected the development of another, or where a second state has contributed, but these points do not detract from a more generalised view that, indeed, on an abstract analysis, conscious climate action by one state which

---

36 See Garibaldi (2009).

actively contributes to limiting the development of others is unconscionable. This much is indeed *common* to all states.

However, it is also true that this concept of *common responsibility* to action cannot be understood to be equal amongst all countries, as climate change action is intimately related to levels of national development. The imperative may be common, but the capability to redirect economies is not within the immediate grasp of all countries. So, as has been expressed in negotiations,[37] universality of application cannot mean uniformity of performance, as the capacity for deviation from pre-established growth pathways, given the economic and developmental constraints faced by developing countries, is clearly different between developed and developing countries. If we look to the components of an effective, global regime under the UNFCCC's principles, and especially since the agreement at the Durban COP17, which sought a more effective and inclusive regime while still remaining faithful to the fundamental principles of the Convention, a pragmatic view should not preclude the fact that CBDR includes elements that are *common* to all nations. This much is clear from the earlier discussion herein on the CBDR principle.

### *F. Differentiated Responsibilities Based on Respective Capabilities*

Since the 1990s, most climate negotiations related to CBDR have centred on the nature of differentiation between Annex I and non-Annex I countries, with the emphasis on burden allocation, support and action, and discussions of conditions for action between different parties. However, it has become evident from domestic political prerogatives and realistic emission projections that this focus is not enough.

How do we move away from a narrative primarily focused on burden allocation to one implying emphasis on climate action at all levels, while still preserving the precepts of CBDR?

37 See e.g. India and China's submissions to the Second Ad Hoc Working Group on the Durban Platform for Enhanced Action (ADP2), Workstream 2, March 2013; see also Opening Statements of ADP1 by India and others; available at http://unfccc.int/bodies/awg/items/7398.php, last accessed 9 May 2013.

Many[38] have argued that, given the seriousness of the global climate change problem, more common action is needed, without necessarily meaning that it would imply application of reciprocity where each country would be treated as having the same quantified obligations. More relaxed requirements from less-developed countries than developed countries should be accepted mainly to treat less-developed countries differently; this by and large has been in the spirit of the development of concepts in the UNFCCC, and is entirely consistent with the principal doctrine of CBDR and respective capabilities.

Over the years, it has become clearer that some countries or groups have more resources in contributing to the objectives of the global climate change action regime. Therefore, a concept which emphasises different treatments for different individuals or groups based on their respective capabilities will be fairer in deciding the treatment of law than a principle which treats everybody equally.

The seriousness of the global climate change problem also means that every country which has the capability for taking action has to do so, and it can be argued that addressing the climate change problem based on respective capabilities brings more justice to the climate change regime. Focusing more on respective capabilities justifies many commonly accepted arguments such as that the poor are the victims of and most vulnerable to environmental degradation without being responsible for it, and that the suffering increases due to their lesser capacity for adaptation to climate change. On the contrary, the fact that the developed countries suffer less from environmental degradation and bear more capacity for responding to environmental problems justifies greater levels of responsibility from them.

Unlike the element of historical responsibility, differentiated treatment based on capacity to respond is less contested by developed countries and much applied by international agreements, yet its articulation through negotiation has become more complex in an era where developing countries now in some cases emit more GHGs than many developed countries do.[39] It also seems that developed countries claim a leadership role in taking response measures to environmental degradation based on their capacity. For exam-

38 See e.g. Submission of 30 September 2008 to UNFCCC by Panama on behalf of Costa Rica, El Salvador, Honduras, Nicaragua and Panama, available at http://unfccc.int/resource/docs/2008/awglca4/eng/misc05.pdf, last accessed 2 May 2013.

39 See UN Statistics, available at http://mdgs.un.org/unsd/mdg/SeriesDetail.aspx?srid=749&crid=, last accessed 15 April 2013.

ple, at Rio, whilst disagreeing with differentiated treatment based on their historical responsibilities, the US agreed to take the leadership. When the US issued an interpretative statement to Principle 7 of the Rio Declaration, it attributed its acceptance of a leadership role to "industrial development", "experience with environmental protection policies and actions", and "wealth, technical expertise and capabilities".[40]

Much discussion during the life of the Convention has focused on the *differentiated* element of the CBDR principle. Yet this focus does little towards encouraging progressive climate change action and policies as it involves a narrative of burden allocation, which naturally results in a much more conservative approach to climate action under the Convention.

This stance cannot be faulted in the multilateral negotiation process because, in terms of international law, tremendous care is afforded to the concept of *precedent*. Under multilateral international law, there are few sources of custom; and, as noted under Article 38(1) of the 1946 Statute of the UN Chartered International Court of Justice,[41] custom is a principal aid in the interpretation of the proper and legal application of international rules.

Thus, it is not surprising that the preponderance of practical interpretation of the principle of CBDR has been conservative, emphasising the differentiation of the responsibility rather than on the more proactive *common* element which needs to include some degree of domestic climate action initiatives within the global context. Ostensibly, the concern would be that, with practice and custom advancing largely the *common* elements of the concept of an interpretation, the differentiation of the responsibility may become diluted.

Under this sequence of logic, neither group of actors, i.e. neither Annex I nor non-Annex I countries, has an incentive to act at more than an absolute minimum, as it is clear under the aegis of *customary application* that any actions which are not reciprocal could be construed as becoming a baseline for the application of any legally binding regime under the Convention. Developed countries, whilst responsible for historical emissions, could not condone a regime where the lion's share of current emissions would continue unchecked, as developing countries could not accept an interpretation of climate action which has shifted away from the differentiation inherent in the original concept.

40 1992 Rio Declaration.

41 Available at http://www.icj-cij.org/documents/index.php?p1=4&p2=2&p3=0#CHAPTER_II, last accessed 18 April 2013.

Today, because of its particular architecture of differentiation of action and because of domestic political reasons arguably emanating from this architecture,[42] the Kyoto Protocol covers less than 20% of projected global emissions and its effect has not tallied to a net reduction on global emissions.

The issue arising, as is clear by now, is that the political conditions for ambitious movement to low emission growth trajectories will not happen broadly in these circumstances. As is evident from many recent scientific reports,[43] the international community is still not doing enough to change emissions pathways so as to bring atmospheric GHG concentrations to levels projected to bring peak warming within 2°C of pre-industrial levels. How can we return to the original concept of CBDR with an interpretation which maintains the core balances inherent in the original formulation, but that will engender action, promoting hope towards a climate outcome consonant with the agreed intent of limiting global warming below 2°C, and above pre-industrial averages?

If we return to the original formulation of the CBDR concept, thinking positively about the construct and about its logical intent, it is clear that its intent is to limit GHG emissions under an agreed framework. This interpretation may be seen as contrary to arguments of atmospheric space, and equity thereto. However, *ceteris paribus*, the *common* descriptor towards responsibilities can only be interpreted as meaning that no country has a right to emit contaminants indiscriminately. This is part of the transboundary discussion we have seen above, and is, in essence, the source of the concept of *historical responsibilities*: it is unacceptable that any one country be construed as having a superior *right* over any other to emit anthropogenic GHGs, especially as the accumulated effects of those emissions impact all countries.

In this sense, returning to the concept of *respective capability* for the CBDR principle, an element of balance, of fairness, to the commonality of responsibility can be addressed. In this formulation, differentiation of responsibility seeks to attend, as much as possible, to the concept of *historical responsibility*, but *respective capability* seeks to temper the element of com-

42 See Byrd-Hagel Resolution of 25 July 1997, where the US Senate unanimously resolved to prevent the ratification of any international agreement that (1) did not require developing countries to make emission reductions, and (2) "would seriously harm the economy of the United States"; available at http://www.nationalcenter.org/KyotoSenate.html, last accessed 1 February 2013.

43 See IPCC (2007) and World Bank and International Energy Agency reports from 2009 onwards.

mon responsibility to what is just, or fair, within the development context of each country.

This view to action in climate change policy generally, as opposed to a stance of resultant inaction, is not a reinterpretation of the principles of the Convention; rather, it is a reading following the literal meaning of principles of equity and of CBDR and respective capabilities, namely that all countries share a common responsibility to climate action, though there is indeed a differentiation in the degree of response, that countries do have different capabilities, and that action needs to be undertaken accordingly.[44]

With these ideas in mind, negotiations for any future agreement need to try to move from a focus on differentiation under CBDR – which could yield a logic of inaction, especially from the perspective of 'free-riding', domestic political objections, and a possibly narrow emphasis of differentiation aimed at burden transfer and action subject to conditions – to one building on common action based on respective capability, without losing the elements of differentiation so essential for a just approach to multilateral climate action.

Over the past few years, there have been numerous examples of valuable and ambitious climate action in circumstances far beyond the expected or anticipated ambit of a given nation. The underlying motivations are diverse, complex and subject to more development than can be discussed here, but two immediate, contrasting and perhaps surprising examples can be cited.

The first example to note was the pledge by the Dominican Republic at Doha COP18, being a commitment sanctioned under national law to reduce their absolute GHG emissions by 25% or more by 2030, while projecting per capita income growth by 140% in the same period, without a requirement or condition of support.

It should be noted that this commitment by a developing country – which is a vulnerable small island state – is enacted under national law, with no obligation to do so under international law.

A second example is in the State of California in the US, individually one of the top ten economies in the world, and the largest state in the US. In 2006,

44 See the submission to the UNFCCC's Ad Hoc Working Group on the Durban Platform by Colombia, Costa Rica, the Dominican Republic and Peru, May 2012, available at http://unfccc.int/resource/docs/2012/adp1/eng/misc03a01.pdf, last accessed 1 February 2013.

California enacted legislation to cap its emissions at 1990 levels by 2020, with a long-term reduction goal of 80% from that figure by 2050.[45]

Again, it has to be noted that this is a commitment to a developed country's state, and was enacted under domestic law, with no obligation to do so under any applicable federal law or international obligations, and indeed contrasting with the US's negotiating stance on climate action, which projects distance from commitments to climate change action policy.

To the extent that domestic political inertia, concerns over 'free-riding' and burden-transfer mentalities in a conservative reading of the current climate action regime tend to limit climate change action programmes, these examples are exceptional inasmuch as they overcame inertia and delivered truly ambitious and firm commitments to action. However, if we look at the CBDR principle with a view to action, why are the benefits of the Dominican Republic's plan so particular to that country, and why are more developing countries not moving this way? Moreover, in the case of California, why are more developed countries not moving this way? Whilst there are many other such examples, it is clear that the multilateral regime we seek to nurture has to function in a way that promotes these types of initiatives without deviating from the necessary global actions and fundamental principles of the Convention. These examples of ambitious climate action, beyond the prevalent legal requirements, should not be outlier examples.

In terms of what is *common* to all, but within the *respective capability* of each, without losing sight of the differentiated context of the overall responsibility for action, the construct of *common responsibilities* can be understood to mean that small economies do what they can, medium-sized economies do more, and large economies do the most of all – but all countries contribute to the purpose of the Convention, and all countries have to be part of the solution.

Moreover, even in the simplest commercial sense, a vital issue at stake is clearly that the impact costs of climate change under business-as-usual scenarios will far outstrip economic growth[46] affecting all countries, especially the poorest. Thus, even on a bare economic appraisal, it makes no sense to follow an interpretation which subsidises the principal costs which the Convention seeks to avoid; this would be especially unconscionable for least-

45 See the 2006 Global Warming Solutions Act, available at http://www.arb.ca.gov/cc/ab32/ab32.htm, last accessed 24 March 2013.

46 See Garibaldi (2009).

developed countries and small island states, whose impact costs would proportionately be highest.

It is fair to argue that all nations need to adopt actions in line with their intrinsic capabilities to climate action, moving in a global environment of evolving higher levels of climate action. In other words, all countries have the capacity to act on climate change with support, so our aim should also be to tap the intrinsic capacities of all countries to act, and build on such intrinsic capacities with subsidised capacities. It is true, as was noted above, that capacities differ between Annex I countries and non-Annex I countries, and it has been a well-established argument that the high emitters are obliged to take stronger actions to limit their emissions. However, it is also true that capacities within non-Annex I countries vary greatly; the argument here is that these different capacities are to be taken into account in a future regime. As such, the *common* element of the CBDR principle may also mean that all countries act according to their respective capabilities towards an agreed aim of limiting concentrations of GHGs, starting from their intrinsic capacity for action, and building on that over time.

Naturally, the poorest countries will have little intrinsic capacity to act. Despite this, their intrinsic capacity may be geared to allow for the easier uptake of low carbon development pathways, for example, which in this case may come through subsidised capacity from wealthier countries[47] with higher emissions levels. In this case, the poor country is doing what its capacity allows it to do; but, even here, the country would be putting itself in a position not to follow high carbon-intensity development pathways and instil higher levels of climate action programmes from wealthier countries.

If one returns to the question of fairness in the system, then it is evident that a multilateral system espousing increased action will ultimately be the most *fair*. Yet, climate action, in a literal interpretation of the fundamental principles of the Convention, is required to be tempered with the concept of *respective capabilities*. In this view, all countries are obliged to act towards climate change – because all countries need to be part of the solution – in accordance with their intrinsic capabilities. More capable countries can act more, and the most capable can not only act with the most ambition, but can also support other countries and unlock further subsidised climate action

47 Some poor countries have started doing so; see e.g. Ethiopia's well-established Green Economy Strategy, available at http://www.epa.gov.et/Download/Climate/Ethiopia's%20Climate-Resilient%20Green%20economy%20strategy.pdf, last accessed 22 April 2013.

capacities. Such contributions will encourage even the less-developed countries to contribute towards more progressive climate change policies.

## G. Conclusions

If we revisit the ultimate objective of the UNFCCC, and bear in mind the irrefutable illustrations of global impact inherent in the problem of anthropogenic climate change, it is clear that the imperative towards a regime of positive global climate action gathers many arguments beyond a conservative zero-sum or burden-allocation narrative, which imperative has to include differentiated action by all. It has been argued here that the transboundary nature of climate change action and impacts includes both environmental and developmental repercussions for all countries, derived from impact costs which affect all countries and, by the same token, that climate action derives benefits for all countries.

From these considerations it is proposed that the imperative for climate change action has to exist in an environment of real yet differentiated action by all, and that a more progressive paradigm, where domestic action – both intrinsic and supported – can be aligned to cross-border benefits, should be conceived as an important component for a new regime.[48]

However, movement towards this new understanding needs to preserve the fundamentals of the UNFCCC consensual regime for climate change. Innovation in the regime has to include returning to these concepts with a view to fostering cooperation, solidarity and action among states so as to move to a view of *common* action where national intrinsic capacity can place every country as part of a solution towards a coordinated and mutually cooperative global climate regime, with supported action leading to more ambition and even more capacity for low carbon development and climate action.

Yet, this new regime should not be conceived as promoting a uniform application of action on all countries, as this would not encompass a solution that is consonant with the underlying CBDR philosophy. The argument is that, whilst the regime will be universally applicable, respective capabilities

48 Roberts & Edwards (2012).

will dictate the extent of *common* efforts, and the differentiation of responsibilities will support even greater efforts.[49]

In the new regime, there will be several challenges to overcome, not the least of which will be an equitable and fair view of the intrinsic capabilities of countries, which is essential not only for delivering more climate action and ensuring that climate action is mainstreamed in all countries, but also to overcome arguments of 'free-riding' which would jeopardise domestic political support for a broad agreement on action.

The new regime also needs to deliver on maintaining and supporting action and progress not only in developing countries, but also in supporting the leadership of developed countries in action to the level that is commensurate with their differentiation. Indeed, further action by developed countries, over and above their high intrinsic capabilities, be it through public funding, direct investment, capacity-building or market-based initiatives, has to work to incentivise more capacity for low carbon development in developing countries, as well as driving down their own emission pathways as required by science and the ultimate objective of the Convention.

Examples such as California should be mainstreamed into global action for all countries. This would herald a multilateral regime which truly deals with what science has required: the most drastic de-carbonisation in history.[50] Many questions will remain to be discussed over the coming years as a new regime is crafted, but it is clear that the new regime will need to both sustain CBDR and invigorate it with a narrative which incentivises climate action by all countries.

## *References*

Birnie, Patricia, Alan Boyle & Catherine Redgwell, 2009, *International Law and the Environment* (Third Edition), Oxford, Oxford University Press.

CISDL/Centre for International Sustainable Development Law, 2002, The Principle of Common but Differentiated Responsibilities: Origins and Scope, CISDL Legal Brief for World Summit on Sustainable Development, Johannesburg, 26 August 2002, available at http://www.cisdl.org/pdf/brief_common.pdf, last accessed 20 April 2009.

Cullet, Phillipe, 1999, Differential Treatment of International Law: Towards a New Paradigm of Inter-state Relations, *European Journal of International Law* 10, 3.

49 Garibaldi et al. (2012).

50 See PwC (2012); Sachs (2012).

Garibaldi, José A., Monica Araya & Guy Edwards, 2012, Shaping the Durban Platform: Latin America and the Caribbean in a Future High Ambition Deal, Policy Brief, Energeia, Climate and Development Knowledge Network & Fundación Futuro Latinoamericano, available at http://www.intercambioclimatico.com/wp-content/uploads/Shaping-Durban-Platform_Final-April_2012.pdf, last accessed 9 May 2013.

Garibaldi, José, 2009, *The Economics of Boldness*, London/Lima, Energeia, available in Spanish at http://libelula.com.pe/IMG/pdf/La_Economia_de_la_Audacia.pdf, last accessed 15 April 2013.

Hey, Ellen, 2009, Common but Differentiated Responsibilities, in: Wolfrum, Rüdiger (Ed.), *Max Planck Encyclopaedia of International Law*, Oxford, Oxford University Press.

Honkonen, Tuula, 2009, The Principle of Common but Differentiated Responsibility in Post-2012 Climate Negotiations, *RECEIL* 18 (3), 257–268.

IEA/International Energy Agency, 2012, *Turn Down the Heat: Why a 4°C Warmer World Must be Avoided – A Report for the World Bank by the Potsdam Institute for Climate Impact Research and Climate Analytics*, Washington, DC, International Bank for Reconstruction and Development/The World Bank, available at http://climatechange.worldbank.org/sites/default/files/Turn_Down_the_heat_Why_a_4_degree_centrigrade_warmer_world_must_be_avoided.pdf, last accessed 15 April 2013.

IPCC/Intergovernmental Panel on Climate Change, 2007, *Climate Change 2007: Synthesis Report. Contribution of Working Groups I, II and III to the Fourth Assessment Report of the Intergovernmental Panel on Climate Change*, Core Writing Team: Pachauri, Rajendra K. & Andy Reisinger (Eds), Geneva, IPCC, available at http://www.ipcc.ch/publications_and_data/publications_ipcc_fourth_assessment_report_synthesis_report.htm, last accessed 13 January 2013.

Joyner, Christopher, 2002, Common but Differentiated Responsibilities, *Proceedings of the American Society of International Law*, Washington, DC, ASIL 96, 358–368.

PwC/PricewaterhouseCoopers, 2012, *Too Late for Two Degrees?: Low Carbon Economy Index 2012*, London, PwC, available at http://www.pwc.com/en_GX/gx/low-carbon-economy-index/assets/pwc-low-carbon-economy-index-2012.pdf, last accessed 8 May 2013.

Rajamani, Lavanya, 2005, *Differential Treatment in International Environmental Law*, Oxford, Oxford University Press.

Roberts, Timmons & Guy Edwards, 2012, A New Latin American Negotiating Group: The Greenest Shoots in the Doha Desert, 13 December 2012, available at http://www.intercambioclimatico.com/en/2012/12/13/a-new-latin-american-climate-negotiating-group-the-greenest-shoots-in-the-doha-desert/#more-5684, last accessed 15 April 2013.

Sachs, Jeffrey D., 2012, Secretariat on Climate Change: First Workshop on Long-term Finance, available at http://unfccc.int/files/cooperation_support/financial_mechanism/long-term_finance/application/pdf/sachs9july2012.pdf, last accessed 9 May 2013.

Sands, Philip, 2003, *Principles of International Environmental Law*, Cambridge, Cambridge University Press.

Stone, Christopher, 2004, Common but Differentiated Responsibilities in International Law, *American Journal of International Law* 98, 276–281.

UNDP & ECLAC/United Nations Development Programme and Economic Commission for Latin America and the Caribbean, 2012, *Sustainable Development 20 Years on from the Earth Summit: Progress, Gaps and Strategic Guidelines for Latin America and the Caribbean*, New York, United Nations.

UNEP/United Nations Environment Programme, 2012, *UNEP Gap Report 2012*, available at http://www.unep.org/pdf/2012gapreport.pdf, last accessed 15 April 2013.

UNEP/United Nations Environment Programme & ECLAC/United Nations Environment Programme and Economic Commission for Latin America and the Caribbean, 2010, *Vital Climate Change Graphics for Latin America and the Caribbean*, available at http://www.grida.no/publications/vg/lac2/, last accessed 8 May 2013.

Weisslitz, Michael, 2002, Rethinking the Equitable Principle of Common but Differentiated Responsibility: Differential Versus Absolute Norms of Compliance and Contribution in the Global Climate Change Context, *Colombia Journal of International Environmental Law and Policy* 13, 473–483.

7

# Beyond International Climate Negotiations: Climate Diplomacy from a Foreign Policy Perspective*

*Dennis Tänzler & Alexander Carius*

*Abstract*

Flooding, droughts, a shift of climate zones, and increasingly frequent and intense extreme weather hazards will have serious economic and social consequences for entire regions. Countries with low adaptation capacities are likely to be hit the hardest by these climate changes, among them many of the so-called fragile states. To address this challenge, a new profile of climate diplomacy is evolving using the full range of available policies, including development cooperation, conflict prevention, and humanitarian assistance, as well as climate change adaptation and mitigation. The aim is to move from risk analysis of climate-related threats to preventive action. This also asks for ways to integrate climate change concerns into development, foreign, and security policies. Based on the discussions of major developments within the European Union, the United Nations and Germany, we outline some of the potentially key tasks for climate diplomacy in the future that are needed to complement the negotiations for a comprehensive, global climate agreement.

## *A. The Climate Security Challenge*

The slow progress in further developing the international climate regime shows that urgent action is needed that complements and stretches beyond international climate negotiations. In recent years, climate change has gained increasing prominence among foreign policymakers.[1] This can partly be ex-

---

* This article is based on insights gained during the project "Climate Diplomacy 2012" supported by a grant from the German Federal Foreign Office.

1 For an overview, see Tänzler & Carius (2012).

plained by the fact that climate change represents a vital challenge for international politics. Flooding, droughts, a shift of climate zones, and increasingly frequent and intense extreme weather hazards will have serious economic and social consequences for entire regions. In addition, there is a broad consensus that countries with low adaptation capacities will be hit the hardest, among them many of the so-called fragile states.[2] Starting in 2007, a number of analyses reveal a growing potential for conflict and an increase in social tension as a result of the impending changes in the climate.[3] Conflicts may arise as a result of water and food shortages, in turn caused by an increase in extreme weather events and climate-change-induced mass migration. Weak and fragile states are considered particularly vulnerable because of their already limited political capacities. The main assumption is that a further weakening of the key services provided by the public sector is likely to lead to national and regional destabilisation, with societal and political tensions potentially developing into violent conflict. Seen in this light, it is not surprising that the foreign policy community is concerned about the slow progress of the international climate negotiations and the decision to agree on an outcome with legal force by 2015 which is to enter into force only by 2020.

However, the role of foreign policies in a changing climate is complex. When assessing whether or not there will be an increase of violent conflicts related to the distribution of natural resources such as water and land, one should avoid one-dimensional causal explanations.[4] Possible conflicts will not be caused by climate change alone; rather, climate change is seen as a factor that multiplies the deficits in other areas such as poverty, a lack of the rule of law, and social and economic injustice.[5] In addition, a worsening of conflict situations as a result of climate change is only one possible pathway. Another is the peaceful avoidance of new conflict situations through early action and cooperation. The latter interpretation is based on research findings about how environmental cooperation toward common challenges could support confidence-building as well as peace-building efforts between former antagonists.[6]

2 Corendea et al. (2012).
3 Campbell et al. (2007); CNA (2007); WBGU (2007).
4 See e.g. Harris (2012).
5 Carius et al. (2008).
6 See e.g. Conca & Dabelko (2002); UNEP (2009).

This opens a different point of entry for foreign policy engagement. In other words, there is a need not only to reflect appropriately the potential security-related impacts of climate change, but also to design appropriate policy measures which are timely enough to avoid a further destabilisation of already weak or fragile states. It seems more than obvious that such approaches have to go beyond traditional climate policy as we have known it for some time. By encompassing the full range of available policies, including development cooperation, conflict prevention, and humanitarian assistance, as well as climate change adaptation and mitigation, a new profile of climate diplomacy is evolving. This new profile most likely requires new strategic alliances beyond the conference halls of Copenhagen, Doha or Durban. In the following section, we discuss selected political processes initiated in recent years on climate change, international security and foreign policies. These processes illustrate how to move from risk analysis to preventive action and how to integrate climate change concerns into development, foreign, and security policies. To this end, in order to address the challenges of climate security, we first highlight major developments within the European Union (EU), the United Nations (UN) and Germany. Based on this discussion, we then outline some of the potentially key tasks for climate diplomacy in the future that are needed to complement the negotiations for a comprehensive, global climate agreement.

## B. *Foreign Policy Perspectives on Climate Change*

### I. *The EU on the Search for International Partners for Climate Security*

An early approach to address the potential security implications of climate changes was initiated by the EU. Under the 2007 German EU presidency, the European Council and the European Commission were asked to prepare a joint paper on climate change and international security. This report, published in March 2008, summarised potential security risks associated with climate change.[7] Broadly, the report outlines that climate change has the potential of becoming a "threat multiplier", exacerbating existing tensions and potentially creating new ones over time.[8] Among the main security-relevant threats of climate change that the EU identified were conflicts over

7 EU (2008).
8 (ibid.).

depleting resources such as water and food, and the economic damage and risks caused by an increase in sea levels and in the strength and frequency of extreme weather events. According to the report, fragile and radicalised situations may be exacerbated owing to the amount of environmental stress and a lack of coping capacity.

Against the backdrop of these risks, the Council stated in its Conclusions of December 2009 that climate change and its international security implications were part of the wider EU agenda for climate, energy, and its Common Foreign and Security Policy.[9] The Council stressed the need to strengthen the EU's comprehensive efforts to reduce emissions as one aspect of conflict prevention.

In the aftermath of the Council's Conclusions, the main focus of the EU's activities has been directed towards enhancing EU capacities for early warning, on the one hand, and towards fostering international cooperation with the aim to creating dialogue and a common awareness in relevant international forums, including the UN, on the other. However, owing to the establishment process of the European External Action Service (EEAS), the initiatives to address climate change and security have only progressed very slowly. In July 2011, however, the EEAS and the services of the Commission presented a conceptual outline of what should be considered as a climate diplomacy blueprint: the Joint Reflection Paper.[10] Most importantly, the Joint Reflection Paper outlined three "strands for action" action on EU climate diplomacy:[11]

- The promotion of ambitious climate action
- The support of implementation of climate policies and measures, and
- Activities in the area of climate change and international security.

Among the 13 recommendations outlined in the Joint Reflection Paper, there are some with immediate implications if they are implemented. For example, the capacities of the EEAS to engage in climate diplomacy should be strengthened "by establishing a focal point in the Service for Climate change issues"[12] as well as local climate change working groups in strategic partner countries to improve the relevant reporting on climate-change-related developments. In addition, by suggesting the mainstreaming of climate action

9 Council of the European Union (2009).
10 EEAS & EC (2011).
11 (ibid.:1).
12 (ibid.:5).

in the multiannual country and regional strategy papers, long-term processes can be initiated that may help to make climate change a cross-cutting issue on EU foreign affairs agendas.

In principle, the EEAS approach can be interpreted as building a bridge between further improvement of early warning capacities on climate-change-related security threats and the diplomatic efforts needed to contribute to a global negotiation deal. The practical relevance of this approach, however, remains to be seen.

## *II. The United Nations Arena*

The UN took the climate change issue seriously right after its appearance on the international agenda. The 1992 United Nations Framework Convention on Climate Change (UNFCCC) as well as the 1997 Kyoto Protocol were highly influential in establishing a portfolio of policy innovations worldwide in respect of mitigating greenhouse gas (GHG) emissions and to adapt to unavoidable climate changes.[13] However, an increasingly complex and slow process of international negotiations has caused some concerns that the UNFCCC cannot achieve its main objective stated in its Article 2, namely to avoid "dangerous anthropogenic interference with the climate system". This is especially understandable from the perspective of small island developing states (SIDSs) as well as other least-developed countries with low-lying coastlines, like Bangladesh, which are already today witnessing the severe impacts of climate change.[14] As a result, the issue of climate security also has gained increasing attention in recent years at UN level. In 2007, the UN Security Council held its first debate on the impact of climate change on global peace and security. The discussions among UN member states revealed broad uncertainty regarding the question of an appropriate international framework for action on responding to the security risks related to climate change. The UN General Assembly, on 3 June 2009, adopted a res-

13 See for an early discussion of the accomplishments, Oberthür & Ott (1999); Yamin & Depledge (2004).

14 See the contributions of the Minister for Environment and Forests of the People's Republic of Bangladesh, Hasran Mahmud (2012:25–29), and the then Minister of State for Housing and Environment of the Maldives, Mohammed Shareef (2012:31–32).

olution on climate change and its possible security implications,[15] which had been proposed by Pacific SIDSs. The resolution was adopted by consensus, and 101 states supported it. For the first time in the history of the UN, the United States (US) co-sponsored a climate protection resolution. The resolution urged UN bodies to strengthen their efforts to combat climate change and to avoid intensifying potential security risks. This was also the first time that a UN resolution had established a direct link between climate change on the one hand, and international peace and security on the other.

On the basis of 35 contributions from member states and relevant regional and international organisations, the UN Executive Committee on Economic and Social Affairs (ECESA) published a comprehensive report in September 2009.[16] The report defined *security* in a broader sense, where vulnerable individuals and communities were the primary concern, and *security* was understood in terms of protection from a range of threats, i.e. disease, unemployment, political repression, disasters, and violence. The report further acknowledged that the security of individuals and communities was important in shaping the security of nation states, which is typically framed in terms of threats of external aggression. The most important aspect of the report was its strong focus on potential threat minimisers, such as –

- climate mitigation and adaptation
- economic development
- democratic governance and strong institutions
- international cooperation, and
- preventive diplomacy and mediation.

In addition, the report highlighted the importance of timely availability of information and increased support for research and analysis in order to improve the understanding of links between climate change and security, and to build up early warning capacities.[17]

Despite the report's clear mandate, however, it only received minor attention as a reference for further activities. It took until July 2011, when the German government, under its Security Council presidency, brought up the topic of climate change and security on the agenda of high-level discussions, namely at the UN Security Council. The open debate in the Security Council

15 Resolution A/63/281.

16 Climate Change and its Possible Security Implications, A/64/350, 11 September 2009, New York.

17 (ibid.).

resulted in a presidential statement not only confirming the concern of climate change affecting security, but also asking for a systematic and regular review of and reporting by the UN Secretary-General to the UN Security Council on the likely security implications of climate change.[18] This result is remarkable, since China, Russia and several countries among the Group of 77 (G77) expressed their concern about linking climate change to security.[19] The unanimous adoption of the Presidential Statement on climate change and security at the Security Council meeting of 20 July 2011, however, revived the spirit of climate diplomacy at international level in particular, because the UNFCCC was also endorsed as the major UN forum for discussing comprehensive climate policy actions by all participating representatives. In the aftermath of the Security Council meeting, the German government took steps to design preventive climate diplomacy – as did the United Kingdom and other governments, who are now involved in entrenching climate change as a key issue in foreign policy.[20]

### *III. German Foreign Policy as regards Climate Change*

German foreign policy has, in recent years, constantly pushed the EU as well as the UN to address the security risks of climate change and to make it a priority in the foreign policy community. This engagement started with asking the EU, under the German EU presidency in 2007, to prepare a report on the security dimensions of climate change. After the report was published, Germany was not only part of the informal steering group on this topic, but also started to develop its own initiatives to actively enter into discussions with partner countries and regions on climate change challenges. For example, in 2008, the German Federal Foreign Office designed and launched the initiative entitled "Water Unites" with the governments of Central Asia to jointly address the challenges of increased water scarcity in Central Asia through –

- promoting transboundary water management
- strengthening research on joint utilisation approaches

---

18 Statement by the President of the Security Council, S /PRST/2011/15. 20 July 2011, New York.

19 The authors' own observation during the open debate of the Security Council, 20 July 2011, New York.

20 See Thölken & Börner (2012:7–8).

- forming a network among experts from Central Asia, the EU and Germany, and
- knowledge transfer and investments in the water sector.

The initiative is also meant to contribute to the implementation of the EU Central Asia strategy, entitled The EU and Central Asia: Strategy for a New Partnership, issued in 2007, and other international initiatives in the region of Central Asia.

In addition, and as a follow-up to the Security Council meeting, the Federal Foreign Office has been implementing a number of climate diplomacy activities since 2011. These include a large international conference in Berlin.[21] More than 100 participants from the foreign policy community discussed how the conclusions from the Security Council could be further operationalised. More concretely, the challenges of water scarcity, food insecurity and coastal instability were discussed, and the prospects of geopolitical change management examined.

According to the statements made during the Berlin conference, and during further regional dialogue initiatives in southern and East Africa, southern Asia and Latin America, climate diplomacy activities are specifically aimed at supporting communication with partner institutions in partner countries; promoting capacity- and network-building in affected regions in particular; and analysing the scientific fundamentals in a bid to identify climate policy options that will prevent conflicts. One of the key aspects within this multilevel effort of climate diplomacy is the integration of regional perspectives – especially from developing countries and emerging economies – into current international policy processes. To this end, it is imperative to raise awareness among key actors in the relevant regions regarding the need to cooperate regionally and globally in respect to climate issues, and this has been addressed by the German Federal Foreign Office and German embassies around the world by a number of public diplomacy means, including exhibitions, information platforms, round tables and conferences. In the following section, we will turn our attention to the question of how these initiatives and activities may be translated into a coherent climate diplomacy agenda that can be addressed with concrete policies and measures.

21 See adelphi (2012) as well as the documentation of the regional consultations, available at http:/www.climate-diplomacy.org, last accessed 9 January 2013.

## C. Towards a Strategy on Climate Diplomacy

The EU and the UN have made considerable progress in describing how a climate diplomacy framework would look, conceptually, in order to support international climate negotiations and strengthen conflict and crisis prevention capacities. German foreign policy has been a key driver behind these discussions. The identification of available threat minimisers, as outlined by the UN Secretary General in 2009, opens the door in principle to move from the stage of risk analysis to one of policy formulation and implementation, for which the parallel processes on this issue at the UN and EU levels can be used. Both levels offer other governments the opportunity to engage in strategy formulation in respect of dealing with the climate security challenge. Again, this matter will hardly be restricted to international climate negotiations: it requires the involvement of a broad spectrum of partners. Three potential areas of engagement for these partnerships are outlined in the following subsections.

### I. Building Transformative Pathways

The concept of a low-carbon economy is relevant for the climate and security debate because it aims to address different political key priorities: climate protection, energy security, and economic and social development. The expansion of renewable energies is also an important element of debate today within the security and defence community: a 2010 report by the Center for Naval Analyses (CNA) outlined the potential opportunities for US national security that could result from the transition to an economy based on clean-energy technology.[22] According to the CNA, innovation and commercialisation of clean, low-carbon energy would contribute directly to the US's future economic competitiveness and would bolster national security.[23] Comprehensive actions to mitigate GHG emissions in industrialised and developing countries are also needed to limit the risk of climate-induced conflicts and allow the global economy to shift towards lower emissions. Such transformative pathways should not only ensure compliance with ambitious

22 CNA (2010).
23 (ibid.).

climate change targets, but also support sustainable growth and the creation of new employment opportunities.[24]

Thus, well-designed mitigation policies have the potential to link climate protection, development and conflict prevention, allowing them to serve together as threat minimisers. To this end, however, some of the key mitigation questions need to be answered, e.g. –

- How will mitigation efforts be distributed among the various countries, above all with respect to the key emitters?, and
- How can poorer countries be supported to link technological progress in strategic key areas such as energy supply, infrastructure development, or transportation with a low-carbon development pathway?

The development of sustainable energy options is especially important to avoid locking in high-carbon technologies while the demand for energy rises and, in turn, often leads to costly energy import dependency. In addition, decentralised grids are likely to offer co-benefits between sustainable energy production and improved access to energy. The impact of mitigation policies will vary significantly by country owing to varying sectoral composition, such as energy supply or transportation infrastructures. Accordingly, there is no silver bullet: ongoing consultations are needed – not least on how to involve the private sector.

### *II. Designing Conflict-sensitive Climate Policies*

The discussion about appropriate policy frameworks is of strategic value. Accordingly, the development of low-carbon growth strategies needs further guidance and international cooperation. One possible option in supporting countries who are entering such a strategy discourse is to use the revenue generated from auctioning emission permits in carbon-trading programmes. At the same time, a conflict-sensitive approach requires that international donors and recipient countries ensure funding is spent transparently and effectively in order to avoid an increase in governance pitfalls such as corruption.[25]

Apart from the energy sector, land use and forest protection have received increasing attention and can serve as an example of how climate mitigation

24 Ellis et al. (2009).
25 See for a general reflection, Hammill et al. (2009).

may be linked to development and stability. Efforts to systemically address the cost-effective emission reduction potential in the forest sector have led to various approaches to conceptualise the UN's Collaborative Programme on Reducing Emissions from Deforestation and Forest Degradation in Developing Countries (UN-REDD), a UN initiative to reduce emissions from deforestation and forest degradation. UN-REDD can, in principle, contribute to economic recovery by generating new sources of income in the forest sector for often-marginalised social groups.

Depending on the concrete design of benefit-sharing agreements, central governments as well as local communities can receive income and use it, for example, to build infrastructure and services. Additional employment opportunities may also be created for forest monitoring and law enforcement.

However, whether sustainable forest management and extractive logging are compatible with UN-REDD regulations will only be seen after an associated international agreement has been adopted. In addition, implementation of UN-REDD requires excellent governance capacities. Governments, communities, and project implementers need to develop sound concepts and implementation capacities to address the drivers of deforestation. When it comes to compliance with any future international agreement, countries need to enforce forest protection (e.g. curb illegal logging) and build up sufficient capacity to measure, report on and verify their commitments. Last but not least, sophisticated benefit-sharing mechanisms are needed in order to avoid conflicts on the national and local levels concerning the distribution of revenues generated through any kind of UN-REDD mechanism.[26]

### *III. Learning to Adapt*

The UNFCCC defines *adaptation* as "... adjustment in natural or human systems in response to actual or expected climatic stimuli or their effects, which moderates harm or exploits beneficial opportunities."

Seen through a more political lens, adaptation requires people to be empowered, their livelihoods to be secured, and their resilience to be strengthened by building appropriate institutions. Adaptation will require both effective local activities and national and regional coordination for the design

26 For a more comprehensive debate on potential risks, see the Rights and Resources Initiative, available at http://www.rightsandresources.org/documents/files/doc_1400.pdf, last accessed 27 December 2012. See also Tänzler & Ries (2012:695–706).

and implementation of appropriate action. To this end, international cooperation is needed, especially in the case of the most vulnerable developing countries, to provide for adequate resources.[27]

The idea of adaptation has taken centre stage in the debate on the security-related implications of climate change – in part because GHG emissions to date have already triggered irreversible global warming. Adapting to a changing environment should help avoid negative effects such as water or food scarcity and, consequently, social and political tensions. Ongoing activities have already made some progress in creating strategic support for future adaptation processes – including in some conflict-prone countries. As at the end of 2010, for example, 45 National Action Plans for Adaptation (NAPAs) for least-developed countries had been submitted to the UNFCCC. Of these, 21 were developed in countries considered to be states at high risk of destabilisation, and 19 in countries at increased risk of destabilisation.[28] Hence, so-called fragile states are also influenced by international support to initiate processes of adaptation.

However, there is only a slow initiation of concrete projects. This not only illustrates the as yet insufficient funding, but also contributes to an increasing loss of credibility for international climate protection measures in those countries most severely affected by climate change.

A coherent implementation of adaptation measures is likely to be facilitated by an institutionalisation of responsibilities. If an appropriate national authority does not exist, this jeopardises the integration of adaptation measures into other development processes, and makes it extremely difficult to incorporate conflict-sensitive considerations into national planning processes. As we learn from the research on environment and security, cooperation over scarce resources such as shared waters harnesses great potential to facilitate sustainable development and political stability between riparian nations as well as within such countries. One key factor for success is the establishment of strong institutions such as river commissions and other transboundary institutional arrangements.[29] Cooperation between countries with bordering watersheds has long been a focus of the international donor community. As a result, it is often possible to make use of existing structures – also to address future adaptation needs. However, the stabilising and trust-

27 For a comprehensive discussion, see Tänzler et al. (2010:741–750); Corendea et al. (2012).
28 See Fund for Peace (2011).
29 See e.g. Houdret et al. (2010).

building potential often demonstrated by transboundary cooperation in the water sector is not yet reflected prominently in existing national adaptation activities. This suggests the need exists to link and coordinate national and regional processes more systematically in order to provide for climate security, which may also be facilitated by appropriate institutions.

### *D. Doha and Beyond: Prospects for Climate Security and Climate Diplomacy*

The impacts of global climate change will be felt differently across the world, but no region will be able to avoid all of them. Moreover, feedback loops between different threats across regions, converging trends, and global interconnectedness requires concerted and global action. The options available to foreign policymakers in respect of addressing climate-related security concerns are not limited to the UN climate negotiations. However, the ongoing debate on 'targets and timetables' cannot delay the establishment of a comprehensive framework for adaptation governance, and support for initiating the development of low-carbon growth strategies. These elements are likely to benefit from a re-energised global process in order to facilitate the mainstreaming of these issues in relevant national and regional processes and to provide a basis for further activities to ensure climate-related security. Beyond the international climate change process, there are further entry points to ensure that the responses to climate change are designed in a conflict-sensitive way. Here representatives from the fields of development, foreign and security policy should engage in a strategic partnership to address the following issues:

- Governments and non-governmental stakeholders should use ongoing risk analysis processes to identify sectors critically affected by climate change, especially in conflict-prone areas. This will also help to ensure coherency and coordination with other planning processes. One possible means would be to expand the use of peace and conflict assessments to consider the impacts of climate mitigation and adaptation activities.
- Aid agencies active in the transatlantic context should initiate conflict-sensitive mitigation and adaptation processes using a multi-dimensional system that incorporates administrative and societal perspectives. Involving representatives from partner countries in risk analysis and strategy formulation will probably increase acceptance for the transformation

processes necessary to secure the supply of food, water, and sustainable energy, and to improve disaster preparedness.

- The establishment of national and regional steering committees in conflict-prone regions can support the monitoring of mitigation and adaptation programmes, coordinating public authorities and external stakeholders such as donor organisations, and establishing mediation bodies. To this end, a substantial increase of capacities on a national and regional level is needed that can be supported not only by the EU but also by relevant UN agencies, and
- The support for adaptation and mitigation processes, especially in already fragile countries, should be integrated into the larger regional context. The further development of the EEAS offers a chance to expand international cooperation with third countries to commence dialogue, create awareness, share analysis, and cooperatively address the challenges of climate change.

## *References*

adelphi, 2012, *Climate Diplomacy: Reducing Risks for Security*, Berlin, adelphi, also available at http:/www.climate-diplomacy.org, last accessed 9 January 2013.

Campbell, Kurt M., Jay Gulledge, John R. McNeill, John Podesta, Peter Ogden, Leon Fuerth, R. James Woolsey, Alexander T.J. Lennon, Julianne Smith, Richard Weitz & Derek Mix, 2007, *The Age of Consequences: The Foreign Policy and National Security Implications of Global Climate Change*, available at http://www.csis.org/media/csis/pubs/071105_ageofconsequences.pdf, last accessed 3 January 2013.

Carius, Alexander, Dennis Tänzler & Achim Maas, 2008, *Climate Change and Security – Challenges for German Development Cooperation*, Eschborn, Gesellschaft für Technische Zusammenarbeit.

CNA/Center for Naval Analyses, 2007, *National Security and the Threat of Climate Change*, Alexandria, VA, CNA.

CNA/Center for Naval Analyses, 2010, *Powering America's Economy: Energy Innovation at the Crossroads of National Security Challenges*, Alexandria, VA, CNA.

Conca, Ken & Geoffrey Dabelko (Eds), 2002, *Environmental Peacemaking*, Baltimore, MD, Johns Hopkins University Press.

Corendea, Cosmin, Koko Warner & Kristina Yuzva, 2012, Social Vulnerability and Adaptation in Fragile States, *InterSecTions Report 11/2012*, Bonn, United Nations University Institute for Environment and Human Security.

Council of the European Union, 2009, *Council Conclusions on Climate Change and Security, 2985th Foreign Affairs Council Meeting, Brussels, 8 December 2009*, Brussels, Council of the European Union.

EEAS & EC/European External Action Service and Services of the European Commission, 2009, *Towards a Renewed and Strengthened EU Climate Diplomacy: Joint Reflection Paper, Brussels, 9 July 2011*, available at http://eeas.europa.eu/environment/docs/2011_joint_paper_euclimate_diplomacy_en.pdf, last accessed 9 April 2013.

Ellis, Karen, Bryn Baker & Alberto Lemma, 2009, *Policies for Low Carbon Growth*, London, Overseas Development Institute.

EU/European Union, 2008, *Climate Change and International Security: Paper from the High Representative and the European Commission to the European Council*, available at http://www.consilium.europa.eu/uedocs/cms_data/docs/pressdata/en/reports/99387.pdf, last accessed 9 April 2013.

Fund for Peace, 2011, Failed State Index Scores, available at http://www.fundforpeace.org, last accessed 29 December 2012.

Hammill, Anne, Alec Crawford, Robert Craig, Robert Malpas & Richard Matthew, 2009, *Conflict-sensitive Conservation: Practitioners' Manual*, Winnipeg, International Institute for Sustainable Development.

Harris, Katie, 2012, Climate Change in UK Security Policy: Implications for Development Assistance?, Working Paper 342, London, Overseas Development Institute.

Houdret, Annabele, Annika Kramer & Alexander Carius, 2010, *The Water Security Nexus: Challenges and Opportunities for Development Cooperation*, Eschborn, Gesellschaft für Technische Zusammenarbeit.

Mahmud, Hasran, 2012, Climate Change and Security: The Challenges for Bangladesh, in: Tänzler, Dennis & Alexander Carius (Eds), *Climate Diplomacy in Perspective: From Early Warning to Early Action*, Berlin, Wissenschaftsverlag, 25–30.

Oberthür, Sebastian & Hermann E. Ott, 1999, *The Kyoto Protocol: International Climate Policy for the 21st Century*, Berlin, Springer.

Rights and Resources Initiative, 2010, *The End of the Hinterland: Forests, Conflict, and Climate Change*, available at http://www.rightsandresources.org/documents/files/doc_1400.pdf, last accessed 27 December 2012.

Shareef, Mohammed, 2012, Climate Change and Security: The Challenges for Small Island States, in: Tänzler, Dennis & Alexander Carius (Eds), *Climate Diplomacy in Perspective: From Early Warning to Early Action*, Berlin, Wissenschaftsverlag, 31–34.

Tänzler, Dennis & Alexander Carius (Eds), 2012, *Climate Diplomacy in Perspective: From Early Warning to Early Action*, Berlin, Wissenschaftsverlag.

Tänzler, Dennis & Felix Ries, 2012, International Climate Change Policies: The Potential Relevance of REDD+ For Peace and Stability, in: Scheffran, Jürgen, Michael Broszka, Hans Günter Brauch, Peter Michael Link & Janpeter Schilling (Eds), *Climate Change, Human Security and Violent Conflict: Challenges for Societal Stability*, Berlin/Heidelberg, Springer Verlag, 695–706.

Tänzler, Dennis, Achim Maas & Alexander Carius, 2010, Climate Change Adaptation and Peace, *Wiley Interdisciplinary Reviews Climate Change* 1(5), 741–750.

Thölken, Heinrich & Weert Börner, 2012, Climate Change and International Security: Why Foreign Policy Must Meet the Challenge, in: Tänzler, Dennis & Alexander Carius (Eds), *Climate Diplomacy in Perspective: From Early Warning to Early Action*, Berlin, Wissenschaftsverlag, 7–8.

UNEP/United Nations Environment Programme, 2009, *From Conflict to Peacebuilding: The Role of Natural Resources and the Environment*, Nairobi, UNEP.

WBGU/Wissenschaftlicher Beirat der Bundesregierung Globale Umweltveränderungen (German Advisory Council on Global Change), 2007, *World in Transition – Climate Change as a Security Risk*, Berlin/Heidelberg, Springer.

Yamin, Farhana & Joanna Depledge, 2004, *The International Climate Change Regime: A Guide to Rules, Institutions and Procedures*, Cambridge, Cambridge University Press.

8

# Confronting Complex Global Challenges: Comparing the Climate Change and Law of the Sea Negotiations

*Gregg B. Walker*


*Abstract*

The United Nations Framework Convention on Climate Change (UNFCCC) features the most important and complex negotiations that the global community has yet addressed. Climate scientists have reached consensus on the significance of climate change, its impacts, and anthropogenic causes. Political leaders and negotiators, though, have yet to achieve consensus agreements on any of the major climate change policy areas, such as extending the Kyoto Protocol, setting and adhering to clear mitigation goals, providing the resources needed to adapt, and developing new institutions, such as the Green Climate Fund.[1]

In contrast, consensus on both science and policy was achieved during an earlier international conference that, for its time, was called "one of the most important negotiations to have ever taken place".[2] Negotiators at the Third United Nations Conference on the Law of the Sea (UNCLOS III) worked on 25 issues over most of a decade to develop a comprehensive consensus agreement.

This essay compares Law of the Sea (LOS) negotiations with the climate change negotiations. Lauded for its innovative negotiation approach and leadership, UNCLOS III may offer some important insights that climate change negotiators may find relevant to the challenges they face.

To compare the ongoing climate change negotiations with the LOS talks, this essay employs the *Progress Triangle* framework. The commentary examines Climate Change and LOS negotiations in the Progress Triangle areas of substance, relationship, and procedure. The conclusion of the essay fea-


1 Jacobs (2013).
2 Raiffa (1982:276).

tures key findings that emerge from the comparison and possible lessons learned.

## A. Introduction

In December 2012, near the close of the two-week Conference of the Parties (COP) climate change negotiations in Doha, Qatar, the co-chairs of the ADP – the Ad hoc Working Group on the Durban Platform for Enhanced Action – presented a draft decision on what the ADP had accomplished and what lay ahead. The Durban Platform had emerged a year earlier at the end of the COP17 climate change negotiations in South Africa as an important compromise among all parties to continue the Kyoto Protocol and to establish a 2015 deadline for a comprehensive climate change agreement.

The Durban Platform co-chairs introduced their draft decision by stating –[3]

> *Recalling* decision 1/CP.17, which *recognized* that climate change presents an urgent and potentially irreversible threat to human societies and the planet and thus requires to be urgently addressed by all Parties, and acknowledged that the global nature of climate change calls for the widest possible cooperation by all countries and their participation in effective and appropriate international response, with a view to accelerating the reduction of global greenhouse gas emissions, *noted with grave concern* the significant gap between the aggregate effect of Parties' mitigation pledges in terms of global annual emissions reductions of greenhouse gases by 2020 and aggregate emission pathways consistent with having a likely chance of holding the increase in global average temperature below 2°C or 1.5°C above pre-industrial levels, and *recognized* that fulfilling the ultimate objective of the Convention will require strengthening of the multilateral, rules-based regime under the Convention …

The draft decision document subsequently highlighted the two workstreams designated to address a wide range of issues related to both mitigation and adaptation, accounting for matters that had been on the agendas of the Ad hoc Working Group on Long-term Cooperative Action and the Ad hoc Working Group on the Kyoto Protocol.

The Durban Platform for Enhanced Action added to the complexity of the climate change negotiations at an international convention that was already the most complex and controversial international negotiation the world community had yet experienced – the United Nations Framework Conven-

3 UNFCCC (2012).

tion on Climate Change (UNFCCC). While other international negotiations – on trade, natural resources, and human rights issues – feature degrees of complexity and controversy, none seem as involved and as challenging as the climate change talks.

Are any other international negotiations comparable to the climate change meetings in terms of their complexity and/or controversy? This essay offers the Third United Nations Conference Law of the Sea (UNCLOS III) as an international negotiation for comparison. The Law of the Sea (LOS) negotiations, convened by the United Nations (UN) General Assembly in 1973, generated a draft treaty by 1981 and was ready for signature in 1982. This treaty, hailed by former US secretary of state Henry Kissinger as "one of the most important international negotiations to have ever taken place", emerged from consensus agreements among over 160 parties.[4] Lauded for its innovative negotiation approach and leadership, UNCLOS III may offer some important insights that climate change negotiators may find relevant to the challenges they face.

To compare the ongoing climate change negotiations with the LOS talks, this essay employs the *Progress Triangle* framework. After presenting the Progress Triangle, the commentary examines the two negotiations in terms of its three dimensions.

## *B. Comparing Climate Change and Law of the Sea Negotiations*

### *I. The Climate Change Negotiations (UNFCCC)*

The nations of the world have been negotiating climate change for the past two decades. Negotiations began formally during June 1992 as part of the UN Conference on the Environment and Development (UNCED), popularly known as the Rio Summit. Over 150 government delegations participated, and they produced the UNFCCC. While some critics have claimed that the Convention was 'watered down' to gain the support of the United States (US),[5] the Convention, through its 26 articles, established an international organisation and template for negotiating specific agreements to combat climate change. In October 1992 the US Senate (the government body for treaty ratification) "voted unanimously to ratify the treaty and commit the

4 Raiffa (1982:276f.).
5 Flannery (2005).

U.S. to join the global effort to "stabilize greenhouse gas concentrations in the atmosphere at a level that would prevent dangerous anthropogenic interference with the climate system".[6]

In the 20 years since its inception, the UNFCCC has added parties (new independent nations) and issues. For example, during the May 2012 climate change negotiation meetings in Bonn, the UN Climate Change Secretariat sponsored a "gender picnic". UNFCCC executive secretary Christiana Figueres invited delegates (negotiators, observers, media) to the event to recognise the critical roles and contributions of women in the development and implementation of climate policy. As the Women for Climate Justice website reported:[7]

> Apart from being a very lively networking event, it [the picnic] also was an excellent opportunity to communicate the various ideas on how to improve gender recognition in the negotiations, in addition to the inclusion of gender and women references in the [negotiation] text. Christiana [Figueres, secretary] asked for ideas on how references can be trickled down to national and local levels and called for the support of non-government organizations (NGOs) to put pressure on governments to integrate gender.

While gender issues were visible at the Bonn session (as they were at the December 2011 17$^{th}$ Conference of the Parties meetings or COP17 in Durban, South Africa), their salience symbolises how the climate change negotiations have changed in the almost two decades of UNFCCC work. Since the first COP and its Berlin Mandate, climate change negotiations have become increasingly complex and controversial. The volume and variety of issues have proliferated, and the number of parties, observer organisations, and media representatives has increased significantly as well.

COP3 in Kyoto, for example, expanded the Berlin Mandate's call for developed countries' commitments for mitigation to control carbon emissions by constructing the Kyoto Protocol. The Protocol broadened commitments to include legally binding commitments and mitigation practices, such as clean development mechanisms and emissions trading. The next three COPs refined the Kyoto Protocol by focusing on finance issues and what constituted appropriate mitigation and credit strategies (e.g. carbon sinks from forests and agricultural lands). Parties at the COP7 in Marrakech, Morocco, agreed to establish the Adaptation Fund to help developing countries

6 Moomaw & Hamel (2013).

7 See http://www.gendercc.net/policy/conferences/road-to-doha.html?L=2, last accessed 16 April 2013.

cope with the impacts of climate change. This marked a significant expansion of the negotiation agenda since adaptation discussions would include technology transfer, capacity building, finance, and governance issues. At this point the US delegation, although representing the foremost producer of greenhouse gases, was participating only as an observer. President George W. Bush had rejected the Kyoto Protocol and the United States of America was no longer a negotiating party.

Over the past decade the complexity of climate change negotiations has increased through the addition of topics such as REDD (reduction in emissions due to deforestation and degradation), loss and damage, and LULUCF (land use, land use change, and forestry). Procedural reforms have been added as well, such as the Ad hoc Working Group on Long-term Cooperative Action (AWG-LCA, created at COP13 in Bali, Indonesia, and terminated at COP18 in Doha, Qatar) and the Ad hoc Working Group on the Durban Platform for Enhanced Action (ADP, initiated at COP17 in Durban, South Africa).

Consequently, at the May 2012 Bonn, Germany, intersession meetings between COP17 and COP18 in Doha, Qatar, UNFCCC, the parties interacted in five major groups. Along with the AWG-LCA and ADP, the Bonn gathering included meetings of the AWG-KP (Ad hoc Working Group on the Kyoto Protocol), the SBI (Subsidiary Body on Implementation) and the SBSTA (Subsidiary Body on Scientific and Technical Advice). COP18 in Doha added two more negotiating bodies: the Conference of the Parties (COP, including all negotiating countries) and the Meeting of the Parties (CMP, created after the Kyoto Protocol took effect in 2005 and including all parties that had ratified the Kyoto Protocol).

As the substantive and procedural complexity of the climate change negotiations increased, so, too, did controversy. At the 2001 Marrakech, Morocco COP, for example, many parties were very upset with the US presidential administration's (George H.W. Bush) dismissal of the climate negotiations and the influence that the actions of the US could have on other major carbon emitters. During the 2009 Copenhagen, Denmark COP, talks about extending the Kyoto Protocol (via a second commitment period) experienced gridlock and the Copenhagen Accord that US president Barack Obama and other select leaders developed was criticised widely by many parties for emerging outside of the UN framework. The subsequent three-day meeting in Bonn (April 2010) focused solely on procedure in an effort to reaffirm the Framework process and restore confidence in it.[8]

The UNFCCC, when established in 1992, emphasised the principle that the nations of the world should negotiate climate change issues and seek agreement "on the basis of equity and in accordance with their common but differentiated responsibilities and respective capabilities".[9] The Convention established a procedure or process – "an agreement to negotiate" – for negotiating policies to manage constructively and comprehensively the causes and impacts of climate change. The UNFCCC provides "the overarching international law framework for intergovernmental efforts to address climate change".[10] Although more than 190 nations endorse the Framework and participate in its negotiations, agreements on specific climate change policies remain elusive.

As COP19 in Warsaw, Poland, approaches, many issues not imagined when the UNFCCC was established in 1992 or when the first COP met in Berlin in 1995 are visible in the climate change negotiations agenda. Matters of financial accountability and transparency, indigenous peoples' rights, loss and damage, gender, governance, technology transfer, and other specific concerns are now being negotiated, as are the issues related directly to mitigation goals and adaptation mechanisms. The substantive agenda can seem overwhelming to delegates and observers alike, and the procedural details can appear confounding.

## *II. The Law of the Sea Negotiations (UNCLOS III)*

The Third UN Law of the Sea Conference (UNCLOS III) is likely to endure as one of the most significant international diplomatic events in modern history. The seeds of UNCLOS III were planted in 1967, when Dr Arvid Pardo, Maltese representative to the UN, addressed the UN General Assembly about ocean policy. He proposed that the resources of the international seabed and ocean floor should be considered "the common heritage of mankind", in effect, calling for a new international order of the sea. Pardo contended further that a constitution or charter was required to guarantee that ocean space

8 See IISD (2010).
9 Article 3(1) UNFCCC; Wirth (2009:xxv); Mace (2010:221).
10 Mace (2010:221).

was treated as an ecological whole and used exclusively for peaceful purposes.[11]

In response to Pardo's vision and concerns for a new economic order, the UN General Assembly called for a new oceans conference (two took place in the 1950s). With much media attention worldwide and high expectations, the UNCLOS III convened in Caracas, Venezuela, in 1974. Almost nine years later, after additional conference sessions in Geneva and New York, a new, comprehensive LOS emerged. On 10 December 1982 in Montego Bay, Jamaica, the United Nations Convention on the Law of the Sea was opened for signature, with 119 nations signing initially. Many additional nations (not including the US) have since signed the treaty convention.

UNCLOS III addressed a complex array of maritime problems. Via three principal committees, representatives of the world's nations considered over 25 substantive agenda items. Major issues included the international regime for the non-national sea bed and ocean floor, the territorial sea, the contiguous zone, international navigation of straits, the continental shelf, the exclusive economic zone, coastal state preferential rights, land-locked nations' rights, preservation of the marine environment, scientific research, technology transfer, archipelagos, and dispute settlement.[12] Among the most volatile conference issues at Caracas and beyond were the related matters of the nature of the international authority for control of the deep seabed and how the resources of the deep seabed should be exploited. In fact, these international sea issues lay at the heart of why the US did not sign the LOS treaty, causing some international leaders to charge that the US bargained in bad faith.[13]

Despite the failure of the US and a few other nations to sign the UNCLOS III Convention, the negotiations stand as one of the most significant international negotiations;[14] an excellent example of international collaborative work and consensus. In his opening remarks to the 17th LOS Institute Conference, Willy Ostreng of Norway's Fridtjof Nansen Institute observed that the completion of the Third UN Conference on the Law of the Sea (UNCLOS III) was "the culmination of the longest, largest, and most ambitious collective effort ever undertaken to promote peace and prevent conflict by agreeing

11 Sebenius (1984); Koh (1986).
12 UN (1974:7–10).
13 Pardo (1983).
14 Raiffa (1982).

on the precise distribution and effective limitation of power among all nations of the world".[15]

British analyst R.P. Barston called UNCLOS III "the most ambitious and complex of contemporary attempts at multilateral diplomacy".[16] According to Cameroon's UN ambassador Paul Bamela Engo, the Conference was a "stimulating, ambitious exercise" that represented "the widening of the scope of dialogue on matters of global interest, providing opportunity for effective participation by all nations, large and small".[17] International legal scholar Elisabeth Mann Borgese commented that the emerging new international economic order would rely heavily on the new LOS and its foundation on the principles of ownership, participation, equity, and peace.[18]

The Convention resulted from a treaty-making process that involved innovative procedural rules, guidelines that emphasised open communication, and a commitment to fairness. Adopted in 1973, UNCLOS III rules of procedure featured the use of single negotiation texts (SNTs), package deals, and decision by consensus. The Conference structure organised representatives in formal and informal committees to consider specific issues.[19]

Consensus decision-making and the package deal approach emphasised interest-based, mutual gains negotiation.[20] Adoption of treaty articles and provisions mandated discussion to achieve general agreement without resorting to a vote. By a procedural "gentleman's (sic) agreement", voting only occurred as a last resort. The decision rules required the conferees to exhaust all efforts to achieve consensus before voting on any substantive matters. Prior to a particular vote, a cooling off period allowed negotiators to continue to work toward consensus, either through direct or backchannel means. By delaying voting as long as possible, divergent aspirations could be reconciled, obviating any need for a vote.[21] During the almost decade-long negotiations, no substantive issues required a majority vote. As legal scholar Milner Ball (1982) has noted, UNCLOS III illustrates "a productive labo-

15 Koers & Oxman (1983:xv).
16 Barston (1980:154).
17 Engo (1985:21).
18 Borgese (1986:131).
19 Zuleta (1983).
20 Fisher et al. (1991); Susskind & Field (1996); Raiffa (1982).
21 Zuleta (1983:xxi).

ratory, working experiments in the form of negotiation, multinational decisions, and transcultural discourse".[22]

UNCLOS III involved a number of unique, changing, and sometimes overlapping coalitions, referred to as "interest groups" in some of the LOS literature. These included the maritime group, the coastal group, the landlocked and geographically disadvantaged group, the Group of 77, regional groups such as the Western European and Others Group, the Group of Five, the environmental group, the territorialists, a boundary limitations group, and so on.[23] These groups were referred to in the statements of delegates published in the Conference minutes, and some of the groups (e.g. the Group of Landlocked and Geographically Disadvantaged States and the Group of 77) produced official position papers and other documents.

The UNCLOS III negotiation process included public and private debate.[24] As a process of consensus and conciliation, its success depended "upon the power of persuasion and the willingness to be persuaded".[25] UNCLOS III generated a comprehensive treaty in part through persuasive argumentation. As Ball has noted, "above all, [consensus] depended upon the tentative mutual trust among those who actually believed that arguments count".[26] "The Conference has proven the possibility for accommodating mutual trust, good faith and proleptic belief in the efficacy of argument … [it] has been a means for impressive multicultural discourse".[27]

While the climate science and policy library expands as the UNFCCC negotiations endure, detailed commentaries on the climate negotiations equivalent to the LOS analyses remain to be written. Still, by reviewing UNFCCC and non-government documents, observing UN climate negotiation sessions, talking with negotiators, and examining media accounts, one can compare climate change negotiations with those of the LOS. This comparison can draw on the Progress Triangle framework for areas of substance, procedure, and relationship as a means for doing so.[28]

22 Ball (1982:463).
23 Beesley (1983:187–188).
24 Ogley (1984).
25 Ball (1982:471).
26 Ball (1985:60).
27 Ball (1982:472).
28 Walker & Daniels (2005); Daniels & Walker (2001).

## *III. Resolving Conflicts and Negotiating Decisions – The Progress Triangle*

Negotiation can be viewed as a process that generates tangible improvements in a conflict or decision-making situation, improvements that can be implemented and evaluated.[29] Improvements represent progress and, in the case of international negotiations, constitute components of a comprehensive agreement. Therefore, conflict resolution, negotiation, and decision-making on matters of climate or oceans can be thought of as 'making progress'. As part of improving the situation, progress can include such ideas as reaching consensus, developing mutual gains, learning, resolving a dispute, achieving agreement, and laying a foundation for future negotiations. Progress is a way of thinking about conflict, negotiation, and decision situations that recognises that conflicts are inevitable and ongoing, and that the competent management of those conflicts comes from continual improvements in areas of substance, procedure, and relationship.

Constructive conflict management, then, involves making progress on these three fundamental dimensions of a conflict situation: the substantive, procedural, and relationship dimensions. These dimensions may be viewed as points of a conflict management Progress Triangle, as presented in Figure 1.

***Figure 1. The Progress Triangle***

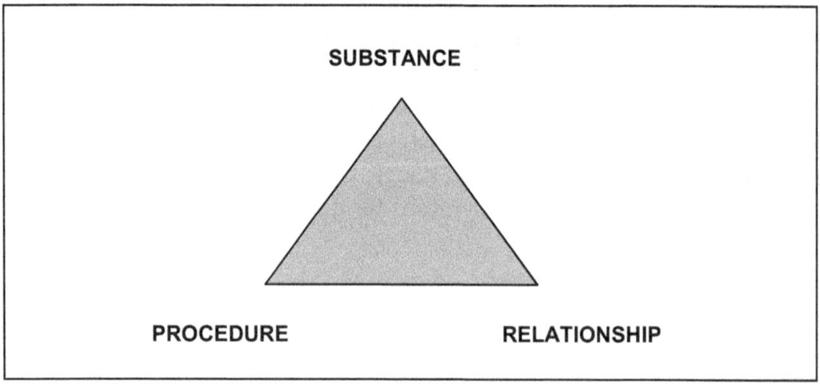


29 Daniels & Walker (2001).

Portraying conflict and decision situations as a triangle of three interrelated dimensions – substance, procedure, and relationship – illustrates a number of things about making progress in those situations. First, any conflict or decision situation includes substantive, procedural, and relationship dimensions. Second, the three dimensions overlap and affect one another. A procedural element such as jurisdiction, for example, may become a substantive issue. Third, one can address the conflict or decision situation initially through any of the three dimensions. An education reform policy conflict situation, for example, might feature substantive concerns related to teacher certification and student test scores. A natural resource conflict situation such as salmon recovery might emphasise procedural and relationship factors related to the sovereign status of native peoples. Fourth, progress on one dimension is likely to contribute to progress on the other dimensions.[30]

Comparing the UN climate change negotiations with the LOS negotiations begins by noting a fundamental structural difference. The climate change negotiations occur within a *framework* that parties have endorsed as an international agreement. That framework, the UNFCCC, was developed in negotiations prior to the Rio Summit in 1992 and signed at that conference. The framework established a secretariat to lead the negotiations and a set of articles that guides the work of the parties. The UN General Assembly created the LOS negotiations by establishing a *conference* to work though ocean and maritime issues. That conference negotiated procedural matters before substantive issues, procedures that guided the eight years of discussions.

### *IV. Comparing Substance Factors*

Negotiations are about substance: the visible issues of a conflict, dispute, or decision situation. Both the climate change and LOS negotiations have addressed a myriad of substantive matters. Table 1 presents a number of substantive areas related to issues, information and texts, for comparing the two international negotiations.

30 Daniels & Walker (2001); Walker & Daniels (2005).

### 1. Issues

Substantive issues are the tangible or 'concrete' topics for negotiation. Climate issues and ocean issues are obviously different (although with some overlap, such as ocean acidification[31]). Climate negotiators work on a wide range of issues related to mitigation and adaptation[32], while LOS parties negotiated issues related to resource development, navigation, and ocean health.

Just as the climate change negotiations involve more issues, they represent more complexity than the LOS issues exhibit. Both negotiations have been complex, but the climate issues have changed as new scientific and technical information has emerged, as the number of parties has increased, and as developing countries have become more vocal and better organised. For example, during the mid-1990s the UNFCCC negotiators were concerned primarily with greenhouse gas emissions (GHGs). Both the Berlin Mandate (COP1) and the Kyoto Protocol (COP3) focused on establishing emission reduction targets. The Kyoto Protocol specifically sets legally binding commitments for Annex I (developed) countries. As the UNFCCC website states:[33]

> During the first commitment period, 37 industrialized countries and the European Community committed to reduce GHG emissions to an average of five percent against 1990 levels. During the second commitment period, Parties committed to reduce GHG emissions by at least 18 percent below 1990 levels in the eight-year period from 2013 to 2020; however, the composition of Parties in the second commitment period is different from the first.

While concerns about GHGs endure and mitigation issues persist, negotiators at more recent COPs (Conferences of the Parties) have confronted an array of issues that were not on the early COP agendas. For example, the decisions coming out of COP16 in Cancun, Mexico, (known as the Cancun Agreements), advanced the following objectives –[34]

- establish clear objectives for reducing human-generated greenhouse gas emissions over time to keep the global average temperature rise below two degrees

31 See Gonzalez (2010).
32 See IISD (2010).
33 See http://unfccc.int/kyoto_protocol/items/2830.php, last accessed 20 May 2013.
34 See http://cancun.unfccc.int/cancun-agreements/main-objectives-of-the-agreements/#c33, last accessed 20 May 2013.

- encourage the participation of all countries in reducing these emissions, in accordance with each country's different responsibilities and capabilities to do so
- ensure the international transparency of the actions which are taken by countries and ensure that global progress towards the long-term goal is reviewed in a timely way
- mobilize the development and transfer of clean technology to boost efforts to address climate change, getting it to the right place at the right time and for the best effect
- mobilize and provide scaled-up funds in the short and long term to enable developing countries to take greater and effective action
- assist the particularly vulnerable people in the world to adapt to the inevitable impacts of climate change
- protect the world's forests, which are a major repository of carbon
- build up global capacity, especially in developing countries, to meet the overall challenge
- establish effective institutions and systems which will ensure these objectives are implemented successfully.

These objectives illustrate the complexity and fluidity of climate change issues. In contrast, the law of the sea negotiators worked on a set of issues that remained stable throughout the eight years of meetings. The primary issues of the three LOS committees – international regime for the deep ocean floor, the territorial sea, the contiguous zone, the exclusive economic zone, international military and commercial navigation of straits, the continental shelf, coastal state preferential rights, rights of land-locked nations, the preservation of the marine environment, scientific research, and technology transfer – were established at the outset of the LOS negotiations and remained stable and constant.[35] Most of these issues were settled within the first four years of the LOS meetings and new issues were not added.

Issue salience also differs between the two international negotiations. Although a limited number of nations produce significant greenhouse gases, all countries are affected by climate change. Consequently, every delegation has a compelling interest to track issues related to mitigation and adaptation even if some delegations may not participate actively in negotiating all issues.

Such was not the case at the law of the sea negotiations. Some issues – such as military navigation rights, territorial sea designation, and the exclusive economic zone – were not important to land-locked countries. These countries though, along with coastal states, were concerned with the devel-

35 UN (1974:7–10).

opment of resources from the deep ocean floor and the health of the marine environment. Issue jurisdiction is both a substantive and procedural issue. UNCLOS III distributed issues among three substantive committees. As reported on the UN Law of the Sea website:[36]

> The Conference allocated to the First Committee the topic of the international regime of the seabed and ocean floor beyond national jurisdiction, and to the Second Committee the topics of the territorial sea, the contiguous zone, the continental shelf, the exclusive economic zone, the high seas, land-locked countries, shelf-locked States and States with narrow shelves or short coastlines and the transmission from the high seas, while the topic of the preservation of the marine environment was allocated to the Third Committee.

While the work of the First Committee took the longest, none of the committees added issues or negotiated on matters assigned to another group. By comparison, significant climate change issues have been discussed by more than one negotiation body. For example, after the parties at COP13 in Bali, Indonesia, created the Ad hoc Working Group on Long-Term Cooperative Action (AWG-LCA), this new group took on issues related to mitigation of greenhouse gas emissions. So, too, did the Ad hoc Working Group on the Kyoto Protocol (AWG-KP). Delegates, both informally and in plenary sessions, voiced concerns about how the work of these two negotiation bodies would be reconciled and what body had jurisdiction or greater influence. When the Ad hoc Working Group on the Durban Platform (ADP) was created at COP17, parties wondered how its work would be related to the efforts of the AWG-LCA and AWG-KP (both of which ended at COP18 in Doha, Qatar).

### 2. *Information and Texts*

Technical information and the use of negotiation texts have been significant for both the climate change and LOS meetings. In both negotiations, parties have accounted for the best scientific, technical, and financial information available. While the recent climate change COPs have featured side events on the latest scientific information regarding climate change impacts and climate models, there is little debate in these areas. Most of the parties en-

36 See http://untreaty.un.org/cod/diplomaticconferences/lawofthesea-1982/lawofthesea-1982.html, last accessed 17 April 2013.

dorse the scientific consensus that exists in support of the work of the Intergovernmental Panel on Climate Change (IPCC).

While the IPCC provides fundamental climate science knowledge, negotiators look to universities, intergovernmental organisations (IGOs), and civil society organisations (CSOs or NGOs) for information in such areas as finance, capacity building, technology transfer, sustainable forestry, and human dimensions. Similarly, the law of the sea negotiators turned to non-government organisations and universities to better understand the range of options, particularly in the area of deep seabed development. For example, during the law of the sea negotiations, parties struggled to determine what policy was fair and appropriate for the development of the resources of the deep ocean floor. In 1976, researchers at the Massachusetts Institute of Technology (MIT) began to develop a model about the technology of ocean mining. They presented their report in 1978. Not long thereafter, the soon to be UNCLOS III president, Singapore diplomat Tommy Koh, convened a panel of financial experts to address monetary issues of seabed development. The work of these technical groups was critical to UNCLOS III progress.[37]

Both the climate change and LOS negotiations have made extensive use of *negotiation texts*. UNCLOS III began without preparatory documents;[38] the chairs of the three committees authored single negotiation texts (SNTs). As Hodgson and Smith reported in 1976, "from the second session of the third UN Law of the Sea Conference came single-text documents from the chairmen of each of the three main committees plus an 'informative paper' on the settlement of disputes".[39] The single negotiating text was an innovative tool in the international negotiation arena; it provided LOS Conference delegates with a common starting point for discussion.[40]

Similarly, negotiation texts play an essential role in the climate negotiations. The chairs of the subsidiary bodies, the ad hoc working groups, and the contact groups prepare texts on the issues within their domain to focus and guide the negotiation. The facilitators of informal consultative groups will prepare *facilitator notes* for the same purpose. While the parties may disagree over the language in these texts, the texts themselves provide the negotiators with a common reference point.

---

37 Antrim & Sebenius (1994); Raiffa (1982).
38 Koh (2009).
39 Hodgson & Smith (1976:225).
40 Buzan (1981).

***Table 1. Substance Factors***

| Factor | Climate Change (UNFCCC) | Law of the Sea (UNCLOS III) |
|---|---|---|
| Issue Areas | Scientific, technical, financial, legal, human dimensions | Scientific, technical, financial, security, legal |
| Issue complexity | Multi-faceted and fluid | Multi-faceted and finite |
| Issue salience or relevance | High for all parties | High for some parties |
| Issue jurisdiction | Different bodies considering the same issue | Clear issue assignment and differentiation |
| Technical information (e.g. role of science) | Significant: Intergovernmental Panel on Climate Change | Varied: Massachusetts Institute of Technology model |
| Texts | Substantial: prepared by chairs and facilitators of working groups, subsidiary bodies, contact groups, and informals | Substantial: use of the single *negotiation text* method prepared by committee chairs |

## *V. Comparing Relationship Factors*

As dispute resolution scholars have noted, in conflict, negotiation, and decision situations relationships matter.[41] The relationships of parties are influenced by a variety of factors, such as history, culture, experience, skill, personality, constituent expectations and home government instructions.[42] All of these relationship factors can potentially affect any complex international negotiation. The salient relationship factors, though, that warrant climate change and LOS comparisons involve the parties: who they are and how they work together.

### *1. Coalitions*

Both the climate change negotiations and LOS meetings have involved most of the world's national actors: currently 195 countries are involved in the UNFCCC and 160 nations participated in UNCLOS III. While each delegation acts on its national interest, the parties have organised into coalitions. Important negotiations occur within the coalitions – and the coalitions wield significant influence.

41 E.g. Wilmot & Hocker (2010); Lewicki et al. (2011).
42 Lewicki et al. (2011); Wilmot & Hocker (2010); Brett (2007).

UNCLOS III involved a number of unique, changing, and sometimes overlapping coalitions, referred to as *interest groups* in some of the law of the sea literature. These included the maritime group, the coastal group, the landlocked and geographically disadvantaged group, the Group of 77, regional groups such as the Western European and Others Group, the Group of Five, the environmental group, the territorialists, a boundary limitations group, and so on.[43] These groups were referred to in the statements of delegates published in the Conference minutes, and some of the groups (e.g. the Group of Landlocked and Geographically Disadvantaged States and the Group of 77) produced official position papers and other documents.

Koh has referred to these coalitions as "interest groups" and noted that they emerged and formed at UNCLOS III.[44] Although the Group of 77 had formed years earlier as a loosely coupled organisation of non-aligned nations, it figured prominently in the law of the sea negotiations.[45] Similarly, the G77 and China coalition at the climate change meetings has served as an essential organisation and voice. And like UNCLOS III, the UNFCCC involves numerous coalitions, such as the African Group, the Association of Small Island States (AOSIS), the Least Developed Countries, the Umbrella Group, the European Union, and more. The coalitions have been very influential at both international meetings. They meet regularly at the climate change talks and generally speak as one voice in the plenary sessions of the major negotiation bodies. At the LOS negotiations the coalitions were visible within and across the major committees. At both conferences, national delegations have participated in more than one coalition, such as the land-locked states and G77 at the UNCLOS III, and the Least Developed Countries and the African Group at the UNFCCC. But the coalitions at the climate change talks have been more emergent and fluid. For example, a new coalition of developing country mountain states has emerged at recent COPs, and all of these countries are part of the G77 and China coalition as well. This illustrates that, at the climate change negotiations, coalitions also operate *within* coalitions, with internal coalition negotiation becoming layered. Not surprisingly, reaching consensus on an issue within a coalition, particularly the G77 and China with its 130 plus members, can be as difficult as reaching agreement among all the parties.

43 Beesley (1983:187–188).
44 Koh (2009).
45 Raiffa (1982).

## 2. Media

In his seminal book, *Conflict Regulation*, Paul Wehr distinguishes between primary and secondary parties.[46] Media organisations constitute important secondary parties and have been prominent at the climate change negotiations throughout the two decades of talks. Since COP13 in Bali, Indonesia, the number of media credentials approved for the COPs has averaged over 1,000, with about 3,500 media badges given out at COP15 in Copenhagen.[47] The media report the status and progress of the negotiations daily to regions and countries globally. Media personnel write blogs, conduct interviews, and talk with parties and civil society delegates.

The media presence at the LOS negotiations was, by comparison, much more limited. There was substantial media coverage when negotiators met in Caracas, Venezuela, in 1974, but media interest waned when journalists realised that the negotiations involved a wide range of issues, significant detail, and a slow, incremental, consensus-based approach. Two other media distinctions warrant comment. First, while both the climate change and LOS negotiations have restricted public access and observation, the climate change negotiations are taking place in a media and technology environment unknown in the 1970s. Satellite television, 24-hour news channels and internet saturation combine with issue salience[48] to produce continual reports about the climate change talks. Second, civil society interest in climate change issues is compelling, and while some non-government organisations played an important lobby and consultation role in UNCLOS III, broad general interest in ocean, coastal, and maritime issues did not exist.

## 3. Civil Society

While civil society organisations (CSOs, with some referred to as NGOs or non-government organisations) were active in the LOS negotiations, they have been very prominent in the climate change talks. At UNCLOS III, the non-government actors, though not numerous, made important contributions. Koh has noted that NGOs at the law of the sea meetings (1) brought independent experts to meet delegates; (2) helped developing countries to

46 Wehr (1979).
47 See http://unfccc.int/meetings/cop_15/items/5214.php, last accessed 20 May 2013.
48 Sebenius (1993).

close the knowledge gap; (3) afforded delegates opportunities to meet outside the conference; and (4) influenced the domestic positions of countries and their delegations.[49]

The UNFCCC meetings involved a significant number of non-state actors, both NGOs and intergovernmental organisations (IGOs). The UNFCCC website provides the most recent data:[50]

> Over 1598 NGOs and 99 IGOs are admitted as observers. The NGOs represent a broad spectrum of interests, and embrace representatives from business and industry, environmental groups, farming and agriculture, indigenous populations, local governments and municipal authorities, research and academic institutes, labour unions, women and gender and youth groups. Constituency groupings have emerged from the above groups to facilitate interaction.

Many of the NGOs affiliate with the constituency groups, something that did not exist in the law of the sea talks. The constituency groups function, in effect, as coalitions of NGOs. For example, the business and industry NGOs participate in BINGO, the environmental NGOs comprise ENGO, and the youth NGOs belong to YOUNGO. Each constituency receives invitations to send representatives to topic-specific meetings (e.g. the Adaptation Fund Board) and to speak at select plenary sessions of the negotiation bodies. By comparison, the civil society organisations have made a much larger and deeper footprint at the climate change talks than they did at UNCLOS III.

## *4. Leadership*

Competent and credible leadership is an essential part of progressive and effective work in the public policy arena,[51] whether domestic or international. Leadership approaches, though, can vary culturally. Consequently, leaders in complex international negotiations need to reflect the cultural community of diplomacy,[52] as well as their own local and national cultures. At both UNCLOS III and the UNFCCC meetings, many leaders have displayed skill and diplomacy. But the leadership positions of the law of the

49 Koh (2009).

50 See http://unfccc.int/parties_and_observers/items/2704.php, last accessed 17 April 2013.

51 Walker & Daniels (2012).

52 Fisher (1989) and (1990).

sea negotiations were more constant and stable than those of the climate change talks. Tommy Koh, president of UNCLOS III from 1980 to 1982, has noted that an unprecedented degree of authority was vested in the four conference leaders – i.e. the UNCLOS III president, and the chairpersons of the three primary committees.[53] LOS scholars Lance Antrim and James Sebenius have observed that responsibility for issuing new versions of the negotiating text added a significant new power to the limited authority of the presiding officer. The authority to revise the draft text was, in effect, the power to define the issues to be addressed as the negotiation process.[54]

Although the executive secretary is the chief administrative officer of the UNFCCC, its president changes annually, depending on the nation that hosts the Conference of the Parties. The COP president is typically a cabinet minister from the host country, such as the minister for the Environment or the minister of Climate and Energy. The UNFCCC works closely with the COP presidency and, in doing so, defers some decisions to the COP host. For example, at COP16 in Cancun, Mexico, the COP presidency made the decision to pair ministers from developing and developed countries to facilitate consultations on specific and challenging issues (e.g. finance). This technique had not been done before; some delegations liked it and others were critical, claiming that the Mexican presidency was asserting too much control and was not transparent.

In the UNFCCC process the chairs of the major negotiation groups – the two subsidiary bodies and the working groups – change regularly. All serve at the pleasure of the parties themselves. Some chairs are very skilled: they provide well-received negotiation texts and facilitate meetings competently. Others struggle to maintain the perception of impartiality and fairness. For example, one working group chair at COP18 in Doha was criticised by developed country delegates for marginalising some of the parties and favouring others.

53 Koh (2009).

54 Antrim & Sebenius (1992:101).

***Table 2. Relationship Factors***

| Factor | Climate Change (UNFCCC) | Law of the Sea (UNCLOS III) |
|---|---|---|
| Role of Media | Significant and increasing as the talks have continued over time; Copenhagen (COP15 in 2009) being the high point | Significant at the beginning; decreasing coverage over time |
| Coalitions | Fluid, emerging, overlapping, multiple memberships; essential and influential | Stable and distinct; overlapping, multiple memberships; essential and influential |
| Civil Society Organisations (CSOs/NGOs) | Numerous and increasing; active participation, integration with some parties, key advisory work, lobbying | Limited numbers; some key advisory work |
| Leadership | Critical and uneven | Critical and consistent |

## *VI. Comparing Procedure Factors*

International negotiations are about substance; parties come together to seek agreements on complex and challenging issues. The ability to generate agreement relies significantly on procedures. Not surprisingly, procedural matters have loomed large at both the LOS and climate change meetings and provide some interesting areas for comparison, as shown in Table 3.

### *1. Consensus Decision-making and Procedural Rules*

Prior to considering the substantive issues, the UNCLOS III delegates negotiated the rules of procedure. Most importantly, the parties resolved to make decisions by consensus. "Reflecting the desire to obtain wide (ideally universal) acceptance of the results of the LOS conference, the decision-making system was designed to avoid votes on matters of substance as much as possible," Antrim and Sebenius[55] explain, noting that "committee chairmen were responsible for identifying opportunities for consensus solutions, with the authority to prepare draft texts that, in their judgment, represented a step toward consensus". UNCLOS III delegates recognised that any treaty needed the support of the major powers, but the developing countries would not accept veto authority akin to the UN Security Council, Consequently,

55 Antrim & Sebenius (1992:99-100).

the UN General Assembly adopted a Gentlemen's Agreement at the outset of the LOS negotiations in November 1973, that "the Conference should make every effort to reach agreement on substantive matters until all efforts at consensus have been exhausted." Antrim and Sebenius clarify that "consensus, in the context of the LOS Conference, implied an absence of explicit disagreement rather than total agreement on all of the issues".[56]

Koh has described the Gentlemen's Agreement in more detail, noting that it involved the following: (1) Before taking a vote, the Conference parties must decide that all efforts at reaching agreement have been exhausted; (2) Parties can consider options during a cooling off period; (3) The LOS Conference president will attempt, with assistance from the General Committee, to achieve an agreement; and (4) the parties will receive two days' notice before voting.[57] Buzan considers the LOS approach to be innovative: "a major international experiment in decision making by consensus".[58] He has emphasised that UNCLOS III is noteworthy procedurally because if formalised "active consensus" (in contrast to passive consensus); that is, consensus that emerged from active, direct discussion.

The UNFCCC process also operates according to a principle of consensus decisions, but no consensus procedure has ever been adopted formally. Article 7.2.k of the UNFCCC charter does state that the parties "agree upon and adopt, by consensus, rules of procedure and financial rules for itself and for any subsidiary bodies". Consensus decision-making is the norm, but Article 7.3 language indicates the possibility of voting:

> The Conference of the Parties shall, at its first session, adopt its own rules of procedure as well as those of the subsidiary bodies established by the Convention, which shall include decision-making procedures for matters not already covered by decision-making procedures stipulated in the Convention. Such procedures may include specified majorities required for the adoption of particular decisions.

Although a number of analysts believe that the consensus standard is either outdated or an obstacle and that other decision systems warrant review,[59] proposals for a consensus/voting combination has not gained much traction (e.g. the current Mexico and Papua New Guinea plan). Annto Vihma of the Finnish Institute of International Affairs points out that "The Cancún meet-

56 (ibid.:100).
57 Koh (2009).
58 Buzan (1981:324).
59 E.g. Schwarte & Wei (2011); Schroeder et al. (2012).

ing [COP16 in 2010] also brought to everyone's attention the inconvenient truth that decision-making in the UNFCCC exists in a legal vacuum. The Conference of the Parties has never agreed its Rules of Procedure, and has during its 17-year history operated with draft Rules of Procedure without voting rules, under a general agreement that decisions are taken by 'consensus'".[60] Similarly, a recent report from the Foundation for International Law and Development (FIELD) contends that "the COP and the CMP have not formally adopted their rules of procedure. However, at all sessions since 1995 a set of draft rules have been applied consistently – with the exception of the disputed rule 42 on voting …",[61] and some form of voting remains possible if the parties conclude that they have exhausted all means of achieving consensus.

Reflecting on the LOS negotiations, legal scholar Milner Ball has noted that "the most striking characteristic of the Conference is that is has proceeded by consensus. The text has been assembled without a vote".[62] It remains to be seen if the climate change consensus process can achieve a successful outcome similar to the law of the sea.

### 2. *Structure*

The UN Framework Convention on Climate Change, the subsequent Kyoto Protocol, and major COP decisions such as the Bali Action Plan have provided structure for the climate change negotiations. The UNFCCC charter established three negotiation bodies: the Subsidiary Body on Implementation (SBI), the Subsidiary Body on Scientific and Technical Advice (SBSTA), and the Conference of the Parties (COP). After the Kyoto Protocol came into force in 2005 (after enough country ratifications) the CMP or Meeting of the Parties (of Kyoto Protocol nations) was established. The Bali Action Plan included the formation of two working groups: the Ad hoc Working Group on the Kyoto Protocol (AWG-KP) and the Ad Hoc Working Group on Long-term Cooperative Action (AWG-LCA). These two negotiation bodies were terminated at COP18 in Doha (December 2012) as the Ad Hoc Working Group on the Durban Platform for Enhanced Action (ADP) emerged.

60 Vihma (2011:2).
61 Schwarte & Wei (2011).
62 Ball (1982:463).

The UNFCCC negotiation structure can be characterised as fluid, evolving, or unstable, depending on one's point of view. The creation of the AWG-LCA at Bali (COP13) was noteworthy; it provided a forum in which the US could participate actively (the US could only observe at the Meeting of the Parties or CMP since it did not ratify the Kyoto Protocol). The ADP, created at COP17 in Durban, has offered a forum for comprise; this new structure has provided a way for the idea of a second commitment period (or extension of the Kyoto Protocol) to endure.

According to some critics, this structure has favoured large UNFCCC delegations, typically from developed nations, over small delegations. Schroeder and colleagues make the case –[63]

> Different delegation sizes to negotiations reflect different priorities, with some countries less interested than others to push or stall a climate change agreement. It also reflects different capacities; poor countries cannot afford to send large delegations and their level of expertise usually remains significantly below that of wealthier countries. This 'capacity gap' – only partly mitigated through assistance from non-state actors (NSAs) such as the Climate Action Network – limits poor countries' negotiating power and makes their participation in each session less effective. Furthermore, many sessions take place in parallel, span a wide range of issue areas and continue into the night during the final 'push' for agreement at the end of a conference. As a result 'negotiation by exhaustion' constrains smaller delegations much more severely than larger ones.

In contrast, the structure of the LOS Conference remained relatively constant throughout its lifespan. The first session opened in Caracas, Venezuela, with three committees; those committees continued until their work was done. By 1977, most of the work of two committees had been completed, with consensus reached on the majority of the 25 LOS agenda issues. Although deep seabed issues remained, negotiators addressed these issues through the stable committee structure.

Structural modifications did occur, though, within the three-committee design. Between 1977 and 1980 the primary unresolved issues pertained to the deep seabed: access to mining areas, technology, and finances.[64] Following the 1977 LOS session, Committee I chair Paul Engo of Cameroon prepared an informal composite negotiation text (ICNT) that addressed the seabed issues. A number of developed nations objected to the text and agreed to continue to participate in the negotiations only if (1) the power of the

63 Schroeder et al. (2012:835).
64 Antrim & Sebenius (1992).

committee chairs was reduced, (2) subgroups were formed to work through the remaining issues, and (3) a new LOS president was elected and given significant authority to lead the negotiations (Antrim and Sebenius, 1992). When ambassador Tommy Koh of Singapore assumed the role of LOS president in 1980, he acted as an informal mediator and guided the LOS negotiations to a consensus agreement.

### 3. *Agendas*

The LOS Conference began without a single preparatory document.[65] Consequently, the general UNCLOS III agenda was established along with procedural rules when the parties met for the first time. Twenty-five primary issues were distributed among the three committees. The agenda items reflected, according to LOS president Tommy Koh, "the theory of inter-relatedness". The agenda items were connected in such a way that they needed to be negotiated as part of one Convention.[66] Koh has observed that the wide-ranging agenda and goal of a single comprehensive treaty or convention combined to generate a lengthy conference.[67] The agenda – both substantively and procedurally – remained relatively stable throughout UNCLOS III. Some issues were refined and "fractionated" – divided up into specific "sub- issues",[68] but no new significant agenda items were added during the middle or later stages of the negotiations.

The reasonably firm and clear LOS agenda enhanced the viability of package deals. The concept of the package deal was salient throughout the LOS discussions. It dictated that the various parts of the Convention (treaty) be considered an entity, "as a single negotiated package, where the laws of give and take presumably had struck a reasonable balance between the participated states considered as a whole".[69] Proposed package deals "worked in practice because governments are made up of people" who required persuasion "to accept ideas and principles they had long resisted. That, as much as anything, was what this conference was about".[70]

---

65 Koh (2009).
66 (ibid.).
67 (ibid.).
68 Fisher (1969).
69 Evensen (1985:27–28).
70 Sanger (1987:40).

For example, developing countries, led by the G77 coalition, and developed, industrialised nations had different interests and objectives regarding the development and use of the oceans and seas. Consequently, the respective parties worked to reconcile their interests in a "package deal" that featured a "trade off" – trading navigational freedom for concessions in the deep sea bed regime.[71]

While the general UNFCCC approach has remained intact, the overall agenda and the more particular agendas of the negotiating bodies have changed frequently throughout the two decades of climate change negotiations. The agenda has expanded, both across and within negotiation bodies. The early COPs focused on mitigation matters, while adaptation issues have become prominent in recent years.

The changing agendas reflect the strength of coalitions, particularly groups like the G77 and China and AOSIS (the Alliance of Small Island States). As voices of developing countries, these and other coalitions have advocated for policies that will help the countries that produce little greenhouse gas, but experience significant impacts from GHG emissions.

The increasingly complex agendas, though, make package deals and consensus harder to achieve. At times parties and observers alike are not sure in what negotiation body a specific issue may be addressed, or an issue appears on more than one agenda. Furthermore, the agendas themselves become the focus of debate, delaying discussion on matters of substance. For example, at the June 2011 two-week intersession meeting in Bonn, Germany, parties debated about the nature of the Subsidiary Body agendas for most of the first week. Until the agenda dispute was settled, no other significant discussions took place.[72]

Although the package deal was central to the UNCLOS III negotiations, packages and trade-offs, while possible at the climate change talks, are harder to construct and discern. To illustrate: in the aftermath of COP15 in Copenhagen, a climate change meeting with both high expectations and harsh criticism, the parties met in Bonn, Germany, in April 2010 to discuss next steps and procedural reforms. During a briefing meeting with NGOs during the Bonn session, UNFCCC executive secretary Yves DeBoer fielded questions about the possibility of breaking the apparent gridlock over a

71 Kikugawa (1999); Friedheim (1993).

72 See the archival reports of the *Earth Negotiations Bulletin* of the International Institute for Sustainable Development (IISD), available at http://www.iisd.ca/vol12/, last accessed 20 May 2013.

number of major issues, such as a second commitment period, finance, and extending GHG decisions to all nations (not just Annex I). Secretary DeBoer was asked if the Parties might combine issues (package) or divide out issues or specific packages (fractionate) so that agreements could be reached on specific items, such as REDD (Reduction in Emissions from Deforestation and Degradation). Secretary DeBoer replied that most of the parties, particularly the developing nations, would not accept such action. For them, he explained, a climate change agreement was "all or nothing".[73] When asked about this "all of nothing" view, a negotiator from an African nation delegation explained that many parties believed that such "fractionation" would allow developed nations to pick and choose their issues, agree on the easiest ones, and then claim to have acted on climate change in the best interest of the global community.[74]

### *4. Technology*

The LOS talks took place throughout the 1970s and into the early 1980s, before the advent of personal computers, cell phones and tablets. The parties corresponded by phone, letter, and in face-to-face interaction. During UNCLOS III sessions, negotiators in one committee or subgroup would find out about the negotiations in another committee after the fact and not in real time. LOS sessions were recorded and transcribed, but in the negotiation sessions delegates took notes with pen and paper.

The UNFCCC sessions and the negotiators make use of a variety of technologies, many of which have emerged during the 20 years of meetings and were obviously not available for the LOS talks. During informal consultations, contact group meetings and plenary sessions, delegates can be seen regularly texting on their cell phones or examining a document on their laptop computer or tablet. Monitors throughout the venue show plenary session speakers in real time, and many plenary sessions are streamed live for access on computers.

These technologies have changed the ways in which the delegates negotiate, both at UNFCCC meetings and between sessions. While participating in a UNFCCC event, negotiators can stay in contact with members of their

73 Walker (Fieldnotes 2010).
74 (ibid.).

delegation or coalition via text and access information immediately. When encountering difficult or uncertain situations junior negotiators can receive immediate instructions from senior or lead negotiators on the team.

Tablets, smart phones and laptop computers are now the norm. The UNFCCC meetings have become paperless, with draft texts, the daily programme, facilitators' notes, and other documents (e.g. the daily *Earth Negotiations Bulletin*[75] by the International Institute for Sustainable Development (IISD)) accessible to parties and observers in a timely fashion.

### 5. *Size and Site*

The law of the sea negotiations involved approximately 1,000 representatives from 164 nations.[76] The number of parties and negotiators remained relatively constant for the duration of UNCLOS III. Observer and media participation was modest, with media coverage decreasing as the LOS talks continued.

In contrast, the number of parties, observers and media has increased through the UNFCCC's two decades of work. As new countries have gained independence (e.g. in southeastern Europe), more parties have sent representatives to climate change meetings. Schroeder and colleagues report that –[77]

> attendance at the international negotiations under the United Nations Framework Convention on Climate Change (UNFCCC) has changed both in terms of the number and diversity of 'expert agents'. Overall, the number of delegates went from 757 individuals representing 170 countries at the first Conference of the Parties (COP) in 1995 to 10,591 individuals from 194 countries attending COP15 in 2009 (13,482 representatives from 937 observer organizations were able to register for COP15 but many more had been nominated). This is a 14-fold increase (1,400%) in attendees over this time period.

Correspondingly, the number of observer organisers and their representatives has increased.

The two international negotiations differ by site as well. The UNFCCC, following the dictates of the Convention, changes the site of the Conference of the Parties (COP) annually, while holding the majority of its intersession

75 Available at http://www.iisd.ca/vol12/, last accessed 20 May 2013.

76 Antrim & Sebenius (1994); Koh (2009).

77 Schroeder et al. (2012:835).

meetings in Bonn, Germany, where the UNFCCC Secretariat is housed. Furthermore, the UNFCCC meetings, particularly the COPs, have become climate multidimensional climate change expositions – with the negotiations, side events, exhibits, and off-site events (such as NGO climate forums) occurring simultaneously.

UNCLOS III was convened in New York City and held its initial meetings in Caracas, Venezuela, where most of the negotiated issues were resolved. The seabed issues, which dominated the talks in the late 1970s, were negotiated in Geneva, Switzerland and New York City. The meetings featured negotiation sessions, but nothing comparable to the climate change gatherings in terms of non-negotiation activities.

***Table 3. Procedure Factors***

| Factor | Climate Change (UNFCCC) | Law of the Sea (UNCLOS III) |
|---|---|---|
| Decision-making | Consensus | Consensus |
| Procedural rules | Negotiated as part of the creation of the UNFCCC; significant modifications throughout; no formal adoption of consensus requirement | A pre-negotiation period; negotiated at the beginning of the UNCLOS III process prior to substantive issues; reasonably consistent throughout |
| Structure | Negotiation bodies; fluid, emerging, changing; seven negotiating bodies at COP18 in Doha | Negotiation committees, constant and stable; use of some ad hoc subsidiary groups |
| Agenda | Changing, expanding | Stable, fixed |
| Technology | Critical to progress on substance and procedure | Limited to what was available during the negotiation period |
| Size and Site | Increased participation, COP locations change annually | Stable participation, multi-year sites |

## C. Conclusion

The climate change negotiations of the past two decades and the LOS negotiations of the 1970s endure as the two most complex international negotiations the global community has undertaken. When comparing the two international negotiations, one claim, not surprisingly, is conclusive: The climate change negotiations are more complex than the LOS proceedings overall, as well as in the three areas of substance, relationship, and procedure. More specifically, a number of points stand out:

- Climate change negotiations involve more issues than LOS.
- Climate change negotiations issue agendas, with issues being added over time, and are thus more fluid than LOS.
- At both the climate change and LOS negotiations, the parties rely on technical expertise. This expertise came from delegation staff, UN organisations, and observer organisations and individuals. The climate change negotiators generally trust the IPCC reports; the LOS participants relied on the MIT model.
- The UNFCCC process has involved multilateral negotiations on a wide range of issues pursuant to the Convention that was established in 1992. UNCLOS III was a treaty-making multilateral negotiation.
- Both the climate change and LOS negotiations emphasise consensus. The LOS consensus rule was formalised with a Gentlemen's Agreement that voting would occur only as last resort. The consensus standard has not been adopted formally by the UNFCCC, but consensus is the accepted norm.
- Consensus worked at UNCLOS III, coupled with trade-offs and package deals. Consensus has proved difficult at the UNFCCC meetings, with limited agreement on substantive matters since COP2 at Kyoto, Japan, where the Kyoto Protocol, which set GHG emission targets for Annex I countries, was established.
- UNCLOS III established a clear and stable structure that did not change substantially over eight years. The UNFCCC structure has expanded, adding negotiation bodies that have generated limited agreements which affect the negotiations both substantively and procedurally (e.g. the Bali Action Plan, the Bali Road Map, Cancun Agreements, and the Durban Platform).
- The UNCLOS III issue agenda was set at the beginning of the LOS negotiation process. While issues were refined, no substantive issues

were added. The UNFCCC process has been addressing an ever expanding agenda.

- Leadership was centralised and constant at UNCLOS III, with committee chairs serving multiple years. Committee chairs had significant power and leeway to draft negotiation texts. Working group, subsidiary body, and contact group chairs at UNFCCC meetings may serve more than one year, but many do not. Conflicts have arisen over the chair position, such as the almost two weeks of negotiation to select the co-chairs of the Ad hoc Working Group on the Durban Platform (ADP) at the 2012 intersession meeting in Bonn.
- The climate change negotiations are very public, with significant public interest and media coverage throughout the world. Consequently, the UNFCCC process is subject to scrutiny from both within and without the negotiations. The LOS negotiations generated limited media attention after the 1974 Caracas session. While governments followed LOS progress closely, the general public did not.

This last point was predicted some twenty years ago. Shortly after the creation of the UNFCCC in 1992, Harvard University professor James Sebenius reflected on the challenges climate change negotiators would face. Drawing on his knowledge of UNCLOS III and analysis from his 1984 book, *Negotiating the Law of the Sea*, Sebenius noted that "climate change issues are far more publicly salient" than LOS issues, which he labelled as "obscure".[78]

In his1992 essay, Sebenius has offered some ideas as lessons for climate change negotiators to consider. "Given the current and future diplomatic activities dealing with climate change," Sebenius wrote in 1993, "it becomes more important to explore the deeper implications of the intensive and precedential experience for negotiated responses to the prospect of greenhouse warming".[79]

A number of lessons from Sebenius remain relevant to understanding the current state of climate change negotiations and what UNFCCC parties might learn from the LOS experience. Sebenius advised climate change negotiators to "expect great pressure to combine issues," and "since any action

78 Sebenius (1993:197).
79 (ibid.:190).

on climate change will involve shared and parallel sacrifice, it is probably only by linking issues... that many countries will be induced to join".[80]

Professor Sebenius offered further advice on linkages. "Link issues into packages that promise that sufficient joint gain is attractive to a large number of parties", and "link with caution", for "it can be extraordinarily difficult to 'unpackage' issues once they have been combined for bargaining purposes". While encouraging linkages and packaging, Sebenius also cautioned that packages should not be "so broadly comprehensive as to risk excessive complexity and delay".[81]

Sebenius voiced other cautions as well, including the risk of encouraging blocking coalitions. Noting the emphasis on protocols when the UNFCCC was created, Sebenius reflects –[82]

> [P]rotocols have been suggested, seemingly without much explicit analysis of their implications for negotiating success, on a virtually endless number of issues .... A good way to guarantee an endless negotiating impasse would be to handle all or many of the ... protocols in a comprehensive Law of the Atmosphere package to be agreed upon by consensus .... Despite joint gains from trades across disparate issues ... a comprehensive climate-change convention might well energize and unify a large set of otherwise separate opposing interests.

Some critics may speculate that "an endless negotiation impasse" has become reality in the UNFCCC process, one that features a significant North-South divide.[83] Notwithstanding the passage of two decades, Sebenius' insights are consistent with conclusions drawn from this essay's comparative analysis.

International multilateral policy negotiations on any global issue are difficult, none more so than those on climate change. The problems of climate change will never be resolved, nor will all climate changed conflicts be settled through negotiated agreements, but the problems and the conflicts can be managed productively and the climate change situation continually improved. Management and improvement will only occur if the UNFCCC parties can move beyond procedural wrangling and self-interest and find ways to achieve shared goals. Such shared goals and the actions to achieve them need not ignore salient national interests; they can and should respect and

---

80 (ibid.:200).
81 (ibid.:200f.).
82 (ibid.:206–207).
83 Khor (2012).

incorporate them in ways discovered through collaboration and compromise. Although not as complex as the climate change negotiations, the LOS Conference experience is pertinent. If 164 nations can work together to achieve a multi-issue agreement that establishes an international LOS, hopefully the negotiations on climate change among 194 nations can garner similar success.

In a 2008 Brookings Institution report on climate change, foreign policy, and national security, Campbell and Weitz conclude that the global community –[84]

> can expect that climate change will exacerbate already existing North-South tensions, dramatically increase global migration both inside and between nations …, lead to increasingly serious public health problems, heighten interstate tensions and possibly conflict over resources, collapse agricultural markets and global fisheries, challenge the institutions of global governance, cause potentially destabilizing domestic political and social repercussions, and spur unpredictable shifts in the global balance of power.

While this long list of possible impacts may seem extreme, the effects of climate change on migration, agriculture, and public health are already apparent. Just as UNCLOS III convened to tackle a crisis of the oceans, so, too, has the UNFCCC been convened to confront the climate crisis. The law of the sea negotiators reached agreement in time to manage the problems of the oceans. Hopefully, the climate change negotiators will, too, achieve agreements in time.

## *References*

Antrim, Lance N., & James K. Sebenius, 1992, Formal Individual Mediation and the Negotiator's Dilemma: Tommy Koh at the Law of the Sea Conference, in: Bercovitch, Jacob & Jeffrey Z. Rubin (Eds.), *Mediation in International Relations*, Basingstoke, Palgrave Macmillan, 97–130.

Ball, Milner S., 1982, Law of the Sea: Expression of Solidarity, *San Diego Law Review* 19, 461–474.

Ball, Milner S., 1985, *Lying Down Together: Law, Metaphor, and Theology*, Madison, WI, University of Wisconsin Press.

Barston, Ronald P., 1980, The Law of the Sea Conference: The Search for New Regimes, in: Barston Ronald P. & Patricia Birnie (Eds.), *The Maritime Dimension*, London, Allen & Unwin, 154–168.

84 Campbell & Weitz (2008:213–214).

Beesley, Alan, 1983. The Negotiating Strategy of UNCLOS III: Developing and Developed Countries as Partners – a Pattern for Future Multilateral International Conferences? *Law and Contemporary Problems* 46, 183–194.

Borgese, Elisabeth, 1986, *The Future of the Oceans*. Montreal, Harvest House.

Brett, Jeanne M., 2007, *Negotiating Globally* (Second Edition), San Francisco, CA, Jossey-Bass.

Buzan, Barry, 1981, Negotiating by Consensus: Developments in Technique at the United Nations Conference on the Law of the Sea, *The American Journal of International Law* 75 (2), 324–348.

Campbell, Kurt M. & Weitz, Richard, 2008, Conclusion: The Clear Implications of Global Climate Change, in: Campbell Kurt M. (Ed.), *Climatic Cataclysm: The Foreign Policy and National Security Implications of Climate Change*. Washington D.C., Brookings Institution Press, 213–223.

Caron, David D., 1983, Reconciling Domestic Principles and International Cooperation, in: Oxman, Bernard H., David D. Caron, & Charles L.O. Buderi (Eds.), *Law of the Sea: U.S. Policy Dilemma*, San Francisco, ICS Press, 3–10.

Daniels, Steven E., & Walker, Gregg B. (2001). *Working Through Conflict: The Collaborative Learning Approach*, Westport, CT, Praeger.

Edwards, Paul N., 2010, *A Vast Machine: Computer Models, Climate Data, and the Politics of Global Warming*, Cambridge MA, The MIT Press.

Engo, Paul, 1985, An Evaluation of the Deep Seabed Mining Controversy, in: Buderi, Charles L.O. & David D. Caron (Eds.), *Perspectives on U.S. Policy Toward the Law of the Sea*, Honolulu, HI, The Law of the Sea Institute.

Evensen, Jens, 1985, The Effect of the Law of the Sea Conference upon the Process of the Formation of International Law, in: Krueger, Robert B. & Stefan A. Riesenfeld (Eds.), *The Developing Order of the Oceans*, Honolulu, HI, Law of the Sea Institute, 23–40.

Fisher, Glen, 1989, Diplomacy, in: Asante, Molefi K. & William B. Gudykunst (Eds.), *Handbook of Intercultural and International Communication*, Newbury Park, CA, Sage, 407–422.

Fisher, Glen, 1990, Addressing the Communication Dimension of Diplomacy and Negotiation: The Intellectual Agenda, in: Korzenny, Felipe & Stella Ting-Toomey (Eds.), *Communicating for Peace: Diplomacy and Negotiation*, Newbury Park, CA, Sage, 237ff.

Fisher, Roger, 1969, *International Conflict for Beginners*, New York, Harper and Row.

Fisher, Roger, William L. Ury & Bruce M. Patton, 1991, *Getting to Yes: Negotiating Agreement Without Giving In* (Second Edition), Boston, Houghton Mifflin.

Flannery, Tim, 2005, The Weather Makers: How Man is Changing the Climate and What it Means for Life on Earth, New York, Grove Press.

Friedheim, Robert L., 1987, The Third United Nations Conference of the Law of the Sea: North-South Bargaining on Ocean Issues, in: Zartman, I. William (Ed.), *Positive Sum: Improving North-South Negotiations*, New Brunswick, NJ, Transaction Books, 73–114.

Friedheim, Robert L., 1993, *Negotiating the New Ocean Regime*, Columbia, SC, University of South Carolina Press.

González, Verónica, 2012, An Alternative Approach for Addressing CO2-driven Ocean Acidification, *Sustainable Development Law & Policy* 12 (2), 45–69.

Hodgson, Robert D., & Robert W. Smith, 1976, The Informal Single Negotiating Text (Committee II): A Geographical Perspective. *Ocean Development and International Law Journal*, 3 (3), 225–259.

IISD/International Institute for Sustainable Development, 2010, Summary of the Bonn Climate Change Talks: 9–11 April 2010, *Earth Negotiations Bulletin* 12 (460), 14 April 2010, available at http://www.iisd.ca/download/pdf/enb12460e.pdf, last accessed 20 May 2013.

Jacobs, Michael, 2013, What is the State of the International Climate Talks?, *The Guardian*, 06 February 2013, available at http://www.guardian.co.uk/environment/2012/sep/17/internattional-climate-talks-faq, last accessed 16 April 2013.

Khor, Martin, 2012, A Clash of Paradigms: UN Climate Negotiations at a Crossroads, *Development Dialogue* 61, *What Next: Volume III, Climate, Development, and Equity,* Uppsala, The Dag Hammarskjold Foundation, available at http://www.dhf.uu.se/wordpress/wp-content/uploads/2012/10/Climate-Development-and-Equity_single_pages.pdf, last accessed 16 April 2013, 76–105.

Kikugawa, Tomofumi, 1999, *A Theoretical Analysis of the Law of the Sea Negotiation in the Context of International Relations and Negotiation Theory*, PhD Dissertation, Department of Politics, Stirling, Univerrsity of Stirling.

Koers, Albert W. & Bernard H. Oxman (Eds), 1983, *The 1982 Convention on the Law of the Sea*, Honolulu, HI, The Law of the Sea Institute.

Koh, Tommy, 1986, Negotiating a New World Order for the Sea, in: Henrikson, Alan (Ed.), *Negotiating World Order*, Wilmington, DE, Scholarly Resources, 33–45.

Koh, Tommy, 2009, The Negotiating Process of UNCLOS III, Video Lecture and Notes, 02 January 2009, Audiovisual Library of International Law, United Nations Office of Legal Affairs, New York, available at http://untreaty.un.org/cod/avl/ls/Koh_T_LS.html, last accessed 16 April 2013.

Laursen, Finn, 1982, Security versus Access to Resources: Explaining a Decade of Ocean Policy, *World Politics* 34, 206–213.

Lewicki, Roy, J., David M. Saunders & Bruce Barry, 2011, *Essentials of Negotiation* (Fifth Edition), New York, McGraw Hill-Irwin.

Mace, M. J., 2010, International Treaties, in: Schneider, Stephen. H., Armin Rosencranz, Michael D. Mastandrea & Kristin Kuntz-Duriseti (Eds), *Climate Change Science and Policy*, Washington D.C., Island Press, 231–244.

Moomaw, Bill & Sonia Hamel, Climate Change Series: The Governance Challenge, *Cognoscenti: Thinking that Matters*, 23 January 2013, available at http://cognoscenti.wbur.org/2013/01/23/climate-politics-moomaw-hamel, last accessed 16 April 2013.

Ogley, Roderick, 1984, *Internationalizing the Seabed*, Hampshire, Gower.

Oxman, Bernard H., 1983, Summary of the Law of the Sea Convention, in: Oxman, Bernard H., David D. Caron, & Charles L.O. Buderi (Eds.), *Law of the Sea: U.S. Policy Dilemma*, San Francisco, ICS Press, 146–161.

Pardo, Arvid, 1983, An Opportunity Lost, in: Oxman, Bernard H., David D. Caron, & Charles L.O. Buderi (Eds.), *Law of the Sea: U.S. Policy Dilemma*, San Francisco, ICS Press, 13–25.

Raiffa, Howard, 1982, *The Art and Science of Negotiation*, Cambridge, MA, Harvard University Press.

Sanger, Clyde, 1987, *Ordering the Oceans*, Toronto, University of Toronto Press.

Schroeder, Heike, Maxwell T. Boykoff & Laura Spiers, 2012, Commentary: Equity and State Representations in Climate Negotiations, *Nature Climate Change* 2, 834–836.

Schwarte, Christoph & David Wei, 2011, The Limitations of Consensus, Foundation for International Law and Development, available at http://www.field.org.uk/files/field_limitationconcensus_web.pdf, last accessed 16 April 2013.

Sebenius, James K., 1984, *Negotiating the Law of the Sea*. Cambridge, MA, Harvard University Press.

Sebenius, James K., 1991, Designing Negotiations Towards a New Regime: The Case of Global Warming, *International Security* 15 (4), 110–148.

Sebenius, James K., 1993, The Law of the Sea Conference: Lessons for Negotiations to Control Global Warming, in: Gunnar Sjöstedt (Ed.), *International Environmental Negotiation*, Newbury Park, CA, Sage, 189–216.

Soltau, Friedrich, 2009, *Fairness in International Climate Change Law and Policy*, Cambridge, Cambridge University Press.

Susskind, Lawrence & Patrick Field, 1996, *Negotiating Environmental Agreements: The Mutual Gains Approach*, Washington D.C., Island Press.

UN/United Nations, 1974, *United Nations Source Documents on the Third U.N. Law of the Sea Conference: Caracas '74 (LOS-3)*, Washington D.C., Nautilus Press.

UNFCCC/United Nations Framework Convention on Climate Change, 2012, Draft Text on ADP Agenda Item 3 Version of 18:30, 06 December 2012 – Draft Decision of the Conference of the Parties at its Eighteenth Session Proposed by the Co-chairs of the ADP, available at http://unfccc.int/files/documentation/submissions_from_parties/adp/application/pdf/draftadp36dec1830.pdf, last accessed 16 April 2013.

UNFCCC/United Nations Framework Convention on Climate Change, 2010a, The Cancun Agreements, available at http://unfccc.int/documentation/decisions/items/3597.php?such=j&volltext=%22cancun%20agreements%22#beg, last accessed 16 April 2013.

UNFCCC/United Nations Framework Convention on Climate Change, 2010b, Participation breakdown, available at http://unfccc.int/files/parties_and_observers/ngo/application/pdf/participation_breakdown_cop1-18.pdf, last accessed 16 April 2013.

Vihma, Antto, 2011, A Climate of Consensus: The UNFCCC Faces Challenges of Legitimacy and Effectiveness, *FIIA Briefing Paper 75*, Helsinki, The Finish Institute of International Affairs.

Walker, Gregg B., 1987, Communicating Across Cultures: Argument and International Negotiation, in: van Eemeren, Frans H., Rob Grootendorst, J. Anthony Blair & Charles A. Willard (Eds), *Argumentation: Analysis and Practices*, Dordrecht, Foris, 238–250.

Walker, Gregg B., 1989, Assessing Multicultural Argument in the Law of the Sea Negotiations: A Rationale and Analytical Framework, in: Gronbeck, Bruce E. (Ed.), *Spheres of Argument*, Annandale, VA, Speech Communication Association, 599–607.

Walker, Gregg B., 1990, Cultural Orientations of Argument in International Disputes: Negotiating the Law of the Sea, Korzenny, Felipe & Stella Ting-Toomey (Eds.), *Communicating for Peace: Diplomacy and Negotiation*, Newbury Park, CA, Sage, 96–117.

Walker, Gregg B., Fieldnotes 2010, 2011 and 2012, Field Notes from UNFCCC Meetings in Bonn (2010, 2011) and Doha (2012), on file with the author.

Walker, Gregg B. & Steven E. Daniels, 2005, Assessing the Promise and Potential for Collaboration: The Progress Triangle Framework, in: Walker, Gregg B. & Williams J. Kinsella (Eds.), *Finding our Way(s) in Environmental Communication: Proceedings of the Seventh Biennial Conference on Communication and the Environment*, Corvallis, OR, Oregon State University Department of Speech Communication, 188–201.

Walker, Gregg B. & Steven E. Daniels, 2012, The Nature and Role of Agency Leadership: Building and Sustaining Collaboration in Natural Resource Management and Environmental Policy Decision-Making, in: Rigling Gallagher, Deborah (Ed.), *Environmental Leadership*, Thousand Oaks, CA, Sage.

Wehr, Paul, 1979, *Conflict Regulation*, Boulder CO, Westview.

Wilmot, William W. & Joyce L. Hocker, 2010, *Interpersonal Conflict* (Eighth Edition), Boston, MA, McGraw-Hill.

Wirth, Timothy E., 2009, Foreword, in: Aldy, Joseph E. & Robert N. Stavins (Eds.), *Post-Kyoto International Climate Policy: Summary for Policy Makers*, Cambridge, Cambridge University Press, xxiii–xxviii.

Zuleta, Bernardo, 1983, *The Law of the Sea*, New York, United Nations.

# 9
# UNEP plus X? A Critical Assessment of Reform: Proposals and Implications for the International Climate Regime*

*Dirk Hanschel*

*Abstract*

The author discusses recent proposals to upgrade the United Nations Environment Programme in order to deal with its alleged deficiencies. Conceding that moderate reforms of global environmental governance may be useful, he views the creation of a fully-fledged international environmental organisation with scepticism, especially where implying a marginalisation of the Commission on Sustainable Development. Stressing the advantages of institutional flexibility, he views the implementation of existing substantive rules as stipulated by Agenda 21 and by issue-specific international regimes (such as the climate regime) as current main challenges of global environmental governance. They may only be tackled successfully if negotiators make sure they do not overemphasise procedure to the possible detriment of substance, the latter involving the quest for suitable models of long-term distribution of international environmental costs in times of economic instability.

## *A. Introduction*

Recent attempts to reform the United Nations Environment Programme (UNEP) have stirred a lively academic debate.[1] While scholars disagree as to the ideal design of the potentially emerging new institution, many seem to suggest that the transformation as such is a useful and necessary pro-

---

* The author wishes to thank Lars Borchardt for editorial advice.

1 For an overview, see e.g. Biermann (2011). According to Biermann (2012:1f.), "proposals to create an international agency on environmental protection have been debated for over 40 years now". He also sketches a brief history of the conduct of political and academic debate since then, including the more sceptical voices.

cess.[2] The author wishes to challenge this notion to a certain extent by maintaining that, while a better fine-tuning of the existing institutional set-up is certainly useful, at least the more ambitious of the current reform efforts may run the danger of actually hampering implementation of Agenda 21[3] and issue-specific environmental efforts, such as those envisaged by the international climate change regime.[4] There can be no doubt that negotiations on institutional reform are important in order to address current issues such as fragmentation, duplication, lack of funding, and lack of political support.[5]

---

2 See e.g. Najam et al. (2006:3), who say that "the GEG [Global Environmental Governance] system has outgrown its original design and intent. The system's high maintenance needs, its internal redundancies and its inherent inefficiencies have combined to have the perverse effect of distracting from the most important GEG goal of all – improved environmental performance"; or (ibid.:4): "Much like children who outgrow their clothes as they mature, the GEG system needs to be rethought so that it can meet the challenges of its own growth, respond to future issues and move from its current emphasis on awareness-raising and treaty creation to actual environmental action and implementation". To the extent that institutional reform may actually spur the latter two aspects, it is hard to disagree with such reform; the question is whether it really will do so.

3 For the text of this non-binding action plan adopted at the UN Conference on Environment and Development in Rio de Janeiro in 1992, see http://www.unep.org/documents.multilingual/default.asp?documentid=52, last accessed 20 April 2013.

4 For a cautious account regarding major UNEP reforms, see also Ivanova (2012a:566), who argues that "granting UNEP specialized agency status is not a panacea for the difficulties besetting global environmental governance". She (ibid.) goes on to say that "the reasoning and exceptional foresight of UNEP's designers in creating a small, agile subsidiary body, to catalyse [sic] and coordinate environmental action remains valid today". Similarly, Young (2008:15ff.) views organisational reforms as a less pressing issue, while distinguishing between organisational and institutional (i.e. regime) reform: "Without a doubt, UNEP suffers from a number of weaknesses. But reorganising UNEP cannot serve as an effective substitute for more fundamental changes in the system of rights, rules, and decision-making procedures covering matters ranging from the use of environmental services... to the destruction of major ecosystems". He adds that such "institutional arrangements need to be well-matched to the defining problems they address" (ibid.:20).

5 See e.g. Chulkov & Zhang (2008:3): "The current framework of international environmental governance is weakened by institutional fragmentation and specialization and the lack of a holistic approach to environmental issues and sustainable development. The duplication and fragmentation of the work of United Nations system organizations stem principally from a blurred distinction in their work programmes between environmental protection and sustainable development and the absence of a single strategic planning framework". On the existing challenges, see also Simon (2011:7ff.).

However, in terms of strategy, one should be aware of the risk that efforts regarding the reorganisation of international environmental governance that go beyond a mere fine-tuning might absorb energy urgently needed for negotiations on substance.

The underlying assumption guiding this hypothesis is that recent setbacks with regard to the solution of international environmental problems may be less due to actual governance deficits than to a lack of political will to move forward. In order to at least maintain the pressure on the pivotal actors, i.e. the governments negotiating issue-specific regimes (e.g. regarding climate) and cross-cutting processes such as the recent Rio+20 Conference, one should be cautious to avoid opening new playing fields that may be used in order to demonstrate progress without getting to grips with the substantive issues, e.g. emission reduction, funding, and technology transfer. In addition, to the extent that current institutional efforts envisage a bypassing of the Commission on Sustainable Development (CSD), they may endanger the continued support of any future environmental negotiation processes from the side of the developing countries; this support, however, is a clear precondition for success – as was aptly conceptualised by the Rio Conference 1992.

## *B. The Genesis and Status Quo of UNEP*

UNEP resulted from the Stockholm Conference in 1972 which successfully called upon the United Nations General Assembly (UNGA) to establish such an institution for purposes of action and coordination.[6] The initial idea was to avoid a large bureaucracy and, hence, to keep UNEP small.[7] Since the institution was not set up by a treaty, it lacks legal personality and merely

6 See Sands (2008:60); Sands & Peel (2003:60ff.); see also United Nations General Assembly (UNGA) Res. 2997 and UNGA Res. 3004 (XXVII), 1972, and Pushkareva (2011:para. 1). On the history and development of UNEP, see also Ivanova (2007). Furthermore, Ivanova (2005a:iii) describes UNEP as an "anchor institution" that was created as a "lean, flexible, and agile entity to gather and transmit information, catalyze action, and coordinate environmental activities in the UN system".

7 Pushkareva (2011:para. 1); Ivanova (2007:347ff.).

constitutes a subsidiary body of the UN.[8] This means it cannot enter into international agreements, claim damages, pass binding secondary law, etc.[9] Further shortcomings include limitations with regard to its mandate, budget and political support.[10]

Nevertheless, UNEP has shown remarkable success in many fields, e.g. with regard to its coordinating and financing role.[11] Furthermore, despite lacking legal personality, it has sponsored intensive lawmaking activities through its Division of Environmental Law and Conventions.[12] In terms of the Programme for the Development and Periodic Review of Environmental Law (the *Montevideo Programme*), it aims at –[13]

> … catalysing progressive development of environmental law aimed at sustainable development; providing legal and technical assistance; and capacity-building training to developing countries and countries with economies in transition to strengthen their capacity to develop and enforce environmental law.

An important precondition for lawmaking is setting agendas.[14] In this regard, UNEP has been particularly successful in setting up its regional seas programme, which created a blueprint for effective environmental rule-making in many instances.[15] It has contributed to a number of multilateral environmental agreements, including –

- the Convention on the Conservation of Migratory Species of Wild Animals (1979)
- the Montreal Protocol on Substances that Deplete the Ozone Layer (1987)
- the Basel Convention on the Control of Transboundary Movements of Hazardous Waste and their Disposal (1989)

---

8 Pushkareva (2011:para. 2); Ivanova (2007:349): "The United Nations Environment Programme was ultimately established as a subsidiary body to both the General Assembly and ECOSOC [Economic and Social Council] reporting to the General Assembly through ECOSOC".

9 Pushkareva (2011:para. 2).

10 (ibid.:para. 3).

11 (ibid.:para. 16ff.). For a thorough assessment of successes and failures of UNEP, see Ivanova (2010); on the controversial scholarly assessment of UNEP's achievements, see Simon (2011:8).

12 Pushkareva (2011:para. 19): see also Ivanova (2005a:7ff.).

13 Pushkareva (2011:para. 12).

14 Ivanova (2005a:10).

15 Pushkareva (2011:para. 25).

- the UN Framework Convention on Climate Change (1992), and
- the Convention on Biological Diversity (1992).[16]

In addition, UNEP has become engaged in extensive environmental assessment activities, in particular through the Global Environmental Outlook.[17]

## *C. Reform Debate*

However, one needs to concede that due to its lacking legal personality, UNEP's activities are limited, particularly because the institution cannot create binding secondary law or even adopt treaties by its members according to its own rules.[18] Furthermore, while UNEP has played the role of an "anchor institution", it has found it hard to coordinate existing international Conventions.[19] After the failure of the World Summit on Sustainable Development in 2002, which also displayed the deficits of the CSD, reforms of international environmental governance were, therefore, vividly discussed, most of them focusing on UNEP, not the CSD. Ultimately, more than 50 governments, including the member states of the European Union (EU), proposed the creation of a UN Environment Organization, while others pleaded for a less ambitious solution.[20]

The various suggestions were pursued and gradually put into more concrete terms in the run-up to the Rio+20 Summit in 2012.[21] Hence, the institutional design of future environmental governance became a central topic of this long-awaited conference, which meant that about half the time would be devoted to institutional matters and not to substance. The other half was indeed reserved for the implementation of Agenda 21, reframed under the

16 (ibid.:para. 25); see furthermore Ivanova (2005a:7).

17 Ivanova (2005a:9).

18 Pushkareva (2011:para. 26).

19 Ivanova (2005a:12).

20 Biermann (2012:1); Pushkareva (2011:para. 27). For an overview of the various opinions, see DGVN (n.d.); see also Biermann (2012:6ff.); Fauchald (2010), and Najam et al. (2006). According to Fauchald (2010:iii), the three major reform models are "1. strengthening UNEP within its current mandate, combined with enhanced cooperation and coordination within groups of MEAs [multilateral environmental agreements]; strengthening UNEP by adding new elements to its mandate; [and] the establishment of a World Environment Organization".

21 See e.g. Simon (2011). On the following analysis of the Rio+20 Summit, see also Hanschel (Forthcoming: 253ff., 262ff.).

term *green economy*.[22] The Summit resulted in a 49-page, non-binding document entitled *The Future We Want*,[23] which had been circulated prior to the Summit as a draft resolution by its President and was ultimately endorsed by the heads of state or government.[24] The resulting General Assembly Resolution stresses the need to strengthen the institutional framework for sustainable development (IFSD).[25] On the one hand, a High-level Political Forum would be created to build on the "strengths, experiences, resources and inclusive participation modalities" of the CSD, ultimately replacing it.[26] This forum is meant to drive the further implementation of sustainable development in a cost-effective manner while avoiding an overlap with pre-existing institutions.[27] The main intention is, however, an enhancement of UNEP, as illustrated by "D: Environmental Pillar in the Context of Sustainable Development". The UNGA is asked to pass a resolution for the strengthening of UNEP, with a view towards achieving universal membership, stable resources (from the regular UN budget), an upgrading and better linkages with other existing institutions, a strengthening of its competencies with regard to technology transfer and capacity-building, an improved transparency, and extensive inclusion of civil society.[28] On 20 December 2012, the UNGA passed a decision following up on these suggestions.[29]

While this may be seen as a step forward, it also means that the parties clearly stopped short of setting up a UN Specialised Agency as some drafters

22 On these two topics, see UNGA, A/66/L.56, No. 12: "We express our determination to address the themes of the United Nations Conference on Sustainable Development, namely a green economy in the context of sustainable development and poverty eradication, and the institutional framework for sustainable development"; on the Rio+20 Summit, also see Martens (2012:27).

23 See http://www.uncsd2012.org/content/documents/727The%20Future%20We%20Want%2019%20June%201230pm.pdf, last accessed 20 April 2013.

24 See Beisheim et al. (2012).

25 UNGA, A/RES 66/288, Annex, IV; see also Martens (2012:24ff.).

26 UNGA, A/RES 66/288, Annex, IV, No. 84.

27 (ibid.); for the current state of discussion see furthermore http://sustainabledevelopment.un.org/index.php?menu=1556, last accessed 18 May 2013. To what extent this High Level Political Forum might manage to maintain the assets of the CSD whilst avoiding its weaknesses, remains to be seen.

28 (ibid.:No. 88); see also DGVN (n.d.).

29 See http://www.unep.org/newscentre/Default.aspx?DocumentID=2700&ArticleID=9363&l=en, last accessed 16 January 2013; see also UNGA, A/RES/67/203.

had suggested beforehand.[30] Instead of a fully-fledged "United Nations Environment Organization (UNEO)" as envisaged by some states,[31] the Rio +20 Declaration called for a more low-key solution that would require neither a treaty nor a fully-fledged international organisation with legal personality. The advantage of this construction is that UNEP may still be sponsored through the regular UN budget.[32]

One may assume that this compromise will not have discouraged the proponents of a still larger solution. A full upgrading of UNEP towards a UNEO, IEO or WEO sounds charming indeed, since it may allow this institution to meet other international organisations such as the World Trade Organization or the International Labour Organization (ILO) at eye level. Theoretically, this could also provide the climate negotiations with a new thrust. One way of achieving this could be, as some have suggested, modelling the new institution according to the tripartite system employed by the ILO, i.e. to engage not only states, but also environmental as well as business non-governmental organisations in the process.[33] However, doubts remain not only as to whether this would increase institutional effectiveness, but also whether this is a feasible option in the first place.

Generally, focussing on the nuts and bolts of institutional design can be very useful with regard to environmental issues, since it may often be easier to achieve consensus on this than when tackling the intricate substantive questions, such as the implementation of Agenda 21. The framework-protocol approach operates precisely on that assumption by focusing on procedure (the framework) before substance (which then follows in the protocols). The drawback, however, is that this may delay the process of agreeing on substance and reduce the pressure, since procedural success may be used as a fig leaf for a lack of substantial progress. Once a suitable institutional and

30 On this idea, see UNEP, Issues, Brief #4; for an overview, see Ivanova (2012a: 567ff.); on the discussion of whether a UNEO as a UN Specialised Agency is needed or merely an upgrading of UNEP (UNEP +) within the existing framework, see Martens (2012:26ff.).

31 See http://www.uncsd2012.org/rio20/content/documents/german-inst-frame.pdf, last accessed 18 September 2012; http://www.uncsd2012.org/rio20/content/documents/EU-Commission-1st-Intersessional-11Jan.pdf, last accessed 18 September 2012; on the German perspective, see Umweltbundesamt (2011), suggesting a participation of civil society as well as the economic actors, following the ILO model of tripartism.

32 See UNEP, Issues, Brief #4; Hanschel (Forthcoming:262ff.).

33 See Umweltbundesamt (2011:6ff.); Hanschel (Forthcoming:262ff.).

procedural framework is in place, the right way forward would, therefore, be to focus primarily on substance rather than procedure. Going back to the drawing board might otherwise distract from the solution of the actual political conflicts of distribution of environmental costs that have been imminent in environmental negotiation processes conducted since the Rio Conference in 1992.[34]

In addition, one may doubt whether the developing countries will, in the long run, accept the strengthening of UNEP to the detriment of the CSD. Even if the latter may not have proved very effective so far, the intrinsic blending of environmental and development topics as envisaged by the original Rio Conference in 1992 has, in fact, been a major stepping stone towards a truly global engagement regarding the pressing issues such as climate change, loss of biodiversity, and desertification.[35] One should concede that a skilfully designed tripartite model might appeal to the interests of developing countries and at the same time engage civil society. But such effects may also be achieved by a more loosely knit institutional set-up that continues to view UNEP as a power broker rather than a power centre, while maintaining the CSD in parallel instead of working towards its abolishment. This may work if the respective competencies are clarified and strengthened,

34 On the framework–protocol approach, see Gollnisch (1995:89ff.); Kelly (1997:481ff.); Susskind (1994:32); and Weiss (1996:276); on the whole discussion, see also Hanschel (2003:260ff.). With regard to the current debate on UNEP, Ivanova (2005b:46), aptly asserts that "[t]he institutional recommendations... need not add a new layer of international bureaucracy. Quite to the contrary, they entail consolidation of the existing panoply of international environmental institutions and a shift towards a more modern 'virtual' environmental regime"; see also Hanschel (Forthcoming:263ff.).

35 As a consequence, China and the Group of 77 (G77), a coalition of developing nations at the UN, stressed the following in their declaration on institutional reform in 2011: "We need to keep in mind that nowadays the Commission is currently the only forum that addresses the three pillars of sustainable development and for that reason we believe that we need to review the CSD in order to strengthen it and to make it more efficient"; available at http://www.g77.org/statement/getstatement.php?id=11030 8c, last accessed 20 April 2013.Furthermore, the work of UNEP is acknowledged and a better coordination with other UN institutions deemed necessary; available at http://www.uncsd2012.org/rio20/content/documents/g77-inst-frame.pdf, last accessed 20 April 2013. Martens (2012:26) points out the rather reluctant reaction of the developing countries with regard to a weakening of the CSD, as well.

and if coordination, cooperation and funding of the two organs improve.[36] There is some reason to hope that the recent UNGA Resolution referred to above[37] is a step in that direction. One might add that the 'either/or' solution may fuel a rather difficult and potentially fruitless debate about the right seat of the remaining environmental institution – be it New York or Nairobi. Hence, the debate on the 'right' institution might, to some extent, disguise a hidden controversy between nations or regions struggling for the maintenance of the status quo.[38] At the same time, this reveals a certain path dependency with regard to the creation of international institutions by showing that initial decisions are hard to change once they have become entrenched by long-standing practice and corresponding interests.[39]

More importantly, however, one may doubt whether a UNEO would really display more power, e.g. with regard to the climate negotiations. In spite of the current deficits[40] (which may call for minor revisions of the existing format), the current informal structure has its own advantages: it has allowed for a less noisy, but not less effective, coordination of interests and the placement of new topics on the international agenda of cross-cutting environmental negotiations since 1992. This, in turn, is neatly linked to the treaty-based, issue-specific approach regarding areas such as long-range transboundary air pollution, ozone depletion, and biodiversity – even if this approach has shown certain limitations with regard to the climate negotiations, which have proven particularly difficult.[41]

---

36 Some of these elements are also contained in the suggestions of the Head of the German delegation, Stephan Contius, available at http://www.uncsd2012.org/rio20/content/documents/german-inst-frame.pdf, last accessed 20 April 2013.

37 UNGA, A/RES/67/203.

38 One should concede, however, that UNEP's location away from the main hosts of UN institutions, such as New York or Geneva, has, in fact, restrained its influence; for more on this see Ivanova (2005a:19); on the historical decision-making process with regard to Nairobi, see Ivanova (2007:355ff.).

39 On these suggestions, see Hanschel (Forthcoming:262ff.).

40 See UNEP, Issue Brief #4:3f.), which points out the relative weakness of the environmental pillar of the concept of sustainable development in comparison with the economic and social pillars; on the normative contents of these pillars, see Gehne (2011:107ff.).

41 With regard to this paragraph, see also Ivanova (2012a:584): "The need for a strong, legitimate and credible authority for the environment is undeniable, but the causal link between specialized agency status and the possession of such authority is unclear at best". On the success of UNEP in its current shape, see Sands & Peel (2003:60f.). For a focus on the existing set-up rather than major new solutions, also see Najam

## D. Conclusion

Nobody knows whether the strong focus on institutional design at the Rio +20 Conference has actually prevented a more effective deal with regard to the burning issues of Agenda 21. But one may hope that it has not opened a Pandora's box that will be difficult to close. Otherwise, attention might be diverted from an admittedly painful, but unavoidable, struggle regarding questions such as the distribution of environmental costs and, potentially, certain limitations on short-term economic growth for the sake of long-term benefits. Reviving the former Club of Rome debate on limitations to growth may sound illusionary for many, but could in the long run prove to be unavoidable in light of increasing environmental damage occurring throughout the world. The weak substantial outcome of Rio+20 shows that, against the background of the global financial crisis and other pressing issues, it has become much more difficult to find common ground in matters of international environmental protection.[42] This should not be a reason to defer matters and to focus too much on institutions instead of substance. To be fair, one should stress that the institutional results of the Rio+20 Summit may constitute some progress, and that the other half of the Summit was, in fact, devoted to substance – even though the relabelling of sustainable development by the term *green economy* may not be entirely satisfactory in that it apparently weakens the human (rights) dimension.[43]

The sectoral approach as employed in the climate regime has the advantage of usually producing binding results that display long-term effectiveness.[44] UNEP should continue to serve as an interface between these issue-specific regimes. What it needs in order to improve its work is primarily money and political acceptance rather than a major institutional trim. At the same time, one main function of UNEP will be to set the agenda for future negotiations and to provide and promote often non-binding, but nevertheless authoritative, principles which may later turn into binding law and serve as a focal point of orientation. Fragmentation resulting from the issue-specific approach might be reduced by a UNEO, e.g. by centralising the functions of

et al. (2006:iii): "... rather than proposing grand institutional reform, the study proposes to work with the existing pieces". On the intricacies of the climate negotiations, see e.g. Droege (2010:11ff.), Hanschel (2013a:277ff.) and Hanschel (2013b:11ff.).

42 See Hanschel (Forthcoming:265).

43 (ibid.:259ff.).

44 See e.g. Hanschel (2003).

secretariats or even Conferences of the Parties.[45] But this might also endanger the advantage of pinpointed bargaining processes as well as specialised expertise evolving from long-term engagement on a particular topic – apart from the fact that such centralisation would be very difficult to achieve, as the Rio+20 negotiations have shown.[46] Environmental regimes are dependent on the coordination that UNEP can provide in its current (or recently amended) format on condition that the institution is properly funded.[47] The most critical issue is its relationship to the CSD, the intricacies of which should, however, not be solved by the latter's abolishment. At the same time, UNEP is linked to many fully-fledged international organisations that provide the legal status that it lacks.[48] Strengthening these linkages may in fact be one of the major steps towards more effective international climate governance.

## *References*

Beisheim, Marianne, Birgit Lode & Nils Simon, 2012, Rio Plus 20 Realpolitik and its Implications for "The Future We Want", *Stiftung Wissenschaft & Politk (SWP) Comments* 25, available at http://www.swp-berlin.org/fileadmin/contents/products/comments/2012C25_bsh_lod_sin.pdf, last accessed 10 May 2013.

Biermann, Frank, 2011, Reforming Global Environmental Governance: The Case for a United Nations Environment Organization (UNEO), available at http://www.stakeholderforum.org/fileadmin/files/WEO%20Biermann%20FINAL.pdf, last accessed 20 April 2013.

Biermann, Frank, 2012, World Environment Organization, in: Ritzer, Georg (Ed.), *The Wiley-Blackwell Encyclopedia of Globalisation*, Hoboken, John Wiley & Sons.

Chulkov, Nikolav & Yishan Zhang, 2008, *Management Review of Environmental Governance within the United Nations System*, Geneva, United Nations, Joint Inspection Unit, available at https://www.unjiu.org/en/reports-notes/archive/JIU_REP_2008_3_English.pdf, last accessed 20 April 2013.

45 On the problem of fragmentation in the existing system, see UNEP Issues Brief #2.

46 Compare UNEP Issues Brief #4.

47 On the crucial issue of funding, see Ivanova (2012a:583f.); for reform proposals, see e.g. Gerstaetter et al. (2012); on further proposals for reform within the UNEP framework, see Ivanova (2005b: 44ff.); see also Ivanova (2012b).

48 See e.g. the report of the Governing Council of the United Nations Environment Programme, Environment in the United Nations System, UNEP/GC.26/INF/23; see also the table displayed by Simon (2011:11).

DGVN/Deutsche Gesellschaft für die Vereinten Nationen e.V., (n.d.), Reform der UN-Umweltinstitutionen, available at http://www.dgvn.de/index.php?id=1186&no_cache=1&type=98, last accessed 20 April 2013.

Droege, Susanne, 2010, The International Climate Policy Negotiations – Objections, Themes, and Prospects for Success, in: Droege, Susanne (Ed.), *International Climate Policy – Priorities of Key Negotiation Parties*, Berlin, Stiftung für Wissenschaft und Politik (SWP), Research Paper 3, 11–29, available at http://www.swp-berlin.org/fileadmin/contents/products/research_papers/2010_RP02_dge_ks.pdf, last accessed 20 April 2013.

Fauchald, Ole Kristian, 2010, International Environmental Governance – A Legal Analysis of Selected Option, *Fridtjof Nansen Institute Report 16/2010*, Lysaker, Fridtjof Nansen Institute.

Gehne, Katja, 2011, *Nachhaltige Entwicklung als Rechtsprinzip*, Tübingen, Mohr Siebeck.

Gerstaetter, Christiane, Benjamin Görlach, Susanah Stoessel, Maria Ivanova, Sandra Cavalieri, Elizabeth Tedsen & Haran Bar-On, 2012, *Rio Plus 20: Financial Resources for Improved International Environmental Governance*, Berlin, Federal Environment Agency, available at http://www.ecologic.eu/files/projects/08/13/2012%20-%2016:50/Gerstetter_12_IEG_Finance_Study_RIO_plus_20.pdf, last accessed 20 April 2013.

Gollnisch, Peter, 1995, *Entwicklungstendenzen im internationalen Umweltrecht*, Herzogenrath, Shaker.

Hanschel, Dirk, 2003, *Verhandlungslösungen im Umweltvölkerrecht,* Stuttgart, Boorberg.

Hanschel, Dirk, 2013a, Developing a Legal Toolkit: Institutional Options to Remove Stumbling Blocks in the Climate Negotiations, in: Sjoestedt, Gunnar & Ariel Macaspac Penetrante (Eds), *Climate Change Negotiations: A Guide for Resolving Disputes and Facilitating Multilateral Cooperation*, London, Routledge Publications, 277–321.

Hanschel, Dirk, 2013b, Institutional Options for the Climate Negotiations in Times of Crisis, in: Rodi, Michael (Ed.), *Opportunities and Drivers on a Way to a Low-carbon Society*, Berlin, Lexxion, 11–24.

Hanschel, Dirk, (Forthcoming), Die Institutionalisierung internationaler Verhandlungslösungen im Umweltvölkerrecht – Rio Plus 20 und die Zukunft des internationalen Klimaregimes, in: Hanschel, Dirk, Sebastian von Kielmansegg, Uwe Kischel, Christian Koenig & Alexander Lorz (Eds), *Mensch und Recht – Festschrift für Eibe Riedel*, Berlin, Duncker & Humblot.

Ivanova, Maria, 2005a, Assessing UNEP as Anchor Institution for the Global Environment: Lessons for the UNEP Debate, *Yale Center for Environmental Law and Policy Working Paper Series*, Working Paper No. 5/01, available at www.yale.edu/gegproject/uneo-wp.pdf, last accessed 20 April 2013.

Ivanova, Maria, 2005b, *Can the Anchor Hold? Rethinking the United Nations Environment Programme for the 21st Century*, Yale, Yale F&ES Publication Series.

Ivanova, Maria, 2007, Designing the United Nations Environment Programme: A Story of Compromise and Confrontation, *International Environmental Agreements* 7, 337–361.

Ivanova, Maria, 2010, UNEP in Global Environmental Governance: Design, Leadership, Location, *Global Environmental Politics* 10 (1), 30–59.

Ivanova, Maria, 2012a, Institutional Design and UNEP Reform: Historical Insights on Form, Function, and Financing, *International Affairs* 88 (3), 565–585.

Ivanova, Maria, 2012b, Global Governance in the 21st Century – Rethinking the Environmental Pillar, available at http://www.stakeholderforum.org/fileadmin/files/IEG%20Paper-Ivanova-Final%20_2_.pdf, last accessed 20 April 2013.

Kelly, Michael J., 1997, Overcoming Obstacles to the Effective Implementation of International Environmental Agreements, *The Georgetown International Environmental Law Review* 9, 447–488.

Martens, Jens, 2012, *Rio+20 – Die UN-Konferenz für nachhaltige Entwicklung 2012*, Bonn, Global Policy Forum Europe.

Najam, Adil, Mihaela Papa & Nadaa Taiyab, 2006, *Global Environmental Governance*, Winnipeg, International Institute for Sustainable Development (IISD), available at http://www.iisd.org/pdf/2006/geg.pdf, last accessed 20 April 2013.

Pushkareva, Elvira, 2011, United Nations Environment Programme (UNEP), in: Wolfrum, Rüdiger (Ed.), *The Max Planck Encyclopedia of Public International Law*, available at www.mpepil.com, last accessed 16 January 2013.

Sands, Peter H., 2008, The Evolution of International Law, in: Bodansky, Daniel, Jutta Brunnee & Ellen Hey (Eds), *The Oxford Handbook of International Environmental Law*, Oxford, Oxford University Press.

Sands, Phillipe & Jacqueline Peel, 2003, *Principles of International Environmental Law* (Third Edition), Cambridge, Cambridge University Press.

Simon, Nils, 2011, International Environmental Governance for the 21st Century – Challenges, Reform Processes and Option for Action on the Way to Rio 2012, *Stiftung für Wissenschaft und Politik (SWP) Research Paper Series*, Research Paper 01, available at http://www.swp-berlin.org/fileadmin/contents/products/research_papers/2011_RP01_sin_ks.pdf, last accessed 20 April 2013.

Susskind, Lawrence E., 1994, *Environmental Diplomacy: Negotiating More Effective Global Agreements*, Oxford, Oxford University Press.

Umweltbundesamt, 2011, *Arbeitspapier United Nations Environment Organisation – Kurzzusammenfassung*, Dessau-Roßlau, UBA-Projektgruppe International Environmental Governance.

UNEP/United Nations Environment Programme, (n.d.), The Environmental Dimension of IFSD, Issues Brief #2, available at http://www.unep.org/environmentalgovernance/Portals/8/InstitutionalFrameworkforSustainabledevPAPER2.pdf, last accessed 20 April 2013.

UNEP/United Nations Environment Programme, (n.d.), The Environmental Dimension of IFSD, Issues Brief #4, available at http://www.unep.org/environmentalgovernance/Portals/8/InstitutionalFrameworkforSustainabledevPAPER4.pdf, last accessed 20 April 2012.

Weiss, Edith Brown, 1996, New Directions in International Environmental Law, in: United Nations (Eds), *International Law as a Language for International Relations*, New York, United Nations.

Young, Oran R., 2008, The Architecture of Global Environmental Governance: Bringing Science to Bear on Policy, *Global Environmental Politics* 8 (1), 14–32.

# PART III:
# A TRANSDISCIPLINARY MIX: CLIMATE CHANGE, POLITICS AND FINANCE

## 10
# Transdisciplinarity: Theory and Visions on Global Transdisciplinary Processes for Adapting to Climate Change*

*Roland W. Scholz*

*Abstract*

Transdisciplinarity, which has become a third mode of using science, supplements disciplinarity and interdisciplinarity. This contribution discusses different notions of transdisciplinarity and introduces into theory, methodology and practice of transdisciplinarity. The core of transdisciplinarity is the integration of different types of epistemics (i.e. ways of knowing) from science and societal actors and systems. Ideally, transdisciplinarity relates abstract, analytic academic rigor with contextualised, experiential intuitive knowledge/wisdom of different key actors from practice. Transdisciplinary is a key methodology of sustainable transitioning and the coping with ill-defined problems such as coping with climate change. The article elaborates how transdisciplinary processes differ from other forms of applied research and participatory research such as action research, public participation, participatory research, or consultation. It is shown how different methods of case/system representation, case/system evaluation and case/system transformation may serve for different types of knowledge integration in transdisciplinary processes The functions of transdisciplinarity, i.e. capacity building, consensus building, mediation and legitimisation are identified and discussed.

## *A. Goals and Variants of Transdisciplinarity*

This paper reviews how *transdisciplinarity* is distinguished from both *interdisciplinarity* and *disciplinarity*. The distinction between *transdisci-*

---

* This paper is based on Scholz (2011:Chapter 15), used with permission, and has been partly reworked.

*plinarity research* and *transdisciplinary processes* is introduced. While there have been different interpretations of transdisciplinarity, this article advocates transdisciplinary processes that involve authentic collaboration between science and society, including representatives from industry, government, administration, different stakeholder groups, and the public at large. Such collaborations, as described here, emphasise mutual learning, joint problem definition and knowledge integration. Transdisciplinary processes should produce relevant, socially robust knowledge, i.e. knowledge that empowers society to cope with societal relevant problems, and which feeds back to scientific knowledge-generation and theory-building.

To initiate a transdisciplinary process, facilitators can employ techniques such as embedded case study methods to structure and organise work. These methods support problem representation and modelling, problem evaluation, development, and transition of the real-world problem. After a brief introduction of these methods, a step-by-step example is presented of a transdisciplinary process in Switzerland – one which has used these methods to develop a sustainable business future in harmony with the environment. Finally, the functions of transdisciplinary processes are described, such as capacity-/competence-building, consensus-building, analytic mediation, and legitimisation of public policy.

There is a practice of more than 20 years of transdisciplinary processes on sustainable transitions of regional, urban and organisational processes.[1] In the case of nuclear waste disposal, transdisciplinary processes have been applied to sustainable policy transformations.[2] The *why* and *how* of global transdisciplinary processes are presented, in which people involved in transdisciplinary case studies in different parts of the world may learn about how to adapt to climate change, and in what way social adaptation processes can inform each other.

## *I. Why Transdisciplinarity?*

Collaboration between science and society is often requested if uncertainty arises about substantial changes in human-environment systems, such as the introduction of a new technology or new medical pharmaceuticals, diag-

1 Scholz (2011); Scholz et al. (2006).

2 Krütli et al. (2010); Scholz et al. (2007).

noses or therapies. Other problems that could benefit from collaborative processes between science and society are the finding of mitigation, adaptation, policy or decision strategies when facing fundamental changes to the natural or social environment, for instance, such as those due to natural hazards, climate change, resource scarcity or changing cultural settings.

It is argued here that, from the perspective of society, transdisciplinarity provides an efficient use of knowledge for coping with complex, socially relevant problems; it provides societal capacity-building and bridges the growing gulf between many areas of research and the public. This equips society with a better understanding of how technologies or the natural environment work, and how the natural environment interacts with man-made systems. Consequently, transdisciplinarity can permit us to master the new and cope more adequately with the unknown both from a scientific and a societal perspective, e.g. with regard to inventions such as nanotech articles. At the same time, transdisciplinarity stimulates academic research by highlighting phenomena, issues and emerging questions that require scientific reflection, and by feeding experiential knowledge into the research process. Furthermore, it frees science from the cumbersome implementation problem. In other words, instead of science having to face the challenge of gaining public understanding, acceptance or appraisal of something ingenious, but which is then rejected for 'non-academic' reasons, transdisciplinary processes put science into practice from the very beginning. Transdisciplinarity is, as will be elaborated, an efficient means of using knowledge in decision-making – at least in certain types of pro-democratic, civic societies.[3]

## *II. Definition and Notions of Transdisciplinarity*

The term transdisciplinarity is occasionally referred to as *perfected interdisciplinarity*, or as the transfer of concepts or methods from one discipline to another. In the first definition of transdisciplinarity by Jantsch, as the "multi-level coordination of [an] entire education/innovation system",[4] the 'beyond science' notion of 'trans'-disciplinarity is highlighted. Since there is some confusion about the distinction between interdisciplinarity and transdisciplinarity, these concepts are briefly defined below.

---

3 Almond (2000).
4 Jantsch (1972:221).

*Disciplines* are characterised by objects and (core) methods by which certain problems are approached. For example, the discipline of mathematics deals with relationships between symbols and numbers by the method of proof. Similarly, the purview of the discipline of pharmacy is to investigate the impact of certain chemicals (called *pharmaceutical drugs*) on diseases by the use of laboratory experiments and clinical trials.

*Interdisciplinarity* is established by the fusion of concepts and methods from different disciplines. A metaphorical example of fusion is the saxophone, which emerged from the clarinet and the trumpet. Biochemistry, the study of chemical processes in living organisms, e.g. investigating reactions between proteins and other molecules, can serve as an example of an interdisciplinary field – at least, that is, before it became established as a discipline. The experimental method is a pillar of this field. The term *industrial food web*, used in the emerging domain of industrial ecology, can also be taken as an integrated concept.

*Transdisciplinarity* is fundamentally different from interdisciplinarity. Most of today's definitions of *transdisciplinarity* include the fact that it goes beyond science, in the sense that it "... deals with relevant, complex societal problems and organizes processes [that relate knowledge and values of] agents from the scientific and the non-scientific world."[5]

There are different notions of transdisciplinarity. This contribution refers to the Zurich 200 definition.[6]

Disciplines efficiently organise the methods and systematised knowledge about the material-biophysical-technological world, as well as the social-cultural-epistemic world. Interdisciplinarity merges concepts and methods from different disciplines for a better understanding and to better explain certain issues, phenomena and processes that cannot be sufficiently explained from a single disciplinary perspective. Transdisciplinarity organises processes that link scientific, theoretical and abstract epistemics with real-world-based experiential knowledge, outside academia; and it relates human wisdom to the analytical rigour of science and academic methodology.

5 Scholz et al. (2000:447).
6 See Scholz (2011:377).

***Figure 1: Transdisciplinary Processes as Interface and Mutual Learning among Three Types of Actors***

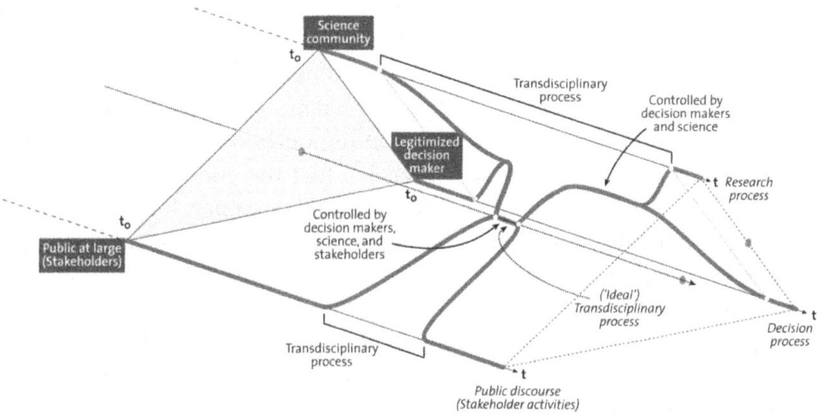


In Figure 1, actors from the science community, the public at large, and legitimised decision-makers are involved in research processes, public discourses or stakeholder activities, and decision processes, respectively. When the actors leave their primary processes (indicated by bold grey action lines in Figure 1) and join in on a collaborative, power-balanced effort, we can call this a *transdisciplinary process*.

Distinguishing transdisciplinary *research* from transdisciplinary *processes* is important. Transdisciplinarity, according to the Zurich 2000 definition described in the next section, organises mutual learning among science and society that can generate socially robust knowledge. For the most part, this mutual learning takes place in transdisciplinary processes in which members from the science community interact with decision-makers, stakeholders or the public at large. Transdisciplinary processes differ from consultancy and contracted research with respect to power-sharing and direction of involvement, i.e. in terms of who participates in whose process. In consultancy and contracted research, scientists operate in the action space of the legitimised decision-maker, who allows science to participate. Members, knowledge and results from the science community become part of the decision process. Here, the legitimised decision-maker ultimately decides how the skills and knowledge of scientists are used during the process, and how outcomes and results are communicated and utilised.

Usually, a transdisciplinary process emerges if a legitimised decision-maker and members from the science community notice that they have a joint interest in a complex, relevant phenomenon that can be better understood and dealt with if knowledge from practice and from science is integrated. Typical examples are sustainable transitions of regional systems caused by, for example, the overexploitation of natural resources (including pollution), or by adapting to changing environmental conditions.

The ways in which scientists, legitimised decision-makers and the public at large can collaborate are illustrated in Figure 1, which shows four horizontal time axes. At the bottom front, we find the public at large, whose priority is to sustain and organise life. This public opinion-building activity, shown as public discourse (in the foreground), represents the cultural-social side of human systems. The top arrow represents the activities of the scientific community, i.e. teaching, in-service training, investigating, etc. Together these make up the research process. At the bottom rear, we find the legitimised decision-maker, whose activities make up decision processes. A legitimised decision-maker can be a national or local government, a local environmental agency that has certain responsibilities, or a property owner such as a landowner or a company director who makes or plans decisions. If a legitimised decision-maker and scientists participate in a transdisciplinary process, they leave their action spaces for a certain period and collaborate. A situation of equal control between decision-makers and scientists – i.e. the transdisciplinary process – is represented by the plane bordered by the research process and the decision process lines in Figure 1.

Here, scientists and decision-makers leave their primary domain (research and decision processes, respectively) and establish a joint transdisciplinary process. In this process, they jointly agree on the topic or specific system to be investigated, e.g. defining the spatial and temporal system boundaries, leading to a joint problem definition.

Members of the public at large can be affected or feel concerned by the decision process and may organise themselves into interest or stakeholder groups. Often, these stakeholders or members of the public at large then participate in the transdisciplinary process (in a limited time period: see middle bold line of Figure 1). However, sometimes these groups can formally – or even informally – control the process. An ideal transdisciplinary process involves all three groups – researchers, decision-makers and stakeholders – in a collaborative, power-balanced relationship. In Figure 1, this ideal transdisciplinary process takes place at the transdisciplinary process line that runs through the centre of the triangle. It should also be noted that industry, as a specific stakeholder in society, may become a key player in a transdisciplinary process.

Transdisciplinary research takes place before, during or after a collaborative process, to provide preparatory research, support information, or follow-up research, respectively. The transdisciplinary *process* provides important input for transdisciplinary *research*, which is controlled by scientists, who can in turn ensure that results are produced through rigorous research methods.

### *III. Different Interpretations of Transdisciplinarity*

Transdisciplinarity and interdisciplinarity have been defined differently at different times and by different scientific domains and disciplines. Table 1 refers to the relevant discussion in the environmental and sustainability sciences. The first, moderate interpretation of 'going beyond science' was suggested by Mittelstrass (1996), who argues that "… transdisciplinarity is primarily a form of research addressing and reflecting on issues in the life-world."[7]

Referring to a distinction of scientific activities suggested by Gibbons et al. (1994), we call this *Mode 1* transdisciplinarity, as such scientific work can be found in traditional disciplines.

As an example of Mode 1 transdisciplinarity, we can take Nobel Laureate Amartya Sen's work on famine and poverty. Sen's work is characterised by classical disciplinary economics; however, many of his papers include ethical ideas – and, thus, are rather of an interdisciplinary nature – and deal, for instance, with inequality and child survival in Bangladesh, India, Pakistan, and North Africa, among other countries.[8] Though he is as a classical economist, Sen's work is interdisciplinary and is, according to Swedberg, "… ultimate[ly] concern[ed] … about the lives we can or cannot lead; about issues of the real world."[9]

Mode 1 transdisciplinarity differs from isolated thinking about the environment (and applied sciences), as it includes interperspectivity[10] and empathy, but not necessarily collaboration with others. Sometimes, pure problem-oriented research is denoted as being transdisciplinary.[11] Factually then,

7 Cited in Hirsch-Hadorn et al. (2008:28).
8 Sen (1999).
9 Swedberg (1990:339).
10 Giri (2002).
11 Jaeger & Scheringer (1998).

Mode 1 transdisciplinarity does not significantly differ from applied research, and is often linked to a 'truth to power' theory–practice relationship.[12]

The second definition of transdisciplinarity (Mode 2), which is here referred to as the *Zurich 2000* definition, resulted from the participation of 500 researchers and 300 practitioners at a conference in Zurich in 2000. This definition reads as follows: "Transdisciplinary research takes up concrete problems of society and works out solutions through cooperation between actors and scientists."[13]

From the perspective of scientific research, this conception of transdisciplinarity –

- organises processes of mutual learning among science and society
- integrates knowledge and values from society into research,[14]
- provides an appropriate research paradigm that better reflects the complexity and multidimensionality of sustainable development.

Following the Zurich 2000 definition, transdisciplinarity has been declared as the appropriate methodology by which sustainable development should be investigated and promoted.[15]

Transdisciplinary processes are periods of cooperation between scientists and practitioners to develop socially robust knowledge for coping with ill-defined, socially relevant problems, as described above. Transdisciplinary research deals with questions emerging from these processes. For the most part, this research is conducted without the participation of actors from the non-academic world who would normally be participating in the transdisciplinary process. However, we speak about transdisciplinary research only if it results from joint problem definition and a transdisciplinary process. Transdisciplinary research provides results that can be fed back to the body of science. Furthermore, transdisciplinary research is an element of the mutual learning process that takes place to develop robust orientations. This is the foundation for solving socially relevant problems, such as groundwater management, soil protection, or sea level rise. Thus, according to the Zurich 2000 definition, transdisciplinary research is simply research that is directed to, and conducted in the context of, transdisciplinary processes.

12 Pohl (2008).
13 Häberli et al. (2001:6).
14 Scholz et al. (2000).
15 Scholz & Marks (2001).

A third conception of transdisciplinarity can be called the *post-normal science approach*, as exemplified by Funtowicz and Ravetz.[16] A key to this approach is the assumption that real-world issues are so complex that the scientific results become just another voice or one vote within the agora of arguments in the real-world discourse. This is reflected in statements such as the following: "The faith in the truth and objectivity of science, established by Descartes and Galileo, is overthrown."[17]

The post-normal view stresses that science has lost credibility, and that "… any genuine understanding of a technology must now take into account malevolence, … also of those applying it to anti-human ends."[18]

It should be noted that the Zurich 2000 definition and the post-normal science understanding show many commonalities. Both approaches assume, for instance, that science incorporates values and world views in its investigations. However, it is important to distinguish between the different types of knowledge (*epistemics*) and constraints of producing knowledge in practice and in science. In the view presented here, experiential knowledge and scientific knowledge differ in their foundations, epistemological status, and the roles they play in different types of real-world problems. Taking the climate change example, we see that models of sea level rise are highly uncertain. Furthermore, they include unknown and unknowable assumptions about the development of the human population, prospective greenhouse gas emissions, technological development, or unknown future natural systems dynamics such as volcanic activities, El Niño-like cycles, or interactions with the biosphere, and there are huge uncertainties in many contextualised environmental research questions.[19] Nevertheless, there is a community of natural scientists[20] developing models, theories and predictions about climate change that, in their genesis, status, precision, and validation strategies, differ from statements uttered by politicians, members of construction companies, etc. In addition, social scientists provide knowledge about drivers of and obstacles to human behaviour that are relevant to understanding adaptation processes. Thus, one should also be careful when stating that "science takes place in the agora"[21] of ideas, though it is acknowledged that scientific

16 Funtowicz & Ravetz (1993, 2008).
17 Funtowicz & Ravetz (2008:364).
18 (ibid.).
19 Van de Kerkhof & Leroy (2000).
20 Waert (2003).
21 Gibbons & Nowotny (2001:79).

statements become relative in complex contextualisation. Thus, both the Zurich 2000 definition and the post-normal science approach assume that science only provides one type of epistemic, and it is vital to be aware that science can err.

One should also note that the Zurich 2000 definition is compatible with a constructivist perspective: it not only acknowledges that truth has a 'social nature', but it is also understood that the science community decides what is considered true or valid. As stated earlier herein,[22] this even holds true for mathematics, for example, as there are many cases where no single scientist is able to verify all prerequisites included in a complex proof of a theorem.[23] However, in the Zurich 2000 definition, *science* refers to a normal science view – i.e. approaching a valid or 'true' description of reality as a reference system – that is lost in its post-normal variant.

In both the Zurich 2000 and the post-normal science definitions, scientists can play a double role: they can contribute as facilitators or moderators to establish an appropriate process, and they can contribute as scientific experts.

A fourth definition of transdisciplinarity has been shaped by Nicolescu[24] and the Charter of Transdisciplinarity.[25] This approach to transdisciplinarity shares many aspects of the Zurich 2000 and the post-normal science approaches. One such aspect is "acknowledging different types of logic".[26] What makes this approach unique is that it "constitutes a personal moral commitment" against the "spiritual and material self-destruction of human species",[27] challenging that the "dignity of the human being is of both planetary and cosmic dimensions".[28] Furthermore, the approach addresses the unity of knowledge that targets the integration of scientific, religious, transcendent, and other forms of knowledge. In the following sections, we refer to the Zurich 2000 definition of transdisciplinarity.

Table 1 offers a schematic representation of the four variants of transdisciplinarity discussed above.

22 See section A.II above.
23 Scholz (1998).
24 Nicolescu (2002).
25 (ibid.); De Freitas et al. (1994).
26 De Freitas et al. (1994:Article 2).
27 (ibid.:Preamble).
28 (ibid.:Article 8).

***Table 1: Variants of Transdisciplinarity***

| **Variant** | **Essentials** |
|---|---|
| Mode 1 transdisciplinarity | Science becomes transdisciplinary if it reflects on real-life problems. |
| Zurich 2000 definition | There are transdisciplinary processes organising mutual learning between the science and the non-science communities, and transdisciplinary research which integrates knowledge and values from practice and science. Both processes deal with tangible, socially relevant, real-world problems. Practice and theory (science) have different reference systems. The transdisciplinary process features joint problem definition, representation and transformation ('problem-solving'). |
| The post-normal science approach | Science becomes just one vote in an agora (*forum, marketplace*) of arguments solving real-world problems because of the uncertainties and incompleteness of knowledge, multitude of logics, etc. |
| The 'Charter of Transdisciplinary' approach | In addition to many aspects of Modes 2 and 3, a personal moral commitment and 'unity of knowledge' is needed. |

## *IV. Sustainability Learning for Generating Socially Robust Knowledge*

Facilitating transdisciplinary processes presents an opportunity for scientists to promote sustainability learning as a process outcome and, in turn, to use process outcomes to generate socially relevant, robust knowledge. In this section, we discuss the scientist's dual epistemic role in transdisciplinary processes. Transdisciplinarity endeavours to use scientific knowledge efficiently to cope with socially relevant problems. Whether and, if so, how this can be done depends on various constraints, in particular on the specific problem at hand and the given sociopolitical context. Here, sustainability, as a widely shared way of regulating ideas from many contemporary societies, is essential. Transdisciplinarity can be linked to processes where society can learn about sustainability. This is known as *sustainability learn-*

*ing*. How sustainability learning is related to transdisciplinarity is explained in the following way:[29]

> Transdisciplinarity can be said to evolve from special types of problems, i.e. real, complex, socially relevant problems, which ask for the integration of the knowledge of science and society (Burger & Kamber, 2003; Scholz et al., 2000; Thompson Klein et al., 2001). Most of these problems are strongly related to sustainable development (Blättel-Mink & Kastenholz, 2005). It can be said that planning and learning processes for sustainable development require transdisciplinarity as an approach (Meppem & Gill, 1998). This holds particularly true if the development and implementation of policies and mutual learning processes are targeted by the behavior of individuals, industries, organizations, and governments. We refer to the corresponding process as "sustainability learning".

With respect to 'ill-defined problems', sustainability learning requires processes that go beyond traditional consultation (i.e. transfer of information from society to science) and knowledge transfer (from science to society), which are common ways of using scientific knowledge. Here it is suggested that developing socially robust knowledge in transdisciplinary processes is a key element of societal capacity-building.

Generating socially robust knowledge involves a form of epistemics, which –[30]

- meets state-of-the-art scientific knowledge
- has the potential to attract consensus, and thus must be understandable by all stakeholder groups
- acknowledges the uncertainties and incompleteness inherent in any type of knowledge about processes of the universe
- generates processes of knowledge integration of different types of epistemics, e.g. scientific and experiential knowledge, utilising and relating disciplinary knowledge from the social, natural, and engineering sciences, and
- considers the constraints imposed by the context both of generating and utilising knowledge.

Here, "mutual learning between science and society"[31] is considered a key characteristic of transdisciplinary processes and sustainability learning. These processes should be characterised by –

29 Scholz et al. (2006:231).

30 Gibbons & Nowotny (2001); Nowotny et al. (2001).

31 Scholz (2000).

- joint problem definition, i.e. determining the actual problem that is being targeted by various stakeholders, and how the sometimes diverging views of these stakeholders can be integrated and agreed on
- joint problem representation, i.e. developing a language that provides a medium of representation for describing the object, content and changes (dynamics) of the object (this step includes problem structuring),[32] and
- jointly initiating a process of problem-solving (perhaps more properly expressed as *problem transition*), which is cost-effective, socially acceptable, scientifically sound, and competitive in the marketplace.[33]

Thus, mutual learning can be viewed as a process that can generate socially robust knowledge that can contribute to coping with challenging societal transitions.

Through the example of global warming, the nature of the problems involved in these transitions and the generation of socially robust knowledge can briefly be looked at. Global warming and its environmental impacts require significant adaptations. Today, it is unclear how these adaptation processes should proceed, or what they should look like. It is unclear which measures are technically feasible and socially acceptable, and how reasonable goals can be presented. Making a country more resilient and less vulnerable to climate change, for example, and making preparations that will help a country to avoid uncontrollable damage[34] are typically ill-defined problems. This holds true for lowland countries and islands such as Bangladesh, the Maldives or the Netherlands having to cope with rising sea levels, or in the case of adapting agriculture in semi-arid regions. Nobody knows exactly what target state can or should be attained, or what barriers need to be overcome to reach this state. Neither is it known which barriers – economic, environmental or social – are the most severe. Problem definition is, therefore, a particularly difficult task, especially with respect to financial payouts that correlate with who is affected (positively or negatively) by environmental impacts. Clearly, all this is linked to the ontology of ill-defined problems.

32 Checkland (2000).
33 Gibbons et al. (1994).
34 KfC (2008).

## *V. Key Messages*

- Transcendisciplinarity is essentially different from interdisciplinarity.
- There are different definitions of transdisciplinarity, ranging from reflecting on society (Mode 1 transdisciplinarity – see Table 1) via theory–practice collaboration- based definitions referring to capacity-building and knowledge integration in transdisciplinarity processes (which underlie this book), to definitions requiring a personal moral commitment.
- Transdisciplinarity is considered a powerful and efficient means of using knowledge from science and society with different epistemics serving societal capacity-building under certain political cultures.
- Transdisciplinarity can stimulate science to identify challenging research questions, and feed experiential knowledge into the research process.
- Transdisciplinarity organises and effects collaboration and mutual learning between theory and practice.
- Transdisciplinary processes, which are jointly controlled processes that involve scientists, decision-makers and stakeholders, should be distinguished from transdisciplinary research, which is controlled by researchers.
- Transdisciplinarity is a means of coping with complex, ill-defined (wicked), contextualised and socially relevant problems that today often suffer from a framework of uncertainty and ambiguity. Transdisciplinary processes can organise sustainability learning and capacity-building in society, and are essential for environmental literacy.
- Generating socially robust knowledge in transdisciplinary processes can be seen as a major goal of transdisciplinarity.

## *B. Implementing Transdisciplinary Processes*

### *I. Sustainability Learning*

If one considers transdisciplinarity as a procedure for sustainability learning or establishing socially robust knowledge, a critical question is what specific methods can be used to implement transdisciplinary processes.[35] This section describes a suite of embedded case study methods for supporting and

35 Reeger & Bunders (2009); Scholz et al. (2006); Scholz & Tietje (2002).

organising transdisciplinary processes. Sometimes acting as key facilitators during these processes, scientists draw on facilitation methods such as capacity-building, consensus-building and analytical mediation.

### *II. Methods for problem representation*

Common issues that community members, practitioners and scientists jointly assess include the following examples:

- Representing how an infectious disease spreads[36]
- Identifying the causes of malnutrition, and
- Understanding the changes that have occurred in a certain system.[37]

A representation of a system or problem structure is often conceptualised by way of iconic mapping. Flow diagrams between concepts or pictures can represent cause–impact or means–end relationships. In principle, these types of representations are mental models. They are simple tools, well known from planning studies[38] or soft systems methodology.[39] Visualisations reveal much that is masked by verbal communication alone.[40] Visual literacy is ubiquitous and universal and, in some cases, a visualisation offers the only way of building a shared external representation of an issue as seen by both researchers and practitioners. The concept of "rich pictures" has been used to explore conscious and unconscious perception of problems and cases.[41]

Constructing a representation of a mental model includes many elements of an analytic process, as both the conceptual representation and the 'arrows' connecting these concepts ask for abstraction. The concept of "future workshops"[42] is a well-known approach that typically involves a two-day meeting that includes community members, researchers and administrators. These workshops result in consensus-building in respect of the current and future state of an urban setting, an institution, etc.[43] Under certain conditions, participants can become part of a transdisciplinary process. One goal of the

36 Kruse et al. (2003).
37 Hellier et al. (1999).
38 Schnelle (1979).
39 Checkland (1981); Checkland & Scholes (1990).
40 Cornwall & Jewkes (1995:1671).
41 Bell & Morse (2010a; 2010b).
42 Jungk & Müllert (1994).
43 Scholz & Tietje (2002).

process is to develop a joint problem representation of complex human–environment relations that are both understandable and acknowledged by the participants, and also compatible with system-theoretic scientific approaches. Facilitating the identification of a joint problem representation is an art that draws on methods which include verbal and system graphs from scenario analysis, system dynamics, material-flow analysis, supply chain analysis, etc. Creating a joint system representation that meets the problem at hand and that adequately deals with the different types of human-environment system complexities is an important part of the transdisciplinary process.

### *III. Methods of Process Management*

Transdisciplinary processes require facilitation of the process of knowledge integration. What are needed here are aspects of moderation, balancing the power between different participants, and methods of integrating knowledge.[44] Process management includes the social processes of interaction between the participants and the content level. For the latter, certain types of integrated modelling such as the joint construction of scenarios can also play a role in a transdisciplinary process.

### *IV. Embedded Case Study Methods*

*Embedded case study methods* are a specific set of methods that have been modified, advanced or newly developed in transdisciplinary case studies of sustainable urban development,[45] as well as regional[46] and organisational transitions.[47] Embedded case study methods were also applied to sustainable transitions of policy processes. In principle, these methods fall into four classes, namely A–D, as shown in Table 2. Transdisciplinary processes often start with case study team methods, which equip the group with strategies that allow them to work together effectively using approaches such as the experiential case encounter. We consider empathy and side change – i.e.

44 Hoffmann et al. (2009); Reeger & Bunders (2009); Scholz & Tietje (2002).
45 Scholz et al. (1996, 1997, 2004, 2005).
46 Scholz et al. (1995, 2002).
47 Mieg et al. (2001); Scholz et al. (2001); Scholz & Stauffacher (2007).

living or working in a real-world setting – a valuable way for researchers to gain a better understanding of what case agents know and understand. Empathy and side change also enhance the development of mutual trust, as the practitioner notices that the researcher is willing to leave the 'ivory tower'.

## *Table 2: Embedded Case Study Methods for Transdisciplinary Processes*[48]

| Case study methods | Key questions | The four types of knowledge integration<br>Disciplines | Systems | Modes of thought | Interests |
|---|---|---|---|---|---|
| ***(A) Case representation and modeling methods*** | | | | | |
| Formative scenario analysis (FSA) | What variables are crucial to the state of a system and to its change?<br>What can be? What ought to be? What can happen? | XX | X | X | |
| System dynamics (SD) | Which variables are the most decisive in temporal dynamics?<br>Which (counterintuitive) system change, outcomes, and feedback loops result from the dynamic interactions of the variables? | XX | X | | |
| Material flux analysis (MFA) | What are critical fluxes in materials for the case?<br>What are the sources and sinks of the system/case? | X | XX | | |
| ***(B) Case evaluation methods*** | | | | | |
| Multi-attribute utility theory (MAUT) | How can different evaluation criteria be integrated?<br>Which (mis)perceptions are inherent in an integral evaluation?<br>What are the preferences of different stakeholder groups? | X | X | | X |
| Integrated risk management (IRM) | How can/shall I cope with uncertainty?<br>Which of a set of different alternatives are the least risky ones? | X | X | | |
| Life cycle assessment (LCA) | How can the main environmental impacts (on a global level) be evaluated? | X | XX | | |
| Bioecological potential analysis (BEPA) | How can the bioecological quality and potential of a case site be evaluated? | | XX | X | |
| ***(C) Case development and transition methods*** | | | | | |
| Mediation: Area development negotiations (ADN) | What causes the conflicts among the principal-agents/key players of the case?<br>Which (mis)perceptions do the case agents have?<br>How can we attain Paretooptimal solutions? | | X | X | XX |
| Future workshops | Which ideas may guide the questions:<br>What can be?<br>What ought to be? | | | XX | X |
| ***(D) Case study team methods*** | | | | | |
| Experiential case encounter | What does the case look like from the case member's perspective? | | | XX | X |
| Synthesis moderation | How can I optimize teamwork to improve the synthesis process?<br>How can I find the right method of synthesis? | | | X | X |

48 All methods are presented and discussed in detail in Scholz & Tietje (2002).

In Table 2, *X* indicates that this type of knowledge integration is established by an embedded case study method, while *XX* indicates that such integration is strongly established by the method.

### *C. New Frontiers for Science–Society Cooperation*

This section discusses how theory-practice or science-society cooperation may continue to evolve. The first part presents a review of how, after nearly two centuries in their 'ivory towers', many scientists are now switching to problem-oriented, Mode 2 science, focusing on real-world problems. The constraints under which ideal transdisciplinary processes can take place are also discussed. The second part elaborates, from an inner-science perspective, on how transdisciplinarity can become a third mode of research, complementing disciplinary and interdisciplinary research. Thirdly, knowledge integration is distinguished from a normal and from a post-normal perspective. The fourth part presents a discussion on discuss how the stance of realism enables scientists to evaluate the quality of scientific assertions of causation, allowing verification of hypotheses.

#### *I. Mode 2: Bringing Science to Society*

During most of the past two centuries, the principle of the division of labour by disciplinary differentiation as well as by specialisation at universities has been dominant. To fulfil the goals of science and engineering disciplines to generate consistent and cohesive theories and methods, knowledge production at universities was intended to be academic, free, curiosity-driven, socially and politically neutral, and shaped by specialty and cryptic language. This type of science has been called Mode 1.[49] However, only a relatively small number of universities – primarily 'top' Western European and North American ones – came close to realising this traditional 'ideal'. As a consequence, research came to be seen as an isolated, elitist endeavour, conducted from an ivory tower.

As Gibbons, Nowotny and other proponents of the 'new production of knowledge' sociology proposed,[50] the role of the university changed dra-

49 Gibbons et al. (1994).
50 Gibbons (1999, 2000); Gibbons et al. (1994); Nowotny et al. (2001).

matically in the 1970s, taking on new aspects that went far beyond a narrow, elitist perspective: "[T]he modern university has become a hybrid institution, with multiple and sometimes incommensurable missions."[51]

Universities became an important participant in and driver of industry and regional development. Today, this holds true not only for top universities but even for those in second or third ranking. Already in 1968, about seven million students (43% within the 18–21 age group) were enrolled in institutions of higher learning in the United States.[52] Some countries endeavour to have more than half of their youth earn a university degree. For instance, in Germany, the number of universities increased from 34 in 1949 (25 in West Germany and nine in East Germany) to 350 in 2000.[53] Many former vocational training institutions in remote areas became universities of applied sciences. Although traditionally operating under the label *application and development*, they are moving towards the label of *research*, cooperating with local business and administration on regional developmental problems. However, the production of knowledge has also changed at the top universities. Scientific and technical work is increasingly performed by temporary teams dealing with specific, real-world problems rather than with theory:[54]

> Mode 2 science and technology includes cognitive science, computing, environment studies, biotechnology and aviation. It [Mode 2] is non-academic in the sense that its ties are with society and social issues. Society determines which problems are to be explored and resolved.

According to Gibbons et al.,[55] Mode 2 science provides a new epistemology and asks for a rethinking of science. Mode 2 proponents look at four principles that govern the new form of sciences research:[56]

- The coevolution of science and society
- Contextualisation
- The production of socially robust knowledge, and
- The construction of narratives of expertise.

Such research brings science out of the ivory tower to work with industry, government and laypeople in order to generate socially robust knowledge,

51 Scott (2007:214).
52 Ben-David (1974/1981).
53 Kehm (2004).
54 Shinn (2005:742–743).
55 Gibbons et al. (1994).
56 Gibbons (1999:3).

or sociotechnically robust solutions. At the same time, scientists can use data from the narrative of experiential expertise which emerges from transdisciplinary processes for knowledge integration and theory-building. This will probably operate at all levels: from small, regional colleges to universities which operate globally. Mode 2 shares commonality with the Zurich 2000 definition of transdisciplinarity, and Gibbons et al. state that "Mode 2 is transdisciplinary".[57] Scientists are required to enter into open theory-practice discourses, which require them to cope with complexity and contextualisation. Here, they meet the following situation:[58]

> Collective narratives of expertise need to be constructed to deal with the complexity and the uncertainty generated by this fragmentation. … Experts must respond to issues and questions that are never merely scientific and technical, and must never address audiences that only consist of other experts.

### *II. Transdisciplinarity as a Third Type of Research*

Based on the views outlined above, it is purported that transdisciplinarity becomes, or should become, a third form of academic activity. It is proposed here that many problems ask for knowledge of the disciplinary, interdisciplinary and transdisciplinary type.

At least four arguments supporting this proposal are evident. The first is that of disciplinary communities serving as a clearing-house. Secondly, there is the social contamination of science, which is closely related to the freedom of research. Two additional reasons relate to differentiating between disciplines, and distinguishing between scientific causation and non-scientific causation, as follows:[59]

> Many antidifferentiationists refuse cognitive and social differentiation, and beyond. They deny the division between nature and culture, science and society, science and technology, and between research and enterprise.

Specifically, the third and fourth arguments are described here.

It is considered problematic not to distinguish between knowledge generated in practice, and knowledge generated through scientific processes. An important issue related to the third argument is that, in addition, sciences

57 Gibbons et al. (1994:11).
58 Gibbons (1999:C83).
59 Shinn (2005:744).

include different rationales in reasoning and validation. There are, for instance, different reference schemes for validation, with practical efficacy on the one side and scientific coherence on the other.

An example of differentiating and relating different types of knowledge is provided in area development negotiations,[60] when the science-based assessment (which relies on disciplinary data and methods) is compared with one based on stakeholder judgments. Without doubt, there are different epistemics at work in these different evaluations.

The current study advocates against abandoning the divisions between the modes of thought of science/theory and society/practice, as well as between different disciplines. Transdisciplinary processes require knowledge integration, both for quantitative and qualitative knowledge, for instance, intuitive versus quantitative knowledge (Table 2). An interesting argument against the antidifferentiationist approach, which partly underlies Mode 2 thinking, is that the abandonment of the differentiation between and among science causation and social-agent causation finally pulls everything to pieces. This has been pointed out in the following statement: "The New Production of Knowledge [i.e. the variant of Mode 2, as suggested by Gibbons et al., 1994] posits atomistic learning and social interaction."[61]

### *III. Key Messages*

- Transdisciplinary research does not (and should not) substitute disciplinary and interdisciplinary research, but complements these types of research. It is a third mode of scientific activity that is based on transdisciplinary processes.
- Transdisciplinarity should avoid antidifferentiationist approaches, at least with respect to two dimensions: firstly, the different rationales that are at work in the intuitive, experience-based judgments of practical experts; and secondly, the different types of causation and statements inherent in different sciences. Also, the specific role of individuals is of relevance as regards their reconstruction and evaluation.

60 Scholz (2011:382–384).
61 Shinn (2005:744).

Knowledge integration is key to transdisciplinarity. Such integration should acknowledge the different epistemics from various participants, and is best carried out using method-driven procedures (see Table 2).

## *References*

Almond, Gabriel A., 2000, The Study of Political Culture, in: Crothers, Lane and Charles Lockhart (Eds), *Culture and Politics*, New York, Palgrave MacMillan.

Bell, Simon & Stephen Morse, 2010a, Rich Pictures: A Means to Explore the 'Sustainable Mind'?, *Sustainable Development* 21(1), 30–47.

Bell, Simon & Stephen Morse, 2010b, Triple Task (TT) Method: Systemic, Reflective Action Research, *Systemic Practice and Action Research* 23(6), 443–452.

Ben-David, Joseph, 1974/1981, *Trends in American Higher Education*, Chicago, The University of Chicago Press.

Blättel-Mink, Birgit & Hans Kastenholz, 2005, Transdisciplinarity in Sustainability Research: Diffusion Conditions of an Institutional Innovation, *Journal of Sustainability and World Ecology* 12(1), 1–12.

Burger, Paul & Rainer Kamber, 2003, Cognitive Integration in Transdisciplinary Science: Knowledge as a Key Notion, *Issues in Integrative Studies* 21, 43–73.

Checkland, Peter, 1981, *Systems Thinking, Systems Practice*, Chichester, Wiley and Sons.

Checkland, Peter & Jim Scholes, 1990, *Soft Systems Methodology in Action*, Chichester, Wiley and Sons.

Checkland, Peter, 2000, Soft Systems Methodology: A Thirty Year Retrospective, *Systems Research and Behavioral Science* 17(1), S11–S58.

Cornwall, Andrea & Rachel Jewkes, 1995, What is Participatory Research?, *Social Science & Medicine* 41(12), 1667–1676.

De Freitas, Lima D., Edgar Morin & Basarab Nicolescu, 1994, Charter of Transdisciplinarity, Unpublished paper presented at the Convento da Arrabida, November 1994. [The Charter of Transdisciplinarity is available at http://www.inters.org/Freitas-Morin-Nicolescu-Transdisciplinarity, last accessed 23 April 2013.]

Funtowicz, Silvio O. & Jerome R. Ravetz, 1993, Science for the Post-Normal Age, *Futures* 7(25), 735–755.

Funtowicz, Silvio O. & Jerome R. Ravetz, 2008, Values and Uncertainties, in: Hirsch Hadorn, Gertrude, Holger Hoffmann-Riem, Susette Biber-Klemm, Walter Grossenbacher-Mansuy, Dominique Joye, Christian Pohl, Urs Wiesmann & Elisabeth Zemp (Eds), *Handbook of Transdisciplinary Research: A Proposition by the Swiss Academies of Arts and Sciences*, Heidelberg, Springer.

Gibbons, Michael, Camille Limoges, Helga Nowotny, Simon Schwartzman, Peter Scott & Martin Trow, 1994, *The New Production of Knowledge*, London, Sage.

Gibbons, Michael, 1999, Science's New Social Contract with Society, *Nature* 402(6761), C81–C84.

Gibbons, Michael, 2000, Changing Patterns of University-industry Relations, *Minerva* 38(3), 352–361.

Gibbons, Michael & Helga Nowotny, 2001, The Potential of Transdisciplinarity, in: Thompson Klein, Julie, Walter Grossenbacher-Mansuy, Rudolf Häberli, Alain Bill, Roland W. Scholz & Myrtha Welti (Eds), *Transdisciplinarity: Joint Problem Solving among Science, Technology, and Society, An Effective Way for Managing Complexity*, Basel, Birkhäuser.

Giri, Ananta K., 2011, The Calling of a Creative Transdisciplinarity, *Futures* 34(1), 103–115.

Häberli, Rudolf, Alain Bill & Walter Grossenbacher-Mansuy, 2001, Synthesis, in: Thompson Klein, Julie, Walter Grossenbacher-Mansuy, Rudolf Häberli, Alain Bill, Roland W. Scholz & Myrtha Welti (Eds), *Transdisciplinarity: Joint Problem Solving among Science, Technology, and Society: An Effective Way for Managing Complexity*, Basel, Birkhäuser.

Hellier, Augustine, Adrian C. Newton & Susana O. Gaona, 1999, Use of Indigenous Knowledge for Rapidly Assessing Trends in Biodiversity: A Case Study from Chiapas, Mexico, *Biodiversity and Conservation* 8(7), 869–889.

Hirsch Hadorn, Gertrude, Susette Biber-Klemm, Walter Grossenbacher-Mansuy, Holger Hoffmann-Riem, Dominique Joye, Christian Pohl, Urs Wiesmann & Elisabeth Zemp, 2008, The Emergence of Transdisciplinarity as a Form of Research, in: Hirsch Hadorn, Gertrude, Holger Hoffmann-Riem, Susette Biber-Klemm, Walter Grossenbacher-Mansuy, Dominique Joye, Christian Pohl, Urs Wiesmann & Elisabeth Zemp (Eds), *Handbook of Transdisciplinary Research: A Proposition by the Swiss Academies of Arts and Sciences*, Heidelberg, Springer.

Hoffmann, Volker, Angelika Thomas & Alexander Gerber (Eds), 2009, *Transdisziplinäre Umweltforschung: Methodenhandbuch*, München, Oekom.

Jaeger, Jochen & Martin Scheringer, 1998, Transdisziplinarität: Problemorientierung ohne Methodenzwang, *GAIA: Ecological Perspectives in Science, Humanities, and Economics* 7(1), 10–25.

Jantsch, Erich, 1972, *Towards Interdisciplinarity and Transdisciplinarity in Education and Innovation*, Paris, Organisation for Economic Co-operation and Development.

Jungk, Robert & Norbert R. Müllert, 1994, *Zukunftswerkstätten: Mit Phantasie gegen Routine und Resignation*, München, Heyne.

Kehm, Barbara M., 2004, Hochschulen in Deutschland: Entwicklung, Probleme und Perspektiven, *Aus Politik und Zeitgeschichte* 25, 6–17.

KfC/Knowledge for Cimate, 2008, *Executive Summary of the Dutch National Research Program*, Utrecht, Knowledge for Climate.

Kruse, Natalie, Freida M. Behets, Georgina Vaovola, Gillian Burkhardt, Texina Barivelo, X. Amida & Gina A. Dallabetta, 2003, Participatory Mapping of Sex Trade and Enumeration of Sex Workers Using Capture–recapture Methodology in Diego-Suarez, Madagascar, *Sexually Transmitted Diseases* 30(8), 664–670.

Krütli, Pius, Thomas Flüeler, Michael Stauffacher, Arnim Wiek & Roland W. Scholz, 2010, Technical Safety vs. Public Involvement? A Case Study on the Unrealized Project for the Disposal of Nuclear Waste at Wellenberg (Switzerland), *Journal of Integrative Environmental Sciences* 7(3), 229–244.

Meppem, Tony & Roderic Gill, 1998, Planning for Sustainability as a Learning Concept, *Ecological Economics* 26(2), 121–137 (17).

Mieg, Harald A., Peter Hübner, Michael Stauffacher, Marianne E. Balmer & Sandro Bösch (Eds), 2001, *Zukunft Schiene Schweiz 2 – Ökologisches Potenzial des SBB-Bahngüterverkehrs für die Region Zugersee, ETH-UNS Fallstudie 2000*, Zürich, Ruegger.

Mittelstrass, Jürgen, 1996, Transdisziplinarität, in: Mittelstrass, Jürgen (Ed.), *Enzyklopädie Philosophie und Wissenschaftstheorie*, Band 4, Stuttgart, Metzler.

Nicolescu, Basarab, 2002, *Manifest of Transdisciplinarity*, Albany, State University of New York.

Nowotny, Helga, Peter Scott & Michael Gibbons, 2001, *Re-thinking Science: Knowledge and the Public in an Age of Uncertainty*, London, Polity Press.

Pohl, Christian, 2008, From Science to Policy Through Transdisciplinary Research, *Environmental Science and Policy* 11(1), 46–53.

Reeger, Barbara J. & Joske F.G. Bunders, 2009, *Knowledge Co-creation: Interaction between Science and Society – A Transdisciplinary Approach to Complex Societal Issues*, Den Haag, Raad voor Ruimtelijk, Milieu- en Natuuronderzoek.

Schnelle, Eberhard, 1979, The Metaplan-method: Communication Tools for Planning and Learning Groups, *Metaplan Series 7*, Quickborn, Metaplan.

Scholz, Roland W. 1998, Umweltforschung zwischen Formalwissenschaft und Verständnis: Muss man den Formalismus beherrschen, um die Formalisten zu schlagen?, in: Daschkeit, Achim & Winfried Schröder (Eds), *Umweltforschung quergedacht: Perspektiven integrativer Umweltforschung und -lehre*, Berlin, Springer.

Scholz, Roland W., 2000, Mutual Learning as a Basic Principle of Transdisciplinarity, in: Scholz, Roland W., Rudolf Häberli, Alain Bill & Myrtha Welti (Eds), *Transdisciplinarity: Joint Problem-Solving among Science, Technology and Society. Workbook II: Mutual Learning Sessions*, Zürich, Haffmans Sachbuch.

Scholz, Roland W., 2011, *Environmental Literacy in Science and Society: From Knowledge to Decisions*, Cambridge, Cambridge University Press.

Scholz, Roland W. & David Marks, 2001, Learning about Transdisciplinarity: Where Are We? Where Have We Been? Where Should We Go?, Thompson Klein, Julie, Walter Grossenbacher-Mansuy, Rudolf Häberli, Alain Bill, Roland W. Scholz & Myrtha Welti (Eds), *Transdisciplinarity: Joint Problem Solving among Science, Technology, and Society, An Effective Way for Managing Complexity*, Basel, Birkhäuser.

Scholz, Roland W. & Michael Stauffacher, 2007, Managing Transition in Clusters: Area Development Negotiations as a Tool for Sustaining Traditional Industries in a Swiss Prealpine Region, *Environment and Planning* A39(10), 2518–2539.

Scholz, Roland W. & Olaf Tietje, 2002, *Embedded Case Study Methods: Integrating Quantitative and Qualitative Knowledge*, Thousand Oaks, Sage.

Scholz, Roland W., Harald A. Mieg & Jenny Oswald, 2000, Transdisciplinarity in Groundwater Management: Towards Mutual Learning of Science and Society, *Water, Air, and Soil Pollution* 123(1–4), 477–487.

Scholz, Roland W., Sandro Bosch, Harald A. Mieg & Jörg Stünzi (Eds), 1997, *Zentrum Zürich Nord: Stadt im Aufbruch: Bausteine für eine nachhaltige Stadtentwicklung*, Zürich, vdf Hochschulverlag.

Scholz, Roland W., Sandro Bösch, Michael Stauffacher & Jenny Oswald (Eds), 2001, *Zukunft Schiene Schweiz 1: Ökoeffizientes Handeln der SBB*, Zürich, Ruegger.

Scholz, Roland W., Theo Koller, Harald A. Mieg & Corinne Schmidlin, 1995, *Perspektive "Grosses Moos" – Wege zu einer nachhaltigen Landwirtschaft*, Zürich, vdf Hochschulverlag.

Scholz, Roland W., Michael Stauffacher, Sandro Bösch & Pius Krütli, 2004, *Mobilität und zukunftsfähige Stadtentwicklung: Freizeit in der Stadt Basel*, Zürich, Ruegger und Pabst.

Scholz, Roland W., Michael Stauffacher, Sandro Bösch & Pius Krütli, 2005, *Nachhaltige Bahnhofs- und Stadtentwicklung in der trinationalen Agglomeration: Bahnhöfe in der Stadt Basel*, Zürich, Ruegger.

Scholz, Roland W., Michael Stauffacher, Sandro Bösch & Arnim Wiek (Eds), 2002, *Landschaftsnutzung für die Zukunft: Der Fall Appenzell Ausserrhoden*, Zürich, Ruegger und Pabst.

Scholz, Roland W., Sandro Bösch, Theo Koller, Harald A. Mieg & Jörg Stünzi (Eds), 1996, *Industrieareal Sulzer-Escher Wyss: Umwelt und Bauen – Wertschöpfung durch Umnutzung*, Zürich, vdf Hochschulverlag.

Scholz, Roland W., Daniel J. Lang, Arnim Wiek, Alexander I. Walter & Michael Stauffacher, 2006, Transdisciplinary Case Studies as a Means of Sustainability Learning: Historical Framework and Theory, *International Journal of Sustainability in Higher Education* 7(3), 226–251.

Scholz, Roland W., Micahel Stauffacher, Sandro Bösch, Pius Krütli & Arnim Wiek, 2007, *Entscheidungsprozesse Wellenberg – Lagerung radioaktiver Abfälle in der Schweiz. ETH-UNS Fallstudie 2006*, Zürich, Rüegger.

Scott, Peter, 2007, From Professor to 'Knowledge Worker': Profiles of the Academic Profession, *Minerva* 45(2), 205–215.

Sen, Amartya, 1999, *Development as Freedom*, New York, Knopf.

Shinn, Terry, 2005, New Sources of Radical Innovation: Research-technologies, Transversality and Distributed Learning in a Post-industrial Order, *Social Science Information* 44(4), 731–764.

Swedberg, Richard, 1990, *Economics and Sociology: Redefining Their Boundaries: Conversations with Economists and Sociologists*, Princeton, Princeton University Press.

Thompson Klein, Julie & William H. Newell, 1997, Advancing Interdisciplinary Studies, in: Gaff, Jerry G., James L. Radcliff & Associates (Eds), *Handbook of the Undergraduate Curriculum*, San Francisco, Jossey-Bass.

Van de Kerkhof, Marleen & Pieter Leroy, 2000, Recent Environmental Research in the Netherlands: Towards Post-normal Science?, *Futures* 32(9–10), 899–911.

Waert, Spencer, 2003, *The Discovery of Global Warming*, Cambridge, MA, Harvard University Press.

# 11
# The Politics of Climate Change: Review and Future Challenges

*Holger Haibach & Kathrin Schneider*

*Abstract*

Issues of climate change – its far-reaching effects and urgently needed countermeasures in particular – are becoming an unprecedented challenge for nowadays society. Hence, climate change politics are at the top of the political agenda for all countries around the world. Starting with a short introduction on the history of climate change politics – from the first major conference on the environment in Stockholm, Sweden in 1972, to the negotiations on the United Nations Framework Convention on Climate Change (UNFCCC) in Rio de Janeiro, Brazil in 1992, the Kyoto-Protocol in 1997 and to the recent Conference of the Parties to the UNFCCC in Doha in 2012 – this article critically questions the effectiveness of the UNFCCC, its Protocol and the annual meetings of the parties. Furthermore, the reasons why it appears to be impossible to reach a global agreement on fighting climate change are discussed. The article concludes that the essential burdens lie in the different interests of developing and developed countries, of polluters and those who are mostly affected. On the basis of theses global disagreements, the article finally outlines possible future challenges for global climate politics and suggests some deeply needed provisions, which can only succeed through a trustful and intense cooperation among all countries of the world.

## *A. Climate Change as a Global Political Issue*

Climate change has become an unprecedented challenge for humankind due to its severe consequences for our environment. Especially in the last three decades, global warming has been increasing rapidly, predominantly caused by the combustion of fossil fuels such as coal, oil and gas, as well as by continuing deforestation.[1]

The frequency and strength of environmental disasters like floods, hurricanes, and long-lasting periods of heat and drought have to be taken as serious warnings. The famine in the Horn of Africa, recent floods in Thailand and Central America, Hurricane Sandy in the Northeastern United States (US)and the Caribbean in October 2012 are only a few shocking examples of how our planet is affected by human intervention.

Consequently, the issues of climate change, its far-reaching effects and the urgently needed countermeasures nowadays top the political agendas in many countries around the world.

What makes climate change different from any other political topic is the mere fact that the climate crisis cannot be solved by a single world power or a group of states. The whole of humankind has to cope with the consequences on a planetary scale. Hence, adequate measures to restrict the effects of climate change are needed, and those can only succeed via a global approach.[2] Yet, although everyone seems to be aware of the necessity of global cooperation, finding a supranational consensus is practically impossible, as the political and economic interests of countries around the world interfere in finding an adequate solution. Thus, the progress in international negotiations is very slow – if discernible at all.

Consequently, over the last few years, the scepticism regarding insufficient progress in climate politics has reached a new level. Especially in the post-Kyoto process, it has become obvious that the different areas of interest of developing and developed countries, of polluters and those who are mostly affected, as well as of the new emerging powers have made it nearly impossible to reach a consensus on a common strategy, let alone effective and efficient countermeasures. Some critics even argue that the ecological footprint and the costs involved have far outweighed the outcome of conferences on climate change in recent years. In order to understand the current situation – which may be described as a deadlock – it is necessary to briefly reconsider the history of international climate change negotiations. The different spheres of interest of all parties involved also need to be assessed.

1 See Hirsch (2012:2).
2 (ibid.:2).

### B. From Polluting the Environment to the First Steps of Action: A Long Way

Ever since the beginning of industrialisation roughly 150 years ago, humankind has affected the global climate. These effects on the environment can no longer be reversed. Hence, many states, and developing countries in particular, have to face the principal environmental challenges of deteriorating air and water quality, inadequate soil conservation, and insufficient food supply. With increased frequency, even more challenging problems appear, including acid precipitation, Arctic haze, depletion of the stratospheric ozone, species extinctions and global warming. Especially in areas like sub-Saharan Africa, the existing problems like migration and disease have been severely aggravated.

The scarcity of natural resources such as oil and their rapidly increasing prices are critical reasons for seriously considering the question as to whether economic growth is *infinite* by definition. At the same time, the idea of sustainability emerged, as did the concept of the fair distribution of wealth amongst the countries and continents of the world. All of this culminated in the Club of Rome initiative, which stated that growth had to have its limits and that it came with a price tag. Economic growth cannot happen at the expense of sustainability. Whatever political stance one takes regarding the findings of the Club of Rome in 1968 and their political reactionism, the serious environmental response to extensive industrial development served the purpose of putting the issue of climate change high on the international agenda. In particular, the disaster at the nuclear power plant in Chernobyl in 1986 and its effects on flora and fauna throughout Europe made it abundantly clear that man-made interventions and their consequences do not take into account territorial or ideological boundaries, but call for an international response.

However, a conceptual approach of the United Nations (UN) was, even back then, still not under way. Initiatives tackling climate change were only taken in the 1980s when the issue emerged on the international political agenda,[3] although they could build on efforts made in the early 1970s. At the first major conference on the environment, namely the UN Conference on the Human Environment (UNCHE) held in Stockholm, Sweden, in 1972, representatives of 113 states underpinned that further environmental challenges could only be overcome through extensive international cooperation.

3 See Mayr (2009:10).

Consequently, participants agreed on a Declaration which incorporated not only 26 common principles,[4] but also provided further guidelines for action concerning issues on environment and development. Twenty years later, in June 1992, heads of state or government convened at the UN Conference on Environment and Development (UNCED) in Rio de Janeiro to pass the UN Framework Convention on Climate Change (UNFCCC). This Convention is still the definitive international treaty whose object is the long-term aim to stem climate change – and the stabilisation of greenhouse gas (GHG) concentrations in the atmosphere in particular – by means of international cooperation through the principle of common but differentiated responsibilities and respective capabilities.[5] Hence, the Rio Conference, also known as the Earth Summit, marked the institutional heyday of global climate politics.

## *C. Post-Rio: Down the Road to Nowhere*

Since the ratification of the UNFCCC in 1992, its signatories have been meeting annually at Conferences of the Parties (COPs) to discuss progress and further steps in tackling climate change. The first such conference (COP1) took place in Berlin, Germany, in 1995. COP1 saw the adoption of the Berlin Mandate, which incorporated a review of initial negotiations as well as new commitments. With hindsight as to the effectiveness of the first two COPs – the second having taken place in Geneva, Switzerland, in 1996, one criticism is that both merely emphasised the necessity of action rather than elaborating on ground-breaking commitments to combat serious climate change.[6] The first legally binding obligations concerning the reduction of GHG emissions were set at COP3 in Kyoto, Japan, in 1997, when the national representatives in attendance signed what became the Kyoto Protocol. Under the Protocol, the 37 leading industrial countries committed themselves to decreasing GHG emissions by an average of 5% within a set

4 The principles focus, amongst other things, on the assertion of human rights, the prevention of oceanic pollution, and the financial support of developing countries. For more details, see http://www.unep.org/Documents.Multilingual/Default.asp?DocumentID=97&ArticleID=1503, last accessed 19 November 2012.

5 See http://unfccc.int/essential_background/convention/background/items/1355.php, last accessed 17 November 2012.

6 For the complete COP1 Report, see http://unfccc.int/resource/docs/cop1/07a01.pdf, last accessed 17 November 2012.

target period between 2008 and 2012, compared with the level in 1990. To date, the Kyoto Protocol is "... generally seen as an important first step towards a truly global emission reduction regime that will stabilise GHG emissions, and provides the essential architecture for any future international agreement on climate change."[7]

As it was not clear how many states would ratify the Protocol, the ensuing annual COPs – whether in Buenos Aires in 1998, The Hague in 2000, or Bonn in 2001 – primarily targeted filling gaps in the Protocol and preparing for ratification, which was originally envisaged for 2000. Unforeseen complications arose shortly before the Sixth Climate Change Conference started in Bonn, in 2001, when the US, the biggest emitter of GHGs worldwide at the time, rejected the ratification of the Kyoto Protocol. The then newly elected President, George W. Bush, justified the US's decision by raising concerns regarding the potential damage to the US economy due to the Protocol's requirements. During the Bonn Conference, UN member states nonetheless tried to find a satisfactory outcome, despite having to cope with refusal or indifference from a large number of states parties. The Kyoto Protocol finally entered into force in February 2005, and had been ratified by 191 states parties by September 2011.[8]

Ten years after Rio, the World Summit on Sustainable Development – informally known as *Rio+10* – took place. Representatives of 192 countries met in Johannesburg, South Africa, not only to recapitulate Agenda 21 – an unprecedented global plan of action for sustainable development decided during the Rio Summit, but also to discuss further steps and targets for better implementation. Moreover, states parties which had already ratified the Kyoto Protocol again called on others to do likewise. Canada announced it would ratify the treaty, while other countries such as Australia, China and Russia reaffirmed they would give it their consideration.[9]

In addition to the Earth Summit, UNFCCC member states continued their meetings at annual COPs. Although a failure with respect to being a follow-up on Kyoto, COP15 in Copenhagen, Denmark, in 2009 made some progress referring to "negotiations on the infrastructure needed for effective global climate change cooperation" and "improvements to the Clean Development

7 See http://unfccc.int/kyoto_protocol/items/2830.php, last accessed 19 November 2012.
8 (ibid.).
9 See http://www.earthsummit2002.org/, last accessed 19 November 2012.

Mechanism of the Kyoto Protocol".[10] Moreover, participants decided on the Copenhagen Accord, which clearly expressed a political intent to constrain carbon dioxide emissions and respond to climate change in both the short and long term. States parties committed to limiting the increase of the global average temperature by 2°C. However, no sufficiently legally binding commitments were made regarding the reduction of GHGs in order to reach that goal.

Additional progress was achieved after COP16 in Cancun, Mexico, in 2010, and during COP17 in Durban, South Africa, in 2011, when participants agreed to adopt a universal legal instrument on climate change. The instrument was to be instituted by 2015 and take effect in 2020.[11] Earlier considerations to expand the existing framework, namely the Kyoto Protocol, could not be agreed on as extensive resistance was put up by emerging countries as well as the US.

A further decision taken in Durban was to set up a Green Climate Fund to assist developing countries with implementing mitigating measures. The target was an amount of 100 billion US$ by the year 2020.[12] Six countries/cities applied to host the Fund: Warsaw (Poland), Mexico City (Mexico), Geneva (Switzerland), Namibia (Windhoek), Germany (Bonn), and South Korea (Seoul), whereby the latter proved to be the successful applicant.

COP18 in Doha, Qatar, continued the ongoing discussions on commitments with very little – if any – success. Only 37 of the nearly 200 participating countries were willing to bind themselves to reducing their GHG emissions in the framework of the second Kyoto phase until 2020. These 37 countries currently emit only 15% of the world's GHGs. The biggest polluters, amongst them Canada, China, Japan and the US, were not willing to further reduce their emissions.[13] After 18 annual COPs[14] to date, it is time to put all the commitments made to a reality test.

---

10 See http://unfccc.int/meetings/copenhagen_dec_2009/meeting/6295.php, last accessed 19 November 2012.

11 See http://unfccc.int/meetings/durban_nov_2011/meeting/6245.php, last accessed 19 November 2012.

12 See http://cancun.unfccc.int/financial-technology-and-capacity-building-support/new-long-term-funding-arrangements/, last accessed 8 January 2013.

13 See Ruppel (2013).

14 2001: COP7, Marrakech, Morocco; 2002: COP8, New Delhi, India; 2003: COP9, Milan, Italy; 2004: COP10, Buenos Aires, Argentina; 2005: COP11/MOP 1, Montreal, Canada; 2006: COP12/MOP2, Nairobi, Kenya; 2007: COP13/MOP3, Bali, Indonesia; 2008: COP14/MOP4, Poznań, Poland; 2009: COP15/MOP5, Copen-

### D. Setbacks and Achievements of the Global Action against Climate Change: The Real World

Taking into account the mounting criticism on formal climate change negotiations, one could easily conclude that not much has been achieved over the last 20 years. Moreover, the prospect of anything further being achieved in the next 20 seems equally problematic, since there are other, allegedly more urgent issues to which the international community needs to attend. Ever since the world plunged into a financial and economic crisis in 2008/2009, the attention has tended to move away from environmental issues and shift to short-term troubleshooting in the areas of finance and the economy. Huge amounts of money are spent on stabilising markets and the economies of entire countries. Furthermore, economic strongholds like the European Union (EU) are struggling to maintain their stability. The effect of the crisis is twofold: the EU is not only experiencing a scarcity of resources to appropriate towards climate change matters, but it also – having been one of the driving forces behind negotiations on the issue – now finds itself in a precarious situation by the debates prevailing on its future financial framework.

Thus, in spite of a continued interest in climate action by non-governmental organisations and environmental movements, the world appears to be moving away from a safe and equitable climate future faster than ever before. Or at least so it seems. Political leaders are fully engaged with the global financial crisis and economic recession. There is a decrease in public interest. Furthermore, there is no master plan on the horizon to somehow break up the situation of deadlock that the negotiations are in right now. The million dollar question remains unresolved, however: how can one streamline the divergent interests of the various parties to the process?

To address this question, it may be worthwhile to explore the origin of these diverging interests. COP participants, for instance in Doha, failed to set legally binding commitments to reduce atmospheric concentrations of GHGs for individual countries: they merely agreed on guidelines that had no legal force. Hence, industrial countries did not really consider these targets as obligatory, but accepted them in a more moral sense. This poses questions regarding the annual COPs and their effectiveness. Although, with

---

hagen, Denmark; 2010: COP16/MOP6, Cancún, Mexico; 2011: COP17/MOP7, Durban, South Africa; 2012: COP18/MOP8, Doha, Qatar. For more detail see http://unfccc.int/meetings/items/6240.php, last accessed 19 November 2012.

the Kyoto Protocol, the UNFCCC states parties for the first time decided on a binding international treaty regarding limits of GHGs emissions, non-ratification or delayed ratification by quite a large number of countries challenged the effectiveness of the Protocol from the very outset.

In particular, the rather surprising decision by the US not to ratify the Kyoto Protocol in 2001 made it difficult to proceed, as it became questionable whether the treaty would still make sense without the participation of the world's biggest polluter and one of the most influential leaders in economic issues.[15] Predictably, the US's withdrawal from the Protocol set the example for other countries not to ratify the treaty. It was only in 2005, with Russia's ratification – ten years after Kyoto – that the treaty finally became effective. It took another three years for the Protocol's first commitment period to start, i.e. the one ending in 2012. So it was to be expected that the Protocol's long-term vision to keep global warming below the threshold of 2°C, or even 1.5°C, could simply not be reached within the treaty's existing framework.[16] The situation at hand is aggravated by the fact that emerging powers and big polluters like China never signed the Kyoto Protocol, and have yet to be convinced that globally decided commitments are indispensable to find a common way out of the climate crisis.

### *E. Essential Burdens: The Gap between the Haves and the Have-Nots*

Although the idea of annual COPs in which all states parties participate may be considered a positive development, the results of these meetings have been relatively disappointing. Hence, massive conflicts of interest and an absence of political will on the part of governments to agree to a fair, far-reaching and binding commitment for all countries explain why especially COP15 in Copenhagen in December 2009 failed so spectacularly. Powerful states like China, the US, and further emerging economies again opposed common, binding emission reduction obligations, and insisted on voluntary commitments in a so-called pledge-and-review arrangement. Once again, industrial states considered the demands as economically harmful, and believed this justified their position not to commit. Rather predictably, the voluntary provisions have so far failed to address global warming, and in

15 Cf. Böhringer (2001:4).
16 (ibid.:9).

addition the principle of common but differentiated responsibilities and respective capabilities, which was emphasised in the Bali Action Plan of 2007, lost its importance.[17]

The more the world has to cope with the consequences of global warming, the more this issue moves from an economic and environmental problem to a social challenge, as economic injustice and global poverty collide with the overconsumption and the desire for ever-increasing wealth. The biggest contrast can be found in the fragile social balance between industrialised and developing countries, or the gap between the rich North and the poor South. Ironically, the poorest, who bear little or no responsibility for the climate problem, have to face the most dire, even catastrophic, consequences of climate change and are unable to cope with them. Tackling the climate crisis seems to have changed to a system of the 'survival of the fittest', in which winners and loser are preordained. Developed countries stand to lose in the short term; but from a long-term perspective, we might all lose: causes are often generated locally, but the harm is felt on a planetary scale.

Despite the recognition that combating climate change can only succeed via global cooperation, it seems almost impossible to realise as developed countries, in particular, focus on personal interests and benefits. Most of them fail to take on the double responsibility of not only reducing their own emissions, but also providing adequate, reliable financial measures for emission reduction and adaption in poorer countries.

### *F. Things Happen for a Reason: Key Players and their Motives*

As the world's second largest single emitter of GHGs after China, the US has a considerable potential for action and should take the lead in reducing their domestic emissions by accepting reduction targets and pledging financial support for developing countries.[18]

However, instead of taking a step towards global climate protection, the US still refuses to accept the commitments of the Kyoto Protocol. The US Senate justifies its position by saying that treaty negotiations were held under unfair conditions because developing countries – among others China and India – were excluded from emissions obligations within the Protocol.[19]

17 See Fuhr et al. (2011:9).
18 See Donner et al. (2007:4).
19 (ibid.:5).

These circumstances also raised the Senate's fear that economic growth in the US would stagnate, and that the threshold countries mentioned – which made no commitments to reduce their emissions at all – were unfairly advantaged by the treaty.[20]

Thus, instead of striving for a more climate-friendly industrial economy, the US focused on the promotion of research and innovation for climate-friendly technologies. The approach to develop more such technologies and types of energy generation is also supported in the Asia-Pacific Partnership on Clean Development and Climate (APP) signed in 2005, which aims at pushing climate-friendly technologies through intergovernmental cooperation.

Further typical strategies of US climate politics are tax concessions, voluntary partnerships between the private and public sectors, and international cooperation, the latter being integrated into federal strategies such as the Climate Change Technology Program and the Climate Change Science Program, both initiated in 2002.[21]

Nevertheless, new approaches referring climate politics of the US government were not made over the last years. Indeed, during a speech in 2008, then President George W. Bush hinted at a new climate protection strategy, whereby GHG emissions should not increase until 2025. This position could also be interpreted as the US's first acceptance in principle of an emissions limitation. Nonetheless, the country remains steadfast in refusing to commit to concrete, legally binding requirements, at least until countries like China and India are similarly obliged to reduce their GHG emissions.[22]

Against the relatively regressive attitude of the US as a political and economic unit, various US states – most notably the State of California – agreed to pursue an active climate change policy. Indeed, legally binding caps for GHG emissions have been set in 17 US states. More and more frequently, these legal changes in domestic politics are reflected in cross-party drafts in the US Congress that aim to reduce emissions through binding caps.[23]

When one focuses on current actions in the US, the 2012 presidential elections and, especially, the topics of their campaign debates, it is clear that the climate change crisis – the biggest global issue of all – still seems to

20 See http://www.oekosystem-erde.de/html/klimapolitik.html, last accessed 27 November 2012.
21 See Donner et al. (2007:5–6).
22 (ibid.:7).
23 (ibid.:4).

enjoy scant priority in US politics. In his election campaign in 2008, President Barack Obama highlighted climate change as one of his priorities; in the 2012 presidential race, almost no reference was made to the issue by the incumbent President or his opponent, Mitt Romney. The only exception was the impact of Hurricane Sandy, where President Obama described climate change as one of its causal factors. Moreover, the Obama Administration did not establish new policies in this regard, so it is yet to be seen whether the re-elected President will make good on his first-term promises during his second term.[24]

Equally unwilling to commit themselves to reduce GHGs in the framework of an international, legally binding treaty are emerging countries and rising powers, including China and India. If one compares the climate policies of China and the US, both countries display similar behaviour towards binding agreements to reduce GHG emissions. The Chinese Government opposed any binding cuts in GHGs during the COP in Copenhagen in 2009, and opted to take its own measures domestically against GHG emissions, predominantly to become more energy efficient.

One of the voluntary commitments offered by China in Copenhagen was to reduce the country's carbon dioxide concentrations to 40–45% by 2020. Taking into account that China's total emission of $CO_2$ doubled in the 15-year period between 1991 and 2006, it is foreseeable that the set long-term objective to decrease emissions will not be achieved by 2020.[25]

However, China has developed measures to decrease emissions. These include integrating climate change mitigation and adaption in national sustainability strategies and initiatives. Although these guidelines imply that China's political leaders may take climate change seriously, the issue takes no priority over objectives such as national economic development and growth, which aim to lift it out of poverty.[26] On the international front, China has advocated for common but differentiated responsibilities and respective capabilities, proposing that developed countries be frontrunners in reducing GHG emissions, and that they technically and financially support developing countries. This approach seems to be problematic in a sense that China still

24 See http://www.guardian.co.uk/commentisfree/2012/nov/05/obama-romney-remain-silent-climate-change, last accessed 6 November 2012.

25 See Harris (2010:2).

26 (ibid.:2). More information on China's policies and actions for addressing climate change are available at http://english.gov.cn/2008-10/29/content_1134544.htm, last accessed 6 November 2012.

considers itself as a developing country and urges other stakeholders to take action, whereas it replaced the US as the world's largest polluter in 2006. Furthermore, on the grounds of national sovereignty, China is not willing to allow its emissions to be monitored and evaluated in an attempt to reduce them.[27]

Like China, other emerging countries have not as yet committed themselves to the Kyoto process or a respective follow-up. One of the underlying arguments is the fear of losing national sovereignty. Another, perhaps more important one, is that the process as a whole is viewed with suspicion by some countries, because it is seen as an attempt by the previously dominant world economies to maintain their dominance. Reducing GHGs will only be possible at the expense of economic growth, so the argument goes. Closely connected to this is the issue of development. While industrialised countries have had the opportunity to grow without any concern as to the damage they were doing to the world's climate, the prevailing feeling among developing countries is that they should be given the same chance to do so.

For quite some time, the EU has been trying to act as a mediator between conflicting interests and as a driver for progress. Although it is relatively difficult to coordinate the positions of nearly 30 member countries, the EU has managed to come up with a common approach during the various COPs. However, the EU has not always succeeded in making use of its political leverage. During the sometimes chaotic COP15 in Copenhagen, final decisions were taken and final deals struck between China and the US without the EU being involved. COP17 in Durban, on the other hand, showed that the EU and the developing countries have some influence on the process. Had these latter two groups not teamed up and followed a common approach, even the little that was achieved back then would not have been possible.

## *G. Least Culpable, but Most Affected: The Role of Developing Countries*

If one looks back on the latest disasters as a result of climate change, developing countries have been the ones to suffer most from the consequences. Hence, it is more than understandable that countries in Africa and Latin America, in particular, not only urge the industrialised world to assume their responsibility for global warming, but also insist on their financial support

---

27 Harris (2010:5).

for devising measures against climate change. However, with budgets being strained by financial turmoil, climate aid for least-developed countries has become a more challenging task.

In Copenhagen in 2009, industrialised countries agreed to provide nearly US$30 billion in grants and loans to developing countries to enable them to counteract the effects of climate change, but those commitments expired in 2012. Later on, the Green Climate Fund was set up to raise US$100 billion annually by 2020 for the same purpose. However, this Fund is not yet operational, and the money has only been promised. Furthermore, it is to be seen if the money pledged is really 'fresh' money or, as happens frequently in development cooperation, money that was pledged to other, earlier commitments. Where does this leave the developing world? It is to be expected that new financial resources may be available to mitigate the consequences of climate change, in spite of some sentiments in developed countries which question whether developing countries in fact do have a solid strategy to tackle the issue.[28]

One can only hope that the problems experienced by many developing countries (including corruption, weak statehood and infrastructure) will not negatively affect the success of measures to mitigate climate change. In addition, developing countries will have to come up with their own initiatives in respect of becoming less vulnerable to exterior political influences, wherever this is possible. A good example in this respect may be Namibia, a country heavily affected by the impact of climate change.[29] Although not a big polluter, Namibia has the opportunity to reduce its carbon footprint by making use of its most abundant source of energy: the sun. Solar installations, combined with other sources such as wind or invasive bush, could make the country a role model for self-sufficient, eco-friendly, decentralised energy supply. Yet, in spite of many private initiatives, Namibia still relies heavily on fossil fuel energy sources, which – to make matters worse – are mostly imported. Furthermore, the country lacks a coherent policy and legislation that could endorse the use of renewable energies.

It is precisely these gaps that the support of developed countries could usefully fill by means of knowledge transfer and financial assistance. However, it would be too simplistic to see developing countries only as being on the receiving end or as passive recipients of financial support. The COP17

28 See http://www.huffingtonpost.com/2012/11/25/2012-un-climate-talks-qatar_n_2188048.html, last accessed 4 December 2012.

29 See Mapaure (2011:289ff.).

in Durban showed that, if such countries speak with one voice and team up with allies, they could have political leverage and bring some pressure to bear on the largest polluters. This applies to both the US and China. Moreover, since China has embarked on a series of interventions in the quest for new suppliers of natural resources – usually in developing countries – to meet the ever-growing needs of its economy, such countries also have ways and means of taking part actively in negotiations.

## *H. Future Challenges for Global Climate Politics: Doha and Beyond*

According to a climate study recently published by the World Bank, we are moving directly towards a situation in which the global average temperature will rise by 4°C by the end of the 21st Century.[30] This greatly exceeds the official aim of limiting the rise in global temperature to below 2°C by 2012, to which the states agreed at a UN climate conference in the Copenhagen Accord some years ago.[31]

As UNFCCC states parties have not complied with the set target, political leaders have been sharply criticised for their ignorance of deeply needed global cooperation, and the progress in counteracting global warming has completely stagnated. Hence, the participants at the COP18 in Doha in 2012 – the first UN climate conference ever in the Arab region – were under particularly intense pressure.[32]

In the run-up to COP18 negotiations, the 195 states parties set themselves key objectives according to which the most urgent task lay in extending the Kyoto Protocol, following the expiry in 2012 of its first commitment period. Other key issues on the agenda concentrated, amongst other things, on strengthening the adaptive capacities of the most vulnerable, producing the financial support pledged to developing countries' climate change mitigation actions, the call for further measures against deforestation, and developing more eco-efficient technologies.[33]

---

30 See http://www.zeit.de/wirtschaft/2012-11/klimawandel-weltbank-bericht, last accessed 4 December 2012.

31 (ibid.).

32 For more information, see http://unfccc.int/meetings/doha_nov_2012/meeting/6815.php, last accessed 4 December 2012.

33 Cf. http://unfccc.int/files/press/press_releases_advisories/application/pdf/pr20102611_cop18_open.pdf, last accessed 4 December 2012.

As we now know, the outcome was rather poor. The one success – keeping Kyoto legally binding until 2020 – has been diluted by the low number of countries committed to the Kyoto process. Furthermore, the biggest polluters are again not part of the negotiated solution. Moreover, no progress has been made on a new treaty that is supposed to be finalised by 2015. On the other hand, time is running out. In order to limit the rise in global temperatures to below 2°C in the 21st Century, global emissions have to be reduced significantly more than originally assumed. Fortunately, this aim still seems feasible if the rate of emissions does not exceed the maximum of 44 Gt of carbon dioxide in 2020.[34]

Otherwise, if emissions continue to increase as before, the target of a projected 56 Gt in 2020 would be missed by 12 Gt – also described as the *Gt gap*. Even if all states honoured their commitments to climate protection, the original aim would be exceeded by 5 Gt.[35] So what measures have to be taken to meet the original target of 44 Gt? It is now more necessary than ever to set higher national objectives regarding emission reduction; these can only be realised through internationally binding treaties such as the Kyoto Protocol. Therefore, consistent and robust ways of measurement, reporting and verification of GHG emissions have to be developed. This automatically leads back to the issue of global cooperation and states parties' common but differentiated responsibilities and respective capabilities.[36] It will take a strong and inclusive effort to address the issue of climate change properly and finally. Thus, the current deadlock has to be overcome. A coalition of the willing is all very well, but without the support of the big polluters, the process is bound to fail. So the challenge for future negotiations and negotiators is threefold:

- There are only two years remaining to reach a consensus on a new treaty. Taking into account the very slow progress from 1990 to 2010, some scepticism may be justified as to whether this is sufficient.
- For such a treaty to achieve its aim, it will be necessary to successfully include countries such as Canada, China, India and the US not only in treaty negotiations, but also in their commitment to signing up, and

34 See http://www.unep.org/annualreport/2011/#, last accessed 4 December 2012.
35 See Hirsch (2012:15f.).
36 (ibid.:3).

- Effective and efficient mechanisms will have to be found regarding technical and financial assistance both for developing countries and those most affected by climate change.

## *I. Outlook: Is There a Way Forward?*

Climate change can no longer be considered as a political issue that can still be put off until tomorrow. It is all too present, and with its complex multitude of devastating consequences, it is a daily reality for many people – especially the poor. While the international climate debate has, for years, centred on implementing plans and concluding adequate agreements to take into account the interests of all countries – with rather disappointing outcomes – it is high time to rethink this approach. In view of the dramatic increase of disasters caused by the impact of climate change, the focus now has to be shifted to issues such as crisis management and plans for preventative measures to combat further natural disasters, among other things.

It seems clear there is no way other than to initiate a global innovation process aiming to reform our current economic models and come up with new technological and social solutions to the problem of climate change. This has to be done with a long-term perspective. In order to be able to concentrate on these issues it will be necessary to create a climate of trust in which cooperation on an international level is easier than it is presently. So how can we overcome the clashing interests of emerging economies and developing countries in this process?

Developed countries will have to acknowledge their responsibility for the current situation, and they have to financially and technically assist other countries to redress the harm they have done. On the contrary, developing countries will have to realise that their ever-growing economies and the understandable desire of their populations for better living conditions can only be environmentally sustainable if they do not repeat the same mistakes developed and industrialised countries have made over a long period. Developing countries will have to be willing to play an active part in international negotiations while implementing respective measures to counter the impacts of climate change at home. The road to change is by no means an easy route to take. It requires the acknowledgment of two basic facts: we have no time, and climate change and its impacts know no boundaries. These two facts apply to us all. There will also need to be renewed focus on climate change as a highly prioritised political topic. Financial and economic crises are serious problems and they certainly require our attention, but the impact of climate change will have more far-reaching and long-lasting effects. It is not rocket science to predict that the road to any solution will be a long and winding one. Many frustrations will have to be overcome in the process which will leave a lot of stakeholders dissatisfied.

Perhaps the most important factor remains the individual. The way we behave towards our environment determines not only our future, but also the future of generations to come. The same goes for the political process. As long as voters regard other issues as being more important than climate change, the pressure on political leaders to act and react will not be very significant. So before we point a finger at negotiators and deplore the state of negotiations, we have to ask ourselves whether we, too, have our priorities right. The history of climate change negotiations does not give us much reason to be overly optimistic as far as a comprehensive and speedy solution is concerned. However, the small margin of progress that has been achieved so far proves that success is possible.

## *References*

Böhringer, Christoph, 2001, Climate Politics from Kyoto to Bonn: From Little to Nothing, *ZEW Discussion Papers, No. 01–4*, Mannheim, Centre for European Economic Research (ZEW), University of Oldenburg, available at http://hdl.handle.ne/10419/24470, last accessed 10 November 2012.

Donner, Susanne & Felix Faltin, 2007, *Klimapolitische Entwicklungen in den USA: Initiativen auf Bundesstaatlicher und Regionaler Ebene*, Wiesbaden, Deutscher Bundestag, available at http://www.bundestag.de/dokumente/analysen/2008/Klimapolitische_Entwicklung_USA.pdf, last accessed 12 November 2012.

Fuhr, Lili, Barbara Unmüßig & Hans J.H. Verolme, 2011, *A Future for International Climate Politics – Durban and Beyond*, Berlin, Heinrich-Böll-Stiftung.

Harris, Paul G., 2010, *China and Climate Change: From Copenhagen to Cancun*, Washington, D.C., Environmental Law Institute, available at http://www.epa.gov/ogc/china/harris.pdf, last accessed 12 November 2012.

Hirsch, Thomas, 2012, *Internationale Klimapolitik: Rückblick und künftige Herausforderungen*, Bonn, Friedrich-Ebert-Stiftung, Online-Akademie.

Mapaure, Isaac, 2011, Climate Change in Namibia: Projected Trends and Effects, in: Ruppel, Oliver C. & Ruppel-Schlichting, Katharina (Eds), *Environmental Law and Policy in Namibia*, Windhoek/Essen, Orumbonde Press & Welwitschia Verlag Dr A. Eckl, 289–307.

Mayr, Christoph, 2009, *Erklärungshilfen zur Entwicklung der internationalen Klimapolitik: Spieltheorie und Public Choice Theorie*, Hamburg, Igel Verlag.

Monbiot, George, 2012, Obama and Romney Remain Silent on Climate Change, the Biggest Issue of All, *The Guardian*, 5 November 2012, available at http://www.guardian.co.uk/commentisfree/2012/nov/05/obama-romney-remain-silent-climate-change, last accessed 6 November 2012.

Ritter, Karl, 2012, 2012 UN Climate Talks in Doha, Qatar Face Multiple Challenges, *Huffington Post*, 25 November 2012, available at http://www.huffingtonpost.com/2012/11/25/2012-un-climate-talks-qatar_n_2188048.html, last accessed 4 December 2012.

Ruppel, Oliver C., 2013, Gassünder ohne Verantwortung – Der 18. Weltklimagipfel: Verlängerung des Kyoto-Protokolls – Nur Minimalerfolg, *Allgemeine Zeitung* (Namibia), 08 January 2013, available at http://www.az.com.na/umwelt/umweltsnder-ohne-verantwortung.161903.php, last accessed 8 January 2013.

# 12
# International Climate Finance: Policies, Structures and Challenges

*Cord Lüdemann & Oliver C. Ruppel*

*Abstract*

Continuing negotiations in the climate change debate have prominently exposed the topic of climate financing. It is broadly accepted that mitigation and adaptation activities require large volumes of capital. The International Energy Agency estimates that the total cost of investment to meet climate goals may amount to US$ 220 billion per year between 2010 and 2020 and to almost US$ 1 trillion per year between 2020 and 2030. Being aware of the financial needs, the international community has come up with a variety of climate financing instruments and mechanisms over the last two decades.

After familiarising the reader with the basic financing provisions of the climate change framework and the guiding principles of climate finance, the article presents recent international developments with regard to climate finance including the latest financial commitments of the international community and the establishment of the Green Climate Fund (GCF) as a new and comprehensive funding mechanism. The article subsequently gives an overview over the various public and private sources of climate financing and over the multitude of channels, mechanisms and actors involved in disbursing climate finance. In this context, the contribution focuses on those disbursement entities specifically designated for mitigation and adaptation issues within the international climate change framework. A separate part deals with the topic of accountability, transparency and corruption as both the large volumes of financial transfers and the complexity of the present climate finance architecture give rise to concern. Since the ambitious commitments of the international community and current estimates of future climate-related investment require a substantial increase of private sector capital, the article finally elaborates on the topical issue of private climate finance. After stressing the importance of the investment climate for climate investments, current barriers for private investment and the measures required for overcoming those barriers and mobilising private sector engagement are described. The success of the new GCF which is supposed to play a key role in channelling new and additional financial means will also depend on whether the instrument is able to mobilise additional private investment on a substantial scale.

## A. Introduction

Mitigation and adaptation activities require large volumes of finance and innovative financial mechanisms. For these purposes, the term *climate finance* is commonly used, though not clearly defined. In a nutshell, climate finance covers financial support for mitigation and adaptation activities, including capacity-building, research and development and broader efforts to facilitate transition towards low-carbon, climate resilient development.[1]

Historically, only a very small share of climate finance has gone to adaptation efforts: the larger proportion being used for mitigation measures[2] – arguably because it is rational to invest more in mitigation as long as the negative effects of climate change can still be reduced by enhancing GHG reductions. Buchner et al.[3] estimate that climate finance flows in 2012 have added up to US$364 billion, of which only US$14 billion have been used for adaptation.

Latest estimates of the total investment needed to tackle climate change draw a clear picture about the situation: the International Energy Agency estimates that the total cost of investment to meet climate goals may amount to US$220 billion per year between 2010 and 2020 and to almost US$1 trillion per year between 2020 and 2030.[4] With regard to adaptation, the World Bank's World Development Report estimates costs to range from US $75 billion to 100 billion per year.[5] A United Nations Framework Convention on Climate Change (UNFCCC) review concluded that "the additional investment and financial flows in 2030 to address climate change amounts to 0.3 to 0.5% of global domestic product in 2030 and 1.1 to 1.7% of global investment in 2030".[6]

1 Buchner et al. (2011:1).
2 (ibid.:7).
3 (ibid.:1).
4 International Energy Agency (2010).
5 World Bank (2010).
6 UNFCCC (2009:1).

In this light, the question that needs to be addressed is how to generate sufficient funds to address climate-change-related challenges. A variety of proposals has been suggested to generate climate finance. Global funding is derived from the public and the private sector,[7] while the amount of private finance is almost three times greater than that of public finance.[8] Main approaches towards generating capital include international taxation or international carbon markets. Official Development Assistance, as one source of climate finance, is unlikely to reach the scale necessary to meet high-level international climate finance commitments.[9] The UN Secretary General's High Level Advisory Committee has thus considered private finance to meet the targets. However, doubts remain on how private sector financing can be effectively mobilised and channelled, especially towards adaptation in developing countries.[10]

This article gives an overview of the climate financing framework. In its first parts, it presents the relevant provisions within the legal framework as well as the underlying basic principles. The chapter then focuses on the sources of climate finance and mechanisms, before finally highlighting the crucial issue of private climate finance.

7 Trying to classify climate finance offers several options. One indicator can be the dividing line between public funds and private investor commitment. This division based on the origin of the funds, however, remains cursory and imprecise. A second classification could be based on the primary purpose of the funds. The dividing line can be drawn between financial instruments that originate from the climate change framework such as the Clean Development Mechanism (CDM), official development assistance (ODA) related to climate change projects and other national and international funding dedicated to climate change mitigation and adaptation.This attempt to classify climate finance also remains vague. Moreover, it adds to the already controversial debate on the relation between climate funding and ODA. For any classification of climate finance, it will be necessary to emphasise that it consists of public funds as well as private sector capital and that it can be differentiated from ODA without losing an integrated perspective, which is particularly important for developing countries.

8 Buchner et al. (2011:III).

9 Energy and Resources Institute (2012:25).

10 Atteridge (2011:25).

### *B. The Legal Framework for Climate Finance*

The UNFCCC provides for basic rules on climate financing. When the UNFCCC was adopted, the parties acknowledged that the climate change phenomenon calls for the "widest possible cooperation between the countries and their participation in an effective and appropriate international response, in accordance with their common but differentiated responsibilities and respective capabilities and their social and economic conditions."[11] These words from the preamble already link cooperation and participation to specific responsibilities and respective capabilities, as well as to economic conditions of the countries involved. Responses to climate change have to consider financial implications for and financial responsibilities of different state actors. Financing matters therefore play an important role in combatting climate change. Accordingly, the legal framework provides financial assistance for developing countries to support the implementation of adaptation and mitigation programmes and projects under the UNFCCC.

The main provisions in the UNFCCC regarding climate finance are Article 4 paragraph 3 and 4 and Article 11 UNFCCC. Additional regulations are stipulated in Article 7 paragraph 2(h) and Article 21 paragraph 3 UNFCCC. The Kyoto Protocol also provides for the mobilisation of financial resources.[12] Article 4 UNFCCC contains the commitments of the parties under the climate change regime. In terms of financial commitments Article 4 paragraph 3 of the UNFCCC stipulates a comprehensive framework for financial assistance by developed country parties. Firstly, developed country parties and other developed parties included in Annex II are required to provide new and additional financial resources to meet the agreed full costs of developing country parties related to the required communication of information under Article 12 paragraph 1 UNFCCC.[13] Secondly, and more broadly, those parties are also asked to provide such financial resources as are needed by the developing country parties to meet the agreed full incremental costs of implementing measures that fall under Article 4 paragraph 1 UNFCCC.[14] These measures comprehensively cover mitigation and adaptation approaches and policies. As a further requirement the measures must be agreed between the developing country party and the entity running the

11 Preample to the UNFCCC.
12 Article 11 Kyoto Protocol.
13 Article 4 para. 3 UNFCCC.
14 Article 4 para. 3 UNFCCC.

financial mechanism, currently the Global Environment Facility (GEF).[15] Article 4 paragraph 4 UNFCCC specifically emphasises the financial support for those developing country parties that are particularly vulnerable to the adverse effects of climate change. Developed country parties are required to assist the countries concerned in meeting the costs of adaptation to those adverse effects.[16] Financial assistance under Article 4 UNFCCC therefore is intended as full financial support of developed countries for developing countries living up to their commitments under the climate change regime. The need for financial assistance is reaffirmed in Article 11 Kyoto Protocol with regard to its specific regulatory content.

The organisation and management of the financial mechanism is regulated in Article 11 UNFCCC. This provision defines a mechanism for the provision of financial resources on a grant or concessional basis, which is supposed to function under the guidance of, and be accountable to, the Conference of the parties (COP).[17] While the COP decides on the policies of the financial mechanism as well as its programme priorities and eligibility criteria for funding, Article 11 paragraph 1 UNFCCC requires the operation of the financial mechanism to be carried out by one or more existing international entities. For an interim period, the GEF, the United Nations Environment Programme (UNEP) and the International Bank for Reconstruction and Development (IDBR) were defined as the international entities entrusted with the operation of the financial mechanism.[18] At present, the operation of the financial mechanism is still entrusted to the GEF. The financial mechanism as such and the commissioning of the international entity is under review every four years.[19] Article 11 paragraph 2 UNFCCC requires the financial mechanism to "have an equitable and balanced representation of all Parties within a transparent system of governance." Accordingly this requirement has to be met by the international entity to be entrusted with the operation of the financial mechanism. For the GEF, this was specifically stipulated in the interim arrangements in Article 21 paragraph 3 UNFCCC.

15 The Global Environment Facility (GEF) was established as a programme in the World Bank and later restructured as a separate institution. Presently, the GEF is a financial mechanism for several global environmental conventions, including the Convention on Biodiversity and the UNFCCC.
16 Article 4 para. 4 UNFCCC.
17 Article 11 para. 1 UNFCCC.
18 Article 21 para. 3 UNFCCC.
19 Article 11 para. 4 UNFCCC.

In their interaction, the COP and the GEF are responsible for arrangements that give effect to financing activities of climate change mitigation and adaptation. These arrangements include modalities to ensure that the funded projects are in conformity with the policies, programme priorities and eligibility criteria for funding, as well as modalities by which a particular funding decision may be reconsidered in light of these policies, programme priorities and eligibility criteria.[20] Initially, the interaction between the COP and the GEF was determined in a memorandum of understanding.[21] Further communication involves regular decisions of the COP, providing additional guidance to the GEF, as well as the GEF's annual reports to the COP, in order to meet the accountability requirement laid down in Article 11 paragraph 1 and 3 UNFCCC.

Finally, Article 11 paragraph 5 UNFCCC and Article 11 paragraph 3 Kyoto Protocol clarify that, in addition to the envisaged financial mechanism, financing can also be provided through bilateral, regional or other multilateral channels. This provision offers many opportunities for state actors and other stakeholders to play an active role in climate financing. Thus, a variety of programmes and activities are carried out through channels other than that of the official financial mechanism provided for by the UNFCCC.[22] Selected financing mechanisms under the UNFCCC framework will be presented below.

## C. *Guiding Principles of Climate Finance*

As indicated in the Preamble to the UNFCCC, the topic of climate finance is linked to the guiding principle of the climate change regime, the concept of common but differentiated responsibilities and respective capabilities. This principle reveals that the climate change debate is strongly influenced by ethical considerations around responsibility, justice and fairness.[23] While responsibilities for climate impacts can be attributed to the developed world

20 Article 11 para. 3 UNFCCC.

21 Decision 12/CP.2 and decision 12/CP.3, available at http://unfccc.int/cooperation_support/financial_mechanism/guidance/items/3655.php, last accessed 24 January 2013.

22 See http://unfccc.int/cooperation_and_support/financial_mechanism/bilateral_and_multilateral_funding/items/2822.php, last accessed 24 January 2013.

23 For an elaborate discussion on ethical considerations regarding negotiations on climate finance, see Grasso (2011:361–377).

to a large extent, climate vulnerabilities are unevenly distributed and predominantly feature in developing countries. Correspondingly, Article 4 paragraph 4 UNFCCC contains a specific value of the principle of common but differentiated responsibilities and respective capabilities in that developed countries are obliged to assist the developing country parties – which are particularly vulnerable to the adverse effects of climate change – in meeting adaptation costs. Climate financing, therefore, is guided by the question on how to share burdens fairly and to distribute costs related to climate change.

The main indicator for a distribution of climate change costs has been the responsibility of developed countries for historical greenhouse gas emissions which have accumulated since the beginning of carbon-based industrial activity. These emissions have contributed to adverse changes in the climate system and weather patterns. In the light of common environmental law principles, like the polluter pays principle and the no-harm principle, the climate change regime follows up on the responsibility of developed countries for past and present greenhouse gas emissions.[24] This responsibility not only involves obligations concerning emissions reductions and limitations, but is also a basis for financial obligations. Accordingly, the industrialised countries, which have primarily contributed to climate impacts, are also held financially responsible for them. Under the climate change regime, they are obliged to give financial assistance to the more vulnerable developing countries, which face the major damage caused by changes in weather and climate patterns. Consequently, financial assistance particularly needs to support adaptation efforts in countries which do not have the resources to adapt to the impacts of climate change.

The principle of common but differentiated responsibilities and respective capabilities is the central concept within the climate change framework to strike a balance towards a fair and just allocation of financial responsibilities. Aspects of justice and fairness play an increasingly important role in negotiations on the further development of climate financing within the climate change regime owing to the growing awareness that strengthening adaptation is crucial. In the past, by far the largest share of financial means has been spent on mitigation efforts in industrialised and large developing coun-

24 While the polluter pays principle is not explicitly mentioned, the no-harm principle features in the Preamble to the UNFCCC and the precautionary principle are included in Article 3 para. 3 UNFCCC.

tries such as China, Brazil and India.[25] The diagram presented in the introduction indicates the small share of climate finance which supports adaptation measures. According to newest findings, in 2010 to 2011, mitigation activities attracted US$350 billion, predominantly focusing on renewable energies and energy efficiency, while financing for adaptation covered US $12.3 billion to 15.8 billion.[26] The figures in this report also reveal that, in 2010 to 2011, the private sector contributed 74% of the finance for mitigation measures, while in adaptation finance, public financial institutions were the predominant sources, accounting for 77.5 % of the total.[27]

Recent negotiations of the climate change regime have focused on increasing the financial capacities and have also strengthened the case of adaptation.[28] The way of administering adaptation funding, especially, has become a crucial element for the development of international climate policy.[29] As soon as discussions concern increased funding for adaptation measures in developing countries, they touch on issues around responsibility, justice and fairness. Whereas mitigation is still mostly a topic in industrialised countries and larger developing countries, adaptation measures are needed particularly in less developed regions.[30] Moreover, developing and especially least developed countries are most vulnerable to climate change impacts, although they contributed least to historical greenhouse gas emissions that are responsible for climate change at present.[31] Fairness, therefore, demands an increased transfer of financial resources from industrialised to developing countries. Accordingly, the challenge for the parties and stakeholders involved in the climate negotiations is to develop secure, adequate and predictable funding streams for the financing of adaptation needs in poorer, more vulnerable countries with least adaptive capacity.[32]

---

25 Von Bassewitz (2011:316–318). Von Bassewitz states that the Clean Development Mechanism (CDM), established under the Kyoto Protocol, has been the largest mechanism to transfer funding from industrialised to developing countries. The CDM has however often been criticised as favouring projects in large developing countries, while being unattractive for projects in smaller developing and least developed countries.

26 Buchner et al. (2012:49).

27 (ibid.:50–54).

28 For the recent developments, see the following part of this chapter.

29 Grasso (2011:362).

30 Dellink (2009:411).

31 (ibid.).

32 Grasso (2011:362).

### D. Recent International Developments Concerning Climate Finance

Since the 1990s growing awareness in the climate change discussion has also exposed the topic of climate financing. The permanent discussions and negotiations in the climate change debate have led to a variety of climate financing instruments and mechanisms. Adjustments of old and introduction of new climate-related funds have been on the agenda at regular intervals. With the decisions taken at the 16th COP to the UNFCCC in Cancun, the international community embarked on the development of a new funding framework, stating that a scaled up, new and additional, predictable and adequate funding is envisaged.[33]

Developed countries committed themselves to fast-tracking the provision of funding in the amount of US$30 billion for the period 2010 to 2012 in the Cancun Agreement. The decision refers to new and additional financial resources and aims at "a balanced allocation between adaptation and mitigation".[34] For the most vulnerable developing countries, the commitment states that funding for adaptation is regarded as a priority. Beyond that fast-track pledge, the Cancun Agreement also contains a commitment of the developed country parties to "a goal of mobilizing jointly US$100 billion per year by 2020 to address the needs of developing countries". This pledge of the international community, in principle, represents one of the largest development programmes in history.[35] However, the international debate also acknowledged that there are "no individual sources that can simultaneously deliver the US$100 billion target and meet the full range of end-use requirements".[36] Correspondingly, the Cancun Agreement also reaffirmed that funding may derive from multiple sources, including public and private, multilateral and bilateral, as well as alternative sources. In this context, the Cancun Agreement acts on the specific financing provisions in the Bali Action Plan. These provisions call upon enhanced action on the provision of financial resources, including, inter alia, improved access to adequate, predictable and sustainable financial resources, the provision of new and addi-

33 UNFCCC, Decision 1/CP.16 The Cancun Agreements, Outcome of the Work of the Ad Hoc Working Group on Long-term Cooperative Action under the Convention, 16, available at http://unfccc.int/resource/docs/2010/cop16/eng/07a01.pdf, last accessed 24 January 2013.

34 (ibid.).

35 Donner (2011:908).

36 UN Secretary General (2010:35).

tional resources, as well as the mobilisation of public- and private-sector funding and the facilitation of climate-friendly investment choices.[37] The ambitious US$100 billion target will require that the envisaged funding sources are secured well in advance of 2020 "in order to allow for sufficient time to develop both the capacity to deliver and the capacity to use wisely the flow of funds made available".[38]

Another important decision in the Cancun agreement is the establishment of the Green Climate Fund (GCF). This new financial instrument will channel both the initial US$30 billion and a substantial fraction of the envisaged US$100 billion per year.[39] The Cancun agreement also points out that the GCF will be in charge of a significant share of new funding for adaptation.[40] According to the Cancun agreement, funding is proposed to flow through multiple channels including public, private, bilateral, and multilateral sources.[41]

The implementation of the GCF under the guidance of – and accountable to – the COP with a balanced and comprehensive governing instrument as well as an intermediary process to get the Fund up and running as quickly as possible has been one of the results of the climate negotiations held at the 17th COP in Durban, South Africa.

Recent decisions taken at the 18th COP in Doha, Qatar, (dubbed the Doha Climate Gateway) emphasise the importance of financing mechanisms in the field of climate change. In the Work Programme on Long-Term Finance it has, for example, been decided:[42]

> to extend the work programme on long-term finance for one year to the end of 2013, with the aim of informing developed country Parties in their efforts to identify pathways for mobilising the scaling up of climate finance to USD 100 billion per year by 2020 from public, private and alternative sources in the context of meaningful mitigation actions and transparency on implementation, and informing Parties in enhancing their enabling environments and policy frameworks to facilitate the mobilization and effective deployment of climate finance in developing countries.

37 UNFCCC, Decision 1/CP.13 *Bali Action Plan*, para. 1 (e), p. 5, available at http://unfccc.int/resource/docs/2007/cop13/eng/06a01.pdf, last accessed 24 January 2013.

38 UN Secretary-General (2010:35).

39 Donner (2011:908).

40 Van Kerkhoff et al.(2011:19).

41 Donner (2011:908);Van Kerkhoff et al. (2011:19).

42 UNFCCC draft decision -/CP.18 Work Programme on Long-term Finance, Advance unedited version, available at http://unfccc.int/files/meetings/doha_nov_2012/decisions/application/pdf/cop18_long_term_finance.pdf, last accessed 24 January 2013.

The agreement also encourages developed countries to increase efforts to provide finance between 2013 and 2015 at least to the average annual level at which they provided funds during the 2010 to 2012 fast-track finance period. This is to ensure that there is no gap in continued finance support while efforts are being scaled up. Furthermore, governments will continue a work programme on long-term finance during 2013 to contribute to the on-going efforts to scale up mobilisation of climate finance and report to the next COP on pathways to reach that target. Germany, the United Kingdom, France, Denmark, Sweden and the European Union Commission announced concrete finance pledges in Doha for the period up to 2015, totalling approximately US$6 billion.

COP18 has also taken note of the first annual report of the Board of the Green Climate Fund to the Conference of the parties and endorsed the consensus decision of the Board of the Green Climate Fund to select Songdo, Incheon, Republic of Korea, as the host of the Green Climate Fund, on the basis of an open and transparent process.[43]

Moreover, the UN Climate Change Secretariat and World Economic Forum have launched an initiative called Momentum for Change: Innovative Financing for Climate-friendly Investment, which aims at identifying and highlighting creative financing models that enable adaptation and mitigation activities in developing countries.

### *E. Sources of Climate Finance and Selected Disbursement Channels*

The landscape of climate finance is complex and manifold. Buchner et al.[44] have described and surveyed the difficult material in detail, establishing the diagram presented in the introduction. A closer look at this diagram reveals the complex structure. The following part aims at giving an overview of the different sources of climate finance and the specific disbursement channels implemented under the UNFCCC and the Kyoto Protocol.

43 UNFCCC Draft decision -/CP.18 Report of the Green Climate Fund to the Conference of the Parties and guidance to the Green Climate Fund, Advance unedited version, available at http://unfccc.int/files/meetings/doha_nov_2012/decisions/application/pdf/cop18_report_gcf.pdf, last accessed 24 January 2013.

44 Buchner et al. (2012).

### *I. Sources of Climate Finance*

Funding for climate change mitigation and adaptation can derive from a variety of sources, including public and private, bilateral and multiple, as well as alternative sources.[45]

In the debate on climate finance, the main differentiation is made between public and private sources. In the period 2010 to 2011, public sources such as government budgets and development finance institutions provided 26% of total climate finance of US$343 billion to 385 billion.[46] In this context, the specific contribution of government budgets ranged between US$16 billion and 22.6 billion (5% of total climate finance), including direct public investments and north-south aid flows.[47] Public funding through development finance institutions accounted for US$76.8 billion.[48]

In the public sector, funding through government budget particularly derives from domestic revenues through direct budget contributions.[49] Scaling-up public finance requires stakeholders to break new ground to increase funding through government budget. In addition to general tax revenues provided for climate funds, new public instruments need to be introduced or, if already applied, expanded. The introduction of instruments based on carbon pricing not only raises revenue, but also provides incentives for mitigation actions.[50] These carbon-related instruments may vary in their design and follow different approaches, e.g. tax-based approaches or the introduction of carbon markets. According to new findings for the period 2010 to 2011, carbon taxes account for US$7.3 billion, while carbon market revenues contribute US$2 billion.[51] The largest amount of the latter derives from the European Union Emissions Trading System, accounting for US$1.62 billion.[52] The further expansion of carbon market instruments, such as international or domestic auctioning of emissions allowances and emissions

---

45 For example, UNFCCC, Decision 1/CP.16 The Cancun Agreements: Outcome of the Work of the Ad Hoc Working Group on Long-term Cooperative Action under the Convention, p 16, available at http://unfccc.int/resource/docs/2010/cop16/eng/07a01.pdf, last accessed 24 January 2013.

46 Buchner et al. (2012:18).

47 (ibid.).

48 (ibid.).

49 UN Secretary-General (2010:7).

50 (ibid.:12).

51 Buchner et al. (2012:22).

52 (ibid.:20).

trading schemes, as well as carbon taxes, was regarded to have the potential to contribute significantly to raising the envisaged US$100 billion per year.[53] Other prospective instruments for raising revenue include, inter alia, taxes on international aviation and shipping, charges on electricity generation, fossil fuel extraction royalties, the removal of fossil fuel subsidies and an international financial transaction tax.[54]

Main contributors in the public sector are development finance institutions providing 21% of total climate funding. The multilateral, bilateral, subregional and national finance institutions falling under this category are generally dominated by national or member governments who fund their capital base.[55] They operate as a crucial link between public and private finance in that they add significant value to financing packages by subsidising interest rates, transforming the maturity of loans to long-term, and absorbing a share of the risks of the loans handed out to the private sector.[56] Thus, development finance institutions have the capacity to leverage large additional amounts of public and private investment in a way that integrates climate action into development programmes.[57]

The private sector, including, inter alia, corporate actors, project developers, commercial financial institutions and households, is the dominant source, providing US$250 billion to 286 billion out of the total climate finance of US$343 billion to 385 billion in the 2010 to 2011 period.[58] Project developers account for the largest contribution in the private sector (34% to climate investments flows in total).[59] In the period 2010 to 2011, they provided US$115 billion to 129.3 billion for designing, commissioning, operating and maintaining emissions reduction projects.[60] The second largest contribution in the private sector is made by corporate actors. In the period 2010 to 2011, they provided 21% of global climate finance by investing in emissions reduction projects such as renewable energies or energy efficiency measures, as well as by financing of own technologies in the case of manu-

53 UN Secretary-General (2010:12).
54 (ibid.:20). The report (at page 15) particularly highlights "carbon pricing of international transport as an important potential source for climate financing (and mitigation) that could contribute substantially towards mobilizing US$100 billion".
55 Buchner et al. (2012:24).
56 (ibid.).
57 UN Secretary-General (2010:32).
58 Buchner et al. (2012:18).
59 (ibid.:29).
60 (ibid.).

facturers of renewable energy systems.[61] Commercial financial institutions and households accounted for 10% and 9%, respectively, of total climate finance in the years 2010 to 2011.[62] These figures reveal that private investment is already quite substantial. However, findings also indicate that most of the private finance is currently invested in mitigation projects in developed countries.[63] Consequently, it will be necessary to scale-up private capital flows in order to meet the goal of mobilising US$100 billion per year by 2020 to address the needs in developing countries.[64] Part G of this chapter will focus on this challenge and the important role of the private sector in scaling-up climate finance, especially for adaptation needs in developing countries.

### *II. Selected Disbursement Channels*

The actors and channels involved in disbursing climate finance are manifold. It is beyond the scope of this chapter to give a detailed insight into the multitude of organisations and funding mechanisms. Therefore, the following part focuses on the entities specifically designated for climate change mitigation and adaptation issues within the UNFCCC and Kyoto Protocol framework.

In general, specific funding for mitigation and adaptation can be based on both international mechanisms and funding instruments under the UNFCCC framework, as well as bi- or multilateral arrangements. The general financing mechanism provided for in Article 11 UNFCCC is complemented by several specific funding instruments under the UNFCCC and/or the Kyoto Protocol, including, inter alia, the Climate Investment Funds (CIF), the Adaptation Fund, as well as the new Green Climate Fund (GCF), established at COP 16 in Cancun, Mexico. Corresponding to the multitude of funding mechanisms, a variety of organisations is involved in disbursing climate funds. They may be local, regional, national, or international organisations from the public and private sector, including, for example, public-private

61 (ibid.:26).
62 (ibid.:18).
63 (ibid.:26ff.).
64 UN Secretary-General (2010:5).

partnerships (PPPs), local financial institutions and banks, multilateral organisations, non-governmental organisations, and civil society.[65]

The financial mechanism under Article 11 UNFCCC promotes projects in energy efficiency, renewable energy, sustainable urban transport and sustainable management of land use, land-use change, and forestry.[66] The operating entity of the financial mechanism is the GEF.[67] The GEF was set up to provide grants for global environmental benefits and, presently, is the world's largest grant funding source dedicated to multilateral environmental agreements and public goods.[68] The daily business of the GEF is run by the World Bank serving as the facility's trustee. With regard to operating the financial mechanism of the UNFCCC, the main task of the GEF is to facilitate multilateral financial support for developing countries, including support for adaptation measures.[69] In doing so, the GEF allocates hundreds of millions of dollars per year for mitigation and adaptation projects in developing countries and economies in transition.[70] Concerning adaptation measures, the GEF initially focused on covering costs for capacity-building and research.[71] In 2005, the GEF expanded its portfolio and introduced the Strategic Priority on Adaptation (SPA), which marks a shift from focusing on planning and capacity-building towards practical adaptation measures.[72]

Several other mechanisms have been introduced to support the GEF's efforts in adaptation funding. In 2001, COP 16 in Bonn recognised the increasing importance of adaptation measures and building adaptive capacity and, consequently, introduced two special funds dedicated to adaptation: the Special Climate Change Fund (SCCF) and the Least Developed Country Fund (LDCF).[73] For both funds, the GEF is the designated operating entity.

The SCCF aims at supporting adaptation and technology transfer in all developing country parties to the UNFCCC.[74] The fund assists developing countries in diversifying their economies, in preparing their national communications to the UNFCCC and in strengthening implementation of adap-

65 Buchner et al. (2012:49).
66 See http://www.thegef.org/gef/climate_change, last accessed 24 January 2013.
67 For history and legal background, see part B. of this chapter.
68 Di Leva (2010:373).
69 McGoldrick (2007:52).
70 See http://www.thegef.org/gef/climate_change, last accessed 24 January 2013.
71 Bouwer & Aerts (2006:53).
72 McGoldrick (2007:52).
73 Bouwer & Aerts (2006:51).
74 See http://www.thegef.org/gef/SCCF, last accessed 24 January 2013.

tation activities related to their national communications.[75] Projects supported by the SCCF include both long-term and short-term adaptation activities in water resources management, land management, energy, agriculture and health, as well as in infrastructure development and fragile ecosystems.[76] According to the latest figures (June 2012), the SCCF adaptation programme had mobilised US$162.24 million for projects and programmes in non-Annex I countries, while the technology transfer window has supported six projects in total, accounting for US$26.64 million.[77]

The LDCF was established to support activities in least developed countries, drawing on financial contributions from developed countries.[78] The fund is specifically designated for the financing of the preparation and implementation of National Adaptation Programmes of Action (NAPAs) in least developed countries.[79] Latest figures (June 2012) reveal that, since its inception, the LDCF has funded the preparation of 48 NAPAs and, subsequent to adopting the NAPA, 46 countries have officially submitted NAPA implementation projects for approval.[80] Altogether, the LDCF now supports 74 projects and 1 programme in 44 countries, accounting for US$334.6 million in total and leveraging another US$1.59 billion in co-financing.[81]

Another fund specifically dedicated to adaptation measures in developing countries is the Adaptation Fund. Unlike the other UNFCCC funds for adaptation, the Adaptation Fund is not regulated by the Convention but by the Kyoto Protocol.[82] The Adaptation Fund is still a young financing instrument. Although already provided for in the 1997 Kyoto Protocol and being established at COP 7 (Marrakesh) in 2001, the final negotiations and decisions on the management and governance of the Adaptation Fund took place at COP12 (Nairobi) and COP13 (Bali) in 2006 and 2007, respectively.[83] The Fund therefore could only start operating in 2008. The operating entity of the Adaptation Fund is the GEF. Funding is provided for "the implementation of concrete adaptation projects in Non-Annex I countries, including activities aimed at avoiding forest degradation and combating land degra-

75 Bouwer & Aerts (2006:51).
76 See http://www.thegef.org/gef/SCCF, last accessed 24 January 2013.
77 (ibid.).
78 McGoldrick (2007:52).
79 See http://www.thegef.org/gef/LDCF, last accessed 24 January 2013.
80 (ibid.).
81 (ibid.).
82 Grasso (2011:363).
83 (ibid.: 363f.).

dation and desertification."[84] Since 2010, the Adaptation Fund has dedicated more than US$165 million to increase climate resilience in 25 countries around the world.[85] Provided for by Article 12 paragraph 8 of the Kyoto Protocol, resources for the Adaptation Fund partly come from a share of 2% of certified emissions reductions issued for Clean Development Mechanism (CDM) projects.[86] Further resources for funding are contributions from governments, the private sector, and individuals.[87]

Additionally, two Climate Investment Funds (CIF) established in 2008 address both mitigation and adaptation needs in developing countries through combining grants with highly concessional funding and risk reduction instruments.[88] The CIF consist of a pair of funds, the Clean Technology Fund (CTF) and the Strategic Climate Fund (SCF), which are channelled through the African Development Bank, the Asian Development Bank, the European Bank for Reconstruction and Development, the Inter-American Development Bank, and the World Bank Group. The CTF focuses on mitigation actions and specifically supports low-carbon technologies in developing countries.[89] While the CTF therefore promotes concrete national investment plans for demonstration, deployment and transfer of clean technologies, the SCF serves as an overarching fund to support targeted programmes such as the Forest Investment Program (FIP) and the Program for Scaling-Up Renewable Energy in Low Income Countries (SREP).[90] Another programme promoted by the SCF is the Pilot Program for Climate Resilience (PPCR), which addresses adaptation needs of developing countries.[91] The financial resources for both funds amount to US$6.5 billion, pledged by 14 contributors.[92]

The youngest climate-related disbursement mechanism is the GCF.[93] This fund was established within the UNFCCC framework. Its purpose is to con-

84 Bouwer & Aerts (2006:52).
85 See https://www.adaptation-fund.org/about, last accessed 24 January 2013.
86 Grasso (2011:363).
87 See https://www.adaptation-fund.org/about, last accessed 24 January 2013.
88 Di Leva (2010:378).
89 (ibid.:378).
90 See http://www.climateinvestmentfunds.org/cif/designprocess, last accessed 24 January 2013.
91 Di Leva (2010:378).
92 See http://www.climateinvestmentfunds.org/cif/funding-basics, last accessed 24 January 2013.
93 For the relevant negotiations, see part D. of this article.

tribute to the achievement of the ultimate objective of the UNFCCC. The GCF will be an operating entity of the financial mechanism under Article 11 UNFCCC and will be governed and supervised by a board with full responsibility for funding decisions.[94] An interim secretariat runs the daily business for the Board of the GCF and, as an interim trustee, the World Bank manages the financial assets of the Fund. The main task of the GCF is to support projects, programmes, policies and other activities in developing countries relating to climate change by using thematic funding windows.[95] As the important role of the GCF with regard to mobilising new funding sources was emphasised during the climate change negotiations, the prospects of this new instrument will be evaluated separately at the end of the following part of this chapter, focusing on the need for increased private sector finance.

## *F. Corruption Risks in Climate Finance*[96]

The previous section gave an overview of the substantial volumes of funding and the multitude of mechanisms involved in the climate finance arena. The overall climate finance architecture is becoming a giant platform for financial resources being shifted from developed countries to developing countries. The large volumes of funding, as well as the complexity of the climate finance architecture, increasingly raise questions about accountability, transparency and corruption.

Deficiencies in transparency and accountability are manifold. They start with the lack of clear definitions of what exactly constitutes climate change funding. Many COP decisions within the UNFCCC refer to new and additional funding. Subsequent to those decisions, it has often been disputed whether funding of climate change mitigation and adaptation is additional to the existing development aid architecture or whether it is possible for developed countries to label existing climate change induced activities as

94 UNFCCC, Decision 3/CP.17 Annex II. A. + B., available at http://unfccc.int/resource/docs/2011/cop17/eng/09a01.pdf#page=58, last accessed 24 January 2013.

95 UNFCCC, Decision 1/CP.16 The Cancun Agreements: Outcome of the Work of the Ad Hoc Working Group on Long-term Cooperative Action under the Convention, 17, available at http://unfccc.int/resource/docs/2010/cop16/eng/07a01.pdf, last accessed 24 January 2013.

96 Parts of the following section are based on Ruppel (2013:308f.).

developing aid. Although the Cancun agreement now specifically states that the new climate change funding must be additional to existing aid commitments, the definition of additionality of climate change financing is still vague.[97] Beyond difficulties around definitions, other deficiencies result from the lack of consistent transparency requirements within the UNFCCC framework, as well as the existing multitude of funding flows, disbursement channels and actors involved. This complexity has led to a patchwork of incomplete, inconsistent, multiple and overlapping data standards and repositories, so that even where data is provided it is difficult to track, analyse and use.[98]

Lack of transparency and accountability within the complex architecture of climate financing increases the danger of the abuse of entrusted power for private gain. The *2011 Global Corruption Report: Climate Change*[99] addressed this danger and stated that corruption was indeed a risk in addressing climate change, since a risk of corruption always exists where "huge amounts of money flow through new and untested financial markets and mechanisms" – the latter sentiment being particularly true for recent, current and future financial flows related to climate change finance, technology and capacity-building meant to support developing countries according to the principle of equity. Indigenous and rural poor communities in remote locations, the urban poor living in precarious settlements, and displaced persons, especially women and children, are especially adversely affected by climate change and they are actually meant to be the main beneficiaries of adaptive action. However, corruption eventually puts at risk the rights of those most vulnerable to the negative effects of climate change. The reasons for the high risk of corruption with regard to climate finance are rooted in the level of complexity, uncertainty and novelty that surrounds many climate issues. A multitude of regulatory grey zones and loopholes exist that are at risk of being exploited by those with corrupt interests.

The report states that "US$250 billion per annum will eventually flow through new, relatively uncoordinated and untested channels" and that "[s]ome estimate total climate change investments in mitigation efforts alone at almost US$700 billion by 2020". Furthermore, carbon markets have been adopted in a number of regions and countries as one method of reducing GHG emissions and it is estimated that the value of leading carbon markets

97 Donner (2011:908).
98 Forstater & Rank (2012:23).
99 Transparency International (2011).

has now reached some US$144 billion. In order to ensure that the investments by the public and private sectors are properly and equitably managed, a system of good climate governance[100], with participatory, accountable, transparent, inclusive and responsive policy development and decisions and the respect of the rule of law, is essential.

### *G. Private Climate Finance*[101]

Despite the substantial amounts of climate finance that are already spent, particularly in developed countries, the Cancun target of an additional US $100 billion per year for developing countries by 2020 remains an ambitious goal. Stakeholders in the international climate change debate, thus, have focused on the challenges connected to this target during the last COP meetings.[102] One of the major challenges in the near future will be the need successfully to leverage private investment in developing countries.

#### *I. Climate Finance and the Role of the Private Sector*

Previous to the Cancun COP, four groups of potential sources of finance have been identified: public sources for grants and highly concessional loans (including, inter alia, carbon taxation, auctioning of emission allowances and removal of fossil fuel subsidies); development bank-type instruments; carbon market finance; and private capital.[103] Accordingly, the Cancun agreements expressly include private investment as one of the sources providing funds to developing countries. The UN Secretary-General's High-Level Advisory Group on Climate Change Financing stated that private investment "in mitigation and adaptation activities will depend on a mix of Government policies, including regulation, standards, support for new technologies, implicit and/or explicit carbon pricing, improved investment cli-

---

100 *Climate governance* can be understood as the processes that currently exist at the international, national, corporate and local levels to address the causes and effects of climate change. See Transparency International (2011).

101 Parts of the following sections are based on Ruppel & Lüdemann (2013).

102 See part D. of this article.

103 UN Secretary-General (2010:9).

mate and the availability of risk-sharing instruments".[104] There is a large potential of sources originating from the private sector. In 2010, for example, private flows of development aid amounted to US$300 billion.

***Figure 1: Development Assistance in 2010 in Millions of US Dollars***

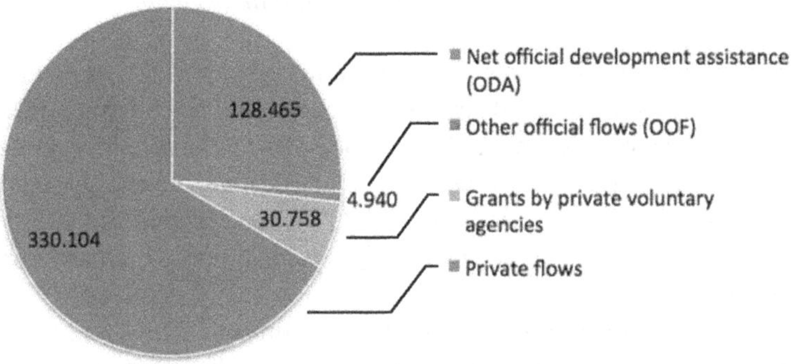


Figures from
*Developement: Key tables from OECD*
available at http://www.oecd-ilibrary.org/development/development-key-tables-from-oecd_20743866;jsessi%20onid=1ovg6qen403kx.delta, last accessed 10 May 2013

In the climate finance sector, private funding is in the form of debt investments and private equity. Further climate finance instruments include policy incentives, risk management facilities, carbon offset flows and grants. Innovative mechanisms to activate private capital need to be identified continuously. Ideas to tap private sources for climate finance have emerged, such as guarantees, funds of funds, project aggregation mechanisms, climate bonds and public-private funds.[105] All the aforementioned ideas require, above all, a reliable regulatory framework for attracting private sector capital to tackle climate change, particularly in developing countries. Political instabilities and financial flaws are major barriers for private investors. However, the ambitious Cancun commitments and the estimates of international institutions concerning required climate-related investment require a substantial increase of private sector capital. The fulfilment of climate change targets – be they financial commitments; be they the limitation of further

104 (ibid.:35).
105 See the more detailed discussion below under IV. 2. and Energy and Resources Institute, (2012:28ff.).

temperature increases – depend highly on the success in mobilising private capital. Therefore, it remains crucial to overcome barriers and create an enabling environment for private investor capital.

## *II. The Role of the Investment Climate for Climate Investment*

Addressing the impacts of climate change requires substantial investment in new technologies, processes and services. Global new investment in clean energy is a good example of the high relevance of a favourable investment climate for climate change: new investment in the sustainable energy sector set a new record in 2010 by reaching US$211 billion, an increase of 32% from a revised US$160 billion in 2009, and more than 600% increase from 2004 (US$33 billion).[106]

Generating and allocating the investment and financial flows needed to attain the levels of growth necessary for job creation and poverty reduction and thus to meet the Millennium Development Goals (MDGs) and at the same time to finance significant climate change mitigation is not an easy task. Taking that the private sector is the major investor in renewable energy and energy efficiency worldwide and in developing countries, a country's investment climate is one essential factor for increased climate investment. A favourable investment climate is pivotal for investments, particularly from the private sector in clean and climate-resilient technologies and renewable energy. Innovative solutions and technologies can however only be implemented, if adequate framework conditions for inclusive climate investment, leveraging private sector resources, and seizing opportunities for innovation exist.

A number of instruments to improve the investment climate have emerged at global, regional, national and sub-national levels. Various factors result in an unfavourable investment climate, including poor governance, institutional failures, macroeconomic policy imperfections and inadequate infrastructure, as well as rampant corruption, bureaucratic red tape, weak legal systems and a lack of transparency in government departments. The World Bank's *Doing Business Report* is one of the instruments to analyse the business climate by tracking a set of indicators, for ranking purposes combined in nine topics, namely starting a business, dealing with construction permits,

106 UNEP and Bloomberg New Energy Finance (2011:12).

registering property, getting credit, protecting investors, paying taxes, trading across borders, enforcing contracts, and closing a business. In the past five years, about 85% of the world's economies have made it easier for local entrepreneurs to operate, through improvements to business regulation. The rankings for 185 countries in 2012[107], however, reveal that of the 33 countries classified as low-income countries only two are within the rankings and lie between 50 to 100 (Rwanda 52nd and Kyrgyz Republic 70th); 17 of the 33 low-income countries rank among the last 50 of 185 countries in total. Of the 50 lowest ranking countries, 32 are in Africa, a continent most vulnerable to the negative effects of climate change. When comparing the World Bank's African 'Ease of Doing Business' ranks of 2011 and the previous year, it can be observed that 10 African countries retained the same ranking they had received in 2010, 24 African countries have been downgraded, while 17 African countries could obtain a higher rank as a result of policy reforms and initiatives with positive impact on the investment climate.

The figures above correspond to those on Foreign Direct Investment (FDI) in Africa, as contained in the United Nations Conference on Trade and Development's World Investment Report (2012). Having reached a peak in 2008, the FDI inflows to Africa continued to decline in 2010, with divergent trends among subregions. "The fall in FDI flows to Africa seen in 2009 and 2010 continued into 2011, though at a much slower rate. The 2011 decline in flows to the continent was due largely to divestments from North Africa. In contrast, inflows to sub-Saharan Africa recovered to $37 billion, close to their historic peak."[108] Although it remains difficult for the African continent to attract foreign capital and mobilise adequate and sustained levels of domestic private investment, some African countries including Mauritius, Botswana, Ghana and Tunisia have made progress and could achieve higher levels of investment.[109]

### *III. Investment Barriers*

Investment barriers have to be evaluated according to the respective country specifics. Several attempts have been made to categorise investment barriers. A survey of those attempts reveals that the barriers are interrelated and,

107 See http://www.doingbusiness.org/ranking, last accessed 24 January 2013.
108 UNCTAD, World Investment Report (2012).
109 Ruppel (2011).

therefore, cannot be strictly divided into groups. However, certain features allow for a categorisation into political/regulatory, project-related and financing barriers, bearing in mind that single risks correlate and especially the group of financing risks to a certain extent results from political and project-related barriers.[110]

With regard to a first category, key risks for private sector investors evolve from political and/or regulatory instabilities. This group of barriers includes, inter alia, political instability, insecurity of property rights, lack of knowledge of legal systems, currency risks, as well as the instability and uncertainty of the regulatory and policy environment, including, for example, the duration of incentives programmes.[111] Another group of barriers is connected to the respective project. In this group, technology risks such as limited performance track records or limited market penetration may have negative effects.[112] Technology risks usually come with high initial costs for the developer. Other project-related risks include execution inefficiency and unfamiliarity risks based on insecurity concerning the capacities and experiences of local project developers, but are often also based on the lack of investor experience in an unknown field.[113] The third group of barriers is related to financing risks and partly results from regulatory and/or project-related barriers and partly from original risks. This category, particularly, features technology cost gaps between high- and low-emission alternatives.[114] Although some renewable energy technologies nowadays expand fast, they are still in their infancy in respect of their market performance. Like with any new technology, project developers are confronted with higher market volatility. Consequently, a market entry entails capital intensity. In addition to this specific technology cost gap, the financial challenges are substantially increased by market distortions based on the market maturity of conventional high-emission technologies and subsidies for the fossil fuel sector, which fall under the first group of regulatory barriers and have to be addressed by the policy maker. Further financial risks include, inter alia, debt availability, reasonable debt terms and equity availability.[115] Particularly

110 This categorisation is based on and reflects Patel (2011:7). See also Brown & Jacobs (2011:2); Sierra (2011:7f.).

111 Brown & Jacobs (2011:2).

112 Patel (2011:7).

113 Brown & Jacobs (2011:2).

114 Sierra (2011:8).

115 Patel (2011:7).

developing countries often feature incomplete financial markets, which makes reliable estimates for risk-adjusted returns difficult and results in a lack of financial instruments to diversify risk over long-term projects.[116]

### *IV. Targeting Investment Barriers*

Mobilising private sector engagement in climate change mitigation and adaptation requires political and financial programmes to overcome substantial barriers on different levels. A catalogue of coordinated and integrated measures must aim at developing a supportive and enabling environment for climate change-related investments. Support policies have to be identified for each category of barriers, and implemented at different levels. While designing strategies and programmes generally emanates from the policy level, project developers and private investors are inclined rather to demand concrete financial instruments to support engagement in climate change-related activities.

#### *1. Policy Reform towards Climate Resilience*

At the policy level, governments have to design and implement strategies and policies for low-emission development to enhance an enabling investment environment.[117] Strategies and policies for low-emission development include, inter alia, measures like reforms of fossil fuel subsidies, renewable energy feed-in tariffs and energy efficiency programmes. The policy maker has to coordinate these measures and integrate them in a coherent policy framework. Without government intervention working in that direction, low-emission alternatives will not be competitive. Removing fossil fuel subsidies and pricing the carbon externality adequately will alleviate pricing distortions that currently work against renewable energies and energy efficiency and contribute to creating a level playing field between energy sources.[118] However, the extent to which policy support measures can contribute to market transformation depends on the strength of the leadership

116 Sierra (2011:8).
117 (ibid.:10).
118 Patel (2011:8).

and the reform programme.[119] Support measures will only reach maximum efficiency and vigour if the regulatory framework is strict and transforms markets according to climate change necessities. Only markets that provide a level playing field between energy sources will attract private investors on a large scale. Regulatory measures therefore have to apply market-wide, instead of being directed to single projects or technology solutions in particular.[120]

The requirement of an integrated policy reform and a coherent regulatory framework poses great challenges on developing countries. Climate change actions and strategies cannot be separated from each country's broader economic and social development programmes, but have to be closely integrated with development strategies and investment plans.[121] Attempts to coordinate climate change and development strategies are ambitious and test the already existing financial constraints. Correspondingly, climate finance already plays an important role at this early stage. In this context, financial assistance does not provide direct funding for private sector activities, but goes into national government budget accounts to support policy reform.[122]

### 2. *Financial Instruments to Leverage Private Investments*

In addition to budget support to create an enabling regulatory framework, it remains mandatory to leverage private investments in projects supporting a low-emission development. Only a stable and competitive risk-return profile of climate investments will mobilise private sector capital and thus contribute to achieving the significant investment volumes required in international climate finance.[123] Thus, for the design of any climate finance architecture, it remains crucial to ensure that scarce public funds are applied to mobilise and leverage private sector investments.[124]

A number of financial tools and initiatives are discussed to address investment risks and potential barriers. They follow different approaches in that they leverage either debt or equity through involving direct public fi-

119 Sierra (2011:11).
120 Patel (2011:10f.).
121 De Nevers (2011: 4).
122 Sierra (2011:10).
123 UNEP Finance Initiative (2011).
124 De Nevers (2011:3).

nancing or providing public guarantees. The different designs of the financial tools facilitate a flexible utilisation, depending on the specific conditions of the project or the specific needs in the particular country. In general, the financial instruments aim at strengthening the role of the private sector as an investor and focus on providing new sources of capital for developing countries.[125]

Among the financial tools leveraging debt, loan guarantees and policy risk insurances are most prominent. Both tools protect private capital investors against risks of default. By using loan guarantees, governments and other public finance institutions underwrite loans to projects and, in doing so, ensure that the loan will be repaid if the borrower is not able to pay.[126] Similar instruments which decrease risk of default for private investors are cash grants and concessional financing. Policy risk insurances are used for climate investments in developing countries to reduce political and even currency and legal risks in order to ensure private investors adequate returns.[127] This financing instrument can involve conventional insurances which cover the risk of policy change, e.g. the risk of abandoning or reducing an existing feed-in tariff supporting renewable energy projects.[128] Policy risk insurances are able to reduce certain risks included in the regulatory framework and provide investors with certainty. However, this option might not be feasible for every developing country. The insurance sector will factor the risks involved in every single country so that this financing instrument "is most likely to succeed in countries with strong regulatory systems and institutions, and where certain policies are already in place or under development".[129]

Equity-leveraging tools are either structured as funds, directly investing in companies and projects, or as a fund of funds that invests in commercially managed funds, which then invest in concrete projects.[130] Pledge funds are one of the instruments used for leveraging private equity. In this model, governments or international financial institutions act as public finance sponsors in that they provide an initial amount of equity to mobilise much

125 Sierra (2011:12).
126 Brown & Jacobs (2011:2).
127 UNEP Finance Initiative (2011).
128 Brown & Jacobs (2011:2).
129 (ibid.:2).
130 De Nevers (2011:24).

larger amounts of private capital.[131] Pledge funds are an interesting financing option in cases where projects have difficulties to access sufficient equity because capital investors are reluctant to invest owing to geographic, country and execution risks.[132] Pledge funds can also be applied for projects that usually have a strong rate of return, but still have limited access to equity because they are too small for private investors to be considered.[133]

Fund of funds approaches are an attractive solution for institutional investors as they allow for diversification of risks and greater investment scales.[134] In this model, a public funder invests as a limited partner into a private fund, which, in turn, invests in other private investment funds.[135] The selection of the second stage funds is supposed to offer different levels of risk profiles reflecting country or technology sector specificities. If managed successfully, the fund of funds model offers investors access to countries or sectors which they might otherwise not have considered owing to insufficient expertise to evaluate the risks of financial commitments.[136]

Another method of leveraging equity is the provision of subordinated equity. In this model, public finance is used under the condition that private equity investors have priority over public funds in the reimbursement. Thus, the so-called subordinated equity funds contribute to increasing the risk-adjusted returns of private equity investors by ensuring that they have first claim on the distribution of profits.[137]

### *V. Role of the GCF in Mobilising Private Sector Finance*

In addition to the already diverse mix of funding mechanisms presented in part E II. of this chapter, the Cancun Agreements established a new funding instrument, the GCF. This fund is supposed to play a central role with regard to the ambitious US$100 billion funding target. Consequently, the GCF will also be an important player when it comes to mobilising private sector capital.

131 Brown & Jacobs (2011:2).
132 Sierra (2011:12).
133 Brown & Jacobs (2011:2).
134 De Nevers (2011:24).
135 Sierra (2011:12).
136 De Nevers (2011:24).
137 Brown & Jacobs (2011:2).

*1. Current Framework*

According to the Cancun decision, the GCF is supposed to channel a significant share of new multilateral funding for adaptation.[138] With regard to the substantial pledges of the developed countries in the Cancun agreement, the GCF will take over a central role in the climate change financing system. This central role was highlighted at COP 17 in Durban, when the GCF was officially launched and its governing instrument was approved. According to the latter, the purpose of the GCF is to make a significant and ambitious contribution towards achieving the goals of the international community in fighting the climate change challenge.[139] The self-conception of the GCF is to play a key part in channelling new, additional, adequate and predictable financial means from both public and private sources at the international and national level. Correspondingly, the governing instrument states that the Fund will receive financial inputs from developed country parties and, beyond that, is open to funding from a variety of other public and private sources.[140] In the long run, it is envisaged that the GCF is to become "the main global fund for climate change finance".[141]

As a consequence of the ambitious targets set out for the GCF, the governing instrument for the GCF provides for specific regulations concerning the integration of private financial resources. A private sector facility is established to finance private sector mitigation and adaptation activities directly and indirectly at the national and international level.[142] Furthermore, the facility will particularly support private sector actors engaging in developing countries.

The GCF governing instrument lists grants and concessional lending as financial instruments. Financing can also be provided through other modalities, instruments or facilities after approval by the Board. According to the governing instrument, financing of concrete projects has to cover the iden-

138 UNFCCC, Decision 1/CP.16 The Cancun Agreements: Outcome of the Work of the Ad Hoc Working Group on Long-term Cooperative Action under the Convention, p 17, available at http://unfccc.int/resource/docs/2010/cop16/eng/07a01.pdf, last accessed 24 January 2013.

139 UNFCCC, Decision 3/CP.17 Launching the Green Climate Fund, available at http://unfccc.int/resource/docs/2011/cop17/eng/09a01.pdf, last accessed 24 January 2013, Annex para. 1.

140 (ibid.:Annex para. 29f.).

141 (ibid.:Annex para. 32).

142 (ibid.:Annex para. 41).

tifiable additional costs of the investment, which are regarded as necessary to make the project viable.[143]

## 2. Key Issues

One of the main challenges of the GCF will be to find its place in the already diverse climate financing architecture. It will be interesting to follow how both relation to and delineation towards the other funding mechanisms mentioned above will develop. The GCF will only become the envisaged key financing mechanism if it manages to operate on a large scale. This development depends on the level of public funds contributed by the developed states, as well as "on the attractiveness of the vehicle, particularly as a catalyst for private sector investment".[144] Regarding the different levels of barriers, the GCF will have several opportunities for engagement. The support for public sector projects and policy reform programmes through tools like budget support will be a crucial element to build a consistent and reliable enabling environment for private investment.[145] In addition to the support of enabling policy and regulatory environments, the GCF will also have directly to leverage public climate funds through risk reduction instruments and new climate instruments to attract private investment.[146] Tools have been mentioned above and include, inter alia, risk guarantees and pledge funds or fund of funds. Correspondingly, it will be necessary that the design of the GCF incorporates ways of leveraging private capital by means of direct investments and by supporting the necessary enabling frameworks in developing countries.[147]

Another key issue follows from the GCF's envisaged role in channelling a significant share of new adaptation funding. For the GCF, the task of strengthening adaptation activities will translate into specifically focusing on private sector engagement. The most vulnerable countries are developing countries with low country creditworthiness and thus least able to attract private investment as they require adaptation investments (e.g. related to water or agriculture), which are less attractive to private investors than mit-

143 (ibid.:Annex para. 54).
144 Sierra (2011:16).
145 (ibid.).
146 De Nevers (2011:25).
147 UNEP Finance Initiative (2011).

igation activities, where large investments in infrastructure are needed (e.g. energy or transport).[148] This dilemma is illustrated by the fact that, in terms of pledges, mitigation globally receives ten times the resources of adaptation. And as mitigation finance is rather spent on fast developing economies, Africa, consequently, receives the lowest level of funding.[149] In this light, the GCF will need to break up the hitherto existing climate financing structures and make a strong case for adaptation. As has been discussed at earlier stages when introducing the Adaptation Fund, it remains crucial to develop secure, adequate and predictable funding streams for the financing of adaptation needs in poorer, more vulnerable countries.[150] The GCF, therefore, needs to develop structures and methods, ensuring that priority in the use of public funds is given to funding adaptation costs, particularly in the most vulnerable countries.[151] However, it will also be mandatory for the GCF to increase private sector engagement in adaptation activities. In order to attract private investments it is necessary to understand the role that private sector finance can play in the most vulnerable countries.[152] If the GCF manages to prioritise public funds for adaptation and to mobilise additional private investment on a substantial scale, it can make a strong case for adaptation. It will be a challenge for the GCF to rather complement and further than to duplicate and impede structures and activities of the Adaptation Fund.

### *H. Concluding Remarks*

The International Energy Agency estimates that the total cost of investment to meet climate goals may amount to US$220 billion per year between 2010 and 2020 and to almost US$1 trillion dollars per year between 2020 and 2030. Mitigation and adaptation activities require large volumes of capital, innovative financial mechanisms and long-term commitment. Therefore both public and private funds have to be increased substantially.

Considering that the amount of private finance is almost three times greater than public finance, it is imperative to mobilise private sector finance for mitigation and adaptation continuously. To this end, it is pivotal to create

148 De Nevers (2011:9).
149 Whande & Reddy (2011:2).
150 Grasso (2011: 362).
151 De Nevers (2011:9).
152 Bird et al. (2011:6).

a favourable investment climate for investments, particularly from the private sector in clean and climate-resilient technologies and renewable energy. Only a stable and competitive risk-return profile of climate investments will mobilise private sector capital and thus contribute to achieving the significant investment volumes required in international climate finance.

Mobilising private sector engagement in climate change mitigation and adaptation requires political and financial programmes to overcome substantial barriers on different levels. Ideas to tap private sources for climate finance have emerged, such as guarantees, fund of funds, project aggregation mechanisms, climate bonds and public-private funds. Further approaches for tapping private capital will need to be designed in order to meet future climate-change-related challenges. Particularly with regard to the most vulnerable regions in the world, it will be important for countries which are affected most by the negative effects of climate change to address the split between financial resources spent on mitigation measures (approximately 95%) and those spent on adaptation. It is hoped that the Green Climate Fund will play a key part in channelling new, additional, adequate and predictable financial means from both public and private sources at the international and national level.

## *References*

Atteridge, Aaron, 2011, Will Private Finance Support Climate Change Adaptation in Developing Countries? Historical Investment Patterns as a Window on Future Private Climate Finance, Stockholm Environment Institute, available at http://www.sei-international.org/mediamanager/documents/Publications/Climate/SEI-WP-2011-05-Private-Sector-Adaptation-Finance-ES.pdf, last accessed 24 January 2013.

Bird, Neil, Jessica Brown & Liane Schalatek, 2011, Design Challenges for the Green Climate Fund, Climate Finance Policy Brief No. 4 (2011), 6, available at www.odi.org.uk/resources/docs/6457.pdf, last accessed 24 January 2013.

Bouwer, Laurens M. & Jeroen C.J.H. Aerts, 2006, Financing Climate Change Adaptation, *Disasters* 30, 49–63.

Brown, Jessica & Michael Jacobs, 2011, Leveraging Private Investment: The Role of Public Sector Climate Finance, Overseas Development Institute (ODI) Background Paper, available at http://www.odi.org.uk/sites/odi.org.uk/files/odi-assets/publications-opinion-files/7082.pdf, last accessed 24 January 2013.

Buchner, Barbara, Angela Falconer, Morgan Hervé-Mignucci, Chiara Trabacchi & Marcel Brinkmann, 2011, The Landscape of Climate Finance, Climate Policy Initiative, available at http://climatepolicyinitiative.org/publication/the-landscape-of-climate-finance/, last accessed 24 January 2013.

Buchner, Barbara, Angela Falconer, Morgan Hervé-Mignucci & Chiara Trabacchi, 2012, The Landscape of Climate Finance 2012, Climate Policy Initiative, available at http://climatepolicyinitiative.org/publication/global-landscape-of-climate-finance-2012/, last accessed 24 January 2013.

Dellink, Rob, Michel den Elzen, Harry Aiking, Emmy Bergsma, Frans Berkhout, Thijs Dekker & Joyeeta Gupta, 2009, Sharing the Burden of Financing Adaptation to Climate Change, *Global Environmental Change* 19, 411–421.

De Nevers, Michele, 2011, Climate Finance – Mobilizing Private Investment to Transform Development, Global Economic Governance Programme (GEG) Working Paper, available at http://www.globaleconomicgovernance.org/wp-content/uploads/Climate-Finance-for-Development_deNevers.pdf. last accessed 24 January 2013.

Di Leva, Charles E., 2010, International Environmental Law, the World Bank, and International Financial Institutions, in: Daniel D. Bradlow & David B. Hunter, *International Financial Institutions and International Law*, The Hague, Kluwer Law International, 343–385.

Donner, Simon D., Milind Kandlikar & Hisham Zerriffi, 2011, Preparing to Manage Climate Change Financing, *Science* 334, 908–909.

Energy and Resources Institute, 2012, The Climate and Development Research Review, Synthesis Report (2012), available at http://www.teriin.org/pdf/Synthesis_Report.pdf, last accessed 24 January 2013.

Forstater, Maya & Rachel Rank, 2012, Towards Climate Finance Transparency, Paper by aidinfo and Publish What You Fund, available at http://www.publishwhatyoufund.org/files/Towards-Climate-Finance-Transparency_Final.pdf, last accessed 24 January 2013.

Grasso, Marco, 2011, The Role of Justice in the North-South Conflict in Climate Change: The Case of Negotiations on the Adaptation Fund, *International Environmental Agreements*, 361–377.

International Energy Agency, 2010, World Energy Outlook 2010, available at http://www.iea.org/publications/freepublications/publication/name,27324,en.html, last accessed 22 February 2013.

McGoldrick, Will, 2007, Financing Adaptation in Pacific Island Countries: Prospects for the Post-2012 Climate Change Regime, *Australian International Law Journal* 14, 45–69.

Patel, Shilpa, 2011, Climate Finance: Engaging the Private Sector, International Finance Corporation (IFC) Background Paper, available at http://www1.ifc.org/wps/wcm/connect/5d659a804b28afee9978f908d0338960/ClimateFinance_G20Report.pdf?MOD=AJPERES, last accessed 24 January 2013.

Ruppel, Oliver C., 2011, Financing for Climate Change in Africa, in: AMCEN/African Union and UNEP/United Nations Environment Programme, *Addressing Climate Change Challenges in Africa,* Nairobi, AMCEN 127-169, available at http://www.unep.org/roa/amcen/docs/publications/guidebook_CLimateChange.pdf, last accessed 24 January 2013.

Ruppel, Oliver C., 2013, International Legal Climate Change Regimes and Climate Finance, in: Ruppel, Oliver C. & Katharina Ruppel-Schlichting (Eds), *Environmental Law and Policy in Namibia*, Windhoek/Essen, Orumbonde Press/Welwitschia Verlag Dr. A. Eckl, 287–308.

Ruppel, Oliver C. & Cord Lüdemann, 2013, Climate Finance – Mobilising Private Sector Finance for Mitigation and Adaptation, Institute for Security Studies (ISS) Situation Report, available at http://www.issafrica.org/uploads/SitRep2013_6May.pdf, last accessed 21 May 2013.

Sierra, Katrin, 2011, The Green Climate Fund: Options for Mobilizing the Private Sector, available at http://cdkn.org/wp-content/uploads/2012/03/The-Green-Climate-Fund_options-for-mobilizing-the-private-sector.pdf, last accessed 24 January 2013.

Transparency International, 2011, *World Corruption Report: Climate Change*, London, Earthscan, available at http://archive.transparency.org/publications/gcr/gcr_climate_change2, last accessed 24 January 2012.

UN Secretary-General, 2010, Report of the Secretary-General's High-level Advisory Group on Climate Change Financing, available at http://www.un.org/wcm/webdav/site/climatechange/shared/Documents/AGF_reports/AGF_Final_Report.pdf, last accessed 24 January 2013.

UNCTAD, World Investment Report, 2012, Towards a New Generation of Investment Policies 4 (2012), available at http://www.unctad-docs.org/files/UNCTAD-WIR2012-Full-en.pdf, last accessed 24 January 2013.

UNEP & Bloomberg New Energy Finance, 2011, Global Trends in Renewable Energy Investment, available at http://fs-unep-centre.org/publications/global-trends-renewable-energy-investment-2011, last accessed 24 January 2013.

UNEP Finance Initiative, 2011, The Green Climate Fund (GCF): The Private Financial Sector's Perspective, available at http://unfccc.int/files/cancun_agreements/green_climate_fund/application/pdf/unep_fi_submission_on_private_sector.pdf, last accessed 24 January 2013.

UNFCCC/United Nations Framework Convention on Climate Change, 2009, Fact Sheet: Financing Climate Change Action, available at http://unfccc.int/files/press/backgrounders/application/pdf/fact_sheet_financing_climate_change.pdf, last accessed 24 January 2013.

Van Kerkhoff, Lorrae, Imran Habib Ahmad, Jamie Pittock & Will Steffen, 2011, Designing the Green Climate Fund: How to Spend $100 Billion Sensibly, *Environment* 53, 18–30.

Whande, Webster & Trusha Reddy, 2011, The Governance of Climate Finance in Africa, Asia and Latin America: Some Critical Reflections, Institute for Security Studies (ISS) Policy Brief No. 24, available at http://www.issafrica.org/uploads/No24Jun2011.pdf, last accessed 24 January 2013.

World Bank, World Development Report 2010, Development and Climate Change, available at http://web.worldbank.org/WBSITE/EXTERNAL/EXTDEC/EXTRESEARCH/EXTWDRS/0,,contentMDK:23062354~pagePK:478093~piPK:477627~theSitePK:477624,00.html, last accessed 22 February 2013.

# PART IV: CLIMATE CHANGE POLICY, COOPERATION AND PROTECTION EFFORTS

13

# Climate Change Law and Policy Positions in the African Union and Related Developments in Selected African Countries*

*Oliver C. Ruppel*

*Abstract*

The African Union (AU) has embarked on new policy pathways to accommodate climate change more effectively in future. These pathways, which are described in this article, are not only deemed to become increasingly relevant in a changing climate, but promise to unfold potential and new opportunities for economic and sustainable development in Africa on regional and sub-regional levels. It is the objective of this article to discuss existing laws and new policy pathways in the AU and related legal developments in selected African countries and, where possible, to assess their potential benefit. While doing so, the article reflects on legal and institutional structures, some relevant cases, contemporary and future challenges, and developmental perspectives pertinent to the issue of climate change and the environment on the African continent.

## *A. Introduction*

Despite Africa's relatively low contribution to the world's total greenhouse gas emissions, it is one of the most vulnerable continents to climate change.[1] Africa emits far less carbon than other continents. Africa's carbon dioxide ($CO_2$) emissions per year represent only a small fraction, 3.6%, of global emissions, yet 14% of the population of the world lives here.[2] The African continent accounts for only 3% of world energy consumption, and the average energy consumption of an African inhabitant is six times less

---

* This updated and amended article is based on Ruppel (2012a).

1 Boko et al. (2007).

2 See http://www.grida.no/graphicslib/detail/emissions-of-carbon-dioxide-in-africa-and-selected-oecd-countries_1400, last accessed 17 January 2013.

than that recorded for individual inhabitants across the populations of the world. About 600 million people in Africa currently lack access to electricity.[3]

In the same light, Africa is particularly vulnerable as a consequence of a combination of stresses, and especially owing to poverty. The complexity of climate change will require the involvement of a diverse range of institutions.[4] It is expected that climate change will generate varied and significant impacts on national, regional and global economies; and it is also not unlikely that this will result in increased local and international conflict.[5]

Climate change poses an enormous threat to Africa's economic growth (through its harmful effects on natural systems and resources), long-term prosperity, and the survival of its already vulnerable populations. Climate change, variability and associated increased disaster risks are an additional burden to sustainable development in Africa, as well as a threat and impediment to achieving the Millennium Development Goals.[6]

In 2011, the South African city of Durban was in the international limelight as the host of the global climate negotiations.[7] The goal of these discussions was to advance the implementation of the United Nations Framework Convention on Climate Change (UNFCCC) and the Kyoto Protocol, and to make progress on previously agreed action plans (Bali) and agreements (Cancun). The 17th Conference of the Parties (COP 17) to the UNFCCC and the 7th Session of the COP serving as the Meeting of the Parties (MOP 7) to the Kyoto Protocol were, however, especially from an African viewpoint, only partly successful. Not only were China and the United States reluctant to sign binding targets, each waiting for the other to move first, but this reluctance by the two biggest polluters had repercussions for the negotiations as a whole, namely that Canada, Japan and Russia refused to enter into a second commitment period for the Kyoto Protocol owing to the lack of legal restriction on the world's largest polluters.[8] The sense of dampened success applies even more to the outcomes of the 18th Conference of the Parties (COP 18) to the UNFCCC and the 8th Session of the COP serving

3 See IRENA (2012).
4 Keohane & Victor (2011).
5 Scholtz (2010).
6 AMCEN (2011).
7 See also http://www.dieburger.com/Suid-Afrika/Nuus/Krisis-raak-net-erger-20111129, last accessed 23 December 2011.
8 Ruppel et al. (2011).

as the Meeting of the Parties (MOP 8) to the Kyoto Protocol held in Doha, Qatar, in 2012, where a number of decisions were adopted (The Doha Climate Gateway).[9] A second commitment period under the Kyoto Protocol has been launched, with the end date being 2020.[10] It has been agreed to work towards a universal climate change agreement covering all countries from 2020. Such agreement is to be adopted by 2015. Countries have furthermore agreed on ways and means to deliver scaled-up climate finance and technology to developing countries and COP18 has also taken note of the first annual report of the Board of the Green Climate Fund to the Conference of the Parties and endorsed the consensus decision of the Board of the Green Climate Fund to select Songdo, Incheon, in the Republic of Korea as the host city of the Green Climate Fund, on the basis of an open and transparent process.[11] Further key elements of the outcome include an agreement to consider loss and damage in developing countries that are particularly vulnerable to the adverse effects of climate change.

At the Doha conference, "Germany, the UK, France, Denmark, Sweden and the EU Commission announced financial pledges for the period up to 2015 totalling approximately $6 billion. Most developed countries did, however, not make pledges! African countries thus left Doha with little more than they already had."[12] This means for Africa that climate change continues to prompt significant challenges in future[13] and it is therefore noteworthy that the African Union (AU) and a number of African countries have embarked on new policy pathways to accommodate climate change more effectively.

## *B. The African Union*

The historical foundations of the African Union (AU) originated in the Union of African States, an early confederation that was established in the 1960s.

---

9 All decisions adopted by COP 18 and CMP 8 can be accessed at http://unfccc.int/meetings /doha_nov_2012/meeting/6815.php#decisions, last accessed 16 January 2013.

10 However, some previously participating countries in the Kyoto Protocol have not joined the second period, namely Russia, Canada, New Zealand and Japan.

11 See UNFCCC (2012).

12 See http://www.un.org/africarenewal/web-features/what-does-'doha-climate-gateway'-mean-africa, last accessed 19 January 2013.

13 Ruppel & Van Wyk (2011).

The Organisation of African Unity (OAU) was established on 25 May 1963. On 9 September 1999, the heads of state and governments of the OAU issued the Sirte Declaration,[14] calling for the establishment of an African Union. The Declaration was followed by summits in Lomé in 2000, when the Constitutive Act of the African Union was adopted, and in Lusaka in 2001, when the Plan for the Implementation of the African Union was adopted. During the same period, the initiative for the establishment of the New Partnership for Africa's Development (NEPAD) was also instituted.[15] The AU was launched in Durban on 9 July 2002 by the then South African president, Thabo Mbeki,[16] at the First Session of the Assembly of the AU. The Union's administrative centre is in Addis Ababa, Ethiopia, and its working languages are Arabic, English, French, Portuguese and Swahili. The AU has 54[17] member states, with Morocco being the only African state that is not a member. Geographically, the AU covers an area of 29,757,900 km$^2$ and, for 2010, the United Nations Population Division estimated a population total of 990,283,000.[18]

Article 3 of the Constitutive Act of the African Union contains the objectives of the AU, which include the promotion of sustainable development, international cooperation and continental integration, as well as the promotion of scientific and technological research to advance the continent's development. In the Protocol relating to the Establishment of the Peace and Security Council (PSC) of the AU, member states committed themselves to various guiding principles (Article 4), including "early responses to contain crisis situations", and the recognition of the "interdependence between socio-economic development and the security of peoples and States". Moreover, in Article 6 of the Constitutive Act, the functions of the PSC are outlined as, among other things, the promotion of peace, security and stability in Africa; early warning and preventive diplomacy; peace-making; humanitarian action; and disaster management. All the aforementioned provisions

14 Named after Sirte in Libya.

15 Ruppel (2011a: 43).

16 Thabo Mbeki was also the AU's first president.

17 Including South Sudan, which ratified the Constitutive Act of the African Union on 15 August 2011; see http://www.au.int/en/sites/default/files/Constitutive_Act_0.pdf, last accessed 28 March 2012.

18 Africa's entire population was estimated to be 1,022,234,000. As Morocco is not a member of the AU, its population – estimated at 31,951,000 – has to be deducted. Data collected from http://esa.un.org/unpd/wpp/unpp/panel_population.htm, last accessed 28 March 2012.

give a clear mandate to address climate change as a matter of priority within the AU.

### C. The New Partnership for Africa's Development (NEPAD)

NEPAD was adopted in 2001 in Lusaka, Zambia, by heads of state and government of the OAU in 2001. NEPAD was ratified by the AU in 2002. Partnership and cooperation between Africa and the developed world are envisaged by this development initiative aimed at the economic and social revival of Africa. NEPAD is –[19]

> ... a pledge by African leaders, based on a common vision and a firm and shared conviction, that they have a pressing duty to eradicate poverty and to place their countries, both individually and collectively, on a path of sustainable growth and development, and at the same time to participate actively in the world economy and body politic. The Programme is anchored on the determination of Africans to extricate themselves and the continent from the malaise of under-development and exclusion in a globalising world.

NEPAD has emphasised Africa's important role in respect of the critical issue of environmental protection. Of the six main thematic areas on which NEPAD focuses, two are of particular relevance here –[20]

- Agriculture and food security, and
- Climate change and natural resource management.

NEPAD's Climate Change and Natural Resource Management Programme focuses on three key areas: environment, water and energy. The Programme aims to assist countries to integrate climate change responses with their national development processes. The Programme also aims to strengthen skills in adaptation, mitigation, technology and finance in order to combat environmental change. In order to achieve these aims, an Action Plan for the Environment Initiative was released in 2003, paving the way for the first decade of the 21st century.[21]

---

19 See NEPAD's founding document, available at http://www.dfa.gov.za/au.nepad/nepad.pdf, last accessed 12 February 2012.

20 Further thematic areas within the NEPAD framework are Regional Integration and Infrastructure; Human Development; Economic and Corporate Governance; and Crosscutting Issues (Gender and Capacity Management).

21 Available at http://www.nepad.org/system/files/Environment%20Action%20-Plan.pdf, last accessed 12 February 2012.

## D. The AU's Judicial System and Consideration of Environmental Rights

In 1998, the African Court on Human and Peoples' Rights (AfCHPR) was established by the Protocol to the African Charter on Human and Peoples' Rights on the Establishment of an African Court on Human and Peoples' Rights, which came into force in 2004. The AfCHPR is situated in Arusha, in the United Republic of Tanzania, and has received cases since June 2008.

In 2003, the African Court of Justice – as the ultimate organ of jurisdiction in the AU – was established by the Protocol of the Court of Justice of the African Union, which came into force in February 2009. However, the Protocol on the Statute of the African Court of Justice and Human Rights adopted in 2008 during the AU Summit of Heads of State and Government in Sharm El Sheikh in the Arab Republic of Egypt provides that the 1998 and 2003 Protocols be replaced, and that the AfCHPR and the AU's Court of Justice be merged into a single court to become what is now known as the African Court of Justice and Human Rights. However, the 2008 Protocol on the merger of the courts has so far only been ratified by five[22] member states, but ratification by a minimum of 15 is required for the Protocol to come into force. Once operational, the merged court will have two sections: a General Affairs Section, and a Human Rights Section, both composed of eight Judges. The court will have jurisdiction over all disputes and applications referred to it which, among other things, relate to the interpretation and application of the AU Constitutive Act or the interpretation, application and validity of Union Treaties, as well as human rights violations.

The African Commission on Human and Peoples' Rights (hereafter *African Commission*) is a quasi-judicial body established by the 1981 African (Banjul) Charter on Human and Peoples' Rights (hereafter *African Charter*) and is responsible for monitoring compliance with the African Charter. The African Charter is a human rights treaty that already proclaims environmental rights in broadly qualitative terms. It protects the right of peoples both to the "best attainable state of physical and mental health" (Article 16) and to a "general satisfactory environment favorable to their development" (Article 24). Article 24 of the African Charter further establishes

22 As of 01 March 2013, the Protocol was ratified by Benin, Burkina Faso, Congo, Libya, and Mali. See http://www.au.int/en/sites/default/files/Protocol%20on%20Statute%20of%20the%20African%20Court%20of%20Justice%20and%20HR.pdf, last accessed 07 April 2013.

a binding human-rights-based approach to environmental protection, linking the right to environment to the right to development.[23]

In the *Endorois* case,[24] the African Commission concluded that several Articles of the African Charter had been violated in the course of the dispossession of the land of Kenya's indigenous Endorois through the creation of the Lake Hannington Game Reserve in 1973, and a subsequent re-gazetting of the Lake Bogoria Game Reserve in 1978 by the Kenyan Government. Among the Endorois' rights found to be violated were their right to culture (Article 17 (1) and (2)) and their right to the free disposition of natural resources (Article 21), as they were unable to access the vital resources in the Lake Bogoria region after their eviction from the game reserve. Moreover, the African Commission held that the Endorois' right to development (Article 22) had been violated, as the respondent state had failed adequately to involve the Endorois in the development process.[25] The decision of the African Commission in the *Endorois* case[26] was influenced by provisions of Convention No. 169 of the International Labour Organisation on Indigenous and Tribal Peoples in Independent Countries.[27] Among other things, the Convention –[28]

23 Van der Linde & Louw (2003).

24 Communication 276/03, *Centre for Minority Rights Development (Kenya) and Minority Rights Group International on behalf of Endorois Welfare Council v Kenya*, available at http://www.achpr.org/english/Decison_Communication/Kenya/Comm.%20276-03.pdf, last accessed 12 February 2012.

25 The Commission's recommendation was to recognise the Endorois' rights of ownership and to restore their ancestral land; to ensure that the Endorois community had unrestricted access to Lake Bogoria and surrounding sites for religious and cultural rites and for grazing their cattle; to pay adequate compensation to the community for all the loss they had suffered; to pay royalties to the Endorois from existing economic activities and ensure that they benefited from employment possibilities within the Reserve; to grant registration to the Endorois Welfare Committee; and to engage in dialogue with the complainants for the effective implementation of these aforementioned recommendations and to report on their implementation.

26 Communication 276/03, *Centre for Minority Rights Development (Kenya) and Minority Rights Group International on behalf of Endorois Welfare Council v Kenya*, available at http://www.achpr.org/english/Decison_Communication/Kenya/Comm.%20276-03.pdf, last accessed 28 March 2012.

27 The Convention came into force on 5 September 1991 and is available at http://www.ilo.org/ilolex/cgi-lex/convde.pl?C169, last accessed 12 February 2012.

28 It should be noted that of the 22 states that have ratified ILO Convention No. 169, as of February 2012, only one – the Central African Republic – is from the African continent.

- provides criteria for describing the peoples it aims to protect
- makes provision regarding the principle of non-discrimination
- calls for special measures to be adopted to safeguard the persons, institutions, property, labour, cultures and environment of indigenous and tribal peoples
- recognises cultural and other specificities of indigenous and tribal peoples, and
- requires that, on all issues that affect them, indigenous and tribal peoples are consulted and that these peoples are able to engage in free, prior and informed participation in policy and development processes.

In the *Ogoni* case, the African Commission held, among other things, that Article 24 of the African Charter imposed an obligation on the state to take reasonable measures to "prevent pollution and ecological degradation, to promote conservation, and to secure ecologically sustainable development and use of natural resources."[29]

The *Ogoni* case led to a landmark decision with regard to the effective protection of economic, social and cultural rights in Africa, particularly the protection of the right of peoples to a satisfactory environment.[30] Article 24 of the African Charter should be viewed together with the Bamako Convention and the first OAU treaty on the environment, namely the Convention on the Conservation of Nature and Natural Resources, which predates the African Charter. The Revised African Convention on the Conservation of Nature and Natural Resources was adopted by the Second Ordinary Session of the AU Assembly of Heads of State and Government in Maputo, Mozambique, in July 2003. However, the latter Convention has not yet come into force.

The recognition in the African Charter of a right to a satisfactory environment and the progressive jurisprudence by the African Commission take up the issue of environmental protection from a human rights perspective, and underline the linkage between climate change and human rights in a modern, holistic approach to one of today's burning issues.[31] The impacts of climate change on human rights have been explicitly recognised by the

29 Communication 155/96, *The Social and Economic Rights Action Center (SERAC) & the Center for Economic and Social Rights (CESR) v Nigeria*, available at http://www.achpr.org/english/_
info/decision_article_24.html, last accessed 28 March 2012.

30 Ruppel (2011b).

31 Ruppel (2010a).

African Commission: in its AU Resolution 153, the African Commission calls on the Assembly of Heads of State and Government to ensure that –[32]

> human rights standards safeguards, such as the principle of free, prior and informed consent, be included into any adopted legal text on climate change as preventive measures against forced relocation, unfair dispossession of properties, loss of livelihoods and similar human rights violations;

and "to take all necessary measures to ensure that the African Commission on Human and Peoples' Rights is included in the African Union's negotiating team on climate change". In the same communication, the African Commission resolves to carry out a study on the impact of climate change on human rights in Africa.[33]

### *E. AU Climate Change Policy and Related Developments*

Many African governments have made progress in addressing climate change and related issues. The AU itself has succeeded in presenting a more cohesive African position on climate change. Although gaps may remain, it was clearly reflected during COP 17 in Durban that Africa maintained a common position in spite of pressure from developed countries. In fact, the African Group "spoke with one voice", according to Seyni Nafo, spokesperson for the African Group and lead negotiator on Mitigation.[34] It has since become apparent, however, that divergent priorities among African countries threaten the potential of the AU to influence international climate politics.[35]

At its 46th Ordinary Session, held in Banjul, The Gambia, from 11 to 25 November 2009, the African Commission urged –[36]

> … the Assembly of Heads of State and Government of the African Union to ensure that human rights standards safeguards, such as the principle of free, prior and informed consent, be included into any adopted legal text on climate

32 AfCHPR/Res. 153 (XLV09).

33 See http://www.achpr.org/english/resolutions/resolution153_en.htm, last accessed 14 February 2012.

34 See http://www.dailytrust.com.ng/index.php?option=com_content&view=article&id=149432:africa-maintains-common-position-in-durban&catid=10:environment&Itemid=11, last accessed 20 February 2012.

35 Hoste (2010).

36 See http://www.achpr.org/english/resolutions/resolution153_en.htm, last accessed 14 February 2012.

change as preventive measures against forced relocation, unfair dispossession of properties, loss of livelihoods and similar human rights violations.

### *I. Climate Change Induced Displacement*

One legal instrument which explicitly deals with the potential impacts of climate change is the African Union Convention for the Protection and Assistance of Internally Displaced Persons in Africa (hereafter *Kampala Convention*),[37] which was adopted in Kampala on 23 October 2009. It is the first regional legal instrument in the world to contain legal obligations for states with regard to the protection and assistance of internally displaced persons (IDPs).[38] So far, the Kampala Convention has 39 signatories, and 17 countries[39] have ratified it. The Convention entered into force on 6 December 2012.[40] Article 1(k) of the Kampala Convention defines IDPs as follows:

> [P]ersons or groups of persons who have been forced or obliged to flee or to leave their homes or places of habitual residence, in particular as a result of or in order to avoid the effects of armed conflict, situations of generalized violence, violations of human rights or natural or human-made disasters, and who have not crossed an internationally recognized State border.

The Kampala Convention explicitly recognises climate change as one of the possible reasons for internal displacement: Article 5 states that "States Parties shall take measures to protect and assist persons who have been internally displaced due to natural or human made disasters, including climate change."

However, the Kampala Convention applies to all situations of internal displacement regardless of its causes (Article 15), which makes sense:

---

37 Text available at http://www.au.int/en/sites/default/files/AFRICAN_UNION_CONVENTION_FOR_THE_PROTECTION_AND_ASSISTANCE_OF_INTERNALLY_DISPLACED_PERSONS_IN_AFRICA_(KAMPALA_CONVENTION).pdf, last accessed 30 January 2012.

38 See Kidane (2011).

39 As of 07 April 2013, the following 17 member states had ratified the Convention: Benin, Burkina Faso, Central African Republic, Chad, Gabon, Gambia, Guinea-Bissau, Lesotho, Mali, Nigeria, Niger, Rwanda, Sierra Leone, Swaziland, Togo, Uganda and Zambia, see http://www.au.int/en/sites/default/files/Convention%20on%20IDPs%20-%20displaced....pdf, last accessed 07 April 2013.

40 Ratification of 15 member states was required for the Convention to come into force.

drivers of migration in general are not mono-causal, but are influenced by multiple factors.

Several obligations are imposed on states parties by the Kampala Convention. For example, the Convention addresses the need to prevent displacement from happening, e.g. by establishing early warning systems and adopting disaster preparedness and management measures to prevent displacement caused by natural disaster. The Convention also requires States Parties to, among other things, protect people against displacement resulting from conflict and violence, discriminatory policies, or human rights violations. Neither should displacement be used as a method of warfare, nor as collective punishment. Forced evacuations should only take place for reasons of health and safety, and, once people have been displaced, the Kampala Convention provides that they are to be protected and assisted (Article 4). According to Article 5, States Parties are obliged to assess the needs and vulnerabilities of displaced persons, as well as those of the host communities, and to provide adequate assistance, if need be with assistance from relevant local and international agencies. One objective of the Kampala Convention is to provide for durable solutions with respect to IDPs, who retain the right to make a free and informed choice on whether to return, integrate or relocate elsewhere in the country (Article 11). Furthermore, States Parties are responsible for establishing an effective legal framework to provide just and fair compensation, and other forms of reparations for damage incurred as a result of displacement (Article 12).

In light of the aforementioned, it is also worthwhile noting that various sub-regional organisations, that is Regional Economic Communities (RECs), have "at the insistence of the AU" established certrain mechanisms.[41] At the seventh ordinary session of the AU's Assembly of Heads of State and Government in Banjul, The Gambia, in July 2006, the AU officially recognised eight such communities.[42] Alphabetically listed, these are as follows:[43]

- The Arab Maghreb Union (AMU)
- The Community of Sahel-Saharan States (CEN-SAD)

41 See Eze (2012:516).

42 See the decision relating to the recognition of RECs, Assembly/AU/Dec.112 (VII) Doc. EX.CL/278 (IX), text in French, available at http://www.africa-union.org/Official_documents/Assemblee %20fr/ASS06b.pdf, last accessed 22 December 2012.

43 Ruppel (2009a:276).

- The Common Market for Eastern and Southern Africa (COMESA)
- The East African Community (EAC)
- The Economic Community of Central African States (ECCAS)
- The Economic Community of West African States (ECOWAS)
- The Intergovernmental Authority on Development (IGAD), and
- The Southern African Development Community (SADC).

All AU member states are affiliated to one or more of these RECs, as tabulated below:

***Table 1: State members of RECs officially recognised by the AU***[44]

| AMU | CEN-SAD | COMESA | EAC | ECCAS | ECOWAS | IGAD | SADC |
|---|---|---|---|---|---|---|---|
| Algeria | Benin | Burundi | Burundi | Angola | Benin | Djibouti | Angola |
| Libya | Burkina Faso | Comoros | Kenya | Burundi | Burkina Faso | Ethiopia | Botswana |
| Mauritania | Central African Republic | DRC | Rwanda | Cameroon | Cape Verde | Kenya | DRC |
| Morocco | Chad | Djibouti | Tanzania | Central African Republic | Cote d'Ivoire | Somalia | Lesotho |
| Tunisia | Comoros | Egypt | Uganda | Chad | Gambia | Sudan | Madagascar |
| | Cote d'Ivoire | Eritrea | | Congo | Ghana | Uganda | Malawi |
| | Djibouti | Ethiopia | | DRC | Guinea | | Mauritius |
| | Egypt | Kenya | | Gabon | Guinea-Bissau | | Mozambique |
| | Eritrea | Libya | | Guinea | Liberia | | Namibia |
| | Gambia | Madagascar | | São Tomé and Príncipe | Mali | | Seychelles |
| | Ghana | Malawi | | | Niger | | South Africa |
| | Guinea-Bissau | Mauritius | | | Nigeria | | Swaziland |
| | Kenya | Rwanda | | | Senegal | | Tanzania |
| | Liberia | Seychelles | | | Sierra Leone | | Zambia |
| | Libya | Sudan | | | Togo | | Zimbabwe |
| | Mali | Swaziland | | | | | |
| | Mauritania | Uganda | | | | | |
| | Morocco | Zambia | | | | | |
| | Niger | Zimbabwe | | | | | |
| | Nigeria | | | | | | |
| | São Tomé and Príncipe | | | | | | |
| | Senegal | | | | | | |
| | Sierra Leone | | | | | | |
| | Somalia | | | | | | |
| | Sudan | | | | | | |
| | Togo | | | | | | |
| | Tunisia | | | | | | |

44 Ruppel (2009a:278).

At first glance it appears that the promotion and protection of displaced persons is not within the RECs' focal range. However, it will still have to be seen how effective these mechanisms actually are. Regional integration, which is the primary engine of RECs, certainly has the potential to provide –[45]

> … an opportunity to enhance political stability by establishing regional organisations which play an increasing role in defusing conflicts within and between countries and in promoting human rights. In terms of climate change related matters, such organisations are of the utmost relevance, especially when it comes to climate change related disaster management and environmentally induced migration.

## *II. Climate Change and Vulnerable Groups*

The African Commission in 2009 urged –[46]

> … the Assembly of Heads of State and Government to ensure that special measure of protection for vulnerable groups such as children, women, the elderly, indigenous communities and victims of natural disasters and conflicts are included in any international agreement or instruments on climate change.

This call is very much in line with Article 3 of the International Covenant on Economic, Social and Cultural Rights,[47] the Convention on the Elimination of All Forms of Discrimination against Women, and the latter's Optional Protocol.[48] The vulnerability of women to climate change and natural disasters is more severe than for other groups for a number of reasons. Women are usually at higher risk of being placed in unsafe, overcrowded shelters owing to a lack of assets such as savings, property or land. In the context of droughts, floods and other disasters that require mobility, cultural constraints on women's movements may hinder their timely escape, access to shelter, or access to health care. These effects are exacerbated when women avoid using shelters out of fear of domestic and sexual violence, and become even less mobile as primary family caregivers. Poor women and

45 See Ruppel & Ruppel-Schlichting (2012:41).

46 See http://www.achpr.org/english/resolutions/resolution153_en.htm, last accessed 14 February 2012.

47 Article 3 encourages States Parties to ensure the equal right of men and women to the enjoyment of all economic, social and cultural rights as set forth by the Covenant.

48 See Ruppel (2008); Ruppel (2010b).

those in countries with greater gender inequality appear to be most at risk.[49]

The impact of climate change on the realisation of child rights[50] as a whole obviously also includes multiple effects on basic rights such as water, food and health, especially in African countries that are vulnerable to temperature and precipitation change. Children are vulnerable to climate change – and policy makers need to adhere to the Convention on the Rights of the Child that requires national governments to ensure that children's specific needs are given due consideration in adaptation and mitigation policy.[51]

It is worth noting that Cancun Decision 1/CP.16 also recognises indigenous peoples as a vulnerable group, alongside women, children and other vulnerable groups. In Africa, indigenous peoples are vulnerable to the actual and potential detrimental impacts of climate change. In line with the Declaration on the Rights of Indigenous Peoples, which was adopted by the United Nations General Assembly in 2007,[52] indigenous peoples should be free from discrimination of any kind – including that in the context of climate change.

### *III. Climate Change, Peace and Collective Security*

The AU has a clear mandate regarding the maintenance of peace and security in Africa. Yet, Africa remains a continent blighted by conflict, where "millions of human beings remain at the mercy of civil wars, insurgencies, state repression and state collapse".[53] But these are not the only challenges. Climate change overlays these ravages, intensifying existing problems as a magnifying glass would. Although climate change is, in the first instance, an environmental issue, it is also political, with "far-reaching economic, societal and political ramifications"[54], which cannot be neglected.

In fact, framing climate change as more of a security issue in Africa could serve to enhance existing policy response mechanisms. In 2011, the United

49 UN Womenwatch (2009).
50 See Ruppel (2009b).
51 UNICEF (2009).
52 Resolution 61/295.
53 Report of the International Commission on Intervention and State Sovereignty (2011).
54 See Scott (2012:220).

Nations Security Council expressed concern that the possible adverse effects of climate change could, in the long run, intensify certain existing threats to international peace and security and that the loss of territory in some states could have possible security implications.[55]

The Constitutive Act of the African Union, which was adopted in July 2000,[56] stipulates the Union's right of intervention in a member state in cases of "grave circumstances", i.e. crimes against humanity, war crimes and genocide. This raises the question of how such grave circumstances relate to climate change? Notwithstanding the controversy around this question, Achim Steiner, when addressing the UN Security Council in the 2011 Debate on Climate Change and Security, clearly stated the following:[57]

> Climate change as a threat multiplier: The scale and pace of climate change acts as a multiplier which could result in simultaneous and unprecedented impacts on where we can settle, grow food, maintain our built-up infrastructure, or rely on functioning ecosystems. Managing the potential disruption, displacement and adaptation to phenomena such as sea-level rise or extreme weather events, represents a profound challenge to sustainable development at the local, national and international level – both in economic and geopolitical terms. ...In 2010, over 90 per cent of disaster displacement within countries was caused by climate-related hazards, primarily floods and storms. Climate scenarios expect such weather events to increase and or intensify as a result of accelerating climate change. (...) Competition over scarce water and land, exacerbated by regional changes in climate, are already a key factor in local-level conflicts in Darfur, the Central African Republic, northern Kenya, and Chad, for example-when livelihoods are threatened by declining natural resources, people either innovate, flee or can be brought into conflict.

The aforementioned statements clearly pinpoint climate change as a potential source of conflict, and a potential threat to national and international peace and human security. One cornerstone of the United Nations Charter paradigm is the notion of collective security which is perhaps the first and most obvious manifestation of the principle of solidarity in the post-World

55 Security Council Meeting SC/10332 document available at http://www.un.org/News/Press/docs/2011/sc10332.doc.htm, last accessed 18 February 2013.

56 Constitutive Act of the African Union, 11 July 2000.

57 Address by UN Under-Secretary-General and UNEP Executive Director Achim Steiner at UN Security Council Debate on the impact of climate change on maintaining international peace and security, 20 July 2011, available at http://www.unep.org/Documents.Multilingual/Default.Print.asp?DocumentID=2646&ArticleID=8817&l=en, last accessed 18 February 2013.

War II era.[58] In fact, it forms the political and legal foundation for the collective security system established by the UN Charter. A still controversial manifestation of the notion of solidarity in international law is the emerging doctrine of the responsibility to protect. This concept was developed by the International Commission on Intervention and State Sovereignty in September 2000, after UN Secretary-General Kofi Annan emphasised the grave failure of the international community to handle gross and systematic violations of human rights such as those perpetrated in Rwanda and others.[59] The aforementioned concept has gained growing attention in the context of the notion of global solidarity and collective security as it aims at addressing legal and political dilemmas for intervention to stop or prevent human suffering and crimes against humanity.[60]

The Peace and Security Council (PSC) of the African Union in its recent report[61] documented its activities and the state of peace and security in Africa pursuant to Article 7(q) of the Protocol Relating to the Establishment of the PSC of the AU. It provides an overview of the state of peace and security on the continent from July 2012 to January 2013. During this reporting period, the PSC considered the crisis and conflict situations in the following states: Central African Republic, Sudan, Democratic Republic of Congo, Guinea Bissau, Mali and Somalia, as well as the situation between Sudan and South Sudan. During the reporting period, the PSC considered several thematic issues relating to the promotion of peace, security and stability in Africa, such the following:[62]

- Capacity-building for effective response to humanitarian assistance and disasters in Africa
- Challenges related to peace and security in Africa from the perspective of delivering humanitarian assistance in situations of crisis and conflict
- The need to mainstream gender in all development, peace and security efforts on the continent, and

58 See Koroma (2012).

59 Report of the Secretary-General on the Work of the Organization, document A/54/1, at 48, 31 August 1999.

60 See Koroma (2012).

61 Assembly of the African Union, 20th Ordinary Session, 27–28 January 2013, Addis Ababa, Ethiopia, Assembly/AU/3(XX), available at http://www.peaceau.org/uploads/assemblyau-3-xx-e.pdf, last accessed 05 April 2013.

62 (ibid.).

- The importance of international and transitional justice in the promotion of peace and security in Africa.

The PSC in its report furthermore –[63]

- reiterated the commitment of the AU to the fight against impunity, and stressed the importance of international and transitional justice in the promotion of peace and security in Africa, and the need, in the context of the search for solutions to crises and conflicts and in view of the fragility of the peace and reconciliation processes on the continent, to ensure that they are mutually reinforcing; and
- underscored the fact that the primary responsibility for the protection of human rights rests with Member States;
- emphasized the need for a close working relationship with the African Commission on Human and Peoples' Rights, as well as with the African Court on Human and Peoples' Rights.

Under Article 52 of the UN Charter, regional organisations may undertake actions aimed at the maintenance of international peace and security. Article 53(I) of the UN Charter specifically provides that such regional organisations may undertake enforcement measures, provided that they have the authorisation of the UN Security Council. As mentioned earlier, Article 4(h) of the Constitutive Act of the African Union provides for such intervention in the event of grave circumstances and it also gives the Assembly of the African Union the authority to decide over it. Moreover, Article 13(3)(c) of the African Union Peace and Security Protocol provides a mandate to the African Standby Force to intervene in the activities of a member state, when grave circumstances demand such intervention, in accordance with Article 4(h) of the Constitutive Act of the African Union.[64]

Most obviously the responsibility to protect concept strongly revolves around the dilemma between state sovereignty and intervention for humanity. In light of this, current discussions focus on the duty of the international community and the territorial state in cases of natural disasters, which raises the question whether the doctrine of the responsibility to protect can actually be extended to the international law relating to disaster relief and in particular to cases of grave circumstances, such as severe human suffering during times of natural disasters.

63 (ibid.).
64 See Kabau (2012).

Unfortunately, international law and politics thus far still seem to attach great significance to whether human suffering is the result of a natural disaster or of an (international) armed conflict.[65] However, when responding to the question whether the doctrine of the responsibility to protect should be extended to international law relating to disaster relief, one should argue in the interest of the African people. Once again, consider the words of Achim Steiner:[66]

> There is no reason why the international community cannot avoid escalating conflicts, tensions and insecurity related to a changing climate if a deliberate, focused and collective response can be catalyzed that tackles the root causes, scale, potential volatility and velocity of the challenges emerging. In bringing forward a response that enhances global security and cooperation on the climate challenge, the world can perhaps also better manage risk from numerous other challenges and in doing so diminish tensions between nations and lay the foundations and possibilities of a more sustainable and equitable peace.

## F. *The African Ministerial Conference on the Environment*

The African Ministerial Conference on the Environment (AMCEN), which has so far played a prominent role in the African response to climate change,[67] has a strong regional and sub-regional focus. Thus, AMCEN builds on the potential for regional economic communities (RECs) to integrate adaptation measures into regional policies and socio-economic development.[68] AMCEN is a permanent forum where African ministers of the environment discuss mainly matters of relevance to the environment on the continent. The forum was established in Egypt in 1985, the year in which the Cairo Programme for African Co-operation has been adopted. AMCEN has convened every second year since then. In 2010, at its Thirteenth Session, AMCEN adopted the Bamako Declaration on the Environment for Sustain-

65 See Thielbörger & Liburd (2012).

66 Address by UN Under-Secretary-General and UNEP Executive Director Achim Steiner at the UN Security Council Debate on the impact of climate change on maintaining international peace and security, 20 July 2011, available at http://www.unep.org/Documents.Multilingual/Default.Print.asp?DocumentID=2646&ArticleID=8817&l=en, last accessed 19 February 2013.

67 2009 Nairobi Declaration on the African Process for Combating Climate Change, UNEP/12/9.

68 Scholtz (2010).

able Development as its contribution towards providing political guidance and leadership on environmental management to Africa. AMCEN was established to –

- provide advocacy for environmental protection in Africa
- ensure that basic human needs are met adequately and in a sustainable manner
- ensure that social and economic development is realised at all levels, and
- ensure that agricultural activities and practices meet the food security needs of the region.

The adequate response to these challenges needs to be aligned with national and regional strategies for development, poverty alleviation, economic growth, and the enhancement of human well-being, while increasing resilience to the physical impacts of climate change. Several bodies have identified opportunities and challenges in the transition to a green economy, with links to the achievement of the United Nations Millennium Development Goals, climate change, and sustainable development. These bodies include the Meeting of African Heads of State and Government at the Seventeenth Session of the AU Summit held in Malabo, Equatorial Guinea, in July 2011; AMCEN's Fourth Special Session held in Bamako, Mali, in September 2011; and, most recently, the Seventh Session of the Committee on Food Security and Sustainable Development, as well as the Africa Regional Preparatory Conference on Sustainable Development (Rio+20) held in Addis Ababa, Ethiopia, in October 2011. In recognition of AMCEN's mandate, which includes guidance in respect of key issues related to multilateral environmental agreements, African governments asked AMCEN to facilitate the provision of information to countries that would assist them with translating available climate science and current international climate policies in their effort to move towards their practical implementation in the context of sustainable development. To this end, AMCEN prepared a guide book with information on climate change matters including science, governance, technological, financial and capacity-building needs, as well as opportunities for effective action that would lead towards sustainable development.[69]

69 See ACMEN (2011).

## G. The African Climate Policy Centre

The African Climate Policy Centre (ACPC) was established in 2010 as an integral part of the Climate for Development in Africa Programme, which is a joint initiative of the African Union Commission, the United Nations Economic Commission for Africa, and the African Development Bank.

The ACPC has been established to develop into a hub for a demand-led knowledge base on climate change in Africa to address the impact of climate change by assisting member states to elevate climate change into mainstream deliberations in their development strategies and programmes. To this end, the ACPC hosted the inaugural Climate Change and Development in Africa (CCDA) Conference between 17 and 19 October 2011 at the United Nations Conference Centre in Addis Ababa, Ethiopia. The theme for the conference was "Development First: Addressing Climate Change in Africa", which reflected the need for integrating development and climate policies, and emphasised the importance of African ownership of policy formulation and decision-making processes. The inaugural conference built directly on the African Development Forum VII, and many other forums, initiatives, activities, and outcomes of initiatives, including AMCEN; the Conference of African Heads of State and Government on Climate Change; the UNFCCC and related instruments; the UN Secretary-General's High-level Advisory Group on Climate Change Financing; the Global Climate Observation System and its sub-regional climate programme; and the Africa-EU Climate Change Partnership. The CCDA Conference helped to position the Climate for Development in Africa Programme within this ever-broadening knowledge and institutional terrain, and ascertained how best it could facilitate the interaction between policy, research and practice. The overall objective of the Conference was to establish a forum for dialogue, enhance awareness-raising, and mobilise effective commitment and action by bringing together policy makers, academicians and practising stakeholders, with the aim of effectively mainstreaming climate change concerns into development policies, strategies, programmes and practices in Africa. The CCDA also aimed at strengthening Africa's position and participation in international climate change negotiations with a view to ensuring the continent's concerns and priorities are adequately reflected in a post-2012 international climate change regime.[70]

70 See http://www.uneca.org/acpc/ccda/ccda1/index.htm, last accessed 13 January 2013.

All the aforementioned developments are laudable, and should be seen as being in the overall interest of the AU. These developments, although not always clearly concerted, reflect that the AU and its subsidiary bodies have acknowledged that climate change and its impacts constitute a pressing policy priority.

### *H. National Climate-change-related Legal Developments*

In light of the aforementioned AU policy pathways and related developments, accommodating national constitutional and legislative stipulations as well as climate change policy developments is critical because these are the linchpin between global, continental and domestic action. When countries enact policies responsive to climate change, it is more likely than not that investment will follow, which in turn can open opportunities for development and political space for international cooperation. Interaction around policy responses at national and international level generates incremental, structural and transformational change. As there is no universal solution to environmental climate change, coordination, participation and cooperation are critical for jointly achieving internationally agreed-upon goals and targets, while also addressing national capacity deficits.[71]

### *I. The Democratic Republic of Congo* [72]

Noteworthy legal developments in environmental protection in the Democratic Republic of Congo (DRC) commenced with the adoption of a new constitution on 18 February 2006.[73] During the 45 years preceding its 2006

71 See http://www.unep.org/geo/pdfs/GEO5_SPM_English.pdf, last accessed 19 January 2013.

72 Based on Ruppel & Bwiza (2013).

73 The adoption of a new constitution was a result of the overthrow of president Mobutu's power in 1996 by the opposition coalesced in the *Alliance des Forces Democratiques pour la Liberation du Congo-Zaire*. After a political impasse that lasted from 1997 to 2003, and the adoption of a government of national unity that ruled the country between 2003 and 2006, the DRC adopted a new constitution, which was followed by presidential elections.

Constitution, the DRC introduced eight previous constitutions.[74] The 2006 Constitution introduced clear environmental rights and obligations as follows:

- Right to clean drinking water (Article 48)
- Right to a healthy environment and the duty to protect the environment (Article 53)
- Obligation of the State to protect the environment and to ensure health of populations (Article 54), and
- Obligatory control of domestic and international toxic waste resulting from economic activities (Article 55).

Article 123 of the Constitution provides for laws to be made concerning, among others, the protection of the environment and tourism. To protect, among others, the environment, natural sites and landscapes, Article 203 allows for cooperative governance by central government and the provincial administrations. The DRC alone accounts for one-fifth of Africa's total forest area and as rainforests play a key role in the Earth's carbon cycle,[75] it is seen to be particularly important to preserve these forests in the fight against climate change.[76] Besides many other International Environmental Agreements, the DRC is a party to the UNFCCC (since 1995) and the Kyoto Protocol (since 2005). A National Climate Change Adaptation Plan (NAPA) was adopted in 2006, in order to develop a concrete priority action programme to guide the DRC's adaptation to climate change.[77] The NAPA is intended to provide the DRC with a framework to guide the coordination and implementation of climate change adaptation initiatives in the country, using a participative approach. The need to build synergies with other environmental programmes, such as the Biodiversty Action Plan is emphasised. A national plan to combat desertification was finalised in 2012.[78]

74 Loi fondamentale du 19 juin 1960, Constitution du 1er août 1964, Constitution du 24 juin 1967, Loi constitutionnelle du 15 août 1974, Acte constitutionnel harmonisé du 2 avril 1993, Acte Constitutionnel de la transition du 9 avril 1994, Décret-loi constitutionnel du 27 mai 1997 et Constitution de la Transition du 3 avril 2003.

75 Food and Agricultural Organization of the United Nations (2011).

76 UNEP (2008).

77 Government of the Democratic Republic of Congo (2006).

78 See http://www.riddac.org/content/view/174/, last accessed 15 November 2012.

Various national environmental laws also relevant to the field of climate change have been enacted during the past years.[79]

## *II. Ethiopia*[80]

The Constitution of the Federal Democratic Republic of Ethiopia in Article 44 provides that –

1. [a]ll persons have the right to a clean and healthy environment.
2. [a]ll persons who have been displaced or whose livelihoods have been adversely affected as a result of State programmes have the right to commensurate monetary or alternative means of compensation, including relocation with adequate State assistance.

Moreover, the Constitution stipulates the Environmental Objectives in Article 92 that –

1. Government shall endeavour to ensure that all Ethiopians live in a clean and healthy environment.
2. The design and implementation of programmes and projects of development shall not damage or destroy the environment.
3. People have the right to full consultation and to the expression of views-in the planning and implementation of environmental policies and projects that affect them directly.
4. Government and citizens shall have the duty to protect the environment.

Ethiopia is a non-Annex I member to the UNFCCC and the Kyoto Protocol, which it ratified in 2005. It has taken an active role in the recent climate negotiations and in the fight against deforestation in developing countries – mainly the poor and vulnerable ones. Ethiopia's 1995 Constitution includes the principle of environmental rights, including the right to a clean and healthy environment and the principle of government responsibility to en-

79 That is, law no 11/009 of 09 July 2011 on fundamental principles of environmental protection in the DRC on fundamental principles of environmental protection in the DRC; the Forest Code no 011/2002 of 29 August 2002, which sets out the law applicable to the conservation, exploitation and development of forestry resources; the Mining Code law no 007/2002 of 11 July 2002, which specifies the need for environmental impact assessments to be conducted for certain activities, such as prior feasibility studies of environmental, social and economic impact of projects, mitigation and rehabilitation plans, and environmental management plans for specific projects.

80 Based largely on GLOBE International (2013:134–141).

sure this right. In 2011, the Ethiopian Government finalised the Climate Resilient Green Economy[81] – the first of its kind in Africa. The strategy seeks to achieve ambitious economic development goals in a sustainable way by building a climate-resilient green economy

Ethiopia's NAPA[82] was finalised in June 2007 by the Ministry of Water Resources and the Meteorological Service. The NAPA identifies high-priority adaptation projects, for example, promoting drought and crop insurance programmes; strengthening or enhancing drought and flood early warning systems; conserving and using wisely selected wetlands to promote the adaptation capacity of the rural community for climate shocks; enhancing the Community Based Carbon Sequestration Project in the Rift Valley System; establishing a national research and development (R&D) centre for climate change; strengthening the malaria containment programme; and promoting farm and homestead forestry and agroforestry practices in arid, semiarid and dry sub-humid parts of Ethiopia. The NAPA was updated and replaced in 2010 by the Ethiopian Programme of Adaptation to Climate Change, which is tasked to identify climate-change-related risks and oversee institutions in charge of mitigating these risks.

### *III. Kenya*[83]

The 2010 Kenyan Constitution[84] provides a modern framework for environmental rights. In its preamble it states that the people of Kenya are respectful of the environment, "which is our heritage, and determined to sustain it for the benefit of future generations". In the section on rights and fundamental freedoms, the Constitution in Article 42 recognises that everyone has a "right to a clean and healthy environment", which includes the right –

(a) to have the environment protected for the benefit of present and future generations through legislative and other measures, particularly those contemplated in Article 69; and
(b) to have obligations relating to the environment fulfilled under Article 70.

---

81 Federal Democratic Republic of Ethiopia (2011).
82 Available at http://unfccc.int/resource/docs/napa/eth01.pdf, last accessed 06 April 2013.
83 Based largely on GLOBE International (2013:269–277).
84 See Glinz (2011:60–80).

Under Chapter 5 (Land and Environment) of the Constitution, the enforcement of environmental rights is regulated in Article 70(1), where it states that if a person alleges that a right to a clean and healthy environment recognised and protected under Article 42 has been, is being or is likely to be, denied, violated, infringed or threatened, the person may apply to a court for redress in addition to any other legal remedies that are available in respect to the same matter. Kenya signed the UNFCCC in 1992, ratifying it in 1994, along with the Kyoto Protocol in 2005. Kenya (a non-Annex I country) has placed great importance on issues of climate change, for example its 2008 Draft National Environment Policy, which recognises that climate change involves many ministries, gives climate change a higher profile at a national level to help address climate-related issues in respect of the relevant ministries.[85]

Kenya has developed a Climate Change Authority Act that is making progress through parliament and is likely to be voted into law in early 2013. In November 2012 a public validation process paved the way for government to approve a land launch in early 2013 of a complementary Climate Change Action Plan that defines clear measures on adaptation and mitigation, including nationally appropriate mitigation actions, a low carbon development strategy, knowledge management and capacity development, financing mechanisms and the creation of an institutional structure to ensure effective coordination. Developed by the Kenyan Government, through the Ministry of Environment and Mineral Resources, and in conjunction with donor partners, the Action Plan provides a platform for the implementation of the 2010 National Climate Change Response Strategy. When it is passed, the Action Plan could qualify to be considered Kenya's flagship legislation on climate change. Another process under way is the formulation of the National Environment Policy, also expected to be adopted soon. The Climate Change Authority Bill was introduced in the National Assembly in 2012. In addition to establishing a Climate Change authority, the Bill sets out to provide for the development of strategies to address the effects of climate change, as well as to forge a framework for mitigation of and adaptation to climate.

85 See Kameri-Mbote & Odote (2012:296–318).

## *IV. Namibia*[86]

The 1990 Namibian Constitution, many international treaties, as well as a multitude of statutory enactments and policies provide for a wide field of environmental protection in Namibia. Over the past years a bundle of new environmental legislation has been passed, and it thus becomes evident that environmental concern has gained momentum. The Namibian environmental policy framework[87] determines the guiding objectives and the strategies to be used to strengthen the respect for environmental values, taking into account the existing social, cultural and economic situation. The foundation for the Namibian environmental policy framework is Article 95 (l) of the Namibian Constitution of 1990. It stipulates that the state shall actively promote and maintain the welfare of the people by adopting policies which include the "maintenance of ecosystems, essential ecological processes and biological diversity of Namibia and utilisation of living natural resources on a sustainable basis ...".

It is worthwhile mentioning that the Namibian Constitution explicitly incorporates international law and makes it part of the law of the land. *Ab initio*, public international law is part of the law of Namibia. No transformation or subsequent legislative act is needed. A treaty will become binding upon Namibia in terms of Article 144 of the Constitution if the relevant international and constitutional requirements have been met.[88] Namibia ratified the UNFCCC in 1995 and became legally obliged to adopt and implement policies and measures designed to mitigate the effects of climate change and to adapt to such changes. Namibia acceded to the Kyoto Protocol in 2003. Namibia's Initial National Communication to the Conference of Parties of the UNFCCC was submitted in 2002 in accordance with decisions taken at various COPs to the UNFCCC. The Ministry of Environment and Tourism (MET) through the Directorate of Environmental Affairs (DEA) is responsible for overseeing the coordination of climate change issues in Namibia.[89]

Vision 2030, the various National Development Plans, various sector policies and Cabinet directives are in place. Moreover, Namibia in 2008 drew up a draft national climate change strategy and action plan – which was

86 Based largely on Ruppel & Ruppel-Schlichting (2013).
87 (ibid.).
88 See with further references Ruppel (2010a).
89 Mapaure (2013).

introduced in 2009.[90] In June 2011 Namibia's prime minister, Nahas Angula, indicated that a national policy on climate change for Namibia should be placed within the global framework of political, developmental, and technological interests.[91] The then Namibian minister of Environment and Tourism made the following significant statement: "While climate change has the potential to side-rail development processes, the key is to prepare sufficiently and effectively and to use the threats and opportunities of climate change to lay the basis for sustainability and prosperity".[92]

### *V. Rwanda*[93]

Article 49 (I) of the 2003 Rwandan Constitution stipulates that every citizen is entitled to a healthy and satisfying environment. This provision obliges the government to commit itself to ensuring that citizens live in a safe and clean environment. To achieve this obligation, reforestation and terracing are emphasised on hilly areas to prevent soil erosion, while marshlands and low-land vegetation are protected. Article 49 (II) provides that every person has a duty to protect, safeguard and promote the environment, and places the State under the obligation to protect the environment. Article 49 (III) stipulates that the law shall determine the modalities for protecting, safeguarding and promoting the environment.

Rwanda has ratified almost all international instruments related to the protection of the environment. The government provides orientation through the national policy on environment and the national policy on water and sanitation. The Rwanda Environment Management Authority takes the lead in regulating, safeguarding and promoting safe-and-clean environment protection programmes. Rwanda ratified the UNFCCC and the Kyoto Protocol in 1998. In 2005 Rwanda submitted its initial report to the UNFCCC, and in June 2012 its second communication, including a stand-alone mitigation strategy, the Carbon Policy and an updated emissions inventory. The Constitution of 2003 is Rwanda's supreme law. In 2011 the Government pub-

---

90 See Mfune et al. (2009).

91 See http://www.namibian.com.na/news-articles/national/full-story/archive/2011/june/article/policy-on-climate-change-needs-review-pm/, last accessed 18 February 2013.

92 Nandi-Ndaitwah (2011).

93 Based largely on GLOBE International (2013:365–371).

lished the National Climate Change and Low Carbon Development Strategy. The Action Plan for the Ministry of Natural Resources July 2011 – June 2012 sets specific targets for reducing climate change vulnerability. In May 2012, a law establishing a national fund for climate change financing, FONERWA, was passed, and is expected to contribute to Rwanda's existing financing gap. In 2009 a Climate Change and International Obligations Unit was established to coordinate carbon market activities. Rwanda's climate change efforts are supported by various international donors. In 2010, the Government of Rwanda, the Government of Japan, the United Nations Development Programme and the United Nations Environment Programme (UNEP) launched two climate change adaptation programmes, one focusing on reducing vulnerability to climate change by establishing early warning and disaster preparedness systems and support for integrated watershed management in flood-prone areas. The other focuses on building an integrated comprehensive national adaptation approach in Rwanda.

### *VI. South Africa*[94]

Section 24 of the 1996 South African Constitution states, among others, that everyone has the right –

> (a) to an environment that is not harmful to their health or well-being; and
> (b) to have the environment protected, for the benefit of present and future generations, through reasonable legislative and other measures that –
>   (i) prevent pollution and ecological degradation;
>   (ii) promote conservation; and
>   (iii) secure ecologically sustainable development and use of natural resources while promoting justifiable economic and social development.

The Constitution also provides the framework for the administration of environmental laws. It designates the environment as a matter of concurrent national and provincial responsibility. Environmental law is contained in a multitude of statutes and regulations.[95]

94 Based on GLOBE International (2013:373–379).

95 Relevant to the context of climate change are, *inter alia,* the following statutes: National Environmental Management Act, 1998 (Act 107 of 1998); National Water Act, 1998 (Act 36 of 1998); Marine Living Resources Act, 1998 (Act 18 of 1998); Disaster Management Act, 2002 (Act 57 of 2002); National Environmental Management: Biodiversity Act, 2004 (Act 10 of 2004); National Environmental Management:

South Africa signed the UNFCCC in 1993 and ratified it in 1997. It acceded to the Kyoto Protocol in July 2002. Within the Sub-Directorate for Global Climate Change of the Department of Environmental Affairs lies the responsibility to ensure that South Africa complies with its obligations under the UNFCCC. In 2011 South Africa's Second National Communication under the United Nations Framework Convention on Climate Change was published.[96] South Africa has largely dealt with climate change through policies, strategies and regulations. Through these measures South Africa is showing its commitment to tackling climate change, particularly in developing market-based mitigation mechanisms and promoting renewable energy and energy efficiency.

The 2004 National Climate Change Response Strategy represented the first direct recognition of the need for action on climate change. Two years later, the Cabinet commissioned the Long Term Mitigation Scenario study in an attempt to produce sound scientific analysis from which the Government could derive a long-term climate policy. The study produced a series of policy recommendations. In July 2008, the Vision, Strategic Direction and Framework for Climate Policy were announced. The current flagship policy in South Africa is the National Climate Change Response Policy, approved by Cabinet in October 2011. This policy's White Paper presents the South African Government's vision for an effective climate change response and the long-term, just transition to a climate resilient and lower carbon economy and society. It reflects a strategic approach referred to as "climate change resilient development", addressing both adaptation and mitigation. The White Paper accepts the conclusions of the Intergovernmental Panel on Climate Change (IPCC); regards climate change as one of the greatest threats to sustainable development; reaffirms its commitment towards the UNFCCC and the Kyoto Protocol, and undertakes to develop a comprehensive national response plan of which the White Paper is an integral part.

The Taxation Law Amendment Bill of 2009 amends the 1962 Income Tax Act to include, among other things, income tax incentives for participation in Clean Development Mechanism projects as well as for energy efficiency savings. The Clean Development Mechanism projects are run by a desig-

Protected Areas Act, 2003 (Act 57 of 2003); National Environmental Management: Waste Management Act 2008; and National Environmental Management: Air Quality Act, 2004 (Act 39 of 2004).

96 See Glazweski & Collier (2012:319–348).

nated national authority established under the Department of Energy, and they are governed by regulations published under the National Environmental Management Act of 1998. During the 2012–2013 budget discussions in February 2012, the minister of Finance introduced a proposed carbon tax on annual emissions for all sectors, including electricity, petroleum, iron, steel and aluminium.

### *I. Developments and Gaps*

The aforementioned passages reflect that much is already been done in Africa: the rolling out of AU policy pathways and related legal processes, the accommodation of national constitutional and legislative frameworks, and the development of climate change policy are all in progress and are also critical for the continent. Resolution (A/RES/67/1) was recently adopted by the UN General Assembly on the Rule of Law – which underlines the importance of fair, stable and predictable legal frameworks for generating inclusive, sustainable and equitable development and for maintaining peace and security. In this light it is also essential to further the discussions on

- the linkage between climate change, sustainable development and the rule of law
- the mutually supportive relationship between the adherence to the rule of law and the respect for all human rights, including those related to the environment
- access to justice in environmental matters and new and emerging principles, such as the non-regression in environmental law; and
- the importance of country systems and other developments at the national level to improve the effectiveness of environmental and climate change law.

Many African governments have progressed in addressing climate change and related issues. The AU has also succeeded in presenting an increasingly cohesive African position on climate change. However, significant gaps remain.[97]

At the regional level, effectively managing consensus and divergence remains challenging. Divergent priorities among African countries threaten

97 This passage is based on Ruppel & Speranza (2011:199–202).

the potential of the AU to wield influence in international climate politics. The AU thus needs to address divergence factors in lieu of conducting international negotiations.

At the national and sub-regional levels, various policy gaps exist, a major one being the sluggishness in bringing climate change into the mainstream planning of all development sectors. In many African countries, a climate policy is non-existent or still in the making. Development and climate policy run parallel and integrated development-climate policy framework is non-existent, making it difficult to stop the rebranding of Official Development Assistance (ODA) as climate response. Very few sectoral policies consider climate change, and need reviewing to close this gap. [98]

Moreover, African governments should make provisions for integrating climate change responses into national budgetary allocations. Economic planners often lack guidelines on mainstreaming climate change adaptation at the national level. As with other policy spheres, climate-proofing development through integrating climate change in all policy spheres has its costs and trade-offs. Considering the chronic understaffing and underfunding of certain government activities, climate change will bring more work and governments should improve staff skills and provide more resources to address the add-on challenges of climate change. Many operational limitations hinder implementation. These include dysfunctional organisational arrangements causing conflicting and overlapping mandates, overburdening reporting requirements of various international agreements and conventions, and inability to retain skilled staff.

Lastly, designing policies for dealing with climate change offers an opportunity to address the dichotomy between parallel regulatory systems, i.e. the customary and the state laws, e.g. in access to land, the aspects of the management, use of and control over natural resources, and benefit-sharing. However, issues remain on how to develop a national framework for compensating natural resource users for providing environmental services and how to proceed with a low-carbon development, renewable energy schemes and the role of green transformation in these processes.[99]

98 (ibid.).
99 (ibid.).

## J. Conclusion

Although it is primarily the responsibility of developed countries to reduce their greenhouse gas emissions first, in line with the principle of common but differentiated responsibility, developing countries also need to make specific policy commitments. What became very clear during the past few years is that we live in a world that is very different from the context in which the Kyoto Protocol was established. Compared with the mid-1990s, the debates are now taking place in a much-changed world. This applies to Africa, but even more notably to China and other developing countries that have gained more political clout – and greater impact on the world's climate. More effective action is still needed.[100]

For Africa climate change is becoming an increasingly threatening reality, but one which is often too narrowly perceived in economic and sectoral policies, meaning that the severe negative consequences on the people are still largely being neglected.[101]

According to IPCC findings –[102]

> Africa is one of the most vulnerable continents to climate change and climate variability, a situation aggravated by the interaction of 'multiple stresses', occurring at various levels, and low adaptive capacity.... Africa's major economic sectors are vulnerable to current climate sensitivity, with huge economic impacts, and this vulnerability is exacerbated by existing developmental challenges such as endemic poverty, complex governance and institutional dimensions; limited access to capital, including markets, infrastructure and technology; ecosystem degradation; and complex disasters and conflicts. These in turn have contributed to Africa's weak adaptive capacity, increasing the continent's vulnerability to projected climate change.

In response thereto the AU has introduced initiatives aiming to ensure that poor and marginalised communities in Africa, i.e. women, children and indigenous peoples, do not suffer a disproportionate burden associated with climate change. The same applies to various African countries where significant climate-change-policy-related (legal) developments reflect political will towards addressing climate change.[103]

The AU has been decisive on action. Populations whose rights are poorly protected are likely to be less well-equipped to understand or prepare for

100 See Ruppel et al. (2011).
101 Susswein (2003:297).
102 Boko et al. (2007:435).
103 Ruppel & Speranza (2011:199–202).

climate change; they would be less able to enforce their rights or lobby them effectively; and are, thus, more likely to be unable to adapt to the anticipated changes in their environmental and economic situation. What remains essential is that Africa is very clear on the fact that the majority of its people are vulnerable, i.e. subjected to displacement, landlessness, joblessness, homelessness, marginalisation, food insecurity, increased morbidity, loss of access to common property resources, and social disarticulation. These people are particularly vulnerable to extreme events, on the basis of a wide range of social vulnerability characteristics.[104] This needs to be brought more sharply to the attention of the global community in order to develop more international solidarity in terms of a global corporate social responsibility to protect those populations that need it most. In the same light it is hoped that in future we can come up with more global and consolidated approaches when it comes to water, climate, energy, economy, security and development policy. In the interest of Africa and its people this – without doubt – requires still more international cooperation, regional integration and local implementation.[105] In fact, a new global "social contract refers to the necessity of humankind taking collective responsibility for the avoidance of dangerous climate change…to the planet".[106]

Moreover, addressing climate change adequately also requires substantial investment in new technologies, processes and services. Therefore a favourable investment climate is pivotal – and adequate framework conditions for more inclusive climate investment, leveraging private sector resources, and seizing opportunities for innovation will be needed. In light of this, Africa now needs to strengthen its long-term vision of working together on the African continent, with global partners; and, in adapting to climate change, it needs to promote climate investment, sustainable economic development and green economic growth.

A major problem unfortunately prevails: The World Corruption Report 2011[107] states that corruption was a risk in addressing climate change, since a risk of corruption always exists where "huge amounts of money flow through new and untested financial markets and mechanisms". This may be particularly true for recent, current and future financial flows related to climate change finance, technology and capacity-building meant to support

104 IPCC (2012).
105 See Ruppel (2012b) for further references.
106 WBGU (2011:8).
107 Transparency International (2011).

developing countries according to the principle of equity. Corruption eventually puts at risk the rights of those most vulnerable to the negative effects of climate change. The reasons for the high risk of corruption with regard to climate finance are rooted in the level of complexity, uncertainty and novelty that surrounds many climate issues. A multitude of regulatory grey zones and loopholes exist that are at risk of being exploited to satisfy corrupt interests. In order to ensure that the investments by the public and private sectors are properly and equitably managed, a system of good climate governance[108] with participatory, accountable, transparent, inclusive and responsive policy development and decisions, and respect for the rule of law, is essential.

## *References*

AMCEN/African Ministerial Conference on the Environment, 2011, *Addressing Climate Change Challenges in Africa: A Practical Guide towards Sustainable Development*, available at www.unep.org/roa/amcen/docs/.../guidebook_CLimateChange.pdf, last accessed 15 January 2013.

Boko, Michel, Isabelle Niang, Anthony Nyong, Coleen Vogel, Andrew Githeko, Mahmoud Medany, Balgis Osman-Elasha, Ramadjita Tabo & Pius Yanda, 2007, Africa, in: Parry, Martin L., Osvaldo F. Canziani, Jean Palutikof, Paul J. van der Linden & Clair E. Hanson (Eds), *Climate Change 2007: Impacts, Adaptation and Vulnerability. Contribution of Working Group II to the Fourth Assessment Report of the Intergovernmental Panel on Climate Change*, Cambridge, Cambridge University Press, 433–467.

Eze, Osita C., 2012, The Conventions Governing the Specific Aspects of Refugee Problems in Africa, in: Yusuf, Abdulqawi A. & Fatash Ouguergouz (Eds), *The African Union: Legal and Institutional Framework*, Leiden, Boston, Martinus Nijhoff Publishers, 495–518.

FAO/Food and Agricultural Organization of the United Nations, 2011, *State of the World's Forests 2011*. Rome, FAO.

Federal Democratic Republic of Ethiopia, 2011, *Ethiopia's Climate-Resilient Green Economy Strategy*, Addis Ababa, Environmental Protection Agency, available at http://www.uncsd2012.org/content/documents/287CRGE%20Ethiopia%20Green%20Economy_Brochure.pdf, last accessed 06 April 2013.

108 *Climate governance* can be understood as the processes that currently exist at the international, national, corporate and local levels to address the causes and effects of climate change. See Transparency International (2011:3).

Glazewski, Jan & Debbie Collier, 2012, South Africa, in: Lord, Richard, Silke Goldberg, Lavanya Rajamani & Jutta Brunnée (Eds), *Climate Change Liability. Transnational Law and Practice*, Cambridge, Cambridge University Press, 319–348.

Glinz, Cornelia, 2011, Kenya's New Constitution: A Transforming Document or Less Than Meets the Eye? *Verfassung und Recht in Übersee* 44, 60–80.

GLOBE International, 2013, *Climate Legislation Study: A Review of Climate Change Legislation in 33 Countries* (Third Edition), Townshend, Terry, Sam Fankhauser, Rafael Aybar, Murray Collins, Tucker Landesman, Michal Nachmany & Carolina Pavese (Eds), available at http://www.globeinternational.org/images/climate-study/3rd_GLOBE_Report.pdf, last accessed 18 February 2013.

Government of the Democratic Republic of Congo, 2006, *Programme d'Action National d'Adaptation au Changement Climatique de la République Démocratique du Congo*, available at http://unfccc.int/resource/docs/napa/cod01.pdf, last accessed 18 January 2012.

Hoste, Jean-Christophe, 2010, Where Was United Africa in the Climate Change Negotiations?, *Africa Policy Brief* No. 2., Brussels, Egmont Royal Institute for International Relations, available at http://www.edc2020.eu/fileadmin/Textdateien/post_COP_15_briefing/Jean_Christophe_Hoste_-_Where_was_united_Africa_in_the_climate_change_negotiations_-_EDC_2020.pdf, last accessed 13 January 2013.

IPCC/Intergovernmental Panel on Climate Change, 2012, *Managing the Risks of Extreme Events and Disasters to Advance Climate Change Adaptation. A Special Report of Working Groups I and II of the Intergovernmental Panel on Climate Change*, Field, Christopher B., Vicente Barros, Thomas F. Stocker, Dahe Qin, David J. Dokken, Kristie L. Ebi, Michael D. Mastrandrea, Katharine J. Mach, Gian-Kasper Plattner, Simon K. Allen, Melinda Tignor & Pauline M. Midgley (Eds), Cambridge, Cambridge University Press.

IRENA/International Renewable Energy Agency, 2011, *Prospects for the African Power Sector*, available at http://www.irena.org/DocumentDownloads/Publications/Prospects_for_the_African_Power_Sector_abstract.pdf, last accessed 17 January 2013.

Kabau, Tom, 2012, The Responsibility to Protect and the Role of Regional Organizations: An Appraisal of the African Union's Interventions, *Goettingen Journal of International Law* 4, 49–92.

Kameri-Mbote, Patricia & Collins Odote, 2012, Kenya, in: Lord, Richard, Silke Goldberg, Lavanya Rajamani & Jutta Brunnée (Eds), *Climate Change Liability. Transnational Law and Practice*, Cambridge, Cambridge University Press, 296–318.

Keohane, Robert O. & David G. Victor, 2011, The Regime Complex for Climate Change, *Perspective on Politics* 9 (1), 7–23.

Kidane, Won, 2011, Managing Forced Displacement by Law in Africa: The Role of the New African Union IDPs Convention, *Vanderbuilt Journal of Transnational Law* 44 (1), 1–85.

Koroma, Abdul D., 2012, Solidarity: Evidence of an Emerging International Legal Principle, in: Hestermeyer, Holger P., Doris König, Nele Matz-Lück, Volker Röben, Anja Seibert-Fohr, Peter-Tobias Stoll & Silja Vöneki (Eds), *Coexistence, Cooperation and Solidarity: Liber Amicorum Rüdiger Wolfrum,* Leiden, Martinus Nijhoff Publishers, 103–129.

Mapaure, Isaak, 2013, Climate Change in Namibia: Projected Trends and Effects, in: Ruppel, Oliver C. & Katharina Ruppel-Schlichting (Eds), *Environmental Law and Policy in Namibia* (Second Edition), Windhoek, Orumbonde Press; Essen, Welwitschia Verlag Dr. A. Eckl, 245–264.

Mfune, John K., Oliver C. Ruppel, Nico E. Willemse & Alfons W. Mosimane, 2009, *Namibia Climate Change Strategy and Action Plan*, Windhoek, Ministry of Environment and Tourism, available at http://www.met.gov.na/Documents/NAMIBIA-proposed%20Climate%20Change%20Strategy%20and%20Action%20Plan %20(13.pdf, last accessed 13 January 2013.

Ruppel, Oliver C. (Ed.), 2008, *Women and Custom in Namibia: Cultural Practice versus Gender Equality?,* Windhoek, Macmillan Education Namibia, available at http:// www.kas.de/proj/home/pub/8/2/dokument_id-15086/index.html, last accessed 13 January 2012.

Ruppel, Oliver C., 2009a, Regional Economic Communities and Human Rights in East and Southern Africa, in: Bösl, Anton & Joseph Diescho (Eds), *Human Rights in Africa,* Windhoek, Macmillan Publishers Namibia, 273–347.

Ruppel, Oliver C., 2009b, The Protection of Children's Rights under International Law from a Namibian Perspective, in: Ruppel, Oliver C. (Ed.), *Children's Rights in Namibia*, Windhoek, Macmillan Publishers Namibia, 53–100.

Ruppel, Oliver C., 2010a, Environmental Rights and Justice in Namibia, in: Bösl, Anton, Nico Horn & André du Pisani (Eds), *Constitutional Democracy in Namibia: A Critical Analysis after Two Decades*, Windhoek, Macmillan Education Namibia, 323–360.

Ruppel, Oliver C., 2010b, Women's Rights and Customary Law in Namibia: A Conflict Between Human and Cultural Rights?, *Basler Afrika Bibliographien* BAB Working Paper 2010/2, available at http://www.baslerafrika.ch/upload/files/WP_2010_2_Ruppel.pdf, last accessed 20 January 2012.

Ruppel, Oliver C., 2011a, International Environmental Law from a Namibian Perspective, in: Ruppel, Oliver C. & Katharina Ruppel-Schlichting (Eds), *Environmental Law and Policy in Namibia*, Windhoek, Orumbonde Press, 33–78.

Ruppel, Oliver C., 2011b, Human Rights and the Environment, in: Ruppel, Oliver C. & Katharina Ruppel-Schlichting (Eds), *Environmental Law and Policy in Namibia*, Windhoek, Orumbonde Press, 219–237.

Ruppel, Oliver C., 2012a, Climate Change Policy Positions and Related Developments in the AU and SADC, *SADC Law Journal* 2 (1), 14–35.

Ruppel, Oliver C., 2012b, Wasser und Land. Brennpunkte innerhalb der Entwicklungsgemeinschaft des südlichen Afrika (SADC), in: Reder, Michael & Hanna Pfeifer (Eds), *Kampf um Ressourcen. Weltordnung zwischen Konkurrenz und Kooperation* (Globale Solidarität – Schritte zu Einer Neuen Weltkultur, 22), Stuttgart, Kohlhammer, 59–85.

Ruppel, Oliver C. & Dignité Bwiza, 2013, Country Report: Democratic Republic of Congo: Recent Developments in Environmental Protection, *IUCN E-Journal* 2013 (4), 138–147.

Ruppel, Oliver C., Sven Grimm & Sanita van Wyk, 2011, *COP 17: Addressing Climate Change in a Changed World – Chinese and African Positions*, available at http://www.ccs.org.za/wp-content/uploads/2011/12/COP17.pdf, last accessed 7 February 2013.

Ruppel, Oliver C. & Katharina Ruppel-Schlichting, 2012, Climate Change and Human Security: Relevant for Regional Integration in SADC?, in: Hartzenberg, Trudy, Gerhard Erasmus, André du Pisani (Eds), *Monitoring Regional Integration in Southern Africa, 2011 Yearbook for Regional Integration,* Stellenbosch, Trade Law Centre for Southern Africa (TRALAC), 32–71.

Ruppel, Oliver C. & Katharina Ruppel-Schlichting (Eds), 2013, *Environmental Law and Policy in Namibia* (Second Edition), Windhoek, Orumbonde Press; Essen, Welwitschia Verlag Dr. A. Eckl.

Ruppel, Oliver C. & Chinwe Ifejika Speranza, 2011, The Institutional, Legal and Policy Frameworks, in: AMCEN/African Ministerial Conference on the Environment, *Guidebook: Addressing Climate Change Challenges in Africa*, Nairobi, African Union, AMCEN & United Nations Environment Programme, 138–166, available at http://www.unep.org/roa/amcen/docs/publications/guidebook_CLimateChange.pdf, last accessed 23 January 2013.

Ruppel, Oliver C. & Sanita van Wyk, 2011, The Effects of Climate Change on Human Security and Human Rights in Africa, *China Monitor*, available at http://www.ccs.org.za/wp-content/uploads/2011/10/China_Monitor_OCT_2011_final.pdf, last accessed 23 December 2012.

Scott, Shirley V., 2012. The Securitization of Climate Change in World Politics: How Close Have We Come and Would Full Securitization Enhance the Efficacy of Global Climate Change Policy?, *Review of European Community and International Environmental Law* 21 (3), 220–230.

Scholtz, Werner, 2010, The Promotion of Regional Environmental Security and Africa's Common Position on Climate Change, *African Human Rights Law Journal* 10, 1–25.

Susswein, Pendo M., 2003, The Use of Market Based Economic Instruments for Sustainable Environmental Management in Africa: A Case Study of the SADC Region, in: Chaytor, Beatrice & Kevin R. Gray (Eds), *International Environmental Law and Policy in Africa*, Dordrecht, Kluwer.

Thielbörger, Piere & Medgy Liburd, 2012, Haiti Two Years After the Earthquake – A Case for the Responsibility to Protect?, Institute for International Law of Peace and Armed Conflict, *BOFAXE* No. 399E.

Transparency International, 2011, *World Corruption Report: Climate Change*, London, Earthscan.

UN/United Nations Womenwatch, 2009, Fact Sheet: Women and Climate Change, available at http://www.un.org/womenwatch/feature/climate_change/downloads/Women_and_Climate_Change_Factsheet.pdf, last accessed 25 June 2011.

UNEP/United Nations Environmental Programme, 2008, *Africa Atlas of Our Changing Environment*, Nairobi, UNEP, available at http://unep.org/dewa/africa/africaAtlas/, last accessed 28 December 2012.

UNFCCC/United Nations Framework Convention on Climate Change, 2012, Report of the Green Climate Fund to the Conference of the Parties and Guidance to the Green Climate Fund, available at http://unfccc.int/resource/docs/2012/cop18/eng/05.pdf, last accessed 18 February 2013.

UNICEF/United Nations Children's Fund, 2009, *A Brighter Tomorrow: Climate Change, Child Rights and Intergenerational Justice*, available at http://www.unicef.org.uk/Documents/Publications/intergenerationaljustice.pdf, last accessed 25 June 2011.

Van der Linde, Morne & Lorette Louw, 2003, Considering the Interpretation and Implementation of Article 24 of the African Charter on Human and Peoples' Rights in Light of the SERAC Communication, *African Human Rights Law Journal* 3 (1), 167–187.

WBGU/Wissenschaftlicher Beirat der Bundesregierung Globale Umweltveränderungen Flagship Report, 2011, *World in Transition: A Social Contract for Sustainability, German Advisory Council on Global Change,* Berlin, WBGU.

# 14
# Council of Europe Climate Law Standards and Perspectives

*Agata Rogalska-Piechota*

*Abstract*

The Council of Europe (CoE) is an organisation which covers virtually the entire European continent and is anchored in the three pillars (or common values) – human rights, democracy, and the rule of law. It is famous for its developments regarding human rights protection and the promotion of democratic standards, and is focused mainly on actions aiming to protect and promote these three pillars. Nevertheless, its scope of interest is very wide and includes not only human rights protection, democracy-building and combating challenges to the rule of law, but also biodiversity protection, challenges to sports, and youth and culture issues. Among this variety of interests and activities are some which are linked – directly or indirectly – with the protection of the environment.

In this article, selected case law related to the environment from the European Court of Human Rights (ECtHR) and from the European Committee of Social Rights (ECSR), as well as the official publications of the CoE in this field will be outlined, to demonstrate its contribution to the climate change debate. Attention will also be paid to various texts relating to the above subject matter and revealing the real level of commitment of the organisation and its entities to the subject matter, mainly by the Committee of Ministers (CM), the Parliamentary Assembly of the Council of Europe (PACE), the Congress of Local and Regional Authorities (CLRA), and the CoE Human Rights Commissioner. Apart from that, references to the attempts to systematise the internal actions of the CoE bodies in the above field by establishing the Inter-secretariat Group on Climate Change will be included, in order to answer the question of whether the organisation has developed its own set of climate law standards. Subsequently, prospects and possibilities that arise from present achievements will be commented on.

One key observation is that the organisation acknowledges that climate change has become a very important issue and uses the environmental achievements it has already made to join the worldwide climate change debate. At the same time, it is apparent that the climatic subject matter is far from being central to the CoE's interests. As a result, the organisation does not play a standard-setting role in the field of climate law. Nevertheless, it undertakes certain actions to visualise its efforts; in particular, it proudly presents the case law of the ECtHR and the ECSR.

Having observed the above, it is anticipated that the trend to emphasise the environmental case law of the human rights protection bodies will continue. The author does not think that the organisation will be able to overcome its general reluctance to discuss climatic issues. Therefore, it is not expected that the CoE will soon become a key player in the worldwide debate that will contribute to the process of climate law standard-setting.

## *A. Introduction*

Today, the Council of Europe (CoE) consists of 47 member states which are home to approximately 800 million citizens. The organisation, founded on 5 May 1949 by ten countries, now covers virtually the entire European continent. It is said[1] to seek to develop common and democratic principles throughout Europe, based on the European Convention on Human Rights (ECHR, the Convention)[2] and other reference texts on the protection of individuals.[3] In this article, it will be examined whether the CoE – the organisation anchored in the triad of human rights, democracy and the rule of law – has developed its own set of climate law standards. Subsequently, prospects and possibilities that arise from present achievements will be commented on. In particular, it is argued that anthropocentric perspectives

1 See http://www.coe.int/aboutCoe/index.asp?page=quisommesnous&l=en; last accessed 15 December 2012.

2 Convention for the Protection of Human Rights and Fundamental Freedoms (1950) as amended by Protocols No. 11 (1994) and No. 14 (2004), ETS/CETS No.'s 005, 155 and 194.

3 Examples of such reference texts are recommendations of the Committee of Ministers (CM) or the Parliamentary Assembly of the Council of Europe (PACE), respectively, under Article 15 or Article 23 of the Statute of the Council of Europe (1949), ETS No. 001. According to Resolution No. 1 (1953) adopted by the Committee of Ministers at its 8th Session, the conclusions of the Committee may, where appropriate, take the form of a convention or agreement. For the full list of treaties, see the official website of the CoE Treaty Office at http://conventions.coe.int/Treaty/Commun/ListeTraites.asp?CM=8&CL=ENG, last accessed 15 December 2012.

adopted by the organisation in fact self-limit the scope and impact of actions undertaken within the CoE.

### B. The Three Pillars

As stated above, the CoE is based on the triad of human rights, democracy and the rule of law.[4] In the author's opinion, the emphasis put on these common values by the CoE itself is apparent and seems to determine the organisation's range of activities. Therefore, the triad will be used as the starting point for the following text.

The triad of values may be treated as the common denominator in all actions taken within the CoE. Some activities commonly associated with the CoE are, de facto, attributable to one of its statutory organs, i.e. a Council of Ministers (CM), consisting of the Ministers of Foreign Affairs, which usually meets at the level of their deputies, and the Parliamentary Assembly of the Council of Europe (PACE), which represents the political forces within the member states.[5] Other actions associated with the CoE are actually performed by bodies established by the CM, such as the Congress of Local and Regional Authorities (CLRA),[6] or are undertaken by statutory organs of the CoE's treaties, including the European Court of Human Rights (ECtHR)[7] and the European Committee of Social Rights (ECSR).[8] Therefore, the subjective range and material scope of actions taken within the CoE which are associable with the notion of *climate law* are very wide. At the same time, it is not easy to indicate actions that would be devoted to climate without reference to the triad of values. In this regard, it should be pointed out that the capacity limitation of the present text does not allow to go too deeply into detail. Therefore, the actions taken by the statutory organs of the CoE will be indicated in principle. Selected activities by other organs and bodies will be highlighted, especially to mark the flow of ideas, standards and stipulations from one entity to another.

4 See the Statute of the Council of Europe, in particular the Preamble and Article 3.

5 See the Statute of the Council of Europe, in particular Article 10.

6 See the Statutory Resolution 94(3) of the Committee of Ministers of the Council of Europe establishing the Congress of Local and Regional Authorities.

7 See Section II of the ECHR regarding the ECtHR.

8 See Article 25 of the European Social Charter regarding the ECSR.

The triad of values appears to divide the CoE's activities into three segments. At the same time, the values are so similar that it is not always easy to exactly match each activity with only one value. Therefore, to specify the appropriate value, it is advisable to analyse the budget of the CoE, its organisation,[9] the arrangement of its Secretariat, the current terms of reference of its internal structures,[10] and other relevant documents.

It might also be helpful to use the divisions applied on the CoE's website.[11] In this regard, one may observe that the so-called climatic activities of the CoE are mostly performed under the following headings:

- *Democracy*, that is, PACE and CLRA activities
- *Biodiversity*, that is, activities relating to the European and Mediterranean Major Hazards Agreement and the Bern Convention on the Conservation of European Wildlife and Natural Habitats, and
- *Human rights*, that is, some aspects of jurisprudence of the ECtHR, development of human rights law and policy, and social rights guaranteed under the European Social Charter.

At the same time, one should take into account the structure of the Secretariat – the CoE's administrative body – which ensures that the organisation's various offices function properly and fulfil their mandates. In particular, it is noted that a clear division is made between the Directorate General of Human Rights and Rule of Law (DG I) on the one hand, and the Directorate General of Democracy (DG II) on the other.[12] The above structure implies that the activities of DG I are derived either from the notion of *human rights* or from the concept of *the rule of law*,[13] whereas DG II concentrates its work on the area of *democracy*.[14]

---

9 See *Council of Europe Programme and Budget 2012–2013*, pp 3, 223; available at https://wcd.coe.int/com.instranet.InstraServlet?command=com.instranet.CmdBlobGet&InstranetImage=1998892&SecMode=1&DocId=1838802&Usage=2, last accessed 15 December 2012.

10 See Terms of Reference of Intergovernmental Structures 2012–2013 of 23 November 2011 adopted by the CM, Ref. No. CM(2011)131 Final.

11 For the CoE website, see http://hub.coe.int/web/coe-portalaa, last accessed 15 December 2012.

12 See the CoE organisational charts published at http://www.coe.int/t/dgi/OrganigrammeDGI_EN.pdf and http://www.coe.int/t/DEMOCRACY/source/OrganisationChart_DGII_en.pdf, last accessed 15 December 2012.

13 See, in particular, the structure of the Human Rights Directorate with the Secretariat of the European Committee for the Prevention of Torture and Inhuman or Degrading Treatment or Punishment (CPT), the Department for the Execution of Judgments of

Summing up, the CoE examines climatic issues mostly under the headings *Democracy* and *Human rights*, which obviously correspond with the tasks of both DG I and DG II. Generally speaking, climate change is being discussed by various actors in different segments of the CoE's activities. Therefore, it is difficult to associate the climate law standard-setting mission with only one specific organ or body responsible for its further development and application. Nevertheless, the topic of climate change seems to be important enough to generate a need to link all the activities. In particular, a sort of inter-secretariat collaboration in this field has already been established in the shape of an Inter-secretariat Group on Climate Change.[15] In the discussion that follows, a brief analysis of the 'climatic' actions taken by chosen actors on various levels of the CoE's structure will be provided in order to answer the question about climate law standards and the organisation's perspectives.

### *I. Democracy*

As noted above, the vast majority of deliberations within the CoE concerning climatic issues take place under the heading of *Democracy*. In this regard, one may indicate two key actors: PACE (a statutory organ of the CoE) and the CLRA (established with the statutory resolution of the CM). The above forums are mainly platforms for exchanging views, and their standard-setting role in the field of climate law is not preponderant. Nevertheless, they do not hesitate to provide inspiration to other relevant organs, in particular the CM. In this respect, these two entities gradually call for the adoption of

---

the ECtHR, the Department of the European Social Charter and Social Security Code, the Human Rights Policy and Development Department, and the Bioethics Division, as well as the tasks of the Information Society and Action against Crime Directorate, the Justice and Human Dignity Directorate, and the Secretariat of the Enlarged Agreement on Democracy through Law (Venice Commission).

14 See, in particular, the structure of DG II, with the European Directorate for the Quality of Medicines and Healthcare; the Directorate of Democratic Citizenship and Participation; the Directorate of Democratic Governance, Culture and Diversity; and the Directorate of Human Rights and Anti-discrimination.

15 See the CoE website devoted to climate change, entitled *Environment: Climate change, a threat to human rights*, at http://hub.coe.int/what-we-do/culture-and-nature/climate-change, last accessed 15 December 2012.

new regulations in the field of environmental law. Therefore, an overview of their actions will be presented.

## *1. Parliamentary Assembly of the Council of Europe*

PACE is a forum for debate for parliamentarians from all over Europe and represents the political forces (majority and opposition) within the CoE member states. It can adopt three different types of texts: recommendations, resolutions and opinions.[16] PACE recommendations contain proposals addressed to the CM, the implementation of which is within the competence of governments. PACE decisions on questions which it is empowered to put into effect, or expressions of view for which it alone is responsible, take the form of resolutions. PACE also expresses opinions, mostly on questions put to it by the CM.[17]

PACE contains, inter alia, the Committee on the Environment, Agriculture and Local and Regional Affairs. This Committee resulted from a merger that took place in 2001 between the Committee on the Environment, Regional Planning and Local Authorities and the Committee on Agriculture, Rural Development and Food. The Committee on the Environment, Regional Planning and Local Authorities, which was established in 1952 as a special body on municipal and regional affairs, became a general committee of the Assembly in 1956. The subject of "Environment" was added to the Committee's title in 1986.[18] This Committee seems to be concerned with the issue of climate change, and stimulates the debate about climate change within PACE. Among the Committee's recent actions is that, in 2011, it issued a Declaration on climate change entitled "As the world's warmest year ends, time for climate change to be seen as a human rights issue".[19] In this Declaration, the Committee called on the CoE to hold a major transversal conference on "human rights and climate change" to discuss the connections

16 Article 23(a) of the Statute of the Council of Europe.

17 See also http://assembly.coe.int/Main.asp?link=/AboutUs/APCE_Procedure.htm, last accessed 15 December 2012.

18 See the PACE Resolution 1425 (2005) on revising the terms of reference of Assembly Committees.

19 See the Declaration of the PACE Committee on the Environment, Agriculture and Local and Regional Affairs on Climate Change of 27 January 2011; available at http://www.assembly.coe.int/Communication/270111_declarationclimate_E.pdf, last accessed 15 December 2012.

between climate change and human rights in Europe, and whether human rights obligations could strengthen international policy in the field of climate change.[20]

Apparently, since the late 1980s, PACE – aware of the environmental challenges facing the planet – has urged for action in various texts, as follows:

- Resolution 919 (1989) on the Destruction of Tropical Forests – Causes and Remedies
- Resolution 926 (1989) on the Danger of Climatic Changes and the Protection of the Ozone Layer
- Recommendation 1130 (1990) on the Formulation of a European Charter and a European Convention on Environmental Protection and Sustainable Development
- Recommendation 1140 (1991) on Global Environmental Change and the Role of Science and Democracy
- Order 458 (1991) on Climate Change: This invites its Committee on Science and Technology to study scientific and technological measures to minimise climate change and to make recommendations on policy options to stabilise global climate
- Resolution 987 (1992) on Climate Change: This was subsequently adopted, calling on member states to implement certain policy responses to the above phenomenon, such as to –
  - encourage international cooperation and legal instruments
  - stabilise carbon dioxide emissions at their 1990 level by 2000
  - review and adjust energy policies and programmes
  - promote energy conservation and energy efficiency
  - promote energy production methods and fossil fuels that emit less carbon dioxide
  - review the development of nuclear power in the light of waste and safety problems
  - promote research on clean energy and nuclear safety
  - promote research, development and demonstration of renewable and other sustainable energy sources
  - review aid and technology transfer programmes to developing countries
  - promote reforestation and care of forests

20 (ibid.).

- encourage research into the reduction of carbon dioxide emissions by road traffic
- promote international monitoring and coordination on climate change issues
- enhance scientific research, and
- improve information transfer to policymakers.

A decade later, the Assembly was mostly occupied by the challenges and possibilities presented by the Kyoto Protocol to the United Nations Framework Convention on Climate Change (UNFCCC). In its Resolution 1243 (2001) on the Kyoto Protocol on Climate Change: Need for Committed International Solidarity, PACE concentrated on expressing its total support for putting the Kyoto Protocol into practice. Subsequently, it adopted Resolution 1250 (2001) and its Recommendation 1520 (2001) on the technological possibilities for fulfilling the targets of the Kyoto Protocol. In both texts, PACE gave specific advice on how to achieve the Kyoto goals, and emphasised the importance of pro-climatic actions. Additionally, it expressed its further support in –

- Resolution 1292 (2002) on the World Summit on Sustainable Development: Ten Years after Rio
- Recommendation 1594 (2003) on the Follow-up to the World Summit on Sustainable Development: A Common Challenge, and
- Resolution 1406 (2004) on Global Warming: Beyond Kyoto.

The above texts were followed by its Resolution 1552 (2007) on the Capture of Carbon Dioxide as a Means of Fighting Climate Change, in which PACE welcomed the entry into force of the Kyoto Protocol on 16 February 2005 and called for further actions. Subsequently, in its Recommendation 1883 (2009) and Resolution 1682 (2009) on the challenges posed by climate change, the Assembly urged CoE member and observer states to negotiate an ambitious integrated package as part of the new global climate change agreement. Then, in Recommendation 1918 (2010) on Biodiversity and Climate Change, PACE recommended that the CM call on the governments of CoE member and observer states to take into account the opportunities offered by the International Year of Biodiversity in 2010.

It is also important to emphasise PACE's activities outside the CoE. In particular, attention should be devoted to the outcome paper issued by the Inter-parliamentary Union (of which PACE is an associate member), together with the Mexican Congress, with the support of the United Nations

Development Programme (UNDP) on the occasion of the 16th Conference of the Parties to the UNFCCC and the 6th Meeting of the Parties to the Kyoto Protocol (COP16/CMP6) in Cancún, Mexico, on 6 December 2010. In the said outcome paper, parliamentarians from all over the world called on the Inter-parliamentary Union to pursue its efforts to mobilise the global parliamentary community around the issue of climate change.[21]

At CoE level, PACE has also issued many texts committed to protecting the environment[22] which are thematically related to the issue of climate change. Among these texts, it is worth mentioning Recommendation 1885

21 See the Outcome Document of 6 December 2010 adopted by the Parliamentary Meeting on the Occasion of the United Nations Climate Change Conference, Cancún (Mexico); available at http://www.ipu.org/splz-e/cop16/final.pdf, last accessed 15 December 2012.

22 To mention the texts pronounced in the last decade alone, these are Resolution 1869 (2012) on the Environmental Impact of Sunken Shipwrecks; Resolution 1851 (2011) on Armed Conflicts and the Environment; Resolution 1815 (2011) on the Potential Dangers of Electromagnetic Fields and their Effect on the Environment; Resolution 1794 (2011) on Preserving the Environment in the Mediterranean; Resolution 1775 (2010) on Military Waste and the Environment; Recommendation 1946 (2010) on Military Waste and the Environment; Recommendation 1885 (2009) on Drafting an Additional Protocol to the European Convention on Human Rights concerning the Right to a Healthy Environment; Recommendation 1879 (2009) on Renewable Energies and the Environment; Recommendation 1863 (2009) on Environment and Health: Better Prevention of Environment-related Health Hazards; Resolution 1655 (2009) on Environmentally Induced Migration and Displacement: A 21st Century Challenge; Recommendation 1862 (2009) on Environmentally Induced Migration and Displacement: a 21st Century Challenge; Recommendation 1837 (2008) on the Fight against Harm to the Environment in the Black Sea; Resolution 1596 (2008) on Protection of the Environment in the Arctic Region; Resolution 1588 (2007) on Radioactive Waste and Protection of the Environment; Resolution 1542 (2007) on Electronic Waste and the Environment; Resolution 1461 (2005) on the Curonian Spit, Oil and the Environment; Resolution 1449 (2005) on the Environment and the Millennium Development Goals; Resolution 1435 (2005) on Energy Systems and the Environment; Recommendation 1689 (2004) on Hunting and Europe's Environmental Balance; Recommendation 1653 (2004) on Environmental Accounting as a Sustainable Development Tool; Recommendation 1637 (2003) on Pan-European Environmental Co-operation: The Council of Europe's Role after the Kyiv Ministerial Conference and the Johannesburg Summit; Recommendation 1614 (2003) on Environment and Human Rights; Resolution 1296 (2002) on the Change of Name of the Committee on the Environment and Agriculture to Committee on the Environment, Agriculture and Local and Regional Affairs; Resolution 1295 (2002) on the State of the Environment of the Baltic Sea; and Recommendation 1571 (2002) on Reducing Environmental Risks by Destroying Chemical Weapons.

(2009) on Drafting an Additional Protocol to the European Convention on Human Rights Concerning the Right to a Healthy Environment.[23] In this Recommendation, PACE noted, inter alia, the case law in the environmental field developed by the ECtHR. In PACE's view, the above case law had afforded protection for the right to a healthy environment through a 'knock-on effect', by upholding the individual rights in Articles 2 and 8 of the European Convention on Human Rights (ECHR). In this regard, PACE recommended that the CM draw up an Additional Protocol to the Convention, recognising the right to a healthy and viable environment.

In this regard, it is noteworthy that the reply from the CM of 19 June 2010 to Recommendation 1883 (2009) on the Challenges Posed by Climate Change, and to Recommendation 1885 (2009) on Drafting an Additional Protocol to the European Convention on Human Rights Concerning the Right to a Healthy Environment, may be perceived as moderate.[24] The only

23 The close relation between the above Recommendation and the subject matter of climate change is reflected by the fact that the CM decided to give its joint reply to PACE's Recommendation 1883 (2009) on the Challenges Posed by Climate Change, and to PACE's Recommendation 1885 (2009) on Drafting an Additional Protocol to the European Convention on Human Rights Concerning the Right to a Healthy Environment. See Reply of 19 June 2010 to the Recommendation 1883 (2009) on the Challenges Posed by Climate Change, and to the Recommendation 1885 (2009) on Drafting an Additional Protocol to the European Convention on Human Rights Concerning the Right to a Healthy Environment, CM/Cong(2010)Rec271-281 Final.

24 In particular, with regard to climate change, the CM pointed out that climate change may have implications regarding the enjoyment of universally recognised fundamental rights. It also underlined the global scope of challenges posed by climate change and suggested that, should the priorities and the resources of the CoE so permit, a conference be held to examine the issue from various angles, e.g. human rights and legal affairs, the environment, and social cohesion. Finally, it found it appropriate to refer to the appended comments of the Committee of Senior Officials of the CoE CM responsible for Spatial/Regional Planning, the Committee of Permanent Correspondents of the European and Mediterranean Major Hazards Agreement, and the Standing Committee of the Bern Convention, which outline aspects of their work relevant to this issue. At the same time, with regard to PACE Recommendation 1885 (2009) on Drafting an Additional Protocol to the European Convention on Human Rights Concerning the Right to a Healthy Environment, the CM adopted a similar position to its replies to Recommendation 1614 (2003) on the Environment and Human Rights, and Recommendation 1862 (2009) on Environmentally Induced Migration and Displacement: A 21st Century Challenge. In the above replies, the CM did not consider it advisable to draw up an Additional Protocol to the Convention in the environmental domain. At the same time, the CM recalled that PACE Recommendation 1614 (2003) had led to the preparation by the Steering

two real actions envisaged by the CM at that time were to consider the possibility of holding a conference on climate change and human rights, and to continue works on updating and extending the 2006 *Manual on Human Rights and the Environment*[25] in the light of the case law of the ECtHT and the ECSR, of relevant standards set out by other international organisations, and of good practices adopted at national level.[26] This temperate reaction of the CM is typical in this category of matters. More generally, one may notice the usually proactive character of PACE Resolutions and Recommendations, contrasting with the more balanced replies given at governmental level by the CM.

In this regard, it should be stressed that PACE texts have no binding effect on either the member states or the CM, and are merely of an opinion-making and advisory nature. It is also not easy to establish to what extent they reflect the political moods present in the governments and parliaments of Europe. This is because parliamentarians appointed to PACE represent various political forces of their member states, as a requirement exists for a balance of political parties within each national delegation. Therefore, PACE texts may principally be considered as a very interesting voice in the discussion on the European response to climate change phenomenon and the challenges it implies, and may be invoked at various forums. Nevertheless, due to the non-binding character of the texts issued by PACE, as well as the specific way in which PACE delegations are elected, the above voice has no decisive impact on the final solutions adopted at governmental level.

---

Committee for Human Rights (CDDH, Comité Directeur pour les Droits de l'Homme) of a *Manual on Human Rights and the Environment: Principles emerging from the case-law of the European Court of Human Rights* in 2006, and had taken note of the agreement within the CDDH, within the framework of the Committee of Experts for the Development of Human Rights (DH-DEV), to update and extend the Manual, in the light of, notably, the case law of the ECtHR and of the ECSR, of relevant standards set out by other international organisations, and of good practices adopted at national level, in order to give effect to the principles emerging from the ECtHR's case law. Finally, the CM referred to the substantial work already carried out by the CoE in the field of the environment, which has led to the adoption of important legal instruments such as the Convention on the Conservation of European Wildlife and Natural Habitats (ETS No. 104), the Convention on Civil Liability for Damage resulting from Activities Dangerous to the Environment (ETS No. 150) and the Convention on the Protection of the Environment through Criminal Law (ETS No. 172).

25 CoE (2006).

26 Reply from the CM of 19 June 2010.

## 2. *Congress of Local and Regional Authorities*

As mentioned above, the present CLRA is not one of the statutory bodies of the organisation, anchored directly in the Statute of the Council of Europe, but was established on 14 January 1994 by way of Statutory Resolution 94(3) of the CM. It is a consultative organ composed of representatives of local and regional authorities.[27] Its main role is to evaluate the application of the European Charter of Local Self-government and to support the improvements of governance of local and regional authorities.[28]

In a similar way to PACE, the CLRA also pays attention in its texts to the issue of climate change by way of recommendations and resolutions.[29] Three pairs of the most important of these issued since 2009 should be mentioned, namely –

- Recommendation 271 (2009) and Resolution 288 (2009) on the Global Challenge of Climate Change: Local Responses[30]
- Recommendation 281 (2010) and Resolution 298 (2010): After Copenhagen, Cities and Regions take up the Challenge,[31]
- Recommendation 298 (2010) and Resolution 317 (2010): Coastal Towns and Cities tackling Threats from the Sea.[32]

The above texts call for purposive actions in the respective areas and emphasise the need to engage local and/or regional authorities.[33] As is the case

27 See Article 1 of the CM Statutory Resolution CM/Res(2011)2 relating to the Congress of Local and Regional Authorities of the Council of Europe and the revised charter appended thereto.

28 (ibid.:Article 2).

29 As with PACE texts, CLRA Recommendations include proposals addressed to the CM for implementation by governments. They may also be addressed to other international organisations. Resolutions refer to authorities of local and/or regional character, as well as to their associations. For further details, see the CLRA website at http://www.coe.int/t/congress/texts/adopted-texts_en.asp?mytabsmenu=6, last accessed 15 December 2012.

30 Dated 14 October 2009.

31 Dated 18 March 2010.

32 Dated 28 October 2010.

33 It is said that "for the Congress, action against climate change must go beyond reduction of greenhouse gas emissions and the use of renewable energy sources. It is also a question of good governance and the right to a healthy environment". See again the CoE website devoted to climate: *Environment: Climate change, a threat to human rights*.

with PACE Recommendations, CLRA Recommendations are not legally binding instruments. Furthermore, the replies from the CM to the climatic recommendations of the CRLA are as temperate as the replies to those of PACE.[34] In particular, the joint reply to the CLRA Recommendations 271 (2009) and 281 (2010) include reference to the reply of the CM to PACE Recommendations 1883 (2009) and 1885 (2009) on the Challenges Posed by Climate Change and on Drafting an Additional Protocol to the European Convention on Human Rights Concerning the Right to a Healthy Environment.[35] As already pointed out, to date the outcome of the above PACE Recommendations at governmental level has been quite moderate.

The character of the CLRA implies that the meaning of its Recommendations and Resolutions is comparable to those of PACE. They may be treated as a very interesting voice in the discussion. Nevertheless, their impact on governmental activities remains limited. At the same time, one should emphasise a very unique feature of the CLRA texts: they all stress the need to engage local and/or regional authorities in activities concerning climate change.

### *3. Other Environmental Work*

Apart from the work of PACE and the CLRA, the substantial work already carried out by the CoE in the field of the environment is worth mentioning, which has resulted in the adoption of significant legal instruments such as the Convention on the Conservation of European Wildlife and Natural Habitats (Bern Convention),[36] the Convention on Civil Liability for Damage resulting from Activities Dangerous to the Environment,[37] and the Convention

---

34 See the CM Reply of 16 June 2010 to CLRA Recommendation 271 (2009): The Global Challenge of Climate Change: Local Responses, as well as Recommendation 281 (2010): After Copenhagen, Cities and Regions take up the Challenge, Ref. No. CM/Cong(2010)Rec271-281 Final. See also the CM Reply of 7 December 2011 to CLRA Recommendation 298 (2011): Coastal Towns and Cities tackling Threats from the Sea, Ref. No. CM/Cong(2011)Rec298 Final.

35 See in particular para. 3 of the CM Joint Reply of 16 June 2010.

36 Convention on the Conservation of European Wildlife and Natural Habitats (1979), ETS No. 104.

37 Convention on Civil Liability for Damage Resulting from Activities Dangerous to the Environment (1993), ETS No. 150.

on the Protection of the Environment through Criminal Law.[38] The subject matter of the above international treaties is linked with the issue of climate change. In this regard, particularly the Bern Convention, comes to mind, which –[39]

> … implements on the European continent the UN Convention on Biological Diversity … sets European standards in environmental policies related to biodiversity and bio security … [and] aims at the protection and sustainable use of Europe's biological diversity through a monitoring mechanism.

As stated earlier, for the purpose of the present text it is not intended to go into the details of the Bern Convention. As pointed out earlier, the objective of this article is to review the actions undertaken by the statutory organs of the CoE and to augment them by way of information regarding works of other bodies within the CoE. Therefore, as regards this particular point, attention should be drawn to the achievements of the Group of Experts on Biodiversity and Climate Change[40] under Article 14 of the Bern Convention.

The Group of Experts on Biodiversity and Climate Change was set up at the 26th Meeting of the Standing Committee of the Bern Convention.[41] The purpose of creating this Group was to "provide guidance to Parties on understanding climate change impacts and threats, and developing appropriate adaptation measures in national policies regarding the species and habitats protected under the Bern Convention."[42]

38 Convention on the Protection of the Environment through Criminal Law (1998), ETS No. 172.

39 See Council of Europe Programme and Budget 2012–13, 128.

40 The Group of Experts on Biodiversity and Climate Change is one of the expert groups subordinated to the Standing Committee of the Bern Convention. Presently, nine such groups exist, namely on the Conservation of Amphibians and Reptiles – Marine Turtles; the Conservation of Plants; the Conservation of Birds; the Conservation of Invertebrates; Protected Areas and Ecological Networks; Invasive Alien Species; Large Carnivores; Biodiversity and Climate Change; and European Island Biological Diversity. See also the Bern Convention website at http://www.coe.int/t/dg4/culturheritage/nature/experts_EN.asp, last accessed 15 December 2012.

41 See the report on the 26th Meeting of the Standing Committee of the Convention on the Conservation of European Wildlife and Natural Habitats of 27–30 November 2006, Ref. No. T-PVS(2006)24E, 11 December 2006, and the Group's Terms of Reference accessible under "Draft Programme of Activities for 2007 of Standing Committee of the Bern Convention adopted at the 26th Meeting on 27–30 November 2006", Ref. No. T-PVS(2006)14, 8.

42 See the Terms of Reference (ibid.).

The Group has already conducted seven general meetings as well as one special meeting.[43] The results of the Group's work consist of various Recommendations on biodiversity and climate change adopted by the Standing Committee,[44] namely –

- Recommendation 159 (2012) on the Effective Implementation of Guidance for Parties on Biodiversity and Climate Change
- Recommendation 158 (2012) on Conservation Translocations under Changing Climatic Conditions
- Recommendation 152 (2011) on Marine Biodiversity and Climate Change
- Recommendation 147 (2010) on Guidance for Parties on Wildland Fires, Biodiversity and Climate Change
- Recommendation 146 (2010) on Guidance for Parties on Biodiversity and Climate Change in European Islands
- Recommendation 145 (2010) on Guidance for Parties on Biodiversity and Climate Change in Mountain Regions
- Recommendation 143 (2009) on Further Guidance for Parties on Biodiversity and Climate Change
- Recommendation 142 (2009) on Interpreting the CBD Definition of Invasive Alien Species to take into Account Climate Change
- Recommendation 135 (2008) on Addressing the Impacts of Climate Change on Biodiversity, and
- Recommendation 122 (2006) on the Conservation of Biological Diversity in the Context of Climate Change.

Having indicated the above long list of Recommendations, it can be observed that the link between the subject matter of the Bern Convention and climate

43 The meetings took place on 14–15 June 2007 (1st Meeting of the Group of Experts); on 12 October 2007 (Select Committee on Biodiversity and Climate Change); on 13–14 March 2008 (2nd Meeting of the Group of Experts); on 11–12 September 2008 (3rd Meeting of the Group of Experts); on 2–3 July 2009 (4th Meeting of the Group of Experts); on 21–23 June 2010 (5th Meeting of the Group of Experts); on 10–11 October 2011 (6th Meeting of the Group of Experts); and on 1–2 October 2012 (7th Meeting of the Group of Experts). See the Bern Convention website, as well as relevant meeting reports placed therein, at http://www.coe.int/t/dg4/culture heritage/nature/experts_EN.asp, last accessed 15 December 2012.

44 For the texts of Recommendations, see the Bern Convention website at http://www .coe.int/t/dg4/cultureheritage/nature/Bern/recommendations_en.asp, last accessed 15 December 2012.

change is quite evident. Nevertheless, it seems to me that it is the subject matter of the Bern Convention that needs to take into account the changing climatic conditions, rather than the other way around. To be more precise, the Bern Convention is, in principle, a tool for biodiversity conservation. Therefore, any attempts at its implementation make no sense without considering the local circumstances such as climate. At the same time, application of the above treaty may have little or no effect on the process of climate change; or, at the very least, the prevention or slowing down of climate change was not the purpose the drafters had in mind.[45]

## *II. Human Rights*

As already stated, there are deliberations relating to climate law within the CoE that take place under the heading of *human rights*. In this regard, the case law of human rights protection bodies, but also the work of other bodies and organs based on that case law are of particular relevance.

### *1. Case Law of Human Rights Protection Bodies*

In discussing human rights protection bodies, the 'environmental dimension' of jurisprudence of the ECtHR needs to be focussed. Taking into account certain other developments, specific findings by the ECSR are also considered.

Both the ECtHR and the ECSR were designed to judge whether states parties are in conformity in law and in practice with the provisions of the human rights treaties – the ECHR and Additional Protocols to the ECHR,[46] or the European Social Charter, respectively. Therefore, it should

45 The aims of the Bern Convention are to conserve wild flora and fauna and natural habitats, to promote cooperation between states, and to give particular attention to endangered and vulnerable species, including endangered and vulnerable migratory species. See Explanatory Report to the Convention on the Conservation of European Wildlife and Natural Habitats (ETS No. 104); available at http://www.conventions.coe.int/Treaty/en/Reports/Html/104.htm, last accessed 15 December 2012.

46 See again the ECHR. See also the Protocol to the Convention for the Protection of Human Rights and Fundamental Freedoms (1952), ETS No. 009; Protocol No. 4 to the Convention for the Protection of Human Rights and Fundamental Freedoms, Securing Certain Rights and Freedoms other than those Already Included in the

first be emphasised that neither the Convention nor the Charter guarantees the right to a healthy environment as such. Indeed, it is also difficult to think of the above human rights protection instruments as being able to prevent or stop climate change. Nevertheless, it is now indisputable that the gradual evolution of jurisprudence by the ECtHR has resulted in the recognition of environmental human rights,[47] mostly in connection to Article 8 of the Convention, which provides a right to respect for one's "private and family life".[48] The above tendency was followed by the ECSR.[49]

The existing case law of the ECtHR seems to be significant for the overall work of the CoE. In particular, it is often treated as a starting point for further deliberations by various bodies and organs. To emphasise its meaning, the appropriate set of references to the jurisprudence of the human rights protection bodies was collected by government experts from CoE member states and published in 2006 as the *Manual on Human Rights and the Environment*.[50] One may also observe the ECtHR's own proactiveness, which aims to promote the link between human rights and the environment. To strengthen the effect of the dissemination of knowledge and to stress the considerable scale of the ECtHR's achievements, in January 2012, the ECtHR Registry decided to issue a fact sheet on environment-related cases in the ECtHR's case law and to publish it on the ECtHR website.[51] Thanks to the above

---

Convention and in the First Protocol Thereto (1963), ETS No. 046; Protocol No. 6 to the Convention for the Protection of Human Rights and Fundamental Freedoms Concerning the Abolition of the Death Penalty (1983), ETS No. 114; Protocol No. 7 to the Convention for the Protection of Human Rights and Fundamental Freedoms (1984), ETS No. 117; Protocol No. 12 to the Convention for the Protection of Human Rights and Fundamental Freedoms (2000), ETS No. 177; and Protocol No. 13 to the Convention for the Protection of Human Rights and Fundamental Freedoms, Concerning the Abolition of the Death Penalty in all Circumstances (2002), ETS No. 187.

47 See the Council of Europe Commissioner for Human Rights Viewpoint of 19 October 2009, entitled "Climate change is causing an unprecedented, global human rights crisis – and must now be countered by co-ordinated, rights-based action", by Thomas Hammarberg; available at http://www.coe.int/t/commissioner/viewpoints/091019_EN.asp?, last accessed 15 December 2012.

48 See the case law of the ECtHR, referred to below.

49 See the case law of the ECSR, referred to below; see also the findings under Article 11 of the European Social Charter.

50 CoE (2006, 2012).

51 See *Factsheet – Environment-related Cases in the Court's Case Law,* available at http://www.echr.coe.int/ECHR/en/Header/Press/Information+sheets/Factsheets/, last accessed 15 December 2012.

efforts, it is not difficult to indicate the ECtHR case law concerning airport noise;[52] neighbouring noise;[53] industrial pollution posing a danger to people's health;[54] industrial pollution posing other adverse effects on the environment;[55] deforestation and urban development;[56] and passive smoking.[57] Of course, the above indication is illustrative, not exhaustive. It covers cases in which the ECtHR has found the violation of 'environmental' human rights[58] and the ones where no such violation was found.[59]

Similar efforts were made with regard to the environmental findings of the ECSR. In 2008, the *Case Law Digest*[60] was published, presenting the interpretation that the ECSR had made of the various Articles of the Euro-

52 See *Powell & Rayner v United Kingdom*, Application No. 9310/81, ECtHR judgment of 21 February 1990; See *Hatton v United Kingdom*, Application No. 36022/97, ECtHR judgment (GC) of 8 July 2003.

53 See *Moreno Gomez v Spain*, Application No. 4143/02, ECtHR judgment of 16 November 2004; *DEÉS v Hungary*, Application No. 2345/06, ECtHR judgment of 9 November 2010; *Mileva & Others v Bulgaria*, Application No.'s 43449/02 and 21475/04, ECtHR judgment of 25 November 2010; *Dubetska & Others v Ukraine*, Application No. 30499/03, ECtHR judgment of 10 February 2011; *Zammit Maempel & Others v Malta*, Application No. 24202/10, ECtHR judgment of 22 November 2011.

54 See *Öneryıldız v Turkey*, Application No. 48939/99, ECtHR judgment (GC) of 30 November 2004; *Lopez Ostra v Spain*, Application No. 16798/90, ECtHR judgment of 9 December 1994; *Fadeyeva v Russia*, Application No. 55723/00, ECtHR judgment of 9 June 2005; *Giacomelli v Italy*, Application No. 59909/00, ECtHR judgment of 2 November 2006; *Guerra & Others v Italy*, Application No. 14967/89, ECtHR judgment of 19 February 1998; *Taşkın & Others v Turkey*, Application No. 46117/99, ECtHR judgment of 10 November 2004.

55 See *Tatar v Romania*, Application No. 657021/01, ECtHR judgment of 27 January 2009; *l'Erablière v Belgium*, Application No. 49230/07, ECtHR judgment of 24 February 2009; *Mangouras v Spain*, Application No. 12050/04, ECtHR judgment (GC) of 28 September 2010; *Di Sarno & Others v Italy*, Application No. 30765/08, ECtHR judgment of 10 January 2012.

56 See *Hamer v Belgium*, Application No. 21861/03, ECtHR judgment of 27 November 2007; *Kyrtatos v Greece*, Application No. 41666/98, ECtHR judgment of 22 May 2003.

57 See *Florea v Romania*, Application No. 37186/03, ECtHR judgment of 14 September 2010.

58 For example, *DEÉS v Hungary*, *Mileva & Others v Bulgaria*.

59 For example, *Powell & Rayner v United Kingdom* and *Hatton v United Kingdom*.

60 See the *Case Law Digest of the European Committee of Social Rights*; available at http://www.coe.int/t/dghl/monitoring/socialcharter/Digest/DigestSept2008_en.pdf, last accessed 15 December 2012.

pean Social Charter.[61] The Digest, under the heading of "The right to enjoy the highest possible standard of health attainable" (Article 11 of the Charter), contains a whole section on healthy environment, with subsections on food safety, nuclear hazards for communities living in the vicinity of nuclear power plants, risks relating to asbestos, and air pollution. It contains references to the ECSR's findings in the cases *International Federation of Human Rights Leagues (FIDH) v France*[62] and *Marangopoulos Foundation for Human Rights (MFHR) v Greece.*[63]

Other CoE bodies commonly refer to these environmental judgments and decisions in order to support their own position and/or activities in the field of 'environmental' human rights. In this regard, once again PACE's Recommendation 1885 (2009) on Drafting an Additional Protocol to the European Convention on Human Rights Concerning the Right to a Healthy Environment needs to be highlighted.[64] In the above text, PACE referred directly to "case law in the environmental field developed by the European Court of Human Rights" as well as to the *Manual on Human Rights and the Environment*,[65] which serves as a guide to the Strasbourg case law. The author would also like to add a reference to the viewpoint of the CoE Commissioner for Human Rights of 19 October 2009 entitled "Climate change is causing an unprecedented, global human rights crisis – and must now be countered by coordinated, rights-based action", by the then Commissioner Thomas Hammarberg.[66] In the author's opinion, the Commissioner's viewpoint, based on the references to ECtHR and ECSR case law, illustrates the way in which the majority of environment-related texts are being structured

61 See information note at http://www.coe.int/t/dghl/monitoring/socialcharter/Digest/DigestIndex_en.asp, last accessed 15 December 2012.

62 See *International Federation of Human Rights Leagues (FIDH) v France*, Complaint No. 14/2003, ECSR decision on the merits, of 3 November 2004, para. 31.

63 See *Marangopoulos Foundation for Human Rights (MFHR) v Greece*, Complaint No. 30/2005, ECSR decision on the merits, of 6 December 2006, para.'s 194–202.

64 See again the PACE Recommendation 1885 (2009) on Drafting an Additional Protocol to the European Convention on Human Rights Concerning the Right to a Healthy Environment, referred to above.

65 CoE (2006).

66 Thomas Hammarberg was CoE Commissioner for Human Rights from 1 April 2006 to 31 March 2012; see biographical note at http://www.coe.int/t/commissioner/Office/prevCommissioners_en.asp, last accessed 7 February 2012.

within the CoE. He invoked ECtHR case law[67] and emphasised that European human rights standards and principles envisaged safeguards which needed to be integrated into plans and policies to address climate change. The Commissioner also provided a short analysis of ECtHR findings, and observed that, according to the latter's case law, severe environmental pollution could affect the well-being of individuals and prevent them from enjoying their homes in such a way as to adversely affect their private and family life.[68] He noted that the ECtHR had confirmed the obligation of states to conduct proper studies before allowing an activity which could cause environmental damage, and to bring such studies to the attention of the public.[69] Furthermore, he indicated that the ECtHR had found a violation of the right to life in a case where the authorities had not taken preventive action when they had been aware of an increased risk of large-scale mudslides and had not informed the population of the risk.[70] He also emphasised that the European Social Charter provided for the right to health, on the basis of which the ECSR had held states responsible for showing measurable progress in lowering levels of pollution.[71] In addition, he stated that the same ruling would cover nuclear hazards as well as risks related to asbestos and food safety.[72]

### 2. *Development of Human Rights Law and Policy*

As far as the development of human rights law and policy is concerned, a brief comment on the work at governmental level by the Steering Committee for Human Rights (CDDH, Comité Directeur pour les Droits de l'Homme)

67 The Commissioner invoked *Lopez Ostra v Spain*, para. 51; *Taşkın & Others v Turkey*; and *Budayeva & Others v Russia*, Application No. 15339/02, judgment 29 September 2008.

68 See *Lopez Ostra v Spain*, para. 51.

69 See *Taşkın & Others v Turkey.*

70 See *Budayeva & Others v Russia.*

71 See *Marangopoulos Foundation for Human Rights (MFHR) v Greece,* Complaint No. 30/2005, ECSR decision on the merits, of 6 December 2006, para.'s 203 and 205.

72 See CoE Commissioner for Human Rights Viewpoint.

should be made,[73] followed subsequently by the work executed by the CM. To do this, it is necessary to note again that the principal role of the CDDH, under the auspices of the CM, is "to set up standards commonly accepted by the 47 Member States with the aim of developing and promoting human rights in Europe and improving the effectiveness of the control mechanism established by the European Convention on Human Rights."[74]

It can be seen that climate change is not situated in the centre of interests of either the CDDH or the CM. At present, the standard-setting role of the said Steering Committee in the field of climate law is quite narrow, as the CDDH is willing to undertake activities solely of a promotional nature.

A fine and relatively current illustration of the above approach is contained in the CDDH Opinion on PACE Recommendation 1883 (2009) on the Challenges Posed by Climate Change.[75] In this Opinion, the CDDH reiterates the position adopted in its comments to Recommendation 1614 (2003) on the Environment and Human Rights,[76] which is that neither the ECHR nor its Additional Protocols expressly recognise a right to the protection of the environment. At the same time, it noted that the ECHR system had already indirectly contributed to the protection of the environment through existing Convention rights and their interpretation in ECtHR case law. Consequently, the CDDH did not consider it advisable to draft an Additional Protocol to the Convention on the suggested topic. Nonetheless, it expressed its support for the idea to pursue studies on the subject at intergovernmental level by means of regular exchanges of views within the framework of the Committee of Experts for the Development of Human Rights (DH-DEV) and by updating and extending the 2006 *Manual on Human Rights and the Environment*, in the light of the jurisprudence of the

73 The CDDH was set up by the CM under Article 17 of the Statute of the Council of Europe, and operates in accordance with Resolution CM/Res(2011)24 on intergovernmental committees and subordinate bodies, their terms of reference, and their working methods.

74 See the CDDH home page available at http://www.coe.int/t/dghl/standardsetting/cddh/default_en.asp, last accessed 15 December 2012. See also the CDDH Terms of Reference set up by the CM under Article 17 of the Statute of the Council of Europe and in accordance with Resolution CM/Res(2011)241 on intergovernmental committees and subordinate bodies, their terms of reference and working methods.

75 CDDH Opinion on PACE Recommendation 1883 (2009) – The Challenges Posed by Climate Change, Ref. No. CDDH(2009)019.

76 CDDH Opinion on PACE Recommendation 1614 (2003) on Environment and Human Rights, Ref. No. CDDH(2003)026.

ECtHR, of the ECSR, of relevant standards set out by other international organisations, and of good practices adopted at national level. The CDDH also noted the possibility, subject to available funding, of holding a conference to examine the issue of climate change from various angles, e.g. human rights and legal affairs, the environment, and social cohesion. If so, it expressed its availability to contribute to this work through the DH-DEV.

The above Opinion illustrates the CDDH's general approach of refraining from deep engagement into establishing new, legally binding human rights standards in the field of protection of the environment. It is notable that the CDDH concentrates on promoting the existing standards inferred from already adopted international treaties. In particular, the CDDH decided that it would be appropriate to make more explicit the protection indirectly afforded by the ECHR to the environment by updating the 2006 Manual. In the CDDH's view, such an approach would be "a useful way of promoting greater awareness in member states of the implications of their existing obligations under the Convention in environmental matters."[77]

As it transpires from the previous submissions under Subsection I above, the CM subsequently shared the CDDH position. The CM expressed its attitude towards the above idea in its reply of 19 June 2010 to Recommendation 1883 (2009) on the Challenges Posed by Climate Change, and to Recommendation 1885 (2009) on Drafting an Additional Protocol to the European Convention on Human Rights Concerning the Right to a Healthy Environment.

### *C. Environmental Activities of the CoE*

The above short review of recent selected activities of the CoE indicates that the environmental achievements of the organisation are not significant. Its main environmental accomplishment in the latter half of 2012 was the updating of the 2006 *Manual on Human Rights and the Environment*. One may also point out similar actions, such as the publication of the relevant Factsheet on the ECtHR website, and of the Digest on the ECSR website. In this context, it is possible to note a general, visible trend to disseminate and promote the already existing 'environmental' human rights standards. Nev-

77 Final Activity Report of the CDDH on Human Rights and Environment of 29 November 2005, Ref. No. CDDH(2005)016, Addendum II, para. 2.

ertheless, the CoE is talking about human rights, human rights perspectives, human rights standards, and a human rights environmental dimension. The link between the above standards and the issue of climate change is not direct. In particular, the primary aim of the jurisprudence collected in the Manual, the Digest and the relevant Factsheet was not to combat climate change, but to provide an adequate response to human rights violations. In this regard only the general individualistic and anthropocentric approach of the human rights protection bodies to the above-mentioned challenges will be emphasised.

As regards the Manual itself, attention should be paid to the overall process of its preparation. The CDDH received the terms of reference to draft the Manual from the CM in a decision dated 21 January 2004.[78] The CDDH entrusted this task to a subordinate intergovernmental body of experts, namely the DH-DEV. The Manual was published in 2006.[79] In 2009, the CM decided[80] on the said PACE Recommendation 1885 (2009)[81] to update the Manual, and it was duly published in 2012.[82] Apart from activities aimed to increase the public understanding of the relationship between the European system of human rights protection and the environment, some internal coordination work was done within the CoE. In particular, the Inter-secretariat Group on Climate Change was established, which represents many different entities and sectors of the CoE and serves the purpose of exchanging information and discussing possible common initiatives.[83] In other words, it improves the flow of information within different CoE entities. It appears that meetings of the Inter-secretariat Group took place on 11 February 2011, 17 March 2011, 30 May 2011, 5 July 2011, 6 February 2012 and 16 March 2012.[84] The deliberations within this Group resulted in the establishment of a climate change CoE website, which contains the most relevant documentation by the organisation on climate change, and is capable of acting as a knowledge base.[85] The Group has also been responsible for examining the

78 See CM Reply of 23 January 2004 on PACE Recommendation 1614 (2003) on the Environment and Human Rights, Ref. No. CM/AS(2004)Rec1614 Final.

79 See CoE (2006).

80 See Document CDDH(2009)019, para. 19.

81 See again the PACE Recommendation 1885 (2009).

82 See CoE (2012).

83 See CoE website.

84 (ibid.).

85 (ibid.).

possibility of organising the above-mentioned conference to explore the relationship between climate change and human rights in Europe.

Thus far, the discussions concerning a possible conference have not led to concrete results. It was first scheduled for 2012 but was then postponed – largely due to budgetary and organisational constraints.[86] Nevertheless, there seems to be consensus at the level of the Inter-secretariat Group, at least, that such a conference is necessary and will be useful. The conference is also perceived as an opportunity to raise awareness in the CM about climate change. Furthermore, it is expected to act as a catalyst for building on the valuable work on climate change being carried out in different parts of the organisation.[87]

## *D. Standards and Perspectives*

As is shown by the above discussion, the achievements of the CoE in the field of climate law standards are modest. The organisation has not developed any legally binding instrument that could be exclusively devoted to the issue of climate change. In fact, it is due only to the creativity of various bodies and organs that the existing legal texts have been linked with climate change. As a result, certain achievements regarding the protection of the environment and/or human rights are now being used as a starting point for further climatic deliberations.

In this regard, the actions taken under the heading *Biodiversity* (Bern Convention-related actions within the Pillar of Democracy) have to be distinguished from other deliberations around the subject of *Human rights* (within the Pillar of Human Rights and/or Democracy). As stated above, in the case of the *Biodiversity* segment, the link between the subject matter of the Bern Convention and climate change is quite evident. Nevertheless, one should be reluctant to name the above treaty as a source of the climate law and standards.[88] Also, the various climatic recommendations of the Standing Committee of the Bern Convention seem to be technical tools for accurate

86 The conference was initially planned for 2012. It was subsequently decided to postpone it until 2013. See 7th Inter-secretariat Group on Climate Change Meeting Report of 6 February 2012 [unpublished].

87 (ibid.:4).

88 As has already been stated, the author perceives the Bern Convention primarily as a nature-preservation instrument.

implementation of the above treaty rather than a source of climate law as such.

At the opposite extreme, there is a range of human rights treaties, such as the ECHR, Additional Protocols to the ECHR, and the European Social Charter. In this respect, it is important to remember that –[89]

> … [n]either the European Convention on Human Rights nor the European Social Charter protects the environment as such, but various individual rights provided for in these treaties which might be affected by the environment. Hence, it is rather the impact on the individual than the environment that both the Court and the Committee are concerned with.

The above anthropocentric perspective makes the present European system of human rights protection an incommodious tool for developing climate law standards. One may notice that the already existing case law of the human rights protection bodies is always retrospective and involves an element of individual rights violation (the victim status), which is also harmful to the environment (e.g. noise pollution, industrial pollution or passive smoking). In this regard, environmental rights protection under the Convention and the Charter is merely fragmentary. Only a few cases[90] may be directly linked to the issue of climate change. Due to this fact, as well as the subject matter involved, the author is not convinced that existing or future case law may play a decisive role in establishing climate law standards on a regional or global scale. In particular, it seems unlikely that the system, which is individual-complaint-oriented, would be the best possible tool to combat pollution, biodegradation and climate change. Such an advocacy may play an auxiliary and not a leading role in the process of standard-setting.

The above insufficiency was already noticed by PACE, which called for the adoption of a new human rights treaty in the form of an Additional Protocol to the European Convention on the Protection of the Environment. As it transpires from the reply of the CM referred to above, the idea did not gain support at governmental level. It resulted merely in updating the 2006 Manual. The internal discussion about the possible organisation of a conference on human rights and climate change at the level of the CoE is still pending.

It is apparent that the issue of climate change was never a priority for the CM. Also, at least until now, the CDDH has not given priority to this issue in its future planning. It seems that climate law has been perceived more as

89 CoE (2012:16).

90 For example, *Budayeva & Others v Russia*, from 2008.

a global issue that should be discussed from a worldwide perspective at the UN level rather than from a European perspective.[91] At the same time, the topic itself seems to be important enough to command the attention of various bodies and organs within the CoE, such as PACE, the CLRA and the Human Rights Commissioner. In this regard, further multiplication of environmental case law of the human rights protection bodies is to be expected. Furthermore, actions aiming to present and promote the principles flowing from existing case law can be anticipated. Nevertheless, it seems rather unlikely that the organisation becomes a key player that would contribute to the process of climate law standard-setting in the worldwide debate at any time soon.

## *References*

Alston, Philip, Ryan Goodman & Hennry J. Steiner, 2007, *International Human Rights in Context: Law, Politics, Morals* (Third Edition), Oxford, Oxford University Press.

Alfredsson, Gudmundur, 2010, Human Rights and the Environment, in: Leary, David & Balakrishna Pisupati (Eds), *The Future of International Environmental Law*, New York/Paris/Tokyo, United Nations University Press, 127–146.

Anton, Donald K. & Dinah L. Shelton, 2011, *The Environment and Human Rights*, Cambridge, Cambridge University Press.

Birnie, Patricia, Alan Boyle & Catherine Redgwell, 2009, *International Law and the Environment* (Third Edition), Oxford, Oxford University Press.

Boyle, Alan, 2010, Human Rights and the Environment: A Reassessment, United Nations Environment Programme Paper 2010 (Revised), available at http://www.unep.org/environmentalgovernance/Portals/8/documents/Events/HumanRightsEnvironmentRev.pdf, last accessed 7 February 2013.

Bodansky, Daniel, Jutta Brunnee & Elle Hey, 2008, *The Oxford Handbook of International Environmental Law*, Oxford, Oxford University Press.

CoE/Council of Europe, 2006, *Manual on Human Rights and the Environment: Principles Emerging from the Case-law of the European Court of Human Rights*, Strasbourg, Council of Europe Publishing.

CoE/Council of Europe, 2012, *Manual on Human Rights and the Environment* (Second Edition), Strasbourg, Council of Europe Publishing.

91 For similar opinions, see the 7th Inter-secretariat Group on Climate Change Meeting Report of 6 February 2012, 2.

Déjeant-Pons, Maguelonne & Marc Pallemaerts (Eds), 2002, *Human Rights and the Environment: Compendium of Instruments and other International Texts on Individual and Collective Rights Relating to the Environment in the International and European Framework*, Strasbourg, Council of Europe Publishing.

Francioni, Francesco, 2010, International Human Rights in an Environmental Horizon, *European Journal of International Law* 21, 41–55.

Fitzmaurice, Malgosia, 2011, The European Court of Human Rights, Environmental Damage and the Applicability of Article 8 of the European Convention on Human Rights and Fundamental Freedoms, *Environmental Law Review* 13 (2), 107–114.

García San José, Daniel, 2005, Environmental Protection and the European Convention on Human Rights, *Human Rights Files* No. 21, Strasbourg, Council of Europe Publishing.

Glazebrook, Susan, 2009, Human Rights and the Environment, *Victoria University of Wellington Law Review* 40 (1), 293–351.

Gouritin, Armelle, 2011, Potential Liability of European States under the ECHR for Failure to take Appropriate Measures with a View to Adaptation to Climate Change, Institute for European Studies (IES) Commune Workshop, "Environmental Law", 27 November 2009, in: Faure, Michael & Marjan Peeters (Eds), *Climate Change Liability*, Cheltenham, Edward Elgar Publishing, 134–162.

Loucaides, Loukis, 2005, Environmental Protection through the Jurisprudence of the European Convention on Human Rights, *British Yearbook of International Law* 75, 249–267.

MacDonald, Karen E., 2008, A Right to a Healthful Environment – Humans and Habitats: Re-thinking Rights in an Age of Climate Change, *European Energy and Environmental Law Review* 17 (4), 213–226.

Pallemaerts, Marc, 2008, A Human Rights Perspective on Current Environmental Issues and Their Management: Evolving International Legal and Political Discourse on the Human Environment, the Individual and the State, *Human Rights and International Legal Discourse* 2, 149–178.

Pallemaerts, Marc, 2009, Human Rights and Sustainable Spatial Development, in: *Proceedings of the International CEMAT Symposium on "The Spatial Dimension of Human Rights: For a New Culture of the Territory", Yerevan, Armenia, 13-14 October 2008, European Spatial Planning and Landscape Series, No. 91*, Strasbourg, Council of Europe Publishing, 45, 61; also available at www.coe.int/t/dg4/cultureheritage/heritage/Landscape/Publications/ATEP-91Assemble_bil.pdf, last accessed 7 February 2013.

Pedersen, Ole W., 2008, European Environmental Human Rights and Environmental Rights: A Long Time Coming?, *Georgetown International Environmental Law Review* 21, 73–111.

Pedersen, Ole W., 2010, The Ties that Bind: The Environment, the European Convention on Human Rights and the Rule of Law, *European Public Law* 16 (4), 571–595.

Schall, Christian, 2008, Public Interest Litigation Concerning Environmental Matters Before Human Rights Courts: A Promising Future Concept?, *Journal of Environmental Law* 20, 417–453.

Shelton, Dinah L., 2010a, Developing Substantive Environmental Rights, *Journal of Human Rights and the Environment* 1, 89–120.

Shelton, Dinah L., 2010b, International Decision: *Tâtar v Romania, American Journal of International Law* 104, 247–252.

Trilsch, Mirja, 2009, European Committee of Social Rights: The Right to a Healthy Environment, *International Journal of Constitutional Law* 7, 529–538.

15

# Renewable Energy Policy in the European Union: A Contribution to Meeting International Climate Protection Goals?

*Christian Calliess & Christian Hey*

*Abstract*

The legal and political relationships between national and European Union (EU) energy policy competencies and the actual policies are multifaceted. In order to understand those relationships fully, one has to analyse both the formal competencies of the EU as enshrined in the Lisbon Treaty and the actual EU policies with direct and indirect impact on the choice of energy sources. The Treaty grants the EU competence as regards (a) the functioning of the energy market; (b) security of energy supply in the Union; (c) promotion of energy efficiency and energy saving and the development of new and renewable forms of energy; and (d) promotion of the interconnection of energy networks. However, the choice of member states between different energy sources and the general structure of its energy supply remain under national control. Any decision affecting this national competence must be adopted by a unanimous vote of the European Council. EU renewable energy support policy needs to develop within the framework of these mixed and multifaceted competencies. The authors' overall argument is that easy fixes do not work. Considering the different national preferences on the energy mix, it is premature to ask for a full-fledged EU energy competence leading to a harmonised support system for renewables. Nevertheless, the emerging climate and renewables policies could also be a driver for deepened energy integration – rather as a bottom-up than a top-down process. In that sense, a framework for 2030, with clear goals for climate mitigation, renewables shares and efficiency, is of pivotal importance for the transition towards a low-carbon economy by 2050.

## A. Introduction

The legal and political relationships between national and European Union (EU) energy policy competencies and the actual policies are multifaceted. In order to understand those relationships properly one has to analyse both the formal competencies of the EU as enshrined in the Lisbon Treaty and the actual EU policies with their direct and indirect impact on the choice of energy sources.

Member states have some freedom in defining a suitable national energy mix, which however is bound to the EU overall rules in the fields of the internal energy market and environment policies, namely EU climate policies. The Lisbon Treaty has introduced new provisions for an energy competence, which – as we shall show in detail – has only incrementally changed the limited EU role in steering national energy policies directly. The EU impact on the national energy mix is predominantly indirect, yet powerful.

So even if, in the sphere of energy policy, considerable national leeway persists, which can be used for organising a national energy transition towards a renewable-energy-based electricity system like the one in Germany, the success of such an energy transition depends very much upon a supporting EU policy framework, especially as regards climate mitigation, special conditions for renewable energy, and dedicated infrastructure development. Such a supporting EU framework is emerging, but it is far from being stable and consistent in view of the long-term requirements for a low carbon economy.

Our overall argument is that easy fixes do not work. Considering the different national preferences regarding the energy mix, it is premature to ask for a full-fledged EU energy competence leading to a harmonised support system for renewables. Besides which, the emerging climate and renewables policies could also be a driver for deepened energy integration – as a bottom-up rather than a top-down process.

The article is divided into two parts: Part B contains a legal analysis of the new allocation of competence between member states and the EU under the Lisbon Treaty, while Part C contains the analysis of the emerging EU policy framework for decarbonisation and renewable energy.

## *B. Allocation of EU and Member State Energy Policy Competence under the Treaty on European Union*

If EU energy policies – which up to now have chiefly been an outgrowth of European environmental and internal market policies – are poised to take on a life of their own, thanks to the Lisbon Treaty, there is no denying the fact that energy and environmental policies are inextricably bound up with each other, particularly when it comes to climate protection. This situation raises a number of issues concerning horizontal competency overlaps and the attendant issue of vertical competency delimitation in terms of the leeway allowed to member states to set their own energy policies. What this mainly boils down to is where the sphere of responsibility of Brussels leaves off, and where that of Germany starts.

### *I. Spheres of EU Authority in Energy Policy*

#### *1. Introduction*

Whenever the EU exercises authority over a particular matter, the EU's overarching statutory competence principle – known as the subsidiarity principle (pursuant to Article 5 of the Treaty on European Union (ex Article 5 of the Treaty establishing the European Union)) – must be taken into account. This Article lays out the fundamental principles for all actions taken by the EU and is thus the lynchpin of all decisions concerning the exercise of EU authority. The principles of limited authority (paragraphs 1 and 2), subsidiarity (paragraph 3), and proportionality (paragraph 4) in Article 5 of the Treaty on European Union constitute a legal code for all exercise of authority by the EU. It therefore follows that the EU has authority to act only insofar as (a) such authority has been formally vested in the EU, (b) the matter at hand involves a cross-border problem that can best be resolved by the EU, and (c) the measures taken leave the member states as much leeway as possible.[1]

Insofar as one of the rare cases that falls solely within the EU's authority does not come into play (see Articles 2 and 3 of the TFEU), the member states also retain authority for any matter that falls within the purview of the

1 Calliess (1999:69ff. and 240ff.).

EU until such time as the EU exercises its authority by enacting a concrete measure (this is referred to as the prohibitive effect).

The earlier Treaty establishing the European Community (TEC) contained no special provision concerning regulatory authority over the energy sector. The competence to take measures in this regard was based on environmental competence (ex Article 175 TEC), authority over internal market harmonisation (ex Article 95 TEC), and authority over trans-European electricity grids (ex Article 156 TEC). It was only when the new Treaty of Lisbon came into force on 1 December 2009 that the EU gained a special authority in the field of energy policy. Nevertheless the mentioned competences were for the most part carried over to and retained their original meaning in the Treaty on the Functioning of the European Union (TFEU).[2]

### 2. *Environmental Policy Authority Pursuant to Article 192(1) and (2) of the TFEU*

Article 192(1) of the TFEU lays out the spheres of authority for EU actions that aim to realise the goals of its Article 191. The Lisbon Treaty defines "promoting measures at international level to deal with regional or worldwide environmental problems, and in particular combating climate change" as the goal of Community environmental policy, pursuant to Article 191(4) (indent 4) of the TFEU, and contains all other environmental policy provisions of the Lisbon Treaty.

In principle, environmental policy measures require a majority vote of the Council, and are also subject to a European Parliament co-decision procedure. However, in derogation of this practice and on policy-related grounds, Article 192(2) of the TFEU enumerates a series of specific types of actions that are of particular importance to the member states and that are therefore subject to "the Council acting unanimously in accordance with a special legislative procedure".

Article 192(2) of the TFEU is relevant for energy in the following two respects. First, pursuant to Article 192(2)(a), policy instruments that take the form of tax incentives (i.e. "provisions primarily of a fiscal nature") are subject to a unanimous vote of the Council. In line with the narrow interpretation of the concept of "derogation" that prevails in the literature, such

2 Hereinafter referred to as TFEU, or as 'the Treaty'.

instruments here refer solely to taxes in the narrow sense of the term; and thus all other fees, charges and the like, such as eco-fees in the guise of special fees and user charges, fall within the scope of paragraph 1 and are thus not subject to the unanimous vote rule.[3] The word "primarily" means that the environmental measures must have a taxation focus; and thus, for example, the tax deductions for low emission motor vehicles do not fall within the scope of paragraph 2. Against this backdrop, some authors have incorrectly claimed that the greenhouse gas emissions trading directive should have been adopted by a unanimous vote since issuance of the certificates for a fee constitutes a fee regulation within the meaning of paragraph 2(a).[4] However, a unanimous vote was required on a proposed 1992 directive concerning a tax on carbon dioxide emissions and energy harmonisation.

Secondly, pursuant to Article 192(2)(c) of the TFEU, "measures significantly affecting a Member State's choice between different energy sources and the general structure of its energy supply" are subject to a unanimous vote and to an ensuing member state veto. "Significantly" here means that the unanimous vote requirement only applies to final measures that affect the general structure of a member state's energy supply.[5] Hence there was considerable opposition to the envisaged directive concerning government subsidies for renewable energies, as this was regarded as a significant interference in the energy supplies of member states.

Although this wording of Article 192(2) of the TFEU lays down special procedural requirements for energy-related environmental measures, it implicitly states that as a rule such measures fall within the scope of Article 192 of the TFEU. Hence this provision forms the basis for EU authority to adopt environmental policy measures, even in cases where such measures infringe on the freedom of action of member states.[6]

### 3. *Authority over Approximation of Laws Pursuant to Article 114(1) of the TFEU*

Numerous energy policy measures, particularly those concerning the establishment of the European internal electricity market (in this connection, the

3 Kahl (2012:recital 21).
4 Kirchhof & Kemmler (2003).
5 Kahl (2012:recital 34f.).
6 Epiney (2005:60); Pernice (1993:110).

European Parliament recently spoke in terms of full "ownership unbundling", i.e. the separation of power companies' generation assets from their transmission networks in the electricity market), were based on the general harmonisation authority pursuant to Article 95 of the Treaty establishing the European Union (now Article 114 of the TFEU),[7] which stipulates that the relevant proposed legislation must relate to the establishment and functioning of the internal market. This criterion is deemed to be met insofar as a particular measure aims to eliminate either obstacles to basic freedom of action or discernible distortions of competition.[8]

### *4. Trans-European Grid Authority Conferred by Article 172(1) of the TFEU*

The authority of Brussels in the sphere of renewable energies takes on outstanding importance when it comes to trans-European electricity grids. For example, equal amounts of solar energy and hydro power cannot be generated in all member states, owing to differences in climatic and topographical conditions. This, in turn, means that solar energy needs to be generated in southern Europe or North Africa, while hydro power mainly comes from Scandinavian and Alpine countries. But in order for this electricity to reach high-demand regions, an efficient grid structure is necessary; and this is where the energy and environmental policy significance of Article 172 of the TFEU comes in.

The EU's competence concerning the trans-european network (TEN-E) is derived from Articles 170 and 171 of the TFEU, which expands on the application domain of Article 172, which confers the requisite authority; whereby in this context the term 'trans-European' indicates that the networks that are to be established or expanded exhibit a specific cross-border attribute and that, by extension, infrastructure projects of a solely local or regional nature are not the EU's responsibility. Nonetheless, the concept of a trans-European network (TEN) also includes infrastructure projects that solely relate to the specific interests of individual member states.[9]

Article 170 of the TFEU contains a complete list of TEN goals that the EU is authorised to pursue ("promotion"). Contrary to the previous practice whereby member states planned and constructed their networks in accor-

7 Calliess (2008).
8 Kahl (2011:recital 22).
9 Koenig & Scholz (2003:223f.); Bogs (2002:49f.).

dance with national standards, under the TFEU "action by the Union shall aim at promoting the interconnection and interoperability of national networks" – which means that what were once border or peripheral regions are now focal points of the internal market by virtue of not only geographic and economic factors, but also oftentimes owing to national defence or military infrastructure elements. Hence the Treaty also stipulates that the Union (a) "shall take account in particular of the need to link island, landlocked and peripheral regions with the central regions of the Union"; and (b) harmonise the member states' diverse technical standards. The goal here is to establish the interoperable trans-European network called for by Article 170 ff of the Treaty, with a view to enabling the networks of neighbouring states to interconnect, thus filling any gaps resulting from network construction or expansion and efficiently interconnecting autonomous national networks in the interest of the functionality of the system as a whole.

Article 171 of the TFEU enumerates the following measures and other actions that the EU is authorised to undertake in order "to achieve the objectives referred to in Article 170": establishing guidelines; ensuring network interoperability; and providing financial support for projects of common interest. The fact that this constitutes a complete list is signalled in the German version of the treaty, by the absence of the term 'in particular'.[10]

While the EU may or may not provide financial support at its discretion, it is obligated to establish guidelines and ensure network interoperability, although there is no ranking relationship between these latter two types of actions. Hence guidelines can also be established in cases where no interoperability measures have been promulgated.[11]

Viewed in this light, such EU guidelines are legally binding frameworks that the member states are required to implement. Article 4(3) of the Treaty on European Union stipulates that the member states are to "refrain from any measure which could jeopardise the attainment of the Union's objectives".[12] The trans-European network guidelines were initially laid down in Decision No. 1254/96/EC amending Decision No. 1741/1999/EC. In addition, Decision No. 96/391/EC lays down a series of actions aimed at improving the conditions for expansion of the trans-European network in the energy domain. The list of categories defined in this decision and the ensuing Decision No. 1229/2003/EC concerning priority projects of common interest

10 Schäfer & Schröder (2012:recital 3).
11 EuGH, Slg. 1996, I-1689, Rz. 26 – Parlament/Rat.
12 Schäfer & Schröder (2012:recital 7).

that are worthy of support was expanded by Article 8 of Decision No. 1364/2006/EC concerning projects of European interest, which are to be given (a) “appropriate priority” when “selected under the budget for the trans-European networks”; and (b) “particular attention” when “selected under other Community co-financing funds”.

These objectives and priorities are to be supported by harmonised procedural principles aimed at their effective implementation. To this end, Article 8 of Directive 680/2007/EC lays down “general rules for the granting of Community support” that are to be fleshed out by the European Commission via its annual and multi-annual work programmes.[13]

In its Green Paper *“Towards a Secure, Sustainable and Competitive European Energy Network”*,[14] the European Commission calls for far-reaching expansion of support for the trans-European network, in its capacity as a key factor for the achievement of EU climate protection objectives.

## *II. The New EU Authority over Energy Policy Introduced by the Lisbon Treaty*

After the Lisbon Treaty came into force in 2009, the EU’s authority over energy policy discussed above was completed by a specific energy policy competence pursuant to Article 194 of the TFEU, wherein authority to implement the energy policy objectives in Article 194(1) is granted by Article 194(2)(1). Article 194(2)(2) contains derogations concerning the relevant application domain, while Article 194(3) calls for a special legislative procedure for energy taxes.

### *1. EU Energy Policy Objectives, Particularly Those Laid Down in Article 194(1)(c) of the TFEU*

The four energy policy goals laid down in Article 194(1) of the TFEU are to: “(a) ensure the functioning of the energy market; (b) ensure security of energy supply in the Union; (c) promote energy efficiency and energy saving

---

13 Beschluss der Kommission zur Festlegung des Arbeitsprogramms 2008 für Finanzhilfen für transeuropäische Netze – Bereich Energieinfrastrukturen – vom 16.4.2008, K (2008) 1360, ABl. C 160 vom 26.4.2008, 33.

14 European Commission (2008a).

and the development of new and renewable forms of energy; and (d) promote the interconnection of energy networks".

These objectives are subject to the following three guiding principles: EU energy policy is to be carried out (a) "in a spirit of solidarity between the Member States"; (b) "in the context of the establishment and functioning of the internal market"; and (c) "with regard for the need to preserve and improve the environment". These vague objectives are essentially the same as those laid down previously by the EU on the basis of its prior statute law. The objective laid down in Article 194(1)(c) of the TFEU ("[to] promote energy efficiency and energy saving and the development of new and renewable forms of energy") is particularly relevant for energy and environmental policy. However, the extent of the environmental policy authority granted by Article 192(2) of the TFEU (ex Article 175(2) of the Treaty establishing the European Union) is unclear – particularly as to whether all renewable energy matters are now to be governed by Article 194. Most authors who have addressed this matter (albeit in a somewhat cursory manner) have concluded that Article 194 is a *lex specialis.*[15] Although this would theoretically meet the goal – pursuant to the EU's new sphere of authority – of folding the EU's current energy policy competence into a new energy regulation,[16] there are also persuasive arguments against such a reading of the provision, namely the following:

First, Article 194 does not speak in terms of promoting renewable energies but rather of the development of such energies – by which, it is safe to assume, only technological development could possibly be meant.[17] Likewise inconsistent with a blanket *lex specialis* reading of the provision is the stipulation that the EU's authority to act is "[w]ithout prejudice to the application of other provisions of the Treaties". Paragraph 2(2) supports this concept as well in that it limits the EU's energy competence to situations involving a measure's "choice between different energy sources and the general structure of its energy supply", albeit "without prejudice to Article 192(2)(c)" of the TFEU. But this non-prejudice clause only makes sense if Article 192 of the Treaty applies in all cases in conjunction with Article 194.

Hence the EU's newfound authority over energy policy solely empowers it to promote the technological development of renewable energies, whereby

15 Britz (2009:71ff.); Heemeyer (2004:228f.); Trüe (2004:786f.).

16 Draft of the Treaty Establishing a Constitution for Europe: Dok CONV 727/03, Annex VII, 110.

17 Kahl (2009:60).

any economically or ecologically motivated support henceforth is governed by environmental regulations.

## 2. *Authority Granted by Article 194(2) of the TFEU*

Article 194(2)(1) empowers the EU to "establish the measures necessary to achieve the objectives in paragraph 1" – an extremely vague formulation, which, coupled with other EU authority, makes its energy policy jurisdiction seem all-encompassing at first glance, while mandating a far-reaching limitation on this authority to the effect that such policy measures "shall not affect a Member State's right to determine the conditions for exploiting its energy resources, its choice between different energy sources and the general structure of its energy supply, without prejudice to Article 192(2)(c)".

Although this limitation is similar to the aforementioned environmental policy provision pursuant to Article 192(2)(c) of the Treaty, it goes considerably further for the following three reasons:

(1) The requirements laid down in Article 192(2) need not be met cumulatively ('or'), whereby, unlike in Article 194 ('and') they can be met alternatively.

(2) There is no requirement that the measures must have a 'significant' effect on the areas subject to a derogation. Article 192(2)(2) of the TFEU should be interpreted narrowly as a derogation,[18] which thus does not apply across the board irrespective of the intensity of the measure in question.[19] It then follows that a measure can be deemed to affect a member state's energy supply solely in cases where, for example, it relates solely to energy supply related details such as technical matters.[20] Nonetheless, in the absence of an expressly defined significance threshold, the derogation clause grants the member states considerable sovereignty vis-à-vis Community energy policy.

(3) Unlike the procedure stipulated by Article 192(2)(c) of the Treaty,[21] its Article 19(2)(2) lays down a genuine restriction on EU energy policy authority, for the formulation "without prejudice to Article 192(2)(c)" should by no means be regarded as a mere procedural law allusion to the unanimous Council vote provision of paragraph 3. Unlike the environmental

---

18 Calliess (2011a:recital 12); Tiefenthaler (2011:119).
19 Ehricke & Hackländer (2008:599).
20 Neveling (2004:343).
21 Tiefenthaler (2011:128ff.).

policy measures governed by Article 192(2)(c) of the Treaty, energy policy measures with no environmental implications and that could potentially infringe on member states' sovereign right to adopt such measures are not subject to the unanimous Council vote provision of paragraph 3, since in fact the Council has no authority in such matters.[22] This concept is supported by two factors. First, paragraph 3 calls for a unanimous Council vote on energy tax measures only – *and* "without prejudice to paragraph 2".[23] Secondly, such a reading runs counter to the process that gave rise to the provision.[24]

### *3. The Unanimous Council Vote Provision of Article 194(3) of the TFEU*

The derogation in Article 194(2)(2) substantially limits the EU's jurisdiction over Community energy policy, which is further limited by the procedural rule laid down in paragraph 3, which – in keeping with Article 192(2)(a) of the TFEU (ex Article 175 2(a) of the Treaty establishing the European Union) and the tax derogation provisions in other treaties – requires a unanimous Council vote "after consulting the European Parliament" in matters that are "primarily of a fiscal nature". The necessarily narrow reading of this restriction notwithstanding, it shows that the member states still regard energy law as a highly sensitive issue when it comes to their national sovereignty.

### *4. Interplay between Article 194 of the TFEU and Other Areas of EU Jurisdiction*

The relationship between Article 194 of the Treaty and the EU's environmental policy authority was discussed above. Other issues regarding the scope of EU authority in this domain are raised by Articles 114, 122, and 222 of the TFEU.

22 Ehricke & Hackländer (2008:599).
23 (ibid.:579ff.).
24 Draft of the Treaty Establishing a Constitution for Europe: Dok CONV 725/03; Calliess (2010:20ff.).

### a) Interplay between Article 194 TFEU and Article 114 TFEU (concerning Approximation of Laws)

Relative to Article 114 of the TFEU (ex Article 95 of the Treaty establishing the European Union), Article 194 is a *lex specialis*.[25] This reading is supported by the wording of Article 194, whose paragraph 1(a) expressly mentions the energy market, and, historically speaking, by the convention presidium's intention of aggregating energy policy authority.[26] Hence the controversy over the admissibility of future-oriented approximation of laws is superfluous, by dint of the fact that pursuant to Article 194 of the TFEU it is admissible beyond the shadow of a doubt.[27]

### b) Interplay between Article 194 TFEU and EU Authority over the Trans-European Network Pursuant to Article 172 TFEU

It is unclear whether Article 194 of the TFEU (ex Article 156 of the Treaty establishing the European Union) is a priority regulation in its capacity as a more specific regulation.[28] Although the contention that Article 172 is a more specific provision than Article 194 of the TFEU would appear to be plausible at first glance, it is negated by the fact that Articles 170, 171 and 172 of the TFEU relate to all member state networks and access thereto, while Article 194 solely governs energy networks. Hence, in view of the lesser statutory scope and application domain of Article 194, it is in fact the more specific provision. However, the application domain of Article 194 still needs to be determined, since Article 172 remains fully applicable in tandem with Article 194.

The issue here is whether the EU's new authority over support for energy network interconnection measures also includes jurisdiction over support for the trans-European network and interoperability of the various member states' energy networks pursuant to Article 170(2) of the TFEU. This would appear to be the case since interconnection is by definition the umbrella term in this context, i.e. interoperability is a subset of and is subsumed by inter-

25 Kahl (2009:46).

26 Draft of the Treaty Establishing a Constitution for Europe: Dok CONV 727/03, Annex VII, 110.

27 Neveling (2004:343); Kahl (2009:51).

28 Trüe (2004:786f.); Kahl (2009:60).

connection. Interoperability refers to the technical ability of two systems to interact with each other, a process that chiefly involves common or, at a minimum, non-mutually exclusive standards. "Interoperability of national networks" refers to the preconditions for trouble-free interconnection of national networks and the components thereof, particularly when it comes to the establishment of a transnational network.[29]

The purpose of such a network is to compensate for the technical incompatibility of individual national networks (e.g. line voltage differences) by harmonising the relevant technical standards or developing purpose-built technical equipment. In the latter case, it is crucial to ensure from the outset that the relevant technical standards are compatible with each other. Interoperability likewise encompasses the organisational realm, which means that harmonisation measures should also lay the groundwork for economically optimal networks that deliver the best possible security of operation. To this end, both statutory regulations and the applicable EU and industry-organisation standards should be adhered to.[30]

Interconnection (in a technical system) has a broader meaning, on the other hand, in that it refers to the interconnection of physical network structures by establishing the relevant standards and installing the relevant equipment at the interconnector and transfer points. However, in economic terms, interconnection refers to a scenario where technically and logically interconnected networks are also used. Hence the term interconnection covers a broad range of scenarios, in that in a general sense it refers to market-actor access to a network used in common by all such actors. For electricity networks it refers to interconnection of the electricity networks of various states. Hence interconnection is used as a catch-all term – for example in a European Commission communication titled *Recent Progress with Building the Internal Electricity Market*[31], which states the following: "[A]greement has been reached to analyse existing bottlenecks in terms of interconnectors between systems".

Hence the EU's authority to "promote the interconnection of energy networks" pursuant to Article 194(1)(d) of the TFEU goes beyond the scope of that provided by current legislation, since this authority is limited by Article 172 of the Treaty in the following ways:

29 Erdmenger (2003:recital 19).
30 Calliess (2011b:recital 16).
31 European Commission (2000).

(1) Pursuant to Article 171(1)(indent 1) of the Treaty, the EU has the authority to enact mandatory guidelines – which however are solely intended to coordinate the relevant measures[32],

(2) The authority granted by Article 171(1)(indent 2) of the Treaty is limited solely to measures that "may prove necessary to ensure the interoperability of the networks", i.e. existing networks only, and

(3) Pursuant to Article 171(1)(indent 3), the EU is only allowed to "support projects of common interest supported by Member States".[33] In contrast, Article 194 of the Treaty empowers the EU to undertake interconnection projects of its own; it also applies to projects that solely have a bearing on the interests of individual member states. Although the EU can require member states to carry out such projects, it cannot stipulate attendant implementation methods (e.g. specific power line routes) by virtue of the fact that the EU lacks the authority to plan such implementation (Article 5(2) of the Treaty on European Union)[34] and of the subsidiarity principle as well (Article 5(3) of the Treaty on European Union). And thus authority over such matters is left to the member states.[35]

Hence the question arises as to the actual scope of the application domain under Article 172 of the Treaty, since the trans-European network provisions of Article 170(1) of the Treaty still apply to energy policy. It is possible that Article 172 empowers the EU to enact basic general regulations across multiple domains, while Article 194 allows for the adoption of regulations that apply specifically to energy networks. It would also probably be necessary to interconnect with other third state networks (pursuant to Article 172), owing to the fact that, unlike Article 194, Article 171(3) states that "The Union may decide to cooperate with third countries to promote projects of mutual interest and to ensure the interoperability of networks".

### *III. Foreign Policy concerning Energy*

According to European Court of Justice rulings, the EU has implicit authority to enter into international treaties that correspond with EU authority over

32 Härtel (2006:recital 13). Trüe (2002:109).
33 Voet van Vormizeele (2012:recital 9).
34 Tiefenthaler (2011:124f.).
35 Rodi (2012:recital 8).

internal matters.[36] Hence the EU has authority over all foreign relations matters, including the intra-Community aspects of such matters. This means that EU member states are prohibited from entering into any third-state treaty insofar as the EU has assumed its internal responsibility to enact regulations for the matter in question.

Of particular significance in this context is Article 191(1)(d) of the TFEU, which calls for the "promotion of measures at international level to deal with regional or worldwide environmental problems" and aims, according to the Lisbon Treaty, now explicitly to fight global warming in a manner that promotes the achievement of Community environmental goals. In case of uncertainty, this provision also allows for the conclusion of EU energy and environmental policy treaties based on a number of legal principles.

### *IV. Scope of the EU's New Energy Policy Competence under Article 194 of the TFEU*

Opinions in literature vary concerning the EU's new energy policy authority granted by Article 194 of the TFEU. Concerns have been expressed in some quarters that this new authority will prompt the EU to adopt additional regulations, since the vaguely worded objectives of Article 194 appear to grant the EU blanket authority over all energy policy matters.[37] However, most authors feel that the change will merely result in amalgamation of the EU's current authority derived from its authority in the field of environmental policy, infrastructure policy and internal market policy.[38]

As noted above, the coming into force of Article 194 of the TFEU following adoption of the Lisbon Treaty merely expanded the EU's policy-making authority over the interconnection of energy networks. Hence Article 194 grants the EU no genuinely new authority for such interconnection, but instead merely expands the scope of its existing authority.

In view of the fact that, as we have seen, Article 194 of the TFEU does not endow the EU with all-encompassing new authority, its significance is largely political in nature – apart, that is, from the greater legal certainty and clarity created by the measure.[39] Thus from now on EU energy policy will

36 EuGH, Slg. 1971, 263, recital 15f.
37 Jasper (2003:211); Classen (2003:351); Götz (2004:46).
38 Blanke (2004:232); Görlitz (2004:381); Rodi (2012:recital 2); Kahl (2009:51).
39 Kahl (2009:51f.); Neveling (2004:342).

issue forth from "a single source"[40] in a manner that will allow for coherent harmonisation of policy goals and measures.

### *V. Exercise of Energy Policy Authority by the EU*

The manner in which the EU exercises its energy policy authority is governed by the stipulations of the EU energy regulations that are discussed above, as well as the general provisions concerning the exercise of power pursuant to the Lisbon Treaty (Article 5 of the Treaty on European Union).

#### *1. Meaning of the Energy Policy Solidarity Clause under EU Law*

Article 194 of the TFEU stipulates that EU energy policy objectives are to be pursued "in a spirit of solidarity between the member states". This clause is a statutory innovation under EU law, since it makes jurisdiction over energy policy subject to the overarching principle of solidarity among the member states. Under EU law, application of this clause is to be governed by the general EU solidarity principle.

By adopting a solidarity clause concerning energy policy competence, the member states have sent a clear signal that they regard energy as a sector involving their common interests; in other words, the member states have realised that when it comes to energy, they're all in the same boat. This solidarity principle gives rise to the two types of binding solidarity obligations referred to in Articles 194 and 222. First, the member states are enjoined not to take any action in the name of national interest that would interfere with achievement of energy policy goals of common interest – although this applies only to areas that fall within the scope of EU energy policy authority. And secondly member states may be obligated to provide assistance to one or more states that are facing an energy policy emergency, particularly in connection with security of supply.[41] This latter aspect of the solidarity principle represents a mindset shift from one where security of supply, once regarded as a national matter, is now seen as a policy concern for the EU as a whole. The solidarity principle enables a member state that is facing an energy supply shortage – occasioned by domestic policy conflicts or the like

40 Kahl (2009:51).
41 European Commission (2007); Ehricke & Hackländer (2008:595).

– to obtain the assistance of another member state. At the same time, it sets the stage for the application of the EU's general subsidiarity principle, which is a precondition for joint action that the EU is required to demonstrate it has undertaken. The energy policy solidarity clause acts as a corrective to the subsidiarity principle by presupposing that the objectives of energy policy measures cannot be adequately governed at the national level alone and can be governed more efficiently in Brussels. Hence, in effect the solidarity clause shifts the burden of proof to the sphere of a collective procedure.

At first glance, the energy policy solidarity clause has no direct implications for energy and environmental law, since the main focus of the clause is security of supply. But measures in this sphere can also have an impact on environmental policy, one example of this being the EU Commission's *Energy Security and Solidarity Action Plan* (2008), which contains measures aimed at promoting development of the combined heat and power sector.

### *2. Stipulations of Article 11 of the TFEU*

The Treaty's Article 11 – the like of which is not to be found in any member state statute – stipulates that "Environmental protection requirements must be integrated into the definition and implementation of the Union policies and activities, in particular with a view to promoting sustainable development", whose requirements stem from the EU environmental policy objectives and principles laid down in Article 191(1) and (2) of the Treaty. Thus this clause means that all measures that are governed by Article 194 of the Treaty must be realised in a sustainable and environmentally compatible manner.

## *VI. Remaining Competences of the Member States*

The entirety of the EU's energy and environmental policy competence is governed by the principle of shared competences pursuant to Article 4(2)(i) of the TFEU[42], whereby the member states "exercise their competence to the extent that the Union has not exercised its competence" (Article 2(2) of the TFEU) – in which case the member states are free to exercise their own

42 De Sadeleer (2012:63ff.).

policymaking competence, subject to the principle of loyal cooperation with the EU.

### *1. Unilateral Action by Member States*

Like ex Article 176 of the Treaty establishing the European Union, Article 193 of the TFEU allows individual member states to introduce more stringent environmental protection measures under Article 191 of the TFEU. Article 194 of the Treaty contains no such provision in the energy policy realm, and thus not for energy law either. The absence of this provision is regarded in some quarters as a structural shortcoming that works to the detriment of environmental protection in the EU, particularly in the realm of energy efficiency measures and technical development of renewable energies.[43] Financial aid for the furtherance of renewable energies falls within the scope of environmental rather than energy competence, as has always been the case.

It has been suggested, in light of the non-prejudice clause of Article 194(2) of the Treaty, that Article 193 be applied *mutatis mutandis* to energy and environmental law[44] – a dubious proposition, as it would set the stage for an unintended statutory loophole. Such a reading of the non-prejudice clause would also be inadvisable in light of the uniqueness of energy and environmental law, whose limited aims and measures necessitate special reconciliation provisions between EU and national policy measures. The delicate balance of the European energy and environmental policy triad could be upended by national 'go it alone' measures.[45] The abscence of a clause allowing for the adoption of more stringent protective measures can thus be viewed as the embodiment of the target and measure limits imposed by energy and environmental law.

43 Britz (2009:86); Kahl (2009:61).

44 Britz (2009:86).

45 See Gundel (2008:468) for a critical view of market differentiation resulting from such measures.

### 2. Restrictions Imposed by Article 345 of the TFEU (ex Article 295 EGV)

The Treaty's Article 345, which is generally regarded as a provision that imposes limitations on competence,[46] stipulates that "[t]he Treaties shall in no way prejudice the rules in Member States governing the system of property ownership" – which has led some to conclude, for example, that the EU is prohibited from adopting property-related measures.

However, Article 345 of the Treaty was originally promulgated in order to assuage member state fears that EU laws would result in privatisation and/or nationalisation.[47] Hence it follows from a historical reading of Article 345 that it aims to ensure that the EU remains neutral when it comes to basic issues concerning national economies; and thus the current prevailing view refers to the wording of Article 345, which concerns not property rights but rather property ownership[48] – which basically means decisions concerning nationalisation and privatisation.

## C. Advancing the EU Energy Policy Framework in Renewable Energies

The EU has pivotal competences for a number of frameworks that relate to the expansion of renewable energies, to which end the EU has adopted the following interrelated policies and strategies in particular:

- EU climate protection policies in conjunction with mandatory objectives for greenhouse gas reduction; and a broad range of implementation instruments in this regard, notably emissions trading
- EU energy policies, in particular those involving to some extent competing objectives as regards an internal European electricity market and expansion of renewable energy capacity
- EU infrastructure policy, via the trans-European network, and
- European energy research (not discussed in detail in this report).

In all four of these areas, relevant developments and discussions are occurring that improve the chances of successful implementation of renewable energy policies in the various member states. Hence, it is of crucial importance that these EU fields of endeavour unfold in a manner that promotes

46 Kingreen (2011:recital 5).
47 BT-Drs. 2/3440, Anhang C, 154.
48 Calliess (2008:27ff.).

and institutionalises national strategies aimed at an all-renewable-electricity supply. Achievement of ambitious national objectives will be greatly eased if the dynamic expansion path mandated by the Renewable Energy Directive[49] continues to unfold in the post-2020 period. In addition, such an expansion via a coordinated approach between the various member states would be less cost intensive than if each individual member state expands its own renewables.[50] Our analysis of the situation clearly shows that the EU has robustly set the stage for renewable energy expansion; whereby in light of this analysis there is good reason to believe that an EU framework conducive to development of renewable energies will be in place for the period after 2020 as well. This framework needs strengthening.

### *I. Refinement of EU Climate Protection Objectives*

The EU climate package of December 2008 – which calls for a triple target of 20% reduction of greenhouse gas emissions with a 30% contingent option; a 20% share of energy from renewable sources; and 20% greater energy efficiency relative to the current trend – could potentially pave the way for a transition to a climate neutral, and largely or wholly renewable electricity supply. This package, whose elements include a reform of the EU emissions trading system and an amended directive concerning the furtherance of renewable energies, also constitutes a breakthrough after the prior long, drawn-out process of EU integration in energy policy, since the package grants the EU considerably greater climate policy authority than that wielded by the member states.[51] This breakthrough from climate policymaking practices of the past was based on a relatively broad consensus in the EU concerning the importance of European climate policymaking in the realms of security, economic and industrial policy.

However, this consensus has been greatly weakened by the economic crisis and the failure of the UN climate summits since Copenhagen – a phenomenon graphically demonstrated by the fact that the EU has as yet been unable to reach an agreement concerning a unilateral 30% greenhouse gas

49 2009/28/EC.

50 European Climate Foundation (2010b); Czisch (2009).

51 Olivier et al. (2008); Geden & Fischer (2008); Schreurs et al., (2009); Jordan et al. (2010a:3ff.).

emissions reduction by 2020.[52] This goal, whose advisability is demonstrated by the European Commission and other economic analyses,[53] is also seen as a way to revitalise international energy policy,[54] but it no longer commands a majority support within the European Commission nor among the member states – a fact demonstrated by this headline from *Ends Daily* of 10 June 2010: '30% $CO_2$ reduction goal put on the back burner'.

This aside, the benchmark for medium-term EU climate protection policy comprises the often-stated position of the European Council in this regard[55] and the roadmap to 2050[56], both of which place at least an 80% greenhouse gas reduction by 2050 on the EU's policy agenda. In the view of the European Commission, only a minute proportion of these reductions can be achieved through implementation of flexible mechanisms outside the EU.[57] Later the Mobility and Energy General Directorates[58] of the European Commission[59] elaborated on strategies, scenarios and consultation documents, further specifying the sectoral dimension of a low carbon economy.

Those roadmaps for the run-up to 2050, if politically supported and effectively implemented, would enable Europe to achieve the greenhouse gas reductions necessary to adhere to the 2° Celsius goal,[60] and thus be an indispensable yardstick for the climate protection policies of industrialised states. From the perspective of the EU's envisaged unilateral greenhouse gas reduction goal, such roadmaps can be also considered to be sensible instruments that are essential in order to establish guideposts for technological development and above all to avoid technological lock-in effects whose reversal would exact a high economic cost if binding international climate policies came into force aimed at bringing about the requisite reductions.[61]

So far, however, it has been difficult to form the necessary political consensus by member states to anchor the overarching objective or respective

52 Geden & Fischer (2012:43).
53 European Commission (2010c).
54 Wissenschaftlicher Beirat Globale Umweltveränderungen (2010).
55 Council of the European Union (2009).
56 European Commission (2011a).
57 European Commission (2010c:6).
58 European Commission (2011e).
59 European Commission (2011d).
60 Sachverständigenrat für Umweltfragen (2008).
61 Holm-Müller & Weber (2010); Sachverständigenrat für Umweltfragen (2009); Unruh (2000).

sector targets more firmly in EU policy.[62] After difficult negotiations within the Councils of Environment and Energy Ministers, 26 of the 27 member states recognised that "under certain assumptions … that decarbonisation of the energy sector on a EU wide scale is technically and economically feasible".[63] So the roadmap has been accepted as "guidance in the further process" by a strong majority of member states, without firmly incorporating the goal of decarbonisation and intermediate steps into an official and binding strategy.

Meanwhile proposals for sectoral roadmaps for the energy and transport sectors exist, which comply with the overall targets for the Low Carbon Economy Roadmap. It has to be emphasised that reduction targets are differentiated from sector to sector. So in the electricity sector, reduction will have to be higher than that for transport in order to achieve efficient reductions. In the electricity sector, even the 80% goal would make it necessary to aim for full decarbonisation.[64] The case for target differentiation would be less evident for a 95% reduction, but the Commission did not opt for this more ambitious target.[65]

### *II. Roadmap 2030: Additional Expansion Objective for Renewable Energies*

#### *1. A Policy Feedback Approach to Renewable Energy Expansion in the EU*

Various models of energy mixes are available that would achieve the sectoral climate protection goals discussed above – one such path being a massive pan-European expansion of renewable energies beyond the mandated 2020 goal, with the aim of achieving a wholly renewable electricity supply. The different scenarios calculated for underlying the technical and economic feasibility of the Energy Roadmap 2050 all assume a renewables share in the electricity sector in the range of 60% or more. That applies even for a scenario relying strongly on nuclear energy, and another relying more on coal combustion with carbon capture and storage (CCS). The Commission

62 Geden & Fischer (2012:41).

63 Council of the European Union (2012b).

64 European Climate Foundation (2010b); Jones (2010); Edenhofer et al. (2009:7); Öko-Institut & Prognos AG (2009).

65 Hey (2012).

scenarios furthermore conclude that the overall cost of a low carbon energy system (as in the Energy Roadmap 2050) is not significantly higher than that of a business-as-usual scenario. Furthermore, technology choice is not a critical factor as regards cost – most scenarios result in similar cost levels. So next to energy efficiency, strong renewables growth beyond 2020 belongs to the no-regret options of a low carbon energy system. The only exception – owing to a number of methodological shortcomings of the scenarios – is an electricity system completely based upon renewable energy sources.[66]

The EU is already on the way towards such a predominantly renewables-based electricity system. Most member state action plans for implementation of the Renewable Energy Directive call for a very significant renewable energy expansion – an evolution that would result in an EU electricity supply that is more than one-third renewable in 2020. Achieving this will necessitate substantial growth in the renewable energy sector in all member states, as well as the establishment of robust incentives for renewable energy development,[67] grid expansion and other complementary measures. It is also likely that coalitions of economic and political actors will rise to greater prominence in all member states. And thus, spurred by EU climate-friendly economic objectives, we are likely to see an altogether more favourable framework for renewable energy expansion in the post-2020 period.

Other pathways towards decarbonisation, relying more on nuclear energy or coal with CCS, seem to be less realistic. This can be illustrated with examples from a number of scenarios, developed for or in close cooperation with leading power companies, which rely on a massive expansion of nuclear power in the order of 200 GW and coal CCS amounting to some 120 GW and limit the share of renewable electricity to 40%.[68] As such visions imply the massive construction of 100 to 150 new nuclear power plants, they have a limited chance to withstand the opposition in many member states. The European Commission favoured an economically and politically more rational approach with much lower shares of nuclear or coal even in the respective pronuclear or pro-coal scenarios.

That said, we need to bear in mind that the EU's competence when it comes to exercising a direct influence over member state energy source choices is limited, which means that any measures in this regard must stem from the EU's environmental competence pursuant to Article 192(2) of the

66 European Commission (2011d); Hey (2012); Matthes (2012).

67 Rathmann et al. (2009).

68 European Climate Foundation (2010a:9 and 50); EURELECTRIC (2010:61ff.).

TFEU, and must be adopted by unanimous consent of all 27 member states for measures that have a major impact on national energy source policy. Hence any EU effort to fix the putative 2050 energy mix in stone would be premature at this point from both an institutional and political standpoint, regardless of whether a wholly renewable electricity supply (as we advocate for Germany) or a mix of nuclear, fossil and renewable energy is involved.

The relatively few actors that have come out in favour of a wholly renewable electricity supply are mainly found in environmental groups, the renewable energy industry and think tanks – plus the European Parliament, particularly in the parliamentary coalition known as the European Forum for Renewable Energy Sources (EUFORES).[69] Only states such as Germany, Denmark, Spain and Portugal that are in the vanguard of the renewable energy movement are likely to push more strongly for a wholly renewable electricity supply; and the only member state that has thus far recognised the need to establish a widely renewable energy electricity supply over the long term is Germany.[70] States such as Austria, Sweden and Lithuania, with largely conventional renewable energy sources, may also jump on the renewable electricity bandwagon, albeit with only measured enthusiasm – as is evidenced by the relatively slow pace of renewable energy expansion in some of these states.[71] However, we are unlikely to see support emerging for a *wholly* renewable electricity supply any time soon in the majority of member states. Take France, for example. Although the French have decided to ramp up the share of energy from renewable sources in their economy from its current level of 15.5% to 27% by 2020, the nuclear industry is still the major player in the French energy policy arena.[72] Another example is Great Britain, whose energy policy calls for a major off-shore wind farm development programme in conjunction with the construction of nuclear power plants and investments in CCS technology.[73] And, as for most of the Central and Eastern European states, electricity is mainly derived from large centralised nuclear power and/or coal power plants, and renewable energy

69 European Renewable Energy Council (2010); Müller-Kraenner & Langsdorf (2012).

70 Bundesministerium für Wirtschaft und Technologie & Bundesministerium für Umwelt, Naturschutz und Reaktorsicherheit (2010).

71 European Commission (2009).

72 Koopman (2008); Mez et al. (2009); Pellion (2008); Guerry (2012).

73 Department of Energy and Climate Change (2009); HM Government (2009); Helm (2006).

development is still in its infancy.[74] In addition, the major power companies will in all likelihood fiercely oppose efforts to establish a wholly or largely renewable electricity supply.[75]

Against this backdrop, the European Commission's current advocacy of a technology-neutral approach towards decarbonisation would be perfectly understandable. This tendency toward technology neutrality on the part of an EU body, which is often referred to as the 'guardian of the treaties' but which is nonetheless keeping the decarbonisation option open for the member states, is also unavoidable at present, in view of the EU Treaty's restrictions on the EU's energy source policy competence. In brief, the EU is very unlikely to take a system decision in favour of renewables-based electricity in the short run. However, the strategy documents of the Commission, as well as first reactions of the Energy Council[76] suggest that, in the context of a multi-source strategy towards decarbonisation, renewable energy sources receive privileged attention, without making a clear-cut system decision as did Germany.

Instead, the European institutions tend to pursue a strategy, which can be described on the basis of the policy feedback approach.[77] This approach explains radical policy innovation by a sequence of incremental reform steps, which each are suboptimal and insufficient, but which create conditions favourable for the next reform cycle. This strategy engenders a new policy path that grows stronger with the passing years and whose initially inadequate institutional innovations and measures now prompt calls for more extensive reform – thus creating a more robust underpinning for the path *per se*. The policy of incremental self-obligation,[78] as the policy feedback paradigm is also called, has enabled the EU to institute reforms despite their initial unpopularity. The Renewable Energy Directives of 2001 started with legally non-binding goals for renewables, which proved to be insufficient. In the 2009 directive this deficiency has been addressed by making the targets legally binding. It seems that the Commission, supported by the Energy

74 Barbu (2007).

75 EURELECTRIC (2010:61ff.); Lamprecht (2009:22ff.).

76 Council of the European Union (2012a). The Council invited the Commission to prepare "the basis for the discussion for a post-2020 perspective for renewable energy sources" and took note "that any of the scenarios of Europe´s energy supply analysed would require a substantially higher share of renewable energy…beyond 2020, including in 2030.".

77 Pierson (1993); Jordan et al. (2010b); Prittwitz (2007:175f.).

78 Eichener (2000).

Council, now opts for such an incremental step-by-step approach on the way towards decarbonising the energy sector. This offers opportunities for a transition based upon renewable energy – but which also may face backlashes or instability during that transition.

### 2. *A Roadmap for Renewable Energy in 2030*

Against this backdrop, a medium-term European roadmap for the expansion of renewable energies in the run-up to 2030 would be needed in order to stabilise that transition. Also planning and investment stability for German and EU infrastructure development call for a more stable framework for renewables beyond 2020[79]. According to Article 24(9) of Directive 2009/28/EC, the European Commission is planning to issue a renewable energy development roadmap for the post-2020 period as late as 2018, which would not allow sufficient lead time to establish conditions conducive to planning certainty, particularly for network and storage capacity expansion for the post-2020 period. Hence the discussion concerning development objectives should get underway long before 2018. The Energy Ministers Council of December 2012 has invited the Commission to present a proposal for a post-2020 framework for renewable energy sources by 2014.[80]

In order to establish international high-voltage direct current transmission networks or strategic regional networks in the North Sea, it is essential that clearly defined goals and guideposts be laid out concerning renewable energy capacity development, since otherwise the investment risks for such projects will be unduly high. Timely establishment of the requisite transmission grids is a key factor in terms of renewable energy capacity development.[81] Grid planning based solely on scenarios – the approach recommended by the European academies of science, among other actors[82] – will not get the job done in terms of establishing the requisite investment certainty.

A prime example of the importance of timely targets for renewable energy as basis for prospective grid planning is the pilot project for a ten-year plan

79 European Environment and Sustainable Development Advisory Councils (2009); European Climate Foundation (2010b:9 and 28).

80 See footnote 78.

81 European Climate Foundation (2010a:16 and 58).

82 European Academies Science Advisory Council (2009); Wagner (2009:54f.).

(2010–2020) devised by the European Network of Transmission System Operators for Electricity (ENTSO-E),[83] according to which transmission system operators need to undertake investment planning for the 2010–2020 period for more than 42,000 kilometres of transmission lines, half of which will be necessitated by renewable energy capacity expansion. But according to ENTSO-E's own calculations, the scope of the grid build-out will need to be even greater than this, since the national action plans for renewable energies, which had not been submitted as at June 2010, could not be taken into account until the next ten-year plan was issued in 2012. Against this backdrop, ENTSO-E also advocated that grid development objectives be set for a more extended period.[84]

Development objectives are essential for the electricity sector in view of the pivotal importance of transmission networks for load balancing. The groundwork for the requisite planning of such networks can only be laid if sectoral development objectives are set – which, as called for by the Renewable Energy Directive, could also be added to and be one of the outcomes of national action plans. Inasmuch as the share of European electricity from renewables may well reach 35% in 2020, a share to the order of 50–70% in 2030 would appear to be well within reach.[85]

### *III. Subsidiarity and Support Instruments*

The Renewable Energy Directive of 2009 – whose adoption was fraught with conflict from start to finish – represents a conscious decision on the part of the EU to leave renewable energy support policy to the member states or to cooperative arrangements between groups of member states.[86] This solution was preceded by a basic conflict over which support instruments are appropriate. Although a harmonised European quota trading system for renewable-based electricity can be more easily coupled with the internal market, national feed-in tariffs have by and large proved to be the more efficient and robust instrument thus far. The debate on this issue is still ongoing, however. The electricity and hydro power industry association known as Bundesver-

83 ENTSO-E (2010:9ff.).

84 (ibid.:17).

85 European Commission (2006); European Environment and Sustainable Development Advisory Councils (2009); European Renewable Energy Council (2010).

86 Schöpe (2010); Jones (2010).

band der Energie- und Wasserwirtschaft (BDEW), as well as a number of large power companies, is still pushing for a harmonised European quota system of the type described in a 2010 study that was conducted for one such organisation by Cologne University's Department of Energy Studies (EWI).[87] But there have also been calls in recent years for a European approach along the lines of Germany's Renewable Energy Act (EEG) or other feed-in tariff instruments[88] – an approach likewise advocated by EU Energy Commissioner, Guenther Oettinger.[89] Also in that respect the European Commission – certainly in the view of the considerations below – has opted for a very soft approach: it will develop guidance on best practice on cost-effective, predictable and consistent national support systems, promoting cooperation on renewables support between member states and market integration of renewables[90]. This guidance also intends to find a balance between the two partly conflicting European policy approaches: on the one hand, the completion of the internal market for Energy;[91] and, on the other hand, the prevalence of national support schemes, which are necessary to implement the requirements of the renewables directive.

The call for a fully harmonised approach to renewables support holds that (a) such an approach would be a better fit with the internal European electricity market, since divergent national feed-in tariff systems could inhibit or distort cross-border electricity trading;[92] and (b) a large-scale network would also open up relatively cost-efficient load balancing options and would greatly reduce storage capacity investment costs.[93]

But in some quarters it is also felt that the current EU directive arrangements concerning bilateral and multilateral cooperation should remain in force in lieu of striving for European harmonisation.[94] The main argument against a harmonised quota system is the evidence that comparable national

87 Fürsch et al. (2010); for EURELECTRIC's position see Berge & Cross (2010).
88 Czisch & Schmid (2007).
89 See Euractiv, 6 August 2010, "Oettinger Presses for European Green Electricity Subsidies", available at http://www.euractiv.de/energie-und-klimaschutz/artikel/oettinger-drangt-auf-europaische-einspreisevergtung-003476, last accessed 08 March 2013.
90 European Commission (2012a); see also footnote 78.
91 European Commission (2012b).
92 Fürsch et al. (2010); Sensfuß et al. (2007).
93 European Climate Foundation (2010a and b); Czisch (2009).
94 Schöpe (2010); Fouquet & Johansson (2008); Müller-Kraenner & Langsdorf (2012).

systems have enjoyed only limited success.[95] A problem with harmonised European feed-in tariffs is that (a) if they are unduly high, they may engender considerable windfall profits in states with conditions more conducive to electricity generation; or (b) basing the tariffs on the lower costs in regions with better electricity generation conditions could result in a concentration of installations in regions that display such conditions;[96] and (c) this would therefore fail to incentivise the requisite investments in other regions. This could in turn provoke a conflict between EU designating optimised installation sites, on one hand; and possible ambitious expansion plans in individual member states, on the other.

Regionally balanced renewable energy development that also takes account of cost differences is realisable under the current regulation framework based on European objectives and national support instruments, in cases where the development objectives in regions with more favourable site conditions are more ambitious than those in regions with less favourable conditions. Applying such an approach would mean, for example, that Germany would place more emphasis on wind energy development, while Spain would focus more on photovoltaics.

The differences in the renewable energy development phases of the various member states also need to be taken into account, and the attendant support instruments will need to be adapted to the conditions in each state.

A total of 21 member states have instituted feed-in tariffs as a central or partial instrument of their energy mix, although the exact modalities of these instruments differ greatly from one state to another.[97] Any attempt at harmonising these systems would inevitably engender high costs and serious conflicts, as partial modification of well-established long-term investment frameworks would also be involved, whereby switching from member state to EU level policy would set in motion a period of investment uncertainty that would temporarily put the brakes on renewable energy growth. Moreover, the resulting compromise, apart from the extensive negotiations it would undoubtedly entail, would probably result in a support system that is relatively impervious to policy innovation. This same problem of barely resolvable conflicts between the various national support systems and a harmonised European support framework would arise under a harmonised quota

95 Fouquet & Johansson (2008).
96 Sensfuß et al. (2007:54).
97 Rathmann et al. (2009).

system, as it would necessarily replace national feed-in tariffs with flexible quota market prices.

Hence EU support frameworks for renewable energy should recognise the subsidiarity principle and enable EU member states sufficient leeway for action that is also compatible with Community principles.[98] And, in point of fact, a workable compromise for the foreseeable future in this regard was put in place by the Renewable Energy Directive of 2009. Also the more recent communications of the European Commission stick to this basic compromise.[99]

The Directive does two main things:

1. It lays down differentiated national contributions to the EU's 20% share of the renewables goal, based on the extremely heterogeneous baseline electricity generation conditions and potential exhibited by the various member states – a condition that will persist until at least the end of this decade. However, since all member states are required to implement support measures for their renewable energy development goals, the directive stipulates that the gap between the support costs in the various member states is to be kept within reasonable bounds. Against this backdrop, the aforementioned roadmap for 2030 is also indispensable, as it will – at least indirectly and despite any unavoidable cost differences – to some extent balance out the development, promote support cost harmonisation, and thus institute a modicum of convergence among the various member state financing instruments.[100]
2. Under the Directive, the member states retain the right to optimise their support instruments and adapt these instruments to the specific renewable energy development phase the state happens to be in – an approach which, it would seem, makes good sense, particularly in terms of allowing for learning-curve-driven optimisation of support instruments. The Renewable Energy Directive also stipulates that member states may agree on and make arrangements for the statistical transfer of a specified amount of energy from renewable sources from a state that has exceeded its development objectives to one that has not (Article 6), for joint projects between member states (Article 7) or for joint support schemes (Article

98 Scharpf (1999).

99 See footnotes 78 and 93.

100 Hildingsson et al. (2010:115).

11).[101] Competition resulting from electricity price differences can be avoided in particular via regional cooperation between neighbouring member states.

Once an extensive trans-European network has been established – an event unlikely to occur before the 2020s – it will be necessary to consider further medium-term Europeanisation of support instruments in an electricity market where renewables may well be the dominant force by this time.

### *IV. Development of the Trans-European Network*

Key to the expansion of renewable energies in the EU is development of a high-capacity trans-European network, or supergrid,[102] which would be overlaid on the existing grids and interconnectors (which would also need to be optimised) and would be chiefly composed of high-voltage direct current transmission (HVDC) lines, even if other technologies would be viable options. In order to establish this supergrid, it would be essential to expand North Sea grids, and in particular also to be able to leverage Norwegian and Swedish pump storage system potential.[103] According to the *Green Paper – Towards a Secure, Sustainable and Competitive European Energy Network,*[104] an offshore wind farm grid and an energy ring in the Mediterranean region are both crucially important projects for successful expansion of renewable energies.

In order to establish policies for a European infrastructure, or for the more limited trans-regional counterparts, we will need to find answers to the following key questions:

- Are the existing network-like and predominantly private sector cooperative arrangements sufficient; or do EU grid development policies need to be bolstered?
- In view of the growing proportion of wind and solar power being fed into the grid, do the current bottom-up grid planning processes get the job done, or are more robust and strategic planning goals and scenario-based planning processes needed?

---

101 Schöpe (2010); Ragwitz et al. (2012:46ff.).
102 Czisch (2009); Battaglini et al. (2008).
103 Woyte et al. (2008); European Environment Agency (2009); Lilliestam (2007).
104 European Commission (2008a and b).

- To what extent can market-driven grid expansion be stimulated? To what extent is public financing or at least risk mitigation measures necessary for such expansion?

### *1. Grid Development Players in the EU*

Grid planning and development activities fall within the province of transmission system operators, which can be either private sector or public sector enterprises and for which the organisational structures, duties (most of which involve coordination activities) and oversight at the EU level are governed by the internal electricity market directive and by Directive 2009/72/EC (implemented in Germany as the Stromhandelszugangsverordnung (StromhandelZVO)).

The 42 transmission system operators that in December 2008 founded the European Network of Transmission System Operators for Electricity (ENTSO-E) are required under EU law to submit, at two-year intervals, revised ten-year Community grid development plans. These plans are not legally binding and indicate, among other things, scenarios and forecasts concerning the adequacy of electricity generation, as well as areas where investments are needed (Article 8(10) of the StromhandelZVO law). As such plans take their cue from national ten-year plans, they constitute the main national plan coordination instrument.

Organisations such as Nordel (Organisation for the Nordic Transmission System Operators) – one of the ENTSO-E entities in charge of developing a cross-border regional grid investment plan – act as an intermediary instrument in this regard (Article 12 of the StromhandelZVO law), while the Agency for the Cooperation of Energy Regulators (ACER) provides advice and carries out oversight activities (Directive 713/2009/EC; law titled *ACER Verordnung*). A network agency that arose from informal cooperation between national regulatory authorities, ACER, along with its governing board, is composed of political appointees (named by the European Commission, the member states, and the European Parliament). ACER oversees the activities of key regulatory decision makers, provides support and coordination for national regulatory authority measures aimed at implementing the objectives of the internal electricity market, has far reaching competence in areas such as access modalities for cross-border infrastructures and work safety pursuant to Article 8 of the relevant regulation (*ACER Verordnung*), reviews ENTSO-E ten-year plans, and draws up a statement of position containing any changes deemed necessary in such plans (Article 8(11) of StromhandelZVO). These statements of position are not legally binding, and ACER has no say in or veto over their content. Although during the negotiating process concerning the internal European Electricity Market Directive it proved impossible to give ACER greater say in these matters,[105] the European Commission has called for strengthening of ACER's competence in connection with the integrated energy market,[106] and thus ACER's competence in this domain could potentially expand over time. In this regard, the StromhandelZVO empowers the national regulatory authority jointly to delegate decision-making rights to ACER, which in some cases (such as incentives rules for interconnectors) is entitled to draw up proposed decisions for the European Commission. Hence ACER may assume a more important role going forward, particularly if the European Commission begins relying on ACER recommendations.[107]

The EU's trans-European network (TEN-E) policies also constitute a key, albeit weak, grid development policy instrument, whereby the TEN-E guidelines, which the European Council and Parliament adopted at the proposal of the European Commission, comprise the main statutory European infrastructure policy instrument. First adopted in 1996, the guidelines, which were amended in 2003, and in 2006 (via Decision No 1364/2006/EC), with a new proposal for revision pending since late 2011 (COM 2011, 659 final), mainly serve the following purposes: to formulate objectives (Article 3) and selection criteria for Community measures in the field of trans-European energy networks (Article 4); to identify corridors of European interest (Article 6), regulate priority projects (Article 7), and "ensure the interoperability of electricity networks" (Article 4(2)); and to adapt and develop networks "to facilitate the integration and connection of renewable energy production" (Article 4(2a)). The TEN-E guidelines are essentially a coordination and financing instrument for cross-border linkages, although they offer only very limited financial contributions to projects of common interest. According to Articles 6 and 9 of the guidelines, when it comes to projects of common interest it is incumbent upon the member states to facilitate and expedite their realisation (including the attendant approval procedures), to coordinate such projects, to submit completion schedules in their regard, and to report any delays in such completion. In this respect, the TEN-E guidelines mirror

105 Hancher & de Hauteclocque (2010).
106 European Commission (2010a).
107 Hancher & de Hauteclocque (2010:6).

current EU competences as laid down in Articles 170 to 172 of the TFEU (ex Articles 154 to 156 EGV), whose scope is limited to improved and trouble-free coordination of cross-border planning processes. In that regard, the European infrastructure package proposed by the European Commission in October 2011 is a step forward. Among others, this package contains new financing instruments and revised guidelines for the TEN-E, which are based upon Article 172 TFEU.[108] The new guidelines contain a number of new instruments and governance approaches, which intend to improve and speed-up the realisation of interconnectors. Among others, four priority corridors for electricity are identified, which are considered projects of common interest and receive priority status in national permitting procedures (Article 8). A project developer – normally a Transmission System Operator (TSO) or a consortium of TSOs – get management and planning responsibility for the project, including for keeping agreed schedules and reporting (Article 5). Progress is monitored and sanctions established for delays. In case of implementation difficulties a 'European coordinator' will be mandated to overcome any difficulties and hurdles (Article 6). Permitting takes place according to the 'one-stop-shop' principle by one central authority (Article 9). According to Article 10, minimum requirements for public participation and consultation are formulated. The new guidelines also contain rules on how to cover investment costs. As a principle, costs are covered on the basis of the 'key beneficiary pays' principle (Article 13, 1). The different national regulatory authorities are requested to find an agreement on how to share investment costs and revenue among the participating TSOs. Also, provisions are created for projects which are considered to be especially risky (Article 14) or for projects which may receive additional Community support (Article 15). In total, those new governance mechanisms offer an overall framework which may be helpful to speed up investments in interconnectors. Factual implementation, however, will depend upon the way national regulatory authorities and TSOs make use of the new instruments, on how potential conflicts may be settled and which resources and capacity the European Regulator may mobilise to overcome problems.

Despite those improvements in terms of coordination and enforcements of projects of common interest, the EU has relatively little direct control to steer grid development, which, as it is mainly driven by the regulatory framework and the financial interests of transmission system operators, un-

108 European Commission (2011b); Schmitz & Jornitz (2012).

folds primarily as a bottom-up process; and thus only its coordination is under EU control. Hence, grid needs planning at the EU level reflects the incentive and planning frameworks for national grid regulation, including all their strengths and weaknesses. In view of the considerable investment risks and planning uncertainty inherent in the renewable energy development sector, such a bottom-up process is likely to prompt only private investors to plough large amounts of money into the development of high-voltage direct current transmission (HVDC) grids, where national frameworks offer long-term predictability both for renewables deployment and related grid planning. In principle the same applies for the EU framework beyond 2020.

As there are various ways to strengthen the hand of European actors in the electricity grid development arena, expanding ACER's competence would appear to be the best option (in conjunction with a comitology procedure), including when it comes to folding scenarios into a high-capacity transmission network plan.[109] To this end, key grid development needs should be laid down as soon as possible in amended TEN-E guidelines – although the success of this undertaking will be largely contingent on modifying the upstream needs analysis process.

### 2. *Needs of Analysis and Project Selection*

Electricity grid planning in Europe is mainly a needs analysis, project identification and bottom-up process involving information interchange and cross-border interconnection planning on the part of neighbouring states,[110] which, in this process, mainly rely on network development plans devised by transmission system operators;[111] whereby such plans ultimately form the basis for updated TEN-E recommendations. The remaining responsibilities are met by mechanisms of the regulated grid markets, which means that "the construction and maintenance of energy infrastructure should be subject to market principles" and that "Community financial aid for construction and maintenance should therefore remain highly exceptional, and such exceptions should be duly justified" (Recital 4, Decision No 1364/2006/EC); whereby exceptions include, in particular, high-voltage di-

---

109 European Climate Foundation (2010b:29).

110 European Academies Science Advisory Council (2009:5).

111 See StromhandelZVO 2009; UCTE (2009).

rect current transmission (HVDC) lines.[112] Projects are to be selected only insofar as a cost-benefit analysis indicates that they display "potential economic viability" (Article 5, Decision No 1364/2006/EC). The Commission Proposal for TEN-E guidelines contains improvements in the respect that very risky projects and projects with considerable positive externalities receive special treatment on the basis of Article 14 and 15.[113]

By dint of this bottom-up planning process alone, it has been shown that the 2006 TEN-E guidelines were sorely lacking when it comes to the development of grids for renewable energies; one example of this being that the 2006 guidelines do not contain a single mention of a high-voltage direct current transmission (HVDC) project of European interest.[114] According to a European Climate Foundation estimate, grid development between 2004 and 2009, which resulted in an aggregate European capacity increase of 12.6 GW, was considerably below the necessary development rate.[115]

Nonetheless the old TEN-E guidelines, as well as UCTE (Union for the Coordination of Transmission of Electricity, the precursor of ENTSO-E) plans, contain grid development projects that clearly undermine Community objectives, one example being transmission lines linking Tunisia and Sicily that put a coal-fired power station on line that was built mainly for the Italian market[116] (project 4.2.4 in Decision No 1364/2006/EC), with a view to avoiding the carbon certificate costs that would have been incurred had a new power plant been built in the emissions trading zone.

The European Academies Science Advisory Council (EASAC) – which has correctly pointed out that the current grid development planning process is highly unsatisfactory, particularly for the requisite renewable energy expansion process[117] – has recommended that the bottom-up planning process be paired with a scenario-based strategic planning process. If this approach is used, EASAC says, more accurate estimates of network development needs and the robustness of specific future scenarios could be obtained based on various future scenarios. EASAC signals in this regard the exemplary practice of NORDEL (Organisation for the Nordic Transmission System Operators), whose Grid Master Plan 2008 is based on three different sce-

112 See Article 17 of StromhandelZVO.
113 See footnote 110.
114 Holznagel & Schumacher (2009:168 and 170).
115 European Climate Foundation (2010b:28).
116 UCTE (2009:42).
117 European Academies Science Advisory Council (2009:5).

narios – namely business as usual, climate protection and integration, and national focus – which allow for determination of both internal and external grid development needs.[118] In the same vein, the European Climate Foundation *Roadmap 2050* calls for the grid development planning process to encompass a far longer period than is currently the case with a view to harmonising in the medium term presumed renewable energy capacity development and grid development needs.[119] ENTSO-E has also indicated that in the absence of clearly defined long-term climate protection and renewable energy capacity development goals, the organisation's members will simply be unable to elaborate electricity grid planning scenarios.[120] A far stronger and more target-oriented planning paradigm is needed so that the EU can send robust signals that will promote grid development for renewable energies. The cause of strengthening planning certainty and greatly reducing investment risk would be served if the scenarios awaiting elaboration could be largely based on mandatory development targets for renewable energies. Such an approach would also call for the use of scenario design backcasting methods, which appear to be more suitable for target-oriented planning than conventional trend and policy scenarios.

Although amending the TEN-E guidelines[121] is a step in the right direction, it would not do enough to reduce the influence of the major market players on grid planning outcomes. Hence it is essential that the European Commission or a subsidiary body acquires the wherewithal to carry out an independent grid development needs analysis for 2020 and 2030 in light of the policy goal of expanding renewable energies, and that this analysis be harmonised with transmission system operator plans. Inasmuch as transitioning to a wholly or largely renewable electricity supply is a primarily policy-driven undertaking, in keeping with EU Treaty tenets, the EU's governing bodies need to acquire the competence also to evaluate market-driven plans and to amend them in the light of the EU's renewable development policies.

118 Organisation for the Nordic Transmission System Operators (2008).
119 European Climate Foundation (2010b:29).
120 ENTSO-E (2010:9 and 45).
121 Holznagel & Schumacher (2009:170).

### 3. *Financing*

EU subsidies cover only a minute proportion of the cost of electricity grid development for priority projects, as well as possibly risky large-scale projects, such as those involving high-voltage direct current transmission (HVDC) lines. Such finance is particularly meagre for preliminary studies and for undertakings involving common structural policy. The €22 million annual trans-European network (TEN-E) budget for the period 2007 to 2013 can only be described as Lilliputian. Even though the Connecting Europe initiative, as presented by the Commission in October 2011,[122] would mean a major increase of available funds to €9,1 billion for the period 2014 to 2020, it is still minimal compared to the expected €140 billion investments for the high-voltage linkages only. Furthermore, the Commission investment plan is still under scrutiny in the context of the very difficult negotiations on the multi-annual budget for the forthcoming period. European Investment Bank loans amounting to €1,135 million annually for 2007 to 2009 are more generous, however; as is cohesion-policy financial support of €223 million a year. There was also at one time a European economic stimulus programme grant of nearly €4 billion that was partly used for grid infrastructures.[123] Despite the European Commission's view that grid infrastructure investments are mainly incumbent upon private sector network operators (i.e. investment decisions should be primarily market-driven), the Commission nonetheless recognises the need for such investments to be supplemented by public funding for non-commercial objectives in projects such as underground cables for environmental purposes, and the incorporation of renewable energies into the electricity grid.[124] In the same vein, the European Parliament and Council have underlined the importance of robustly promoting investments in large-scale infrastructures, particularly in view of the exceptionally high-risk profile that such investments entail (Recital 23 StromhandelZVO). It is for this reason that the said regulation exempts investors who are willing to invest in high-voltage direct current transmission (HVDC) lines from the differentiation requirements of the internal electricity market directive, subject to review by the agency. However, it is doubtful whether such a derogation – whose aim, of course, is to promote renewable energy capacity expansion investments by large investors – will be a sufficient in-

122 European Commission (2011c).
123 Proprietary calculations, derived from European Commission (2010b).
124 European Commission (2008a:12).

centive.[125] In the view of the European Commission, far more comprehensive public financing instruments and risk mitigation measures will be needed to promote grid expansion, particularly in the renewable energy sphere.

In the interest of establishing a high-voltage overlay network, we recommend that public contracts be awarded, for point-to-point connections, to the bidder that offers the requisite investments in conjunction with the lowest grid charges over a 20-year period. This tendering procedure could also be used for cross-border connections between member states, whereby measures that facilitate cooperation between member states for the cost-sharing arrangements, as suggested by the Commission, would be particularly useful. It should also be determined whether set EU procedures containing a number of standardised elements aimed at expediting joint tenders for key cross-border connection contracts would also be useful and could help to expedite the process.

## *D. Conclusions*

Article 194(1) of the TFEU grants the EU competence as regards the following energy policy goals: (a) ensuring the functioning of the energy market; (b) ensuring security of energy supply in the Union; (c) promoting energy efficiency and energy saving and developing new and renewable forms of energy; and (d) promoting the interconnection of energy networks.

In terms of renewable energies, Article 194(1) expands the scope of EU energy competence solely in respect to promoting technological development, and thus all remaining aspects of renewable energies still fall within the environmental competence laid down in the Treaty's Articles 192(1) and (2) – which are therefore also governed by the "more stringent protective measures" clause of the Treaty's Article 193, thus leaving the member states some leeway to institute measures as they see fit, despite EU legislation.

Thanks to the EU's environmental competence pursuant to Article 192(1) and (2) of the TFEU, the EU is entitled to set requirements for the member states concerning the aspects of renewable electricity expansion capacity, but to the exclusion of the relatively minor and specialised sphere of promoting technological development. EU measures pursuant to Article 192(2) (c) of the TFEU reach their statutory procedural limit insofar as they signifi-

125 Holznagel & Schumacher (2009).

cantly affect "a Member State's choice between different energy sources and the general structure of its energy supply," whereby such measures must be adopted by a unanimous vote of the European Council. This is the key change brought by the EU's new energy policy competence under Article 194(2)(2) – which, unlike the Treaty's purely procedural provisions in Article 192(2)(c), constitutes a genuine competence delineation. Consequently, the EU has no authority over non-environmental energy policy measures that fall within the competence of the member states.

However, it is no easy matter to determine exactly which types of measures are governed by Article 192(2)(c) of the Treaty, particularly when it comes to the share of energy from renewable sources that are mandated for the various member states. But any decision that institutes a durable all-renewables electricity supply would in any case necessitate a unanimous vote. Under the provisions of Article 193 of the Treaty, the member states are entitled to exceed the share of energy from renewable sources stipulated by the EU.

The EU's authority over the electricity transmission network expansion necessary for a wholly renewable electricity supply is expanded on in Article 194 of the Treaty, particularly in terms of the interconnection of energy networks, whose expansion is one of the lynchpins of the internal European electricity market. The EU's competence for the promotion of grid interconnection is reaching further than the trans-European network competence accorded by Article 172 of the TFEU. Nevertheless, the EU's network interconnection financing competence is limited to coordination measures for existing networks or to financing ongoing network projects that are already being subsidised by one or more member states. Hence, save for cross-border network interconnections, the EU is prohibited from imposing on the member states any measure involving transmission network expansion exceeding the scope of that which is in the pipeline in the member states at any given time. However, this restriction also has an upside – namely that the EU can use guidelines as an instrument to coordinate and finance measures aimed at expansion of cross-border networks, and can thus further the cause of expanding such networks to the requisite degree. As a result of this situation, network expansion is mainly the legal responsibility of private transmission system operators. Carrying out such planning at the European level is not mandatory, but instead mainly allows for coordination and consultation, and in some cases information-related revision, of member state transmission network plans from a European perspective. Bolstering EU policies with a view to promoting network expansion will need mainly to focus on successfully interconnecting the various national networks – a goal that will, however, open up considerable member state leeway.

Article 194(1)(c) of the TFEU endows the EU with far-reaching (albeit not new) authority over promoting energy efficiency and saving energy. The extent to which Article 194 of the TFEU empowers the member states to adopt more stringent energy efficiency policies than those mandated by the EU is open to question. In our view, however, the member states are not entitled to adopt "more stringent protective measures" in this regard within the meaning of Article 193 of the TFEU.

The statutory grounds for energy efficiency provisions, measures and programmes have traditionally been Article 175(1) ECT (now Article 192 of the TFEU) or Article 95 ECT (now Article 192 of the TFEU), both of which empower the member states to introduce "more stringent protective measures". However, the member states are not empowered to do so under Article 194 of the TFEU, which lays down the EU's new competence for energy efficiency.

This problem can only be resolved by either applying the more stringent protective measures clause of Article 193 of the TFEU (ex Article 176 of the Treaty establishing the European Union) in accordance with Article 194[126] or incorporating such a clause into future energy efficiency legislation. Such an application of Article 193 would probably be inadmissible, since the existence of a statutory loophole for an area in which the EU intends to find a definitive solution cannot be presumed. Hence EU energy efficiency regulations that are based on Article 194 of the TFEU should expressly empower the member states to enact more stringent protective measures. One example of such a regulation in the realm of energy efficiency is the Directive on energy end-use efficiency and energy services (2006/32/EC), which expressly empowers the member states to set a higher national energy-saving objective than that laid down in the Directive's 13th recital.

EU renewable energy support policy needs to develop within this framework of these competences. The key policy areas that come into play here are climate protection, meeting renewable energy development goals, and adapting the trans-European network in a timely manner to a higher proportion of renewables.

It is essential that renewable energy capacity expansion and the expansion of incentive and subsidy programmes are keyed to statutory medium-term

126 Britz (2009).

EU climate objectives, whose touchstone should be the position taken by the European Council in October 2009 and the European Commission's Decarbonisation Roadmap 2050, according to which greenhouse gas reductions of at least 80% in 2050 compared to 1990 levels are on the EU policy agenda. This is the minimum ambition level that is consistent with the global reduction of greenhouse gases needed to achieve the 2° Celsius objective. In order to implement the reduction path necessary for this objective and at the same time avoid investment missteps in the run-up to 2020, a minimum 30% reduction target will be necessary for 2020.

The Renewable Energy Directive of 2009 will go a long way toward keeping renewable energy capacity expansion on track for the remainder of this decade and achieving partial convergence of renewable energy support schemes. This policy should be extended beyond 2020. A European roadmap that lays down a framework for renewable energy expansion up to 2030 should be developed, particularly in terms of national and European infrastructure development beyond 2030. Moreover, EU support schemes for renewable energy should take account of the subsidiarity principle and should allow EU member states sufficient leeway, but in a manner that is compatible with Community principles. The Renewable Energy Directive sets an overall goal for the share of renewable sources to primary energy consumption, which will effectively lead to a 35% share of electricity from renewable sources in 2020, while allowing for differences in the various member states' contribution to achievement of this goal. In addition, the Directive allows, and indeed encourages, the member states to enter into cooperative regional arrangements that could potentially resolve problems associated with cross-border electricity trading and joint infrastructure projects. Priority should be given to forge such alliances.

Member state grid expansion should be accompanied by intensified needs planning at the EU level. Despite the indisputably key European dimension of grid expansion in general and the development of high-voltage direct current transmission (HVDC) grids or equally high-capacity technologies in particular, EU policy instruments in this domain are in need of being further strengthened. Grid expansion is chiefly market-driven and for the most part is realised by merging national ten-year plans. Those plans mainly mirror national planning systems and the incentive effects of national market regulations and the interests of the various grid operators. Only by way of exception (e.g. in Germany) do they reflect the need to transition to a wholly or largely renewable electricity supply over the long run. And, while this approach to grid expansion planning may suffice for incremental development of the electricity supply, it cannot hope to bring about the requisite long-term, target-oriented transformation. On the other hand, continued renewable energy capacity expansion will make it indispensable to strengthen the policymaking hand of all supranational European players – namely the European Commission, the European Parliament, and the recently established European Agency for the Cooperation of Energy Regulators.

In this regard, member state grid expansion programmes should be strengthened via improved coordination, notably as regards cross-border expansion needs for renewables and high-capacity, long-distance connections, whereby such efforts should focus on the following in particular:

- More tightly intermeshed coordination of renewable energy expansion and grid planning measures for the post-2020 period.
- The European Commission or its subordinate authorities should conduct dedicated needs analyses, based on information from transmission network operators, concerning expansion and optimisation of the trans-European grid, with a view to achieving efficient quality assurance for EU energy policy objectives.
- Cross-border cooperation for public contracts, and notably for new cross-border high-capacity, long-distance connections, should be intensified.
- The groundwork should be laid for regional cooperation among grid operators, notably in the North Sea and Mediterranean.
- national remuneration systems for renewable energies should be further strengthened in the view of the European and national targets.

The policy framework for all those measures is gradually evolving and merits broad political support in the view of the emerging agenda of the EU on climate, renewable and efficiency targets for 2030.

## *References*


Barbu, Anca-Diana, 2007, *Investing in Renewable Energy Sources, Wirtschaftspolitik in Forschung und Praxis*, Hamburg, Verlag Dr. Kovač.

Battaglini Antonella, Nadejda Komendantova, Johan Lilliestam & Anthony Patt, 2008, *Linking Noth Africa's Renewable Energy Resources to Europe*, Potsdam, Potsdam Institute for Climate Impact Research, available at http://www.supersmartgrid.net/wp-content/uploads/2008/11/background_paper.pdf, last accessed 08 March 2013.

Berge, Hans ten & Sam Cross, 2010, Renewable Energy and Beyond: Delivering on the EU Targets and Defining Pathway to a Low Carbon Energy Future, in: Fouquet, Dörte, *EU Energy Law*, Vol. 3 (3), Leuven, Claeys & Casteels Publishers, 101–152.

Blanke, Hermann-Josef, 2004, Die Zuständigkeit der Union, *Zeitschrift für Gesetzgebung* 19, 225–249.

Bogs, Eckhard, 2002, *Die Planung transeuropäischer Netze*, Berlin, Duncker & Humblot.

Britz, Gabriele, 2009, Klimaschutzmaßnahmen der EU und der Mitgliedstaaten, in: Schulze-Fielitz, Helmuth & Thorsten Müller (Eds), *Europäisches Klimaschutzrecht*, Baden-Baden, Nomos Verlagsgesellschaft, 71–90.

Bundesministerium für Wirtschaft und Technologie & Bundesministerium für Umwelt, Naturschutz und Reaktorsicherheit, 2010, *Energiekonzept für eine umweltschonende, zuverlässige und bezahlbare Energieversorgung*, Berlin, Bundesministerium für Wirtschaft und Technologie & Bundesministerium für Umwelt, Naturschutz und Reaktorsicherheit.

Calliess, Christian, 1999, *Subsidiaritäts- und Solidaritätsprinzip in der Europäischen Union* (Second Edition), Baden-Baden, Nomos Verlagsgesellschaft.

Calliess, Christian, 2008, *Entflechtung im europäischen Energiebinnenmarkt*, Stuttgart, Boorberg.

Calliess, Christian, 2010, Sinn, Inhalt und Reichweite einer europäischen Kompetenz zur Energieumweltpolitik, in: Cremer, Wolfram & Johann-Christian Pielow (Eds), *Probleme und Perspektiven im Energieumweltrecht*, Bochumer Beiträge zum Berg- und Energierecht Band 54, Stuttgart, Boorberg.

Calliess, Christian, 2011a, Artikel 192, in: Calliess, Christian & Matthias Ruffert (Eds), *Das Verfassungsrecht der Europäischen Union mit Europäischer Grundrechtecharta – Kommentar* (Fourth Edition), München, C.H. Beck.

Calliess, Christian, 2011b, Artikel 194, in: Calliess, Christian & Matthias Ruffert (Eds), *Das Verfassungsrecht der Europäischen Union mit Europäischer Grundrechtecharta – Kommentar* (Fourth Edition), München, C.H. Beck.

Classen, Claus D., 2003, The Draft Treaty Establishing a Constitution for Europe, *German Yearbook of International Law* 46, 323–352.

Council of the European Union, 2009, *Presidency Conclusions of the Brussels European Council*, (29/30 October 2009), 15265/1/09, available at http://www.consilium.europa.eu/uedocs/cms_data/docs/pressdata/en/ec/110889.pdf, last accessed 08 March 2013.

Council of the European Union, 2012a, *Council Conclusions on Renewable Energy*, 2304[th] Transport, Telecommunications and Energy Council Meeting, Brussels, 3 December 2012.

Council of the European Union, 2012b, *Presidency Conclusions on the Energy Road Map 2050*, ENER 320, Brussels, Council of the European Union, 18 June 2012, available at http://register.consilium.europa.eu/pdf/en/12/st11/st11553.en12.pdf, last accessed 07 March 2013.

Czisch, Gregor & Jürgen Schmid, 2007, *Mitigation Country Study for Germany, Human Development Report*, Geneva, United Nations Development Programme, available at http://hdr.undp.org/en/reports/global/hdr2007-8/papers/Czisch_Schmid.pdf, last accessed 08 March 2013.

Czisch, Gregor, 2009, *Möglichkeiten des großräumigen (transeuropäischen) Ausgleichs von Schwankungen großer Teile intermittierender Elektrizitätseinspeisung aus regenerativen Energiequellen in Deutschland im Rahmen einer 100% regenerativen Stromerzeugung mit dem Zeithorizont 2050*, Berlin, Sachverständigenrat für Umweltfragen, available at http://www.umweltrat.de/SharedDocs/Downloads/DE/03_Materialien/2010_MAT40_Czisch.pdf?__blob=publicationFile, last accessed 07 March 2013.

De Sadeleer, Nicolas, 2012, Principle of Subsidiarity and the EU Environmental Policy, *Journal for European Enviromental and Planning Law*, 63–70.

Department of Energy and Climate Change, 2009, *Draft Overreaching national Policy Statement for Energy (EN-1), Presented to Parliament pursuant to section 5(9b) of the Planning Act 2008*, London, Department of Energy and Climate Change.

Edenhofer, Ottmar, Carlo Carraro, Jean-Charles Hourcade & Karsten Neuhoff, 2009, *Report on Energy and Climate Policy in Europe (RECIPE) – The Economics of Decarbonization*, Potsdam, Potsdam Institut für Klimafolgenforschung, available at http://www.pik-potsdam.de/research/sustainable-solutions/research/ClimatePolicies/recipe-groupspace/working-papers/recipe-synthesis-report, last accessed 08 March 2013.

Ehricke, Ulrich & Daniel Hackländer, 2008, Europäische Energiepolitik auf der Grundlage der neuen Bestimmungen des Vertrags von Lissabon, *Zeitschrift für Europarechtliche Studien*, 11 (4), 579–600.

Eichener, Volker, 2000, *Das Entscheidungssystem der Europäischen Union. Institutionelle Analyse und demokratietheoretische Bewertung*, Opladen, Leske & Budrich Verlag.

ENTSO-E/European Network of Transmission System Operators for Electricity, 2010, *TEN-Year Network Development Plan 2010-2020*, Brussels, ENTSO-E, available at https://www.entsoe.eu/index.php?id=232, last accessed 08 March 2013.

Epiney, Astrid, 2005, *Umweltrecht in der Europäischen Union* (Second Edition), Köln, Heymanns.

Erdmenger, Jürgen, 2003, Artikel 155, in: von der Groeben, Hans & Jürgen Schwarze (Eds), *Kommentar zum Vertrag über die Europäische Union und zur Gründung der Europäischen Gemeinschaft* (Sixth Edition), Baden-Baden, Nomos Verlagsgesellschaft.

EURELECTRIC, 2010, *Power Choices: Pathways to a Carbon-Neutral Electricity in Europe by 2050*, Brussels, EURELECTRIC, available at www.eurelectric.org/powerchoices2050, last accessed 07 March 2013.

European Academies Science Advisory Council, 2009, *Transforming Europe's Electricity Supply, An Infrastructure Strategy for a Reliable, Renewable and Secure Power System*, EASAC Policy Report, London, European Academies Science Advisory Council.

European Climate Foundation, 2010a, *A Practical Guide to a Prosperous, Low-Carbon Europe*, Vol. 1, available at http://www.roadmap2050.eu/attachments/files/Volume1_fullreport_PressPack.pdf, last accessed 07 March 2013.

European Climate Foundation, 2010b, *A Practical Guide to a Prosperous, Low-Carbon Europe*, Vol. 2, available at http://www.roadmap2050.eu/attachments/files/Volume2_Policy.pdf, last accessed 07 March 2013.

European Commission, 2000, *Die jüngsten Fortschritte bei der Schaffung eines Elektrizitätsbinnenmarktes*, Mitteilung der Europäischen Kommission an den Europäischen Rat und das Europäische Parlament, Brussels, European Commission, KOM (2000) 297 final, 16 May 2000, available at http://eur-lex.europa.eu/LexUriServ/LexUriServ.do?uri=CELEX:52000DC0297:DE:HTML, last accessed 07 March 2013.

European Commission, 2006, *European Energy and Transport – Scenarios on Energy Efficiency and Renewables*, Luxembourg, European Commission, available at http://ec.europa.eu/energy/observatory/trends_2030/doc/ee_and_res_scenarios.pdf, last accessed on 08 March 2013.

European Commission, 2007, *An Energy Policy for Europe*, Communication from the Commission to the European Council and the European Parliament, Brussels, European Commission COM (2007) 1 final, 10 January 2007, available at http://ec.europa.eu/energy/energy_policy/doc/01_energy_policy_for_europe_en.pdf, last accessed 07 March 2013.

European Commission, 2008a, *Green Paper: Towards a Secure, Sustainable and Competitive European Energy Network*, Brussels, European Commission, COM (2008) 782 final, 13 November 2008, available at http://eur-lex.europa.eu/LexUriServ/LexUriServ.do?uri=COM:2008:0782:FIN:EN:PDF, last accessed 07 March 2013.

European Commission, 2008b, *Offshore Wind Energy: Action Needed to Deliver on the Energy Policy Objectives for 2020 and Beyond*, Communication from the Commission to the European Parliament, the Council, the European Economic and Social Committee and the Committee of the Regions, Brussels, European Commission COM (2008) 768 final, 13 November 2008.

European Commission, 2009, *The Renewable Energy Progress Report: Commission Report in accordance with Article 3 of Directive 2001/77/EC, Article 4(2) of Directive 2003/30/EC and on the implementation of the EU Biomass Action Plan, COM (2005) 628*, Communication from the Commission to the Council and the European Parliament, Brussels, European Commission, COM (2009) 192 final, 24 April 2009.

European Commission, 2010a, *Stock Taking Document: Towards a New Energy Strategy for Europe 2011-2020*, Brussels, European Commission, available at http://ec.europa.eu/energy/strategies/consultations/doc/2010_07_02/2010_07_02_energy_strategy.pdf, last accessed 08 March 2013.

European Commission, 2010b, *The Implementation of the Trans-European Energy Networks in the Period 2007-2009, Pursuant to Article 17 of Regulation (EC) 680/2007 and Articles 9(2) and 15 of Decision 1264/2006/EC*, Report from the European Commission to the European Parliament, the Council, the European Economic and Social Committee and the Committee of the Regions, Brussels, European Commission, COM (2010) 203 final, 04 May 2010.

European Commission, 2010c, *Analyse der Option zur Verringerung der Treibhausgasemissionen um mehr als 20 % und Bewertung des Risikos der Verlagerung von $CO_2$-Emissionen,* Mitteilung der Europäischen Kommission an das Europäische Parlament und den Rat, und den Wirtschafts- und Sozialausschuss und den Rat der Regionen, Brussels, European Commission, COM (2010) 265 final, 26 May 2010.

European Commission, 2011a, *A Roadmap for Moving to a Competitive Low Carbon Economy in 2050,* Communication from the Commission to the European Parliament, the Council, the European Economic and Social Committee and the Committee of the Regions, Brussels, European Commission COM (2011) 112 final, 8 March 2011.

European Commission, 2011b, *Proposal for a Regulation on Guidelines for Trans-European Energy Infrastructure and Repealing Decision No 1364/2006/EC,* Brussels, European Commission, COM (2011) 658 final, 19 October 2011.

European Commission, 2011c, *Proposal for a Regulation of the European Parliament and the Council Establishing the Connecting Europe Facility,* Brussels, European Commission COM (2011) 665, 19 October 2011.

European Commission, 2011d, *Energy Roadmap 2050,* Communication from the Commission to the European Parliament, the Council, the European Economic and Social Committee and the Committee of the Regions, Brussels, European Commission, COM (2011) 885 final, 15 December 2011.

European Commission, 2011e, *Roadmap to a Single European Transport Area – Towards a competitive and Resource Efficient Transport System,* White Paper, Communication from the Commission to the European Parliament, the Council, the European Economic and Social Committee and the Committee of the Regions, Brussels, European Commission, COM (2011) 144 final, 28 March 2011.

European Commission, 2012a, *Renewable Energy: a Major Player in the European Energy Market,* Communication from the Commission to the European Parliament, the Council, the European Economic and the Social Committee and the Committee of the Regions, Brussels, European Commission, COM (2012) 271 final, 6 June 2012.

European Commission, 2012b, *Making the Internal Market for Energy Work,* Communication from the Commission to the European Parliament, the Council, the European Economic and the Social Committee and the Committee of the Regions, Brussels, European Commission, COM (2912) 663, 15 November 2012.

European Environment Agency, 2009, *Europe's Onshore and Offshore Wind Energy Potential,* Luxembourg, European Environment Agency, available at http://www.eea.europa.eu/publications/europes-onshore-and-offshore-wind-energy-potential, last accessed 08 March 2013.

European Environment and Sustainable Development Advisory Councils, 2009, *Towards Sustainable European Infrastructures,* Brussels, European Environment and Sustainable Development Advisory Councils, available at http://ec.europa.eu/dgs/secretariat_general/eu2020/docs/eeac_en.pdf, last accessed 08 March 2013.

European Renewable Energy Council, 2010, *RE-thinking 2050 – A 100% Renewable Energy Vision for the European Union,* Brussels, European Renewable Energy Council.

Fouquet, Dörte & Thomas B. Johansson, 2008, European Renewable Energy Policy at Crossroads, *Energy Policy,* 4079–4092.

Fürsch, Michaela, Christiane Golling, Marco Nicolosi, Ralf Wissen & Dietmar Lindenberger, 2010, *European RES-E Policy Analysis*, Köln, Institute of Energy Economics at the University of Cologne, available at http://www.ewi.uni-koeln.de/fileadmin/user_upload/Publikationen/Studien/Politik_und_Gesellschaft/2010/EWI_2010-04-26_RES-E-Studie_Teil 1.pdf, last accessed 08 March 2013.

Geden, Oliver & Severin Fischer, 2008, *Die Energie- und Klimapolitik der Europäischen Union*, Baden-Baden, Nomos Verlagsgesellschaft.

Geden, Oliver & Severin Fischer, 2012, EU-Energy Roadmap 2050 – Surrogat für eine ehrgeizige Dekarboniserungspolitik?, *Energiewirtschaftliche Tagesfragen* 10, available at http://www.et-energie-online.de/AktuellesHeft/Topthema/tabid/70/Year/2012/Month/10/NewsModule/423/NewsId/331/EUEnergy-Roadmap-2050--Surrogat-fur-eine-ehrgeizige-Dekarbonisierungspolitik.aspx, last accessed 07 March 2013, 41–44.

Görlitz, Niklas, 2004, Europäischer Verfassungsvertrag und künftige EU-Kompetenzen, *Die Öffentlich Verwaltung*, 374–383.

Götz, Volkmar, 2004, Kompetenzverteilung, in: Schwarze, Jürgen (Ed.), *Der Verfassungsentwurf des Europäischen Konvents*, 2004, Baden-Baden, Nomos Verlagsgesellschaft, 43–61.

Guerry, Anaïs, 2012, Die Wirkung von Fukushima auf die Energiepolitik in Frankreich, *Die Zeitschrift für Neues Energierecht* 3, 244–247.

Gundel, Jörg, 2008, Nachhaltigkeit und Energieversorgung, in: Kahl, Wolfgang (Ed.), *Nachhaltigkeit als Verbundbegriff*, Tübingen, Mohr Siebeck Verlag, 443–470.

Hancher, Leigh & Adrien Hauteclocque, 2010, *Manufacturing the EU Energy Markets: The Current Dynamics of Regulatory Practice*, European University Institute Working Papers, Florence, Robert Schumann Centre for Advanced Studies, available at http://cadmus.eui.eu/bitstream/handle/1814/13077/RSCAS_2010_01.pdf?sequence=1, last accessed 08 March 2013.

Härtel, Ines, 2006, *Handbuch Europäische Rechtsetzung*, Berlin, Springer.

Heemeyer, Uwe, 2004, *Kompetenzordnung eines zukünftigen europäischen Verfassungsvertrags*, Münster, LIT Verlag.

Helm, Dieter R., 2006, Großbritanniens nostalgische Energiepolitik, in: Braml, Josef, Karl Kaiser, Hanns W. Maull, Eberhard Sandschneider & Klaus-Werner Schatz (Eds), *Weltverträgliche Energiesicherheitspolitik*, Jahrbuch Internationale Politik, Oldenburg, Oldenbourg Wissenschaftsverlag, 204–209.

Hey, Christian, 2012, Low-carbon and Energy Strategies for the EU. The European Commisions´s Roadmaps: A Sound Agenda for a Green Economy?, in: *GAIA – Ecological Perspectives for Science and Society* 21 (1), 43–47.

Hildingsson, Roger, Johannes Stripple & Andrew Jordan, 2010, Renewable Energies: A Continuing Balancing Act?, in: Jordan, Andrew, Dave Huitema, Harro van Asselt, Tim Rayner & Frans Berkhout (Eds), *Climate Change Policy in the European Union. Confronting the Dilemmas of Mitigation Policy an Adaptation?*, Cambridge, Cambridge University Press, 103–124.

HM Government, 2009, *The UK Low Carbon Transition Plan. National Strategy for Climate and Energy: Presented to Parliament pursuant to Sections 12 and 14 of the Climate Change Act 2008*, London, HM Government, available at http://centralcontent.fco.gov.uk/central-content/campaigns/act-on-copenhagen/resources/en/pdf/DECC-Low-Carbon-Transition-Plan, last accessed 08 March 2013.

Holm-Müller, Karin & Michael Weber, 2010, *Plädoyer für eine instrumentelle Flankierung des Emissionshandels im Elektrizitätssektor*, available at http://www.umweltrat.de/SharedDocs/Downloads/DE/06_Hintergrundinformationen/2010_06_Emissionshandel_Strom.pdf, last accessed 07 March 2013.

Holznagel, Bernd & Pascal Schumacher, 2009, Wüstenstrom aus der Sahara für die finnische Sauna, *Natur und Recht* 31 (3), 164–170.

Jasper, Maren, 2003, Der Verfassungsentwurf des europäischen Konvents und mögliche Konsequenzen für das Energie- bzw. Atomrecht, *Zeitschrift für Neues Energierecht*, 210–213.

Jones, Christopher, 2010, A Zero Carbon Energy Policy for Europe, in: Fouquet, Dörte, *EU Energy Law*, Vol. 3 (3), Leuven, Claeys & Casteels Publishers, 21–101.

Jordan, Andrew, Dave Huitema & Harro van Asselt, 2010a, Climate Change Policy in the European Union: An Introduction, in: Jordan, Andrew, Dave Huitema, Harro van Asselt, Tim Rayner & Frans Berkhout (Eds), *Climate Change Policy in the European Union. Confronting the Dilemmas of Mitigation Policy an Adaptation?*, Cambridge, Cambridge University Press, 3ff.

Jordan, Andrew, Dave Huitema, Tim Rayner & Harro van Asselt, 2010b, Governing the European Union: An Introduction, in: Jordan, Andrew, Dave Huitema, Harro van Asselt, Tim Rayner & Frans Berkhout (Eds), *Climate Change Policy in the European Union. Confronting the Dilemmas of Mitigation Policy an Adaptation?*, Cambridge, Cambridge University Press, 29ff.

Kahl, Wolfgang, 2009, Energie und Klimaschutz, in: Schulze-Fielitz, Helmuth & Thorsten Müller (Eds), *Europäisches Klimaschutzrecht*, Baden-Baden, Nomos Verlagsgesellschaft, 21–70.

Kahl, Wolfgang, 2011, Artikel 114, in: Calliess, Christian & Matthias Ruffert (Eds), *Das Verfassungsrecht der Europäischen Union mit Europäischer Grundrechtecharta – Kommentar* (Fourth Edition), München, C.H. Beck.

Kahl, Wolfgang, 2012, Artikel 192, in: Streinz, Rudolf (Ed.), *Vertrag über die Europäische Union und Vertrag über die Arbeitsweise der Europäischen Union – Kommentar* (Second Edition), München, C.H. Beck.

Käller, Anja, 2012, Artikel 192, in: Schwarze, Jürgen (Ed.), *EU-Kommentar* (Third Edition), Baden-Baden, Nomos Verlagsgesellschaft.

Kingreen, Thorsten, 2011, Artikel 345, in: Calliess, Christian & Matthias Ruffert (Eds), *Das Verfassungsrecht der Europäischen Union mit Europäischer Grundrechtecharta – Kommentar* (Fourth Edition), München, C.H. Beck.

Kirchhof, Ferdinand & Iris Kemmler, 2003, Einstimmigkeitserfordernis im Rat bei der Beschlussfassung über eine europäische Richtlinie zum Handel mit Treibhausgasemissionsberechtigungen, *Europäisches Wirtschafts- und Steuerrecht* 14 (5), 217–223.

Koenig, Christian & Michael Scholz, 2002, Die Förderung transeuropäischer Netzinfrastrukturen, *Europäisches Wirtschafts- und Steuerrecht* 5, 223-229.

Koopman, Martin, 2008, Französische Perspektiven in der Energie- und Klimapolitik, in: Braml, Josef, Karl Kaiser, Hanns W. Maull, Eberhard Sandschneider & Klaus-Werner Schatz (Eds), *Weltverträgliche Energiesicherheitspolitik*, Jahrbuch Internationale Politik, Oldenburg, Oldenbourg Wissenschaftsverlag, 196–203.

Lamprecht, Franz, 2009, Strommärkte 2050: Smart, integriert und $CO_2$-neutral, *Energiewirtschaftliche Tagesfragen* 59 (9), 22–24.

Lilliestam, Johan, 2007, *The Creation of the Pan-Nordic Electricity Market*, Potsdam, Potsdam-Institut für Klimafolgenforschung, available at http://www.supersmartgrid.net/wp-content/uploads/2008/06/lilliestam-2007-the-creation-of-the-pan-nordic-electricity-market.pdf, last accessed 08 March 2013.

Matthes, Felix C., 2012, Langfristperspektiven der europäischen Energiepolitik- die Energy Roadmap 2050 der Europäischen Union, *Energiewirtschaftliche Tagesfragen* 62 (1-2), 50–53.

Mez, Lutz, Mycle Schneider & Steve Thomas (Eds), 2009, *International Perspectives on Energy Policy and the Role of Nuclear Power*, Brentwood, Multiscience.

Müller-Kraenner, Sascha & Susanne Langsdorf, 2012, *A European Union for Renewable Energy – Policy Options for better Grids and Support Schemes*, Brussels, Heinrich-Böll-Foundation, available at http://www.boell.eu/downloads/HBS-EU_Renewables_web.pdf, last accessed 08 March 2013.

Neveling, Stefanie, 2004, Der Europäische Verfassungsentwurf, Grundlage für eine erweiterte Energiepolitik der EU?, *Energiewirtschaftliche Tagesfragen* 54, 340–345.

Öko-Institut & Prognos AG, 2009, *Modell Deutschland – Klimaschutz bis 2050*, Basel, Freiburg, Öko-Institut & Prognos AG, available at http://www.oeko.de/oekodoc/971/2009-003-de.pdf, last accessed 08 March 2013.

Olivier, Jos G.J., Willemijn Tuinstra, Hans E. Elzenga, Ruud A. van den Wijngaart, Peter R. Bosch, Bas Eickhout & Marcel Visser, 2008, *Consequences of the European Policy Package on Climate and Energy*, Bilthoven, Netherlands Environmental Assessment Agency, available at http://www.pbl.nl/sites/default/files/cms/publicaties/500094009.pdf, last accessed 07 March 2013.

Organisation for the Nordic Transmission System Operators, 2008, *Nordic Grid Master Plan 2008*, Brussels, Organisation for the Nordic Transmission System Operators.

Pellion, Antoine, 2008, *Renouveler la Production d'énergie en Europe: un défi environnemental, industriel et politique*, Paris, Fondation Robert Schuman.

Pernice, Ingolf, 1993, Umweltschutz und Energiepolitik, in: Rengeling, Hans-Werner (Ed.), *Umweltschutz und andere Politiken der EG*, Köln, Heymanns.

Pierson, Paul, 1993, When Effect becomes Cause: Policy Feedback and Political Change, *World Politics* 45 (4), 595–628.

PricewaterhouseCoopers, 2010, *100% Renewable Electricity, A Roadmap to 2050 for Europe and North Africa*, London, PricewaterhouseCoopers et al., available at http://www.pwc.co.uk/pdf/100_per cent_renewable_electricity.pdf, last accessed 19 December 2012.

Ragwitz, Mario, Simone Steinhilber, Barbara Breitschopf, Gustav Resch, Christian Panzer, Andre Ortner, Sebastian Busch, Max Rathmann, Corinna Klessmann, Christian Nabe, Isabelle de Lovinfosse, Karsten Neuhoff, Rodney Boyd, Martin Junginger, Ric Hoefnagels, Niccolò Cusumano, Arturo Lorenzoni, Jitske Burgers, Maroeska Boots, Inga Konstantinaviciute & Botond Weöres, 2012, *RE-Shaping: Shaping an Effective and Efficient European Renewable Energy Market*, Karlsruhe, Fraunhofer Institute for Systems and Innovation Research (ISI), available at http://www.reshaping-res-policy.eu/downloads/Final%20report%20RE-Shaping_Druck_D23.pdf, last accessed 08 March 2013.

Rathmann, Max, Thomas Winkel, Eva Stricker, Mario Ragwitz, Anne Held, Benjamin Pfluger, Gustav Resch, Christian Panzer, Sebastian Busch & Inga Konstantinaviciute (Eds), 2009, *Renewable Energy Policy Country Profiles*, Karlsruhe, Fraunhofer Institute for Systems and Innovation Research, available at http://www.reshaping-res-policy.eu/downloads/RE-SHAPING%20Renewable%20Energy%20Policy%20Country%20profiles%202009.pdf, last accessed 08 March 2013.

Rodi, Michael, 2012, Artikel 194, in: Vedder, Christoph & Wolff Heintschel von Heinegg (Eds), *Europäisches Unionsrecht: EUV, AEUV, Grundrechte-Charta – Handkommentar*, Baden-Baden, Nomos Verlagsgesellschaft.

Sachverständigenrat für Umweltfragen, 2008, *Umweltgutachten 2008 – Umweltschutz im Zeichen des Klimawandels*, Berlin, Erich Schmidt Verlag.

Sachverständigenrat für Umweltfragen, 2009, *Weichenstellungen für eine nachhaltige Stromversorgung*, Berlin, Sachverständigenrat für Umweltfragen, available at http://www.umweltrat.de/SharedDocs/Downloads/DE/06_Hintergrundinformationen/2009_Thesen_Weichenstellungen_Stromversorgung_Hohmeyer.pdf?__blob=publicationFile, last accessed 07 March 2013.

Schäfer, Peter & Meinhard Schröder, 2012a, Artikel 170, in: Streinz, Rudolf (Ed.), *Vertrag über die Europäische Union und Vertrag über die Arbeitsweise der Europäischen Union – Kommentar* (Second Edition), München, C.H. Beck.

Schäfer, Peter & Meinhard Schröder, 2012b, Artikel 171, in: Streinz, Rudolf (Ed.), *Vertrag über die Europäische Union und Vertrag über die Arbeitsweise der Europäischen Union – Kommentar* (Second Edition), München, C.H. Beck.

Scharpf, Fritz W., 1999, *Governing in Europe: Effective and Democratic?*, Oxford, Oxford University Press.

Schmitz, Holger & Philipp Jornitz, 2012, Regulierung des deutschen und des europäischen Energienetzes: Der Bundesgesetzgeber setzt Maßstäbe für den kontinentalen Netzausbau, *Die Neue Zeitschrift für Verwaltungsrecht* 6, 332–337.

Schöpe, Martin, 2010, The New EU Directive on Renewable Energies from the Perspective of a Member State, in: Fouquet, Dörte, *EU Energy Law*, Vol. 3 (3), Leuven, Claeys & Casteels Publishers, 177–198.

Schreurs, Miranda, Henrik Selin & Stacy D. VanDeveer (Eds), 2009, *Transatlantic Environment and Energy Politics*, Surrey, Ashgate.

Sensfuß, Frank, Mario Ragwitz, Marlene Kratzat, Ole Langniß, Carlo Obersteiner, Thorsten Müller, Frank Merten & Manfred Fischedick, 2007, *Fortentwicklung des EEG zur Marktdurchdringung erneuerbarer Energien im deutschen und europäischen Strommarkt*, Karlsruhe, Fraunhofer Institute for Systems and Innovation Research et al., available at http://www.isi.fraunhofer.de/isi-media/docs/e/de/publikationen/Fortentwicklung_EEG_Endbericht_final.pdf?WSESSIONID=5712ff2ca5ffcf0d9590afc8ef7e1486, last accessed 08 March 2013.

Tiefenthaler, Veronika, 2011, Planning in Europe – The Impact of European Union Law on National Planning Systems and Territorial Transnational Cooperation, *Journal for European Enviromental and Planning Law*, 115–140.

Trüe, Christiane, 2002, *Das System der Rechtsetzungskompetenzen*, Baden-Baden, Nomos Verlagsgesellschaft.

Trüe, Christiane, 2004, EU-Kompetenzen für Energierecht, Gesundheitsschutz und Umweltschutz nach dem Verfassungsentwurf, *Juristenzeitung* 59 (15-16) 2004, 779–787.

UCTE/Union for the Cooperation of Transmission of Electricity, 2009, *Transmission Development Plan – Update 2009*, Brussels, UCTE.

Unruh, Gregory C., 2000, Understanding Carbon Lock-In, *Energy Policy*, 817-830.

Voet van Vormizeele, Philipp, 2012, Artikel 172, in: Schwarze, Jürgen (Ed.), *EU-Kommentar* (Third Edition), Baden-Baden, Nomos Verlagsgesellschaft.

von Prittwitz, Volker, 2007, *Vergleichende Politikanalyse*, Stuttgart, Lucius & Lucius.

Wagner, Hermann-Josef, 2009, Erneuerbare Energien erfordern koordinierten Ausbau des europäischen Netzes, *Energiewirtschaftliche Tagesfragen* 59 (11), 54–55.

Wissenschaftlicher Beirat Globale Umweltveränderungen, 2010, *Climate Policy Post-Copenhagen*, Berlin, Wissenschaftlicher Beirat Globale Umweltveränderungen, available at http://www.wbgu.de/fileadmin/templates/dateien/veroeffentlichungen/politikpapiere/pp2010-pp6/wbgu_pp2010_en.pdf, last accessed 07 March 2013.

Woyte, Achim, Jan de Decker & Vu Van Thong, 2008, *A North Sea Electicity Grid [R]evolution*, Brussels, Greenpeace, available at http://www.greenpeace.org/eu-unit/Global/eu-unit/reports-briefings/2009/5/A-North-Sea-electricity-grid-(r)evolution.pdf, last accessed 08 March 2013.

# 16
# China-Africa Cooperation: Joint Engagement in Adaptation to Climate Change

*Wanxin Li, Sven Grimm & Harrie Esterhuyse*


## *Abstract*

China's economic success since the 1990s has been staggering. With its economic growth, the country has experienced new challenges, particularly with regard to environmental sustainability on the one hand, and increasing international demands and responsibilities on the other. Climate change is a key issue for both these challenges. This article explores the rise of environmental concerns, particularly climate change, on the domestic policy agenda in China, and the issue of greening China's growth. Domestic policies and their challenges are the background against which the article investigates the potential and limitations of international engagement in the mitigation of, and adaptation to, climate change. International challenges are discussed in the second part of the article, which looks into the negotiations on actions to mitigate and manage climate change, and specifically the support China has received regarding adaptation to climate change within the China–Africa cooperation platform. In the context of this cooperation, renewable energy, among other things, offers the potential for a viable solution for Africa's energy deficiency, specifically as regards electricity. This co-operation is still in the early stages. Therefore, this article looks at the potential in the relationship rather than at specific actions.


## *A. Introduction*

China's economic development has been one of the big global development success stories of the last generation. China's gross national income has grown tenfold over the past 25 years. Given China's size as well as its demographic weight, this impressive economic growth is a basis for global political weight.[1] After having pursued the Reform and Opening Up Policy[2] proposed by Xiaoping Deng in 1978, China has increasingly been 'going out' since the late 1990s – not least so by engaging with and investing in other Asian countries and Africa. Under President Hu Jintao, the country has been engaging more and more in global politics, including more cooperation with African states. Environmental policies increasingly form part of the country's external political agenda, for both domestic and international reasons.

In recent decades, the rapid development growth path in the major emerging economies such as Brazil, China, India and Russia has been accompanied by severe environmental degradation.[3] In China itself, development is happening at breathtaking speed. The challenges, which include land use, water pollution, air quality and, of course, climate change, are manifold. China's economic growth has led to numerous environmental problems that are being increasingly addressed by policymakers, not least because environmental pollution has reached a level that clearly comes with economic costs.[4] Pollution today is a topic around which Chinese citizens organise and challenge state decisions.[5] It might be a lesson learnt from the fall of the communist regimes in Central Europe that environmental pollution is one of the crystallising points for a dissident civil society if the complaints and immediate threats to citizens' health and well-being are not addressed by the state.

Climate change and the required adaptations to it clearly constitute a double challenge for emerging powers like China, as they step up their global engagement and their struggle with environmental damage at home. China is home to the world's ten most polluted cities. In 2007, 46% of the global coal production was mined in China, and used for the generation of electricity. The 2.5 billion t of coal, however, were not sufficient for China's entire domestic consumption needs, and in 2007 the country became a net coal importer.[6] Energy demand in China is still on a steep increase, with more strain on global resources to be expected, given that China is still in a

1 Kaplinsky & Messner (2008).
2 The nature of the economy has changed from a centrally planned system under rigid political control to a more market-oriented economy that has a rapidly growing private sector and is a major player in the global economy.
3 Shaw (2012).
4 World Bank (2007); Economy (2007).
5 Li et al. (2012); Chen (2009:xxi); Burgess (2012).
6 Zhu (2010:2).

catching-up phase of development. In this agenda of economic priorities with immediate pollution problems, climate change is a rather complex and abstract issue, yet one that is high on the international agenda. There are, nonetheless, linkages between combating some of the pollution problems in China and working to mitigate climate change. China's 12th Five-year Plan (2011–2015) attributes high importance to environmental investments, such as renewable energy (for mobility and electricity), and biotechnology. Addressing environmental challenges, including climate change, has been rationalised in China as both necessary for solving domestic problems, and important due to its potential to gain new business opportunities and competitive advantages.

At the international level, China has also indicated the importance of sustainable development and renewable energy in its relationship with Africa. New institutions have been created in the interests of China–Africa relations, such as the Forum on China–Africa Cooperation (FOCAC), and economic cooperation has skyrocketed in the last decade.[7] As a Chinese business participant in South Africa enthusiastically stated at a Centre for Chinese Studies workshop in Stellenbosch, South Africa, in August 2011, "China will change the face of this continent".[8]

It is, in fact, very likely that China's economic growth will change the global political, economic and ecological balance. The Chinese government may be more cautious in the presentation of its influence, but there is little doubt that China's developmental rise not only influences the global situation, it also directly impacts on development prospects in Africa.

Immediate environmental challenges exist on the African continent. Large-scale environmental pollution in the Niger Delta is one example for already existing environmental troubles: oil extraction that comes with substantial pollution results in little to no developmental gains for the affected communities. This is one of the prime examples of ignorance towards the ecology and people's livelihoods as a driver for conflict.[9] Compared with the existing local challenges, climate change is more abstract, but no less threatening. The global change in climate is being felt in Africa today already, leading to further development problems on the African continent. The Chinese experiences with steep development gains at the expense of

7 Cissé (2012).
8 Statement during a Centre for Chinese Studies workshop, Stellenbosch University, August 2011.
9 See e.g. Obi (2008).

nature might be an illustration of what is to be expected in the future, if African economic growth accelerates further. Global connectivity has increased in terms of trade and personal exchanges. Changing climate is expected to result in fundamental shifts in the livelihoods of African societies. These shifts will result in more potential for conflicts and challenges, such as the escalation of forced migration.[10] With regard to climate-change-related policy and mitigation of climate change, the global linkages are particularly obvious. Not least due to generally higher per capita carbon dioxide emissions (with the notable exception of South Africa), China has started to accept responsibility for assisting in the adaptation to climate change, as has been repeatedly stated in the declarations at the end of FOCAC meetings. Indeed, the Fifth FOCAC Action Plan (2013–2015) states that China and Africa actively support advancing cooperation in clean energy and renewable resource projects with the "ideals of mutual benefit and sustainable development".[11]

This article first looks at China's domestic policy activities and asks the question: Can China green its own growth? With the domestic policy emphasis and challenges in mind, the thrust and the limitations of China's international engagement in respect of mitigation and adaptation to climate change can be better understood. The focus in the second part of the article is on the international agenda regarding negotiations on measures to mitigate and adapt to climate change. This ultimately leads to questions of adaptation to climate change, specifically in the context of the China–Africa cooperation.

## B. *Climate Issues Climbing in Importance on the Chinese Domestic Political Agenda*

In 2005, having realised the urgent need to improve the quality of China's economic growth, President Hu Jintao called for a scientific approach to development. In the same year, responding to the public call for environmental and social justice, the central government set the national objective of building a harmonious society. In 2006, riding the wave of environmental concern, the then State Environmental Protection Agency[12] announced an

10 Ruppel & Van Wyk (2011).
11 FOCAC (2012:section 4.6.3).
12 Upgraded to the Ministry of Environmental Protection (MEP) in 2008.

environmentally adjusted gross domestic product (GDP) growth rate, aiming to alert provincial governments and the public to losses caused by pollution, as well as the danger of a single-minded pursuit of economic growth in China. Also in 2006, the State Council announced pollution reduction and energy efficiency targets in the 11th Five-year Plan (2006–2010), and local leaders were made accountable for meeting those targets.

In the context of international efforts to reduce carbon emissions, on 25 November 2009 at the 16th Conference of the Parties (COP16) in Copenhagen, Premier Wen Jiabao announced that, by 2020, China aimed to reduce its carbon emissions per unit of GDP by 40–45% from its 2005 level.[13] Continuing the same trend, greater emphasis was placed on pollution control, energy efficiency and carbon reduction in the 12th Five-year Plan. Students of politics and the environment in China interpreted this as a signal that the central government was readjusting its focus from economic growth to environmental and social aspects that are relevant for people and their livelihoods.[14]

As observers have correctly pointed out, one of the leading themes in the 12th Five-year Plan is low-carbon development. The Plan includes an investment of ¥2.37 billion into energy-saving projects, for example. It also aims to reduce energy consumption by 16% per ¥10,000 by 2015, using 2010 as the base year. A 30% increase in clean energy generation year-on-year for each year running is another target. The following objectives are considered indispensable for achieving the ambitious transformation goals:

1. In the short term, adopting low-carbon technologies and controlling carbon emissions
2. In the medium term, upgrading industrial structure, decoupling economic growth and carbon emissions, and cultivating low-carbon consumption and lifestyles, and
3. In the long term, achieving improved resource efficiency, a healthier and more ecologically balanced environment, and harmony between humans and nature.[15]

Much attention has been paid to the co-benefits of carbon mitigation, i.e. upgrading technology, reducing pollution, offering a sustainable supply of energy and other resources, and improving the quality of life. China hosted

13 State Council (2009).
14 Editorial Report (2010).
15 Xiong (2011).

almost half (45.72%) of the global total of 3,447 Clean Development Mechanism (CDM) projects registered with the United Nations Framework Convention on Climate Change (UNFCCC) in 2011. Furthermore, cities such as Baoding have become more competitive in the global market regarding the production and export of solar panels.[16] Despite no consensus having been reached within Chinese society, the central government has worked hard to mainstream the environment and a climate agenda in government decision-making.

Not only have policymakers seen the potential of these co-benefits, the business community has also been trying to develop this competitive edge in almost all sectors of economic life, e.g. in construction, urban planning, new energy technologies and products, and innovative instruments for financing low-carbon initiatives. For instance, engineers have successfully worked with the Guangzhou municipal government to build a zero carbon emission building – the Pearl River Tower. Urban planners have also successfully sold their ideas to cities in order to plan and build low-carbon districts/zones, as follows:

- Lile Island in Hainan Province
- The Shouan township in Chengdu
- The Yujiabao financial district in Tianji
- The Wangjiadun Green Central Business District in Wuhan, and
- Ten low-carbon townships in Huizhou, Guangdong Province.

New energy technologies are being sold both domestically and internationally, such as batteries for electric cars, fast-charging stations and solar panels. However, comparatively fewer companies are engaged in inventing financial instruments for funding low-carbon development. In June 2010, the China Beijing Environment Exchange established an investment platform for contractual power management. With the Chinese government starting to aggressively promote pilot low-carbon cities and counties in the 12th Five-year Plan period, it is expected that the energy industry will quickly seize business opportunities and come up with improved climate mitigation and adaptation policy to further develop competitiveness around these measures.

It is a given that business opportunities in the low-carbon development environment are largely dependent on government regulations and supportive initiatives which encourage the devising and deployment of environ-

16 Zhu (2011b).

mental technologies that address climate change. To this end, the Chinese government has invested in environmental research and development (R&D), e.g. Minister Wan Gang of the Ministry of Science and Technology indicated, on 12 November 2009 at the High-level Forum on Energy Efficiency and Pollution Reduction v Global Climate Change, that the Ministry of Science and Technology had invested over ¥10 billion in developing technologies that addressed climate change.[17]

Furthermore, funds from private sources have been increasingly mobilised to match public investments in R&D. Since 1981, the total number of national key technology R&D projects, as well as R&D funding from private sources, has greatly increased. Yet the share of funds from the central government has declined since 1981. During the 6th Five-year Plan period (1981–1985), the Ministry of Finance (MOF) in the central government contributed up to 60% of the total funds, while during the 10th Five-year Plan period (2001–2005), the share taken up by the MOF dropped to only 8.3%. Thus, the central government has clearly changed its role from major funder to advisor and facilitator.[18] More private investments have entered into energy- and carbon-related R&D in recent years, especially now that a global consensus has been built on the urgency of addressing climate change.

The deployment and adoption of environment-friendly technologies are as important as the development of environmental technologies. Industry experts generally agree that China's new energy industry is limited by the underdeveloped domestic application of new technologies. Official data show China's production of solar cells accounted for 50% of the world's total output in 2010; however, 95% of that output was exported.[19]

Another government priority entailed formulating tax incentives and subsidies at consumer level for those using environment-friendly products. In 2008 and 2009, for example, the MOF and the National Development and Reform Commission (NDRC) offered households a subsidy of up to 50% of the price of an energy-saving light bulb.[20] A national collection system for used electronic home appliances was established in 2009. Logistic companies were employed by government to pick up used appliances and send them to certified waste-disposal companies. Furthermore, households were

17 Xinhua News Agency (2009).
18 OECD (2009).
19 Zhu (2011a).
20 MOF & NDRC (2007).

awarded with coupons which they could use when purchasing new electronic home appliances.[21]

The MOF also subsidises individuals who purchase electric vehicles.[22] The city of Shenzhen, where the headquarters of the largest electric vehicle manufacturer, BYD, is located, provides matching funds of up to ¥50,000 per vehicle to subsidise customers who purchase BYD electric vehicles. Other incentives are more indirect. Currently, a carbon tax is under consideration by the MOF and the NDRC, the latter having acquired green government procurement regulation that is hoped to also greatly boost the demand for green technologies and products.[23]

In addition, the Chinese government has devised measures to assist companies and individuals in making desirable environmental choices through information-sharing and capacity-building. The China Standard Certification Centre, for example, certifies and labels energy-efficient products. The Ministry of Environmental Protection and local governments rate the environmental performance of polluting firms using a colour-coded scale. This information is made publicly available.[24] By implementing the Top-1000 Energy-Consuming Enterprises Program, the NDRC is helping to build capacity in businesses in order to enhance energy efficiency.

### *C. Can China Green its Own Growth? Challenges for China in Addressing Climate Change*

The fact that the Chinese government is paying increased attention to climate-change-related policy is reassuring, but it also begs the question of where obstacles or challenges remain. There are at least four broader areas that require increased attention in order to meet the domestic goals regarding environmental sustainability. These four are discussed individually below.

21 MOF et al. (2009).
22 MOF & MOST (2009).
23 Xinhua News Agency (2009b).
24 Nature (2009).

### *I. Fragmented Government Structure*

China's fragmented government structure means that industrial, energy and innovation policies do not always coincide. There is no doubt that addressing climate change and realising its co-benefits have been widely accepted by the different levels of government, yet at the implementation of the 12th Five-year Plan, there were five different schemes funded by different ministries for developing low-carbon pilot and demonstration projects, as follows:

- 5 provinces and 8 cities chosen by the NDRC
- 108 counties chosen by the MOF, together with the Ministry of Agriculture and the National Energy Bureau
- 100 new energy-demonstration cities chosen by the NDRC
- Low-carbon transportation pilot cities chosen by the Ministry of Communication, and
- A low-carbon demonstration city, jointly built by local government and the Ministry of Housing and Urban–Rural Construction.

When local governments volunteer themselves to be part of one or more of these programmes, much productive time and many productive resources are lost in bureaucracy, e.g. preparing paperwork, or receiving delegations from different ministries that do not necessarily coordinate their travel plans. These examples show how the fragmented government structure is not necessarily conducive to coherence as regards industrial, energy and innovation policies. The challenge is, therefore, to come up with a coherent national strategy to address climate change.

### *II. Weak Environmental Compliance and Enforcement*

Weak environmental compliance and enforcement is also not conducive to technological innovation and/or adoption of environmental policy. Even though China has devoted considerable resources to measures such as treatment facilities aimed at abating and controlling pollution, rapid industrialisation and urbanisation have outpaced efforts to clean up. The situation in this regard is even more complicated with local governments than it is with central government. This is partly due to the fact that provincial governors and county/city mayors are evaluated according to the GDP growth rates of their localities, irrespective of the impact of rapid growth on the local environment. Therefore, they tend not to take environmental or low-carbon growth objectives seriously.[25] This is an example of how weak environmental compliance and enforcement has failed to reinforce technological innovation and/or policy adoption.

### *III. Insufficient Incentives and Disincentives*

Insufficient incentives and disincentives are offered to enterprises to become green. Pollution discharge fees have been so low that it has made more economic sense for polluters to pollute rather than to attempting to avoid pollution or pay costs for treatment of polluted natural resources. According to one account,[26] the operating cost of wastewater treatment in one high-polluting industry on average is around ¥1.2–1.8 per ton. The fixed investment in a wastewater treatment facility is ¥100 million for the 150-t-per-day alkali recycling equipment used in the paper and pulp industry. Yet the maximum fine for wastewater discharge is ¥100,000. Hence, it is a rational business choice to pay the fine rather than treat the pollution.[27]

### *IV. Lack of Social Infrastructure for Addressing Climate Change*

Not only is there a need for trust, consensus-building, social organising and collective action, there is also a lack of social infrastructure for addressing climate change at the local level. Access to environmental information, participation in environmental decision-making, and environmental justice has been limited. Furthermore, the root cause of local environmental degradation can often be attributed to the changing climate. Heightened public awareness of environmental degradation and increasing anxiety over health and property values drive people to fight for more political space to influence decisions that have an impact on the environment. Despite the promises made in the letter of Chinese law, Chinese society lacks a meaningful institutional framework to allow public participation in the area of environmental pro-

25 Tilt (2007); Zhu & Ru (2008).
26 Yang & Ge (2006).
27 (ibid.).

tection.[28] On the whole, the Chinese government only responds to public demands on an ad-hoc basis, with no institutional commitment to proactively engaging the public in environmental issues.[29] Often, the public does not trust the government and/or business, which presents an immense challenge as regards addressing climate change because it requires collective action at the local level.

As far as global climate change is concerned, local and domestic action is urgently needed. On the other hand, the global challenge also requires global cooperation. In the next section we focus on the global dimension of China's climate change agenda.

### D. *Global Engagement on Climate Change – Issues for International Negotiations*

On the global level, China has actively engaged with the climate regime. The country is a signatory of the Kyoto Protocol. On 2 September 2005, a bilateral Memorandum of Understanding on a European Union and China Partnership on Climate Change was signed. The Chinese and United States (US) heads of state initiated a Strategic Dialogue in 2005 as well as a biannual Strategic Economic Dialogue in 2006. In 2009, led by the US Departments of State and Energy and the Chinese NDRC, the two countries negotiated a Memorandum of Understanding to Enhance Cooperation on Climate Change, Energy and the Environment. China has also worked on a project basis with other developed countries. These include Switzerland, in the Sino–Swiss Low Carbon Cities in China project; and the United Kingdom (UK), in the China Climate Adaptation Project co-funded by the UK Department for International Development and the Swiss Agency for Development and Cooperation.

28 For example, the Beijing municipal government only started to make public the local concentration levels of particular matter smaller than 2.5 micrometres ($PM_{2.5}$) after the United States Embassy in Beijing published its monitoring data on $PM_{2.5}$; see http://beijing.usembassy-china.org.cn/aqirecent3.html, last accessed The serious smog in January 2013 in Beijing and other parts of the country alarmed even the politically and economically powerful, as it affected all parts of the population, irrespective of their social status. Thus, a wider consensus was built on addressing environmental degradation and combating climate change.

29 Li et al. (2011).

These activities have all been carried out in line with the following six principles, articulated by Xie Zhenhua of the NDRC:[30]

1. Addressing climate change under the framework of sustainable development
2. Common but differentiated responsibilities
3. Mitigation and adaptation are equally important
4. International Conventions and Protocols are the principal channel for addressing climate change at the global level
5. Emphasis on technological innovation and technology transfer, and
6. Wide public participation and international cooperation.

China has shown respect for the international climate regime and engages as a developing country. Members of the private sector in China have been participating in the CDM as beneficiaries of the transfer of funds and technology. China has hosted twice as many CDM projects as India, which has hosted the second largest number of CDM projects. This has put China in competition with other developing countries for international assistance.

For China, engaging with the post-Kyoto climate change regime is a challenge. Due to its status as a developing country, China is unlikely to accept a national cap on its carbon emissions in the years to come. However, as is stated in Chapter 21 of the 12th Five-year Plan, the country aims to construct a carbon inventory and institutionalise the calculation for developing carbon markets for all relevant activities.[31] There are already five emissions exchanges – in Beijing, Chongqing, Shanghai, Shenzhen and Tianjin. The discussion on the inability of these exchanges to conduct business and launch pilot carbon emissions markets has largely concerned geographic issues, e.g. where to form such a market. Unfortunately, China has not yet formulated sectorial strategies to prepare for a global carbon emissions trading scheme. At a more basic level, for a carbon emissions trading scheme to work, countries have to agree on how many tons of carbon emitted they are responsible for. This introduces the methodological challenge of how to calculate carbon emissions: by production or by consumption. Researchers and policymakers do not agree on these issues, not least because oil production and global production chains for manufacturing result in carbon emissions outside the countries of consumption. China has emitted less

30 Xie et al. (2009).
31 State Council of China (2011).

than a third of the carbon produced per capita, compared with developed countries. It is most likely that China would support the allocation of responsibility for carbon emissions on a consumption basis. However, China has not yet formulated a convincing methodology for calculating the life cycle of carbon emission footprints of resource extraction and industrial outputs, and may, therefore, miss opportunities to engage with other developing countries and the developed world.

### *E. Adaptation to Climate Change in China–Africa Relations*

The relationship between China and Africa is often understood as being driven by China's resource needs, which includes fossil fuels from Africa. While China's dependence on fossil fuel imports will not be easily overcome, its endeavours to promote renewable energy also have an international dimension which is explored in this section with regard to China–Africa relations. As discussed above, for domestic reasons, China needs to upgrade its technology and increase its domestic production of renewable energies in order to reduce environmental pollution and mitigate climate change.

A renewed emphasis is placed on sustainable development and, by extension, renewable energy, in the Fifth FOCAC Action Plan:

- China pledges to increase the size of the China–Africa Development Fund to US$5 billion[32]
- China commits to providing a US$20 billion credit line to African countries for infrastructure development and for boosting sustainable development[33]
- China promises to keep African interests in mind, stating that it wishes to assist African countries in raising the added value of resource products, while "protecting [the] local eco-environment and promoting … sustainable economic and social development",[34] and
- China and Africa agree to strengthen cooperation in the area of infrastructure development, including in the generation of electricity and energy.[35]

32 FOCAC (2012:section 4.2.4).
33 (ibid.:section 4.5.2).
34 (ibid.:section 4.6.2).
35 (ibid.:section 4.6).

China has pledged between US$15 billion and US$20 billion for sustainable development projects in Africa and is already investing US$700 million in Guinea's rural electricity system.[36] China has the expertise and the capital reserves to partner with Africa. The need and the plans for the expansion of electricity supply based on renewable energy are evident in Africa. For example, Ethiopia has plans for 150,000 home solar systems, 300 wind pumps, 300 solar pumps, 3,000 institutional photovoltaic systems, and three million solar lanterns, as well as measures to improve domestic use of biomass.[37] Egypt, which features conditions that are of the best in the world for wind and solar power, is planning to increase energy production from renewable sources, including hydropower, to 20% in 2020. Hydropower generation currently accounts for 12% of the energy mix, but this will decline to 8% by 2020 if no further investment is made.[38] Thus, there is clearly a market for foreign investment in Africa – not least that of Chinese origin.

Strategies for Africa can use both on-grid solutions, such as making use of grid-connected, concentrated solar power stations, wind farms or hydropower stations, or small-scale, off-grid solutions, such as solar photovoltaic modules. The available technology lends itself to both rural and urban scenarios in Africa. Solar photovoltaics, in which technology the Chinese are particularly experienced and skilled, is especially promising as it can easily be installed as an off-grid solution in rural areas, where grid extension is very expensive.

Some African successes are encouraging. Ghana increased its population's electricity access from 25% in 1989 to 66% by 2011, with rural access in particular having risen from 5% to 40% in the same period.[39] South Africa increased its urban electrification from 30% in 1994 to 83% by 2011, and rural electrification from 12% to 57% in the same period.[40] With its Green Economy Accord, South Africa is also slowly turning towards more sustainable solutions. Although it is currently heavily reliant on coal-based power, it aims to increase its production of renewable energy, which has become much more affordable in recent years due to maturing markets and technologies.

36 (ibid.:section 4.5.2); IRENA (2011:7).
37 IRENA (2011:19).
38 (ibid.:18).
39 (ibid.:6).
40 (ibid.).

Expertise is also required in the regulation of energy sectors, as energy and electricity generation faces political and regulatory shortcomings in African countries.[41] Regulatory frameworks need to be coherent and consistent, and need to incentivise the production of renewable energy over fossil fuels, including access to finance. They also need to address technical obstacles for on-grid solutions. Most of Africa's power supply at present is produced by government-backed utilities. Thus, governments will have to engage actively in maintaining and expanding the grid, as well as in promoting off-grid solutions, implementing regulatory frameworks that are conducive to the production and consumption of renewable energy and creating investment incentives. A mere liberalisation of energy production is unlikely to result in increases in the production of renewable energy.

### *F. Conclusions*

Green growth requires conserving and increasing environmental and natural assets, and distributing the benefits of growth fairly throughout society to improve general well-being beyond economic growth, and without causing environmental harm. On 25 November 2009, Premier Wen Jiabao announced at COP16 in Copenhagen that, by 2020, China aimed to reduce its carbon emissions per unit of GDP by 40–45% from their 2005 levels.[42] Continuing in this trend, increased emphasis was placed on pollution control, energy efficiency and carbon reduction in the 12th Five-year Plan. Short-, medium- and long-term objectives for achieving a transformation toward low-carbon development were outlined in Section B of this article. These new policy targets signalled a readjustment in focus by the central government, from economic growth to the environmental and social aspects relevant for people and their livelihoods.[43] Together with a strong emphasis on social development during the 12th Five-year Plan, China is embracing the concept of green growth for a sustainable development path and a harmonious society. Therefore, Chinese society needs to incentivise government agencies, economic agents and civic organisations to articulate their pos-

41 See Lüdemann (2011).

42 State Council of China (2009).

43 Central Committee of the 17th National Congress of the Communist Party of China (2010).

itions more effectively, as well as heed environmental and social interests in their decision-making.

In the context of the China–Africa relationship, including the FOCAC, renewable energy not only provides a viable solution for Africa's energy – and, specifically, electricity – shortage, but it also has the potential to overcome obstacles in infrastructure development in Africa in the electricity sector, and alleviate the infrastructure shortage. In this context, the discussion in this article focused on the potential for action and benefits because cooperation on adaptation to climate change is still in its infancy. Small, off-grid solutions, such as solar photovoltaic modules, are best suited to rural areas. People in remote areas of Africa have long-standing experience with wind energy in the operation of pumps for wells. Current investment in wind energy for electricity is, however, rather an on-grid solution which may be more suitable for urban agglomerations. This is because peaks and valleys in consumption need to be balanced to maintain the stability of the electricity supply.

Small rather than large energy projects should be funded first, and they should be introduced gradually into rural areas, allowing for a learning curve. Decentralised options provided by renewables can provide rapid access to basic levels of electrification. In time, large-scale renewable projects can be introduced, such as concentrated solar and geothermal plants. China has the technology and capital for these investments. For these activities to be successful and to reduce the obstacles to renewable energy implementation, African governments will have to devote attention to national regulation and legislation and, most importantly, its implementation. The policy changes should include lower import tariffs on renewable energy equipment and opening up national electricity production to competition. African governments should also integrate renewable energy into their electricity production plans, which can be done through a tendering process, or by guaranteeing viable feed-in tariffs.

The investments required to foster renewable energy are high. There is, thus, a need for Africa and China, as well as African countries amongst themselves, to recognise a need for cooperation between the private and public sectors. Regional bodies such as the New Partnership for Africa's Development and international bodies such as FOCAC should also play a role in promoting renewable energy, e.g. by facilitating investments and enabling transfrontier solutions, where appropriate. Establishing regional energy bodies and consulting with the private sector in energy and electricity planning are important in this process. Energy trade across borders should be increased as well. Furthermore, stepping up regional integration by way of connected grids will enhance grid reliability and enlarge the available pool of skills in respect of grid maintenance.

As regards cooperative solutions, it is the task of African governments, as well as the responsibility of the China–Africa development partnership, to ensure that not only is technology provided for renewable energy installation, but also that the necessary technology transfer takes place. African states need to upgrade the skills that are essential to post-installation system maintenance and expansion. China–Africa agreements in the field of energy should, thus, include clear references to technology transfer.

## *References*


Burgess, Meryl, 2012, Thick Air in Beijing – Air Pollution as a Political Issue, *CCS Commentary*, Stellenbosch, Centre for Chinese Studies, 2 April, available at http://www.ccs.org.za/wp-content/uploads/2012/04/Beijing-pollution-commentary-MB_final.pdf, last accessed 28 January 2012.

Central Committee of the 17th National Congress of the Communist Party of China, 2010, *Suggestions by the Central Committee of the Communist Party of China on How to Make 12th Five-Year Plan* (中共中央关于制定国民经济和社会发展第十二个五年规划的建议), available at http://www.ndrc.gov.cn/125gh.pdf, last accessed 20 November 2010.

Chen, Gang, 2009, *Politics of China's Environmental Protection: Problems and Progress*, Singapore, World Scientific Publishing Co. Pte. Ltd.

Cissé, Daouda, 2012, FOCAC: Trade, Investments and Aid in China–Africa Relations, *CCS Policy Brief*, Stellenbosch, Centre for Chinese Studies, available at http://www.ccs.org.za/wp-content/uploads/2012/05/FOCAC_Policy-Briefing_tradeinvest_final.pdf, last accessed 28 January 2012.

Economy, Elizabeth, 2007, Environmental Governance: The Emerging Economic Dimension, in: Carter, Neil and Arthur P.J Mol (Eds), *Environmental Governance in China*, London/New York, Routledge, 23–41.

FOCAC/Forum on China Africa Cooperation, 2012, *The Fifth Ministerial Conference of the Forum on China–Africa Cooperation Beijing Action Plan (2013–2015)*, available at http://www.focac.org/eng/z, last accessed 23 January 2013.

Hu, Jintao, 2006, Development Cannot be Pursued at the Cost of Ecological Environment and Human Life, *Xinhua Net*, 29 March.

IRENA/International Renewable Energy Agency, 2011, Scenarios and Strategies for Africa, paper presented at the IRENA–Africa High-level Consultations, Abu Dhabi, 8–9 July 2011.

Kaplinsky, Raphael & Dirk Messner, 2008, Introduction: The Impact of Asian Drivers on the Developing World, *World Development* 36 (2), 197–209.

Li, Wanxin, Jieyan Liu & Duoduo Li, 2012, Getting Their Voices Heard: Three Cases of Public Participation in Environmental Protection in China, *Journal of Environmental Management* 98, 65–72.

Liu, Yi & Sun Xiuyan, 2006, Interview with Zhou Shengxian: A Good Administrative Chief Would Better to Do a Good Job in Environmental Protection and Not to Pollute, *Beijing News*, *Xinhua Net*, *People's Daily*, 20 April 2006.

Lüdemann, Cord, 2011, Renewable Energy Promotion in Developing Countries – The Case of South Africa, *The China Monitor* 67, 10–17.

MOF/Ministry of Finance, Ministry of Commerce, National Development and Reform Commission, Ministry of Industry and Information, Ministry of Environmental Protection, State Administration for Industry and Commerce, and General Administration of Quality Supervision and Quarantine, 2009, *Measures on Replacing the Used Electronic Home Appliances with the New Ones* (家电以旧换新实施办法), Beijing, Government of the People's Republic of China.

MOF/Ministry of Finance & MOST/Ministry of Science and Technology, 2009, *Guidelines on the Management of Fiscal Subsidies on Energy Efficient Vehicles and Vehicles Using New Energy Sources (Trial Implementation)* (节能与新能源汽车示范推广财政补助资金管理暂行办法), Beijing, Government of the People's Republic of China.

MOF/Ministry of Finance & NDRC/Ministry of Finance and National Development and Reform Commission, 2007, *Guidelines on the Management of Fiscal Subsidies on Promoting LED Products (Trial Implementation)*, 高效照明产品推广财政补贴资金管理暂行办法, Beijing, Government of the People's Republic of China.

*Nature*, 2009, Raising the Standards, *Nature* 459, 1033–1034.

Obi, Cyril, 2008, Enter the Dragon? Chinese Oil Companies and Resistance in the Niger Delta, *Review of African Political Economy*,35, 417–434.

OECD/Organisation for Economic Co-operation and Development, 2009, *Ensuring Environmental Compliance: Trends and Good Practices*, Paris, OECD.

Ruppel, Oliver & Sanita van Wyk, 2011, Climate Change Effects on Human Security in Africa: Environmental Migration – A Case for Human Rights?, *The China Monitor* 66, 10–17.

Shaw, William, 2012, Will Emerging Economies Repeat Environment Mistakes of their Rich Cousins?, *Carnegie Endowment Online*, available at http://carnegieendowment.org/2012/03/01/will-emerging-economies-repeat-environmental-mistakes-of-their-rich-cousins/9zni, last accessed 10 September 2012.

State Council of China, 2009, State Council Decided on Carbon Emission Reduction Targets at the General Committee Meeting, Beijing, State Council of China.

State Council of China, 2011, The 12th Five-year Plan for the Economic and Social Development of the People's Republic of China, Beijing, Xinhua News Agency.

Tilt, Bryan, 2007, The Political Ecology of Pollution Enforcement in China: A Case from Sichuan's Rural Industrial Sector, *The China Quarterly* 192, 915–932.

World Bank, 2007, *Cost of Pollution in China – Economic Estimates of Physical Damage*, Washington, DC, World Bank.

Xie, Dengke, Wenjing Yum & Zhenhua Xie, 2009, China Insists on Six Principles in Addressing Climate Change, *Xinhua News Agency*, 26 June 2009.

Xiong, Yan, 2011, Objectives and Principles of the Transformation to Low Carbon Development, *21st Century Economy*, 21 February 2011.

Xinhua News Agency, 2009a, China Invested over 10 Billion in Technologies Addressing Climate Change during the 11th Five-year Plan Period, *Xinhua News*, 12 November 2009.

Xinhua News Agency, 2009b, Xie Zhenhua: Five Major Achievements by China in Addressing Climate Change, *Xinhua Net*, 26 June 2009.

Yang, Zhangli & Yongde Ge, 2006, *Capacity Building in Local Environmental Governance: Case Study of Nantong City*, Beijing, Centre for Industrial Development and Environmental Governance, available at http://166.111.100.15:8081/cms/Reports/560.jhtml, last accessed 2 May 2013.

Zhu, Da & Jiang Ru, 2008, Strategic Environmental Assessment in China: Motivations, Politics, and Effectiveness, *Journal of Environmental Management* 88 (4), 615–626.

Zhu, Jianhua, 2011a, Low Carbon City – Baoding Pursued a Different Path for Development and Led the Trend of Green Growth of Cities, *People's Daily*, 25 August 2011.

Zhu, Tony, 2011b, China Plans to Have 100 New Energy Demonstration Cities by 2015, *Business China*, 27 April 2011.

Zhu, Zhiqun, 2010, *China's New Diplomacy: Rationale, Strategies and Significance*, Farnham, Ashgate.

17

# The BRICS Partnership: Development and Climate Change Policy from an African Perspective

*Oliver C. Ruppel & Katharina Ruppel-Schlichting*

*Abstract*

As a group of leading emerging economies, the BRICS partnership consisting of Brazil, the Russian Federation, India, China and South Africa has embarked on new policy pathways to exploit potential and new opportunities for economic and sustainable development. BRICS is gaining international importance, not only because its land area is more than a quarter of that of the world and because almost half of the world's population lives in its territories, but also because of its ever-growing share in the world economy. Although economic development may be at the heart of the BRICS partnership, environmental protection and climate change prevention are topics that have been on the agenda of BRICS since its formal inception in 2009. Having recognised that "climate change is one of the global threats challenging the livelihood of communities and countries", BRICS aims at a constructive dialogue on how to deal with climate change based on the principle of common but differentiated responsibilities. This article intends to shed some light on the development and climate change policy of BRICS from a predominantly southern African perspective.

## *A. Introduction*

The BRICS partnership is a grouping of leading emerging economies, namely Brazil, the Russian Federation, India, China and South Africa, playing a growing role in the world economy.[1] BRICS considers itself to be "a plat-

---

1 Prior to South Africa's first attendance of the Summit in 2011, the group was named BRIC (Brazil, Russian Federation, India and China). South Africa had received a formal invitation to join from China in 2010.

form for dialogue and cooperation amongst countries that represent 43% of the world's population, for the promotion of peace, security and development in a multi-polar, inter-dependent and increasingly complex, globalising world".[2] Looking at the enormous potential of BRICS, it seems that in this globalising world the "centre of economic gravity is shifting. The rise of Asia and emerging economies in other parts of the world is ushering in a new configuration of economic power and influence. This is a continuing process and we are now beginning to see the emergence of Africa".[3]

The global economy is undergoing a shift in economic power from North to South and West to East. The BRICS partners are fast emerging as the new sources of global economic growth, trade and investment flows, reshaping the global economy and driving recovery from the 2008/2009 Great Recession. Advanced economies are projected to grow only 1.2% in 2013, compared to 5.3% for emerging and developing economies as a group. By 2014 BRICS is expected to account for 61% of global growth. Developing countries' share of world trade will double over the next 40 years, to 69% by 2050. Hence there is a strategic importance to strengthening South–South trade, investment and development cooperation.[4]

In its early years the four initial BRIC countries (Brazil, Russian Federation, India and China) predominantly used the term BRIC in a much narrower sense, mostly in the context of companies wanting to expand into the emerging markets and banking groups wanting to provide new financial products. Eventually the four countries developed the idea further to foster closer links and the dialogue on the political level started in 2006.[5] Since 2009, annually and on a rotational basis, BRIC nations have held their summits where the heads of state as well as ministers and key business people come together. The first BRIC Summit was held in 2009 in Yekaterinburg, Russia; and the second took place in Brasília, Brazil, in 2010. In 2010, South Africa received a formal invitation from China to join the partnership. Thereafter BRIC became BRICS in 2011[6] and with South Africa joining the

2 See para. 3 of the 4th BRICS Summit Delhi Declaration, available at http://www.brics.utoronto.ca/docs/120329-delhi-declaration.html, last accessed 01 May 2013.
3 WTO (2013:17).
4 Projections in this paragraph have been given by Davies (2013).
5 See Mostert (2013) with further references.
6 See para. 2 of the Sanya Decalaration http://english.sanya.gov.cn/publicfiles/business/htmlfiles/englishsite/sanya%20brics/201109/2398.html, last accessed 01 May 2013.

partnership its commitments suddenly also became relevant for the entire African continent – at least indirectly.[7] With and after South Africa's inclusion, three further BRICS summits have been held: 2011 in Sanya, China, 2012 in New Delhi, India, and 2013 in Durban, South Africa.

## *B. The Nature of BRICS*

A study on the European Union (EU) foreign policy towards BRICS argues that the –[8]

> BRICS countries do not form a bloc and should thus not be approached ... as a coherent bloc, but rather as an increasingly dense and influential network which, together with other overlapping multilateral settings, has a growing impact on international political and economic governance.

The impact of the BRICS countries can only be correctly understood if it is "seen as part of a wider shift in the international balance of power, both politically and economically".[9] BRICS is emerging as an intergovernmental network – somewhat comparable to, for instance, the G20.[10] It functions on agenda-setting, consensus-building,[11] policy coordination[12] and as a platform for knowledge production and information exchange. So far BRICS consists of five states with no founding document (formal charter or treaty). This means that there is actually no formal structure, voting procedure or central secretariat. Moreover, BRICS so far fails to provide for any mechanism to come up with legally binding decisions, nor does it have a dispute settlement procedure or mechanism in place.[13] However, the BRICS leaders have issued several joint statements and declarations. Of particular impor-

7 See Ruppel (2013a).

8 Keukeleire (2011:32).

9 (ibid.).

10 For information on the G20 see http://www.g20.org/docs/about/about_G20.html, last accessed 28 April 2013.

11 The wide range of topics that the countries cover at the BRICS summit which translates into a declaration where the BRICS collectively state on which issues they formed consensus on; see the various BRICS Summit Declarations.

12 For example on climate change; the countries are committed to finding sustainable outcomes; see para. 22 of the Sanya Declaration http://english.sanya.gov.cn/publicfiles/business/htmlfiles/englishsite/sanya%20brics/201109/2398.html, last accessed 01 May 2013.

13 See Mostert (2013).

tance are the official documents that have resulted from the BRIC and BRICS summits, namely:[14]

- the 2009 Joint Statement, Yekaterinburg, Russia
- the 2010 Joint Statement, Brasília, Brazil
- the 2011 Sanya Declaration, Sanya, China
- the 2012 Delhi Declaration, New Delhi, India, and
- the 2013 eThekwini Declaration, Durban, South Africa.

Several other official documents have been produced by the Summits and on the BRICS ministerial level, such as the 2011 BRICS Agriculture Ministers Declaration or the 2011 BRICS Finance Ministers Communiqué. Yet, BRICS does not constitute an international organisation in the strict sense of public international law and it will have to be seen whether it will still develop as such in future.[15] BRICS is neither an international organisation nor a trade bloc in terms of a regional (or preferential) economic community. It refers to itself as a "partnership",[16] which comprises "a non-hierarchical governance structure in which relations among actors are repeated and enduring, but where no one has the power to arbitrate and resolve disputes among the members".[17] BRICS is merely a networking forum that has the potential for more (i.e., to become an economic powerhouse in future), considering that it covers an enormous population (more than 40% of the world's population) and landmass (over a quarter of the world's land area). It has been stated that while China and India can be considered to be the 'world's factory' and the 'world's office', respectively,[18] Russia has come to be known as the 'world's gas station' and Brazil as the 'world's farm'. In this context, South Africa can either serve as a 'gateway to Africa' or as 'a gatekeeper to Africa' and its rich natural resources.[19]

14 These and other official documents are available from the University of Toronto BRICS Information Centre at http://www.brics.utoronto.ca/docs/index.html, last accessed 08 May 2013.
15 See ITAR-TASS (2012).
16 Boisson de Chazournes (2009:657).
17 Podolny & Page (1998).
18 Fujita (2006).
19 Van den Bosch (2011).

In the 2013 Durban eThekwini Declaration,[20] BRICS reaffirmed its commitment to the promotion of international law, multilateralism and the central role of the United Nations (UN). The discussions also reflected a growing intra-BRICS solidarity, as well as the shared goal to contribute positively to global peace, stability, development and cooperation based on an inclusive approach of shared solidarity and cooperation towards all nations and peoples:[21]

> We aim at progressively developing BRICS into a full-fledged mechanism of current and long-term coordination on a wide range of key issues of the world economy and politics. The prevailing global governance architecture is regulated by institutions which were conceived in circumstances when the international landscape in all its aspects was characterised by very different challenges and opportunities. As the global economy is being reshaped, we are committed to exploring new models and approaches towards more equitable development and inclusive global growth by emphasising complementarities and building on our respective economic strengths.

## *C. The BRICS Development Agenda – A Brief Overview*

The BRICS agenda is calling for a comprehensive reform of the UN Security Council[22] and has identified other common themes such as the control of climate change, the combatting of terrorism, and the reform of the International Monetary Fund (IMF) and the World Bank.[23] The Sanya BRICS Declaration on the global economy, international financial issues and developmental affairs emphasises that "[I]n the economic, financial and development fields BRICS serves as a major platform for dialogue and cooperation" and the group has agreed to continue further expanding and deepening of economic, trade and investment cooperation among BRICS countries. Furthermore, BRICS countries, by signing the declaration, have committed themselves "to assure that the BRICS countries will continue to enjoy strong and sustained economic growth supported by our increased cooperation in

---

20 2013 Durban 5th BRICS summit eThekwini Declaration and Action Plan available http://www.brics5.co.za/fifth-brics-summit-declaration-and-action-plan/, last accessed 01 May 2013.

21 Para. 2 of the 2013 eThekwini Declaration.

22 Para. 8 of the Sanya Declaration.

23 Para. 9 of the Delhi Declaration.

economic, finance and trade matters, which will contribute to the long-term steady, sound and balanced growth of the world economy".[24]

On the challenges related to the aforementioned balanced growth envisaged by the BRICS countries, a recently completed report on *Economic Policy and Social Affairs in the BRICS* concludes as follows:[25]

> Sustaining growth and addressing the equity challenge in the long run will require the BRICS countries to strengthen capacities for production and innovation, while improving links with the global knowledge economy. This will be possible only if countries take an integrated approach to development. Identifying priorities, aligning activities aimed at scientific and technological development, and upgrading production capacities will need to be accompanied by targeted actions addressing skills-based, financial and infrastructural obstacles. Barriers and incentives differ from country to country; there will be no single, unique response. Each country has to develop its particular, innovative way of shaping its own development trajectory.

According to the Sanya Declaration, BRICS is generally committed to supporting a strong, open, rules-based multilateral trading system embodied in the World Trade Organisation (WTO) and a successful, comprehensive and balanced conclusion of the Doha Development Round.[26] After more than 10 years of repeated negotiation failures of the Doha Development Round, rumours have become louder that the Doha negotiations are unlikely to be concluded in the near future. Some even contend that the "WTO risks its future by keeping Doha alive".[27] During the recent process of selecting a new director general for the WTO, BRICS countries have been standing closely together[28] in support of the application of the Brazilian candidate Roberto Azevêdo (the United States and the European Union had supported Mexican trade minister Herminio Blanco). Azevêdo was approved by WTO members as the new director-general at the general council meeting on 14 May 2013 and is the first Latin-American holding the post.[29] Expectations are high that the new head of the WTO may help to ease global trade talks and to conclude the Doha Development Round successfully.

24 Para. 13 of the Sanya Declaration.
25 Reisen (2013).
26 Para. 26 of the Sanya Declaration.
27 Miles (2011).
28 With strong support from Africa, which has the most WTO members by continent.
29 Süddeutsche Zeitung (2013).

In the 2013 eThekwini Declaration, BRICS reaffirmed its support for an open, transparent and rules-based multilateral trading system:[30]

> We will continue in our efforts for the successful conclusion of the Doha Round, based on the progress made and in keeping with its mandate, while upholding the principles of transparency, inclusiveness and multilateralism. We are committed to ensure that new proposals and approaches to the Doha Round negotiations will reinforce the core principles and the developmental mandate of the Doha Round. We look forward to significant and meaningful deliverables that are balanced and address key development concerns of the poorest and most vulnerable WTO members, at the ninth Ministerial Conference of the WTO in Bali. We note that the process is underway for the selection of a new WTO Director-General in 2013. We concur that the WTO requires a new leader who demonstrates a commitment to multilateralism and to enhancing the effectiveness of the WTO including through a commitment to support efforts that will lead to an expeditious conclusion of the DDA.

Concerning excessive volatility in commodity prices, BRICS supports the international community in strengthening cooperation to ensure stability and development of physical markets by reducing distortions and further regulating financial markets. In the 2013 eThekwini Declaration, BRICS also expressed its openness to increasing engagement and cooperation with non-BRICS countries, in particular emerging markets and developing countries (EMDCs), and relevant international and regional organisations, as envisioned in the Sanya Declaration.[31]

BRICS considers safe nuclear energy as an important element in future energy supply and supports the development of renewable energy as being a part of a future energy mix in and for BRICS countries. "Russia and Brazil are core oil and gas exporters, China and India are consummate consumers; South Africa can unlock marginal production from the Gulf of Guinea to the deep waters of Mozambique, adding extra energy…to the BRICS bloc". So far BRICS has no common vision on the energy front. And it will most probably be "the energy question that exemplifies the BRICS overall economic potential …".[32]

30 Paras 15 and 16 of the 2013 eThekwini Declaration.

31 Para. 3 of the 2013 eThekwini Declaration.

32 Hulbert & Brütsch (2012).

### D. *BRICS-Africa Relations*

The Africa relationship of BRICS is far more complex, internally divergent and perhaps precarious than it may seem. This does not necessarily take away from the fact that all five of the BRICS countries will increase in global influence and that partnerships with one another and African nations will be significant, but these dynamics will not be straightforward, and could become increasingly complex with the rise of other developing economies. Nevertheless, it is important to understand that BRICS is intent on cooperating with Africa to enhance its overall influence. Though it may not be the magic fix that many desire, and though it will not be the straightforward relationship sometimes suggested, the BRICS-Africa relationship provides potential for positive cooperation in the future.[33]

In the 2013 eThekwini Declaration, BRICS plans to hold a retreat together with African leaders with the theme, *Unlocking Africa's potential: BRICS and Africa Cooperation on Infrastructure*. The retreat will be an opportunity for BRICS and African leaders to discuss how to strengthen cooperation between the BRICS countries and the African continent:[34]

> Recognising the importance of regional integration for Africa's sustainable growth, development and poverty eradication, we reaffirm our support for the Continent's integration processes. Within the framework of the New Partnership for Africa's Development (NEPAD), we support African countries in their industrialisation process through stimulating foreign direct investment, knowledge exchange, capacity-building and diversification of imports from Africa. We acknowledge that infrastructure development in Africa is important and recognise the strides made by the African Union to identify and address the continent's infrastructure challenges .... We will seek to stimulate infrastructure investment on the basis of mutual benefit to support industrial development, job-creation, skills development, food and nutrition security and poverty eradication and sustainable development in Africa. We therefore, reaffirm our support for sustainable infrastructure development in Africa.

Thus far Africa remains poor regardless of its high concentration of natural resources.[35] Africa's economic outlook may look bright in GDP terms, but this conceals structural deficiencies and deep imbalances. First of all, African economies are heavily dependent on exports (mainly commodities and raw materials), which means that the current high international com-

33 Khanna (2013).

34 Para. 5 of the 2013 eThekwini Declaration.

35 See Ruppel (2013b and c).

modity prices are inflating growth statistics.[36] Most foreign direct investment attracted by African states relates to natural resources.[37]

Moreover, the global –[38]

> thirst for energy has generated exceptional conditions for African economies to continue exploiting their natural resources (all of which are depleting at an ever accelerating rate) and focus only on infrastructure that serves the purpose of taking 'stuff' away from the African continent (the Chinese, for instance, have invested in ports, railway systems and other 'corridors' to ease the transportation of commodities towards the East).

Two of the three sub-Saharan states currently receiving the largest amount of investment from China include South Africa and the Democratic Republic of Congo.[39] South Africa is China's second largest source of imports dominated by mineral resources.[40] Since Chinese and African populations together constitute about one fourth of the global community,[41] aspects of this relationship have a major impact on global sustainable development. It is vital for Africa to have the policies, laws and regulations effectively in place to propel development, but also provide protection against overuse, abuse and pollution of non-renewable natural resources by foreign investors.[42]

In an assessment of whether a relationship with BRIC will prove to be South Africa's way ahead, it has been summarised as follows:[43]

> Poverty, poor infrastructure, lack of productive capacity and transfer of technology, the emerging threats associated with climate change as well as the food, energy, financial and economic crises, have been identified as areas where Africa can enhance its capacity by cooperating with other developing countries. Furthermore, the increased bargaining power of developing countries in multilateral negotiations, as reflected in the current Doha negotiations of the World Trade Organisation (WTO), has been cited as another reason for cooperation. It is against this background that since 2000, African countries have entered into new partnerships and arrangements with the South, increasingly driven by economics rather than politics. The new partnerships are often based on formal frameworks with dialogue forums and action plans. The agenda behind the renewed and increased global economic interest in Africa, a continent that was

36 Fioramonti (2013).
37 Marafa (2009:13).
38 Fioramonti (2013).
39 Scissors (2011); see also Van der Lugt et al. (2011).
40 Marafa (2009:10).
41 (ibid.:8).
42 Susswein (2003:309).
43 Sandrey (2013:6).

> once dubbed 'hopeless' should be considered. In historical representations Africa has been regarded as underdeveloped and poor, but of late Africa has been regarded as a continent brimming with potential and opportunities. The need of boom economies, like China and India, for raw materials, is generating valuable new opportunities. Can this be regarded as a new scramble for Africa in a post-colonial era?

China indeed attaches ever-increasing importance to BRICS and its Africa relations. The reasons lie mainly in two aspects. The first is economic, namely to promote China's influence and interest in the global economy and international financial institutions, and to share China's development achievement with other developing partners. The second is to place political and security affairs on the BRICS agenda. The fast development of Sino-African relationships in the past decade is one of the most significant international advancements.[44] Sino-African relations have significantly upgraded from a linear and single-dimension relationship to a multilayer and multidimension one, with the support of institutionalisation efforts, especially the establishment and development of the Forum on China-Africa Cooperation (FOCAC).[45]

## E. South Africa: The BRICS Development Partner in Africa?

When compared to its BRICS partners, South Africa's role of is somewhat different, as its economy is much smaller than that of the other four countries and, strictly speaking, South Africa does not comply with all the characteristics generally associated with the country group: (1) the outstanding size of their economies, (2) strong growth rates, leading to increasing significance in world economy, and (3) the demand for a stronger political voice in international governance structures, which corresponds to their economic status. Nevertheless, BRICS represents a heterogeneous group, with individual countries also forming other coalitions. Beside the differentiation made above for South Africa, China is in an exceptional position concerning most aspects of economic cooperation and Russia stands out as a former superpower.[46]

---

44 49 Countries out of the continent's 54 countries have established diplomatic ties with China, see Chun (2013).

45 See with further references CCS (2013).

46 Morazán et al. (2012).

There had been other contenders for a spot in the former BRIC, but South Africa was chosen and not necessarily for economic reasons per se. South Africa has a relatively small economy as well as a much smaller population and a slower growth rate compared to the BRIC countries. However, from global governance and network-strengthening perspectives, the next natural partner was a country such as South Africa. South Africa is highly interested in support for its African agenda, through which it hopes, inter alia, to foster the growth of infrastructure on the continent, as well as to promote development that will advance governance, peace and security in Africa. So far, South Africa is the only African BRICS member and it thus has specific stakes in the African continent. South Africa's role within BRICS has been formulated as follows:[47]

> The strong stance of South Africa on the African continent has limitations in a setting with 54 states. South Africa has been publicly denounced as 'not representing Africa' in the past – not least so by the African Union Commission. The country's government is cautious to be a team player in Africa and has a strong liking for multilateral processes. … The engagement within the BRICS is thus complementing other South African activities meant to foster development on the African continent. This certainly includes debates on peace and security and structural reform of global governance. And, not to forget, this also includes an element of competition, as South African enterprises are facing competition by other BRICS actors in what they consider their natural area for economic expansion. This competition has to be managed by all BRICS countries. With the particularities of South Africa, however, trade is not following the flag (nor vice versa); rather, both seem to operate alongside each other. The African agenda is emphasised as the underlying tune of all these endeavors.

During the Public Dialogue on *South Africa and the BRICS: Prospects for the Future*, held at the Centre for Conflict Resolution in Cape Town on 24 April 2013, the South African Minister for Trade and Industry, Dr Rob Davies inter alia mentioned the following policy positions:[48] South Africa is currently the rotating chair for BRICS for the next year and BRICS is a flagship for South African trade and foreign policy. Major structural changes in the international environment affect the world economy where the rise of important emerging economies is contributing towards an East–West shift in development, moving away from the North-South divide of the past. South Africa is a member of the 'club of strong emerging economies' because it is the most industrialised and largest economy on the African continent. Fur-

47 Grimm (2013:39 and 43).
48 Davies (2013).

thermore, Africa – apart from Asia – is the fastest growing continent and it is ready for industrialisation now. In this light the commodity cycle in minerals trade will be overcome: in order to industrialise the continent it is important to promote more value added trade (especially in the context of minerals) and improved agriculture. Regional integration efforts from Cape to Cairo, such as envisaged by the initiative consisting of the Common Market for Eastern and Southern Africa, the Eastern African Community and the Southern African Development Community (COMESA-EAC-SADC tripartite initiative) and industrial infrastructure development, are crucial. In terms of South Africa-BRICS trade this means that, while traditional developed country partners remain important, South Africa needs to diversify and strengthen its economic relations with other emerging economies. South Africa aims at championing a new paradigm for equitable development that focuses on mutually beneficial complementary trade, not destructive competition. In the same light South Africa is prepared to render increasing support for Africa's development agenda (especially regional integration, industrial capacity and infrastructure development).[49]

### *F. BRICS Development Bank*

At the 4th BRICS Summit in Delhi, India, in 2012, the BRICS countries called for the various finance ministers to examine the feasibility and viability of a new development bank to mobilise resources for infrastructure and sustainable development projects in BRICS countries, as well as in other emerging and developing economies. At the 5th BRICS Summit, held in Durban, South Africa, in March 2013, BRICS leaders expressed their intention to set up a BRICS development bank, which would finance infrastructure and sustainable development projects across BRICS nations and other developing countries. The theme of the Summit was *BRICS and Africa Partnership for Development, Integration and Industrialisation*.

The eThekwini Declaration considers a range of international issues, including the global economy, sustainable development, climate change and trade. The Declaration states that the initial contribution to the development bank should be substantial and sufficient for it to be effective in financing infrastructure. At the same time BRICS calls for the reform of international

49 (ibid.).

financial institutions to make them more representative and to reflect the growing weight of BRICS and other developing countries –[50]

> We remain concerned with the slow pace of the reform of the IMF. We see an urgent need to implement, as agreed, the 2010 International Monetary Fund (IMF) Governance and Quota Reform. We urge all members to take all necessary steps to achieve an agreement on the quota formula and complete the next general quota review by January 2014. The reform of the IMF should strengthen the voice and representation of the poorest members of the IMF, including Sub-Saharan Africa. All options should be explored, with an open mind, to achieve this. We support the reform and improvement of the international monetary system, with a broad-based international reserve currency system providing stability and certainty. We welcome the discussion about the role of the SDR in the existing international monetary system including the composition of SDR's basket of currencies. We support the IMF to make its surveillance framework more integrated and even-handed. The leadership selection of IFIs should be through an open, transparent and merit-based process and truly open to candidates from the emerging market economies and developing countries. We emphasise the importance of ensuring steady, adequate and predictable access to long term finance for developing countries from a variety of sources. We would like to see concerted global effort towards infrastructure financing and investment through the instrumentality of adequately resourced Multilateral Development Banks (MDBs) and Regional Development Banks (RDBs).

Regarding the establishment of a BRICS development bank, an "irreversible decision has been taken in Durban in March 2013",[51] namely to go ahead with the bank. What this exactly means, however, still needs to be determined, as a variety of questions are still unresolved.[52] These include the modalities of the bank, the financial contribution required from each BRICS country, how voting rights will be allocated and the bank's physical location, etc. The World Bank has welcomed the BRICS plan for a development bank: "The World Bank Group believes partnership is central to its development mission and would naturally welcome a strong working relationship with a new BRICS Development Bank."[53] It is expected that such a BRICS development bank will only start functioning in a couple of years after all the unresolved details have been worked out.[54]

50 Paras 13 and 14 of the 2013 eThekwini Declaration.

51 Davies (2013).

52 Coleman (2013).

53 Donelly (2013).

54 Further discussions on the bank are set to take place alongside the G20 Summit in Russia in September 2013; see Viljoen (2013).

## G. BRICS and Climate Change

Accelerating sustainable growth of developing countries is one of the major world challenges and BRICS sees growth and development as central to addressing poverty and to achieving the United Nations Millennium Development Goals (MDGs). Infrastructure development in Africa and the industrialisation of the continent within the framework of the New Partnership for Africa`s Development (NEPAD) is a major objective of the BRICS members. Therefore, in its fourth Declaration signed in New Delhi in March 2012,[55] BRICS states again emphasised their strong commitment to enhancing sustainable development by also focusing on environmental protection. In the Declaration, the BRICS states affirm their commitment to the implementation of the United Nations Convention on Climate Change (UNFCCC) and the Convention on Biological Diversity, among others. It is specifically pointed out that "sustainable development should be the main paradigm in environmental issues, as well as for economic and social strategies".[56] It is noteworthy that the Delhi Declaration spells out the commitment of the BRICS states towards environmental protection and respective multilateral environmental Agreements, and the commitment towards the achievement of the Millennium Development Goals, but the Declaration is also very clear regarding the responsibilities of developed nations and the need for them to ensure that growth in non-developed countries is not affected.[57]

The Sanya Declaration refers to environmental matters, with climate change leading the way, as climate change is considered to be "one of the global threats challenging the livelihood of communities and countries".[58] In this regard, and highlighting the principle of equity and common but differentiated responsibilities, BRICS leaders have committed themselves "to work towards a comprehensive, balanced and binding outcome to strengthen the implementation of the United Nations Framework Convention on Climate Change and its Kyoto Protocol"[59] and to enhance "practical cooperation in adapting our economy and society to climate change".[60] Moreover, cooperation has been envisaged in order to "reach new political commitment

55 Paras 28–35 of the Delhi Declaration.
56 (ibid.:para. 32).
57 Paras 29–35 of the Delhi Declaration.
58 Para. 22 of the Sanya Declaration.
59 (ibid.).
60 (ibid.).

and achieve positive and practical results in areas of economic growth, social development and environmental protection under the framework of sustainable development".[61]

With the fourth BRICS Declaration signed in New Delhi in March 2012,[62] BRICS states have welcomed the significant outcomes of the 17th Conference of Parties to the United Nations Framework Convention on Climate Change and the 7th Conference of the Parties serving as the Meeting of the Parties to the Kyoto Protocol (COP17/CMP7) held in Durban, South Africa, in December 2011. It was declared that the BRICS states "are ready to work with the international community to implement its decisions in accordance with the principles of equity and common but differentiated responsibilities and respective capabilities". Furthermore, the following commitment with regard to climate change was laid down in the Delhi Declaration:[63]

> We are fully committed to playing our part in the global fight against climate change and will contribute to the global effort in dealing with climate change issues through sustainable and inclusive growth and not by capping development. We emphasise that developed country Parties to the UNFCCC shall provide enhanced financial, technology and capacity-building support for the preparation and implementation of nationally appropriate mitigation actions of developing countries.

In the context of BRICS and climate change, it is also worth mentioning that the so-called BASIC Group – Brazil, South Africa, India and China – which was formed in 2009, has, as a group with increasing economic and geopolitical influence, embarked on international climate change negotiations by, amongst others, drafting the Copenhagen Accord together with the United States for the 15th Session of the Conference of the Parties to the UNFCCC (COP15).[64] The role of the BASIC countries for climate-related issues has been described as follows:[65]

> BASIC countries – both individually and collectively – are rapidly moving to the centre stage of international politics. Their increased influence in international climate diplomacy was clearly seen for the first time at the Copenhagen Climate Conference (COP15). … The foundation of the BASIC cooperation can be traced back to a common 'third world' identity formed during several decades

61 (ibid.:para. 23).
62 Paras 29 and 30 of the Delhi Declaration.
63 Para. 30 of the Delhi.
64 Hallding et al. (2011).
65 (ibid.:13).

> in the G77 group of developing countries, in which the four BASIC countries have each played leading roles. … BASIC ministers responsible for climate-related issues have met quarterly, suggesting that the group is committed to continued cooperation. From the joint statements issued after the meetings, two clear patterns emerge: first, a call for a second commitment period of the Kyoto Protocol; and, second, an effort to build bridges with and show support for the rest of the G77. The discussions thus seem to generate agreement on a few broad principles, but have so far been unable to settle the finer details needed to articulate concrete contributions for the international negotiating process.

In the 2013 eThekwini Declaration, the BRICS indicate that the 5th BRICS Summit "concluded the first cycle of BRICS Summits". The countries reaffirmed their "commitment to the promotion of international law, multilateralism and the central role of the United Nations (UN)". BRICS representatives acknowledge that the current global governance architecture is ruled by entities conceived when the international landscape presented very different challenges and opportunities:[66]

> We are committed to building a harmonious world of lasting peace and common prosperity and reaffirm that the 21st century should be marked by peace, security, development, and cooperation. It is the overarching objective and strong shared desire for peace, security, development and cooperation that brought together BRICS countries. We welcome the twentieth Anniversary of the World Conference on Human Rights and of the Vienna Declaration and Programme of Action and agree to explore cooperation in the field of human rights.

In 2013 BRICS called on all parties to build on the decisions adopted in the 18th Session of the Conference of the Parties (COP18) to the UNFCCC held in Doha, Qatar, at the end of 2012, in order to reach a successful conclusion by 2015 of negotiations on the development of a protocol, another legal instrument or an agreed outcome "with legal force, under the Convention applicable to all Parties, and guided by its principles and provisions".[67] Lastly, the eThekwini Action Plan furthermore envisages consultative meetings of BRICS senior officials in the fields of sustainable development, environment and climate change.

---

66 Paras 20–23 of the 2013 eThekwini Declaration.

67 Para. 37 of the 2013 eThekwini Declaration.

## H. Concluding Remarks

BRICS provides a new climate for development with high potential in a world of changing power balances. BRICS countries now constitute the largest trading partners of Africa and the largest investors too. No BRICS country is an Organisation for Economic Co-operation and Development (OECD) member and, more than ever before, non-OECD member economies seem to be transforming into world economic forces in the global economy. At the same time, BRICS is fast emerging as the new source of global economic development. Although at present BRICS does not constitute an international organisation in the strict sense of public international law, BRICS countries are forming an increasingly influential network with a growing impact on international political and economic governance. In the 2013 Durban eThekwini Declaration, BRICS reaffirmed its commitment to the promotion of international law, multilateralism and the central role of the United Nations.

The cooperation of BRICS members with one another and with African nations provides an enormous potential for development in the future. China attaches ever-increasing importance to BRICS and its Africa relations for economic and other reasons. South Africa is highly interested in support for its African agenda, through which it hopes, inter alia, to foster the growth of infrastructure on the continent, as well as to promote development that will advance governance, peace and security on the continent. South Africa is so far the only African BRICS member and it thus has specific stakes in the African continent. A new BRICS development bank may become very useful for this purpose, especially once the unresolved details have been worked out.

Africa must select its partners carefully as it remains poor regardless of its high concentration of natural resources. This fact should place particular focus on its sustainable development agenda in future. Although Africa's economic outlook looks bright, it is vital for Africa to have adequate policies and the rule of law in place to provide protection against exploitation by foreign investors. This seems to be in line with the following message of United Nations secretary-general, Ban Ki-moon (May 2011):[68]

> For most of the last century, economic growth was fuelled by what seemed a certain truth: the abundance of natural resources. The world mined its way to

68 Ki-moon (2011).

> growth and burned its way to prosperity. Those days are gone. In the twenty-first century, supplies are running short and the global thermostat is running high. Climate change is showing us that the old model is more than obsolete. It is in fact extremely dangerous. How do we lay the foundation for future growth? How do we lift people out of poverty while protecting the planet and ecosystems that support economic growth? How do we regain the balance? All of this requires rethinking. We have to be prepared to make major changes – in our lifestyles, our economic models, our social organisation, and our political life. We have to connect the dots between climate change and issues such as water, energy and food. The challenge is great – but, so too, is the opportunity. The sustainable development agenda is the growth agenda for the twenty-first century.

BRICS proclaims to be committed to playing its part in the global fight against climate change and to contribute to the global effort in dealing with climate change issues through sustainable and inclusive growth, and not by capping development. Although energy derived from fossil fuels still dominates its energy mix for the foreseeable future, BRICS states intend to expand sourcing of clean and renewable energy, as well as the use of energy-efficient and alternative technologies, to meet the increasing demand of their economies and people, and to respond to climate concerns as well. The commitment of BRICS to climate concerns is laudable: by calling on all parties to build on the decisions adopted in the 18th Session of the Conference of the Parties (COP18) to the UNFCCC held in Doha, Qatar, at the end of 2012, BRICS shows ambition to reach a successful conclusion by 2015 of negotiations on the development of a protocol, another legal instrument or an agreed outcome "with legal force, under the Convention applicable to all Parties, and guided by its principles and provisions".[69]

It will, however, have to be seen whether the BRICS partners – who are at the same time economic competitors – will be able to achieve their ambitious (climate-change-related) goals in the absence of a more structured formal environment.

## *References*

Boisson de Chazournes, Laurence, 2009, Changing Roles of International Organizations: Global Administrative Law and the Interplay of Legitimacies, *International Organizations Law Review* 6 (2), 655–666.

69 Para. 37 of the 2013 eThekwini Declaration.

CCS/Centre for Chinese Studies, 2013, The BRICS Summit in 2013: Is the Road from Durban Leading into Africa?, *The China Monitor*, Special Edition, Stellenbosch, Centre for Chinese Studies, available at http://www.ccs.org.za/wp-content/uploads/2013/03/China_Monitor_BRICS_Special-Edition_March-2013_Website.pdf, last accessed 07 May 2013.

Chun, Zhang, 2013, A Promising Partnership between BRICS and Africa: A Chinese Perspective, *The China Monitor*, Special Edition, Stellenbosch, Centre for Chinese Studies, 30–37, available at http://www.ccs.org.za/wp-content/uploads/2013/03/China_Monitor_BRICS_Special-Edition_March-2013_Website.pdf, last accessed 07 May 2013.

Coleman, Isobel, 2013, Ten Questions for the New BRICS Bank, *Foreign Policy*, 09 April 2013, available at http://www.foreignpolicy.com/articles/2013/04/09/ten_questions_for_the_new_brics_bank, last accessed 07 May 2013.

Davies, Rob, 2013, Speech given by the South African Minister for Trade and Industry, Dr Rob Davies during the Public Dialogue on South Africa and the BRICS: Prospects for the Future, held at the Centre for Conflict Resolution (CCR) in Cape Town on 24 April 2013, audio file available at http://www.ccr.org.za/index.php?option=com_phocadownload&view=category&id=164:sa-brics-group&Itemid=92, last accessed 07 May 2013.

Donelly, Lynley, 2013, World Bank Welcomes Brics Plan for Development Bank, *Mail and Guardian* 25 January 2013, available at http://mg.co.za/article/2013-01-25-world-bank-welcomes-brics-plan-for-development-bank, last accessed 24 April 2013.

Fioramonti, Lorenzo, 2013, Africa: The Brics and Africa's Growth Dilemma, *AllAfrica*, 26 March 2013, available at http://allafrica.com/stories/201303271114.html, last accessed 30 April 2013.

Fujita, Masahisa, 2006, The Coming Age of China-plus-One: The Japanese Perspective on East Asian Production Networks, available at http://siteresources.worldbank.org/INTCHIINDG-LOECO/Resources/Fujita_Hamaguchi_draft2_ACCEPT_ALL.doc, last accessed 19 October 2011.

Grimm, Sven, 2013, South Africa: BRICS Member and Development Partner in Africa, *The China Monitor*, Special Edition, The BRICS Summit in 2013: Is the Road from Durban Leading into Africa?, 38–44, available at http://www.ccs.org.za/wp-content/uploads/2013/03/China_Monitor_BRICS_Special-Edition_March-2013_Website.pdf, last accessed 07 May 2013.

Hallding, Karl, Marie Olsson, Aaron Atteridge, Antto Vihma, Marcus Carson & Mikael Román, 2011, *Together Alone. BASIC Countries and the Climate Change Conundrum*, Copenhagen, Nordic Council of Ministers.

Hulbert, Matthew & Christian Brütsch, 2012, Charging the BRICS to Power the World, Global Policy Essay, January 2012, available at http://www.globalpolicyjournal.com/sites/default/files/pdf/Hulbert%20and%20Brutsch%20Essay%20Jan2012%20Charging%20the%20BRICS%20to%20Power%20the%20World.pdf, last accessed 07 May 2013.

ITAR-TASS, 2012, BRICS is Not Full-fledged International Organisation So Far, *Russia and India Report* 14 September 2012, available at http://indrus.in/articles/2012/09/14/brics_is_not_full-fledged_international_organization_so_far_-_lavrov_17659.html, last accessed 07 May 2013.

Joel M. Podolny & Karen L. Page, 1998, Network Forms of Organisations, *Annual Review of Sociology* 24 (1), 57–76.

Keukeleire, Stephan, Mikael Mattlin, Bas Hooijmaaijers, Timo Behr, Juha Jokela, Mikael Wigell & Vadim Kononenko, 2011, The EU Foreign Policy Towards the BRICS and other Emerging Powers: Objectives and Strategies, Brussels, European Parliament, Directorate-General for External Policies, available at http://www.europarl.europa.eu/committees/en/studiesdownload.html?languageDocument=EN&file=49151, last accessed 07 May 2013.

Khanna, Kartikeya, 2013, BRICS in Africa: A Development Dream?, *Think Africa Press*, 25 March 2013, available at http://thinkafricapress.com/development/brics-africa-living-development-dream; last accessed 07 May 2013.

Ki-moon, Ban, 2011, Message (delivered by Jean Christophe Bouvier) at the Fourth Nevsky International Ecological Congress, Saint Petersburg, Russian Federation, 16 May 2011, available at http://www.un.org/News/Press/docs/2011/sgsm13576.doc.htm, last accessed 12 April 2013.

Marafa, Lawal M., 2009, Africa's Business and Development Relationship with China: Seeking Moral and Capital Values of the Last Economic Frontier, NAI Discussion Paper No. 44, Uppsala, Nordiska Afrikainstitutet.

Miles, Tom, 2011, WTO Risks Future by Keeping Doha Alive, *Taipei Times*, 31 December 2011, available at http://www.taipeitimes.com/News/editorials/archives/2011/12/31/2003522031, last accessed 07 May 2013.

Morazán, Pedro, Irene Knoke, Doris Knoblauch & Thobias Schäfer, 2012, The Role of BRICS in the Developing World, Brussels, European Parliament, Directorate-General for External Policies, Policy Department, available at http://www.ab.gov.tr/files/ardb/evt/1_avrupa_birligi/1_9_politikalar/1_9_8_dis_politika/The_role_of_BRICS_in_the_developing_world.pdf, last accessed 07 May 2013.

Mostert, Werner, 2013, *BRICS as Intergovernmental Network and a Proposal for the Mandate and Shareholding of the Envisaged BRICS Development Bank*, Unpublished thesis submitted in partial fulfilment oft the degree of Master of Laws (LLM), University of Stellenbosch, on file with the author.

Reisen, Helmut, 2013, Economic Policy and Social Affairs in the BRICS, Gütersloh, Bertelsmann Stiftung, available at http://www.sgi-network.org/brics/pdf/BRICS_Economy_and_Social_Affairs.pdf, last accessed 07 May 2013.

Ruppel, Oliver C., 2013a, Environmental Law in the African Union (AU), in: Ruppel, Oliver C. & Katharina Ruppel-Schlichting, *Environmental Law and Policy in Namibia* (Second Edition), Windhoek and Essen, Orumbonde Press and Welwitschia Verlag Dr. A. Eckl, 53–68.

Ruppel, Oliver C., 2013b, International Trade Law and the Environment, in: Glazewski, Jan (Ed.) *Environmental Law in South Africa*, Cape Town, LexisNexis.

Ruppel, Oliver C., 2013c, Trade, Environment and Sustainable Development, in: Ruppel, Oliver C. & Katharina Ruppel-Schlichting, *Environmental Law and Policy in Namibia* (Second Edition), Windhoek and Essen, Orumbonde Press and Welwitschia Verlag Dr. A. Eckl, 421–470.

Sandrey, Ron, 2013, BRICS: South Africa's Way Ahead? Summary, Stellenbosch, tralac, available at http://www.tralac.org/files/2013/03/BRICS-South-Africas-way-ahead-Summary-20130306.pdf, last accessed 07 May 2013.

Scissors, Derek, 2011, China's Investment Overseas in 2010, The Heritage Foundation, available at http://report.heritage.org/wm3133, last accessed 02 June 2011.

Süddeutsche Zeitung, 2013, Brasilianer Azevêdo wird neuer WTO-Chef, *Süddeutsche Zeitung*, 08 May 2013, available at http://www.sueddeutsche.de/wirtschaft/welthandelsorganisation-brasilianer-azevdo-wird-neuer-wto-chef-1.1668065, last accessed 08 May 2013.

Susswein, Pendo M. 2003, The Use of Market Based Economic Instruments for Sustainable Environmental Management in Africa: A Case Study of the SADC Region, in: Chaytor, Beatrice & Kevin R. Gray (Eds), *International Environmental Law and Policy in Africa*, Dordrecht, Kluwer Academic Publishers, 297–326.

Van den Bosch, Servaas, 2011, Development: African LDCs Won't Benefit Much from BRICS Arrival, *AllAfrica* 31 January 2011, available at http://allafrica.com/stories/201101311879.html, last accessed 19 October 2011.

Van der Lugt, Sanne, Victoria Hamblin, Meryl Burgess & Elizabeth Schickerling, 2011, Assessing China's Role in Foreign Direct Investment in Southern Africa, Stellenbosch, Centre for Chinese Studies, available at http://www.ccs.org.za/wp-content/uploads/2011/03/Final-report-CCS-March-2011-CCS.pdf, last accessed 07 May 2013.

Viljoen, Willemien, 2013, The Fifth BRICS Summit, 26-27 March 2013, tralac hot seat comment available at http://www.tralac.org/2013/03/27/the-fifth-brics-summit-26-27-march-2013/?utm_source=Weekly+tralac+Newsletter&utm_campaign=f65ff63003-NL20130327&utm_medium=email, last accessed 28 April 2013.

WTO/World Trade Organization, 2013, *The Future of Trade: The Challenges of Convergence*, Report of the Panel on Defining the Future of Trade convened by WTO Director-General Pascal Lamy, Geneva, WTO, available at http://www.wto.org/english/thewto_e/dg_e/dft_panel_e/future_of_trade_report_e.pdf, last accessed 07 May 2013.

# PART V: CLIMATE CHANGE RESPONSES, EQUITY AND SUSTAINABLE DEVELOPMENT

18

# Sustainable and Inclusive Adaptation to Climate Change and Public Policy Challenges in Central America*

*Julie G. Lennox*

*Abstract*

Highly vulnerable, small developing countries have to develop public policy responses to climate change in the face of fiscal constraints, the current economic recession, excruciatingly slow international negotiations, and their immediate and long-term development challenges. This article presents findings regarding potential climate change impacts and policy response discussions in Central America. The separation between adaptation and mitigation, so prevalent in international negotiations and finance, is probably not appropriate for policymaking in small developing countries. An alternative is to prioritise adaptation that is both sustainable and inclusive. Measures to transition to economies that are more sustainable and low-carbon-oriented and that improve intra- and intergenerational equity would be integrated within this basic approach. Sector-specific knowledge and policies will have to be developed, while identifying intersectoral co-benefits and adverse effects, and aligning them with fundamental development challenges. Countries that are already increasingly exposed to extreme climate events could prioritise measures to reduce vulnerability to these events, given their higher profile on the political agenda, while ensuring that these

---

* This chapter is a translated version of the key messages of ECLAC et al. (2012f) published by the Economic Commission for Latin America and the Caribbean (ECLAC), the Ministries of Treasury/Finance and Environment of the seven Central American countries, through their Council of Finance or Treasury Ministers (COSEFIN) and their Commission for Environment and Development (CCAD) of the Central American Integration System (SICA) and the Secretariat for Central American Economic Integration (SIECA), with funding from UKAID/DFID of the Government of the United Kingdom and the Government of Denmark´s development cooperation programme, DANIDA. The Coordinator of this initiative, Julie Gail Lennox, is responsible for the introduction and final considerations, and these are not necessarily those of the partner institutions of this initiative.

measures serve as first steps along a sustainable and inclusive adaptation pathway.

## A. Introduction

One of the great challenges of public policy development and governance for small, highly vulnerable developing countries is their response to climate change, given both immediate and long-term development challenges, the economic recession in developed economies with spill-over effects for their countries, and excruciatingly slow international negotiations aimed at a global agreement. This article explores this challenge in the light of findings from an initiative in Central America that aims to develop knowledge of potential impacts of climate change, to increase policymakers' awareness across multiple sectors, and to encourage knowledge-based dialogue and policymaking. The initiative addresses the following questions: Is the separation between adaptation and mitigation that is so prevalent in the international arena appropriate for policymaking in small developing countries with limited public budgets? How can knowledge of sector-specific potential impacts and responses, and their costs and benefits, be developed, while ensuring that the response of one sector does not undo efforts or have adverse effects on another? Can the response to the additional challenges generated by climate change be made in ways that bring about progress in fundamental development challenges?

## B. Potential Climate Change Impacts in Central America

Central America is one of the regions of the world most exposed to the effects of climate change, although it produces a minimum part of greenhouse gas (GHG) emissions, viz. less than 0.8% of global gross emissions.[1] This narrow isthmus that serves as a land bridge between the two continents of the Americas and is surrounded by the Pacific and Atlantic Oceans is frequently

1 Equivalent to less than 0.3% of emissions without land-use change. Estimates based on national inventories for 2000 and global figures from IPCC (2007b) and the Climate Analysis Indicators Tool data base from the World Resources Institute (http://www.wri.org/tools/cait). It is important to note the high uncertainties around land-use change emissions and absorptions.

hit by droughts, cyclones and the El Niño Southern Oscillation. Over the past three decades there has been a trend of reduced rainfall, especially in the western part of the isthmus, and of temperature increases ranging from 0.7°C to 1°C. Given that diverse economic activities, such as agriculture and production of hydroelectricity, are climate dependent, climatic changes could increasingly affect the region's economic evolution over the course of the current century. In fiscal terms, climate change constitutes a contingent public liability that will have an increasingly detrimental effect on public finance.

At the same time, the region has valuable natural reserves that must be preserved for their contribution to the support and development of current and future generations. Ecosystems rich in biodiversity, including forests, coral reefs and mangroves, among others, all provide the population with multiple products and services, such as food, shelter, medicines, pollination and pest control, and regulation of local climate, water and humidity. These ecosystems are already suffering the ravages of unsustainable exploitation and will be further affected by climate change. The population of these countries should also be regarded as a treasure, given its youth and cultural, ethnic, linguistic and lifestyle diversities. This population requires not only investment in its development, but also a greater recognition of the value of the knowledge of its local communities and indigenous peoples. These assets could make important contributions to climate change responses, but they are also probably the most vulnerable to its effects and are already suffering from the consequences of increased extreme events such as hurricanes, intense rainfall, floods and drought.

The Global Climate Risk Index prepared by German Watch ranks over 180 countries according to the impact of events such as storms, floods and temperature extremes. In the cumulative ranking from 1992 to 2011, Honduras was placed first as the most affected country, Nicaragua third, Guatemala eleventh, El Salvador fifteenth and Belize twenty-sixth. In more recent years, these countries have often appeared in the first ten places: Guatemala was first and Honduras seventh in 2005, Nicaragua third in 2007, Belize ninth in 2008, El Salvador first in 2009, Guatemala second and Honduras fifth in 2010, and El Salvador fourth and Guatemala ninth in 2011.[2]

In its Fourth Assessment Report, the Intergovernmental Panel on Climate Change (IPCC) notes that the frequency of intense rainfall has increased over

2 Harmeling & Eckstein (2012:28).

most land masses, consistent with global warming and increases in water vapour. It reports that there is a medium certainty that anthropogenic influences have contributed to intensification of extreme precipitation on a global scale and of droughts in some areas, including Central America, due to rainfall reductions and/or increases in evapotranspiration. The report also warns that droughts and landslides can result from a set of accumulative events that individually are not considered as extreme.[3]

Just over 290 major climate-related extreme events have been registered for Central America in the Emergency Events Database, EM-DAT, between 1930 and 2011, with an estimated increase of 7% annually in the last three decades relative to levels recorded in the 1970s. The most frequently registered events are hydrological, with 86% associated with floods, storms, landslides and mudslides, and 9% with droughts. Many more such events have occurred on a lesser scale with cumulative effects that have yet to be assessed.

Hurricanes generate the greatest measured costs to date, with the Atlantic coast being most exposed. At the same time, hurricanes originating in the Caribbean force the Intertropical Convergence Zone northwards, provoking intense rainfall, landslides and mudslides over a far wider territory. This was the effect of Hurricane Mitch in 1998, which generated estimated costs of close to $8,000 million in five countries of the region. In the last few decades, storms and hurricanes originating in the Pacific have also started to make landfall in Central America, while previously they did so further north in Mexican territory.[4] In addition, tropical depressions and storms are being recorded with more intense rainfall and destruction, such as Tropical Depression 12E of 2011, which caused extensive damage and loss in El Salvador and parts of Guatemala, Honduras and Nicaragua. Although climatic events in Eastern Pacific have not been studied in similar detail, National Oceanic and Atmospheric Administration records for the Atlantic Ocean have provided evidence of an increased frequency of storms of short duration (less than two days), especially since the 1960s.[5]

The frequency of moderate duration storms has also increased since 1980, but has historically fluctuated in a cycle of about three decades. The relationship between frequency of these events and climate change may become clearer as we see if this oscillation changes its historic pattern in the coming

3 IPCC (2007b) and (2012).
4 MARN (2012).
5 ECLAC et al. (2010).

years. Regarding the relationship between intensity of these events and climate change, the evidence is stronger. It is estimated that the oceans have absorbed about 20 times more heat than the atmosphere during the last half century, causing higher temperatures in shallow and deep waters – which are factors contributing to the increased intensity of tropical cyclones. The surface temperatures of the Eastern Pacific Ocean and Caribbean Sea, whose waters affect the climate of Central America, have increased over the last hundred years: the time series indicates that the area of the Pacific Ocean associated with El Niño Southern Oscillation (20N–20S and 90W–120W) has experienced a temperature rise in this century and the Caribbean has suffered an acceleration of warming since the mid-nineties.[6] A review of the international literature suggests that the intensity of hurricanes could increase by between 5% and 10% during this century.[7]

Another line of analysis is based on the laws of thermodynamics, which suggest that elevated temperatures generate higher rates of evaporation, evapotranspiration and water vapour in the atmosphere and an acceleration or destabilisation of the water cycle. Stott of the Hadley Centre for Climate Prediction and Research has estimated that for every 1°C increase in temperature there should be an increase of 7% global average moisture in the atmosphere, which would cause more intense rainfall events.[8]

Historical climate databases indicate that Central America has seen a rise in average temperature of about 0.5°C over the past 50 years. The climate change scenarios for temperature and precipitation developed by the Economics of Climate Change in Central America (ECCCA) initiative used emissions scenarios and climate models recommended by the IPCC.[9] In an emissions scenario to 2100 that is lower than the current trend (IPCC scenario B2), temperature could climb between 2.2°C and 2.7°C, depending on the country, with a 2.5°C regional average increase over the 1980–2000 average. A continuation of the current trend of rising emissions (IPCC scenario A2) could result in temperatures rising between 3.6°C and 4.7°C, depending on the country, and a regional average of 4.2°C. With this scenario,

6 Jury (2011).

7 ECLAC et al. (2011).

8 Carey (2011).

9 The Intergovernmental Panel on Climate Change (IPCC) has developed four families of developmental and emissions scenarios. The ECCCA initiative primarily used scenarios A2 and B2 with four general circulation models, three of which were used for this reported average.

the temperature rise limit of 2°C could be reached by a number of Central American countries in the next 40 years.

The estimates for future precipitation levels involve even greater uncertainty. In the B2 emissions scenario, precipitation could fall by 3% in Panama, 7% in Guatemala, between 10% and 13% in Costa Rica, Belize, El Salvador and Honduras, and 17% in Nicaragua by 2100. The average reduction for the region could be 11% by that year. Using scenario A2, precipitation could be reduced in the order of 18% in Panama, 35% in Nicaragua and between 27% and 32% in Costa Rica, Belize, El Salvador, Guatemala and Honduras. On a region-wide basis, the decrease could average 28% by 2100.[10]

However, even with a lower reduction in precipitation under B2 scenario, rising temperatures will have their own effect on evapotranspiration, and result in reduced availability of water, especially in the second half of the century. In the more pessimistic scenario (A2), this multiplier effect would be greater. The ECCCA analysis of aridity patterns found that levels of temperature and precipitation for the period 1950-2000 generated an aridity index of 1.6 for Central America, varying between the Western Highlands of Guatemala with higher humidity (aridity index of 1.96) and parts of the Dry Corridor of Central America (index results between 0.91 and 1.25). This study estimated that by the end of the century the region could experience conditions associated with an aridity index of 1.4 in the least pessimistic scenario (B2) and 1.2 in the most pessimistic scenario (A2), with a general prevalence of conditions similar to those experienced by the driest part of the Dry Corridor in the period between 1950 and 2000.[11]

Central America is privileged in terms of the average availability of water in the region, but there is a very uneven distribution of this resource between countries and regions at the subnational level. This situation often leads to alternating periods of floods and of severe droughts. Population and economic growth alone could cause water demand to grow by almost 300% by 2050 and more than 1,600% by 2100 in a baseline scenario without climate change and without improvements in efficiency of water use. With climate change, demand may expand 20% more than in this baseline scenario in the case of B2 and 24% more with A2. The total availability of renewable water could fall 35% by 2100 compared to current levels under B2 and 63% with

10 ECLAC et al. (2011).

11 ECLAC et al. (2012a).

A2. In these scenarios, El Salvador would be the most affected, followed by Honduras and Nicaragua. The combination of changes in water demand and supply could result in a regional intensity of water use of 36% by 2100 in a scenario free of climate change, 140% with B2 and more than 370% with A2, if adaptation and efficiency measures are not adopted. These levels would be greater than the 20% threshold internationally accepted as critical for water stress, and are similar to current levels of intensity found in Egypt and some countries on the Arabian Peninsula.

These scenarios suggest greater risks and uncertainty for activities such as hydroelectricity production. The combined effect of the rise in temperature and precipitation changes would affect evapotranspiration in watersheds, river flows and evaporation in dam reservoirs. In a pilot study of two hydroelectricity plants (Chixoy of Guatemala and Cerron Grande of El Salvador), this chain of effects results in reductions in electricity generation of over 20% in the two plants for 2020 in the most pessimistic scenario (A2) relative to average generation during reference periods (1979–2008 for Chixoy and 1984–2009 for Cerron Grande). By 2050, the reductions would be above 40% in both plants, and would reach more than 80% for Chixoy and 70% for Cerron Grande at the end of this century. In the less pessimistic scenario (B2) there could be an increase between 4% and 6% in the plants for around 2020, but from then on production is reduced, with up to a 26% decrease in Chixoy and 17% in Cerron Grande by 2100. The study recommended making a more detailed analysis of possible changes in the next two decades and reviewing the operating models of the reservoirs.[12]

The agricultural sector is a driver of the region's economy. It represents 18% of total GDP when agro-industry is included, and it will be one of the sectors most affected by climate change. According to initial estimates, the regional agricultural index could register a reduction of approximately 9% under scenario A2 by 2100, if no adaptation measures are taken. Maize yields could grow in the near term with levels slightly greater than 2 tonnes per hectare, but would then begin to decline, possibly falling to as little as 1.4 tonnes per hectare around 2100. Average bean yields may decline from more than 0.7 to less than 0.1 tonnes per hectare by the end of the century. Rice production could fall from the historical average of 3.5 tonnes per hectare to between 2 and 1 tonnes per hectare.[13]

12 ECLAC et al. (2012c).
13 Ramirez et al. (2009).

The importance of these crops goes beyond their economic value, because they are staple food crops for large segments of the population. For example, bean production represents less than 4% of agricultural GDP but, combined with corn or rice, is an invaluable source of vegetable proteins and iron. With differences between the various countries and crops, low-income small-scale farmers produce a significant proportion of these staples. Climate change will have a significant impact on food security of these rural producers by reducing their production-based direct access to these staples and could cause shortages and price increases to urban consumers, depending upon import possibilities. Thus, the implications for food security and poverty are serious.

Central America is home to 7% of the planet's biodiversity and exhibits great geological, geographic, climatic and biotic diversity. In a business-as-usual scenario of land-use change (without climate change), one measure of biodiversity, the Potential Biodiversity Index (PBI),[14] could decrease by approximately 13% during the current century, especially in the period before 2050. With climate change, under scenarios B2 and A2, the PBI could decline by 33% and 58% respectively by 2100. Guatemala, Nicaragua, El Salvador and Honduras would be the countries hardest hit with PBI reductions ranging between 75% and 70% under scenario A2.[15] Another ECCCA study on forests used the Holdridge life zones (HLZ) classification. Under this approach, the surface of natural cover would decrease under a land-use change scenario without climate from approximately 28.5 million hectares to 16.3 in 2050 and 16.9 million hectares by 2100. However, the proportion of each of the six predominant HLZ forest types in Central America would not change significantly. If drivers of this land-use change scenario occurred with the B2 emissions scenario, it was estimated that the humid tropical forest could increase its cover from 44% in 2005 to just over 70% by 2100. In contrast, with scenario A2, the largest increase in surface would go to the dry tropical forest, rising from 11% to 39% of the total. Thus, both scenarios suggest changes, one to drier HLZ (in A2) and the other toward more humid ones (in B2). Both scenarios estimate initial increases in areas with conditions appropriate for humid forests up to 2020 and a reduction in the diversity

14 The Potential Biodiversity Index includes species and ecosystems and makes an inference about the probability of encountering greater diversity in function of a series of relevant variables. It does not necessarily coincide with the present-day recorded number of species and ecosystems.

15 ECLAC et al. (2011).

of HLZ as the century progresses. This study evaluates the conditions of temperature and precipitation associated with the various HLZ, but it remains to be analysed whether their ecosystems could successfully make these transitions, especially considering the speed of the projected changes in precipitation and temperature coupled with the pressures of land use changes. These results confirm that reducing deforestation and increasing protection and restoration of natural ecosystems is a development challenge in itself, and that climate change, especially in the more pessimistic scenario (A2), could bring greater loss of forests and their ecosystem services.[16]

Of the approximately 41 million people in America, two-thirds live in settlements that combine poverty with poor sanitation and health services.[17] These adverse conditions make the population vulnerable to climate change, because poverty-related diseases, such as malaria and dengue, are also associated with changes in climate.[18] In addition, these populations face direct and indirect effects of extreme events on health, including crop destruction and relocation in overcrowded and unsanitary spaces.[19] Central America has a mosaic of ecological niches favourable for the transmission of diseases associated with climate change, owing to its varied topography and proximity to two oceans. A recent ECCCA study has identified valuable efforts in the region to assess climate variability and its influence on health and the epidemiology of certain diseases. It proposes priorities for a future agenda on the relationship between temperature and precipitation and the incidence of disease, the potential climate change impacts on human health, and the design of adaptation measures.[20]

The analysis of the many direct and indirect impacts of climate change on the vulnerability of certain populations requires consideration of the multiple dimensions of the condition of poverty, such as that of "capabilities and opportunities".[21] This requires an analysis of the ability of people to adapt to climate change, not only in terms of the availability of financial and natural resources, education and health, but also their ability to use these assets. The IPCC states that the causes, problems and solutions related to climate change are laden with equity issues, since the countries that contribute least to

16 ECLAC et al. (2012b).
17 FAO & ETEA (2008).
18 Hotez et al. (2008).
19 Noji & Toole (1997).
20 ECLAC et al. (2012e).
21 Sen (1999).

greenhouse gas emissions are the most vulnerable and have less capacity to adapt.[22] The Stern Report[23] states that "climate change is a great threat to the developing world and a major obstacle to continued poverty reduction across its many dimensions". Therefore, there is a need to integrate adaptation strategies with those being taken to reduce poverty and inequality.[24]

About half of the population of Central America lives in poverty, and about a third in extreme poverty, especially in rural areas. There are still high levels of socioeconomic, ethnic and gender inequalities, as reflected in several indicators, including the relatively high Gini index (0.53 in 2010), and high rates of child mortality and morbidity, maternal mortality, and malnutrition, and well as limited access to food, clean water, health services, education, social security, capital and productive credit. A significant part of this population, especially in rural areas, depends directly on the environment for access to water, food, shelter, medicines and energy, among other needs. The lack of capital and livelihood options has sometimes led to overexploitation of the environment by these populations. The general pattern of unsustainable and inequitable development and weak risk management has created a vicious circle of human impoverishment and environmental degradation, which further complicates the response to climate change.

Another part of the population living in poverty inhabits marginal urban areas, is often dependent on the informal urban economy and accesses most of its goods and services through the market. These communities will face the economic instabilities that climate change could cause with serious handicaps. The reduction and instability of water availability and crop yields can affect labour markets, the supply and price of commodities, and the migration to urban areas.

The already existing challenges of social investment are related to the prevalence of informal work and the limited coverage and quality of education and social protection in most countries. Only those who are employed in the formal sector have access to this coverage, including pensions, unemployment insurance and health services. Low per capita social spending, although relatively higher in Panama and Costa Rica, also limits resilience and adaptive capacity. Recent decades have seen the implementation of conditional cash transfer programmes for poor families in several countries to supplement income and encourage the use of basic health and education

22 IPCC (2007a).
23 Stern (2007).
24 ECLAC (2009) and (2010); IPCC (2007a); UNDP (2007); AfDB et al. (2007).

services. These programmes have the advantage of focusing on poor households with children, adolescents, and female heads of households and the unemployed and economically inactive population.[25] Cecchini and Madriaga state:[26]

> In some countries they facilitate access to social services for the population in greatest need. However, one should not lose sight that these programs do not replace the functions meant to be provided for by other policy instruments and their effectiveness depends largely on the presence of strong health systems and universal education.

Strengthening the supply of these services remains a priority need in the region and it would be advisable to consider that climate change may worsen the intergenerational cycle of impoverishment and that such measures should be made sooner rather than later.

Participation and political representation are important and will be key to successful climate change adaption. Although the countries of the region have democratic electoral systems, there is still a way to go to for marginalised groups, such as women, small-scale farmers, indigenous peoples and communities of African origin to achieve effective participation.

An initial estimate prepared by the ECCCA initiative of the measurable accumulative cost to 2100 for the impact on agriculture, water resources, biodiversity and the intensity of hurricanes, storms and floods under scenario A2 could be equivalent to about 54% of the regional GDP of 2008 at net present value (NPV) with a discount rate of 0.5%. With a discount rate of approximately 4%, the equivalent value is 9% of the 2008 regional GDP at NPV, underscoring the importance of the rate applied. The measurable accumulative cost to 2100 of the same sectors under scenario B2 could be equivalent to about 32% of the regional gross domestic product (GDP) of 2008 at NPV with a discount rate of 0.5%. With a 4% discount rate the equivalent value is 6% of regional GDP of 2008 at NPV. The current dollar cost under B2 is equivalent to 60% of the same cost under A2. It is important to note that the greatest increase in costs could occur during the second half of the century, and in general costs will be extremely high at the end of the

25 ECLAC (2012).
26 Cecchini & Madriaga (2011).

century in a scenario of inaction.[27] There is considerable uncertainty involved in such long-term scenarios and in the integration of various analytical *layers*, such as climate and macroeconomic scenarios with impact studies for different sectors and their economic valuation. In addition, there are notable methodological challenges in the various sectors and areas of concern. In this sense, these results should be regarded as an indication of relative trends and magnitudes, not as exact figures. In the future, it will be necessary to explore how changes in one sector influence what may occur in other sectors.

In conclusion, a scenario of rising emissions such as IPCC A2 will bring significant and growing impacts and costs to Central America, with a certain degree of variation between countries. This result confirms the asymmetrical nature of climate change with the most polluting developed countries probably experiencing the least effects and having a greater ability to adapt, while the countries that have contributed least to the problem will suffer greater impacts and have less resilience. It lends weight to the concern that the costs of climate change in a scenario of global inaction, particularly on the part of major emitting countries, would be higher than those in a scenario with an equitable and inclusive international agreement that significantly lowers emissions with shared yet differentiated responsibilities between countries. This second scenario would need to ensure adequate support for the most

27 The initiative first defined various baseline or business-as-usual scenarios for the macroeconomy, demographics, land use and energy consumption without climate change, against which the phenomenon's impacts and costs were measured. A bottom-up analysis was used to analyse impact in key sectors and areas of concern such as agriculture, water resources, extreme events and biodiversity services, and then an economic valuation of these impacts was generated in relation to the projected GDP baseline. Analysis and policy-oriented discussions were held regarding both adaptation and low-carbon economies. The initiative developed long-term future impact and cost scenarios to 2100, with cut-offs at 2020, 2030, 2050 and 2070 so as to uncover potential risks that could grow over time, particularly in the second half of the current century. On the other hand, mitigation scenarios were made only to 2030 owing to uncertainty surrounding technological changes, with cut-offs at 2010 and 2020. Lastly, a common focus was agreed upon for the treatment of discount rates for the economic valuation. The costs described are initial estimates and relate to the impacts analysed for the agricultural sector, water resources (availability and municipal and agricultural demand), biodiversity (direct costs registered in economic statistics and indirect impact on agriculture), and the increasing intensity of hurricanes, storms and floods (not including their increased frequency nor the costs of other extreme events). Thus, these results represent a conservative and initial estimate of the costs of economic impact.

exposed and vulnerable countries, such as those of Central America, to adopt adaptation and mitigation measures in the context of sustainable and inclusive development.

## *C. An Exploration of Climate Change Policy Options*

From an economic standpoint, it is more cost effective to act now than to leave the matter to future generations. It is also more ethical to do so. The initial cost estimates of the ECCCA initiative suggest that climate change impacts will become progressively higher if ambitious and immediate emission reduction measures are not taken. The work also confirms that climate change is the greatest market failure to date for not having internalised the value of climate as a global public good and not properly registering its social impact and effects on environmental services. This failure implies the need to make ethical decisions that go beyond the realm of economics regarding the implicit inequalities within and between current and future generations.

Climate change could be considered a phenomenon that will only affect us in the distant future, not worthy of significant current investment given budget constraints deepened by the current global recession. But the growing impact of extreme events suggests that urgent action must be taken regarding climate resilience. In addition to the growing threat of major impacts and costs in the future, current risk reduction and reconstruction efforts need to be reoriented and strengthened by adopting more climate resilient infrastructure, housing and land use standards, more efficient water management, stepped-up protection of forests and watersheds, and natural coastal barriers such as mangroves. This investment should reduce vulnerability to upcoming extreme events and generate a greater ability to cope with the impacts of climate change.

The challenge of adaptation will be highly onerous for Central America because it demands a redoubling of efforts to reduce poverty, inequality and both socio-economic and environmental vulnerability, while heightening the resilience and adaptive capacity of these societies, especially high-risk populations and related ecosystems. There will be limits to what adaptation can achieve in the face of increasingly irreparable losses and damages, even if abundant financing were to be available, and especially in a business-as-usual, high-carbon scenario. This analysis demonstrates that the present value of the long-term costs of climate change impacts will prove to be too high if we do not take ambitious and immediate measures. Given that this is a market failure, climate change cannot be treated as if it were the exclusive responsibility of environmental institutions, but instead must be recognised as a central and cross-cutting economic problem with serious fiscal implications.

Central American societies will need to avoid ad hoc strategies with a business-as-usual logic that might respond to emergencies, but only heighten risks. In such a logic, climate change might be regarded as important, but not a matter that could be fully addressed owing to existing budgetary restrictions exacerbated by the current global recession and the need to address urgent social and economic issues in a conventional manner.

We will have to address these challenges in a period in which the model of market self-regulation has demonstrated limitations. As Barcena has noted, we are experiencing a new era, which demands profound structural changes made necessary by climate change and other externalities caused by industrialisation and the hydrocarbon-based economy – changes on a scale equivalent to the industrial revolution.[28] Other elements to take into consideration are the significant population increase still to come before reaching its stabilisation point in the second half of this century, the demographic transition as the population ages, and the migration between countries and from the countryside to the city. The challenge of achieving inclusive development with better quality of life and opportunities is made greater when we consider our responsibility to future generations in the context of climate change. The globalisation of communications has promoted deregulation, self-regulation and increased information flows, which favour democracy. But it has also strengthened the role of market forces in the definition of identities. Reversing the consequences of market self-regulation and excessive dependence on hydrocarbons requires greater collective consciousness of global public goods. This will require transforming the processes and structures of global and national governance.[29]

National, regional and international agreements should be oriented towards sustainable and inclusive adaptation strategies that integrate vulnerability and poverty reduction with adaptation actions and measures for the transition to more sustainable and lower-carbon economies. This should include mitigation actions designed to generate adaptation co-benefits in a range of instruments directed at sustainable and equitable development. This

28 ECLAC (2010).
29 ECLAC (2010); ECLAC (2012).

requires a strategic vision to focus on inclusion and sustainability in its many dimensions and to maximise co-benefits and minimise costs across sectors and between adaptation and mitigation, and all within the development agenda. For example, improved protection and restoration of forests and energy efficiency are part of a sustainable development agenda, and, if well designed, co-benefits could result in better adaptation of these ecosystems, reduced emissions and improved wellbeing of populations living in these ecosystems, including indigenous peoples.

In this scenario, the global economic recession and climate change risks would be used as an opportunity to review thoroughly the current productive specialisation of these economies, including their linkages to regional and global markets, the ties between their energy patterns and negative externalities from conventional pollution and GHG, losses to public health and harvests, weaknesses in rural and urban infrastructure, degradation of ecosystems and loss of their services.

Public policies aimed at sustainable and inclusive adaptation could be designed to take into account intra- and inter-sectoral synergies in the following major policy clusters with explicit sectoral and territorial objectives:

- Inclusion and adaptation by human populations as part of policies aimed at the reduction of poverty and inequality, including food security, integrated management of water resources and reduction of extreme event impacts with strengthened land-use and territorial planning
- Transition to sustainable, low-carbon economies that are efficient in the use of natural resources, introducing structural and technological changes especially regarding energy security and efficiency, integrated water management and the curbing of deforestation and pollution
- Protection and restoring of natural ecosystems and rural landscapes, including forests, in order to improve their own adaptation and assure the long-term provision of ecosystem services, as a key policy area for both adaptation and a transition to more sustainable economies, including both effective use of economic incentives and ethical and cultural appreciations
- Far-sighted and proactive fiscal and investment policies as a cross-cutting policy area to establish climate resilient criteria for public investments and economic incentives for risk retention, reduction and transfer, technological innovation and adaptation, and the effective use of national and international resources, and

- Strengthening of the Central American integration process to take advantage of opportunities where regional responses offer added value for challenges such as managing common water resources, food and energy security, competitiveness, trade implications and international negotiations.

Central American societies need to become more audacious managers of their water resources, securing their sustainable and efficient use for the benefit of the population and production. In many ways, the key indicator of adaptation is related to more efficient use of this resource and its judicious distribution among multiple uses, including that needed by ecosystems. Forest conservation and restoration of rural landscapes are essential for the management of watersheds, for reducing erosion, landslides and floods, and for production of hydroelectricity. Extensive efforts are required to make more efficient use of water, reduce pollution and recycle it in domestic use, agriculture, industry and services. A much more effective institutional framework for managing water across sectors and between countries is needed, given that transnational watersheds cover 40% of the territory of Central America.

Protecting food security in the face of climate change, especially access to basic grains, and making the transition toward more sustainable agriculture is a major challenge in order to protect the poorest members of these societies, whether as small-scale producers or urban consumers. With a few notable exceptions, most countries have experienced lower levels of investment in rural areas in recent decades, and the dismantling of rural programmes in land titling, extension, post-harvest loss reduction, market access and capacity building. Much more effort is required for the protection and promotion of native varieties of crops and other local, indigenous and national technological know-how, which can make important contributions to climate resilience and adaptation.

The response to climate change in the agricultural sector will require close coordination with policies to reduce deforestation, protect biodiversity and manage water resources. It will also require recognising and expanding notable experiences in the region that have strengthened the welfare of rural and indigenous populations by establishing more sustainable production processes, such as agroforestry and other activities that combine farming with the protection of ecosystems and systems of payment for environmental services. The region's strategic agrobiodiversity is currently as unprotected as the small farmers and indigenous peoples that have developed it over many generations. Increased access by rural populations to decentralised renewable energy sources, such as solar, wind and small-scale hydroelectric dams, can increase their resilience to climate change and reduce emissions at the same time. In general, rural areas, with their natural and productive resources, will be key to a successful response to climate change.

Active development of appropriate technologies is essential for adaptation and the transition to low-carbon economies, both in terms of using and adapting 'modern' technologies and the recovery of traditional and local knowledge and technologies, especially those of indigenous peoples and small-scale agriculturalists. The region has developed a serious dependence on contaminating and imported hydrocarbon energy sources. The transition to an energy matrix based as much as possible on local, renewable sources would bring multiple benefits, including improved energy security, foreign currency savings and reduced adverse effects of fossil fuels on human health, as well as lower GHG emissions.

Investments in waste management would generate multiple benefits such as reduced pollution, increased raw material for production, power generation by methane capture at landfills, and better drainage of water during extreme hydrometeorological events. There are opportunities to improve energy efficiency and reduce the intensity of GHG emissions and other pollutants with new rules and requirements for motor vehicles and industrial machinery and expansion of safe and efficient public transport systems. The expansion of hydropower, if designed with a focus on sustainability and inclusion, could expand access to electricity for low-income populations and contribute to sustainable production and social development of the surrounding areas.

The energy sector in Central America has designed its Sustainable Energy Strategy 2020,[30] which proposes expanding regional renewable energy sources, including hydro, wind and geothermal sources and importing natural gas. It is the first regional strategy that considered sectoral GHG emissions. It was adopted by the ministers of Energy and the Central American presidents and proposed the following measures:[31]

- Achieve at least 90% of electricity coverage in each country
- Achieve 10% reduction in wood fuel consumption for cooking by introducing more efficient stoves in a million rural households

30 ECLAC & SICA (2007).
31 (ibid.).

- Reduce electricity consumption by 12% in the residential, commercial, industrial sectors and for public lighting with more efficient lighting systems
- Reduce residential electricity use by 35% by replacing obsolete refrigerators for more efficient units in 2.7 million households
- Reduce electricity use in the industrial sector by 10% with more efficient engines
- Bring the level of losses in the electricity systems of the countries down to 12%
- Increase in the regional share of renewable electricity production by 11%, favouring the construction of hydropower plants
- Replace 15% of petroleum consumption with biofuels in public and private transport, and
- Reduce GHG emissions by 20% relative to the 2020 baseline scenario, maximising the use of emissions reduction certificates.

With greater access to technology and funding, Central America could advance further in implementing this regional strategy. The energy sector has shown its ability to carry out long-term coordinated investment projects, such as those conducted to develop the Central American Electrical Interconnection System. It is currently working on the harmonisation of fuel standards in the process of the Customs Union and the implementation of the Action Matrix for the development and integration of the sector in Central America.

Adaptation of societies to climate change is clearly linked to the adaptation of the ecosystems on which we depend. Meeting this challenge will require further assessment of the value and contribution of environmental services and taking non-market measures to create incentives and regulatory frameworks. It is necessary to use the precautionary principle and establish minimum standards of protection and restoration, considering the irreversibility of biological loss, risk and uncertainty.

Reducing deforestation and degradation and restoring rural landscapes will generate benefits in many aspects of the development agenda per se, even without considering climate change. The system of Protected Natural Areas (currently more than 550 in the region) needs strengthening, and biological corridors will have to cover larger-scale biogeographical areas and give greater scope to buffer areas and climate 'shelters'. These efforts could be complemented by programmes for expanding sustainable agriculture, agroforestry and protection of local and endemic varieties of crops and wildlife. Other measures to facilitate the adaptation of forests and rural populations include programmes that involve these communities in the conservation and restoration of the ecosystems on which they depend. This includes adopting technologies for sustainable livelihoods; making full use of traditional knowledge and diversifying livelihoods; improving systems of forest management, such as control of deforestation and forest fires, afforestation and reforestation; and establishing regulations and certification of organic products and ecotourism.

Land-use planning is essential to achieve sustainable development and to improve the distribution of the population, its activities and infrastructure over the landscape in order to reduce damage and loss from extreme events and climate change. Natural ecosystems can reduce a population's vulnerability to extreme weather and serve as complements or substitutes for investment in 'grey' infrastructure, which may have higher costs. For example, forests and coastal mangroves provide protection against storms, floods, hurricanes and tsunamis.

It is advisable to expand and strengthen fiscal and financial policies that encourage a transition to higher energy and water efficiency, sustainable management of forests and the recognition of the economic value of environmental services, including water cycle regulation and carbon sinks. The region has developed programmes such as the National Forestry Financing Fund in Costa Rica, the Fund for the Conservation of Protected Areas in Belize, the Forestry Incentive Certificate in Panama, the Forestry Incentives Programme and the Programme of Incentives for Small Holders Forest Vocation Land and Agroforestry, both in Guatemala. Further assessments are needed of the benefits and costs of voluntary plans to reduce net deforestation at national and regional levels and options for funding with national and international resources, including a future expanded version of the Clean Development Mechanism, national, regional or international carbon markets, or payments for environmental services. National and regional goals for ecological conservation and sustainable use of ecosystems could be linked to those aimed at improving the quality of life of local populations.

Fiscal sustainability is already a serious concern in the region, and the impact of extreme events is putting further pressure on scarce public resources, even before the increasing effects of climate change and the need for an incentives framework for the transition to low-carbon economies are considered. Extreme climate events affect public finances in various ways: directly through increased emergency and reconstruction expenditures, often involving the need for lines of credit, but also by way of lost fiscal income owing to economic losses and damages. Such events can also increase the demands for social services and the relocation of populations and economic activities. With little disaster insurance coverage in place, the cost of responding to the needs of affected populations usually falls on limited public financing or international aid. This incomplete list of climate change pressures on public finances suggests that the fiscal impact should be seen as a serious contingent liability, which in the long run will become far less contingent.

Despite the immediate challenges of the current economic crisis, the Ministries of Finance and Central Banks have begun to pay attention to climate change. These institutions have acquired experience in carrying out debt swaps for funding climate change programmes and creating a system to label expenditures on extreme events, and are now developing proposals for national climate change funds, domestic carbon markets, climate change planning requirements in sectoral public programmes and budgets, disaster contingency funds and investment in infrastructure adaptation.

Because climate change involves a market failure, it cannot be treated as the sole responsibility of environmental institutions, but should be seen as a central economic problem with serious fiscal implications. Climate change presents a complex series of multisectoral challenges that will need the proactive response of many stakeholders, including the public and private sectors, civil society organisations, academia, integration institutions and the international community.

## *D. Final Considerations*

Central American countries are increasingly affected by the rising losses and damages of extreme climate events, and policymakers are more concerned about the rising costs of reconstruction and the need to reduce vulnerability. The effect on the frequency, intensity and patterns of extreme events that can be attributed to climate change is a subject of intense research and debate. Better methods of attribution and increasing evidence are regularly reported. In the region, there is clearly a need to strengthen climate monitoring and analysis in order to establish early warning systems, as well as to determine the additional impact of climate change on these events. From an adaptation point of view, it is becoming clearer that the many urgent steps needed to better protect the population, infrastructure and ecosystems from the ravages of current extreme events are also the first steps for climate change adaptation. In the Central American region, these challenges include agricultural production losses due to droughts, intense rainfall and pest infestations, increased incidence of illnesses such as dengue and those related to pollution, location of dwellings and communities in high-risk areas, productive and social infrastructure badly designed for current climate variability, and increased public budget expenditures and debt to pay for post-disaster reconstruction.

Given that climate change affects multiple sectors and aspects of human activity, one of the challenges for public policy is to ensure that it becomes integrated into the agendas of sectoral ministries and other key actors, such as universities, NGOs and chambers of industry. In Central America, it became clear to many Ministers of the Environment that they rapidly had to change the perception about climate change so that it was no longer seen as a purely environmental issue under the remit of their Ministries, but a major economic and social threat with multisectoral and fiscal impacts and so necessarily involving Ministries of Treasury, Public Works, Health, Agriculture and Education, among others. This process has required developing sector-oriented analysis and arguments in the languages of these sectors, as well as a process of dialogue and consensus-building for new policies. Most countries are presently engaged in this process at the national level.

At the same time, it is becoming apparent that each government needs to develop a capacity to maintain an overarching and integrated coordination of response measures. This effort has been made more difficult as many Ministries of Planning have been downsized or closed in recent decades. Different strategies have emerged, such as leadership being taken up by the Environment Ministries or the Presidency or the signing of bilateral agreements between an Environment Ministry and the Ministries of key sectors such as Treasury, Public Works and Agriculture. Most countries have developed overall climate change strategies, which include proposals both for adaptation and mitigation, and have established national climate change committees. Some are constituted mostly within the national public sector, but others have a broader range of participating institutions – for example, in Honduras this committee has contributed to national public policies and encouraged initiatives by NGOs, academic centres and professional associations. Where sectoral ministries assigned technical staff to this committee, they then became focal points for interinstitutional work on different issues or policy drafts.

For many years, mitigation was the centre of concern in international negotiations and finance mechanisms. Less attention has been given to adaptation and vulnerability reduction. In addition, adaptation and mitigation are still largely treated as separate issues at the international level. While this approach may be necessary for negotiations and financial support based on the principle of shared but differentiated responsibilities, it is not clear that it is useful for public policies, especially for countries that are emission scenario 'takers' with demanding development agendas and limited budgets. Some sectors, such as agriculture and forestry, are usually identified as being a priority for adaptation and for emissions reductions or increased absorptions. Thus, one of the working hypotheses developed is that public policies should consider both aspects in an integrated manner.

In Central America, most countries started developing their climate change policy discussions prioritising adaptation and risk reduction, and there was an intense debate and different positions regarding mitigation. Currently all countries consider that they can at least do their part to reduce emissions on a voluntary basis. Taking a 'double perspective' has come about in different ways. In the case of Costa Rica, national climate change policy aims at both adaptation and mitigation, while the national development goal to achieve carbon neutrality by 2021 has given impetus to mitigation efforts with a focus on improved productivity. On the other hand, El Salvador first designed programmes focused on adaptation and then identified additional measures or co-benefits in emissions reductions. The country has proposed an 'adaptation-based mitigation' approach for its programme to restore forests and rural landscapes and make agriculture more sustainable.

For countries with limited financial resources, separate adaptation and mitigation policies would probably be fiscally onerous and inefficient in terms of both design and implementation efforts, especially considering the many sectors involved. A process that considers both adaptation and reducing GHG emissions could help to identify not only potential co-benefits, but also possible adverse impacts of a measure relative to another within and across sectors. In this regard, national development goals and sustainability, inclusion and protecting public goods would be important criteria. Potential areas for special attention include food security, energy efficiency and security, safe and efficient public transportation, and protection and restoration of forests, rural landscapes and watersheds.

A major concern for many developing countries is water availability and use. Water availability could be affected by a cascade of climate change impacts, including increased temperature and evapotranspiration, and possible declines in rainfall and in water regulation of ecosystems. Meanwhile the efficient and just allocation of water is vital for adaptation of these same ecosystems, the population and its economic activities. Thus, reducing climate change is not just about low-carbon economies, but also about water-efficient economies. In the case of Central America, dependency on imported hydrocarbons and vulnerability to fluctuations in international supplies and prices could be reduced by increasing hydroelectric power capacity in the next decades. Nevertheless, designs, watershed management and operating procedures will have to take climate risks into account.

In addition, new hydroelectric projects will need to improve the track record for protecting the environment and benefiting local populations. Thus, the agenda for water is not only about water efficiency but also about ensuring equitable access to this resource for the population and creating effective social mechanisms for negotiating rights and needs between actors with different demands. The effective management of water resources now needs to include not only how to generate electricity, but how best to regulate water reserves in the face of fluctuating flows and extreme events; to protect local populations in the case of emergencies; to preserve ecosystems in the watershed; and to respond to the increased demand for agriculture and human consumption. This is one example of how complex decisions about options for climate change responses are becoming. This example also suggests that there is an opportunity to generate important advances for sustainable and inclusive development.

The essential nature of climate as a common public good has to be addressed by national policymakers and the international community. Although the principle of shared and differentiated responsibilities is established in the United Nations Framework Convention on Climate Change, those societies that are historically responsible for this externality have yet to assume its collective costs, while the negative impacts and costs have been borne by all, and suffered most by other populations that have not benefited to the same extent by its use. Those in the latter group do not necessarily wield the economic and political power needed to ensure that their needs are respected within countries and in the international arena. The cumulative and rising emissions of GHG have generated one of the greatest 'tragedies of the commons'. In economic terms, climate change is an externality whose costs are not fully accounted for in the current economy. Stern[32] considers it to be the greatest negative externality, or failure of market mechanisms,

32 Stern (2007).

that has ever existed. Thus, efforts to 'internalise' the value of climate as a global public good and register its social impact and effects on environmental services are important. In addition, climate change is riddled with complex equity dilemmas. Climate change is a phenomenon of flows and accumulated stocks of GHG over generations. Actions of the past and present generate impacts far into the future. For this reason, climate change has a negative impact on equity among the world´s current population and between it and future generations.

The equity issue and the failure of the market to value our climate as a common public good require ethical decisions far beyond the realm of economics. In the short term, a serious policy discussion is needed about the effectiveness and sustainability of current social security and poverty reduction mechanisms in the light of increased climate-related impacts on already strained fiscal budgets. There must be an explicit consideration of the value assigned to the needs of future generations and to ecosystems. These systems and their biodiversity provide us with multiple environmental services, which we are at an even higher risk of losing, as market signals of their decline will appear too late in the process.

Public sector institutions responsible for economic policy are essential partners in designing climate change responses, including Ministries of Treasury, Trade, Industry and Central Banks. Many of these institutions have a mandate to undertake analyses of longer-term risk and sustainability, although more traditionally in the fields of debt, exchange rates, international reserves, and trade negotiations. Adaptation not only implies responding adequately to the impacts of climate change, but also requires anticipating changes in the global economy, especially the transition to a low-carbon model. This transition could involve measures such as taxes on the carbon content of products and services and the establishment of barriers or tariffs on the carbon content of imports. Small developing countries may encounter both opportunities and threats in this transition. Identifying and taking advantage of the opportunities and counteracting the threats will take time. If left to the last minute, serious trade discontinuities could be encountered.

Finally, in Central America, one key incentive for public policy formulation was the establishment of climate-change-related mandates for national and regional institutions by the Presidents of these countries, leaders of the Central American Integration System (SICA). This agenda was clearly established during their summit on climate change held in May 2008, and then reiterated and broadened in summits in June 2010, November 2011 and June 2012. SICA has sectoral structures which bring together ministers of the different portfolios, who have the responsibility of responding to the agenda set by their presidents and in which they can discuss strategies with a regional perspective. In some sectors, one or a few ministers have been able to champion climate change concerns and sensitise other colleagues. In addition, many of these structures have technical secretariats and work programmes with the academic community, NGOs, regional and international development agencies and donors. While a few years ago climate change was seen to be the responsibility of the Ministers of Environment and their sectoral integration secretariat, the Ministers of Health and Agriculture, among others, have now established regional programmes of work on potential impacts and policy measures. The Ministries of Environment and Treasury have a joint regional technical committee within the ECCCA Initiative which has provided a forum for developing common analyses and policy proposals.

## *References*

AfDB/African Development Bank et al., 2003, *Poverty and Climate Change: Reducing the Vulnerability of the Poor through Adaptation*, New York, World Bank, available at http://www.unpei.org/PDF/Poverty-and-Climate-Change.pdf, last accessed 15 April 2013.

Carey, John, 2011, Global Warming and the Science of Extreme Weather, *Scientific American*, 29 June 2011, available at http://www.scientificamerican.com/article.cfm?id=global-warming-and-the-science-of-extreme-weather, last accessed 15 April 2013.

Cecchini, Simone & Aldo Madariaga, 2011, *Programas de transferencias condicionadas. Balance de la experiencia reciente en América Latina y el Caribe*, N° 95, (LC/G.2497-P), Santiago de Chile, Economic Commission for Latin America and the Caribbean (ECLAC), available at http://www.cepal.org/publicaciones/xml/6/44126/Programas-transferencias-Condicionadas-ALC-95.pdf, last accessed 15 April 2013.

ECLAC/Economic Commission for Latin America and the Caribbean, 2009, *Cambio climático y desarrollo en América Latina y el Caribe: una reseña*, Project Document, available at http://www.eclac.cl/publicaciones/xml/5/35435/28-W-232-Cambio_Climatico-WEB.pdf, last accessed 15 April 2013.

ECLAC/Economic Commission for Latin America and the Caribbean, 2010, *La hora de la igualdad, Brechas por cerrar, caminos por abrir*, Summary, Thirty-third Session of ECLAC, Brasilia, Santiago de Chile.

ECLAC/Economic Commission for Latin America and the Caribbean, 2012, *Cambio estructural para la igualdad. Una visión integrada del desarrollo*, Thirty-fourth session of ECLAC, San Salvador 27–31 October.

ECLAC/Economic Commission for Latin America and the Caribbean, CCAD/Comisión Centroamericana de Ambiente y Desarrollo, SICA/Sistema de la Integración Centroamericana, UKAID/UK Department for International Development & DANIDA/ Agencia de Cooperación para el Desarrollo de Dinamarca, 2011, *La Economía del Cambio Climático en Centroamérica*, Reporte Técnico 2011, ECLAC, LC/MEX/L. 1016, Mexico D.F., United Nations.

ECLAC/Economic Commission for Latin America and the Caribbean, COSEFIN/Consejo de Ministerios de Hacienda o Finanzas de Centroamérica, Panamá y República Dominicana, CCAD/Comisión Centroaméricana de Ambiente y Desarrollo, SICA/ Sistema de Integración Centroamericana, UKAID/UK Department for International Development & DANIDA/Agencia de Cooperación para el Desarrollo de Dinamarca, 2012a, *La economía del cambio climático en Centroamérica: Impactos potenciales en la aridez y los meses secos, serie técnica* 2012, ECLAC LC/MEX/L.1074, Mexico, D.F., United Nations.

ECLAC/Economic Commission for Latin America and the Caribbean, COSEFIN/Consejo de Ministerios de Hacienda o Finanzas de Centroamérica, Panamá y República Dominicana, CCAD/Comisión Centroaméricana de Ambiente y Desarrollo, SICA/ Sistema de Integración Centroamericana, UKAID/UK Department for International Development & DANIDA/Agencia de Cooperación para el Desarrollo de Dinamarca, 2012b, *La economía del cambio climático en Centroamérica: Impactos potenciales en los ecosistemas*, serie técnica 2012, ECLAC LC/MEX/L.1077, Mexico, D.F., United Nations.

ECLAC/Economic Commission for Latin America and the Caribbean, CEL/Comisión Ejecutiva Hidroeléctrica del Río Lempa de El Salvador, MARN/Ministerio de Ambiente y Recursos Naturales de El Salvador, COSEFIN/Consejo de Ministerios de Hacienda o Finanzas de Centroamérica, Panamá y República Dominicana, CCAD/ Comisión Centroaméricana de Ambiente y Desarrollo SICA/Sistema de Integración Centroamericana, UKAID/UK Department for International Development & DANIDA/Agencia de Cooperación para el Desarrollo de Dinamarca, 2012c, *La economía del cambio climático en Centroamérica: Dos casos de impactos potenciales en la generación de hidroelectricidad*, serie técnica 2012, ECLAC LC/MEX/L.1070, México, D.F., United Nations.

ECLAC/Economic Commission for Latin America and the Caribbean, COSEFIN/Consejo de Ministerios de Hacienda o Finanzas de Centroamérica, Panamá y República Dominicana, CCAD/Comisión Centroaméricana de Ambiente y Desarrollo, SICA/ Sistema de Integración Centroamericana, UKAID/UK Department for International Development & DANIDA/Agencia de Cooperación para el Desarrollo de Dinamarca, 2012d, *La economía del cambio climático en Centroamérica: Impactos potenciales en los patrones intraanaules y espaciales del clima*, serie técnica 2012, ECLAC LC/ MEX/L.1073, Mexico, D.F., United Nations.

ECLAC/Economic Commission for Latin America and the Caribbean, COSEFIN/Consejo de Ministerios de Hacienda o Finanzas de Centroamérica, Panamá y República Dominicana, CCAD/Comisión Centroaméricana de Ambiente y Desarrollo, SICA/Sistema de Integración Centroamericana, UKAID/UK Department for International Development & DANIDA/Agencia de Cooperación para el Desarrollo de Dinamarca, 2012e, *La economía del cambio climático en Centroamérica: Evidencia de las enfermedades sensibles al clima,* serie técnica 2012, ECLAC LC/MEX/L.1069, Mexico, D.F., United Nations.

ECLAC/Economic Commission for Latin America and the Caribbean, COSEFIN/Consejo de Ministerios de Hacienda o Finanzas de Centroamérica, Panamá y República Dominicana, CCAD/Comisión Centroaméricana de Ambiente y Desarrollo, SICA/Sistema de Integración Centroamericana, UKAID/UK Department for International Development & DANIDA/Agencia de Cooperación para el Desarrollo de Dinamarca, 2012f, *La economía del cambio climático en Centroamérica: Sintesis* 2012, ECLAC LC/MEX/L.1076, Mexico, D.F., United Nations

ECLAC/Economic Commission for Latin America and the Caribbean & SICA/Sistema de la Integración Centroamericana, 2007, *Estrategia Energética Sustentable Centroamericana 2020*, (LC/MEX/L.828), Mexico, D.F., Unites Nations.

ECLAC/Economic Commission for Latin America and the Caribbean, UKAID/UK Department for International Development, & CCAD/Comisión Centroamericana de Ambiente y Desarrollo, SICA/Sistema de la Integración Centroamericana, 2010, *The Economic of Climate Change in Central America*: Summary 2010, ECLAC Mexico, United Nations.

FAO/Food and Agriculture Organization & ETEA/Escuela Superior de Técnica Empresarial Agrícola, 2010, *Desarrollo Rural y Seguridad Alimentaria y Nutricional en Centroamérica desde la Perspectiva del Proceso de Integración: Balances y Retos*, Fundación ETEA para el desarrollo y la cooperación y FAO.

Harmeling, Sven & David Eckstein, 2012, Global Climate Risk Index 2013. Who Suffers Most from Extreme Weather Events? Weather-related Loss Events in 2011 and 1992 to 2011, Briefing Paper, Berlin, Germanwatch, available at http://germanwatch.org/en/download/7170.pdf, last accessed 15 April 2013.

Hotez, Peter, Maria E. Bottazzi, Carlos Franco-Paredes, Steven Ault & Mirta R. Periago, 2008, The Neglected Tropical Diseases of Latin America and the Caribbean: A Review of Disease Burden and Distribution and a Roadmap for Control and Elimination, *PLOS Neglected Tropical Diseases* 2 (9), e300.

IPCC/Intergovernmental Panel on Climate Change, 2007a, *Cambio climático 2007: evaluación de la vulnerabilidad e impactos del cambio climático y del potencial de adaptación en América Latina*, Lima, United Nations.

IPCC/Intergovernmental Panel on Climate Change, 2007b, Climate Change 2007: The Physical Science Basis. Contribution of Working Group I to the Fourth Assessment Report of the Intergovernmental Panel on Climate Change, Solomon, Susan, Dahe Qin, Martin Manning, Zhenlin Chen, Melinda Marquis, Kristen B. Averyt, Melinda Tignor & Henry L. Miller (Eds), Cambridge, Cambridge University Press.

IPCC/Intergovernmental Panel on Climate Change, 2012, Managing the Risks of Extreme Events and Disasters to Advance Climate Change Adaptation. A Special Report of Working Groups I and II of the Intergovernmental Panel on Climate Change, Field, Christopher B., Vicente Barros, Thomas F. Stocker, Dahe Qin, David J. Dokken, Kristie L. Ebi, Michael D. Mastrandrea, Katharine J. Mach, Gian-Kasper Plattner, Simon K. Allen, Melinda Tignor & Pauline M. Midgley (Eds), Cambridge, Cambridge University Press.

Jury, Mark R., 2011, Long-term Variability and Trends in the Caribbean Sea, *International Journal of Oceanography*, 1–9.

MARN/Ministry of Environment and Natural Resources Government of El Salvador, 2012, *Forging the Way Forward, Despite the Odds. How El Salvador is Responding to Climate Related Disasters and Climate Change*, El Salvador, MARN.

Noji, Eric & Michael Toole, 1997, The Historical Development of Public Health Responses to Disasters, *Disasters* 21 (4), 366–376.

Ramírez, Diana, Juan L. Ordaz & Jorge Mora, 2009, *Istmo Centroamericano: efectos del cambio climático sobre la agricultura*, LC/MEX/L.924/Rev. 1, México, D.F., Economic Commission for Latin America and the Caribbean (ECLAC).

Sen, Amartya, 1999, *Development as Freedom*, Oxford, Oxford University Press.

Stern, Nicholas, 2007, *The Economics of Climate Change – The Stern Review*, Cambridge, Cambridge University Press.

Stern, Nicholas, 2008, *Key Elements of a Global Deal on Climate Change*, London, London School of Economics and Political Science.

UNDP/United Nations Development Programme, 2007, *Human Development Report 2007-2008: Fighting Climate Change: Human Solidarity in a Divided World*, New York, Palgrave Macmillan.

# 19 Mainstreaming Sustainable Development into National Climate Change Responses: Assessing the Legal Options to Reinforce Equity

*Robert Kibugi*

*Abstract*

Climate change is one of the most pressing challenges of our time as it is a global problem. The impacts, however, are experienced very differently in the developed and developing worlds. Therefore, responsiveness to the impacts of climate change is important to all nations of the world since environment-dependent economic sectors have been adversely affected by climate change. Human livelihoods are increasingly at risk due to a significant rise in extreme weather events that result in more frequent or severe droughts or floods. The adverse impacts of climate change tend to exacerbate or peak due to the vulnerability of systems and people, which in turn increases the risk and susceptibility to adverse effects. The level of vulnerability is impacted by socio-economic status, raising important questions about equity. This article argues that equity is a common goal to be fulfilled in order to have successful climate change responses, and to realise sustainable development. *Sustainable development*, as a legal concept, is advanced as being normatively more equipped with tools to fulfil equity, and such tools could be adapted to address climate change. Thus, the article explores how existing legal tools of sustainable development could have utility in shaping equitable responses to climate change.

## *A. Introduction*

Climate change is one of the most pressing challenges of our time as it is a global problem. The impacts, however, are experienced very differently in the developed and developing worlds.[1] The Intergovernmental Panel on Climate Change (IPCC), in its 2007 assessment report, clearly notes that warming of the climate system is unequivocal, as evidenced by observations of increases in global average air and ocean temperatures, widespread melting of snow and ice, and the rising average sea level.[2] The IPCC further notes that observational evidence from all continents, and most oceans, shows that many natural systems are being affected by regional climate changes, particularly temperature increases.[3] The anthropogenic connection to climate change and global warming has also been clearly pinpointed, with the IPCC further noting that "global GHG [greenhouse gas] emissions due to human activities have grown since pre-industrial times, with an increase of 70% between 1970 and 2004",[4] and that "most of the observed increase in global average temperatures since the mid-20th century is very likely due to the observed increase in anthropogenic GHG concentrations".[5]

Responsiveness to the impacts of climate change is important to all nations of the world, whether developed or developing, because of the implications for the social, economic and environmental dimensions of development. Environment-dependent economic sectors, including water resources and supply, agriculture and forestry have been adversely affected by climate change. Human livelihoods are increasingly at risk due to a significant rise in extreme events that result in more frequent or severe droughts or floods. The increased flood risk poses challenges to human life, livelihoods, physical infrastructure and water quality.[6]

Agricultural production has been under significant pressure due to extreme weather events, but also due to the need to build resilience against changing weather patterns, and to reduce agriculture's contribution to GHG emissions.[7] Water resources are also stressed by climate change, and it is estimated that the number of people projected to experience an increase in water-related stresses is between 0.4 to 1.7 billion for the 2020s and between 1 and 2.7 billion for the 2050s. The degree of water stress increases further

1 Richardson et al. (2009:1).
2 IPCC (2007a:2).
3 (ibid.).
4 (ibid.:5).
5 (ibid.).
6 See Earth Watch Institute (2009).
7 IPCC (2007a:5).

when the minimum water quantities required as environmental flow to sustain integrity of ecosystems are incorporated.[8]

The adverse impacts of climate change tend to exacerbate or peak due to the vulnerability of systems and people. This vulnerability is the degree to which geophysical, biological and socio-economic systems are susceptible to, and unable to cope with, any adverse impacts of climate change.[9] The vulnerability increases the risk and susceptibility to adverse effects due to a lack of basic mechanisms to cope with or adjust to climate-change-induced variations in the environment or in economic circumstances. The level of vulnerability is impacted by socio-economic status, which could result in social stress and environmental damage.[10] Poverty, for instance, is a primary element in causing or raising levels of vulnerability amongst populations in urban and rural settings of both developed and developing countries. Since vulnerability to climate change differs considerably across socio-economic groups, it also raises important questions about equity.[11] Similarly, the equity question arises in discourse regarding climate change effects on countries and their ability to respond to the impacts. Equity is, therefore, a fundamental concern for climate law, and is anticipated in Article 2 of the United Nations Framework Convention on Climate Change (UNFCCC), which stipulates that vulnerability and equity amongst countries is indeed a major consideration when developing legal and other mechanisms for addressing climate change.

One of the consequences of both national and international law failing to adequately resolve challenges of inequity and vulnerability to climate change is impairment of the ability to pursue or attain sustainable development. Notably, in the UNFCCC objectives,[12] control of global warming is sought in part to enable economic development to proceed in a sustainable manner. The increasing prevalence of adverse impacts on people and countries suggests the object of the UNFCCC is far from being realised. The Stern Review supports this reasoning with respect to developing countries, noting that they are –[13]

8 (ibid.).
9 Schneider et al. (2007:783).
10 (ibid.).
11 Schneider et al. (2007:784).
12 Article 2.
13 Stern (2007:Part II, 93).

> … especially vulnerable to the physical impacts of climate change because of their exposure to an already fragile environment, an economic structure that is highly sensitive to an adverse and changing climate, and low incomes that constrain their ability to adapt.

Illustratively, the cost of the 2010 floods that inundated Pakistan resulted in damage to infrastructure, crops and the economy, is estimated at US$43 billion.

Thus, failure to attain equity in response to climate change to reduce peoples' and countries' vulnerability will likely lead to a development deficit and further limit or even negate the ability to realise sustainable development. Normatively, the concept of *sustainable development* revolves around the principles and practice of equity, both among present and future generations. The correlation between climate change and sustainable development is not disputed. According to the IPCC, there is a dual relationship between sustainable development and climate change: climate change influences key natural and human living conditions and, therefore, also the basis for social and economic development; and society's priorities on sustainable development influence both the GHG emissions that are causing climate change and vulnerability to it.[14] The UNFCCC acknowledges that climate change responses should be undertaken within a framework of sustainable development. While this provision has been in place since 1992, the conceptual linkage of sustainable development as a tool to reinforce climate change response mechanisms has been expanded in recent years. The Bali Action Plan (BAP) expanded this linkage further with a decision to enhance national/international action on mitigating climate change with nationally appropriate mitigation actions by developing countries in the context of sustainable development. BAP also urged states parties to pursue policy approaches and positive incentives on issues relating to reducing emissions from deforestation and forest degradation, including the role of conservation, sustainable management of forests and enhancement of forest carbon stocks in developing countries.[15] This approach has been reiterated in subsequent Conferences of Parties (COPs) to the UNFCCC such as the Copenhagen Accord, where countries agreed to enhance long-term cooperation to combat climate change, on the basis of equity and in the context of sustainable de-

14 IPCC (2007b:para. 2.1.3).

15 UNFCCC, Report of the Conference of the Parties on its Thirteenth Session, held in Bali from 3 to 15 December 2007 – Decision 1/CP.13 on Bali Action Plan, para 1.

velopment.[16] Rio+20, the global conference on sustainable development held in June 2012, further enhanced the connection by reaffirming –[17]

> … the necessity to enhance sustainable agriculture, including crops, livestock, forestry, fisheries and aquaculture, that improves food security, eradicates hunger and is economically viable, while conserving land, water, plant and animal genetic resources, biodiversity and ecosystems, and enhancing resilience to climate change and natural disasters.

While the above iteration confirms an existing linkage between climate change and sustainable development, it does not offer any normative content to ensure that attaining sustainable development remains a key objective of climate change responses. This provides a critical moment for law and policy to address climate change and sustainable development simultaneously. Indeed, a key argument by the IPCC is that climate policies can be more effective when consistently embedded within broader sustainable development strategies.[18] Thus, for instance, the challenge of equity, which pervades efforts to address climate change, could be more holistically addressed within the rubric of sustainable development. The core argument made in this article, therefore, is that addressing the adverse impacts and inequities of climate change is a critical ingredient for realising sustainable development. Cyclically, sustainable development provides the requisite tools to respond to climate change.

The article is divided into three parts. The first is an introduction that highlights the evidence and impacts of climate change, and their contribution to inequity. The second part argues that equity is a common goal to be fulfilled in order to have successful climate change responses, and to realise sustainable development. It further argues that *sustainable development*, as a legal concept, is normatively more equipped with tools to fulfil equity, and such tools could be adapted to address climate change. The third part explores how existing legal tools of sustainable development could have utility in shaping equitable responses to climate change.

16 UNFCCC, Copenhagen Accord, Draft Decision -/CP.15, Conference of the Parties Fifteenth Session, Copenhagen, 18 December 2009, 1.

17 United Nations, The Future We Want: Rio+20 Conference on Sustainable Development, Outcome of the Conference A/CONF.216/L.1, Rio de Janeiro, Brazil, 20–22 June 2012, para. 111 (hereinafter *The Future We Want*).

18 (ibid.).

### B. Exploring Commonality of Equity in Climate Change and Sustainable Development

The UNFCCC prominently poses the question of equity in combating climate change. Article 3 stipulates that states parties –

> ... should protect the climate system for the benefit of present and future generations of humankind, on the basis of equity and in accordance with their common but differentiated responsibilities and respective capabilities.

This explicit reference establishes a requirement for countries to protect the climate system –

- while applying a framework of generational equity, and
- guided by the rules of common but differentiated responsibilities, and their respective capabilities.

These frameworks are reflected in discourse on sustainable development in which intergenerational equity is paramount; and special considerations to developing, least-developed and small island states are applied in a similar context as common but differentiated responsibilities and their respective capabilities. This submission focuses on the interaction between climate change and generational equity, and suggests that applying the integration methodology of sustainable development will provide effective mechanisms. Inherently, the pursuit of intergenerational equity necessitates the application of differential treatment even among natural citizens, with the same normative content as applied through common but differentiated responsibilities and respective capabilities amongst states.

#### I. Intergenerational Equity

Equity amongst present and future generations of humankind is also referred to as intra- and intergenerational equity, respectively. According to Edith Brown Weiss, *intergenerational equity* denotes an inherent obligation on the current generation to conserve and utilise the environment without negatively impacting the ability of future generations to meet their own needs.[19] The present generation also enjoys an intergenerational right to enjoy the benefits of a suitable environment bequeathed from earlier genera-

19 Brown Weiss (2008).

tions. In this case, it is necessary to ensure fairness is established within present generations through opportunities to utilise the natural resources and eradicate poverty – hence reducing vulnerability. This exposes intergenerational equity as having an inherent element of intragenerational equity.

This approach is supported in international law through provisions in various Conventions. The Convention on Biological Diversity, for example, has the objective of conserving biodiversity, the sustainable use of its components, and the fair and equitable sharing of benefits amongst present and future generations.[20] This is reiterated by the 2003 Revised African Natural Resources Convention, which requires its provisions to be implemented in the interest of present and future generations.[21] Agenda 21, which also embraces generational equity, argues that a specific anti-poverty strategy is one of the basic conditions for ensuring sustainable development.[22] Such a strategy – to tackle the problems of poverty, development and environment simultaneously – should focus on resources and people, especially on enhanced health care and education; the rights of women; the role of the youth, indigenous people, and local communities; and improved governance. This reinforces the arguments that claim attaining equity is an inherent objective of sustainable development.

## *II. Adapting the Methodology of Sustainable Development to Reinforce Equity in Climate Change*

### *1. Equitable Objectives of Climate Change and Sustainable Development*

The protection of the climate system, which is the target of the UNFCCC, seeks to secure intergenerational equity. The utility of generational equity is reinforced because climate change continues to have devastating impacts on impoverished peoples, largely due to their vulnerability and low capacity to adapt.[23] This vitiates people's resilience to climate change and undermines

20 Article 3.

21 Article 4.

22 *Report of the United Nations Conference on Environment and Development, Rio de Janeiro, 3–14 June 1992*, New York, United Nations, Sales No. E.93.I.8 and corrigenda, Vol. I: Resolutions Adopted by the Conference, Resolution 1, Annexes I and II, Chapter 3, para. 3.2 (hereinafter *Agenda 21*).

23 Brown Weiss (2008).

their ability to cope with changes. Therefore, adaptation to climate change is one mechanism that can demonstrate the application of generational equity in responding to climate change. *Adaptation* can be defined as the process through which people reduce the adverse effects of climate change on their health and well-being, and take advantage of the opportunities that their climatic environment provides.[24] Thus, it is a process that aims to build people's resilience to the unique circumstances posed by climate change.

With respect to building resilience, two possible approaches in respect of adaptation to climate change arise: reactive adaptation, and planned adaptation. *Planned adaptation* is the result of a deliberate policy decision, based on an awareness that conditions have changed or are about to change and that action is required to return to, maintain, or achieve a desired state.[25] It is, therefore, a suitable tool for addressing the vulnerabilities and risks posed by climate change, including improving livelihoods to minimise poverty. Planned adaptations can be reactive or anticipatory, i.e. undertaken before impacts are apparent, and can include potential actions to share losses, modify threats, prevent or decrease effects, or change use.[26]

Enhancing adaptive capacity, particularly through planned adaptation, is a necessary condition for reducing vulnerability and inequity. This is especially the case for the most vulnerable regions, nations, and socio-economic groups: activities required for the enhancement of adaptive capacity are essentially equivalent to those promoting sustainable development.[27] Climate adaptation and equity goals can be jointly pursued by initiatives that promote the welfare of the poorest members of society, e.g. by improving food security, and facilitating access to safe water and other resources.[28] This same objective can be attained where the norms of sustainable development are applied, particularly since many climate change impacts arise from development activities of an anthropocentric nature.

24 Lin (2009:129).
25 (ibid.).
26 IPCC (2001:982).
27 (ibid.:881–885); see particularly Chapter 3, "Adaptation to Climate Change in the Context of Sustainable Development and Equity".
28 (ibid.).

## 2. *Extending the Methodology of Sustainable Development to Climate Change*

Normatively, the link between climate change, equity, and sustainable development addresses the question of justice within and among generations. In the case of sustainable development, this is demonstrated by the classic definition advanced by the Brundtland Commission, namely that it is *development* "that meets the needs of the present without compromising the ability of future generations to meet their own needs."[29]

This definition presents sustainable development as containing –

- a concept whereby the essential needs of the world's poor should be given priority, and
- the idea of limitations of the environment's ability to meet present and future needs.

The concept of *needs* addresses poverty and vulnerability, while the notion of *limitations* suggests an imperative to preserve the quality of the environment. This reveals that both sustainable development and climate change seek to attain the goals of equity to resolve vulnerability, poverty, and degradation.

The actual methodology through which sustainable development is implemented can aid the implementation of equitable climate change mechanisms such as adaptation. This is because implementation of sustainable development centres on the notion of the integration of environmental, social and economic dimensions. As reiterated by the Brundtland Commission's Report, the concept of *sustainable development* provides a framework for the integration of environmental policies and development strategies.[30] *Integration* entails a process for considering and weighing up social, environmental and economic considerations during decision-making. Principle 4 of the 1992 Rio Declaration[31] is notable in this context, providing that, in order to achieve sustainable development, environmental protection has to be an integral part of the development process and cannot be considered in isolation from it. In support of this position, the International Court of Justice has recognised sustainable development as the principle that makes it possible

29 World Commission on Environment and Development (1987:43).

30 (ibid.:40).

31 Rio Declaration on Environment and Development, in UN (1992); hereinafter *Rio Declaration*.

to maintain the balance between environmental and developmental considerations.[32]

With the methodology of integration, the overarching objectives and essential requirements of sustainable development include managing the natural resource base of economic and social production, and eradicating poverty.[33] This approach significantly reinforces climate change, since adaptation mechanisms, for instance, would focus on building or strengthening resilience in social, economic and environmental systems. In order to avoid possible deleterious effects of response mechanisms, adaptation would apply the notion of *integration* to ensure building resilience is holistic by seeking to balance the socio-economic and environmental systems. This is necessary because climate change – as reiterated during the Rio+20 Conference[34] – is a cross-cutting and persistent crisis with grave negative impacts that affect all countries and undermine their ability – particularly that of developing countries – to achieve sustainable development.[35] This requires a concerted effort, therefore; and mainstreaming the methodology of integration will enhance resilience and assure climate systems are protected for present and future generations.

### C. *Assessing the Legal Options*

Sustainable development contains norms and principles that can be deployed to enhance the application of integration as a legal tool to aid the implementation of adaptation programmes necessary to achieve equity. These norms and principles, when applied through law, would ensure adaptation options are considered within the parameters of sustainable development, hence diminishing the likelihood of the deleterious effects of various climate

---

32 *Gabčikovo-Nagymaros Project* (*Hungary v Slovakia*), Judgment, [1997] ICJ Reports, Separate Opinion of Vice-President Weeramantry., 88–116, p. 85.

33 The Future We Want, para. 11.

34 Rio+20, formally known as the *United Nations Conference on Sustainable Development* (UNCSD), was organised in pursuance of General Assembly Resolution 64/236 (A/RES/64/236), and took place in Brazil on 20–22 June 2012. The Conference emphasised the need to further mainstream sustainable development at all levels, integrating economic, social and environmental aspects so as to achieve sustainable development in all its dimensions. Climate change was identified as a major impediment to the realisation of sustainable development.

35 (ibid.:para. 25).

change response mechanisms. They would, thus, provide legal options for policymakers.

### *I. People at the Centre of Sustainable Development*

The adverse impacts of climate change often affect people, especially the most vulnerable, due to weak coping mechanisms or an inability to build resilience. Adaptation responses should, therefore, aim to strengthen the ability of such people to cope with climate variability. This approach resonates with the concept of putting people at the centre of sustainable development. While an important concept, this approach has been termed *anthropocentric*, with a risk of encouraging a strong yet narrow focus on the socio-economic interests of humankind, with insufficient corresponding care for the environment. This is mainly because anthropocentrism confers intrinsic value on people and regards all other forms of life, including the environment, as being only instrumentally valuable, i.e. to the extent that they are or can be useful to serve human beings.[36]

Where people's interests are put first in an anthropocentric setting, it would be the utility of nature and its instrumental value to human beings that would matter most. This is because anthropocentrism revolves around the concept of *value*. Joseph des Jardins explains *value* as instrumental and intrinsic.[37] *Instrumental value* is a function of usefulness such that an object will possess that value because of the use to which such object can be put. By extension, that instrumental value is lost or diminished when the object no longer has a use – as the sense of *value* presupposes the existence of an external valuer or beneficiary, such as a human being.[38] Where anthropocentrism is applied in law and policy, it is only human beings that possess moral value; humans may have responsibilities *regarding* the natural world, but no direct responsibilities *to* the natural world.[39] This reasoning compounds the challenge of integration because, in practical terms, the line to draw between destruction and preservation or conservation is rather vague. This is more so in developing countries with extreme poverty, where the search for survival is desperate, and hope for tomorrow a mirage at best. It

36 Callicot (1984:299).
37 Des Jardins (1997:127–130).
38 Bowman (1996:14).
39 Des Jardins (1997:9–11).

is also in these countries where the need to build resilience through adaptation is highest.

In contrast, people could still be placed at the centre of sustainable development through the ecocentric approach that reinforces integration during decision making. The ecocentric land ethic advanced by US forester Aldo Leopold offers a contrast to anthropocentrism.[40] Leopold suggests that the land ethic reflects the existence of an ecological conscience, which is a conviction of an individual responsibility to attain and retain the health of the land. He clarifies *land health* as the capacity of the land for self-renewal. Therefore, according to Leopold, the ecological conscience involves love, respect and admiration for the land, dedicated to a high regard for its value beyond economic self-interest. The land ethic also examines the role of humankind, with Leopold urging that such a land ethic seeks to alter the role of humans from conqueror of the land community to a citizen of the biotic community. The land ethic conceptualises human beings having an entitlement to utilise environmental resources, but guided by an ecological conscience. Arguably therefore, this approach seeks to ensure the socio-economic and environmental dimensions are addressed. One could, therefore, conclude that, in placing humans at the centre of concerns for sustainable development and, by extension, taking steps to build resilience to climate change, it is the ecocentric approach that provides room to apply integration.

### *II. The Concept of Rights as a Tool for Equity*

Human rights have increasingly become a mechanism for guiding the realisation of sustainable development, particularly the attainment of equity. These rights, especially when they attain the status of a constitutional, or fundamental, right are critical because they acquire superiority in a legal system. Internationally, rights are protected through the binding nature of treaties, and nationally, they could be protected by the supremacy of constitutions.[41] The domain of rights has expanded significantly and incorporates both environmental and development rights, indicating an increasing focus on the realisation of sustainable development. The 2003 African Convention on the Conservation of Nature and Natural Resources (Revised Ver-

40 Leopold (1981/2003:215–224).

41 See e.g. Article 2, Constitution of Kenya, 2010.

sion)[42] incorporates "the right of all peoples to a satisfactory environment favourable to their development".[43] The 1981 African Charter on Human and Peoples' Rights[44] equally guarantees all peoples the right to a "general satisfactory environment favourable to their development".[45] The 2003 Protocol to the African Charter on Human and Peoples' Rights on the Rights of Women in Africa guarantees a right to nutritious and adequate food for women, including provision with "access to clean drinking water".[46] The African Charter on the Rights and Welfare of the Child guarantees every child the right to enjoy the best attainable state of physical, mental and spiritual health, including "provision of adequate nutrition and safe drinking water".[47]

Constitutions, just like treaties, entrench both environmental and socio-economic rights, and in certain cases explicitly set out sustainable development as an output of implementing these rights.[48] Kenya, South Africa and Uganda are illustrative examples where constitutional environmental rights are accompanied by socio-economic rights (to food, water, sanitation, health, etc.) in the Bill of Rights.

Such inclusions in the Bill of rights suggest an obligation to apply integration and balancing of socio-economic and environmental rights when implementing the related entitlements. This argument is supported by the view of Claasen J in *BP Southern Africa (Pty) Ltd v MEC for Agriculture, Conservation, Environment and Land Affairs*,[49] where he argued that the environmental right enshrined in the South African Constitution was on par with other basic rights such as freedom to trade or the right to property, and none should be considered more important than the other. This reasoning supports a conclusion that the framework of environmental and socio-economic human rights seeks equity, and therefore provides a basis on which climate change programmes such as adaptation could be mounted. However, equity will only become an outcome where the methodology of integration is applied to balance the three dimensions of sustainable development.

---

42 11 July 2003, reprinted in Heyns (2010:95).

43 (ibid.:Article 3).

44 27 June 1981, reprinted in Heyns (2010:29).

45 (ibid.:Article 24).

46 Article 18.

47 Article 14(2)(c).

48 See e.g. Article 42, Constitution of Kenya 2010, as read with Article 69 (environmental rights and sustainable development) and Article 43 (socio-economic rights).

49 2004 (5) SA 124 (W).

### III. Local Agenda 21

The role of local authorities such as city councils, municipalities and other devolved governments in sustainable development was introduced, conceptually, by Agenda 21 in 1992. Agenda 21 argued that, since most sustainability problems had their roots in local activities, local authorities were the ones to provide governance closest to the people; for this reason they played a vital role in educating, mobilising and responding to the public to promote sustainable development.[50] They would, thus, be responsible for implementing a 'Local Agenda 21'. This approach of devolved governments taking on such responsibility would work well because local authorities construct, operate and maintain economic, social and environmental infrastructure, oversee planning processes, establish local environmental policies and regulations, and assist in implementing national and subnational environmental policies.[51] Du Plessis argues that Local Agenda 21 translates Agenda 21 into a framework for local authorities to seek local solutions for global challenges through voluntary action.[52]

Local authorities have to develop their own Local Agenda 21 through a process of consultation, consensus-building, and community participation towards preparation of sustainable development strategies. They have to implement and monitor programmes which should apply equity by ensuring that women and the youth are represented in decision-making, planning and implementation processes.[53] The framework for Local Agenda 21 is notable in that it does not provide specific guidance or specify the format for its design and implementation, suggesting that consultations should guide both the design and content to respond to local sustainability concerns. Climate change adaptation is one such sustainability concern inherently suited to response at a local level. As Richardson urges, while the impacts of climate change are commonly widely dispersed, the benefits of adaptation measures are often quite localised.[54] He suggests, therefore, that climate change adaptation invites a legal framework that facilitates capacity-building to develop

50 Agenda 21, Chapter 28, para. 28.1.
51 (ibid.).
52 Du Plessis (2011:48).
53 Agenda 21, Chapter 28, para 28.2.
54 Richardson (2012:7–8).

and implement local adaptation plans that align the work of national and local governments.[55]

Local authorities are empowered by national laws with powers and functions over a defined jurisdiction. They often exercise (quasi-)legislative functions, either making local laws, or devising regulations to implement national, provincial or state law. Many of the functions performed by local authorities impact on sustainability, and could simultaneously be modified to implement climate change adaptation. These include modifying building codes to require the use of solar or geothermal energy sources in buildings, reinforcing infrastructure to withstand climate change impacts, and implementing food security programmes. Certainly, the capacity of local authorities to fully design a local climate change agenda pegged on the rubric of Local Agenda 21 will depend on various factors, including scope of jurisdiction, financial capability, and maturity of planning and enforcement systems. Nonetheless, Local Agenda 21 is a concept of sustainable development that can enable the successful implementation of climate change responses at a local level, with various modifications.

### *IV. Green Economy*

A *green economy* is defined as one that results in improved human well-being and social equity, while significantly reducing environmental risks and ecological scarcities. The United Nations Environment Programme (UNEP) argues that a green economy entails low carbon emissions, resource-efficiency and social inclusiveness.[56] This, according to the Food and Agriculture Organization of the United Nations (FAO), makes a green economy an economic system that is compatible with the natural environment and social concerns. This requires using clean technology and clean energy to provide safer and healthier environments, create alternative green jobs, and safeguard the development of societies. In addition to seeking low carbon emissions, a green economy seeks "green growth" which extends beyond economic output growth, and indicates "sustainable economic progress".[57]

Although the concept of a *green economy* is still evolving, Rio+20 clarified it further that a green economy would promote sustained and inclusive

55 (ibid.:8).
56 UNEP (2011:01–02).
57 FAO (2010:3–4).

economic growth; foster innovation; provide opportunities, benefits and empowerment for all; and respect all human rights. A green economy would also mobilise full and equal contributions by men and women; it would enhance the welfare of women, children, the youth, persons with disabilities, and smallholder and subsistence farmers; and it would improve the livelihoods and empowerment of the poor and of vulnerable groups in developing countries in particular.

An important element of a green economy is the absence of a tailor-made set of rules on how to implement it at national level. Indeed, Rio+20 suggested that each country should be free to choose an appropriate approach in accordance with its own national sustainable development strategies – which would enhance the ability to manage natural resource sustainability and with lower negative impacts.[58] FAO suggested five major elements of a green economy, which will still evolve, providing a normative indicator of the linkage that a green economy brings between sustainable development and climate change:[59]

- Generation and use of renewable energy
- Energy efficiency
- Waste minimisation and management
- Preservation and sustainable use of existing natural resources, and
- Green job creation, offering a decent wage, job security and career prospects.

It would, therefore, be left to individual countries to determine how to apply the concept of a *green economy* as a legal mechanism for unifying sustainable development and climate change.

## *V. Disaster Risk Reduction*

Globally, natural disasters are increasing in frequency and strength. Extreme events such as earthquakes, tsunamis, cyclones, hurricanes, typhoons, floods, droughts and famine are occurring in more frequent cycles, and causing greater adverse effects. The debilitating impacts of disasters are compounded by increasing vulnerabilities related to changing demographic and socio-economic conditions, development within high-risk zones, under-

58 *The Future We Want*, Part III.
59 (ibid.).

development, environmental degradation, climate variability, climate change, geological hazards, and competition for scarce resources.[60] The concept of *disaster risk reduction* (DRR) is concerned with decreasing disaster risks through systematic efforts to analyse and reduce the causal factors of disasters.[61] DRR is concerned with reducing exposure to hazards, minimising the vulnerability of people and property, the wise management of land and the environment, and improving preparedness for adverse events.[62] The absence of DRR in policies significantly vitiates a country or people's ability to make sustainable development a reality. Climate-change-induced extreme events, for instance, have the capacity to reverse sustainable development – hence the need to deploy DRR tools for anticipatory action. These tools include developing early warning systems and a culture of community safety, undertaking hazard and risk assessments, and planning humanitarian relief work ahead of time.

At the global level, the Hyogo Framework on DRR represents a ten-year plan (2005–2015) on measures to reduce the risk of vulnerabilities and hazards turning into disasters. Priority 1 of the Framework urges countries to mainstream DRR into national laws and principles, arguing that, in such instances, countries will have greater capacity to manage risks and to achieve widespread consensus for, engagement in, and compliance with DRR measures across all sectors of society.[63] The strategic action proposed under the Framework establishes a link between the norms and the overarching role of sustainable development in guiding integration when it calls on countries to integrate risk reduction, as appropriate, into development policies and planning at all levels of government, including poverty reduction strategies and sectors, and multisectoral policies and plans. The linkage with sustainability could be strengthened by countries adopting or modifying legislation to support DRR, including regulations and mechanisms that encourage compliance and promote incentives for undertaking DRR and mitigation activities.

The absence of DRR frameworks either aligned to Hyogo or framed to country-specific needs will prevent a country from reducing or eliminating its vulnerability to the adverse effects of climate change, such as extreme

60 UNISDR (2007).

61 See International Strategy for Disaster Reduction (ISDR), available at http://www.unisdr.org/who-we-are, last accessed 11 April 2013.

62 (ibid.).

63 UNISDR (2007:6).

events that amount to disasters. This means that the country and its people will lack the resilience required to withstand disasters, while its ability to achieve sustainable development is further diminished. It is critical, therefore, that DRR be a central concern whenever climate change responses are planned in the context of sustainable development and in pursuit of equity.

## *D. Conclusion*

Normatively, adaptation requires the application of differential treatment, since different people have different coping or resilience deficiencies. The failure to address the equity challenge facing climate change mechanisms will mean that those resilience deficiencies cannot be sufficiently addressed. This would mean that vulnerabilities arising from the negative impacts of climate change will persist. Nations and people who are at high risk due to vulnerability will continue without the capacity or means to realise sustainable development. It is now indisputable that sustainable development and climate change responses are mutually reinforcing. For developing countries whose economies continue to rely on natural resources for macro and small-scale economic productivity, it is critical to view climate change through the lens of sustainable development. A failure to do so will result in a development deficit, because of an exceedingly anthropocentric approach to utilising environmental resources, or because of pervading poverty, which will, cyclically, undermine any adaptation efforts. Applying sustainable development legal concepts such as rights provides a framework within which the law requires decisions – including those on climate change – to be integrated to ensure that the three dimensions of sustainable development are always considered on par. As the outcome from Rio+20 so eloquently states, political will by governments will remain paramount if further progress is to be realised in framing sustainable development as the overall objective of national development.

## *References*

Bowman, Michael, 1996, The Nature, Development and Philosophical Foundations of the Biodiversity Concept in International Law, in: Bowman, Michael & Catherine Redgwell (Eds), *International Law and the Conservation of Biological Diversity*, The Hague, Kluwer Law International, 5–31.

Brown Weiss, Edith, 2008, Climate Change, Intergenerational Equity and International Law, *Vermont Journal of Environmental Law* 9, 615–627.

Callicot, J. Baird, 1984, Non-anthropocentric Value Theory and Environmental Ethics, *American Philosophical Quarterly* 21 (4), 299–309.

Des Jardins, Joseph, 1997, *Environmental Ethics: An Introduction to Environmental Philosophy*, Belmont, CA, Wadsworth Publishing.

Du Plessis, Anel, 2011, Agenda 21: A Rights-Based Approach to Local Environmental Governance, in: Benidickson, Jamie, Ben Boer, Antonio Benjamin & Karen Morrow (Eds), *Environmental Law and Sustainability after Rio,* Cheltenham, Edward Elgar, 47–65.

Earth Watch Institute, 2009, *Climate Change: The Social and Economic Impacts*, available at http://www.earthwatch.org/europe/downloads/Get_Involved/ClimateChange3.pdf, last accessed 10 April 2013.

FAO/Food and Agriculture Organization of the United Nations, 2010, *Stakeholders Consultation: From Payment of Environmental Externalities to Remuneration of Positive Externalities in the Agriculture and Food Sector, Rome, 27–28 September 2010 – Payments for Environmental Services within the Context of the Green Economy*, Rome, FAO.

Heyns, Christopher & Magnus Killander (Eds), 2010, *Compendium of Key Human Rights Documents of the African Union*, Pretoria, Pretoria University Law Press.

IPCC/Intergovernmental Panel on Climate Change, 2001, *Climate Change 2001: Impacts, Adaptation, and Vulnerability. Contribution of Working Group II to the Third Assessment Report of the Intergovernmental Panel on Climate Change*, McCarthy, James, Osvaldo Canziani, Neil A. Leary, David J. Doken & Kasey S. White (Eds), Cambridge, Cambridge University Press.

IPCC/Intergovernmental Panel on Climate Change, 2007a, *Climate Change 2007: Synthesis Report. Contribution of Working Groups I, II and III to the Fourth Assessment Report of the Intergovernmental Panel on Climate Change*, Core Writing Team: Pachauri, Rajendra K. & Andy Reisinger (Eds), Geneva, IPCC, available at http://www.ipcc.ch/publications_and_data/publications_ipcc_fourth_assessment_report_synthesis_report.htm, last accessed 13 January 2013.

IPCC/Intergovernmental Panel on Climate Change, 2007b, *Climate Change 2007: Mitigation of Climate Change. Contribution of Working Group III to the Fourth Assessment Report of the Intergovernmental Panel on Climate Change*, Bert Metz, Ogunlade R. Davidson, Peter R. Bosch, Rutu & Leo A. Meyer (Eds), Cambridge, Cambridge University Press.

Leopold, Aldo, 1981, *A Sand County Almanac with Other Essays on Conservation from Round River*, New York, Oxford University Press, and [reprinted 2003], in: VanDeveer, Donald and Christine Pierce (Eds), *The Environmental Ethics and Policy Book*, Belmont,CA, Thomson/Wadsworth.

Lin, Jolene, 2009, Supporting Adaptation in Developing Countries and the National and Global Levels, in: Richardson, Benjamin & Yves Le Bouthillier, Heather Mcleod-Kilmurray & Stephan Wood (Eds), *Climate Law and Developing Countries: Legal and Policy Challenges for the World Economy,* Cheltenham, Edward Elgar, 127–150.

Richardson, Benjamin, Yves le Bouthillier Heather Mcleod-Kilmurray & Stephan Wood (Eds), 2009, Introduction: Climate Law and Developing Countries, in: Richardson, Benjamin, Yves le Bouthillier, Heather Mcleod-Kilmurray & Stephan Wood (Eds), *Climate Law and Developing Countries: Legal and Policy Challenges for the World Economy,* Cheltenham, Edward Elgar, 1–34.

Richardson, Benjamin, 2009, Local Climate Change Law, in: Richardson, Benjamin (Ed.), *Local Climate Change Law: Environmental Regulations in Cities and Other Localities,* Cheltenham, Edward Elgar, 3–28.

Schneider, Stephen H., Sergey Semenov, Anand Patwardhan, Ian Burton, Chris H.D. Magadza, Michael Oppenheimer, A. Barrie Pittock, Atiq Rahman, Joel B. Smith, Avelino Suarez and Farhana Yamin, 2007, Assessing Key Vulnerabilities and the Risk from Climate Change, in: IPCC/Intergovernmental Panel on Climate Change, 2007, *Climate Change 2007: Impacts, Adaptation and Vulnerability. Contribution of Working Group II to the Fourth Assessment Report of the Intergovernmental Panel on Climate Change*, Parry, Martin L., Osvaldo F. Canziani, Jean Palutikof, Paul J. Van der Linden & Clair E. Hanson (Eds), Cambridge, Cambridge University Press, 779–810.

Stern, Nicholas, 2007, *The Economics of Climate Change: The Stern Review*, Cambridge, Cambridge University Press.

UN/United Nations, 1992, *Report of the United Nations Conference on Environment and Development*, UNGA OR, A/CONF.151/26, Vol. I, New York, UN.

UNEP/United Nations Environment Programme, 2011, *Towards a Green Economy: Pathways to Sustainable Development and Poverty Eradication – A Synthesis for Policy Makers*, Nairobi, UNEP.

UNISDR/United Nations International Strategy for Disaster Reduction, 2007, *Hyogo Framework for Action 2005–2015: Building the Resilience of Nations and Communities to Disasters,* Extract from the final report of the World Conference on Disaster Reduction (A/CONF.206/6), Geneva, UNISDR.

World Commission on Environment and Development, 1987, *Our Common Future*, Oxford, Oxford University Press.

# 20
# Agenda 21 and Climate Protection: The Development of Global and Local Governance for Environment and Development – Observations from Research in Namibia

*Manfred O. Hinz*

*Abstract*

It was due in particular to the difficulty experienced with achieving tangible results in the worldwide efforts to deal with the deterioration of the climate that challenged scholars in political science, legal sociology and related disciplines to brainstorm what is called the *architecture of governance for sustainable development in a globalising world.* Various attempts at using the architectural approach to re-determine the role and function of sub- and suprastatal social levels and institutions have been submitted, leading to theoretical consequences with respect to the role and function of the state and its relationship to the said non-statal levels and institutions. In practical political terms, the architectural approach results in the empowerment of the non-statal levels and institutions, de facto through the recognition of the pertinent role and function these levels and institutions hold, de jure by calls to extend the existing recognition through legal reforms.

In exploring and employing the architectural approach, this chapter relies on research done in Namibia. The focus is on findings from empirical research done within the legal component of the Biodiversity Monitoring Transect Analysis in Africa (BIOTA) and its successor The Future of Okavango (TFO) Projects, both administered under the University of Namibia's Faculty of Law.

## A. The Architecture of Governance for Sustainable Development in a Globalising World

The difficulty experienced with achieving tangible results in the worldwide efforts to deal with the deterioration of the climate[1] challenged scholars in political science, legal sociology and related disciplines to brainstorm what is called the *architecture of governance for sustainable development in a globalising world.*[2] In exploring and employing the architectural approach, empirical research done in Namibia will be used to illustrate special components in the description of the said architecture of governance. The research was conducted within the legal component of the Biodiversity Monitoring Transect Analysis in Africa (BIOTA) and its successor The Future of Okavango (TFO) Projects, both administered under the University of Namibia's Faculty of Law.[3]

---

1 The 1992 Earth Summit adopted the United Nations Framework Convention on Climate Change (UNFCCC). The UNFCCC entered into force on 21 March 1994, with 194 parties having signed it. The UNFCCC was one of three so-called Rio treaties – along with the Convention on Biological Diversity and the Convention to Combat Desertification in Those Countries Experiencing Serious Drought and/or Desertification, Particularly in Africa. In implementing the interest to protect climate, the Kyoto Protocol emerged within the UNFCCC and was adopted at the Third Conference of the Parties of the UNFCCC. With 192 members, the Kyoto Protocol came into force on 11 December 1997. The Protocol set a target to reduce gas emissions by a certain level by the end of 2012, the date on which the Protocol was to expire. For the African implications of the Protocol, see Ruppel (2011). After long negotiations, the Protocol was extended in November 2012 to 2020.

2 Cf. here Biermann et al. (2010); Winter (2006b; 2012).

3 Both sub-projects are under the direction of the author. On the BIOTA Project, see Jürgens et al. (2010). The TFO Project succeeded the BIOTA Project. The TFO Project started in 2010 and will end in 2015. In the words of a TFO document on file with the author, the Project is "dedicated to support sustainable land use and resource management in the Okavango river basin. The Okavango River, connecting the countries Angola, Namibia and Botswana, is a global hydro-political hotspot, a large-scale indicator for climate change and a benchmark for the success of sustainable management. A multitude of different factors make the Okavango basin a trans-boundary study region of high international visibility and high potential transferability of results to other tropical and sub-tropical regions". The following article summarises observations developed in the author's lecture entitled "Sustainable Development: Political Concept and Legal Implications" offered in the Spring semester of 2012 at the Jacobs University of Bremen as part of its Integrated Environmental Studies Programme. The observations were guided by an earlier publication, co-edited by the author, entitled *Biodiversity and the Ancestors* (Hinz & Ruppel 2008). A previous version of this

The orientation towards an architecture of governance for sustainable development in a globalising world denotes more than an enquiry into the functioning of international law, international policies, and their transformation into political and legal instruments of the state. *Architecture* denotes a comprehensive, multilevel construct in which the various levels are interlinked, like the storeys of a building. Frank et al. introduce the concept of *global governance architecture* as follows:[4]

> There is no commonly agreed definition of the term 'global governance architecture'. We define the term in this book as the overarching system of public and private institutions – that is, organisations, regimes and other forms of principles, norms, regulations and decision-making procedures – that are valid or active in a given issue of world politics. Architecture can thus be described as the meta-level of governance.

This would be *governance*, indeed, in the broadest possible manner; and this without any normative focus on public institutions such as the state or what is normally given the centre of attention – the law of the state. The reference to "architecture" is – in this sense, for the authors of the quoted definition of *architecture*, and different from what the reference to international order would be – value-neutral.[5] This specifically allows one to take note of the fragmentation of governance without judging such fragmentation to be negative and/or dysfunctional for the society concerned only because of the fragmentation's existence. According to Biermann et al., "all global gover-

chapter appeared as the author's introduction to *Knowledge lives in the lake* (Hinz et al. 2012), entitled "Agenda 21 – or: The Legal Obligation to Strengthen Local Levels of Societies (including Traditional Authorities) in Support of Sustainable Development". For this publication, the text has been partly rewritten, amended in order to take note of more recent international developments, and enlarged by incorporating parts of an article by Hinz & Mapaure (2012) and a public talk by the author at the Jacobs University in December 2012. Special words of thanks go to my friends and colleagues Senator Councillor Gunther Hilliges, the former Director of the State Office for Development Cooperation of the Free Hanseatic City of Bremen (who, as thc long-standing promoter of the international Towns and Development movement, shared his inexhaustible experience and knowledge in the field of local authorities and development with me) and Prof. Gerd Winter of the University of Bremen's Faculty of Law (who spent time commenting on an earlier version of the text).

4 Biermann et al. (2010b:16).

5 (ibid.:17).

nance architectures are fragmented to some degree; that is[,] they consist of distinct parts that are hardly ever fully interlinked and integrated."[6]

Non-fragmented structures of governance are conceivable, but they most probably do not exist in practice. Biermann et al. distinguish between governmental fragmentations that may be synergistic, cooperative or conflictive.[7] In other words, fragmentation is not bad per se: it may be either good or bad, depending on the repercussions involved.

The architectural approach to global governance as a consortium of governmental entities that extends from public and private international regimes to public and private local actors is very close to what emerged from the work by scholars of legal and political pluralism, which is informed by the, again, basically globally prevailing existence of interwoven (semi-) autonomous social fields. These various (semi-)autonomous social fields have their own systems of governance, including arrangements with the other fields to which they are linked.[8]

In a recent publication on the architecture of global governance with regard to the planet's protection against the harsh effects of climate change, Winter submitted a diagram showing the various levels involved.[9] He places international state organisations at the top of the hierarchy, followed by contractual or non-contractual interstate arrangements. Below that we find state law. Next to state law we see public actors. Transnational public and private governance also appears in the structure, while private actors are located at the base of the diagram. Parallel to this formalised structure of governance, we are informed that the orientation of the various levels will differ as to whether they consume or protect resources.[10]

The sociological functional or legal-normative question in view of architectures of this kind is to what extent the plurality (or fragmentation) contributes to the functioning of the interwoven set of rules and institutions. Philipp Pattberg, one of the co-editors of *Global climate governance beyond 2012*,[11] notes the following in his chapter therein:[12]

---

6 (ibid.).

7 (ibid.); cf. also Fischer-Lescano & Teubner (2006).

8 Cf. Moore (1978:54ff.) and generally, Menski (2006:82ff.). The chapter will address legal pluralism again in Section F.

9 Winter (2012); cf. also Börzel & Risse (2005); Sand (2006); Winter (2006b).

10 Winter (2006b:114ff.).

11 Biermann et al. (2010a).

12 Pattberg (2010:147).

> Recent scholarly debate within the discipline of international relations has focused on the transformation of the global order from a territorial-based one to one of multiple spheres of authority in flexible and issue-specific arrangements.... Reflecting debates about the organisational transformation of the modern nation state, theorists of international relations have begun to reflect on the changing nature of the Westphalian system itself.... One central empirical observation is the emergence of networked forms of organisation that operate under a different logic compared to other types of social organisations, such as markets and hierarchies. Whereas network governance has been discussed as a complementation and gradual innovation of older forms of policy-making (for example[,] corporatism) within the domestic context, networks at the transnational and global level have been largely conceptualised as new forms of governance that potentially overcome the limitation of more traditional approaches.

In his focus on networked climate governance, Pattberg analyses three distinct types of networked climate governance: public non-state governance, public private networks, and private networks.[13] For the purposes of this article, the first type is of particular importance.[14] In looking at the first type, Pattberg refers to two initiatives of local authorities which became most remarkable in networking for climate change and protection, namely the Cities for Climate Protection Programme initiated by Local Governments for Sustainability, and the C40 network initiated by the Large Cities Climate Leadership Group.

These climate-specific initiatives by local authorities, started in 1991 and 2006 respectively,[15] are part of the earlier global movement to engage local authorities in matters of development. The organisation Towns and Development entered the scene and became an international initiative of far-reaching importance.[16] The Cologne Appeal, adopted in 1985, called for charity to be replaced by justice.[17] The Towns and Development Conference

13 (ibid.:149ff.).

14 As will become apparent as the arguments of this chapter develop further.

15 Cf. here also Bulkeley & Betsill (2003); C40 (2008).

16 A Local Authorities Conference held in Florence, Italy, in 1983, gave rise to what became the international organisation now known as Towns and Development; see Shuman (1994:6). See further Gold et al. (2001); Krenzer-Bass (1988); Kussendrager (1988).

17 See Towns and Development (1985); Shuman (1994:5ff.). The Cologne Conference was initiated by Gunther Hilliges from the State Office for Development Cooperation in Bremen, Germany, and Paul van Tongeren from the Dutch National Commission for Development Education in Amsterdam, The Netherlands. From a jurisprudential point of view, it is noteworthy that a certain German tradition was instrumental in

in Cologne launched a process that engaged the linking of local authorities not only among developed and developing countries, but also between the two country categories.[18] After Cologne, the Charter of Berlin was agreed in 1992 by local authorities from developed and developing countries.[19] "Joint Action for Sustainable Development" was chosen as the document's title. With the establishment of the International Council for Local Environmental Initiatives (ICLEI) at the World Congress of Local Governments for a Sustainable Future at the United Nations (UN) in New York in 1990, a world structure emerged that has remained active ever since.[20] Today, ICLEI has more than 1,000 members from 70 countries, representing more than 500 million people.[21]

The political way to the 1992 Rio Conference was, on the one hand, facilitated by movements of this nature; on the other, the broad governmental and non-governmental support for the Earth Summit and its messages, culminating in Agenda 21, was also instrumental for strengthening these movements. Chapter 1.1 of Agenda 21 offers an insight into its vision:

> Humanity stands at a defining moment in history. We are confronted with a perpetuation of disparities between and within nations, a worsening of poverty, hunger, ill health and illiteracy, and the continuing deterioration of the ecosystems on which we depend for our well-being. However, integration of environment and development concerns and greater attention to them will lead to the fulfilment of basic needs, improved living standards for all, better protected and managed ecosystems and a safer, more prosperous future. No nation can achieve this on its own; but together we can – in a global partnership for sustainable development.

Chapter 1.3 adds the following:

> Agenda 21 addresses the pressing problems of today and also aims at preparing the world for the challenges of the next century. It reflects a global consensus

designing the approach to local authorities within the wider political structure and, based on this, to matters of development in a world perspective; cf. Schefold (2011).

18 Cf. Hilliges (1994; 1995a; 1995b; 2001; 2005, 2012); Hilliges & Nitschke (2007).

19 The Charter is contained as Appendix II in Shuman (1994:150ff.). The Berlin Charter was adopted by the German Bundestag in June 1994 (see Servicestelle 2009), calling on the local authorities in Germany to go ahead with efforts to implement Agenda 21. Appendix II (Shuman 1994:144ff.) also contains an African and an Asian Towns and Development document, namely "The Bulawayo Appeal: From Dependence to Justice", which was adopted by the Bulawayo Towns and Development Conference in 1990, and the Sevagram Declaration (India).

20 See http://www.iclei.org; last accessed May 2012.

21 (ibid.).

> and political commitment at the highest level on development and environment cooperation. Its successful implementation is first and foremost the responsibility of Governments. National strategies, plans, policies and processes are crucial in achieving this. International cooperation should support and supplement such national efforts. In this context, the United Nations system has a key role to play. Other international, regional and sub-regional organisations are also called upon to contribute to this effort. The broadest public participation and the active involvement of the non-governmental organisations and other groups should also be encouraged.

At the end of Chapter 1,[22] Agenda 21 is said to be a "dynamic programme":

> It will be carried out by the various actors according to the different situations, capacities and priorities of countries and regions in full respect of all the principles contained in the Rio Declaration on Environment and Development. It could evolve over time in the light of changing needs and circumstances. This process marks the beginning of a new global partnership for sustainable development.

In its first chapter, Section I of Agenda 21, entitled "Social and Economic Dimensions", focuses on international cooperation, states' role in such cooperation, and the expected consequences from these levels of governance for the promotion of sustainable development. Section III, which is devoted to "Major Groups", addresses how to strengthen them and achieve sustainable development. For the purpose of the current discussion, some quotations from Section III are pertinent here. For example, its Preamble states the following:

> 23.1. Critical to the effective implementation of the objectives, policies and mechanisms agreed to by Governments in all programme areas of Agenda 21 will be the commitment and genuine involvement of all social groups.
>
> 23.2. One of the fundamental prerequisites for the achievement of sustainable development is broad public participation in decision-making. Furthermore, in the more specific context of environment and development, the need for new forms of participation has emerged. This includes the need of individuals, groups and organisations to participate in environmental impact assessment procedures and to know about and participate in decisions, particularly those which potentially affect the communities in which they live and work. Individuals, groups and organisations should have access to information relevant to environment and development held by national authorities, including information on products and activities that have or are likely to have a significant impact on the environment, and information on environmental protection measures.

---

22 Chapter 1.6.

Nine "Major Groups" are identified by Agenda 21 as playing a special role on the road towards sustainable development:

- Women (Chapter 24)
- Children (Chapter 25)
- Indigenous peoples and their communities (Chapter 26)
- Non-governmental organisations (Chapter 27)
- Local authorities (Chapter 28)
- Workers and trade unions (Chapter 29)
- Business and industry (Chapter 30)
- The scientific and technological community (Chapter 31), and
- Farmers (Chapter 32).

In view of the special interest of the current discussion in the role of local structures and their placement in the overall governmental architecture, the position of indigenous communities and of local authorities is of particular relevance. As to indigenous peoples, the statements from Agenda 21 cited above can be referred to; as to local authorities, Agenda 21 has the following to say about the basis of action for these bodies:[23]

> Because so many of the problems and solutions being addressed by Agenda 21 have their roots in local activities, the participation and cooperation of local authorities will be a determining factor in fulfilling its objectives. Local authorities construct, operate and maintain economic, social and environmental infrastructure, oversee planning processes, establish local environmental policies and regulations, and assist in implementing national and subnational environmental policies. As the level of governance closest to the people, they play a vital role in educating, mobilising and responding to the public to promote sustainable development.

The Programme for the Further Implementation of Agenda 21 adopted by the UN General Assembly in 1997 – the Rio+5 meeting – took special note of the so-called major groups, as borne out by the following:[24]

> The major groups have demonstrated what can be achieved by taking committed action, sharing resources and building consensus, reflecting grass-roots concern and involvement. The efforts of local authorities are making Agenda 21 and the pursuit of sustainable development a reality at the local level through the implementation of 'local Agenda 21s' and other sustainable development programmes. Non-governmental organisations, educational institutions, the scientific community and the media have increased public awareness and discussion

---

23 Chapter 28.1.
24 UNGA A/RES/S-19/2, paragraph 12.

of the relations between environment and development in all countries. The involvement, role and responsibilities of business and industry, including transnational corporations, are important. Hundreds of small and large businesses have made 'green business' a new operating mode. Workers and trade unions have established partnerships with employers and communities to encourage sustainable development in the workplace. Farmer-led initiatives have resulted in improved agricultural practices contributing to sound resource management. Indigenous people have played an increasing role in addressing issues affecting their interests and particularly concerning their traditional knowledge and practices. Young people and women around the world have played a prominent role in galvanising communities into recognising their responsibilities to future generations. Nevertheless, more opportunities should be created for women to participate effectively in economic, social and political development as equal partners in all sectors of the economy.

The World Summit for Sustainable Development (Rio+10) held in Johannesburg in 2002 also took note of Agenda 21's major groups in its Plan of Action. In the section on strengthening institutional frameworks for sustainable development at national level, the Plan of Action calls for the following:[25]

162. States should:
   (a) Continue to promote coherent and coordinated approaches to institutional frameworks for sustainable development at all national levels, including through, as appropriate, the establishment or strengthening of existing authorities and mechanisms necessary for policy-making, coordination and implementation and enforcement of laws;
   (b) ...
163. Each country has the primary responsibility for its own sustainable development, and the role of national policies and development strategies cannot be overemphasised. All countries should promote sustainable development at the national level by, inter alia, enacting and enforcing clear and effective laws that support sustainable development. All countries should strengthen governmental institutions, including by providing necessary infrastructure and by promoting transparency, accountability and fair administrative and judicial institutions.
164. All countries should also promote public participation, including through measures that provide access to information regarding legislation, regulations, activities, policies and programmes. They should also foster full public participation in sustainable development policy formulation and implementation. Women should be able to participate fully and equally in policy formulation and decision-making.

25 The following is quoted at length as the Johannesburg Plan of Action has – in addition to the decisions of the Rio Conference of 1992 – remained a document of frequent reference in the debate on sustainable development.

165. [States should] [f]urther promote the establishment or enhancement of sustainable development councils and/or coordination structures at the national level, including at the local level, in order to provide a high-level focus on sustainable development policies. In that context, multi-stakeholder participation should be promoted.
166. [States should] [s]upport efforts by all countries, particularly developing countries, as well as countries with economies in transition, to enhance national institutional arrangements for sustainable development, including at the local level. That could include promoting cross-sectoral approaches in the formulation of strategies and plans for sustainable development, such as, where applicable, poverty reduction strategies, aid coordination, encouraging participatory approaches and enhancing policy analysis, management capacity and implementation capacity, including mainstreaming a gender perspective in all those activities.
167. [States should] [e]nhance the role and capacity of local authorities as well as stakeholders in implementing Agenda 21 and the outcomes of the Summit and in strengthening the continuing support for local Agenda 21 programmes and associated initiatives and partnerships and encourage, in particular, partnerships among and between local authorities and other levels of government and stakeholders to advance sustainable development....

The following section of the Plan deals with the participation by major groups, stating that they should –

168. [e]nhance partnerships between governmental and non-governmental actors, including all major groups, as well as volunteer groups, on programmes and activities for the achievement of sustainable development at all levels.
169. [a]cknowledge the consideration being given to the possible relationship between environment and human rights, including the right to development, with full and transparent participation of Member States of the United Nations and observer States.
170. [p]romote and support youth participation in programmes and activities relating to sustainable development through, for example, supporting local youth councils or their equivalent, and by encouraging their establishment where they do not exist.

ICLEI played a special role in the preparation of the Johannesburg Summit. In fact, a survey carried out by ICLEI served as an official background paper for the Summit.[26]

26 The ICLEI survey and a related summary report are available at www.iclei.org/la21survey, last accessed 20 June 2012. The ICLEI website also offers information on Rio+20. See also Servicestelle 2002.

The UN Conference on Sustainable Development of 2012 (the Rio+20 conference) reconfirmed in a special chapter of its final paper the need for "engaging major groups and other stakeholders".[27] The Conference expressed this in the following words:

> 43. We underscore that broad public participation and access to information and judicial and administrative proceedings are essential to the promotion of sustainable development. Sustainable development requires the meaningful involvement and active participation of regional, national and subnational legislatures and judiciaries, and all major groups: women, children and [the] youth, indigenous peoples, non-governmental organisations, local authorities, workers and trade unions, business and industry, the scientific and technological community, and farmers, as well as other stakeholders, including local communities, volunteer groups and foundations, migrants and families, as well as older persons and persons with disabilities. In this regard, we agree to work more closely with the major groups and other stakeholders, and encourage their active participation, as appropriate, in processes that contribute to decision-making, planning and implementation of policies and programmes for sustainable development at all levels.

Agenda 21 is the first comprehensive, internationally sanctioned document that describes the basic framework of global governance – or, more succinctly put, the architecture of global governance with respect to environment and development. This point deserves special emphasis. Agenda 21 has not really been acknowledged as a document of authority that indeed reflects what has been described and analysed as *the architecture of global governance for sustainable development*.[28]

## *B. Traditional Authorities: A Major Group in Terms of Agenda 21?*

In southern Africa, in countries such as Botswana, Namibia and South Africa, traditional authorities perform the functions and tasks of local authorities. They are, using the words of Agenda 21, "the level of governance

---

27 Cf. the final declaration of the Conference: A/Conf.216/L.1 at C.

28 This is true for academic writing in the specific sense, e.g. if one takes the collection of papers on global climate governance beyond 2012 (Biermann et al. (2010a)), but also for action-oriented writing, such as the background paper for Rio+20 by Martens (2012), in which one will search in vain for references to work of the major groups in terms of Agenda 21, particularly that of local authorities.

closest to the people".[29] In many areas of the countries mentioned, local authorities are either remote or non-existent. Thus, the functions of local authorities are taken care of by traditional authorities.[30] In Namibia, the Traditional Authorities Act[31] takes note of the substatal government functions of the various traditional authorities. The Act recognises the overall responsibility of traditional authorities to "promote peace and welfare amongst the members of... [the traditional] community".[32] The responsibility of traditional authorities for environmental matters is provided for in section 3(2)(c) of the Act:

> A member of a traditional authority shall in addition to the function referred to in subsection (1) have the following duties, namely –
>
> (a) ...
> (c) to ensure that the members of his or her traditional community use the natural resources at their disposal on a sustainable basis and in a manner that conserves the environment and maintains the ecosystems for the benefit of all persons in Namibia; ...

Although certainly modelled after Article 95(l) of the Constitution, the environmental obligation of the Traditional Authorities Act is – in legal terms – stronger than the constitutional obligation, which is only one of the principles of state policy and, thus, has limited legal relevance.[33]

The normative translation of the responsibility for environmental matters is special in the sense that it maintains a holistic approach. Customary law follows broadly what Agenda 21 summarises as the holistic approach to land by what it calls "indigenous people":[34]

> Indigenous people and their communities have an historical relationship with their lands and are generally descendants of the original inhabitants of such lands.... [T]he term "lands" is understood to include the environment of the areas which the people concerned traditionally occupy.... They [the indigenous people] have developed over many generations a holistic traditional scientific knowledge of their lands, natural resources and environment.

29 Cf. Agenda 21, Chapter 28.1.
30 Cf. here Hinz & Gatter (2006); Hinz & Katjaerua (1998); Keulder (1997).
31 No. 5 of 2000.
32 See the chapeau of section 3(1) of the Act; cf. Hinz (2009). There are large geographical areas in Namibia which are not under a local authority, meaning that all local authority functions are either under traditional authorities or are administratively facilitated by the latter.
33 See Article 101, Namibian Constitution.
34 See Chapter 26.1, A/CONF.151/26/Rev. 1.

Although the majority of the traditional communities in Namibia are not *indigenous* in terms of the Declaration on the Rights of Indigenous Peoples,[35] their concept of *land* does not differ from that understood by indigenous peoples in terms of Agenda 21.[36] The holistic understanding of *land* is, in a more general sense, particularly true with respect to the holistic ecological cosmologies that are shared by indigenous and traditional communities. These cosmologies are at the foundation of indigenous/traditional environmental ethics and contribute to the very special relationship between indigenous/traditional communities and their environment.[37] As Mary E. Tucker and John Grim state in their Foreword to the publications of the Religions of the World and Ecology Project:[38]

> ... most indigenous peoples have environmental ethics embedded in their worldviews. This is evident in the complex reciprocal obligations surrounding life-taking and resources-gathering which mark a community's relations with the local bioregion. The religious views at the basis of indigenous lifeways involve respect for the sources of food, clothing, and shelter that nature provides. Gratitude to the creator and to the spiritual forces in creation is at the heart of most indigenous traditions. The ritual calendars of many indigenous peoples are carefully coordinated with seasonal events such as the sound of the returning birds....

While the politico-legal discourse has accepted having to deal with indigenous communities, the discourse on traditional authorities is still in its infancy.[39] Although even international organisations have begun to acknowledge that traditional authorities cannot be ignored when it comes to development in rural areas,[40] the conceptual side of this acknowledgment has not

35 UNGA Res. 61/295 and other international instruments on indigenous people, such as the International Labour Organization's Conventions 107 of 1957 and 169 of 1989. For more on the concept of *indigenous people*, see Niezen (2010:105ff.).

36 See Hinz (2008).

37 However, the subproject on Indigenous Traditions and Ecology of the Religions of the World Ecology does not distinguish between *traditional* and *indigenous*. For the subproject, *traditional* is synonymous with *indigenous*.

38 Tucker & Grim (2001:XXVIf.). This Project was conducted by the Centre for the Study of World Religions of the Harvard Divinity School and led to a number of conferences and subsequent publications on the studies of religion and ecology.

39 However, see Hinz & Gatter (2006); Hinz & Ruppel (2008); Ray (1997).

40 Cf. Mapaure (2010); UNDP (2004). The climate debate has, however, opened another avenue for the debate on the role and function of traditional authority. The research on the responses to climate change asks what forms of adaptation societies and communities have in dealing with the consequences of climate change. See here e.g. Kpadonou et al. (2012).

advanced much. The basic obstacle in advancing conceptualisation is that traditional authorities have developed from some kind of sovereignty into systems of integration governed by states – which see themselves as the sole and overall representatives of sovereignty.[41] Before colonialism and in the years of colonial subjugation, the kings and queens of communities were the sovereign holders of authority over such communities. Their authority is not delegated by the state but original, entrusted to them by their ancestors. Modern state centralism would like to see traditional authorities as part of the administration of the state. To date, traditional authorities have seen themselves as bearers of ancestral authority. The different statutory models which the various states have pursued in dealing with traditional authorities[42] point to the unsolved political question as to how to deal with political plurality represented by such indigenous bodies.

Whether observations of this kind were also behind the fact that Agenda 21 did not explicitly consider traditional authorities is worthy of speculation,[43] but be that as it may: the aforementioned fact that traditional authorities share many criteria described by Agenda 21 with indigenous communities prompts the suggestion to include traditional authorities in the list of Agenda 21's major groups. The de facto closeness of traditional authorities to local authorities prompts a step further forward: allowing traditional authorities to be interpreted as falling under local authorities in terms of the Agenda 21 classifications.[44]

## *C. Ekongoro Owns the River, or: The Question of Ownership under Customary Law*

The founder of African law as a distinct academic area of research and teaching, Antony Allott, from the Law Department of the School of Oriental and African Studies, University of London, advised legal scholars many years ago not to use the term *ownership* when it came to land and related matters in Africa.[45] Allott's point was that the use of the term *ownership*

41 Here and in respect of what follows, cf. Hinz (2009:61ff.).

42 Cf. e.g. Keulder (1998).

43 See Hinz (2008:221ff.).

44 With this view, this paper corrects the more cautious approach expressed in Hinz (2008).

45 Personal experience.

would mislead the understanding of the status and function of land in the African approach. The research in Namibia nevertheless asked the *ownership* question, namely *Who owns the land, the wild animals, the trees, the water, the minerals?*

The people themselves used "ownership", but not in the sense of common law to denote the right to do with that which is owned as the owner wishes to do, including the right to dispose of that which is owned. *Ownership* for the ordinary person in rural Namibia stands for "having authority over, being the guardian and custodian". In relation to the Kavango River, therefore, *Who owns the Kavango River? Who owns its water?* were the questions asked.

The interest behind these questions was to establish who has the right to access water, and who is responsible for its control and protection. The interest was to find out how the people – the factual and potential users of the water – would determine their relationship to water. The expectation from the answers was to gain facts that would also lead to consequences in informing policies with respect to the sustainable management of natural resources.

Interviews conducted in several parts of Namibia led to a high percentage of voices who at least implicitly denied the ownership of water by the state. For these voices, water belonged to authorities close to them, to the community, the water point committee, to God, or to an entity called *Ekongoro*. In the area of the Mbunza traditional community of the Kavango Region, out of 31 subjects interviewed in 2009,[46] only 8 held that the state owned the river and its water, while 18 noted ownership close to the local level. One respondent put it as follows: "The Creator is our Creator, the water point committee is our committee; we are the community."

Prompted by this result, the scope of the fieldwork was extended to different places in the Kavango Region. More than 100 interviews were conducted.[47] Wherever the questions touched on the ownership of the Kavango River and its water, many people responded by referring to *Ekongoro*. Listening to and analysing the information collected, the references to *Ekongoro* as the owner of the river and its water appeared to be of much greater

46 Clever Mapaure – research assistant in the BIOTA Project and now reading for his PhD in the Faculty of Law of the University of Namibia – conducted this research; cf. Mapaure (2012b).

47 These interviews were done under the TFO Project by Christian Mukuve and the author.

importance than the references to the Creator, the community and the water point committee.

So what is *Ekongoro*? *Ekongoro* "is similar to a very big snake, it is very long, about 15 to 20 m, it pushes and pulls water...."[48]

*Ekongoro* (as recorded from another interview) –[49]

> … attacked my friends and me once. [In the river], our canoe was [stopped]... we realised that we were held by [*Ekongoro*]. We started talking to it as we had been directed by our elders. We said, *Fumu* [King], what have we done to you? We did nothing to hurt you, we are just passing by. Won't you please release us to go on?
> ... Those who were with me fell out of the canoe and I was left alone... [*Ekongoro*] took me back to the deep water... me, the canoe and all our belongings... I lost my mind... I felt a very bad smell... my mind was off. I was walking around under the water but managed to breathe. It took me almost six hours before [*Ekongoro*] eventually released me. I strongly believe that this [*Ekongoro*] is still around in the place where it attacked my friends and me.

People who live along the Kavango River believe that the way they use the water is closely monitored by this mystical creature called *Ekongoro*, which can punish polluters and abusers of the water resource. People believe that while the (Christian) God created water, *Ekongoro* holds more power in controlling its conservation and utilisation. Because humans are less powerful in comparison with this mystical creature, there is general deterrence regarding potential abuse or pollution of the waters of the Kavango River. It is believed that *Ekongoro* sends messages to water users that they have to manage and utilise water resources in a proper and sustainable way.

One of the researchers in the TFO Project[50] reported to the TFO team that one of his uncles had been 'taken away' by a Zimbabwean *Ekongoro* and had disappeared for years. When he was 'released', he came back with mystical powers on how to heal people using herbs, i.e. traditional healing. As he reported back to his people, he had received immense knowledge on the utilisation of natural resources – including water resources. The knowledge he had received allowed him to advise his community on where best to water their livestock, where to dig a well or drill a borehole in the river basin. In other words, local knowledge had received a spiritual blessing and, with this blessing, had gained a protected status within the customary order of the community.

48 According to one of the interviewees; interview on file with the author.

49 Interview on file with the author.

50 Clever Mapaure.

What has been exemplified with reference to the Kavango River and its water, i.e. the ownership of both, could also be shown with respect to other natural resources. The question *Who owns them?* is answered in very comparable terms: the animals, the trees, plants minerals are owned by "the Chief", "God", "the community", "the ancestors".[51] All the answers have one point in common: the animals, trees, etc. are not fragmented items but are part and parcel of one central item – land – and, thus, fall under the same customary rules as the land itself.

## D. *Statutory Responses to the Customary Law Concept of Ownership*

It is well-established practice in the general law of many countries to have separate laws for animals, trees, plants, mineral resources and water.[52] It is, therefore, of great interest to look not only at the question of ownership of communal land under general law, but also at how the various Namibian statutes handle the accessories to land: animals, trees, plants, mineral resources and water.

Whenever the question *Who owns communal land in Namibia?* is put to legally minded persons, the answers usually point to Article 100 of the Namibian Constitution, which states the following:

> Land, water and natural resources below and above the surface of the land and in the continental shelf and within the territorial waters and the exclusive economic zone of Namibia shall belong to the State if they are not otherwise lawfully owned.

Sometimes, the answers also refer to Schedule 5 of the Constitution, which, in particular, decrees in Sub-Article 1 that –

> [a]ll property of which the ownership or control immediately prior to the date of Independence vested in the Government of the Territory of South West Africa, or in any Representative Authority constituted in terms of the Representative Authorities Proclamation, 1980 (Proclamation AG 8 of 1980), or in the Government of Rehoboth,.... shall vest in or be under the control of the Government of Namibia.

---

51 See in particular Mapaure (2012b) and (2012c), but also already Hinz (1998, 2003a).

52 So it is for Namibia as well; see the Nature Conservation Ordinance 4 of 1975, as amended; the Water Act, 1956 (No. 54 of 1956), as amended; the Water Resources Management Act, 2004 (No. 24 of 2004); the Forest Act, 2001 (No. 12 of 2001); and the Minerals (Prospecting and Mining) Act, 1992 (No. 33 of 1992). Cf. here also Ruppel (2012).

Legal work was developed to trace Namibia's communal land legislation back to the time when the land was occupied and used by the various indigenous communities of the territory that became the colonial (German) South West Africa. Apart from early colonial decisions, such as the Crown Land Disposal Ordinance of 1920,[53] which extended the Transvaal Crown Land Disposal Proclamation of 1903 to Namibia, and the Development Trust and Land Act of 1936,[54] the socio-political transformation enacted in accordance with the Representative Authorities Proclamation of 1980[55] attracted the interest of lawyers in order to find out what happened to the land in the so-called self-governing territories of Namibia.[56] While the customary law regime that had existed up to that point remained basically in place in the various areas designed to be the 'homelands' of the various population groups,[57] a change with far-reaching consequences took place in the Rehoboth *Gebied*.[58] Here, the Rehoboth Government decided to register the land of the Basters in its name. This led to the still ongoing political controversy about communal land in Rehoboth; the Supreme Court decision based on the quoted Schedule to the Constitution led to a confirmation that ownership of Rehoboth communal land vested in the Namibian Government[59] – a decision that has not been accepted by many Basters. Attempts to register the communal land, e.g. of the Owambo communities, in the name of the then second-tier Owambo administration, did not take place, thus leaving communal land under the full administration and control of those communities, as was also the case in the other self-governing territories of the then South West Africa.[60]

After Namibia's Independence in 1990, the debate about land in general and communal land in particular started with the intention to address the land issue. The discussions focused on commercial land, i.e. the land that had lost its communal quality and was now, as a result of many interventions in the

53 Proclamation 13 of 1920.
54 No. 18 of 1936.
55 Proclamation AG 8 of 1980.
56 Cf. Hinz (1998:191ff.).
57 (ibid.).
58 "Territory".
59 *Rehoboth Bastergemeente v The Government of Namibia & Others*, 1996 NR 338 (SC). Cf. here Harring (2012).
60 Pers. comm., Dr. Kuno Budack.

colonial period of the country, privately owned, as well as on communal land that was under the authority of indigenous communities.[61]

The attempt by some Namibian government officials to push through a position whereby the government would own communal land was rejected by various traditional authorities.[62] What we find today in the Communal Land Reform Act[63] is the result of the negotiations between the government and the traditional authorities. In line with the Namibian Constitution, which recognises communal land at least by its more incidental reference to the concept in Article 102(5),[64] the Communal Land Reform Act accepted the inherited communal land tenure system in principle. Section 20 of the Act confirms the allocation of communal land rights under customary law. The primary power to allocate or cancel any customary land right in respect of any portion of land in the communal area of a traditional community vests either in the Chief of that traditional community or, where the Chief so determines, in the traditional authority of that traditional community.

This provision has to be read together with section 17(1) of the Act, which provides as follows:

> Subject to the provisions of this Act, all communal land areas vest in the State in trust for the benefit of the traditional communities residing in those areas and for the purpose of promoting the economic and social development of the people of Namibia, in particular the landless and those with insufficient access to land who are not in formal employment or engaged in non-agriculture business activities.

Although the Act *vests* communal land in the state, the Act avoids any reference to *ownership*. Even if one interpreted the *vesting in* as an indication

61 Cf. here, in particular, the Land Conference of 1991; Republic of Namibia (1991a, 1991b, 1991c).

62 Cf. the proceedings of the Communal Land Conference of 1997; Hinz & Malan (1997).

63 No. 5 of 2002.

64 Article 102(5) reads as follows: "There shall be a Council of Traditional Leaders to be established in terms of an Act of Parliament in order to advise the President on the control and utilisation of communal land and on all such other matters as may be referred to it by the President for advice". With this, the Constitution has taken note of the legal principle *communal land.* Read together with Article 66(1) of the Constitution, one could conclude that communal land and the customary law related to it have been constitutionally confirmed. Article 66(1) states the following: "Both the customary law and the common law of Namibia in force on the date of Independence shall remain valid to the extent to which such customary or common law does not conflict with this Constitution or any other statutory law".

of ownership,[65] such ownership would remain limited to an *entrusted* ownership, meaning that the rules of handling such ownership will not be the rules of general private law, but the rules that apply to trusts, as specified by the Communal Land Reform Act.

Therefore, the quoted sections of the Act appear to be acceptable to the customary law system of land tenure as it leaves operational the authority to administer the rights on communal land by the responsible structures under traditional governance. Ownership of land is not really an issue to customary law as long as the reference to *ownership* does not result in what is understood as limiting the customary law authority over communal land. This is because customary law does not know ownership of land as general law knows it, i.e. as encompassing the right of almost free disposal of the object of ownership.[66]

The way communal conservancies were introduced into the nature conservation law further illustrates the problem that statutory law has with communal land. The policy that led to the 1996 Nature Conservation Amendment Act[67] was driven by the intention to restore rural communities' rights to wildlife. The policy was informed by anthropological evidence which showed that traditional communities had a balanced approach to the use of animals as part of their natural resources, which appeared to be in support of the conservation policy of the state. Why not, therefore, use the societally grounded traditional conservation ethos as a potential contribution to conservation as an overall state-wide conservation policy? The positive answer to this question led to the following provision in section 3 of the Nature Conservation Amendment Act:[68]

> Any group of persons residing on communal land and which desires to have the area which they inhabit, or any part thereof, to be declared a conservancy, shall apply therefor to the Minister in the prescribed manner, and such application shall be accompanied by ….

Leaving aside that the Nature Conservation Amendment Act continued to be based on the premise that the state owns wildlife on communal land,[69]

65 Cf. here Mapaure (2012b:220ff.).

66 Cf. here Hinz (1998:186ff.).

67 No. 5 of 1996, amending Nature Conservation Ordinance 4 of 1975.

68 Section 24A, amended Nature Conservation Ordinance.

69 Cf. here Hinz (2003a:28ff.).

who is "any group of persons" in communal areas?[70] What is the relationship between "any group" and the traditional authority in place? The establishment of a communal conservancy means a change in the existing land tenure system. Giving the administration and management a special place in the overall administration of land in a given area will mean that certain modes of production will be excluded or, at least, limited. Most probably, the common practice of cattle husbandry would not be possible in core areas designated for wildlife. Such a change in the tenure of customary land affects not only customary rights holders, who would be part of "any group", but also the overall responsibility of traditional authorities over the communal land in their jurisdictions. The Nature Conservation Amendment Act did not provide any role for traditional authorities in the process of establishing communal conservancies. However, research has shown that in most – if not all – cases, the relevant traditional authorities have played a role in the establishment of such conservancies. Most – again, if not all – of the Conservancy Committees required by the Nature Conservation Amendment Act have formalised links with the traditional authority in whose territory the conservancy is located.[71] In this sense, traditional authorities have developed a type of customary amendment to the Nature Conservation Amendment Act and, thus, have provided a remedy for the Act's shortcomings in this respect.

Subsequent legislation on other aspects of matters related to land expresses comparable difficulties in dealing with ownership of natural resources on communal land. The legislation on minerals, for example, provides that exercise and control of "any mineral or group of minerals vests, notwithstanding any right of ownership of any person in relation to land in, on or under which any such mineral or group of minerals is found, in the State."[72]

Forest legislation claimed ownership of trees as belonging to the state. As with the Nature Conservation Amendment Act, the possibility of establishing community forests was offered to traditional communities as an option, but, in this case as well, without recognising the authority of traditional au-

70 To this and the following, see Hinz (2011a, 2011b), but also Anyolo (2012:29ff., 41ff.).

71 Cf. Hinz (2003a:82ff.).

72 The Minerals (Prospecting and Mining) Act, 1992 (No. 33 of 1992). A PhD thesis in progress and under the supervision of the author (Renkhoff, Forthcoming) will address questions with respect to minerals.

thorities over communal land.[73] The same eventually happened to water: this resource in communal land is also under the authority of the state.

The water law of Namibia has been in the process of reform for quite some time. New legislation – the Water Resources Management Act,[74] which was to replace the pre-Independence law – was passed by Parliament in 2004. However, the new Act is still awaiting the status of enforceable law because the required notice by the competent Minister to bring the Act into force has not yet been issued. Whether or not the 2004 Act will become law is not known. It is nevertheless interesting to note how this post-Independence Act deals with water matters at the local level, how it deals with the local water law, and how it deals with customary water law.

The Act recognises customary law in general terms, but there is not much follow-up to this recognition. All water resources belong to the state. Section 4 reads as follows:

> Subject to this Act –
>
> (a) ownership of water resources in Namibia below and above the surface of the land belongs to the State;...

The bodies responsible for the administration of customary law, its custodians, the traditional authorities, have no place at all in the Act. The responsibility for managing water at the local levels is regulated in section 16 of the Act, which sets out the details for the establishment of water point user associations and water point committees. The water management structure is not accountable to authorities in place at the local level, namely the traditional authorities, but to the Ministry of Agriculture, Water and Forestry through government-appointed agricultural extension officers.

It is certainly possible for water point user associations to elect traditional leaders as members, but whether such elections will be accepted as recognition of traditional authority in such associations remains to be seen. Indeed, research done in an area to the west, that is, in the Ohangwena Region, supports this scepticism. The traditional authority that has jurisdiction in this area does not recognise the water point committees as they exist. The traditional authority has voiced that water point committees have no legitimacy: they are called "puppets" of the Ministry of Agriculture, Water and Forestry, the water point committees are viewed as being "abused" by the central

---

73 Cf. sections 13, 14 and 15 of the Forest Act; cf. Mapaure (2012a).

74 No. 24 of 2004.

government to impose the rules of the state on the traditional management of water.[75]

## *E. What does Ekongoro Stand for? Or: The Challenge of Legal Pluralism*

The Namibian Constitution reflects a substantial jurisprudential change by confirming customary law as one branch of the land being at the same level of validity as the colonially inherited common law (Article 66(1)), and by recognising the right to culture (Article 19). As the previous section herein has shown, the post-Independence legislator, in its attempts to provide Constitution-inspired law with respect to the administration and management of natural resources, has not really honoured the existing customary law. How can this fact be assessed in legal, but also political, terms?

Article 66(1) of the Namibian Constitution has this to say about the customary law of the various communities that uphold such traditions:

> Both the customary law and the common law of Namibia in force on the date of Independence shall remain valid to the extent to which such customary or common law does not conflict with this Constitution or any other statutory law.

Article 19 reads as follows:

> Every person shall be entitled to enjoy, practise, profess, maintain and promote any culture, language, tradition or religion subject to the terms of this Constitution and further subject to the condition that the rights protected by this Article do not impinge upon the rights of others or the national interest.

When enacted in 1990, Article 66 reflected a revolutionary approach to the African law inherited in the various so-called traditional communities of Africa. The African traditional law was confirmed; it was recognised as law at the same level as the received Roman–Dutch common law. African customary law was not subjected to this common law, but was accepted as having the same level of validity as common law and, like the latter, subject only to the Constitution and to statutory inroads into it by an Act of Parliament.

In more general terms, what are the options a state has to regulate the relationship between the law of the state and African traditional law? Five models are possible:[76]

75 Cf. Mapaure (2012b, 2012c).
76 Cf. Hinz (2009:61ff.).

- Model 1: Abolish African traditional law (strong modern monism)
- Model 2: Leave African traditional law unregulated (unregulated dualism)
- Model 3: Regulate African traditional law (regulated – weak or strong – dualism)
- Model 4: Integrate African traditional law into the modern structures (weak modern monism), and
- Model 5: Make the traditional level the overarching level of governance (strong traditional monism).

The Namibian approach follows Model 3, which, incidentally, is the model applied by many African states. Whether the Namibian approach is strongly or weakly regulated could be debated; however, the fact that the Traditional Authorities Act, the quasi-constitution of traditional governance, even accepts that traditional communities – in the words of the Act – "make customary law",[77] in other words, they have legislative power independent from the supreme lawmaker of the country, Parliament, indicates that the Namibian legal system accepts the quasi-sovereignty of traditional communities.

Although the matter has not been considered by the courts of Namibia, one can, with good reason, hold that, by virtue of the right to culture and the confirmation of customary law, the traditional governance – the traditional authority as an institution with an associated legal framework – enjoys constitutional protection. This means that there are, again with good reasons, limits to parliamentary inroads into traditional governance and its legal framework. Indeed, any inroads will require reasoning that is informed by constitutional requirements, e.g. the reasoning that a particular rule of customary law violates the fundamental rights and freedoms guaranteed in the Constitution.

With these provisions, the Namibian legal system can be seen as having accepted lessons as they were offered by the theory of legal pluralism, but is also facing some of the challenges as they became apparent by the research on legal pluralism. Hence, what does the theory of legal pluralism say?[78]

Legal pluralism has occupied many researchers who have studied the law in the colonies of Africa and elsewhere, but also the legal situation in the post-colonial set-up in former colonies. The research shows that legal plurality is found basically everywhere, even in legal orders which adhere to a

77 See section 3(3)(c) of the Traditional Authorities Act.
78 To the following, see Menski (2006:82ff.).

strong Kelsenian concept of *law*, according to which there is no law apart from the law of the state and all law originates in the state. John Griffiths put it as follows: "Legal pluralism is the fact. Legal centralism is a myth, an ideal, a claim, an illusion."[79]

For the proponents of legal pluralism, legal centralism is a powerful ideology that has dominated the understanding of law by lawyers and social scientists. Sally Falk Moore added to the discussion on legal pluralism by generalising the findings of legal empirical research in her concept of "semi-autonomous social fields".[80] All human beings live in different, sometimes overlapping, social fields, which are run in accordance with their own rules – the "living law" as described by Eugen Ehrlich[81] – generated by the actors in these fields. The social fields are autonomous – at least to some degree. The degree of such autonomy depends on the degree of (exercised and/or accepted) authority of other, neighbouring fields, but also the state.

Referring to legal pluralism as 'fact' focuses on legal pluralism as a theory that interprets the findings of empirical work. Apart from this, the discourse on legal pluralism has added a normative dimension to it. As shown elsewhere,[82] it was in particular the increase in international recognition of the rights of indigenous peoples (in the specific language of the International Labour Organisation and the UN bodies concerned with indigenous peoples) that supported the development of indigenous peoples' "right to one's own law"[83] within the broader context of states in whose territories these peoples settle. Although the extent of the right to one's own law may vary from jurisdiction to jurisdiction, it will encompass the community's right to their own forms of governance, including their own courts, as well as their own –

- family law
- inheritance law
- land law, and
- law dealing with wrongs.

Namibia is not alone: many African countries have accepted traditional governance and their legal frameworks by setting out the duties and functions of traditional authorities and describing (limiting) the competence of African

79 Griffiths (1986:4).
80 Moore (1978:54ff.).
81 See Ehrlich (1967:11ff.).
82 Hinz (2006).
83 Cf. Hinz (2006).

traditional law.[84] They have done so without explicitly recognising the mentioned right to one's own law. The numerous conflicts between traditional stakeholders and their states illustrate that the right to one's own law is part of the living law of traditional stakeholders and their communities, but not necessarily part of the respective legal system.

This article is not the place to argue whether or not the Namibian statutory inroad into inherited customary water law is tantamount to an expropriation with legal consequences in accordance with the constitutional protection of property, which is provided for in Article 16 of the Namibian Constitution.

However, even if we conclude that a constitutional case would not have merit, there are good political reasons for reconsidering the statutory inroads into customary law. The holistic approach of traditional[85] communities to land is certainly in contrast to the fragmenting approach of governments, but, at the same time, also very close to the concept of *sustainable development*, which is said to be equally holistic[86] – although the translation of sustainable development into law is still far from reflecting the holistic nature of such development.[87] This is a challenge for law reform![88]

The statutory inroads into the customary law as noted above are unhealthy, and lead to frustration on the ground. They are dysfunctional, as they do not appreciate the local understanding of water and water supply.

84 Cf. Hinz (2003b:142ff.).

85 Grim (2001) does not distinguish between indigenous communities in terms of the special law of indigenous peoples, e.g. in terms of the aforementioned UN Declaration on the Rights of Indigenous Peoples and other communities called *traditional*.

86 Cf. Gärditz (2008:141).

87 If one looks at an attempt to compile sustainable development law (Cordonier Segger & Khalifan 2004), one sees such law structured in line with categories drawn from the discussion on sustainable development. Further jurisprudential reflections that also go into the jurisprudential foundation of sustainable development (see Ekardt 2005; Jonas 1984; Kahl 2008; Schönherr-Mann 2010) will be needed to shed more light on this.

88 The quoted statutes dealing with land or aspects related to land from a customary law perspective not only differ in the way they do or do not take note of traditional authorities: their administration is also formally fragmented as they fall under the ambit of different Ministries. These are the Ministry of Agriculture, Water and Forestry; the Ministry of Environment and Tourism; the Ministry of Lands and Resettlement; and the Ministry of Mines and Energy; with the Ministry of Regional and Local Government, Housing and Rural Development bearing overall responsibility for implementing the Traditional Authorities Act, while the Ministry of Justice is responsible for traditional courts in terms of the Community Courts Act, 2003 (No. 10 of 2003).

By ignoring the legal folk concepts, the inroads have bypassed local potential for the responsible and sustainable management of water. The seriousness of this circumvention becomes apparent when one goes beneath the surface of the folk philosophy of the concept of *water ownership*.

The British anthropologist Tim Ingold used one of the many recorded nature-related folk concepts according to which the rainforest was understood to be the 'parent' of the community that lived close to it:[89]

> To speak of the forest as a parent is not, then, to model object relations in terms of primary intersubjectivity, but to recognise that at root, the constitutive quality of intimate relations with non-human and human components of the environment is one and the same.

Generalising this, Ingold draws the following conclusion:[90]

> For hunter-gatherers as for the rest of us, life is given in engagement, not in disengagement, and in that very engagement, the real world at once ceases to be 'nature' and is revealed to us as an environment for people. Environments are constituted in life, not just in thought, and *it is only because we live in an environment that we can think at all.* [Emphasis added]

*Ekongoro*, like all other animated parts in the environment, are also 'others' in the sense of the quoted concept. *Ekongoro* represents water, and more so the right of the water to be respected in its created form. With respect to the conceptualisation of *Ekongoro*, there is more than *the right of, the right to* and *respect for*, there is also an element of fear: the fear that *Ekongoro* may swallow somebody despite the fact that the swallowed person will be released to life after a period in the *Ekongoro*'s belly. This element of fear leads to what the philosopher Hans Jonas stated in *Imperative of Responsibility: In Search of an Ethics for the Technological Age* – up to now one of the most important philosophical reasonings on sustainable development:[91]

> ... moral philosophy must consult our fears prior to our wishes to learn what we really cherish. And although what is most feared is not necessarily what most deserves to be feared, and still less so is its opposite the thing most deserving our desire, that is, the highest good... – although, in consequence, the heuristics of fear is surely not the last word in the search for goodness, it is at least an extremely useful first word and should be used to the full of its helpfulness in a sphere where so few unlooked-for words are vouchsafed us.

89 Ingold (1996:129).
90 (ibid.:151).
91 Jonas (1984:27).

### *F. Obligations to Strengthen Agenda 21's Major Groups 21 in Support of Sustainable Development?*

Noting what has been analysed and described as the *architecture of global governance for sustainable development* leads to the fundamental question as to whether or not Agenda 21 contains a normative obligation with respect to the said architecture, and if so, what form that obligation takes. A second and more specific question would be whether or not there is any legal or quasi-legal obligation to strengthen the major groups listed in the Agenda.

The Kyoto Protocol did not deliver what was expected of it.[92] The world is still largely behind the level of emissions reduction the Protocol anticipated; one of the world's biggest emitters, the United States of America, refused to sign the Protocol; Canada, another major industrialised country, withdrew from it; and there is no mechanism in sight for dealing with the level of emissions by the leading developing countries, China and India. This already shows that sustainable development as such, and as a principle, is far from being part of generally applicable international law. Moreover, although sustainable development has entered the field of international and state law, it also does not mean that the concept of *sustainable development* has achieved the goal of binding international law.[93]

The same applies to Agenda 21. In formal terms, it is no more than a resolution; and, like any other UN General Assembly Resolution, it does not achieve the quality of law – irrespective of the number of members of the General Assembly who may vote for it.[94] Nevertheless, even the strongest opponents to giving sustainable development a place in the law have to accept that the amount of effort that has been and is being made in international, regional, national and subnational forums to substantiate the meaning of *sustainable development* have added to the concept a quality which allows it and Agenda 21 to be called *soft law*.[95] The Rio+10 Conference in Johannesburg in 2002 was an important event in this respect. Although the Johannesburg Summit broadened the approach to development, as understood by developing countries and as adopted ten years earlier, the principle of

92 Here and to that which follows, cf. von Bassewitz (2011).

93 See Gärditz (2008).

94 Cordonier Segger & Khalifan (2004:21).

95 (ibid.).

sustainable development and Agenda 21 remained the overarching points of reference.[96]

What is *soft law*?[97] The category of *soft law* was developed in order to have a characterisation of norms which had not yet received the quality of law, but which enjoyed wide support by all who were dealing with a specific matter with authority. This new category of law injected parts of that authority into documents whose content was normative but not legally binding.

States are not obliged to implement existing soft law, but they may be confronted with critical questions as to why they have ignored it. The fact that something has the quality of soft law changes the burden of proof to the deviating authority to give good reasons why certain rules equipped with high authority have been ignored.

Although there is no legally binding rule to translate the architecture of governance of Agenda 21 into state law, there is a clear indication that states would be well-advised to consider such a translation. However, it will be particularly difficult for states that have achieved their independence – and with this, their statehood, after long anticolonial struggles and the loss of many lives – to accept that their newly acquired sovereignty will be limited by a structure of governance in which the Westphalian model of statehood and sovereignty has been overcome. The globalising world had its input everywhere and *sovereignty*, as it used to be defined, is a thing of the past. The focus today is increasingly on participation, local empowerment, supporting local responsibility, and accepting existing local structures.

In other words, Agenda 21 is not only the political anticipation of what researchers have described as the architecture of global governance with respect to sustainable development: it is also a normative, quasi-constitutional political charter of good governance in matters of sustainable development, demanding not only that the structures already reflecting the global

96 (ibid.:25ff.).

97 See Paech & Stuby (2001:482ff.); Thürer (1985). Paech & Stuby (ibid.) refer to Agenda 21 in their deliberations on soft law. They observe in this context another result of the Rio Summit of 1992, namely the one that deals with the conservation of forests. The Rio Conference adopted a document that was originally meant to be a Convention, but it did not achieve that status. The document is entitled *A Non-legally Binding Authoritative Statement of Principles for a Global Consensus on the Management, Conservation and Sustainable Development of all Types of Forests*, in short, the Forest Principles (A/CONF.151/26 (Vol. III). The combination of "non-legally binding" and "authoritative" as qualities of one and the same element is as contradictory as the combination of *soft* and *law*.

architecture of governance be respected, but also that their extension and strengthening be supported.[98]

In the early days of the Towns and Development movement,[99] German officials from the central level of government raised their concern about local authority activities in the field of development cooperation. The concern was that such activities would interfere with the competence of the central government. Meetings were held, negotiations took place, and, eventually, legal scholars took it upon themselves to show that their concerns had no foundation in Germany's constitutional law.[100]

In southern Africa, one finds discourses comparable to those prevailing in Germany on the legal status and place of traditional authorities – government entities that are usually not elected in accordance with the rules applied to democracy at the level of state. But how can a traditional authority have government functions which it has not been properly delegated by Acts of Parliament? How can traditional authorities "make customary law", as it is stated in section 3(3)(c) of the Traditional Authorities Act of Namibia? The proponent of Kelsenian centralism may wish away the existence of the rule that traditional authorities "make law", but even the Kelsenists will at least suggest that the quoted phrase from section 3(3)(c) implies a delegation of authority from the state to the traditional authority. Such a delegation could even be challenged because it does not set any framework in which the delegation can be used. Legal pluralism will instead observe what traditional authorities do when making use of their authority, and will – with empirical evidence, if so wished – assist in enhancing the required jurisprudential reflections on what is being observed.[101]

The multilayered architecture of globalised environmental governance requires new approaches to legitimacy. Only new approaches can provide answers to the new questions that are being raised by the new dispensation. Since the discovery of 'living law', we know that the demands for unrestricted priority of state law lose their power where people insist on their own ways of doing things. In view of this, the issue on the table is how a society will deal with the realities in which such living law is applied.

---

98 By doing so, also providing input as to the interpretation and further development of (human) rights; cf. Gärditz (2008:159ff.).

99 As described above.

100 Here and to what follows, cf. Hinz (1985); Von Schwanenflügel (1993).

101 For example, this is what Namibia's Customary Law Ascertainment Project does; cf. Hinz (2010).

This is a political issue to which basically two answers are possible. One would be to go the easy route, i.e. to reject the results of the non-statal/non-official formations of governance as not legitimised by Parliament. However, research on legal pluralism has shown that decisions of this kind will not end the matter, but rather lead to conflict.[102] The other answer is to open up not only space for negotiation, but also for remedies within the parameters set for a specific legal environment.[103] The Agenda 21 Preamble declares that the document is a "dynamic programme".[104] Its dynamism requires political creativity. The challenge of the architecture of global environmental governance is to overcome inherited concepts that are unable to respond to changing demands over time.

Legal and legal anthropological research shows that traditional authorities take their environmental responsibility seriously. The research shows where traditional authorities experience very practical limits in pursuing action to support the sustainable maintenance of their resources.[105] The research also shows that some of these limits result in statutory deficits and related, politically motivated reluctance to accept the dynamics in traditional communities. The need to accept the dynamics of traditional authorities led the introduction to this paper to focus on legal problems originating in the described tensions and contradictions between statutory and customary law, including the fragmentation in the administration of matters which, at the local level, are under the power of traditional authorities.

Law reform is called for to address these issues. The attempt followed in this paper could be an additional stimulus for law reform, namely to place the operation of traditional authorities within a wider governmental framework. The construction of this wider framework would be guided by the call to reconsider the various actors relevant to sustainable development and their place in the architecture of global governance in support of sustainable development. Managerial reasons support this attempt, but so do legal ones that refer to recognising and confirming customary law and traditional governance as provided for in the Namibian Constitution. The reference to suggestions to improve the architecture of global governance in support of sustainable development will also more adequately expose the nature of frag-

102 Which one finds, for example, where African states have tried to abolish customary law!

103 See Winter (2012:145).

104 Chapter 1, section 1, paragraph 1.6.

105 See here the research assembled in Hinz & Ruppel (2008) and in Hinz et al. (2012).

mentation in the architecture. Thus, transforming traditional authorities into subnational administrative units of the state will not meet the very special legitimacy of traditional governance; providing for better state responses to traditional governance will enhance local ownership and, with this, strengthen local responsibility.

Apart from this, it also has to be noted that the area under research in the TFO Project is occupied with special international arrangements which contribute important aspects to the very concrete architecture of governance that go beyond the borders of states. In addition to general public international law that binds, in 1994, Angola, Namibia, and Botswana entered into an agreement concerning the Kavango River Basin,[106] by means of which they want to set rules that take note of the interests of the three countries and, in particular, the people who depend on the water of the river. According to the agreement, the objective of the Permanent Water Commission made up by these three countries and known as *OKACOM* is –

> … to act as technical advisor to the Contracting Parties on matters relating to the conservation, development and utilisation of the resources of common interest to the Contracting Parties (basin member states)....

OKACOM's mandate is as follows:[107]

- Determine the long-term safe yield of the river basin;
- Estimate reasonable demand from the consumers;
- Prepare criteria for conservation, equitable allocation and sustainable utilisation of water;
- Conduct investigations related to water infrastructure;
- Recommend pollution prevention measures;
- Develop measures for the alleviation of short[-]term difficulties, such as temporary droughts …

On 7 December 2006, the Governments of Angola, Botswana, Namibia, Zambia and Zimbabwe signed a Memorandum of Understanding to establish the Kavango–Zambezi Transfrontier Conservation Area (KAZA

106 Agreement between the Governments of the Republic of Angola, the Republic of Botswana and the Republic of Namibia on the Establishment of a Permanent Okavango River Basin Water Commission (OKACOM), concluded in Windhoek on 15 September 1994; available at www.okacom.org, last accessed 15 January 2013.

107 See Article 4 of the Agreement.

TFCA).[108] The relevant treaty was concluded on 18 August 2011 and states as its purpose the following:[109]

> By this Treaty, the Partner States establish the Kavango Zambezi Transfrontier Conservation Area... for the primary purpose of harmonising policies, strategies and practices for managing shared Natural Resources that straddle the international borders of the five... Partner States and deriving equitable socioeconomic benefits through the sustainable use and development of their natural and cultural heritage resources.

OKACOM and the KAZA TFCA have overlapping jurisdiction and many goals in common. The analysis of both these international arrangements, their policies, the implementation of the policies, and their effects on the various societal levels, including the input of the sub-international levels in the formulation of those policies and their implementation, will contribute further to the understanding of the architecture of governance for sustainable development in a globalising world.

## *References*

Adams, Fiona & Wolfgang Werner, 1990, *The Land Issue in Namibia: An Inquiry*, Windhoek, Namibian Institute for Social and Economic Research, University of Namibia.

Anyolo, Prisca N., 2012, Conservancies in Namibia: Tools for Sustainable Development, in: Hinz, Manfred O., Oliver C. Ruppel & Clever Mapaure (Eds), *Knowledge Lives in the Lake: Case Studies in Environmental and Customary Law from Southern Africa*, Windhoek, Namibia Scientific Society, 29–39.

Biermann, Frank, Philipp Pattberg & Fariborz Zelli (Eds), 2010a, *Global Climate Governance Beyond 2012: Architecture, Agency and Adaptation*, Cambridge, Cambridge University Press.

Biermann, Frank, Fariborz Zelli, Philipp Pattberg & Harro von Asselet, 2010b, The Architecture of Global Climate Governance: Setting the Stage, in: Biermann, Frank, Philipp Pattberg and Fariborz Zelli (Eds), *Global Climate Governance Beyond 2012: Architecture, Agency and Adaptation*, Cambridge, Cambridge University Press, 15–24.

108 Treaty between the Government of the Republic of Angola, the Government of the Republic of Botswana, the Government of the Republic of Namibia, the Government of the Republic of Zambia, and the Government of the Republic of Zimbabwe on the Establishment of the Kavango–Zambezi Transfrontier Conservation Area; available at www.kavangozambezi.org, last accessed 15 January 2013.

109 See Article 2(1) of the Treaty.

Börzel, Tanja A. & Thomas Risse, 2005, Public–private Partnership: Effect and Legitimate Tools of International Governance, in: Grande, Edgar & Louis W. Pauly (Eds), *Reconstituting Political Authority: Complex Sovereignty and the Foundation of Global Governance*, Toronto, University of Toronto Press, 195–216.

Bulkeley, Harriet & Michele M. Betsill, 2003, *Cities and Climate Change: Urban Sustainability and Global Environmental Governance*, London, Routledge.

C40, 2008, History of the C40, available at www.c40cities.org, last accessed May 2012.

Cordonier Segger, Marie-Claire & Ashfaq Khalifan, 2004, *Sustainable Development Law: Principles, Practices and Prospects*, Oxford, Oxford University Press.

D'Engelbronner-Kolff, Marina F., Manfred O. Hinz & John L. Sindano (Eds), 1998, *Traditional Authority and Democracy in Southern Africa*, Windhoek, New Namibia Books.

Ehrlich, Eugen (edited by Manfred Rehbinder), 1967, *Recht und Leben: Gesammelte Schriften zur Rechtstatsachenforschung und zur Freirechtslehre*, Berlin, Duncker & Humblot.

Ekardt, Felix, 2005, *Das Prinzip Nachhaltigkeit: Generationsgerechtigkeit und globale Gerechtigkeit*, München, C.H. Beck.

Fischer-Lescano, Andreas & Gunther Teubner, 2006, *Zur Fragmentierung des globalen Rechts*, Frankfurt/M, Suhrkamp Verlag.

Gärditz, Klaus F., 2008, Nachhaltigkeit und Völkerrecht, in: Kahl, Wolfgang (Ed.), *Nachhaltigkeit als Verbundbegriff*, Tübingen, Mohr Siebeck, 137–179.

Gold, Jane, Anna Muller & Diana Mitlin, 2001, *The Principles of Local Agenda 21 in Windhoek: Collective Action and the Urban Poor*, London, International Institute for Environment and Development.

Griffiths, John, 1986, What is Legal Pluralism?, *Journal of Legal Pluralism and Unofficial Law* 24, 1–55.

Grim, John A. (Ed.), 2001, *Indigenous Tradition and Ecology: The Interbeing of Cosmology and Community*, Cambridge, MA, Harvard University Press.

Harring, Sidney, 1998, Inconsistencies in the Constitution of Namibia Regarding the Land Question, in: D'Engelbronner-Kolff, Marina F., Manfred O. Hinz & John L. Sindano (Eds), *Traditional Authority and Democracy in Southern Africa*, Windhoek, New Namibia Books, 155–182.

Harring, Sidney, 2012, The *Baster Community of Rehoboth* Case in the Supreme Court of Namibia: Courtroom Observations and Reflections on the Decision, in: Limpricht, Cornelia (Ed.), *Rehoboth, Namibia – Past and Present*, Windhoek, Solitaire Press, 386–392.

Hilliges, Gunther, 1994, Towns and Development entwickelt weltweite Strukturen: Kommunen und NRO gründen deutsche Arbeitsgruppe, *Terre des hommes*, Intern 9/94, 20–21.

Hilliges, Gunther, 1995a, Große Ereignisse werfen ihre Schatten voraus: 32. IULA-Weltkongress in Den Haag, 03.–7.9.1995, Zum Thema "Eine Welt der Kommunen: Lokale Innovation in internationaler Zusammenarbeit", *Nord–Süd Info-Dienst* 71, 12–13.

Hilliges, Gunther, 1995b, Auf dem Weg in ein neues Jahrtausend, *Nord–Süd Info-Dienst* 72, 19.

Hilliges, Gunther, 2001, Internationale Vernetzung im Agenda-21-Prozess, in: Berndt, Michael & Detlef Sack (Eds), *Global Governance? Voraussetzungen und Formen demokratischer Beteiligung im Zeichen der Globalisierung*, Wiesbaden, Westdeutscher Verlag, 193–194.

Hilliges, Gunther, 2005, Inhalte und Ergebnisse der bisherigen Bundeskonferenzen, in: Servicestelle Kommunen in der einen Welt (Eds), *Globalisierung gestaltet Kommunen – Kommunen gestalten Globalisierung: 9. Bundeskonferenz der Kommunen und Initiativen, Magdeburg, 2004*, Bonn, Servicestelle Kommunen in der einen Welt, 6–21.

Hilliges, Gunther, 2012, Ein schwieriger Weg bis zur Gründung der Servicestelle Kommunen in der einen Welt, Mimeo.

Hilliges, Gunther & Ulrich Nitschke, 2007, Städte als Partner für nachhaltige Entwicklung von Rio de Janeiro: Material No. 15, Bonn, Servicestelle Kommunen in der einen Welt.

Hinz, Manfred O., 1985, Notes on Possible Development Policies for Local Authorities, in: Towns and Development (Eds), *Report on the European Conference on NGO and Local Community Joint Action for North–South Cooperation*, Cologne, Towns and Development, 13–14.

Hinz, Manfred O., 1998, Communal Land, Natural Resources and Traditional Authority, in: D'Engelbronner-Kolff, Marina F., Manfred O. Hinz & John L. Sindano (Eds), *Traditional Authority and Democracy in Southern Africa*, Windhoek, New Namibia Books, 183–227.

Hinz, Manfred O., 2002, Traditional Authorities: Sub-central Governmental Agents, in: Hinz, Manfred O., Sam K. Amoo & Dawid van Wyk (Eds), *The Constitution at Work: Ten Years of Namibian Nationhood*, Pretoria, University of South Africa, 81–94.

Hinz, Manfred O., 2003a. *Without Chiefs There Would be no Game: Customary Law and Nature Conservation*, Windhoek, Out of Africa.

Hinz, Manfred O., 2003b, *Customary Law in Namibia: Development and Perspective* (Eighth Edition), Windhoek, Centre for Applied Social Sciences.

Hinz, Manfred O., 2006, Legal Pluralism in Jurisprudential Perspective, in: Hinz, Manfred O. (Ed., in collaboration with Helgard K. Patemann), *The Shade of New Leaves: Governance in Traditional Authority: A Southern African Perspective*, Münster, Lit Verlag, 29–45.

Hinz, Manfred O., 2008, Findings and the Way Forward, in: Hinz, Manfred O. & Oliver C. Ruppel (Eds), *Biodiversity and the Ancestors: Challenges to Customary and Environmental Law – Case Studies from Namibia*, Windhoek, Namibia Scientific Society, 211–225.

Hinz, Manfred O., 2009, Traditional Governance and African Customary Law: Comparative Observations from a Namibian Perspective, in: Horn, Nico & Anton Bösl (Eds), *Human Rights and the Rule of Law in Namibia* (Second Edition), Windhoek, Macmillan Namibia, 59–87.

Hinz, Manfred O. (Ed., assisted by Ndateelela E. Namwoonde), 2010, *Customary Law Ascertained, Volume 1: The Customary Law of the Owambo, Kavango and Caprivi Communities in Namibia*, Windhoek, Namibia Scientific Society.

Hinz, Manfred O., 2011a, Customary Law and the Environment, in: Ruppel, Oliver C. & Katharina Ruppel-Schlichting (Eds), *Environmental Law and Policy in Namibia*, Windhoek/Essen, Orumbonde Press/Welwitschia Verlag Dr. A. Eckl, 169–199.

Hinz, Manfred O., 2011b, Traditional Governance and Communal Conservancies, *Namibia Law Journal* 3 (2), 115–124.

Hinz, Manfred O. & Frank Thomas Gatter (Eds), 2006, *Global Responsibility – Local Agenda: The Legitimacy of Modern Self-determination and African Traditional Authority*, Berlin, Lit Verlag.

Hinz, Manfred O. & Brian Katjaerua (assisted by Monica Kalondo), 1998, *To Promote Peace and Welfare: The Political and Economic Sustainability of Traditional Authorities*, Windhoek, Centre for Applied Social Sciences, University of Namibia.

Hinz, Manfred O. & John Malan (Eds), 1997, Communal Land Administration: Consultative Conference on Communal Land Administration, 26–28 September 1997, Proceedings, Windhoek, Centre for Applied Social Sciences.

Hinz, Manfred O. & Clever Mapaure, 2012, Water is Life: Customary and Statutory Water Law – A Problematic Relationship, Ongoing Research in the Kavango River basin, *Namibia Law Journal* 4 (1), 189–196.

Hinz, Manfred O. & Oliver C. Ruppel (Eds), 2008, *Biodiversity and the Ancestors: Challenges to Customary and Environmental Law – Case Studies from Namibia*, Windhoek, Namibia Scientific Society.

Hinz, Manfred O., Oliver C. Ruppel & Clever Mapaure (Eds), 2012, *Knowledge Lives in the Lake: Case Studies in Environmental and Customary Law from Southern Africa*, Windhoek, Namibia Scientific Society.

Ingold, Tim, 1996, Hunting & Gathering as Ways of Perceiving the Environment, in: Ellen, Roy and Katsuyoshi Fukui (Eds), *Redefining Nature: Ecology, Culture and Domestication*, Oxford/Washington, Berg Publishers, 117–155.

Jürgens, Norbert, Ute Schmiedel & M. Timm Hoffman (Eds), 2010, *Biodiversity in Southern Africa, Volume 1: Patterns at Local Scale – The BIOTA Observatories*, Göttingen/Windhoek, Klaus Hess Publishers.

Jonas, Hans, 1984, *The Imperative of Responsibility. In Search of an Ethics for the Technological Age*, Chicago/London, The University of Chicago Press.

Kahl, Wolfgang (Ed.), 2008, *Nachhaltigkeit als Verbundbegriff*, Tübingen, Mohr Siebeck.

Keulder, Chris, 1997, *Traditional Authorities and Regional Councils in Southern Namibia*, Windhoek, Friedrich Ebert Foundation.

Keulder, Chris, 1998, *Traditional Leaders and Local Government in Africa: Lessons for South Africa*, Pretoria, Human Sciences Research Council.

Kpadonou, Rivaldo A.B., Patrice Y. Adégbola & Silvère D. Tovignan, 2012, Local Knowledge and Adaptation to Climate Change in Ouémé Valley, Benin, *African Crop Science Journal* 20 (Supplement), 181–192.

Krenzer-Bass, Angelika, 1988, Städte und Entwicklungszusammenarbeit, *Entwicklungspolitik* 88, 1.

Kussendrager, Nico, 1988, *NGO and Local Authority Joint Action for North–South Co-operation: Case Studies from Belgium, Germany, the Netherlands and the United Kingdom* (Second Edition), Cologne, Towns and Development.

Mapaure, Clever, 2009, Jurisprudential Aspects of Proclaiming Towns in Communal Areas, *Namibia Law Journal* 1 (2), 23–48.

Mapaure, Clever, 2010, Accused and Cherished: Traditional and Informal Justice Systems as seen by International Organisations, in: Hinz, Manfred O. (Ed., in cooperation with Clever Mapaure), *In Search of Justice and Peace: Traditional and Informal Justice*, 25–49.

Mapaure, Clever, 2012a, Trees at the Crossroads: Internal Conflict of Laws in the Ownership of Forests in the Kavango Region, in: Hinz, Manfred O., Oliver C. Ruppel & Clever Mapaure (Eds), *Knowledge Lives in the Lake: Case Studies in Environmental and Customary Law from Southern Africa*, Windhoek, Namibia Scientific Society, 131–148.

Mapaure, Clever, 2012b, Who Legally Owns Water in Namibia's Communal Areas? A Critical Analysis of Observations from Empirical Experience, in: Hinz, Manfred O., Oliver C. Ruppel & Clever Mapaure (Eds), *Knowledge Lives in the Lake: Case Studies in Environmental and Customary Law from Southern Africa*, Windhoek, Namibia Scientific Society, 207–232.

Mapaure, Clever, 2012c, The Politics of Rural Water Supply in Namibia: Exploring the Dynamics of Multi-layered Institutional Interactions, in: Hinz, Manfred O., Oliver C. Ruppel & Clever Mapaure (Eds), *Knowledge Lives in the Lake: Case Studies in Environmental and Customary Law from Southern Africa*, Windhoek, Namibia Scientific Society, 233–258.

Martens, Jens, 2012, *Rio+20: Die UN-Konferenz für nachhaltige Entwicklung 2012: Hintergründe – Konflikte – Perspektiven*. Bonn/Osnabrück, Global Policy Forum Europe.

Menski, Werner, 2006, *Comparative Law in a Global Context: The Legal Systems of Asia and Africa* (Second Edition), Cambridge, Cambridge University Press.

Moore, Sally F., 1978, *Law as Process: An Anthropological Approach*, London, Routledge.

Niezen, Ronald, 2010, *Public Justice and the Anthropology of Law*, Cambridge, Cambridge University Press.

Paech, Norman & Gerhard Stuby, 2001, *Völkerrecht und Machtpolitik in den internationalen Beziehungen*, Hamburg, VSA.

Pattberg, Philipp, 2010, The Role and Relevance of Networked Climate Governance, in: Biermann, Frank, Philipp Pattberg & Fariborz Zelli (Eds), *Global Climate Governance Beyond 2012: Architecture, Agency and Adaptation*, Cambridge, Cambridge University Press, 146–164.

Ray, Donald I. (in collaboration with Keshav C. Sharma and Ibikunle May-Parker), 1997, Symposium on Traditional Leadership and Local Government: Report, Mimeo.

Renkhoff, Nathalie, (Forthcoming), Das Konzept der nachhaltigen Entwicklung bei Bergbauprojekten auf kommunalem Land in Namibia.

Republic of Namibia, 1991a, *National Conference on Land Reform and the Land Question, Volume 1*, Windhoek, Office of the Prime Minister.

Republic of Namibia, 1991b, National Conference on Land Reform and the Land Question, Conference Brief, Windhoek, Office of the Prime Minister.

Republic of Namibia, 1991c, National Conference on Land Reform and the Land Question, Consensus Document, Windhoek, Office of the Prime Minister.

Ruppel, Oliver C., 2011, The International, African and Regional Institutional, Legal and Policy Framework of Climate Change, in: African Ministerial Conference on the Environment (AMCEN) (Eds), *Addressing Climate Change Challenges in Africa: A Practical Guide Towards Sustainable Ddevelopment*, Nairobi, AMCEN, 169–202, also available at www.unep.org/roa/amcen/docs/publications/guidebook_ClimateChange.pdf, last accessed 5 August 2012.

Ruppel, Oliver C., 2012, Wasser und Land: Brennpunkte innerhalb der Entwicklungsgemeinschaft des südlichen Afrika (SADC), in: Reder, Michael & Hanna Pfeifer (Eds), *Kampf um Ressourcen: Weltordnung zwischen Konkurrenz und Kooperation, Globale Solidarität – Schritte zu einer neuen Weltkultur,* Vol. 22, Stuttgart, Kohlhammer, 59–85.

Ruppel, Oliver C. & Katharina Ruppel-Schlichting (Eds), 2011, *Environmental Law and Policy in Namibia*, Windhoek/Essen, Orumbonde Press/Welwitschia Verlag Dr. A. Eckl.

Sand, Peter H., 2006, Global Environmental Change and the Nation State: Sovereignty Bounded, in: Winter, Gerd (Ed.), *Multilevel Governance of Global Environmental Change: Perspectives from Science, Sociology and the Law*, Cambridge, Cambridge University Press, 519–538.

Schefold, Dian, 2011, Von der Kolonialpolitik zur Selbstverwaltung: Entwicklungen im Denken des jungen Hugo Preuβ, in: Ruppel, Oliver C. & Gerd Winter (Eds), *Justice from Within: Legal Pluralism in Africa and Beyond, Liber Amicorum Manfred O. Hinz in Celebration of his 75th Birthday*, Hamburg, Verlag Dr Kovač, 105–116.

Schönherr-Mann, Hans-Martin, 2010, *Die Macht der Verantwortung*, München, Karl Alber Freiburg.

Servicestelle/Servicestelle Kommunen in der einen Welt, 2002, Local Government Declaration to the World Summit on Sustainable Development and Johannesburg Call (*Material – Info Nr. 2*), Bonn, Servicestelle Kommunen in der einen Welt.

Servicestelle/Servicestelle Kommunen in der einen Welt, 2009, Die Bundeskonferenzen der Kommunen und Initiativen 1988–2009 *(Material – Info Nr. 37)*, Bonn, Servicestelle Kommunen in der einen Welt.

Shuman, Michael, 1994, Towards a Global Village: International Community Development Initiatives, London/Boulder, CO, Pluto Press.

Thürer, Daniel, 1985, 'Soft Law' – Eine neue Form des Völkerrechts?, *Zeitschrift für Schweizerisches Recht (Neue Fassung)*, 429–453.

Towns and Development, 1985, Report on the European Conference on NGO and Local Community Joint Action for North–South Cooperation, Cologne, Towns and Development.

Tucker, Mary E. & John A. Grim, 2001, Series Foreword, in: Grim, John A. (Ed.), *Indigenous Tradition and Ecology: The Interbeing of Cosmology and Community*, Cambridge, MA, Harvard University Press, XV–XXXII.

UNCED/United Nations Conference on Environment and Development, 1992, *Agenda 21*, Rio de Janeiro, United Nations.

UNDP/United Nations Development Programme, 2004, *Human Development Report: Cultural Diversity in Today's Diverse World*, New York, UNDP.

Van der Byl, Pieter C., 1992, *Legal Position to Land Occupied in Namibia on a Communal Basis: Report*, Windhoek, Ministry of Lands, Resettlement and Rehabilitation.

Von Bassewitz, Nadia, 2011, International Climate Change Policy and Legislation: Where do we Stand?, in: Ruppel, Oliver C. & Katharina Ruppel-Schlichting (Eds), *Environmental Law and Policy in Namibia*, Windhoek/Essen, Orumbonde Press/ Welwitschia Verlag Dr. A. Eckl, 316–352.

Von Schwanenflügel, Mathias, 1993, *Entwicklungszusammenarbeit als Aufgabe der Gemeinden und Kreise*, Berlin, Duncker & Humblot.

Winter, Gerd (Ed.), 2006a, *Multilevel Governance of Global Environmental Change: Perspectives from Science, Sociology and the Law*, Cambridge, Cambridge University Press.

Winter, Gerd, 2006b, Introduction, in: Winter, Gerd (Ed.), *Multilevel Governance of Global Environmental Change: Perspectives from Science, Sociology and the Law*, Cambridge, Cambridge University Press, 1–33.

Winter, Gerd, 2012, Zur Architektur globaler Governance des Klimaschutzes, *Zeitschrift für ausländisches öffentliches Recht und Völkerrecht* 72 (1), 103–145.

21

# Ethics for Climate Justice and Sustainability through Value-based Approaches: A New Tanzanian Model and Paradigm Shift

*Aidan G. Msafiri*

*Abstract*

This article has a threefold purpose. Firstly, it underscores the unique importance of ethics and values in responding to the current challenges of climate justice and resource depletion in sub-Saharan Africa – and Tanzania in particular. These ethics and values include food insecurity, health hazards, prolonged and severe drought spells, hydroelectric power crises, destruction of human settlements, rapid extinction of wildlife, and invisible psychological consequences to humans.

Secondly, behind the ever-increasing threat of the climate crisis is a plethora of deep-seated philosophical, ideological and policy root causes of today's wanton climate injustices and unsustainable lifestyles. These include the radical anthropocentric, biocentric, cosmocentric, pathocentric, and hyper-post-modern consumerist lifestyles, as well as the weaknesses of present climate policies, locally and globally.

Thirdly, on a more formative and methodological note, today more than ever before it is becoming increasingly necessary (and evident) that humans need to use value-based, qualitative approaches in responding to the issues of climate justice and sustainability, i.e. to go beyond the hitherto quantitative (mathematical), functionalist and legalist approaches, which are neither exhaustive nor sustainable. In short, the major argument is that climate justice and sustainability are mainly value-based issues which first and foremost call for a rethinking of the value of values, beyond empirical and legal solutions and methods. The present situation calls for a radical paradigm shift to viable and sustainable ethics for climate rights if humanity and the earth are to survive.

## *A. Introduction*

### *I. Terminology*

#### *1. Climate Justice*

The broadest and deepest meaning of *climate justice* refers to the short- and long-term rights and abilities of the earth to regenerate and support all life forms, human and non-human, in a sustainable and dignified manner, with a fair distribution of resources and environmental burdens.

Briefly, in sub-Saharan Africa, *climate justice* simply means real commitment to the 1992 Rio Declaration[1] which was endorsed by 160 countries worldwide. The Declaration affirms that "human beings are entitled to a healthy and productive life in harmony with nature" (Principle 1).[2] The right to development must be fulfilled so as to equitably meet the needs of present and future generations (Principle 3).[3] Indeed, climate justice encapsulates a myriad of socio-human, economic, environmental, political, cultural, technological and ethical trajectories, using such mechanisms as rights and duties.

#### *2. Sustainability/Sustainable Development*

I have argued that *sustainability* or *sustainable development* is a highly complex, new and fluid concept which is often ambiguous and confusing.[4] It is a new paradigm based on continued use of the earth's natural resources in such a manner and degree that they are not overstressed, exhausted or used to the detriment of future human and non-human generations.[5] Admittedly, in the last few decades, the word *sustainability* has widely and synonymously been used with concepts such as *sustainable development*, *integral development*, *sustainable economy*, and *sustainable growth* – but the list goes on and on. Fundamentally, *sustainability* is part and parcel of climate justice, both as a means and an end.

1 UN (1994:Articles 1–3).
2 (ibid.).
3 (ibid.).
4 Msafiri (2007:105).
5 UN (1992:1–20).

## *II. The Key Issues at Stake*

Today, facts, figures, observations and feelings indicate that climate change is real and not fictitious. This is evidenced by floods, drought spells, stress on health systems, melting of ice, food insecurity, etc. Unfortunately, climate justice does not figure as prominently today on global media as, for instance, financial crises, global politics and elections. Yet, climate injustices pose direct threats to the planet and to all life forms, today and in the future, locally and globally. The earth has an intrinsic right and mechanism to sustain all life forms in a dignified way. Ever-worsening climate injustice dilemmas are a 'wake-up call' for humanity generally, and governments in particular, to urgently rediscover the inviolable dignity, sanctity and well-being of all life forms on earth.

Consequently, climate protection should be the first and foremost human obligation, i.e. the human as *homo conservator* and *homo intergralis* (responsible stewards of creation) as opposed to *homo faber* or *homo consumerismus*. There is an urgent need to go beyond the common view of 'business as usual'. Our planet has inviolable rights. If they are infringed, all forms of life are subjected to potentially irreversible risks and consequences. As Archbishop Desmond Tutu aptly put it at an inter-faith rally prior to an international conference on climate change, "We only have this one planet. We do not have planet B".[6]

## *III. Justifications/Rationale*

Today, more than ever before, the entire world – and Africa in particular – continues to witness increased tragic life scenarios largely resulting from unsustainable socio-economic, technological, cultural and political worldviews and lifestyles which threaten the very existence of humans, present and future. It is true that climate injustices and climate change are a result of human *habitus* (lifestyle) and activity. Climate injustices are essentially ethical crises which need ethical solutions. This requires deep and radical transformation within, and not outside, humans. Admittedly, the problems of climate change and wanton climate injustices cannot be solved in the cabins of large aeroplanes flying to international conferences on climate, or

6 Seventeenth Conference of the Parties (COP17) to the United Nations Framework Convention on Climate Change, November 2011.

in the luxurious hotels which accommodate climate change experts and activists, but rather in the hearts and consciences of each and every human being on this planet.

### *IV. Structure*

There are three main sections to this Pre-COP18 Doha reflection on the ethics of climate justice and sustainability. Section B attempts to identify the magnitude of climate injustices, both from a local (Tanzanian) and a global perspective. Section C offers a wide range of philosophical and ideological policy, as well as life views behind climate injustices and sustainable living. Section D unveils key value-based principles for a new model and paradigm shift, in sub-Saharan Africa in particular, but also in the world in general.

## *B. Climate Injustice and Vulnerability Scenarios: Local and Global Realities*

### *I. A Tanzanian View*

#### *1. Negative Agricultural Impacts*

Due to unpredictable rainfall patterns in Tanzania, average food and cash-crop production has decreased very significantly. This has not only threatened national food and health security (nutritional well-being), but has also caused a dramatic drop in foreign income.[7] It was recently confirmed that, with increased mean annual temperatures and reduced rainfall, the production of maize – the staple food crop for most Tanzanians – has decreased by 34%.[8] Furthermore, recent research by the Tanzania Meteorological Agency (TMA) indicate that some of the previously highly productive regions, especially the Southern and Northern Highlands, are badly affected year after year, especially by declining rainfall, acute droughts and massive rainfall variability.[9]

7 URT (2011:20).
8 (ibid.:21).
9 (ibid.).

As far as livestock activity is concerned, recent severe drought spells are reducing water availability and pastures for rearing livestock,[10] causing increasing potential threats and risks to keepers of livestock and agro-based communities in Tanzania. Moreover, it has been claimed that, due to severe drought spells, savannah grassland is encroaching on natural forests and woodlands countrywide.

### 2. *Health Hazards and Risks*

Increased and fluctuating temperatures have a direct influence and impact, particularly as regards the survival of lethal vectors, pathogens and hosts for new, climate-driven habitats. These health threats include malaria, meningitis, dysentery, cholera, plague, Rift Valley fever, and schistosomiasis. Today, for instance, there are several potential incidences of epidemic malaria, particularly in the previously cold highland regions of Tanzania such as Iringa, Kagera, Kilimanjaro, Mbeya and Njombe.

Besides being a major killer disease in Tanzania, affecting especially children, malaria has recently been estimated to cost Tanzania potentially about US$20–100 million a year by 2030, increasing to US$25–160 million a year by 2050.[11]

### 3. *Negative Effects on Freshwater Resources and Reliability of Water*

The levels of many lakes and basins in Tanzania are decreasing significantly. Recent predictions by the Tanzania National Climate Change Strategy and Action Plan indicate that the actual water levels of most Tanzanian lakes, particularly Lake Victoria, but also Lakes Eyasi, Mayara and Rukwa, are decreasing, some at a rate of 0.6–5.0% annually.[12] This is mainly because many ecosystems, wetlands, groundwater aquifers and rivers in Tanzania and in neighbouring East African countries have been drastically affected by severe and recurrent droughts, coupled with immense water evaporation.

10 (ibid.:2–7).
11 Global Climate Adaptation Partnership (2011:30).
12 URT (2011:16).

Water is a finite resource. Due to increased pressure on water catchments, increased industrialisation and urbanisation, both the quantity and quality of clean water sources and resources in Tanzania have decreased noticeably.

### *4. The Hydroelectric Power Crisis*

Recent data from the Tanzanian Ministry of Energy and Minerals shows that, due to prolonged severe droughts, the water levels in most of the country's hydroelectric power (HEP) stations have declined to their lowest ever.[13] Indeed, 54% of electricity in Tanzania comes from water sources which have been exposed to prolonged dry periods. These include the Mtera HEP Dam, the Nyumba ya Mungu HEP Dam, and the Hale Pangani HEP dam.

From a socio-economic point of view, intermittent power blackouts and rationing have become commonplace, particularly in the large cities such as Arusha, Dar es Salaam, Morogoro, Mwanza, and Tanga. Prolonged power rationing for both domestic and industrial use definitely threatens the well-being of the Tanzanian people and exposes them and their economy to greater risks of poverty and social instability.

### *5. Negative Effects on Human Settlements*

Reports indicate that there are several instances of sea-level rise and prolonged coastal erosion, particularly around the Island of Zanzibar and Dar es Salaam.[14] This is spectacular in the northern part of Ras Mkumbuu Peninsula and New Chake-Chake, where several ancient commercial and religious centres have been completely abandoned. In Dar es Salaam, the threats of increased coastal erosion are escalating at a very rapid pace. Most of the beaches, such as those at Mbezi, Msasani and Upanga, are clear examples of this.

Unexpected heavy rainfalls caused by climate change have caused huge disruption, damage and loss, including loss of human lives and the destruction of homes, schools, industries, railways, roads, power lines, sewage systems, bridges, etc. The El Niño spells in Dar es Salaam in December 2011 were the worst Tanzania has ever seen.

---

13 (ibid.).
14 (ibid.:31).

### 6. *Negative Effects on Wildlife and Tourism*

Due to increased water shortages, large habitats, ecosystems and national parks which sustain diverse fauna and flora are under increased risk, including animals being threatened by extinction. This especially affects water-dependent animals such as hippopotami, crocodiles, buffalos and elephants, as well as bird species such as flamingos, which are either migrating to other countries in search of water or are vanishing. This situation causes endless conflict in human–wildlife relationships, especially around national parks and wildlife zones.

Climate change is a constant threat to the beauty of many places in Tanzania, including mountains, lakes, craters, hills, beaches, and coral reefs. It is also a threat to the tourism industry. It has been scientifically proven that Mount Kilimanjaro lost 80% of its ice cover between 1912 and 2005. It is envisaged that the melting of the ice on this mountain, coupled with sea-level rise which will submerge small islands and destroy beaches and coastal infrastructure such as hotels, will reach catastrophic proportions in the near future.[15]

### 7. *Long-term Negative Invisible Effects*

Besides the quantifiable effects of climate change on humans, there are also deep-seated psychological and socio-pathological effects which are just as real, but are often overlooked. These range from severe sound pollution to long and unpleasant traffic jams in most of the big cities of Tanzania, particularly in Dar es Salaam. Most city dwellers spend more time commuting to and from their place of work than they do sleeping. As a result, many people suffer from psychological disturbance and are becoming increasingly tense, impatient, selfish and unsympathetic. They are also becoming less careful. This has a direct and indirect negative impact on day-to-day relationships in the familial, societal, medical, professional, academic and religious spheres.

15 (ibid.:23).

## II. Regional and Global Outlook and Effects

Recent findings indicate that, in East Africa, atmospheric dust is one of the key factors contributing to the subregion's climate variability and change:[16]

> Dust storms over the eastern plains of Somalia, northern Kenya, northern Sudan and Ethiopia are common phenomena through most of the year. Dust is one of the least understood components of the Earth's atmosphere and it may have a greater importance for climate change than has been realised up until now.

Increased global temperatures are causing the world's fauna and flora to vanish more rapidly than ever before. Professor John van Klinken of the University of Groningen in the Netherlands claims that, between 1880 and 1950, one animal species became extinct per year; in 1989 it was one species per day; by 2000, it was one species per hour.[17] Worse still, he shows that within 50 years from now, 25% of animal and plant species will have disappeared due to the effects of global warming and climate change.[18]

Franz Alt et al. argue that, due to wanton global climate injustices, the availability of quality water for human domestic and industrial use is at stake.[19] It is affirmed that, today, chemical industries worldwide produce more than 116,000 types of chemicals, including agro-chemicals and pesticides. These have far-reaching toxic effects on our groundwater and on all forms of life.[20]

From the perspective of the developing world, Franz Alt observes that, every year, due to water pollution and severe restrictions on fresh water, about two million people suffer from malaria, 4.6 million children under five suffer from acute cholera, 50 million Africans are confronted with the potential risk of contracting river blindness, 200 million people worldwide suffer from bilharziasis, and one billion people suffer from an acute pandemic of dysentery.[21] All these are the direct causes, effects and consequences of climate injustices. The diverse visible and invisible consequences of climate change and injustices should serve as a 'wake-up call' for all of humanity. We are one common human species. Consequently, we are called to live in a just and sustainable relationship with each other for the benefit

16 UNEP (2006:647).
17 Cited in WCC (2005:41).
18 (ibid.).
19 Alt et al. (2002:148).
20 (ibid.).
21 (ibid.).

of all humans and non-humans today, tomorrow and in the future. This clearly makes climate change and sustainability justice issues par excellence. In Section C below, the deep-seated philosophical, ideological, political, psycho-sociological, socio-economic and technological root causes behind climate change and injustice as a whole are identified.

## *C. Anthropological Perspectives*

### *I. Philosophical-ideological and Policy Root Causes and Lifestyles*

#### *1. The Anthropocentric Life View*

*Anthropocentric* is derived from the Greek *anthropos*, which connotes the human person or being. Consequently, *anthropocentrism* is a radical philosophical life view which considers the human person as the nucleus, centre and key determinant of life, human actions and everything.[22] This life view exploits the rights of creation (or the planet) and places human goals and whims at the forefront. Among others, Immanuel Kant (1724–1804) is considered an ardent proponent of this life view.

#### *2. The Biocentric Life View*

According to this life view, which is based on the term *bios*, which is Greek for "life", animal and plant rights are accentuated and safeguarded. Nobel Prize winner Albert Schweitzer (1875–1965) pioneered this view. As a radical life view centred on the *cosmos* (Greek for "natural", thus "physical world" or "creation"), biocentrism places the rights of the natural world above the rights and dignity of humans (*anthropos*).

#### *3. The Cosmocentric Life View*

This life view places the rights of the physical world (*cosmos*) above all, while downplaying those of humans, plants and animals. A moderate, rational, anthropocentric life view would take into account the intrinsic inter-

22 Msafiri (2007:60).

relationship and interdependence between humans (*anthropos*) and other life forms (*bios*), including nature (*cosmos*).

### 4. The Pathocentric Life View

*Pathocentric* originates from the Greek *pathos*, meaning "pain" or "suffering". The Australian scholar Peter Singer is the main proponent of this radical life view, which exaggerates the right and roles of animals over those of humans, plants and nature Singer considers a purely carnivorous life style as one of the key factors behind ecological destruction. Briefly, it has far-reaching impacts, both direct and indirect, on the environment as a whole.[23]

### 5. The Hyper-post-modern Consumerist Life View and Lifestyle

As far as the insatiable craving for the endless production and consumption of material things is concerned, humans are persistently being oriented into more of a 'having' than a 'being' culture. The Cartesian philosophical life view of *Cogito ergo sum* ("I think, therefore I am") has radically been displaced by "I consume, therefore I am", or "I buy, therefore I am".[24] This has led to massive post-modern environmental risks and threats.

This post-modern consumerist syndrome is embedded in powerful theories, principles and views, of which I have identified the following:

- the You-Need-More-Than-One-Fashion Theory
- the Use-Once-and-Throw-Away Principle
- the Limitless-Technological-Advancement Life View. This entails a Constant-Commodity-Transformation-And-Betterment Life View, e.g. smartphones, ipads, high-definition TVs and hybrid cars.

A hyper-consumerist society is unsustainable, and it exacerbates ecological and climate injustices and burdens. It is deeply rooted in human selfishness, self-interest and greed, which drives the insatiable craving for material possessions.

23 (ibid.:70).
24 Msafiri (2008:54–55).

*6. The Curse of the Champagne Glass Economy[25] Model and Analogy*

In the so-called Champagne Glass Model of the world economy, Figure 1 shows that, as far as global resources (both human and natural) are concerned, 20% of the world's population (i.e. the so-called First World) has amassed 83% of the world's resources, while 20% of the world's population (i.e. the so-called Second World) has access to 10% of those resources, and the remaining 60% (i.e. the so-called Third World) has access to about 7% only.

***Figure 1: The Champagne Glass Model of the World Economy***

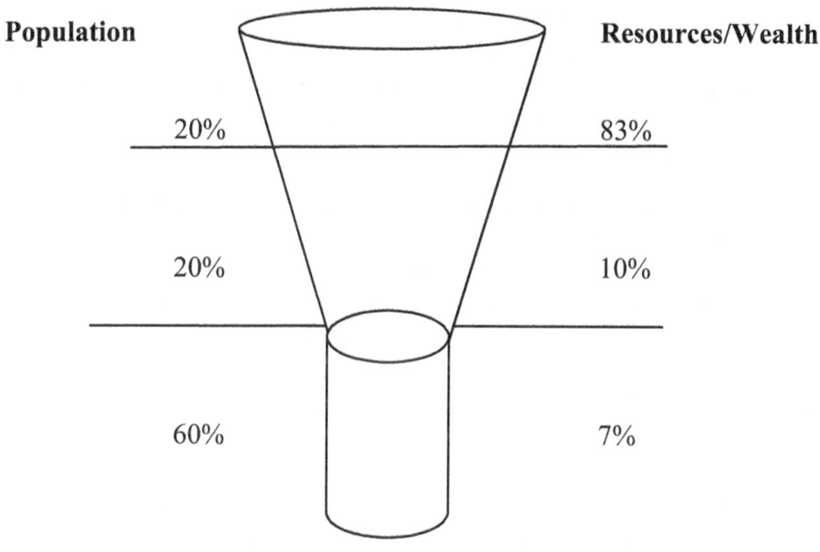


Source: Justice, Peace and Creation Commission (WCC 2005:10)

Undoubtedly, the reality behind this model not only systematically manipulates and destroys life forms and resources, but also paralyses the planet's ability and mechanisms to regenerate resources and ensure their long-term sustainability.

25 (ibid.:10).

### *7. The Inherent Weaknesses of Environmental and Climate Policy Paradigms*

These can be summarised in three models which fail to address global climate injustice and sustainability challenges, as follows:

- The 'Greedy Jackal' Climate and Sustainability Policy Model: This represents a landscape with complex issues on climate injustices and sustainability, but exploitation continues regardless of sustainability imperatives.
- The 'Ignorant Ostrich' Climate Policy Model: This model represents exploitative and selfish interests at the expense of the rights of plants, animals and all the planetary resources, at present and in the future.
- The 'Busy Bee' Climate and Sustainability Policy Model: This represents a sensitive policy which is very keen to deal with the challenges of climate injustice and sustainability, but it lacks focus and does not result in real or deep change.

Using the above-mentioned paradigms, the following logical but candid observations can be made regarding eco-policy and sustainability weaknesses.

The global response hitherto largely lacks global synergy, common understanding, vision and true commitment. It does not promote inclusivity or solidarity with nature. For instance, greenhouse gases go beyond their producer and geographical boundaries. They roam from one country to another, from one continent to another, and eventually engulf the entire planet.

Climate justice and sustainability policies, both local and global, lack a holistic and transformative framework, especially in identifying their deeper causes. Furthermore, they lack in-depth and holistic approaches, particularly in analysing and responding to the 'unknown unknowns', or unquantifiable deeper causes, behind climate injustices, such as greed, selfishness, insensitivity and wanton indifference.

There is frequently a lack of common and coordinated, long-term, value-based climate change laws and policies. There is also a lack of strategic rights from the national, regional, continental and global perspectives. Referring to Hans Jonas, Jeffrey Sachs argues that "we need a whole new ethic for the future ... Futurology was once mocked as pseudoscience. Yet we must make it operational, at least within the boundaries of our understanding and capacity."[26]

From a global perspective, it is true that –[27]

> … our response to climate injustice and sustainability challenges is remarkably ignorant and short-sighted, which will undoubtedly lead to disaster. Of course, worse than a death wish has been at play: the greed of powerful vested interests.

## 8. *The Negative Effects of the 'High-speed Maniac' on Climate Integrity and Sustainability*

The so-called high-speed culture of today is essentially energy-intensive: the higher the speed, the more the energy is consumed. For instance, Wolfgang Sachs claims that a "bicycle trip over 16 kilometres needs 350 calories of energy equivalent to a bowl of rice[;] a car trip over the same distance on the other hand may consume up to 18,000 calories."[28]

Speed is considered a virtue, slowness a vice. This 'high-speed' culture translates not only into more energy and non-renewable resources consumed, but also into more greenhouse gas emissions from vehicles, industry, aeroplanes, etc. Ironically, despite the fact that cars remain the major cause of environmental pollution, people continue to buy them.

Philip Vinod Peacock remarks that the increase of pollution caused by 'high-speed' societies means a decrease of biodiversity, as species are being killed by pollutants that enter the soil, air and water.[29] For instance, penguins have been found to be contaminated by DDT[30] and PCBs,[31] even though neither of these is being used within hundreds of miles of the birds' whereabouts.[32] In the same vein, Jeremy Geedom remarks that the car –[33]

> … is emblematic of the human enterprise that is killing off so many species today. Many scientists are saying that biological diversity is declining at a dangerous rate. Meanwhile the artificial diversity of machines explodes as we humans repopulate with creatures of our own invention.

26 Jonas (1985, cited in Sachs 2011:176).
27 (ibid.:175).
28 Sachs (1995:14).
29 Peacock (2011:77).
30 Dichlorodiphenyl-trichloro-ethene.
31 Polychlorinated biphenyls.
32 Peacock (2011:77).
33 Geedom (1989, cited in Peacock 2011:77).

### 9. *The Quantitative, Mathematically-oriented Mechanistic Approaches to Climate, Development and Sustainability Issues*

Western (classical) formulae and views are profoundly compartmentalised and dualistic. They often overlook the holistic realities and truths inherent in the deep interconnectedness and interdependence of all creation. They cannot measure ethical and value-based humano-ontological altruisms such as dignity, welfare, well-being, and happiness. For this reason, John M. Itty, for instance, calls for a radical rethinking regarding the inherent weaknesses of the current neo-liberal formulae in measuring economic development, particularly by using the gross national product (GNP) and/or gross domestic product (GDP) paradigm.[34]

Itty claims that the GDP increases even when human disasters occur (e.g. money spent to repair the World Trade Center in New York. Ironically, the GDP grows even as the environment is being constantly damaged. This is because it gives rise to a lucrative opportunity for the economically powerful to exploit and oppress the poor, who are at the Bottom of the Pyramid.[35] Hence, a need exists for a radical paradigm shift to value-based, qualitative approaches and indicators.

### 10. *The Commercialisation of Climate Change Non-governmental Organisations and Conferences Worldwide*

As the axiom goes, "The business of business is business".[36] Some organisations and bodies that deal with crises related to climate and sustainability, both locally and globally, put more emphasis on self-interest and profit than on anything else. As I have pointed out in *Globalisation of Concern II*, most local and international conferences and symposia take place in academic halls for scholars, gurus, and special delegates and politicians only.[37] Worse, there is an ever-growing tendency to conduct such 'lucrative' events like the

34 Itty (2007:27–28).

35 (ibid.).

36 "Widely attributed to Friedman, and sometimes cited as being in his work *Capitalism and Freedom* (1962) this is also attributed to Alfred P. Sloan, sometimes with citation of a statement of 1964, but sometimes with attestations to his use of it as a motto as early as 1923"; cited from Wikipedia, http://en.wikiquote.org/wiki/Friedman,_Mi lton, last accessed 29 January 2013.

37 Msafiri (2012:43).

Conferences of the Parties to the United Nations Framework Convention on Climate Change (UNFCCC), the Olympic Games, and other mega-events in five- or six- or ten-star luxury palaces and beach resorts, very far away from the 'real' world of poor people and direct victims. Again, climate and sustainability rights will not be regained in such locations but in the very hearts of the good and willing populace.

The current mathematical, legalistic and mechanistic models and alternatives to mitigation and adaptation, used particularly by the greatest polluters (the West) of the Developing World – particularly sub-Saharan Africa – are neither effective nor exhaustive. These include, to mention a few, the Carbon Development Mechanism (CDM), the Prototype Carbon Fund (PCF), 'The Polluter Pays' Principle (PPP) and Reducing Emissions from Deforestation and Forest Degradation (REDD). All these have their strengths and weaknesses. They place a price tag on carbon, i.e. a quantitative mathematical solution to a qualitative, ethically centred problem. In conclusion, I concur with the Intergovernmental Panel on Climate Change's observation to policymakers that Africa remains –[38]

> … one of the most vulnerable continents to climate change and climate variability, a situation aggravated by the interaction of 'multiple stresses', occurring at various levels, and low adaptive capacity (high confidence). Africa's major economic sectors are vulnerable to current climate sensitivity, with huge economic impacts, and this vulnerability is exacerbated by existing developmental challenges such as endemic poverty, complex governance and institutional dimensions; limited access to capital, including markets, infrastructure and technology; ecosystem degradation; and complex disasters and conflicts.

## *D. Ethics*

### *I. Rediscovering Ethics and the Value of Values for Climate Justice and Sustainability: Key Principles and Norms for Sub-Saharan Africa*

#### *1. The Principle of Care and Compassion*

This value-based, ethical principle calls for us all to avoid the current insensitivity and 'business as usual' attitude. Care and compassion for the planet and awareness of climate change should necessarily "avoid the po-

38 Boko et al. (2007:435); cited in Ruppel (2012:34).

tential dangers of mal-development of the human person".[39] Briefly, the values and virtues of true empathy and moderation need to be rediscovered and practised.

### 2. *The Principle of Globalisation of Concern*

A new ethos and values-based commitment to climate and sustainability is the foundation of this principle. It emphasises the values of human responsibility with and for nature. I advocate this life- and values-based principle which underscores the responsibility of humanity in the collective call or quest to globalise values, virtues and an ethos for human life and the planet using a profound, proactive and preventive approach. It underlines the axiom that "When good people do nothing, evil increases".

### 3. *The Principle of Fairness and Equity*

This is an ethical value which emphasises and demands not only equal treatment regarding the use and distribution of resources, but also proper and viable stewardship of the planet's resources, present and future. *Equity*, however, does not mean *equality*.

### 4. *The Principle of Personality*

This is a fundamental ethical value. It emphasises the inalienable dignity of every human to take into account his/her entire life without threatening the basic and necessary means for a dignified life. Furthermore, it refers to the unique intrinsic dignity, centrality and interrelationship of the human person with the rest of the created world. Hans Kueng aptly summarises this view by affirming that "being human must be the ethical yardstick for all economic actions".[40]

---

39 Hentsch & Shanata (2011:206).
40 Kueng (2009:5).

### 5. *The Precautionary Principle*

This is a key ethical principle which stresses a conscientious and proactive mindfulness in attitude and lifestyle, particularly in avoiding environmental destruction and depletion of the earth's non-renewable resources. It entails the prevention principle and calls for a systemic, in-depth, eco-climate and resource-impact assessment so as to prevent and avoid the worst-case scenarios of eco-destruction and climate change.

### 6. *The Principle of the Common Good*

This is a value-based principle which emphasises holistic approaches in particular, and calls for every individual's well-being. It fosters the new culture of solidarity, inclusivity and care, particularly in promoting socio-economic, cultural, environmental, planetary and technological welfare and integrity. This means fighting against human greed, self-interest and insensitivity to others, to climate, and to the future as a whole. It involves a radical shift from an 'I' culture to a 'we' culture, from exclusivity to inclusivity, and from equality to plurality and diversity.

### 7. *The Principle of the Value of Values*

This is a new ethical paradigm which tries to rediscover the indispensable role of value prioritisation, especially as a qualitatively viable and credible solution to the current climate and sustainability crises. Values do not simply fall from the skies: they are formed and internalised. This principle entails an ensemble of ethical and moral values, particularly faith, hope, prudence, fortitude, agape, justice, communality, integrality, a pro-life stance, peace, trust, solidarity, partnership, subsidiarity, transparency, honesty, moderation, fairness (justice), conversion and forgiveness.

This 'value of values' approach, especially in responding to the climate and sustainability crises, goes beyond the current quantitative, mathematical approaches, solutions and alternatives. It should be prioritised because it provides a more credible and effective alternative response than has been put forward to date.

### *8. The Principle of Deep Change and the "Middle Path"*[41]

This is one of the most important principles in addressing today's climate and sustainability challenges. Deep change goes beyond ordinary arithmetic and/or geometric change. Robert E. Quinn argues that "deep change differs from incremental change in that it requires new ways of thinking and behaving. It is change that is major in scope, discontinuous with the past and generally irreversible."[42]

In Judeo-Christian philosophy, *deep change* refers to the Greek term *metanoia*, or total conversion. Deep change is necessarily based on the virtue of moderation, especially towards climate and sustainability issues as a key to human fulfilment and true happiness.

In this regard, Jeffrey Sachs makes the following value-based observation:[43]

> The essential teaching of both Buddha and Aristotle is the path of moderation pursued through life-long diligence, training and reflection. It is easy to become addicted to hyper consumerism, the search for sensory pleasures, and the indulgence of self-interest, leading to a brief high but long-term unhappiness.

Moderation in the use of the resources of the planet demands mindfulness of self, others, nature, and the future.

### *9. The Principle of Efficiency*

This is a value-based life view which tries to promote the balanced use and reuse of both renewable and non-renewable resources. It supports such approaches as the Reduce, Reuse and Recycle (3Rs) view, and the Planet, People and Profit (3Ps) model. The principle of efficiency is profoundly interconnected with the other value-based ethical approaches to climate justice and sustainability, like the savings, the rotation and the reversibility principles. According to Ernst Ulrich von Weizsaecker et al., the principle of efficiency has to start with a true efficiency revolution. It entails seven key foundations, namely better quality of life, less pollution, ethically based

41 Sachs (2011).
42 Cited in Anderson (2001:156).
43 Sachs (2011:162).

profit, resource reuse mechanisms, international security, justice, and the promotion of job opportunities.[44]

### *10. The Principle of the 'Golden Rule' (Mt. 22:37–39) and the Right to Food and Water (Mt. 25:34–36)*

The 'Golden Rule' remains the centre and zenith of all value-based approaches towards the Creator, fellow humans and the created world. It underscores both the verticalist and horizontalist trajectories, particularly of the human person towards God, environment, climate and sustainable living. Patricia Mische summarises it as follows:[45]

> [The love of] one's neighbor also includes respecting their need for and rights to water, food, shelter and adequate resources. By further extension, one can see that loving one's neighbor includes respect for the rights and need of future generations. Those yet to come depend on our proper stewardship resources on a finite planet ....

Furthermore, analogically, Mt. 25:34–36 encapsulates broad-spectrum ethics of climate justice and sustainable growth. It calls for food for the hungry, clean water for the thirsty, solidarity for and with the alien, empathy and true care for the sick and the marginalised, and total inclusion.[46] In its strictest sense, these ethics include not only the deontological dimensions (duties) of humans to fellow humans and nature, but also the teleological exigencies (results/consequences) of human action and inaction to the present and future generations of humans, plants, animals and the cosmos. The right to food and water helps to foster freedom from fear, want and need. Having identified the key facets which comprise the value-based principles for climate justice and sustainability, I shall now make some concluding remarks.

## *E. Conclusion*

Undoubtedly, the quest for truly value-based ethics of climate justice and sustainability, locally and globally, cannot be exaggerated. The following

44 Von Weizsäcker et al. (1995:21–23).

45 Hessel & Ruether (2000:594).

46 Stueckelberger (2009:20–21).

conclusions need special attention and priority as they constitute new climate-related ethics.

Firstly, climate and sustainability challenges are ethical issues per se. They cannot be adequately addressed through the Western (classical) mechanistic, quantitative approaches. Integral justice, particularly for climate and a sustainable world, must necessarily be built on deep change, ethos and *habitus* (habit), as these pillars remain key to behavioural change, individually and collectively. Therefore, there is an important role of the value of values as a key driver and indicator for present and future climate justice. This calls for a change of heart, mind, lifestyles and priorities. Value-based approaches revive the deeper aspects of human consciousness, which remind us that this earth is our only home, and that its life forces of millions of years cannot simply be allowed to be destroyed in a matter of decades, or even centuries.

Secondly, in the realm of politics, politicians maximise power above all else; in the sphere of economics, entrepreneurs maximise profit, the motto being 'business as usual' or 'the business of business is business'. But in life and ethics, we are obliged to maximise values above all else, i.e. the value of the values for climate justice and sustainable growth against the vices of greed, power, selfishness, short-term consumption lifestyles, lust, etc. Briefly, value based approaches and systems provide moral benchmarks for proper reflection and action.

Thirdly, there is a greater need than ever before to revisit the nexus and interdependency of climate on the one hand, and environmental justice and environmental peace on the other. Indeed, they are two sides of the same coin. Environmental peace necessarily entails not only peace as the fair distribution of natural resources, but also peace as holistic peace – socially, economically, politically, existentially – and that both locally and globally.

Finally, there is an urgent need to reaffirm ethical values, norms and life views for universal responsibility. This demands that every person should live with a deep sense of responsibility towards climate and sustainability issues as we are all citizens of this one world and home, regardless of our different nationalities, genders, cultures or statuses. Therefore, we need to synergise our various abilities and resources – both human and natural – for a life of community and dignity for thousands of years to come.

## *References*


Alt, Franz, Rudolf Bahro & Marco Ferst, 2002, *Wege zur ökologischen Zeitenwende*, Berlin, Edition Zeitsprung.

Anderson, Walter Truett, 2001, *All Connected Now: Life in the First Global Civilisation*, Boulder, CO, Westview Press.

Geedom, Jeremy, 1989, The Fantasy Machine, *New Internationalist*, 195 (May).

Global Climate Adaptation Partnership, 2011, *The Economics of Climate Change in the United Republic of Tanzania*, available at http://www.economics-of-cc-in-tanzania.org/images/Final_report_launch_vs_3.pdf, last accessed 29 January 2013.

Hentsch, Ariane & Premawardhana Shanata (Eds), 2011, *Sharing Values*, Geneva, Globethics.net.

Hessel, Dieter & Rosemary Ruether (Eds), 2000, *Christianity and Ecology*, Cambridge, MA, Harvard University Press.

Itty, John M, 2007, *Illusions About Economic Growth*, Kerala, Vichara Publishers India.

Kueng, Hans, 2009, *Manifesto for a Global Economic Ethic*, Tübingen, Global Ethics Foundation.

Mische, Patricia, 2000, The Integrity of Creation, in: Hessel, Dieter & Rosemary Ruether (Eds), *Christianity and Ecology*, Cambridge, MA, Harvard University Press, 591–602.

Msafiri, Aidan G., 2007, *Towards a Credible Environmental Ethics for Africa: A Tanzanian Perspective*, Nairobi, The Catholic University of Eastern Africa.

Msafiri, Aidan G., 2008, *Globalisation of Concern Vol. I*, Dar es Salaam, Dar es Salaam University Press.

Msafiri, Aidan G., 2012, *Globalisation of Concern Vol. II*, Geneva, Globethics.net Focus.

Peacock, Philip Vinod, 2011, Challenging the Idols of Speed, in: *World Communion of Reformed Churches* 61, 1 (March).

Ruppel, Oliver C., 2012, Climate Change Policy Positions and Related Developments in the AU and SADC, *SADC Law Journal* 2 (1), 14–35.

Sachs, Jeffrey, 2011, *The Price of Civilisation: Economics and Ethics After the Fall*, London, The Bodley Head.

Sachs, Wolfgang, 1995, If Wishes Were Horses: Desire and Democracy in the History of Transport, *The Ecologist* (Asia), January–February.

Stueckelberger, Christoph, 2009, *Das Menschenrecht auf Nahrung und Wasser: Eine ethische Priorität*, Geneva, Globethics.net.

UN/United Nations, 1992, *The Global Partnership for Environment and Development: A Guide to Agenda 21*, New York, UN Publications.

UN/United Nations, 1994, *The United Nations Earth Summit: Agenda 21: The United Nations Programme of Action from Rio*, New York, UN Reproduction.

UNEP/United Nations Environment Programme, 2006, *UNEP, Africa Environmental Outlook 2*, Nairobi, Progress Press Ltd.

URT/United Republic of Tanzania, 2011, *National Climate Change Strategy and Action Plan*, Dar es Salaam, Vice President's Office, Division of Environment, URT.

Von Weizsäcker, Ernst Ulrich, Amory B. Lovins & L. Hunter Lovins, 1995, *Faktor Vier: Doppelter Wohlstand – Halbierter Naturverbrauch*, Munich, Droemer Knaur.

WCC/World Council of Churches, 2005, *Alternative Globalisation Addressing People and Earth (AGAPE)*, Geneva, WCC Publications.

# PART VI: CLIMATE CHANGE, SECURITY AND THE MIGRATION DEBATE

# 22
# Science, Facts and Fears: The Debate on Climate Change and Security

*Marcel Leroy & Fana Gebresenbet*

*Abstract*

A broad range of studies have attempted to identify pathways through which climate change could contribute to conflict. Resource scarcity and climate-induced migration are two avenues which have received ample attention. The link between precipitation and conflict has also been examined, and some excellent work has been done regarding the impact of temperature on conflict. On balance, however, there is mixed evidence about the security implications of climate change. Furthermore, in many writings on the subject, scientific analysis takes a back seat to conjecture.

Studying the consequences of climate change requires examining a broad social context, including the role of governance. Recognising the wide range of social mechanisms and intellectual premises underlying the response of various cultures to new challenges is also essential. The multitude of choices in the process of adaptation reduces the scope for identifying 'standard' mechanisms through which societies react to climate change. Correspondingly, research methods and underlying premises need to encompass the full range of options through which cultures respond.

However, in scientific research – including the physical, social and life sciences – pressures to conform with prevailing thinking may restrict the scope of what is being investigated and, hence, limit the conclusions being advanced. In addition, many scientists cannot escape the ideological bent which shapes their worldviews and ends up pervading their work. Keeping in mind the perspectives of relativity and non-linearity, it is rarely warranted to 'predict' developments over a long-term horizon.

The 'securitisation' of the climate change debate, and predictions of dire consequences for future domestic and international stability, have contributed to raising the debate – and international negotiations on how to address climate change – to the realm of high politics. Identifying climate change as a security problem has encouraged state-sponsored reflection on possible future scenarios, and on measures to be taken to mitigate risks to global security.

The global climate change debate is as much about perception as it is about science. Fear of the unknown has allowed doomsday sayers to dominate headlines, linking climate change with many evils and catastrophes. Alarmist statements have come from world leaders as well as from academics. Rather unusual alliances of interests appear to have been formed, with senior military officers drawing conclusions which sound rather similar to those of environmental activists.

This article places the debate on the security impact of climate change in a broad context, with a critical review of the literature produced by academics, government agencies, think tanks and non-governmental organisations. It analyses the assumptions and conclusions of this work, and makes recommendations on how to advance our understanding of the security implications of climate change.

## *A. Introduction*

Global warming has moved to the top of the international agenda since the early 1990s. Although the warming effects of increasing carbon dioxide levels in the atmosphere had been understood since the late 19th Century,[1] there was no widespread fear that increasing these levels would engender drastic changes – physical or social – in the further development of the so-called blue planet. In fact, following a period of cooling from the 1940s to the early 1970s, some climate scientists drew a linear extrapolation to conclude that the world might be sliding into a new ice age.[2]

This article looks at possible links between climate change and conflict. It addresses the science, perceptions and myths surrounding the current debate on security implications of global climate change. The article also reviews government-sponsored studies, and comments on some academic research dealing with the climate change–conflict nexus. While there is no intention to limit the geographic scope of the analysis, specific examples refer mostly to Africa, which is often seen as the 'canary in the coalmine'

---

1 Arrhenius (1896:237–239).

2 For a good overview of contemporary scientific and popular literature, see Morano (2009).

of climate security – the first continent expected to experience the full effect of climate change on political and economic stability.

### B. The Science of Global Warming

The trend toward global warming is now unmistakable. At the end of 2007, the World Meteorological Organization (WMO) concluded that the 11 years with the highest mean global temperatures – since reliable methods to collect weather data were put in place globally in the second half of the 19th Century – had occurred since 1995.[3] The year 2009 was the second warmest on record, and the decade 2000–2009 the warmest ever.[4] The WMO indicated that nine years of the 2001–2010 period were among the top ten warmest on record. In spite of some anomalies, such as a cold winter in north-western Europe, 2010 was the warmest year on record, closely followed by 2005; 2011 was the 11th warmest, as well as the warmest compared with other La Niña years.[5]

The United States (US) National Oceanic and Atmospheric Administration (NOAA) reported that the first eight months of 2010 tied the same period in 1998 for the warmest combined land and ocean surface temperature on record worldwide.[6] This trend continued into 2012. The US National Snow and Ice Data Center in Boulder, Colorado, reported on 16 September 2012 that Arctic ice covered about 3.4 million km$^2$, which is about three quarters of a million square kilometres below the previous record low set in 2007, and 50% below the 1979–2000 average.[7] It has also been reported that the mean annual temperature of the North Sea rose by 1.7°C from 1962 to 2012, as measured in the vicinity of the German island Helgoland by the Alfred Wegener Institute.[8] NOAA also identified the summer of 2012 as the third hottest on record.[9]

It is now almost universally accepted that human action since the Industrial Revolution has triggered a greenhouse effect. Moreover, it is likely to

---

3 WMO (2007).
4 NASA (2010).
5 WMO (2011).
6 NOAA (2010).
7 New York Times (2012); De Standaard (2012a).
8 De Standaard (2012b).
9 NOAA (2012).

take decades before the release of greenhouse gases (GHGs) – including methane, which accounts for about 20% of the total observed warming[10] – can be stabilised. Even if global efforts succeeded in rolling back GHG emissions, the warming effect of past emissions would linger for decades.

GHGs already released will eventually break down, but their residence time in the atmosphere varies greatly. While methane breaks down after 12 years, the mean breakdown time for nitrous oxide is 114 years.[11] The atmospheric lifetime of the bulk of carbon dioxide released through burning fossil fuels has been estimated at 300 years, but about a quarter of this carbon dioxide lingers for thousands of years.[12] Furthermore, the oceans have absorbed about 84% of the total heating of the earth over the last 40 years,[13] being retained for decades and creating a 'thermal flywheel effect' which could also counteract any cooling tendencies which might occur.

### C. *Facts and Fears*

The global climate change debate is at least as much about perception as it is about science. Fear of the unknown tends to take hold of people's psyche, overtaking reason and caution. At times, alarmist language is used with quasi-religious overtones.[14] Doomsday sayers have dominated the headlines, linking climate change with many evils and catastrophes.

Decision-makers, academics and journalists have made linkages of varying intensity between climate change and conflict. United Nations (UN) Secretary General Ban Ki Moon stated that "the Darfur conflict began as an ecological crisis, arising at least in part from climate change", with a progression from drought to scarcity and violence.[15] French President Nicolas Sarkozy was more explicit when he addressed a meeting of ministers from the world's major economies in April 2008:[16]

> In Darfur, we see this explosive mixture from the impact of climate change, which prompts migration by increasingly impoverished people, which then has

10 Lassey (2007:120).
11 GHG Management Institute (2010).
12 Archer (2005:3).
13 Barnett et al. (2005); Levitus et al. (2005).
14 Hulme (2008:7).
15 Moon (2007).
16 AFP (2008).

> consequences in war. If we keep going down this path, climate change will encourage the migration of people with nothing towards areas where the population do have something, and the Darfur crisis will be only one crisis among dozens of others.

In 2007, former United Kingdom (UK) Foreign Secretary Margaret Beckett, at the first-ever UN Security Council debate on the impact of climate change, stated that –[17]

> [r]ecent scientific evidence has given us a picture of the physical impacts on our world as our climate changes. And those impacts go far beyond the environmental. Their consequences reach to the very heart of the security agenda.

The language of climate change as a security threat has been applied not only to highlight the risks attributed to the phenomenon itself, but also against the polluters themselves. At an African Union debate in early 2007, President Yoweri Museveni of Uganda called GHG emissions an "act of aggression" by the developed world against the developing world.[18] And at the 2007 UN Security Council debate on the impact of climate change, the Namibian representative Kaire Mbuende called GHG emissions tantamount to "low intensity biological or chemical warfare".[19]

It must be acknowledged that the 'securitisation' of the climate change debate – and international negotiations on reducing GHG emissions – has allowed this debate to be invested with a greater sense of urgency, raising climate change to the realm of high politics and creating space for serious commitments. This has assisted climate campaigners to mobilise support for strong action on mitigation and adaptation. Identifying climate change as a security problem suggests that it is an issue that warrants "a policy response commensurate with war, in effort if not in kind".[20]

## D. Climate Change and Security

The ability to reduce complex matters to bite-size morsels is a valuable skill for politicians to possess. By simplifying the questions being addressed, political leaders may be able to offer clear choices and facilitate decision-making. However, governments have increasingly recognised the complexity of

17 UNSCDPI (2007).
18 Brown & Crawford (2008).
19 UNSCDPI (2007).
20 Barnett (2001).

the climate change/security nexus and have also commissioned rather elaborate studies in order to define the implications of climate change.

A study commissioned by the US Department of Defense[21] and carried out by 11 retired military officers came up with some rather sweeping conclusions. These include the contentions that –[22]

- "projected climate change poses a serious risk to America's national security"
- "climate change will provide the conditions that will extend the war on terror", and
- "projected climate change will add to tensions even in stable regions of the world".

The authors of the US study recognise that the impact of climate stress and extreme weather events is likely to be more severe under conditions of weak governance. They also argue, however, that climate change has the potential to trigger "multiple chronic conditions, occurring globally within the same time frame". Their judgement sounds rather ominous:[23]

> Overall, climate change has the potential to disrupt our way of life and force changes in how we keep ourselves safe and secure by adding a new hostile and stressing factor into the national and international security environment.

The German Advisory Council on Global Change, composed of nine scientists from Germany and other European countries, and working under a mandate of the German federal cabinet, was more modest and nuanced in its conclusions.[24] It focused on how governance and political systems are likely to be affected by climate change, concluding that states with weak governance will be particularly vulnerable. The need for addressing the impact of climate change will place additional demands on governments, pushing countries with weak steering and problem-solving capacities further toward fragile statehood. Also, by imposing economic costs, such as reducing agricultural yields and triggering migratory movements, climate change will reinforce obstacles to development.[25]

Furthermore, the German study attempts to identify linkages between the risk for conflict and climate-induced environmental changes. These include

21 CNA (2007).
22 (ibid.:6–7).
23 (ibid.:6).
24 WBGU (2007).
25 (ibid.).

degradation of freshwater resources, declining food production, increases in extreme weather events, and environment-induced migration. There may also be a spillover effect of the social impacts of climate change, transcending borders and expanding the geographic range of conflicts. Such risks are exacerbated when accompanied by weak governance and rapid population growth.[26]

A study commissioned by the Danish Ministry of Foreign Affairs and carried out by Oli Brown and Alec Crawford on behalf of the International Institute for Sustainable Development concentrates on West Africa, in particular the country-level security impacts of climate change in Burkina Faso and Ghana.[27] Although the African continent is least responsible for GHG emissions, it may in the end be the most severely affected by climate change through increasingly scarce water, reduced agricultural yields and encroaching deserts. In his foreword to the report, Ib Petersen, State Secretary for the Ministry of Foreign Affairs, says that these developments may cause destabilising population movements, raise tensions over dwindling resources, and tip fragile states towards failure.[28]

In the UK, economist Sir Nicholas Stern – with a team of UK scientists and research centres – prepared a review on the economics of climate change for the Prime Minister and Chancellor of the Exchequer.[29] The review argues that developing countries are particularly vulnerable because of their tropical geography, high population growth, heavy dependence on agriculture, rapid urbanisation, weak infrastructures and lack of resources. The Stern Review concluded that climate change could exacerbate poverty and endanger development in the poorest countries, forcing millions to migrate internally and across borders in search of food and water, thereby worsening prospects for security, education and gender equality. Rising global temperatures will put growing numbers of people at risk of hunger.[30]

The European Commission and the European Union's High Representative for Common Foreign and Security Policy catalogued the different forms of conflicts which may be driven by climate change, as follows:[31]

26 (ibid.).
27 Brown & Crawford (2008).
28 (ibid.).
29 Stern (2006:11–12).
30 (ibid.).
31 EU (2008).

- Conflict over resources as a result of shortage of water, diminishing fish stocks and drops in agricultural productivity
- Economic damage to megacities and coastal zones resulting from rises in sea level
- Environment-induced migration
- Increased instability by overstretching weak and failing states, and
- Pressures on international governance resulting from resentment between those most responsible for climate change and those most affected by it.

A report by the Swedish Defence Research Agency[32] indicates that research on climate change and armed conflicts largely uses a state-based concept of *security*, while discourse on climate change and vulnerability tends to be based on the concept of *human security*.[33] Security analysis centres on risks which may affect a state's energy security, economic security or environmental security. However, the effects of global climate change form additional risks which may affect individuals as well as the state.[34]

Politicians tend to respond to popular perceptions, while government-sponsored studies proceed more analytically. Nevertheless, policies cannot ignore perceptions, which often acquire a quasi-factual status.

### E. International Concerns

Multilateral agencies and international non-governmental organisations (NGOs) with a strong presence on the ground tend to look more at concrete situations and have generally come to nuanced conclusions about the security impact of climate change.

A study carried out by United Nations Environment Programme (UNEP) on post-conflict Sudan acknowledges that many factors that have little or no link to climate, environment or natural resources contribute to conflict in the country.[35] In addition, where environment and natural resource management issues are important, they are generally not the sole cause for tension, but only contributing factors.[36] The study nevertheless posits a strong link bet-

32 Totalforsvarets Forskningsinstitut, known as *FOI*.
33 Mobjörk et al. (2010).
34 (ibid.).
35 UNEP (2007:77–87).
36 (ibid.).

ween the occurrence of local conflict and environmental degradation of rangeland and rain-fed agricultural land in the drier parts of Sudan. UNEP further notes that struggles over the control of scarce resources – including fertile land and water – have been a factor in many conflicts, including those in Darfur and the Middle East. Moreover, a link to natural resources and environment has been found to double the chance for conflict relapse within the first five years of a peace agreement.[37]

A report on the human impact of climate change by the Global Humanitarian Forum – which was set up in 2007 by former UN Secretary General Kofi Annan and ceased operations in 2010 – treats conflicts as complex emergencies, and views climate change as a catalyst or threat multiplier. The report found that, at the time of its writing, evidence linking climate change to conflict was inconclusive.[38] Whereas it seems plausible that scarcity of water might engender conflict over shared water resources, for instance, the evidence is overwhelming that states have in the vast majority of cases addressed these issues through increased collaboration and resource-sharing agreements, with over 200 international water treaties negotiated in the last 50 years.[39]

The Pan African Climate Justice Alliance, an African civil society coalition on climate change and sustainable development, takes a similar view, stating that the relationship between climate change and security is complex, and that it is difficult to anticipate where conflicts may occur or to attribute conflicts directly to environmental changes.[40] The review by the Alliance further finds that competition for food, water, energy and land are possible pathways for climate change to contribute to conflict, as are unregulated migration, destabilised settlements, and an increased tendency to join armed groups as a result of reduced employment opportunities.

The former President of the International Crisis Group, Gareth Evans, notes that climate-induced changes can exacerbate humanitarian and security strains.[41] Nevertheless, he cautions against leaping "into confident predictions about the impact of climate in generating conflict".[42] The interaction between environmental and climate factors with governance and ethnic is-

37 UNEP (2009).
38 GHF (2009).
39 UNDP (2006).
40 PACJA (2009).
41 Evans (2008).
42 (ibid.:1).

sues complicates any attempts at predicting when and where violence will break out. Thus, while climate change can certainly play a role in deadly conflicts, it is highly unlikely to be the sole or primary cause. Migrations and subsequent inter-group rivalry in areas where migrants settle have played an important role in many conflicts that have been described as 'environmental' or 'climate-induced'. It is particularly important, therefore, to understand the way in which climate change may induce migrations.[43]

Saferworld, an NGO with extensive experience in conflict-related research mostly funded by the European Commission, also sees climate-induced migrations as a possible source of insecurity in the regions of origin, transit and destination because of increased competition over already scarce resources and livelihoods.[44] Migrants face double insecurity: it is hard for them to find employment and provide for their basic needs, but they are also held responsible by already established residents for the increased competition for resources, and face the threat of reprisal from them.[45] Tensions between local and migrant communities over access to resources and employment could result in a breakdown in social cohesion and a rise in crime levels.[46]

This mechanism seems to apply to various cases studied by Saferworld in Bangladesh (where rising sea levels have induced migration to urban centres and across the border to India) as well as studies among pastoralists in northern Kenya (where increased migrations are a coping mechanism in periods of adverse environmental conditions, bringing different groups in greater competition over the same or dwindling resources). However, the form which conflicts may take will vary, and each form may be affected differently by climate change; cattle raiding, for example, seems to occur more frequently in the rainy season.[47] In sum, the relationship between different types of conflicts and climate change merits further investigation.[48]

A study commissioned by International Alert points out that, among the arguments for taking action against climate change, perhaps the most compelling are the potential security implications.[49] Yet there is a risk of over-

43 ICG (2010).
44 Saferworld (2008a).
45 BIISS & Saferworld (2009).
46 Saferworld (2008b).
47 CDC & IISD (2009); Saferworld (2009).
48 CDC et al. (2009).
49 Smith & Vivekananda (2009).

stating the conflict dimension in order to convince a lukewarm public – and the politicians who form their governments. Securitisation may also run the risk of promoting rapid and high-cost responses over more cost-effective and sustainable options. Yet the security dimension cannot be overlooked, as climate change is likely to place additional demands on institutions. Fragile states with weak governance structures may be least able to respond, meaning that climate change will further weaken confidence in the social order and erode the stability of these societies.[50]

It can also be argued that diminishing resources will hinder the capacity of states to respond effectively to the challenges posed by climate change. Saferworld's 2007 human security survey in Bangladesh[51] raised many areas of concern relating to the state's ability to provide basic 'freedom from want' (economic, food, health and environmental security), as well as with respect to the effectiveness of the state security sector to maintain peace and safeguard the population from the risk of crime and violence. As Bangladesh is likely to suffer severely from rising sea levels, additional strains will result.

From a conflict analysis perspective, climate change is not in itself a direct cause of conflict. Analysing how climate change will affect security and conflict dynamics is about understanding the "consequences of consequences".[52]

Looking back at the rapid rise of climate change among global priorities and as a security concern, it is clear that some official institutions as well as civil society groups are not averse to endorsing alarming scenarios. Exaggerations of the nature and intensity of the link between climate change and conflict may, in fact, increase the role of those who may be called to address the emerging risks. This may include some rather unexpected bedfellows, such as the military and some rather activist or left-leaning NGOs. If the maximalist view of climate change risks – with the spectre of an increase in the intensity, scope or frequency of disasters and conflicts triggered by climate change – was to form a basis for policy, this might add to the stature of these organisations, as well as to their ability to raise funds.

50 (ibid.).
51 Saferworld (2008b).
52 International Alert (2007).

## F. Can Academic Research Clear the Mist?

Research on the relevance of environmental issues in the realm of security has pre-dated work on climate change. Throughout the 20th Century, a sizable body of literature has developed on the spatial dimension within which political decisions are made and implemented. Sustained and continually refined analyses from the 1930s to the 1970s by Harold and Margaret Sprout recognised the connections between the decision-maker, the "psychological milieu" (the perceived environment upon which the policymaker's reactions are based), and the "operational environment" – being the real world within which the policy is implemented.[53] This analysis is also relevant today with respect to climate security, where ideologically tinted perception addresses the void which science has – at least as yet – been unable to fill.

At its most basic level, climate change is beyond the reach of political processes. It is, however, impossible to separate 'external' or 'physical' changes from their effects, after they interact with other natural and human processes. Desertification is a case in point: while a reduction in amounts of precipitation is undoubtedly a factor, poor farming practices, overgrazing, deforestation, and poor governance relating to water and resources also contribute to advancing deserts.[54] In this sense, climate change aggravates other environmental and resource use problems and complicates the search for an appropriate human response.

In more general terms, climate shifts aggravate environmental stress which, in conjunction with other factors, may lead to violent conflict. However, environmental stress – manifested through different and often interlinked environmental scarcities – is not a direct cause of violent conflict. Hence, the burden of explanation is simply moved further down the causal chain.

Academic exploration of the link between climate change and violent conflict has been placed mainly within the broader context of how resource scarcity (demand-induced, supply-induced or structural) relates to violent conflict.[55] The physical impacts of climate change, such as: increase in temperature; change in seasonality and amount of rainfall; wind storms; sea-level rise; and increases in the frequency and severity of extreme weather events, e.g. drought and flooding, are likely to contribute toward reducing

53 Sprout & Sprout (1968).
54 Leroy (2009:1–7).
55 See e.g. Homer-Dixon (1999:47–49).

the quantity and quality of various renewable natural resources – thus generating supply-induced scarcity.

Various research efforts have attempted to clarify the pathways and processes through which these impacts – and the supply-induced scarcity they generate – lead to violent conflict. A frequently used method is to look for historical analogies to various effects of climate change – including drought, variability of precipitation, extreme weather events and ecologically induced migration – in order to assess whether these phenomena have in the past been correlated with the onset of violent conflict.[56]

However, findings on the links between climate change and conflict differ quite widely. A climate-change-induced decline in agricultural productivity could reduce the opportunity cost of fighting, thus making it more plausible to fight than to till one's land. Furthermore, increasing supply scarcity coupled with ecological marginalisation of the poor, and potential moves by elites to capture valuable resources, will worsen the economic welfare of the general population.[57] Such outcomes are likely to be exacerbated by pre-existing structural differences in access to resources and by property rights regimes, as well as by weak institutions and governance failures.[58]

Collier et al. single out migration and reduced flow volumes of international rivers as the major pathways to conflict.[59] However, previous studies on river basins have concluded that shared water resources promote international cooperation rather than conflict.[60]

For the northern hemisphere, the historical link between climate and conflict has been investigated rather thoroughly. Colder periods of the previous millennium – such as the Little Ice Age which peaked in Western Europe and China toward the middle of the 17th Century – were associated with declining agricultural productivity and increases in food prices, as well as societal tension and the increased likelihood of war. Colder climates also coincided with a higher incidence of war in both China and Europe, as well as with high depopulation rates.[61] Le Roy Ladurie also noted that excep-

56 Buhaug et al. (2010); Busby (2010).
57 Homer-Dixon (1999:177–180).
58 Leroy (2009:361–370).
59 Collier et al. (2008:337–353).
60 See e.g. Elhance (1999).
61 Zhang et al. (2006); Tol & Wagner (2010); Zhang et al. (2007).

tionally cold winters and poor growing seasons in the 1640s and 1650s were associated with social unrest in France and five other European states.[62]

Academic studies investigating the climate change–conflict nexus use climatic parameters as an independent variable and conflict data as a dependent variable. However, different studies use different parameters as climate change indicators: most use rainfall data,[63] while others use temperature[64] and the occurrence of El Niño events.[65] Even the group which uses rainfall does not use the same attributes of rainfall; for example, Miguel et al. used annual rainfall;[66] Levy et al. used the Weighted Anomaly Standardised Precipitation Index;[67] Hendrix and Glaser used standard deviations from the previous year's rainfall;[68] Hendrix and Salehyan used standardised rainfall deviation;[69] and Theisen et al. used meteorological drought as an independent variable.[70] This range of parameters complicates drawing conclusions which might advance the understanding of links between climate change and conflict.

Both reduction and increase in rainfall could lead to conflict. Dry extremes may lead to slow-onset conflict events, the social impacts of which will first be accentuated by migration and economic impacts. Wet extremes, specifically floods, usually result in the swift onset of conflict, due to their immediate destabilising impacts on lives and livelihoods as well as on social and physical infrastructure.[71] Within Africa, rainfall could show either positive or negative shifts due to climate change, depending on the specific geographic region under consideration, while temperature appears to be increasing continent-wide. For African agriculture, temperature increase is linked to a drop in productivity, in effect increasing conflict risk.[72]

Empirical studies report opposing findings at times: while some find a negative correlation between rainfall and conflict risk, others find a positive

62 Le Roy Ladurie (2005).
63 Hendrix & Glaser (2007); Hendrix & Salehyan (2012); Levy et al. (2005); Meier et al. (2007); Miguel et al. (2004); Smith (2012); Theisen et al. (2009).
64 Burke et al. (2009:20670–20674); Zhang et al. (2006, 2007).
65 Hsiang et al. (2011:438–441).
66 Miguel et al. (2004).
67 Levy et al. (2005).
68 Hendrix & Glaser (2007).
69 Hendrix & Salehyan (2012).
70 Theisen et al. (2009).
71 Smith (2012).
72 Burke et al. (2009).

association and still others do not seem to find a statistically significant correlation at all. For example, Miguel et al. found that, through its impact on economic growth, rainfall is strongly negatively correlated with the risk of civil conflict in Africa.[73] Levy et al. found that the onset of high-intensity conflict is strongly related with rainfall deviations, with a one-year lag.[74] Hendrix and Salehyan found that both negative and positive rainfall deviations increase the likelihood of social conflict, the correlation being strongest for positive deviations.[75] A similar finding of higher conflict likelihood in higher rainfall years is also found by Meier et al. in pastoral areas of the Horn of Africa,[76] and by Smith in the whole Horn of Africa region.[77] In pastoral areas, it makes more sense to steal fatter cattle. Moreover, taller grasses in the rainy season provide cover for the rustled livestock. Thus, cattle rustling and consequent pastoral conflict tend to be more common in higher rainfall years. Conversely, Theisen et al. did not find a statistically discernible increase in the risk of civil war in Africa in years experiencing meteorological drought.[78]

Burke et al. found temperature to be a more important determinant than rainfall in influencing the likelihood of conflict.[79] They further found that an increase of 1°C resulted in a 4.5% rise in the incidence of civil war in the same year, and a 0.9% rise the following year. Hsiang et al. found that the probability of new conflicts doubled in El Niño years relative to La Niña years in the tropics between 1950 and 2004, concluding that El Niño has contributed to 21% of all civil conflicts in that period.[80]

Kevane and Gray investigated the importance of climate change in the Darfur conflict, which has often been seen as a causal explanation by journalists and politicians.[81] They found that there had been no significant decline in annual rainfall in the years prior to the outbreak of large-scale con-

---

73 Miguel et al. (2004).
74 Levy et al. (2005).
75 Moreover, while other studies use the Armed Conflict Database from the International Peace Research Institute of Oslo and the University of Uppsala, Hendrix and Salehyan (2012) and Smith (2012) used the new Social Conflict in Africa Database from the University of Texas at Austin to generate their independent variable.
76 Meier et al. (2007).
77 Smith (2012).
78 Theisen et al. (2009).
79 Burke et al. (2009).
80 Hsiang et al. (2011).
81 Kevane & Gray (2008).

flict in 2003, although there had been a structural break to a lower mean in the early 1970s.[82]

Extreme weather events form one category of climate change that is often attributed to nature alone. Certainly, exceptional deluges like the events in Pakistan in July and August 2010 are bound to cause devastation. It is questionable, however, whether climate change alone is at the basis of an increase in their intensity or frequency. Urbanisation – leaving large areas paved over, thus essentially producing more run-off rather than absorbing moisture into the soil – as well as deforestation and poor land use practices no doubt amplified the dramatic course of events, aggravating the consequences for tens of millions of Pakistanis.[83]

### G. *Cause for Alarm?*

Embarking on a critical analysis of the consequences of global warming requires more than scientific data. At least as important is an understanding of the analyst's perspective on how to approach facts and build understanding. This is by no means a straightforward exercise. We are operating in a multilogical world, in which there are many cultural differences and a rich variety of indigenous knowledge starting from differing premises.[84] Moreover, the globalisation of media flows has made it easier to spread authoritative and diverging views, while also exposing us to more sophisticated propaganda and deliberate distortions.

The observable phenomena that have given rise to the scientific concept of *climate change* are themselves difficult to grasp. Meteorology is a fickle science, with minute variations giving rise to very different outcomes in terms of weather prediction. Meteorological observations by Edward Lorenz formed a significant part in the formulation of chaos theory.[85] Furthermore, the locally observed outcomes of global climatic changes – both in measurable climate variations and in terms of the human response to them – are varied and defy neat categorisation.

Since there is a wide range of social and intellectual premises on which cultures base their responses to new challenges, the scope for identifying

82 (ibid.).
83 Homer-Dixon (2010).
84 Kincheloe (2007).
85 Lorenz (1963).

'standard' mechanisms of how societies adapt to climate change is reduced. In order to allow consideration of the full range of options, social debate and scientific research should not be constrained.

Nevertheless, in the social sciences – as well as in politics and the media – the scope of what is being debated and investigated is often limited through pressures to conform to the dominant creed. This tends to become a form of propaganda to maintain orthodoxy, frequently turning into censorship.[86]

It should also be acknowledged that scientists are not neutral and, however hard they may try, they cannot escape the ideological bent with which they embark on their work.[87] Ideological elements also infiltrate the funding processes for research projects. Donors want to be seen as supporting 'correct' lines of work, which do not run counter to prevailing thinking.

At the same time, researchers working on climate issues should be careful not to overstretch the significance of their data and findings. Whether in the social, physical or life sciences, modesty is an asset in research. Projections made to the second half of the 21st Century and beyond are rarely warranted. Indeed, given our inadequate understanding of current developments, and keeping in mind the perspectives of relativity and non-linearity, it is rather hazardous to move toward such a bold horizon.

Another element in the climate debate relates to the end of the Cold War, which has eased the way toward new power struggles – geographically as well institutionally, including the struggle for domination between the institutions of science, government, the military, and business. Claims of impending climate doom are at least in part a manifestation of this phenomenon.[88]

There is also an ideological element in alarmist statements coming from some scientists and politicians in developing countries, and the echo they receive from some Western NGOs, assigning guilt to the industrialised world as being responsible for yet further threats to their progress. While this may be true in a historical sense, GHG emissions are now far higher than they were a century ago, and they emanate increasingly from growing economies with a non-capitalist past.

Assigning blame and seeking retribution relates more to a struggle for power than to sound analysis about how to address climate-induced ills.

---

86 Herman & Chomsky (1988).

87 For a critique of the ideological element in demographic writings of the 1960s and 1970s, see Leroy (1981:737–743).

88 Hajer (1995).

While it is clear that the use of fossil fuels by the richer economies has tipped the scale in the global warming process, all nations will benefit from shifting development efforts toward renewable energy and conservation. However, international negotiations held in recent years, including Copenhagen, Cancun, Durban and Rio, have demonstrated that there is a continued strong ideological component in the debate on how to respond to climate change.[89]

## *H. Conclusion*

Climate change does indeed have an impact on human security and conflict risk, in combination with other natural and man-made elements. However, governance is the strongest intermediate factor when it comes to explaining the link between environment and climate change on the one hand, and conflict on the other. Problematic land management decisions, for example, such as current trends to allot large tracts for mechanised farming or biofuel production, could endanger local ecosystems and might marginalise small farmers and pastoralists who use such land for grazing, besides placing additional demands on surface water and groundwater reserves. Undesirable outcomes of such practices should not be blamed on climate change.

If climate change is likely to have significant security ramifications, it would of course be important to devise an appropriate response to address these. However, the processes involved are poorly understood. Furthermore, climate scientists and those dealing with climate negotiations generally lack the expertise for taking account of the complex links between climate, development, governance and security. Politicians and climate experts rarely speak the same language; scientific findings, however accurate or widely accepted, will never dictate a specific approach and will always allow a wide range of policy responses. Perceptions, whether based on good climate science or on popular impressions, will continue to have an influence on the policies that are formulated.

Climate change is also likely to have economic consequences which will impose differential burdens – and benefits – on national economies. For example, most infrastructure has been built on the assumption that the physical environment will not change. However, rising sea levels may result in

89 Veening (2012).

increased flooding risks, such as for airports built near the coast. Permafrost is thawing in polar areas; in the Russian Federation, this is already resulting in the degradation of pipelines, railways and buildings, and complicating the transport of timber and minerals which were moved by truck over frozen land or over river and lake ice in winter periods in winter periods.

The impact of natural changes may be exacerbated by human action, such as the massive withdrawal of groundwater, which causes subsidence and will worsen the impact of rising sea levels. Large coastal cities such as Shanghai are particularly at risk: they are important population centres in which housing and productive capacity could be endangered in addition to infrastructure.

It should also be recognised that global warming will create advantages for some states. In Canada, for example, the northern limit of cultivated land may move up, allowing more grain to be produced; the same may happen to the boreal forests, expanding northward into what is currently tundra. Moreover, the thawing of polar ice is likely to result in shortening sea lanes, allowing cargo to be moved between northern Russia and the Far East through Arctic waters. Also, opening of the Northwest Passage through the Canadian Arctic would considerably lower costs and travel time for shipping between Europe and the Far East.

While academic studies offer rather convincing evidence that climate change has historically been associated with conflict in the northern hemisphere,[90] an analogy with the current episode of global warming may not be in order. Rapid and accelerating technological changes since the Industrial Revolution may offer additional avenues for absorbing societal and economic strains introduced by climate change. Coupled with appropriate measures to conserve resources and to complete the shift toward renewable energy, new technologies could go a long way toward reducing tensions associated with resource scarcities and further climatic shifts. Such an approach is admittedly more problematic for countries which face challenges in the fields of technology, ingenuity and creativity.

As it is not clear at this stage which responses might increase conflict or cooperation, addressing the security implications of climate change will require a multifaceted approach, with climate-sensitive development policies and strengthening of institutions, in order to facilitate the implementation of adaptive measures, ensure equitable access to stressed resources, and miti-

90 Tol & Wagner (2010); Zhang et al. (2007).

gate the impact of climate-induced resource scarcities and extreme weather events.

Migration is likely to increase as a response to environmental and climate changes. There is a definite potential for strain to be felt as a result, including ethnic tension and even international conflict. However, if migration can be channelled in a way which allows legal and political controls to prevent an overload in the receiving countries, international migration flows have the potential to mitigate negative impacts in countries which are likely to experience the most severe consequences of climate change. Remittances are already as important as official development assistance in transferring wealth from developed economies to least-developed countries.[91] Further increasing these flows may at least partly offset the unequal economic burdens imposed by climate change. International flows could also reduce the burden on cities in least-developed countries, which would otherwise be the primary destination of environmental migrants from rural areas suffering the consequences of climate change.

In the face of continually rising populations, food production is likely to be the main point of friction globally – including in Africa. New production techniques and shifts in the types of crops produced may bring partial relief for the ill effects of climate change, but there may be increased reliance on grain from a few 'breadbasket' nations.

By mitigating the factors that result in climate change, e.g. cutting the emission of GHGs in addition to carbon dioxide, substantial benefits could be yielded. For example, curbing short-lived GHGs such as hydrofluorocarbons and methane through existing technology could cut the rate of global warming substantially, making it more likely that global warming might be limited to below 2°C. Adding gases like methane to the Montreal Protocol, which was initially adopted in 1987 and has been amended several times since, would be the least cumbersome way to achieve this.[92]

It is clear that all sectors that are major consumers of energy can contribute to achieving climate change mitigation goals. The US military alone consumes US$25 billion worth of fuel annually; also, fuel costs make up 9% of expenses for UN peacekeeping operations. By embracing new technologies and applying renewable energy sources, the military could be part of the solution to climate-change-related problems, rather than potentially having

91 World Bank (2011).

92 Statement by Durwood Zaelke at the "Building Climate Change Institutions" Conference, European Parliament, Brussels, 21 March 2012.

to deal with the consequences. Initiatives such as the Military Green of the European Defence Agency are contributing to making this change.[93]

There is a wide range of possible responses for addressing the challenges raised by climate change. Concrete circumstances on the ground will differ by country and region; policies will need to be designed taking account of these variations. Cultural diversity may limit certain options while opening others; political systems and traditions will guide possible responses. Governance and institutional capabilities are always central to devising appropriate policy responses and ensuring their implementation. The ability of the various levels of government to manage and regulate access to natural resources can limit the ill effects of climate change and environmental degradation. In this sense, conflict over natural resources often reflects failing governance.[94]

Climate change and the food insecurity which may accompany it are likely to amplify tensions over land tenure as well as over access to water and land. Strong institutions and leadership should be able to manage those tensions, and should limit the consequences of drought, famine, extreme weather events, and conflict; weak governance worsens the outcomes. Approaches to land use which allow the bulk of the rural population to improve their living standards while staying put – such as through agro-forestry and short-cycle vegetable and fodder production – would appear to be a first line of defence.[95]

There is obviously room for more research on the security impacts of climate change. Increasingly, this work can be done in the field, assessing the consequences of real cases. Field studies should be based on good climate science, recording changes in climate indicators, and documenting their evolution for as long a period as weather records permit. Furthermore, an effort should be made to disaggregate the effects of climate change from those of other factors – such as population growth – which may result in resource constraints that produce impacts similar to those of climate change. Another complexity is added when one considers that climate change is unlikely to affect different occupational groups within the same region in the same way. Conclusions reached through studies regarding the involvement of pastoralists groups in conflict as a result of climate change are unlikely to have the same validity when applied to sedentary farmers or urban pop-

93 European Defence Agency (2012).
94 Leroy & Gebresenbet (2011:9–15).
95 (ibid.).

ulations in the same region. On balance, comparative and historical analyses will be needed in order to view the issues from a broader perspective. Jumping to conclusions will risk discrediting the many serious efforts that have been directed toward understanding these questions.

Intensifying efforts for research, analysis, monitoring and early warning will yield an exponential growth in our understanding of what is happening to global and local climates, and permit identifying policy options to address the consequences. Understanding how and to what extent environmental and climate changes contribute to security issues will also allow integrating environment, climate variability and natural resources into conflict prevention and peace-building strategies.

The above analysis brings out the rather rudimentary state of our understanding of the links between environment and climate change on the one hand, and security on the other. This underlines the need to approach further reflection with an open mind. Dogmas can only impede further research. Axioms and assumptions should always be stated up front. Ideologies are unavoidable in any social endeavour, and may be useful as guides for further work, but they should not obscure signals coming from other persuasions.

We alluded above to the tendency, in many human pursuits, to become dogmatic and to reject evidence which goes counter to the dominant creed. There is a risk of this evil being repeated in climate security research. If scholars limit themselves to reading and quoting a relatively small group of authors, a consensus may ensue quite easily – though it would be a rather incestuous one. It would be of greater value to conduct research on the social consequences of climate change in different cultural settings, and to benefit from the conclusions arrived at through different perspectives. African oral histories will shed light on human responses to climate fluctuations which have occurred on the continent over past centuries. These can provide clues about steering further scientific research. The writings of French historian Emmanuel Le Roy Ladurie give a detailed view of the impact of climatic changes on Western European history over the past millennium, based on solid analysis and meticulous work carried out over 50 years, though his work is rarely quoted.

Social psychology should also be applied in order to understand perceptions and their link to reality. After looking at literature which investigates links between climate change and violence, Harvard psychologist Steven Pinker agrees with those who are sceptical of the idea that people fight wars over scarce resources and comments that "it is... foolish to let our lurid imaginations determine our sense of the probabilities."[96]

Applying critical views of how human thoughts take form and lead to apparent (though often erroneous) consensuses can add to our ability to grasp present developments regarding climate change and gain insights into what awaits future generations. We should resist the tendency to dig in and ignore work that is based on different premises than those we have become familiar with during our professional training.

Searching for ways of coping with climate change which are compatible with local culture and circumstances is a challenge that will need to be faced many times over. This effort will be successful only if it combines an open search for understanding with indigenous values and experience.

**Acknowledgement:** The authors acknowledge the financial support provided by the Preventive Diplomacy Programme of the Belgian Ministry of Foreign Affairs.

## *References*

AFP/Agence France-Presse, 2008, *Climate Change: Progress at Polluters' Talks, but Obstacles Ahead*, available at http://afp.google.com/article/ALeqM5jbisWIJJP7g0ld htFE1NE5AymCpg, last accessed on 28 July 2009.

Archer, David, 2005, Fate of Fossil Fuel $CO_2$ in Geologic Time, *Journal of Geophysical Research* 110, C09S05, doi:10.1029/2004JC002625.

Arrhenius, Svante, 1896, On the Influence of Carbonic Acid in the Air upon the Temperature of the Ground, *Philosophical Magazine* 41, 237–276.

BIISS & Saferworld/Bangladesh Institute of International & Strategic Studies and Saferworld, 2009, *Climate Change and Security in Bangladesh: A Case Study*, available at http://www.saferworld.org.uk/downloads/pubdocs/Bangladesh_climat_change_June09.pdf, last accessed 6 July 2012.

Barnett, Jon, 2001, Security and Climate Change, *Working Paper 7*, Norwich, Tyndall Centre for Climate Change Research, University of East Anglia.

Barnett, Tim P., David W. Pierce, Krishna M. AchutaRao, Peter J. Gleckler, Benjamin D. Santer, Jonathan M. Gregory & Warren M. Washington, 2005, Penetration of Human-induced Warming into the World's Oceans, *Science* 309, 284–287.

Brown, Oli & Alec Crawford, 2008, *Assessing the Security Implications of Climate Change for West Africa: Country Case Studies of Ghana and Burkina Faso*, Winnipeg, International Institute for Sustainable Development.

96 Pinker (2011:376–377).

Buhaug, Halvard, Nils Petter Gleditsch & Ole Magnus Theisen, 2010, Implications of Climate Change for Armed Conflict, in: Mearns, Robin and Andrew Norton (Eds), *Social Dimensions of Climate Change: Equity and Vulnerability in a Warming World*, Washington, DC, World Bank.

Burke, Marshall B., Edward Miguel, Shanker Satyanath, John A. Dykema & David B. Lobell, 2009, Warming Increases the Risk of Civil War in Africa, *Proceedings of the National Academy of Sciences of the United States of America* 106 (49), 20670–20674.

Busby, Joshua W., 2010, After Copenhagen: Climate Governance and the Road Ahead, *CFR International Institutions and Global Governance Working Paper*, New York, Council on Foreign Relations, 1–25.

CNA/Center for Naval Analyses, 2007, *National Security and the Threat of Climate Change*, Report from a panel of retired senior US military officers, Alexandria, VA, Center for Naval Analyses Corporation, available at http://securityandclimate.cna.org/, last accessed 28 June 2009.

Collier, Paul, Gordon Conway & Tony Venables, 2008, Climate Change and Africa, *Oxford Review of Economic Policy* 24 (2), 337–353.

CDC/Conservation Development Center, International Institute for Sustainable Development and Saferworld, 2009, *Climate Change and Conflict: Lessons from Community Conservancies in Northern Kenya*, available at http://www.saferworld.org.uk/downloads/pubdocs/Climate%20change%20and%20conflict.pdf, last accessed 6 July 2012.

*De Standaard*, 2012a, Hoeveelheid ijs op de Noordpool naar dieptepunt gedaald, available at http://www.standaard.be/Artikel/PrintArtikel.aspx?artikelId=DMF20120920_179l, last accessed 20 September 2012.

*De Standaard*, 2012b, Noordzee warmt sneller op dan andere zeeën, available at http://www.standaard.be/artikel/detail.aspx?artikelid=DMF20120917_178&word=Noordzee+temperatuur, last accessed 17 September 2012.

Elhance, Arun P., 1999, *Hydropolitics in the Third World: Conflict and Cooperation in International River Basins*, New York, United States Institute of Peace.

EU/European Union, 2008, *Climate Change and International Security: Paper from the High Representative and the European Commission to the European Council*, available at http://www.consilium.europa.eu/uedocs/cms_data/docs/pressdata/en/reports/99387.pdf, last accessed 9 April 2013.

European Defence Agency, 2012, *Military Green 2012: Programme – High Level Day – 19 June 2013*, available at http://www.eusew.eu/upload/events/2194_44_military%20green%202012%20-%20programme%20high%20level%20day%20v 2.pdf, last accessed 28 April 2013.

Evans, Gareth, 2008, *Conflict Potential in a World of Climate Change: Address to Bucerius Summer School on Global Governance*, available at http://www.gevans.org/speeches/speech303.html, last accessed 28 April 2013.

GHF/Global Humanitarian Forum, 2009, *The Anatomy of a Silent Crisis*, available at http://www.bb.undp.org/uploads/file/pdfs/energy_environment/CC%20human%20impact%20report.pdf, last accessed 12 November 2009.

GHG Management Institute, 2010, What is a Global Warming Potential? And Which One do I Use?, available at http://ghginstitute.org/2010/06/28/what-is-a-global-warming-potential/, last accessed 30 October 2012.

Hajer, Maarten A., 1995, *The Politics of Environmental Discourses: Ecological Modernisation and the Policy Process*, Oxford, Oxford University Press.

Hendrix, Cullen S. & Sarah M. Glaser, 2007, Trends and Triggers: Climate, Climate Change and Civil Conflict in Sub-Saharan Africa, *Political Geography* 26 (6), 695–715.

Hendrix, Cullen S. & Idean Salehyan, 2012, Climate Change, Rainfall and Social Conflict in Africa, *Journal of Peace Research* 49 (1), 35–50.

Herman, Edward & Noam Chomsky, 1988, *Manufacturing Consent: The Political Economy of the Mass Media*, New York, Pantheon Books.

Homer-Dixon, Thomas F., 1999, *Environment, Scarcity and Violence*, Princeton, Princeton University Press.

Homer-Dixon, Thomas F., 2010, Disaster at the Top of the World, *New York Times*, 22 August 2010, available at http://www.nytimes.com/2010/08/23/opinion/23homer-dixon.html?pagewanted=all, last accessed 6 July 2012.

Hsiang, Solomon M., Kyle C. Meng & Mark A. Cane, 2011, Civil Conflicts are Associated with the Global Climate, *Nature* 476, 438–441.

Hulme, Mike, 2008, The Conquering of Climate: Discourses of Fear and their Dissolution, *The Geographical Journal* 174 (1), 5–16.

International Alert, 2007, *A Climate of Conflict: The Links between Climate Change, Peace and War*, available at http://www.crisisgroup.org/en/publication-type/key-issues/thematic/climate-change-and-conflict.aspx, last accessed 6 July 2012.

ICG/International Crisis Group, 2010, *Climate Change and Conflict*, available at http://www.crisisgroup.org/en/key-issues/climate-change-and-conflict.aspx, last accessed 6 July 2012.

Kevane, Michael & Leslie Gray, 2008, Darfur: Rainfall and Conflict, *Environmental Resource Letters* 3.

Kincheloe, L. Joe, 2007, Critical Pedagogy in the Twenty-first Century: Evolution for Survival, in: McLaren, Peter & Joe L. Kincheloe (Eds), *Critical Pedagogy: Where are we Now?*, New York, Peter Lang Publishers, 9–42.

Lassey, Keith R., 2007, Livestock Methane Emission: From the Individual Grazing Animal through National Inventories to the Global Methane Cycle, *Agricultural and Forest Meteorology* 142, 120–132.

Leroy, Marcel, 1981, Population Ideology in the More Developed Countries in the 1970s and 1980s, in: International Union for the Scientific Study of Population *(Eds), International Population Conference, Manila: Proceedings*, Liège, International Union for the Scientific Study of Population, 737–743.

Leroy, Marcel, 2009, Introduction, in: Leroy, Marcel (Ed.), *Environment and Conflict in Africa: Reflections on Darfur*, University for Peace, Africa Programme, Addis Ababa, 1–7.

Leroy, Marcel & Fana Gebresenbet, 2011, Climate Conflicts in the Horn of Africa?, *Conflict Trends*, Durban, ACCORD (The African Centre for the Constructive Resolution of Discord), 9–15.

Le Roy Ladurie, E., 2005, *Histoire humaine et comparée du climat du dernier millénaire*, paper presented to the Academie des sciences morales et politiques, 4 April 2005, available at http://pratclif.com/climatechange/leroyladurie.htmlast accessed 29 April 2013.

Levitus, Sydney, John I. Antonov & Timothy P. Boyer, 2005, Warming of the World Oceans 1955–2003, *Geophysical Research Letters* 32, L02604, doi: 10.1029/2004GL021592.

Levy, Marc A., Catherine Thorkelson, Charles Vörösmarty, Ellen Douglas & Macartan Humphreys, 2005, Freshwater Availability Anomalies and Outbreak of Internal War: Results from a Global Spatial Time Series Analysis, paper presented at the international workshop Human Security and Climate Change, Holmen, Norway, 21–23 June 2005, available at http://www.ciesin.columbia.edu/documents/waterconflict.pdf, last accessed 28 April 2013.

Lorenz, Edward N., 1963, Deterministic Non-periodic Flow, *Journal of the Atmospheric Sciences* 20, 130–141.

Meier, Patrick, Doug Bond & Joe Bond, 2007, Environmental Influences on Pastoral Conflict in the Horn of Africa, *Political Geography* 26 (6), 716–735.

Miguel, Edward, Shankar Satyanath & Ernest Sergenti, 2004, Economic Shocks and Civil Conflict: An Instrumental Variables Approach, *Journal of Political Economy* 112 (4), 725–753.

Mobjörk, Malin, Mikael Eriksson & Henrik Carlsen, 2010, *On Connecting Climate Change with Security and Armed Conflict: Investigating Knowledge from the Scientific Community*, Stockholm, Totalforsvarets Forskningsinstitut (Swedish Defence Research Agency).

Moon, Ban Ki, 2007, A Climate Culprit in Darfur, *The Washington Post*, available at http://www.washingtonpost.com/wp-dyn/content/article/2007/06/15/AR200706150 1857.html, last accessed 27 February 2009.

Morano, Marc, 2009, Fears of a Coming Ice Age, Showed up in Peer-reviewed Literature, at Scientific Conferences, by Prominent Scientists and Throughout the Media, *Climate Depot*, 6 October 2009, available at http://www.climatedepot.com/a/3213/Dont-Mis s-it-Climate-Depots-Factsheet-on-1970s-Coming-Ice-Age-Claims, last accessed 6 July 2012.

NASA/National Aeronautics and Space Administration, 2010, NASA Research Finds Last Decade was Warmest on Record, 2009 One of Warmest Years, available at http:// www.nasa.gov/home/hqnews/2010/jan/HQ_10-017_Warmest_temps.html, last accessed 6 July 2012.

*New York Times*, 2012, Ending its Summer Melt, Arctic Sea Ice Sets a New Low That Leads to Warnings, posted on 19 September 2012, http://www.nytimes.com/2012/09 /20/science/earth/arctic-sea-ice-stops-melting-but-new-record-low-is-set.html?nl=to daysheadlines&emc=edit_th_20120920&moc.semityn, last accessed 28 April 2013.

NOAA/National Oceanic and Atmospheric Administration, 2010, 2010 Tied with 1998 as Warmest Global Temperature on Record, 15 September 2010, available at http://www.noaanews.noaa.gov/stories2010/20100915_globalstats.html, last accessed 28 April 2012.

NOAA/National Oceanic and Atmospheric Administration, 2010, *State of the Climate*, Asheville, NC, NOAA National Climatic Data Center, available at http://www.ncdc.noaa.gov/sotc/, last accessed 28 April 2013.

PACJA/Pan African Climate Justice Alliance, 2009, *The Economic Cost of Climate Change in Africa*, Nairobi, PACJA.

Pinker, Steven, 2011, *The Better Angels of Our Nature: Why Violence has Declined*, New York, Viking.

Saferworld, 2008a, *Addressing Climate Change and Conflict Challenges through the EU Development Cooperation: Submission to the Climate Change and International Security European Union Roadmap Steering Group*, available at http://www.saferworld.org.uk/downloads/pubdocs/Saferworld%20submission%20to%20EU%2012122008.pdf, last accessed 28 April 2013.

Saferworld, 2008b, *Human Security in Bangladesh*, available at http://www.saferworld.org.uk/downloads/pubdocs/Bangladesh_HS_report.pdf, last accessed 28 April 2013.

Smith, Todd G., 2012, Fighting in the Rain: Monthly Rainfall Extremes and Social Unrest in Africa, in: Berhe, Mulugeta Gebrehiwot & Jean-Bosco Butera (Eds), *Climate Change and Pastoralism: Traditional Coping Mechanisms and Conflict in the Horn of Africa*, Addis Ababa, Institute for Peace and Security Studies, Addis Ababa University.

Smith, Dan & Janani Vivekananda, 2009, Climate Change, Conflict and Fragility: Understanding the Linkages, Shaping Effective Responses, *International Alert*, available at http://www.international-alert.org/sites/default/files/publications/Climate_change_conflict_and_fragility_Nov 09.pdf, last accessed 6 July 2012.

Sprout, Harold H. & Margaret T. Sprout, 1968, *An Ecological Paradigm for the Study of International Politics*, Princeton, Center for International Studies, Woodrow Wilson School of Public and International Affairs, Princeton University.

Stern, Nicholas, 2006, *Stern Review on the Economics of Climate Change*, London, HM Treasury, available at http://www.hm-treasury.gov.uk/sternreview_index.htm, last accessed 6 July 2012.

Theisen, Ole Magnus, Helge Holtermann & Halvard Buhaug, 2009, Drought, Political Exclusion, and Civil War, unpublished paper presented at the conference Climate Change, Social Stress and Violent Conflict: State of the Art and Research Needs, Hamburg University, Hamburg, 19–20 November 2009.

Tol, Richard S.J. & Sebastian Wagner, 2010, Climate Change and Violent Conflict in Europe over the Last Millennium, *Climatic Change* 99, 65–79.

UNEP/United Nations Environment Programme, 2006, *Spectre of 'Water Wars' Distracts from Urgent Need for Cross-border Cooperation*, Nairobi, UNEP.

UNEP/United Nations Environment Programme, 2007, *Sudan: Post-conflict Environmental Assessment*, Nairobi, UNEP.

UNEP/United Nations Environment Programme, 2009, *From Conflict to Peace Building: The Role of Natural Resources and the Environment*, Nairobi, UNEP.

UNSCDPI/United Nations Security Council Department of Public Information, 2007, *United Nations Security Council Holds First Ever Debate on Impact of Climate Change on Peace, Security, Hearing over 50 Speakers*, UN Security Council 5663rd Meeting, 17 April 2007, available at http://www.un.org/News/Press/docs/2007/sc9000.doc.htm, last accessed 17 July 2010.

Veening, Wouter J., 2012, *Summary of Rio +20*, The Hague, Institute for Environmental Security.

WBGU/Wissenschaftlicher Berat der Bundesregierung Globale Umweltveränderungen, 2007, *World in Transition: Climate Change as a Security Threat*, available at http://eeac.hscglab.nl/files/D-WBGU_ClimateChangeRisk_Jun07.pdf, last accessed 28 June 2009.

World Bank, 2011, *Migration and Remittances Factbook 2011* (Second Edition), New York, World Bank, available at http://siteresources.worldbank.org/INTLAC/Resources/Factbook2011-Ebook.pdf, last accessed 6 July 2012.

WMO/World Meteorological Organization, 2007, *The World Meteorological Organization Reports on Extreme Weather and Climate Events*, available at http://www.wmo.int/pages/mediacentre/press_releases/pr_791_en.html, last accessed 28 June 2009.

WMO/World Meteorological Organization, 2011, *WMO Annual Statement Confirms 2011 as 11th Warmest on Record: Climate Change Accelerated in 2001–2010, According to Preliminary Assessment*, available at http://www.wmo.int/pages/mediacentre/press_releases/pr_943_en.html, last accessed 6 July 2012.

Zaehlke, Durwood, 2012, unpublished presentation to Conference on Climate Change and Conflict, European Parliament, Brussels, 21 March 2012.

Zhang, David D., Peter Brecke, Harry F. Lee, Yuan-Qing He & Jane Zhang, 2007, Global Climate Change, War, and Population Decline in Recent Human History, *Proceedings of the National Academy of Sciences of the United States of America* 104, 49.

Zhang, David D., C.Y. Jim, George C.-S. Lin, Yuan-Qing He, James J. Wang & Harry F. Lee, 2006, Climatic Change, Wars and Dynastic Cycles in China over the Last Millennium, *Climatic Change*, 76, 459–477.

# 23
# Dangerous Anthropogenic Climate Change from the Perspective of Adaptation

*Kristie L. Ebi & Ian Burton*

*Abstract*

The stated "ultimate objective" of the United Nations Framework Convention on Climate Change is "stabilisation of greenhouse gas concentrations in the atmosphere at a level that would prevent dangerous anthropogenic interference with the climate system". In the more than 20 years since the Convention was negotiated, understanding has increased of the risks posed to human and natural systems by climate change. The social construction underlying the negotiations is that greenhouse gases (GHGs) are pollutants whose control could prevent serious (and future) impacts. However, research on climate change impacts has led to an enlarged framework: the magnitude and extent of possible risks of climate change depend not only on changes in global average temperature (e.g. global warming), but also the human and natural systems exposed to those changes and their underlying vulnerability. Climate change interacts with other drivers to increase (or decrease) risks, and affects risk through multiple pathways. Adaptation then focuses not just on climate change, but also on addressing underlying exposure and vulnerability.

The determination of what atmospheric concentration of GHGs constitutes dangerous anthropogenic interference with the climate system is a value judgment, which means that science alone cannot provide an answer. To help inform the policy process, the Intergovernmental Panel on Climate Change issued reports in 2001 and 2007, synthesising the literature on climate change impacts and identified reasons of concern to enable readers to evaluate the relationships between increases in global mean temperature and associated impacts. Each concern is consistent with a paradigm that could be used independently or in combination with other reasons to help determine dangerous atmospheric concentrations. The reasons for concern are the relationship between global mean temperature increase and damage to or irreparable loss of unique and threatened systems; distribution of impacts among people and across regions; global aggregate damages; and probability of extreme weather events and of large-scale singular events. Assessments in 2001 and 2007 showed increasing risk with increasing temperatures, with the temperature at which risks become apparent varying across the reasons for concern, and with greater risks in 2007 than in 2001. The assessments addressed only how risks change as global mean temperature increases, and not how risks might change at different levels of warming. The assessments also identified impacts, vulnerabilities, and risks that would merit policy-makers' attention, including food supply, infrastructure, health, water resources, coastal systems, ecosystems, global biogeochemical cycles, ice sheets, and modes of oceanic and atmospheric circulation.

Reasons for concern about the ability to adapt to projected impacts and the likelihood of sustainable adaptation include contractions and uncertainties in the window of opportunity for adaptation; the difference between adaptive capacity and adaptive action; the risk of maladaptation; and the misguided measures of loss.

Effective and efficient adaptation will be critical to increase the resilience of human and natural systems over the next few decades, although it will not be possible to prevent all impacts. Over the longer term, the magnitude and extent of climate change impacts will depend on the mix of adaptation and mitigation, with rapid and successful reductions in GHG emissions reducing how much adaptation will be needed later this century. Slower and less comprehensive mitigation will increase the likelihood of crossing thresholds that will result in dangerous impacts to human and natural systems.

## *A. Introduction*

The United Nations Framework Convention on Climate Change (UNFCCC) lays out its objective in Article 2:[1]

> The ultimate objective of this Convention and any related legal instruments that the Conference of the Parties may adopt is to achieve, in accordance with the relevant provisions of the Convention, stabilisation of greenhouse gas concentrations in the atmosphere at a level that would prevent dangerous anthropogenic interference with the climate system. Such a level should be achieved within a

1 United Nations Framework Convention on Climate Change, Document FCCC/INFORMAL/84, GE.05-62220 (E) 200705 (1992), available at http://unfccc.int/essential_background/convention/items, last accessed 15 January 2013.

> time frame sufficient to allow ecosystems to adapt naturally to climate change, to ensure that food production is not threatened and to enable economic development to proceed in a sustainable manner.

Scientific understanding at the time this paragraph was negotiated in 1991–1992 was insufficient for government negotiators to identify an atmospheric concentration of greenhouse gases (GHGs) that would meet the criteria for preventing dangerous anthropogenic interference with the climate system. Science alone will never be able to provide an answer because the determination of "dangerous" is a value judgment. In addition, the science and policy contexts continue to change with further socio-economic development, a growing knowledge base, and increased climate change. Greater understanding of the breadth and depth of multiple stresses associated with climate change means that climate change has moved from being simply a pollution problem to issues of global development, equity, and ethics (because those who have contributed the least to GHG emissions will experience the most severe consequences).

As detailed later, few impacts of climate change had been observed at the time the UNFCCC was negotiated, so the negotiations assumed impacts were unlikely to occur until later in the 21st Century. However, observations since then show that climate change is altering the mean and variability of temperature, precipitation, and other weather variables, and that the rise in sea level is increasing risk of storm surges, saltwater intrusion into fresh water, and inundation. Impacts are already evident in many sectors and regions with some species extinction, childhood mortality, and changing landscapes at least partially attributed to climate change. This raises the question of whether dangerous interference has already started. Nevertheless, the international policy process adopted a definition of *dangerous* as an increase in global mean surface temperature of +2°C above pre-industrial temperatures, based on interpretation of the scientific evidence and literature.

This article provides text on adaptation from, and offers a historic perspective on, the UNFCCC. The discussion includes a brief review of scientific perspectives on how to estimate dangerous climate change, particularly key vulnerabilities and reasons for concern, as well as issues of relevance to the question of what is "dangerous" anthropogenic climate change from the perspective of adaptation.

## *B. Adaptation in the UNFCCC*

Article 2 is not the only place in the UNFCCC text that mentions adaptation. Article 3 lays out the principles underlying the Convention. As paragraph 3.3 states, the states parties "should take precautionary measures to anticipate, prevent, or minimise the causes of climate change and mitigate its adverse effects."

To achieve this where there are threats of serious or irreversible damage, states parties should implement policies and measures that take into account different socio-economic contexts; are comprehensive; cover all relevant sources, sinks, and reservoirs of GHGs; cover adaptation; and comprise all economic sectors.

Article 4 then lays out the commitments of the states parties and mentions *adaptation* in several paragraphs that have guided negotiations for the UNFCCC's implementation. Article 4 states the following:

> All Parties, taking into account their common but differentiated responsibilities and their specific national and regional development priorities, objectives and circumstances, shall:
>
> …
>
> 4.1 e: Cooperate in preparing for adaptation to the impacts of climate change; develop and elaborate appropriate and integrated plans for coastal zone management, water resources and agriculture, and for the protection and rehabilitation of areas, particularly in Africa, affected by drought and desertification, as well as floods.
>
> 4.1 f: Take climate change considerations into account, to the extent feasible, in their relevant social, economic and environmental policies and actions, and employ appropriate methods, for example impact assessments, formulated and determined nationally, with a view to minimising adverse effects on the economy, on public health and on the quality of the environment, of projects or measures undertaken by them to mitigate or adapt to climate change. …
>
> 4.4: The developed country Parties and other developed Parties included in Annex II shall also assist the developing country Parties that are particularly vulnerable to the adverse effects of climate change in meeting costs of adaptation to those adverse effects. …
>
> 4.8: In the implementation of the commitments in this Article, the Parties shall give full consideration to what actions are necessary under the Convention, including actions related to funding, insurance and the transfer of technology, to meet the specific needs and concerns of developing country Parties arising from the adverse effects of climate change and/or the impact of the implementation of response measures.

4.9: The Parties shall take full account of the specific needs and special situations of the least developed countries in their actions with regard to funding and transfer of technology.

## C. Historical Perspective

National and international organisations began serious consideration of the possible consequences for human and natural systems of increasing GHG emissions in the 1970s. For example, in 1970, the Massachusetts Institute of Technology convened a one-month study of critical environmental problems, focusing on environmental issues whose cumulative effects on ecological systems would be so large and prevalent that they would have worldwide significance.[2] The 50 participants in the study were primarily concerned with the effects of pollution on humans through changes in climate, ocean ecology, and large terrestrial ecosystems. The subjects investigated for climatic effects included the increasing carbon dioxide content of the atmosphere, the particle load of the atmosphere, and contamination of the troposphere and stratosphere by subsonic and supersonic aircraft. This list of topics highlights an important historical perspective: that climate change is an environmental pollutant and reducing GHG emissions through mitigation will solve any negative consequences of exposure.

This perspective is understandable in context of other environmental concerns starting in the 1960s, with the publication in 1962 of *Silent Spring* by Rachel Carson on the environmental hazards of pesticides, particularly on birds.[3] The book is widely credited with helping launch the contemporary environmental movement. Other environmental issues of importance in the 1970s and 1980s included stratospheric ozone depletion and acid rain. Stratospheric ozone depletion went from an unknown issue in early 1970 to a multilateral environmental agreement in 1985 and an international treaty[4] in 1987 that successfully reduced the emissions of ozone-depleting chemicals.[5] Throughout the 1970s and 1980s, there was ongoing scientific and policy debate about the effects of sulphur deposition (*acid rain*) on ecosystem resources in the United States, resulting in the US Congress pass-

2 SCEP (1970).

3 Carson (1962).

4 The Montreal Protocol on Substances that Deplete the Ozone Layer.

5 United Nations Framework Convention on Climate Change, Copenhagen Accord, Document FCCC/CP/2009/L.7 (2009).

ing the Acid Deposition Act 1980, establishing an 18-year assessment and research programme that also was successful in reducing the relevant emissions.[6]

A common thread running through these and similar environmental problems is that they are caused by an agent (pesticides, chemicals that deplete ozone, sulphur compounds) whose release or emission was relatively easily and successfully controlled after overcoming initial resistance. These agents are typically short-lived compared with carbon dioxide, so reducing emissions relatively quickly led to improvements in the impacts of concern. Furthermore, alternatives or substitutes could be made fairly readily available in most cases. A key first step in understanding these issues was risk identification (showing the agents of concern which led to adverse impacts), followed by the scientific determination of a level of exposure that would lead to 'acceptable' risk (where *acceptable* was defined by regulators), usually in terms of risk to human health. This approach – and its success – informed efforts to understand the impacts of and strategies to control climate change. In essence, under this approach, impacts are directly related to emissions. Equally important, the way to manage impacts is to reduce GHG emissions; thus, mitigation is the primary policy task. That perspective is reflected in the language in the UNFCCC and in activities since its negotiation, underscoring the original intention that the treaty should focus on reducing the source of climate change, rather than on adapting to the changes.[7] Adaptive capacity was not regarded as a policy objective but as an indicator of the extent to which societies could tolerate changes in climate. The tension between mitigation and adaptation has strongly characterised the discourse on climate change policy.

Table 1 summarises the historic framing of the climate change debate and adaptation thinking.

6 Lackey & Blair (1997:9–13); Likens & Bormann (1974:1176–1179).
7 Schipper (2006:82–92).

***Table 1: Historic Framing of Climate Change and Adaptation Thinking***[8]

| Time Frame | Forum | Main Questions |
|---|---|---|
| CLIMATE CHANGE DEBATE | | |
| 1960s–1970s | World Meteorological Organization (WMO)<br>Climate scientists | Is climate change an issue of concern?<br>How will climate change affect the weather? |
| Mid-1980s–early 1990s | UNFCCC<br>Intergovernmental Panel on Climate Change (IPCC) | Is climate change occurring?<br>How will climate change affect global human and natural systems?<br>Who should be responsible for reducing emissions? |
| Late 1990s–early 2000s | UNFCCC<br>Regional decision-makers | What are the relative costs of mitigation and adaptation?<br>How vulnerable are communities to climate variability and its consequences? |
| ADAPTATION THINKING | | |
| 1970s–early 1980s | Club of Rome<br>Academics | What are the ecological limits to human development and growth?<br>What are the options to respond to climate change?<br>What sort of impacts can systems sustain?<br>Will systems adapt automatically? |
| Late 1980s | IPCC<br>WMO, International Council of Scientific Unions and UN Environment Programme Advisory Group on Greenhouse Gases (AGGG) | What will the impacts be?<br>How much adaptation are society and ecosystems capable of?<br>How much can adaptation offset the need to mitigate? |
| Early 1990s | IPCC | Is mitigation more important than adaptation for responding to climate change?<br>What is the optimal balance between mitigation and adaptation in responding to climate change? |
| Late 1990s | UNFCCC<br>Research bodies | How can policy support adaptation?<br>Who is vulnerable to climate change and why?<br>How much adaptation will be needed?<br>What are the links between adaptation and development? |
| Early 2000s | IPCC<br>United Nations Development Programme<br>Global Environment Facility<br>The World Bank<br>Donors<br>Research bodies | What constitutes *adaptive capacity*?<br>How can adaptation be integrated into existing sustainable development plans?<br>What is needed to mainstream adaptation?<br>How can adaptation policy be designed? |

8 (ibid.).

### *D. Framework for Impacts and Adaptation*

Research over the past 30 years on the impacts of climate change has led to a more nuanced framework of how climate change could affect human and natural systems. The magnitude and extent of possible risks of climate change depend on –[9]

- changes in temperature, precipitation, and other weather variables
- human and natural systems exposed to these changes, including people and their livelihoods; infrastructure; economic, social, or cultural assets; environmental services and resources, and
- vulnerability of these systems, where vulnerability is defined as the propensity or predisposition to be affected.

Figure 1 illustrates this framework, focusing on extreme weather and climate events. The figure shows the three components of risk, highlighting that realised risk in terms of disasters can influence subsequent development, including through disaster risk management and climate change adaptation, and that development is a driver of the anthropogenic climate change that influences the frequency and intensity of extreme weather and climate events.

9 IPCC (2012:3–21).

***Figure 1: Components of Disaster Risk from Weather and Climate Events***[10]

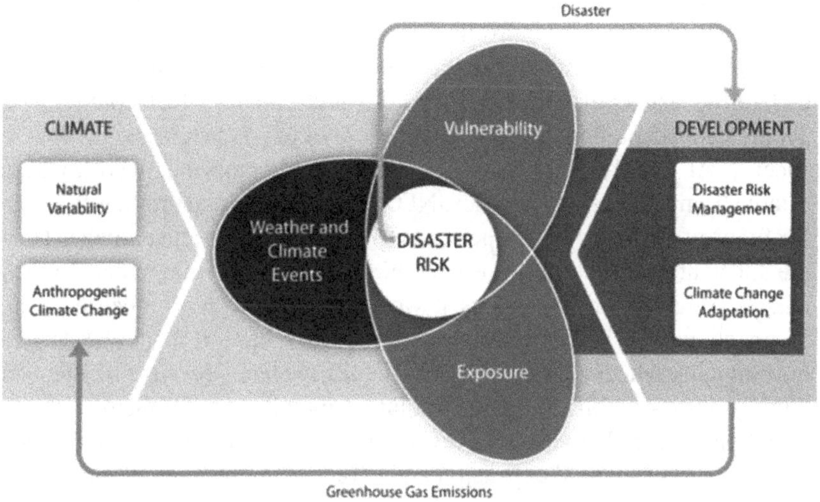


Using this framework, adaptation is understood to increase resilience by decreasing exposure or increasing vulnerability. Climate change could be considered to be dangerous when levels of exposure or vulnerability are deemed unacceptable.

Irrespective of whether climate change is considered primarily as a pollution problem (as at the outset) or as a more complex and nuanced challenge involving questions about, among other things, the costs of strategies, policies, and measures to control, prepare for, respond to, and recover from impacts, it is also about sustainable development, equity, and social justice. Whatever the framing, there has always been an underlying question about the locus of responsibility. Under the UNFCCC, the states parties agreed that climate change was a common responsibility. Article 3.1 states the following:

> The Parties should protect the climate system for the benefit of present and future generations of humankind. On the basis of equity and in accordance with their common but differentiated responsibility and respective capabilities. ... Accordingly the developed countries should take the lead in combatting climate change and the effects thereof.

10 (ibid.:2).

The question of liability was addressed in the 17th UNFCCC Conference of the Parties (COP17) held in Durban in December 2011. At COP17, the developing countries, increasingly dissatisfied with the lack of progress by developed countries in controlling emissions and by the poor results from the Kyoto Protocol, introduced the concept of *loss and damage* that refers to liability and compensation for losses and damages that could not be avoided by adaptation, including loss of land to sea level rise and loss of species. At COP18 in Doha in December 2012, proposals were made to establish a mechanism under the UNFCCC to manage and provide oversight for funds provided by developed countries to cover loss and damage. There are substantial difficulties in establishing and implementing such a mechanism, including on what basis it would be decided how much each developed country should contribute, and on what basis it would be decided how much each developing country could receive. Because of the contentious issues involved, the only solution may be to refer them to the international judicial process. However, before this could happen, there would have to be a substantial expansion and evolution in the international law of the environment.

### *E. Determining Dangerous*

The determination of what atmospheric concentration of GHGs constitutes "dangerous anthropogenic interference with the climate system" is a policy decision, although increasing legal actions suggest that national and international courts may play a role in providing an answer. *Dangerous* is a function of the degree to which impacts are negative and are considered unacceptable. The latter component is a value judgment. There is growing research on impacts to help answer the first component. As indicated previously, scientific understanding of climate change, current and future impacts, and range of policy instruments has increased significantly since 1992, including through assessments prepared by the Intergovernmental Panel on Climate Change (IPCC). The IPCC was established in 1988 by the World Meteorological Organization and the United Nations Environment Programme to provide national governments with a clear scientific view on the current state of knowledge in climate change, its potential environmental and socio-economic impacts, and options to manage the attendant risks. The three working group reports contributing to the IPCC's Fifth Assessment Report will be completed in late 2013 and early 2014.

As the breadth and depth of scientific understanding has increased, so has the social construction of what impacts are unacceptable. The UNFCCC specifies three criteria for *dangerous*: time for ecosystems to adapt naturally; food production not to be threatened; and economic development enabled in a sustainable matter. However, these are not quantifiable criteria that can be measured and monitored, which makes them impractical to operationalise.[11] Also, while these three criteria are clearly important, they are not the only possible impacts of climate change that could have large-scale consequences. For example, there are growing concerns about a wide range of other consequences that could be considered dangerous, including the availability of sufficient quantities of safe water in some regions; the impacts of changing patterns of extreme weather and climate events; changes in the geographic range and incidence of climate-sensitive health outcomes; melting of large ice sheets in Greenland, the Arctic, and Antarctica; sea level rise; and the acidification of the oceans.

Furthermore, determining what concentration of GHGs causes unacceptable harm or injury varies by sector and geographic region. The UNFCCC recognises certain regions are more vulnerable to climate change, including least-developed countries, small island states, and areas with fragile ecosystems. At any particular concentration of atmospheric carbon dioxide, some vulnerable regions and sectors will experience significant impacts that they perceive to be unacceptable, while others will experience little to no impacts. In any one place there is likely to be a combination of adverse and beneficial impacts or opportunities, such as a longer growing season in some high-latitude countries. Scientific evidence overwhelmingly supports the conclusion that adverse effects will far outweigh the beneficial effects, especially as global mean surface temperature rises beyond +2°C above pre-industrial temperatures.

The time frame of impacts also is an important consideration, in both the short and longer term. The UNFCCC text has the implicit assumption that current atmospheric concentrations (or concentrations over the few decades following 1992) are not dangerous, so adaptation and mitigation are future issues. This is in contrast to the large number of scientific publications and empirical observations showing impacts are already being felt from climate change, particularly in developing countries and the high Arctic.[12] Further-

11 Burton et al. (forthcoming 2013).
12 Bernstein et al. (2007:52).

more, there was limited understanding of what is termed the *climate change commitment*: that current atmospheric concentrations of GHGs, particularly carbon dioxide, will take many decades to centuries to come to equilibrium – even without any further increase in concentrations. Sea level rise in particular is expected to continue for several centuries. Therefore, even if global emissions were immediately reduced, the climate would continue to change for several more decades. In other words, the planet is committed to additional warming in the short term no matter whether mitigation activities fail or succeed.

Human and natural systems need to adapt to these changes, which will include increases in the frequency, intensity, spatial extent, and duration of many extreme weather and climate events.[13] Implementation of a wide range of adaptation policies and measures are critical in the short term if human and natural systems are going to cope successfully with the changes built into the climate system. Furthermore, it is not just the changing weather patterns themselves to which adaptation is required, but also to the consequences of those changing patterns, such as increases in the geographic range of insects and other disease vectors, leading to infectious diseases spreading to new regions.

Over the very long term, there is recognition that current atmospheric concentrations of carbon dioxide will continue to drive changes in climate for many hundreds of years.[14] Furthermore, the climate change resulting from carbon dioxide emissions is largely irreversible for 1,000 years – even after emissions stop.[15] Following the cessation of emissions, there will be a slow reduction in the atmospheric concentration of carbon dioxide with a compensating slow loss of heat to the ocean, resulting in global mean surface temperatures not changing significantly for at least 1,000 years. This could result in irreversible impacts such as dry-season rainfall reductions in several regions comparable to those of the 'Dust Bowl' era in North America in the 1930s and continuing sea level rise. Thermal expansion of the warming ocean provides a conservative lower limit for irreversible global average sea level rise of at least 0.4–1.0 m if 21st-Century carbon dioxide concentrations exceed 600 ppmv and 0.6–1.9 m for peak carbon dioxide concentrations exceeding about 1,000 ppmv. There will likely be additional contributions from melting glaciers and ice sheets. Thus, for coastal communities in low-

13 IPCC (2012).

14 Bernstein et al. (2007).

15 Solomon et al. (2009:1704–1709).

lying regions, atmospheric concentrations of carbon dioxide are already dangerous.

Finally, the UNFCCC alludes to – but does provide a framework for – addressing the fact that climate change is a stress multiplier. Changing weather patterns are often not the only driver of impacts, but can exacerbate other stresses to significantly increase risks. Therefore, a determination of *dangerous* will depend on the context. For example, a uniform amount of sea level rise will have very different impacts on coastal communities depending on their vulnerability.

These and other issues make global determination of dangerous anthropogenic influence very challenging. Despite these challenges, the Copenhagen Accord states the "international scientific consensus" that a global mean surface temperature increase of 2°C above pre-industrial levels is the upper limit of what human societies could adapt to, and that anything above that concentration would be dangerous.[16] However, this is more a political than a scientific consensus. Further research indicates the impacts associated with 2°C are greater than previous studies indicated, such that 2°C may now represent the threshold between dangerous and extremely dangerous.[17] Pledges for GHG emission reductions put forward since the Copenhagen Accord could result in a 50:50 chance of a peak global temperature increase of at least 3°C above pre-industrial levels, with some estimates as high as 3.9°C.[18]

### *F. Reasons for Concern*

Based on a growing literature base, the contribution of Working Group II to the IPCC's Third Assessment Report, which addressed impacts, adaptation and vulnerability, included a chapter on vulnerability to climate change and what were termed *reasons for concern.*[19] The chapter set out to synthesise the results of Working Group II's report and assess the state of knowledge relevant to Article 2 of the UNFCCC. The authors specified that it was not their task to determine whether impacts were tolerable or dangerous. They

16 United Nations, United Nations Framework Convention on Climate Change, Copenhagen Accord, Document FCCC/CP/2009/L.7 (2009).
17 Anderson & Bows (2011:20–44).
18 Parry (2010:18–19); Sustainability Institute (2010).
19 Smith et al. (2009:915–967).

synthesised information on climate change impacts to enable readers to evaluate the relationships between increases in global mean temperature and such impacts, and created reasons for concern to aid readers in making their own determination of what constituted dangerous climate change. These reasons were taken from debates and literature about the risks of climate change. Each concern was consistent with a paradigm that could be used independently or in combination with other reasons to help determine what level of climate change was dangerous, and none was considered more important than another. The reasons for concern are the relationship between an increase in the global mean temperature and the –

- ***damage to or irreparable loss of unique and threatened systems:*** This recognised that some systems restricted to a relatively narrow geographic range, but which could affect other entities, might be irreparably harmed by changes in climate beyond certain thresholds. Examples include melting of tropical glaciers, destruction and loss of coral reefs, loss of mangrove ecosystems, loss of biodiversity hotspots, and impacts on indigenous communities.
- ***distribution of impacts among people and across regions:*** Some regions, countries, islands, and cultures might be adversely affected by climate change, while others might or might not have net gains. Within countries, some regions or groups of people could be harmed more than others.
- ***global aggregate damages:*** The authors used a consistent method of measurement to aggregate impacts with global mean temperature increases, assessing whether change would be positive or negative, would occur smoothly or in more complex dynamic patterns, and whether aggregate impacts might mask unequal distribution of impacts.
- ***probability of extreme weather events:*** Increasing mean climate change alters the frequency, intensity, spatial extent, and duration of some extreme weather and climate events, such as heatwaves, extreme floods, droughts, and storms. This reason for concern considers whether the probability and consequences of such events might change as global mean temperature increased.
- ***probability of large-scale singular events:*** These include the breaking up of the West Antarctic Ice Sheet, the collapse of the North Atlantic thermohaline circulation, or destabilisation of international order by environmental refugees and the emergence of conflicts as a result of multiple climate change impacts.

The authors concluded it was not possible to combine the different reasons for concern. They reviewed the associations between temperature and impacts for each reason and drew preliminary conclusions about the potential severity and risk of such impacts. Because of substantial uncertainty with respect to the temperature at which impacts occur, the temperatures are approximate indications of impacts, not absolute thresholds. Furthermore, the authors note that global mean temperature does not describe all relevant aspects of impacts, such as the rate and pattern of change, changes in precipitation and extreme weather and climate events, or latent effects such as rising sea levels. Also, the authors did not factor in the potential role of adaptation.

Figure 2 presents a summary of their findings.[20] Climate change consequences are plotted against increases in global mean temperature (°C) after 1900. Each column corresponds to a specific reason for concern and represents a range of associated outcomes with increasing global mean temperature. The colour scheme represents progressively increasing levels of risk. Global mean temperature increased approximately 0.6°C between 1900 and 2000, which led to some impacts. The figure shows increasing risk with increasing temperatures, and that the temperature at which risks become apparent varies across the reasons for concern. For example, risks were already becoming apparent to unique and threatened systems, but global average temperature would need to increase by about 3–4°C before risks of large-scale discontinuities might become apparent. The figure addresses only how risks change as global mean temperature increases, not how risks might change at different levels of warming. It also does not address when risks might be realised, nor does it account for the effects of different development pathways on vulnerability.

20 Smith et al. (2009:4134).

***Figure 2: Reasons for Concern from the IPCC's Third Assessment Report[21]***

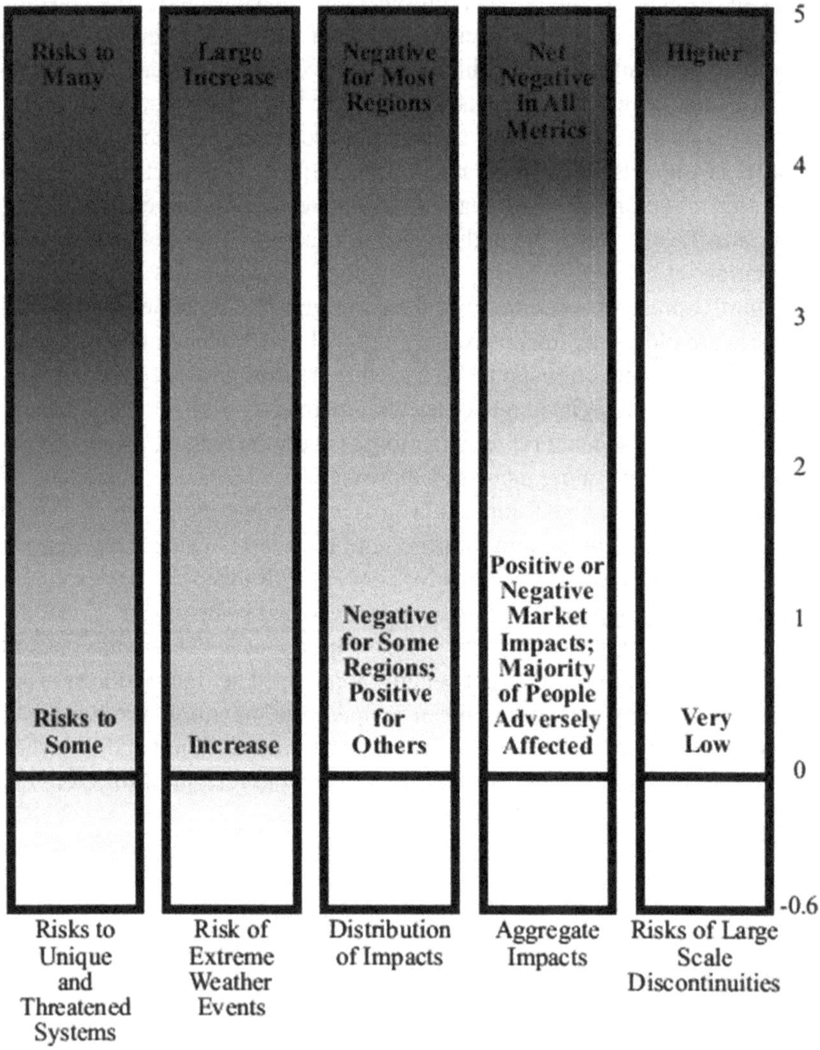


21 (ibid.).

In 2007, the IPCC's Fourth Assessment Report[22] revisited the reasons for concern assessed in the previous report. In addition to updating the assessment on these reasons, the authors also identified what they termed *key vulnerabilities*. Key vulnerabilities were impacts, vulnerabilities and risks that would merit policymakers' attention. Seven criteria were described to identify a key vulnerability: magnitude of impacts; timing of impacts; persistence and reversibility of impacts; likelihood of impacts and vulnerabilities, and confidence in those estimates; potential for adaptation; distributional aspects of impacts and vulnerabilities; and importance of the system(s) at risk. Key vulnerabilities were associated with many climate-sensitive systems, including food supply, infrastructure, health, water resources, coastal systems, ecosystems, global biogeochemical cycles, ice sheets, and modes of oceanic and atmospheric circulation.

The chapter concluded that the following appeared robust across a diverse set of studies:

- A risk-management framework was a useful approach for addressing key vulnerabilities. However, the assignment of probabilities to specific key impacts was often very difficult, due to the large uncertainties involved. Uncertainties were due to factors such as climate sensitivity, regional climate change, vulnerability to climate change, and adaptive capacity and the likelihood of bringing that capacity to bear.
- Mitigating climate change and reducing GHG emissions would reduce the risk associated with most key vulnerabilities. Postponement of such actions generally increased risks.
- Current atmospheric GHG concentrations and the range of projections for future climate change meant that some key impacts (e.g. loss of species, partial deglaciation of major ice sheets) could not be avoided with high confidence. The probability of initiating some large-scale events was very likely to continue to increase as long as GHG concentrations and temperature continued to rise.

Figure 3 summarises the reasons for concern from the IPCC's Fourth Assessment Report,[23] using the same approach as that adopted in Figure 2 (although the figure only appeared in a subsequent publication).[24] In the six

22 Schneider et al. (2007:779–810).
23 (ibid.).
24 Smith et al. (2009:4133–4137).

years between the two assessments, risks increased considerably for all reasons for concern.

***Figure 3: Updated Reasons for Concern from the IPCC's Fourth Assessment Report*[25]**

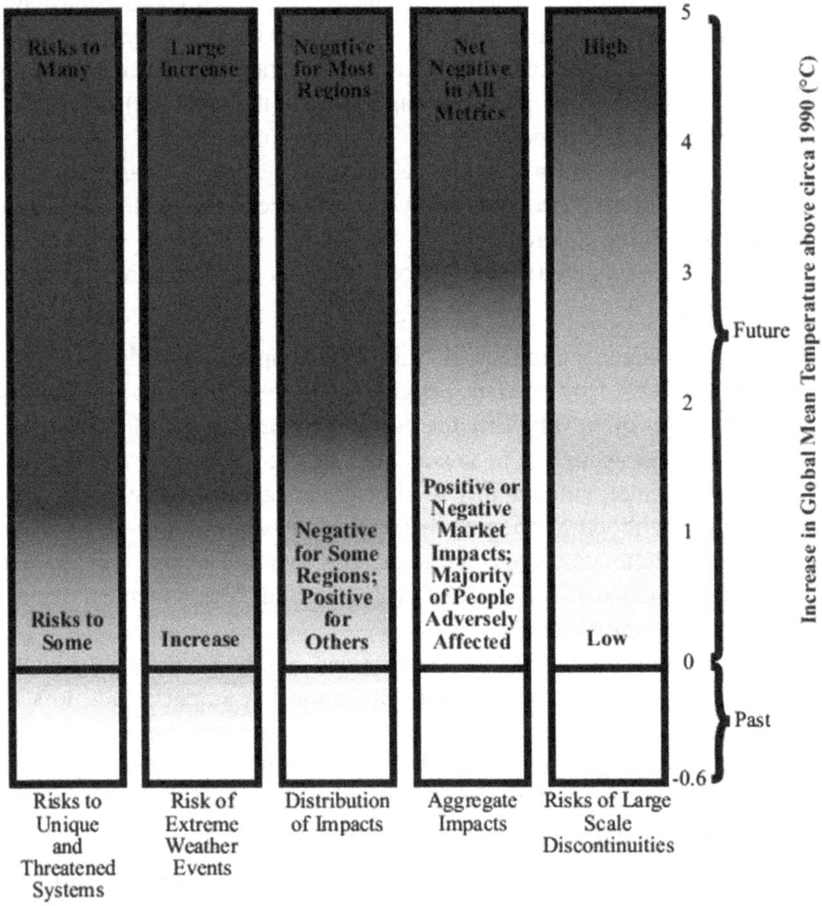


The chapter also concluded that adaptation could significantly reduce many potentially dangerous impacts of climate change and reduce the risk of many key vulnerabilities. However, the technical, financial, and institutional ca-

25 Schneider et al. (2007:779–810).

pacity, and the actual planning and implementation of effective adaptation, is quite limited in many regions. Furthermore, the risk-reducing potential of planned adaptation is either very limited or very costly for some key vulnerabilities, such as loss of biodiversity, melting of mountain glaciers, and disintegration of major ice sheets.

The Australia and New Zealand chapter of the IPCC's Fourth Assessment Report assessed the extent to which adaptation could reduce regional reasons for concern.[26] The left-hand panel shows global temperature change from the IPCC's Third Assessment Report, with the coloured curves representing temperature change associated with stabilisation at different carbon dioxide concentrations from 450 ppm to 1,000 ppm. The year of stabilisation is shown as black dots. The shaded area indicates the range of climate sensitivity across the stabilisation cases. The thin vertical lines next to the stabilisation curves show uncertainty in the year 2300. Crosses indicate warming in 2100. The right-hand panel summarises relative coping range, adaptive capacity, and vulnerability for critical sectors in this region, showing that, for example, the region has limited capacity to cope with further water insecurity, but would likely be able to cope with the impacts of an increase of 2°C on food security.

26 Hennessy et al. (2007:507–540).

***Figure 4: Vulnerability to Climate Change Aggregated for Key Sectors in the Australia and New Zealand Region***[27]

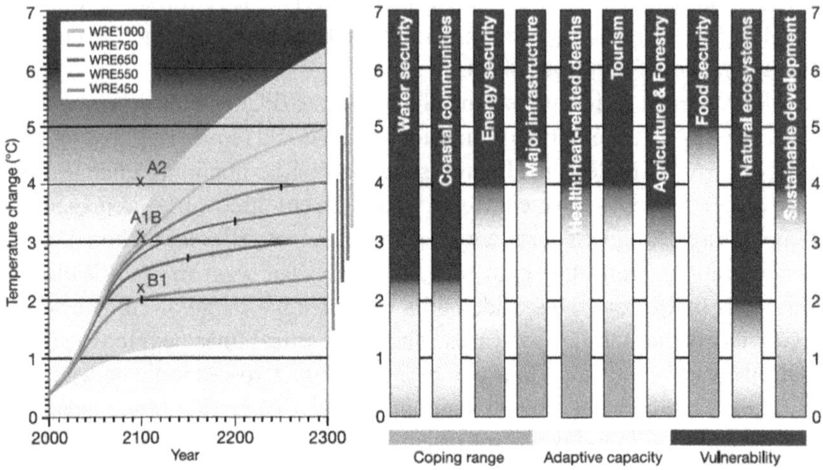


## G. Adaptation Reasons for Concern

Building on the 'reasons for concern' framework, Adger and Barnett[28] identified four reasons for concern regarding the ability to adapt to the identified impacts and the likelihood of sustainable adaptation:

- Contractions and uncertainties in the window of opportunity for adaptation
- The difference between adaptive capacity and adaptive action
- The risk of maladaptation, and
- The misguided measures of loss.

The first reason for concern is that the scale of projected changes and the interconnectedness of impacts mean that the window of opportunity for adaptation may be smaller than assumed. There will have to be major changes in policies and priorities if world governments commit to keeping the global mean temperature increase to less than 2°C above pre-industrial temperature; emissions reductions on the order of 80% or more by 2050

27 (ibid.:529).
28 Adger & Barnett (2009:2800–2805).

would be needed.[29] There is limited confidence that this is achievable, which means adaptation will be needed to ever larger changes in global mean temperature. It is hard to imagine the ability of societies to adapt to the significant impacts projected at temperature increases of 4°C or more, particularly on access to water and food security.[30] Ecosystems will transform into new states that may bear little resemblance to current functioning, and which would have potentially catastrophic consequences on the provision of ecosystem goods and services. Furthermore, these impacts are likely to interact – creating even more surprises.

In addition, as human and natural systems move into new territory with respect to weather patterns, there are increasing concerns about the possibility of crossing thresholds that result in disruptive regime shifts. For example, the Arctic is melting more rapidly than projected, which could lead to much larger increases in sea level over shorter time periods than many countries would be able to manage.[31]

The second reason for concern is that adaptive capacity will not necessarily translate into action. The assumption that it would is called the "adaptation myth".[32] For example, the US assumes the impacts of climate change will be within the limits of its ability to adapt. However, as one example illustrates, the number of annual natural catastrophes is rising faster in North America than anywhere else worldwide, with the increase entirely due to weather events.[33] The potential for weather-related losses in North America continues to rise due to socio-economic factors such as ongoing urbanisation and increasing property values. In addition, new technologies may create further risks.

The third reason for concern is the extent to which implemented adaptation policies and measures are not sustainable. Human choices have created path dependencies that may limit the range of future adaptation options in, for example, managing water resources because of the placement of dams and other infrastructure that may not be in optimal locations under future weather regimes, water rights agreements, etc.

The fourth reason for concern is that approaches to measure the success of adaptation often do not include social and cultural aspects. Adaptations

29 Meinshausen et al. (2009:1158–1162).
30 Parry et al. (2009:111).
31 Oppenheimer (2005:1399–1407).
32 Repetto (2009).
33 Munich (2012:12).

that do not take these perspectives into account may appear successful to policymakers, but may not be legitimate and equitable from the perspective of the communities involved.[34] An obvious example is the issue of managed retreat from small islands. Inhabitants of small islands such as Tuvalu have significant cultural, spiritual, familial, and historical ties with their land, which means that relocation would entail unbearable psychosocial losses.[35]

In addition to the reasons for concern raised by Adger and Barnett, estimates of the costs of adaptation may be much larger than societies are willing to pay. For example, a global estimate of the adaptation costs of just treating diarrhoeal disease and malaria due to climate change in 2010 was US$3–5 billion (in 2005 US$), with the costs expected to decline over time with improvement in basic health services; the estimate also assumed UN population projections and strong economic growth.[36] Aggregating the potential adaptation costs over many sectors (but not over all possible impacts) for the year 2030 leads to upper-end estimates of more than US$150 billion required in annual investment and financial flows to cover the costs of adaptation.[22] It is highly unlikely that governments will be willing or able to pay this amount annually.

Another reason for concern is that adaptation measures seen to be beneficial in the short and medium term may prove to be maladaptive in the longer run. For example, measures to protect communities in the exposed coastal zones of Bangladesh, if successful, could improve living standards and make the region more attractive to additional settlement. However, in the longer run, it seems highly likely that these lands will be inundated with rising sea levels and will have to be permanently abandoned. Thus, good short-term adaptation may only be palliative.

## *H. Discussion*

In the short term, the magnitude and extent of impacts of climate change – and, therefore, what societies may perceive to be dangerous anthropogenic interference with the climate system – will depend not only on the degree and rate of climate change, but also on the vulnerability of natural and social systems to these changes, and on the effectiveness of adaptation options to –

34 Barnett & Campbell (2010:211).
35 Montreux & Barnett (2009:105–112).
36 Pandey (2010).

- reduce exposures to a changing climate
- decrease the susceptibility of individuals, communities, nations, and regions to harm from these exposures, and
- increase their ability to prepare for, cope with, respond to, and recover from the exposures.

Although the UNFCCC indicates its ultimate objective as being to achieve stabilisation of GHG concentrations at a level that will allow time for ecosystems to adapt naturally; ensure that food production is not threatened; and economic development enabled in a sustainable manner, scientists, the general public, and policymakers are now considering a much wider range of impacts as indicating dangerous interference with the climate system, such as increases in the frequency and intensity of extreme weather events that disrupt societies and lead to security threats. Sectoral and regional assessments of the risks of climate change indicate a wide range of subglobal threats to human and natural systems.[37] Effectively anticipating and preparing for these risks requires a wide range of research – from obtaining a better understanding of approaching thresholds, to how to motivate appropriate behavioural change, to modifications of current and implementation of new strategies, policies, and measures addressing the risks of climate change. Adaptation research and practice can raise awareness of the impacts of climate change at local and regional levels, and of the financial and technical assistance required to avoid even more dangerous impacts than have been observed.

Over the longer term, the magnitude and extent of impacts will depend on the mix of adaptation and mitigation; rapid and successful reductions in GHG emissions will reduce how much adaptation will be needed. Slower and less comprehensive mitigation will increase the challenges to which human and natural systems will need to adapt. Effective and efficient adaptation may prevent dangerous impacts in some situations, although there are few studies estimating the trade-offs and associated costs.

Ultimately, the determination of what constitutes "dangerous" interference with the climate systems is a social choice that science can inform, highlighting the risks associated with various levels of climate change, the extent to which adaptation and mitigation can prevent or reduce those risks, and the associated costs and trade-offs that these actions will entail.

---

37 Bernstein et al. (2007).

## *References*

Adger, W. Neil & Jon Barnett, 2009, Four Reasons for Concern about Adaptation to Climate Change, *Environment and Planning* A 41, 2800–2805, doi:10.1068/a42244.

Anderson, Kevin & Alice Bows, 2011, Beyond 'Dangerous' Climate Change: Emission Scenarios for a New World, *Philosophical Transactions of the Royal Society* A 369, 20–44.

Barnett, Jon & John Campbell, 2010, *Climate Change and Small Island States: Power, Knowledge and the South Pacific*, London, Earthscan.

Bernstein, Lenny, Peter Bosch, Osvaldo Canziani, Zhenlin Chen, Renate Christ, Ogunlade Davidson, William Hare, Saleemul Huq, David Karoly, Vladimir Kattsov, Zbigniew Kundzewicz, Jian Liu, Ulrike Lohmann, Martin Manning, Taroh Matsuno, Bettina Menne, Bert Metz, Monirul Mirza, Neville Nicholls, Leonard Nurse, Rajendra Pachauri, Jean Palutikof, Martin Parry, Dahe Qin, Nijavalli Ravindranath, Andy Reisinger, Jiawen Ren, Keywan Riahi, Cynthia Rosenzweig, Matilde Rusticucci, Stephen Schneider, Youba Sokona, Susan Solomon, Peter Stott, Ronald Stouffer, Taishi Sugiyama, Rob Swart, Dennis Tirpak, Coleen Vogel & Gary Yohe, 2007, *Climate Change 2007: Synthesis Report*, Geneva, Intergovernmental Panel on Climate Change.

Burton, Ian, Achala Chandani Abeysighe & Thea Dickinson, forthcoming 2013, UNFCCC Article 2 Revisited, Briefing Paper, International Institute for Environment and Development (IIED).

Carson, Rachel, 1962, *Silent Spring*, New York, Houghton Mifflin.

Hennessy, Kevin, Blair Fitzharris, Bryson C. Bates, Nick Harvey, Mark Howden, Lesley Hughes, Jim Salinger and Richard Warrick, 2007, Australia and New Zealand, in: Parry, Martin L., Osvaldo F. Canziani, Jean P. Palutikof, Paul J. van der Linden & Claire E. Hanson (Eds), *Climate Change 2007: Impacts, Adaptation and Vulnerability. Contribution of Working Group II to the Fourth Assessment Report of the Intergovernmental Panel on Climate Change*, Cambridge, Cambridge University Press, 507–540.

IPCC/Intergovernmental Panel on Climate Change, 2012, Summary for Policymakers, in: Field, Christopher B., Vicente Barros, Thomas F. Stocker, Dahe Qin, David J. Dokken, Kristie L. Ebi, Michael D. Mastrandrea, Katharine J. Mach, Gian-Kasper Plattner, Simon K. Allen, Melinda Tignor & Pauline M. Midgley (Eds), *Managing the Risks of Extreme Events and Disasters to Advance Climate Change Adaptation, A Special Report of Working Groups I and II of the Intergovernmental Panel on Climate Change*, Cambridge/New York, Cambridge University Press, 3–21.

Lackey, Robert T. & Roger L. Blair, 1997, Science, Policy, and Acid Rain: Lessons Learned, *Renewable Resources Journal* 15, 9–13.

Likens, Gene E. & F. Herbert Bormann, 1974, Acid Rain: A Serious Regional Environmental Problem, *Science* 184, 1176–1179.

Meinshausen, Malte, Nicolai Meinshausen, William Hare, Sarah C.B. Raper, Katja Frieler, Reto Knutti, David J. Frame & Myles R. Allen, 2009, Greenhouse-gas Emission Targets for Limiting Global Warming to 2°C, *Nature Letters* 458, 1158–1162.

Montreux, Colette & Jon Barnett, 2009, Climate Change, Migration and Adaptation in Funafuti, Tuvalu, *Global Environmental Change* 19, 105–112.

Munich RE, 2012, *Severe Weather in North America: Perils, Risks, Insurance – Executive Summary*, Munich, Munich RE.

Oppenheimer, Michael, 2005, Defining Dangerous Anthropogenic Interference: The Role of Science, the Limits of Science, *Risk Analysis* 25, 1399–1407.

Pandey, Kiran, 2010, Costs of Adapting to Climate Change for Human Health in Developing Countries, World Bank Discussion Paper 11, December.

Parry, Martin, 2010, Copenhagen Number Crunch, *Nature Reports Climate Change* 4, 18–19.

Parry, Martin, Nigel Arnell, Pam Berry, David Dodman, Samuel Fankhauser, Chris Hope, Sari Kovats, Robert Nicholls, David Satterthwaite, Richard Tiffin & Tim Wheeler, 2009, *Assessing the Costs of Adaptation to Climate Change: A Review of the UNFCCC and Other Recent Estimates*, London, International Institute for Environment and Development.

Repetto, Robert, 2009, The Climate Crisis and the Adaptation Myth, Working Paper 13, New Haven, CT, School of Forestry and Environmental Studies, Yale University.

SCEP/Study of Critical Environmental Problems, 1970, *Man's Impact on the Global Environment: Assessment and Recommendations for Action*, Cambridge, MA, MIT Press, available at http://mitpress.mit.edu/books/mans-impact-global-environment last accessed 15 January 2013.

Schipper, E. Lisa F., 2006, The History of Adaptation in the UNFCCC Process, *Review of European, Comparative and International Environmental Law* 15, 82–92.

Schneider, Stephen H., Sergey Semenov, Anand Patwardhan, Ian Burton, Chris H.D. Magadza, Michael Oppenheimer, A. Barrie Pittock, Atiq Rahman, Joel B. Smith, Avelino Suarez & Farhana Yamin, 2007, Assessing Key Vulnerabilities and the Risk from Climate Change, in: Parry, Martin L., Osvaldo F. Canziani, Jean P. Palutikof, Paul J. van der Linden & Claire E. Hanson (Eds), *Climate Change 2007: Impacts, Adaptation and Vulnerability. Contribution of Working Group II to the Fourth Assessment Report of the Intergovernmental Panel on Climate Change*, Cambridge, Cambridge University Press, 779–810.

Smith, Joel B., Hans-Joachim Schellnhuber, M. Monirul Qader Mirza, Sam Fankhauser, Rik Leemans, Lin Erda, Laban Ogallo, A. Barrie Pittock, Richard Richels, Cynthia Rosenzweig, Uriel N. Safriel, Richard S.J. Tol, John Weyant & Gary Yohe, 2001, Vulnerability to Climate Change and Reasons for Concern: A Synthesis, in: McCarthy, James J., Osvaldo F. Canziani, Neil A. Leary, David J. Dokken & Kasey S. White (Eds), *Climate Change 2001: Impacts, Adaptation, and Vulnerability. Contribution of Working Group II to the Third Assessment Report of the Intergovernmental Panel on Climate Change*, Cambridge, Cambridge University Press, 915–967.

Smith, Joel B., Stephen H. Schneider, Michael Oppenheimer, Gary W. Yohe, William Hare, Michael D. Mastrandrea, Anand Patwardhan, Ian Burton, Jan Corfee-Morlot, Chris H.D. Magadza, Hans-Martin Fussel, A. Barrie Bittock, Atiq Rahman, Avelino Suarez & Jean-Pascal van Ypersele, 2009, Assessing Dangerous Climate Change through an Update of the Intergovernmental Panel on Climate Change (IPCC) "Reasons for Concern", *Proceedings of the National Academies of Sciences* 106, 4133–4137.

Solomon, Susan, Gian-Kasper Plattner, Reto Knutti & Pierre Friedlingstein, 2009, Irreversible Climate Change due to Carbon Dioxide Emissions, *Proceedings of the National Academies of Sciences* 06, 1704–1709.

Sustainability Institute, 2010, Copenhagen Accord Submissions, 4 February, available at http://climateinteractive.org/, last accessed 15 January 2013.

24
# Climate Change, Disasters and Migration: Current Challenges to International Law

*Erika Pires Ramos*

> We must make no mistake. The facts are clear: climate change is real; it is accelerating in a dangerous manner; and it not only exacerbates threats to international peace and security, it is a threat to international peace and security.
>
> Extreme weather events continue to grow more frequent and intense in rich and poor countries alike, not only devastating lives, but also infrastructure, institutions, and budgets – an unholy brew which can create dangerous security vacuums.
>
> *Ban Ki Moon, Security Council 6587th Meeting,* New York, 20 July 2011

*Abstract*

The increasingly frequent occurrence of extreme environmental events and the gradual degradation of essential environmental resources, which seriously compromise the life and safety of individuals, communities and entire populations/nations around the world, can derail the survival of people in their places of origin – leading to new legal situations that need to be regulated by international law. Based on this scenario, two central points guide this study: the first is the emergence of a new category of persons in the international order; the second is the lack of specific legal protection by international instruments. Without intending to isolate the environmental triggers of other related causes, this article proposes a multicausal look into the causes generating such forced displacement. This is done through a brief analysis of the initiatives aimed at the recognition and protection of those affected, especially in situations when environmental and human vulnerability are greater than resilience and responsiveness. In this sense, an integrated legal approach is necessary to deal with migration caused by global environmental changes and the adoption of new commitments founded on global responsibility and international solidarity should be considered.

### A. Introductory Notes: Contextualising a New Emergent Issue

The phenomenon of environmental migration is an age-old and unquestionable reality. Extreme events and major environmental disasters have existed throughout history, forcing individuals and groups to move. However, the increasingly frequent changes in the global environment, caused or accelerated by human action, are already at levels considered intolerable and irreversible. These changes have challenged those involved in the various areas of knowledge to develop effective mechanisms to mitigate negative environmental impacts, restore the deteriorated environment as far as possible, and prevent new threats of degradation.

Migration induced by environmental causes shows a clearly increasing trend – with the escalation seeming to occur parallel with the worsening global environmental crisis. Human beings are at the centre of this process and are doubly exposed because of the progressive destruction of ecosystems and biodiversity that they depend on and the disappearance of the territories where they live. Thus, the processes of degradation of the global environment cannot be considered only as an environmental concern, but should be analysed together with the need to protect the rights of affected people and ensuring dignity and respect for the human being who is in a position of special vulnerability.

In this sense, it seems clear that among the numerous dimensions to be considered in a context of extreme changes in the global environment is the human dimension. Environmental degradation, whether caused by natural phenomena or exacerbated by human action, is a known factor contributing to increased forced migration within the territory of the state or beyond its borders. The reverse is also confirmed: the growing number of people affected by extreme events (natural or man-made) can also be considered an important indicator of the extent and degree of global environmental deterioration.[1] In situations of disaster, the people affected need immediate assistance in the form of food, medicine and shelter, as well as the reconstruction of the environment and assurance of return, or resettlement elsewhere. If the disaster can be managed locally, the aid is usually provided by the government and local organisations. If not, international assistance will be required.[2]

1 Jacobson (1988:7).
2 El-Hinnawi (1985:20).

It is estimated that by 2050 there will be 200 million people who have had to leave their homes owing to degradation processes and environmental disasters, especially as a result of climate change. In 2010, there were already 50 million people in this condition – exceeding the number in the conventional category of refugees.[3] The World Disasters Report released in 2012 by the Red Cross reveals that there were approximately 29.9 million persons internally displaced (IDPs) by environmental factors in the year 2011, including those affected by development projects.[4] Despite the alarming data, this category of people remains without proper recognition or specific protection under international law. The issue of migration driven by environmental factors, which emerges from the debate about climate change, thus reveals novel legal situations not covered by international law, and demanding new institutional strategies, new forms of cooperation and long-term commitments, since the mechanisms and existing international legal instruments were not designed to address this new demand from a global perspective.

This article intends to emphasise the existence of an important gap in international law that must be filled urgently and, therefore, provides tools to assist *stakeholders* and *decision-makers* in the challenge to create a specific system of international legal protection to this new category of people. The point of departure is an integrated approach, suitable to cope with the complexity of the issue: addressing vulnerabilities, establishing and implementing responsibilities.

### B. *People Affected by Global Environmental Pressures: The Dilemma of a Legal Definition*

It is not intended here to detail all possible causes and impacts of environmental change, but only to draw attention not only to their complexity and the need to analyse important aspects that are not currently at the centre of political negotiations, but also their impact on the law and international re-

3 A study of the Institute for Environment and Human Security of the United Nations University (UNU-EHS) mentioned by the Red Cross indicates a greater number of people displaced by environmental disasters due to the conflict, see UNU-EHS (2005).

4 The recently launched World Disasters Report of Red Cross, IFRC (2012), shows an estimated 57.7 million internally displaced persons by environmental factors in 2010, surpassing the figures published in 2005.

lations. The causes generating flows of environmental refugees are fairly broad and have their origin in climate change, natural disasters caused by non-climate factors (with or without human intervention), environmental degradation processes (caused or accelerated by human action) or by the combined action of these factors.

Besides sudden events such as tsunamis, earthquakes, hurricanes, floods, whose magnitude and effects are easy to see, it has been observed that over-exploitation, desertification, pollution and scarcity of environmental resources, whether continuous or progressive (less visible over the short term), may also seriously jeopardize human life and biodiversity in many regions, making them unproductive and uninhabitable in the long term – thus unfit for survival. Likewise, the implementation or inadequate management of projects that have the potential to degrade can also generate negative impacts with irreversible effects to the environment and quality of human life.

On the other hand, it is relevant to highlight that environmental changes will impact differently on regions and locations because of their different geophysical aspects and the varying responsiveness of local structures (social, political and economic) to environmental pressures. While recognising the multiple reasons for human displacement and the difficulty in isolating its causes, environmental migration can be characterised by events (*triggers*) that trigger the migration process. These events may be natural, anthropogenic or mixed, and of such magnitude that they could seriously jeopardize the life and safety of individuals and groups of any particular locality or region. It is also important to note that environmental factors interact with other factors that go beyond the environmental issue. Consider, for example, that the crises that afflict the world economy today are environmental in their origin and involve climate change, pollution, water shortages, loss of biodiversity, decline in arable land, depletion of ocean fishing areas, depletion of oil, pockets of poverty, the threat of pandemics and inequality in the appropriation of resources. It is a mistake to address these issues separately: in order to address them properly, there needs to be an adequate understanding that the issues are connected through a relationship of cause and effect.[5]

Moreover, there are cases where environmental degradation is a consequence rather than a cause of migration. In cases of violent conflicts and wars, for example, the destruction of the environment is hardly a 'natural' consequence and is often adopted as a strategy by the parties in conflict. In

5 Sachs (2008:XII).

this case, therefore, the trigger is the conflict itself, without which the population would not have been compelled to migrate, and not environmental degradation per se. Therefore, the trigger to the displacement or refugee process in this case is not the environment, and thus cannot be called an environmental refugee situation. It is important to note also that the conflict may be a consequence of both the struggle for control of the exploitation of certain natural resources and the environmental degradation processes that led to such a dispute. In this case, the environment will also serve as a trigger for the environmental conflict.

It is true that the enormous complexity and new dimensions of global migration call into question the established categories, migration policies and existing international norms on the subject. In the context of current changes in the global environment, adverse environmental effects of climate change are becoming more visible with the increasing frequency of disasters. Less explored aspects of this process of transformation of the natural and human environment should be examined, such as the protection of human beings in cases of extreme environmental events.

However, the complexity of the interaction between causes that generate migration and questions about the real existence of a direct causal link between environmental change and migration should not represent obstacles to finding solutions. In this context, one enters the debate about the need for recognition by international law of individuals and groups that move because they are driven by environmental degradation, and are understood as a new category of persons worthy of protection and assistance: the so-called environmental refugees. It is important to note that the global environmental changes can affect individuals and groups within states where they occur, as well as neighbouring states or non-affected states (transboundary impacts). In addition to the environmental impacts, it is important to consider the impacts that arise from displacement of people, especially in the case of external displacement, which can affect several states in the migration routes. Thus, there is not always a relationship of cause and effect between the places where the environmental changes occur and locations of origin and destination of forced migration flows induced by environmental causes. The *global* effects of climate change, for example, confirm this thesis.

In this sense, the relevance of the definition of environmental refugees proposed by Essam El-Hinnawi in 1985 lies precisely in analysing the complexity of the phenomenon:[6]

> In a broad sense, all displaced people can be described as environmental refugees, having been forced to leave their original habitat (or having left voluntarily) to protect themselves from harm and/or to seek a better quality of life. However, for the purpose of this book, environmental refugees are defined as those people who have been forced to leave their traditional habitat, temporarily or permanently, because of a marked environmental disruption (natural and/or triggered by people) that jeopardized their existence and/or seriously affected the quality of their life. By "environmental disruption" in this definition is meant any physical, chemical and/or biological changes in the ecosystem (or the resource base) that render it, temporarily or permanently, unsuitable to support human life. According with this definition, people displaced for political reasons or by civil strife and migrants seeking better jobs purely on economics ground are not considered environmental refugees. There are three broad categories of environmental refugees. First, there are those who have been temporarily displaced because of an environmental stress.... The second category of environmental refugees comprises those who have to be permanently displaced and resettled in a new area.... The third category of environmental refugees consists of individuals or groups of people who migrate from their original habitat, temporarily or permanently, to a new one within own national boundaries, or abroad, in search of a better quality of life.

This definition establishes a relationship between the growing number of natural disasters (droughts, floods, cyclones and earthquakes) with the number of people affected because of these and other disturbances or environmental stresses (poverty, hunger, the negative impacts of development, industrial accidents) in developing and underdeveloped countries.

Related to characterising the phenomenon of forced displacement induced by environmental factors and defining its legal name, it seems that the alternatives presented to the use of the term environmental refugees – a name which faces strong resistance when the discussion arises in the framework of International Refugee Law – have also failed to find acceptance. To date, none of these descriptions – *environmentally displaced persons*, *climate refugees*, *environmentally forced migrants*, *eco migrants*, *eco evacuated*, *eco victims* – were able to demonstrate the complexity and breadth of this phenomenon.

Admittedly, the terminology and definition are important elements to assess the feasibility of adaptation and application of existing instruments and

6 El-Hinnawi (1985:4–5).

mechanisms in international law or for the construction of a special protection system. When bringing the debate on the nomenclature and the usefulness of the term environmental refugees to the field of law, the conclusion will depend on the approach that each author intends to adopt.[7] If the conventional use of the term *refugee* is taken as basis, in fact, it will be more difficult to assimilate extended criteria for the granting of refugee status for environmental refugees by international refugee law and the mechanisms already established by the Convention Relating to the Status of Refugees (1951) and its Protocol (1967), especially with regard to the element of "well-founded fear of persecution" and its determinant reasons. However, this argument does not hold when one goes beyond this perspective. It is also noted that this discussion seems to have more sense in theory than in practice. Supposedly, the adoption of the term environmental refugee, either by extending the original concept, or as a new category, is not legally impossible, although it was initially conceived without an immediate concern with the possible legal implications or policies regarding the use of the expression.

At the regional level (Latin America), this debate was prompted in 1989, under the Cartagena Declaration on Refugees (1984), specifically on the topic of implementation of the expanded concept of *refugee* adopted in this Declaration for people affected by environmental disasters.[8] The conclusion was to the effect that natural disasters, unlike climate change, would not be man-made, and this was the reason why one could not consider such events as "other circumstances which have seriously disturbed public order". Until now, the forced displacement induced by environmental pressures was not formally recognised by this declaration.

On one hand, such an approach would create an impermissible discriminatory treatment of individuals and groups in a similar situation as a single category (climate or natural disaster refugees, for example). If such an approach had been adopted at the time, the victims of natural disasters, who also need protection either in domestic law or under international law, would

7 Wijnberg (2007:3) identified multiple views or perceptions about the construction of the term environmental refugees and concluded that only the legal view, being more traditional, does not accept another construction of the term refugee than the conventional.

8 The debate is contained in the document on *Principles and Criteria for the Protection and Assistance of Central Americans Refugees, Returnees and Displaced Persons in Latin America* presented by a committee of legal experts at the International Conference on Central American Refugees held in Guatemala on 29–31 May 1989. See Trindade (1993:133–134).

have remained unprotected. Leaving aside the approach of traditional instruments, and taking into account that categorisations and institutions must follow the dynamics of global transformations and not remain bound to traditional approaches, the use of the term environmental refugee (external mobility) and environmentally displaced (internal mobility) would be acceptable as a definition for new legal categories. In essence, they have the same origin and are still waiting for an effective system of protection at the global level. The need for caution with the legal definition has to be registered, however, in order to avoid the possible consequence that a very comprehensive name may generate an undesired trivialisation of the phenomenon, further hindering the construction of a protection system that effectively fulfils the needs of affected individuals and communities, and thus contributing negatively to maintaining the existing legal gap.

Underlying this debate is the political-normative question of how far the international community and its institutions are willing to commit themselves permanently to the protection of one more category of people and expand their 'mandates'. In this sense, the acceptance or rejection of the terminology influences the decisions made by members of the international community, especially when it comes to multilateral negotiations that require long-term actions to address emerging issues, such as migration flows driven by impacts and environmental pressures that often transcend national borders.

To define the rights of internally displaced persons within their states or of those that had to move abroad owing to drastic changes in the environment, it is therefore necessary to consider the causes and possible scale effects of the displacements in order to identify the needs of affected populations. Thus it would be possible properly to assign responsibilities and formulate strategies for action at the international level, which should also be internalised by states.

Given this, a shift in focus on the theme is essential, detaching it from a classic approach on the topic of refugees and the anthropogenic factors as the only ones capable of generating migration flows in order to overcome the impasse that dealt with in the previous section, especially the terminological barrier. It might also be useful to create a typology or categorisation for the environmental refugees by cause: climate refugees/IDPs, disaster refugees/IDPs, development refugees/IDPs, conservation refugees/IDPs, etc. It is necessary to acknowledge that the different definitions and classifications proposed, although not infallible, are important tools to guide decision-makers and, as such, require flexibility to adapt to the dynamic changes of the global environment which are not restricted to climate change. Thus, the following aspects are essential for an adequate characterisation of environmental refugees: nomenclature; definition; detailed description of the natural and anthropogenic phenomena determinant for the generation of migration and possible interactions with economic, political and social factors; and mapping of environmental and human vulnerability (identification of priority risk areas or already affected) and identifying the needs of affected individuals and groups. It can be concluded that human mobility with environmental motivation, in most cases, is forced, irregular, collective and, depending on the severity and extent of environmental pressures, may be temporary or permanent, internal or external. Environmental migration, in most cases, is a specie of reactive migration, manifesting itself in response to environmental perturbations that threaten human life and safety (sudden onset), but can also be of a proactive nature when the processes of environmental deterioration are gradual (slow onset). The nomenclature or terminology to be used, according to the identified tendencies, depends on the option for the protection regime to be adopted. A new scheme could be developed jointly establishing a nomenclature and definition, without standing as an obstacle to progress in the search for solutions. However, the utmost care must be taken not to adopt an overly broad or restrictive term and definition, which may affect access to a special protection system.

The fact is that the recognition of a new category, regardless of the name that is adopted, has motivated numerous controversies inside and outside the international refugee regime. However, we must emphasise, beyond any controversy, the greater urgency to overcome this issue.

Official recognition of environmental refugees or environmentally displaced persons in international instruments certainly would provide a deeper understanding of the main causes of environmental deterioration and better tools to deal with them. It would therefore be a major step towards finding a lasting solution to the problem, through preventive policies and actions to combat the structural causes of environmental migration at the global, regional and local levels. The legal and environmental dimensions of the issue should be complementary and not mutually exclusive.

### C. The Rule of Law: Are Existing Regimes in International Law Sufficient to Provide Protection?

The recent experience of major environmental disasters worldwide has raised numerous concerns about the situation of human rights protection at the domestic and international level. The main concern of this article are the conditions of individuals, groups and communities in the event of environmental disasters and the need to enforce the protection of their rights in those exceptional situations; where the instability generated as a result of these phenomena causes fundamental rights of affected victims (potential and actual) to be entirely or seriously violated by the environmental deterioration. What is observed is that the treatment of environmental disasters internationally, as a rule, has been dealt with via an operational approach predominantly through the coordination of efforts to obtain financial resources to provide assistance to victims.

It is here that a deeper reflection on the theme of integrating human rights protection with prevention and response strategies to different types of disasters or environmental disturbances is necessary. Once an extensive list of rights not only during but after the occurrence of a disaster has been identified, one should also strengthen the preventive dimension of protection.[9] The absence of regulation and international standards of protection can lead to situations of injustice and discrimination towards people who are in the same circumstances in different states. That is, the protection conferred by the instruments of human rights needs to be amended or new instruments must be created specifically to address this issue and to do so in a uniform manner, so as not to further aggravate the situation of insecurity and human rights violations of the victims affected, especially when the displacement is forced beyond state borders.

Therefore it is possible to contend that the overall protection provided by general instruments – focusing on the fundamental principle of the dignity of the human being – should be the foundation and final rationale for the protection of environmental refugees. In general, it is considered that environmental refugees are already covered by existing universal instruments, but without specific instrumentation, the normative gap subsists. There is no doubt that we are facing flagrant violation of human rights – which must be fought.

9 Prieur (2010).

Thus, a comprehensive legal treatment, which is fair and equitable for that category, should be established with due urgency to provide care, without discrimination, for the special and pressing needs of this category, and to establish a global commitment based on shared responsibility between states and solidarity between states and non-state actors with respect to all people forced to leave their roots in order to save their lives and provide protection outside their places of residence (inside or outside their own country) when such persons are severely compromised by degradation of the environment.

An integrated and sensitive approach to this new global challenge proves to be essential so that we can properly deal with the specificities and multiple facets of the problem. The difficulty regarding the categorisation of forced migration, demonstrated in the previous section, only reinforces the need for a sufficiently broad legal definition to ensure minimum standards of protection at international level. In the case of environmental refugees, an adequate protection depends on the formal recognition of the legal status of individuals and groups severely affected by environmental events whose survival and safety require urgent international assistance, even when they remain within the borders of their own state. Accordingly, although the Guiding Principles on Internal Displacement constitutes an important tool for the development of protective policies at the national level and for the contribution to the development and expansion of a broader scheme itself for the environmental refugees theme, they are not sufficient in themselves for guaranteeing effective protection, since they lack breadth (they apply only in situations of internal displacement) and the necessary binding force. From the point of view of external displacement, the International Refugee Law seems also inadequate to ensure a comprehensive and lasting solution to the issues of environmental refugees, because the system was designed with a more restrictive view and has remained so until the present time.

It is clear, therefore, that environmental refugees, resembling in part classical refugees and in part IDPs, remain doubly exposed, either because they are not legally recognised as refugees or as any other conventional category of people leaving the country, or because there is no specific and binding international instrument that guarantees effective international assistance if the state of origin does not have the capacity to protect its nationals.

This justifies the use of the term environmental refugees to describe a new category, which requires a special legal regime, broader than the conventional regime of refugee protection and the current treatment given to IDPs, which are not binding. A new regime has to be broader than that granted to the conventional refugee, in order to reach individuals and groups in need of international protection because of forced displacement by serious environmental pressures, even when it occurs internally. That said, environmental refugees may be considered, including from a legal point of view, as special kinds of migrants, with classical refugees' features (forced uprooting and need of international protection) as well as distinctive features of their own. A new legal category in the global international order is, therefore, a challenge to be faced. One must remember, however, that the international regime of refugee protection is the product of its time. The social dynamics inaugurated with the process of globalisation indicates that the problem of environmental refugees tends to be increasingly comprehensive and will require extensive planning. The figures presented in the previous section confirm this trend.

Furthermore, political aspects complicate a solution to the issue, especially regarding the practical implications of a possible change in the system of international protection to refugees. Thus, the difficulties of expanding the category of refugees to include those affected by environmental factors remain, as do the blurring of responsibilities for states, both domestically and internationally. However, it is possible to outline several initiatives targeted at obtaining the legal recognition of environmental refugees from different perspectives. The first was proposed by the government of Maldives in 2006. It sought the adoption of a new protocol to the Convention Relating to the Status of Refugees of 1951, with the aim of reducing and preventing losses from disasters caused by natural, anthropogenic or mixed factors, involving human beings, environmental resources, and biodiversity in its multiple dimensions: environmental, economic, social and cultural.[10] It should be noted that the proposed new protocol was not an isolated action of the government of the Maldives, but involved their ministries, UN delegations, participation of interested states,[11] meetings with representatives of inter-

10 Republic of Maldives, Ministry of Environment, Energy and Water (2006). A working draft of this proposal was written by Michael See.

11 Other states (Angola, Argentina, Azerbaijan, Comoros, Ethiopia, Guinea-Bissau, Liberia, Tajikistan, Rwanda, Sri Lanka and Tuvalu) have shown interest in collaborating in the development of the proposal with concrete ideas or simply declaring support for the initiative. Malaysia brought the experience of cooperation assistance for environmental refugees of the 2004 Tsunami. Subsequently joining the initiative were: Bangladesh; Canada; Egypt; Ecuador; Monaco; Asian Development Bank; Environment, Conflict and Cooperation, Germany; European Commission, Environment Directorate General; Foundation for International Environmental Law and Development (FIELD), UK; Friends of the Earth, Australia; Kyoto USA; Tides

national organisations and programmes (IFRC – International Federation of Red Cross and Red Crescent Societies, WHO – World Health Organization, UNICEF – United Nations Children's Fund, UNDP –United Nations Development Programme, the Japanese agencies, JICA – Japan International Cooperation Agency and JOCV – Japan Overseas Cooperation Volunteers, besides the collaboration of NGOs, among which LISER – Living Space for Environmental Refugees), and researchers.[12]

The essence of the Protocol on Environmental Refugees was the review of the key elements of the 1951 convention, expanding the scope of its application from a typology of causes established by the new protocol: "both anthropogenic origins and force majeure so that remedial measures are not disproportionate, but applied in the measures as possible equal to the damage and post-damage scenarios irrespective of origin."

This proposed protocol introduces comprehensive environmental causes as motivation for granting refugee status, ensuring protection even in disaster situations where there is no human interference. The proposal also provides for protection in situations of internal displacement in order to ensure that international aid is always within range of individuals who need it. Despite the pioneering initiative and relevance of the changes suggested in the proposal, it must nevertheless be recognised that the problem of environmental refugees requires more than the recognition of the legal status (which is presently the central occupation of the international protection system), which is the first step towards the formal existence of the category. The current scenario of increasingly restrictive immigration policies hinders any renegotiation or revision of the refugee protection regime. As already demonstrated, climate change has brought the issue of environmental refugees to the international agenda, since global climatic conditions have contributed greatly to increasingly extensive and more frequent forced human mobility. Thus, the proposals under the United Nations Framework Convention on Climate Change (UNFCCC) also deserve mention. In general terms, the international regime on climate change comprises the UNFCCC and the Kyoto Protocol. The convention provides a framework for action and cooperation to the states in regard to climate change. The Kyoto Proto-

Center; KK Chow. Messaging and communications: European Investment Bank, Sweden, Professor Norman Myers.

12 A meeting with the United Nations High Commissioner for Refugees (UNHCR) was planned as part of the agenda of the discussions about the protocol. However, no results have been reported in the document (draft protocol).

col, in turn, imposes quantifiable obligations on member states within their responsibilities and capabilities to reduce their emissions of greenhouse gases.

Until this moment, the main instruments (convention and protocol[13]) and the recent decisions of the Conference of Parties (COP)/Meeting of Parties (MOP) that comprise the current climate change regime do not directly address the issue of legal protection to people in case of forced displacement, while focusing on related issues such as mitigation (COP15), adaptation (COP16), disaster risk reduction (COP17) and loss and damage (COP18). Indeed, the implementation of the climate regime to protect environmental refugees is limited by the structure of the convention itself:

> In our opinion, the UNFCCC – despite its focus on adaptation – has structural limitations to deal with the displacement due to climate change. Displacement is not their focus, their concerns are different. Its structure and institutions are not designed to meet the offset and the problems associated with it. Furthermore, as the climate change conference in Copenhagen reveals the UNFCCC cannot be easily changed to accommodate displaced persons to climate change; deal with the existing provisions is already problematic.
> These arguments are succinctly put by two lawyers from Harvard University who observe that the UNFCCC has legal limitations to deal with the displacement of climate change. As an environmental law treaty, the UNFCCC, they say, is mainly concerned with the relations between states, but do not discuss the rights that states have to grant to individuals or communities, such as those established for human rights or the right of refugees. It is also preventive in nature and less focused on corrective actions that are necessary in the context of refugees. Finally, although the UNFCCC has an initiative to help states with adaptation to climate change, the program does not specifically address the situation of climate change refugees. As the refugee regime, the UNFCCC was not designed for, and so far have not adequately addressed the problem of climate change refugees.

In theory, therefore, the implementation of this scheme for the category of environmental refugees, understood in its broad dimensions (internal and external), could only be made possible if it were modified to include this issue, which can be done during the conferences and meetings of the convention parties taking place annually. However, nothing concrete has been done in this direction so far.

---

13 The UNFCCC was adopted by Brazil on 9 May 1992 and promulgated by Decree No. 2652 of 1 July 1998. The Kyoto Protocol was adopted on 11 December 1997 and promulgated by Decree No. 5445 of 12 May 2005.

About the specific issue of persons displaced by climate change, David Hodgkinson, Tess Burton, Heather Anderson and Lucy Young propose the adoption of a Convention for Persons Displaced by Climate Change,[14] which aims to overcome existing gaps in current systems of protection of human rights, refugee and humanitarian law.

Therefore, such a convention would cover internal and external displacement, establish criteria for the designation of a mass status to climate-change-displaced persons, and obligations of protection and assistance shared between states of origin, host states and the international community, based on the principle of common but differentiated responsibilities.

One cannot fail to recognise the importance of the essence of this proposal, which clearly seeks to integrate elements and mechanisms of various international systems in order to address specifically the human dimension of climate change. However, the absence of a convention draft text hinders further analysis in this article. It is also necessary to recognise that there is no coherent multilateral governance framework that adequately protects climate-change-displaced people. There has been no coordinated response by governments to address human displacement as a result of climate change. Given the nature and magnitude of the problem, ad hoc measures based on existing domestic regimes are likely to lead to inconsistency, confusion and conflict.[15] However, in the experience of the author, such a system would be born with already limited application potential, since it would reach individuals and groups affected by environmental changes caused only by climate factors. In this sense, the construction of an international system of protection only for the category of climate refugees does not seem justifiable from the point of view of human rights protection, since the scope would be restricted to the description of people and communities affected by environmental causes arising from climate change, and exclude all those affected by environmental disasters caused by non-climatic factors but requiring identical protection.

The Draft Convention on International Status of Environmentally-Displaced Persons, drafted in 2008,[16] presents a more comprehensive, innovative and independent proposal, which also has contact points with existing international regimes. Through this convention draft document, states, international and regional organisations, NGOs and local stakeholders are ex-

14 Hodgkinson et al. (2009:155–174).

15 CCDP (undated).

16 CRIDEAU et al. (2010).

pressly urged to raise awareness about the plight of environmentally-displaced persons, in order to recognise and proclaim an international status for that category, act preemptively to combat causes of migration flows, respond to emergency situations and build long-term policies. This project sets out specifically to contemplate the many legal situations related to the condition of this category, in addition to providing the institutional and financial mechanisms in order to achieve such protection.

Regarding the subjective scope of the convention, the project also features innovations: rights to potential victims threatened with displacement in order that they should have access to information on environmental threats prior to occurrences and the participation of potential victims in prevention policymaking, in order to increase preparedness for disaster situations.

The creation of a specific agency, the World Agency for Environmental Displacement, in the proposal deserves mention. The participation of non-state actors is expressly foreseen in the composition of the decision-making body (the High Authority). Also foreseen is a specific fund, the Worldwide Fund for Environmentally Displaced Persons, which will include mandatory contributions by states and voluntary contributions by individuals and corporations.

Furthermore, a system of governance was designed based on cooperation between the institutions of the convention with international regional organisations, and the connection of the convention organs with local authorities (through national committees), as well as cooperation with the secretariats of international conventions on the environment and human rights. With this in view, therefore, the human uprooting caused by global environmental changes can be seen in dual perspective. First is the international responsibility: Everyone, regardless of being part of any specific treaties, is obliged to cooperate in the protection and maintenance of environmental balance. Second is the human rights perspective: Everyone is obliged to uphold and protect issues of nationality, family, work, residence, culture and all rights related to a dignified existence.[17] This proposal recognises that protecting the environment is closely linked to the protection of human rights, and reinforces essential aspects of international environmental law, such as the incorporation of the human dimension on environmental issues, concern for

17 Leão (2010).

future generations, and multidisciplinarity – making this protection system unique.[18]

The temporal dimension of environmental protection, especially as regards the protection of future generations, has significant relevance to the topic under study, given the irreversible or irreparable damage to the environment can assume global proportions. Hence the importance of prevention, preparedness and response to the effects of such global changes that threaten not only the present generation (actual victims), but the very existence of future generations (potential victims).

Beyond the issue of human displacement lies the need to understand its causes and recognise the need to prevent and respond them. The international environmental law allows a viewing of all these aspects with proper sharpness.

Thus, there remains a clear need for a new global commitment founded on a broader basis, balancing the allocation of responsibilities to the states – based on the principle of common but differentiated responsibilities – and the responsibility of the entire international community – based on the principle of solidarity to give adequate protection without discrimination to environmental refugees. The recognition of this new category emerges unequivocally as a new normative and social demand and in the international sphere.

Another important initiative has been developed by the International Law Commission, based on International Humanitarian Law and its principles: the Draft Convention on the Protection of Persons in the Event of Disasters[19] – proposing to regulate the relationships between affected and non-affected states, especially in terms of cooperation, assistance and its limits: the duty to cooperate, duty of the affected state to seek assistance, consent of the affected state to external assistance, and the duty or right of non-affected actors to offer assistance.

It is worth noting that existing regional initiatives, although extremely important, do not eliminate the need for adopting a comprehensive instrument that sets minimum general standards of protection for people displaced internally and externally, and, if necessary, provides access to international assistance. In this sense, the Kampala Convention, which entered into force in December 2012, deserves attention. The Kampala Convention is the first

18 Soares (2001:37).

19 Available at http://untreaty.un.org/ilc/guide/6_3.htm, last accessed 23 November 2012.

international treaty for the protection and assistance of internally displaced persons, being those displaced within their own countries on the African continent. This convention imposes on states the obligation to protect and assist people displaced by natural disasters and man-made actions, such as armed conflicts, mitigation and elimination of the causes that generate displacement and its consequences from a legal framework based on solidarity, cooperation and mutual support between the states parties.[20] This initiative will certainly serve as a model for other continents, but only applies in cases of internal displacement.

Thus, the adoption of a specific international instrument to recognise this emerging category of migrants in a global context is most adequate to ensure an integrated response in the sense of protection of human rights, to provide humanitarian relief and to promote environmental restoration, adaptation and prevention strategies to those forced to leave their places of origin and their livelihoods due to the deterioration of the environment and an incapacity to cope with the adverse effects of environmental change impacts.

These proposals represent an important effort to find a lasting solution to the problem, including preventive policies and actions to combat the structural causes of environmental migration at the global, regional and local levels. As seen, the legal (legal status) and environmental issues are complementary and not mutually exclusive. An integrated approach goes to the root of the problem, namely, the causes which generate the displacements.

### D. *Final Recommendations: The Need for an Integrated Approach*

Even with all the advances made regarding the international protection of human rights and the environment, international law has not responded adequately to ensure sufficient and necessary, comprehensive and adequate protection to individuals and groups forced by the drastic environmental changes to leave their places of origin. This lack of legal recognition has generated unacceptable situations of flagrant violations of human rights – especially the right of all people, without distinction, to social and international order to enable the full realisation of these rights – which is a direct

20 African Union Convention for the Protection and Assistance of Internally Displaced Persons (IDPs) in Africa, Kampala Convention (2009), Article 2.

reflection of the situation of legal vacuum in which these people affected by global environmental changes are.

Whereas many of the extreme environmental events have causes or global effects, the responsibility of the international community concerning the occurrence of such events tends to be increasing, even as the incapacity to deal with problems and conflicts whose causes and effects are not limited to the internal borders of the affected states exposes a growing contingent of individuals and even entire nations to all kind of human rights violations inside and outside its territory. The magnitude and complexity of these violations cannot be underestimated. One must also consider the huge pressure on the environment caused by the rapid growth of the world population, with rates that in general are more significant in localities that are poorest and most vulnerable to degradation and depletion of environmental resources. This is a factor that directly influences the considerable increase in human mobility in those regions where migratory flows tend to be more and more intense. The need for protection of environmental refugees is inserted in this context, which is still in legal vacuum, and reinforces the need for an integrated approach by law. An integrated approach may produce a solution consistent with the multiple peculiarities of the phenomenon in order to prevent its causes and enable an urgent response to its consequences.

This integrated response cannot be limited to humanitarian assistance (material, psychological and legal), but should also include ecological assistance to affected countries, provided by the states, international organisations, civil society and local organisations, regardless of geographic proximity. The ecological assistance must include cooperative efforts in prevention and response – adaptation, resilience, mitigation, and capacity-building of communities and affected populations – to reduce human and environmental vulnerability to disasters and increase safety.

Connect the protection of human beings and the environment is necessary to overcome the fragmented treatment of this issue, limited and inadequate to address the complex dynamics of forced displacement by environmental causes. A broad legal definition of the terms *environmental refugee* or *environmentally displaced persons* could ensure minimum standards and global unified protection for individuals and groups severely affected by environmental events, whose survival and security require also international protection, whether inside or outside the limits of their country of origin or habitual residence. It is important therefore to strengthen the state's role in the institutionalisation of the measures, without which the formal recognition of this new category will not be possible. However, it is necessary to recognise that the signing of agreements and treaties will not be enough to solve the problem if its causes remain unchanged.

International cooperation and the participation of *new actors* are also essential in this process, especially in relation to states which are more sensitive and vulnerable to global environmental changes – usually those with low capacity to respond to such changes, whether in the preventive aspect or in relation to responding and adapting to environmental events. The economic costs of prevention measures will always be a consideration and will depend on the comprehensiveness of the global effort in this direction, but it is also true that such costs are infinitely smaller than the cost of human, material and environmental losses and damages, as well as the cost of material and legal assistance to the victims and for the reconstruction of the environment. The lack of preparation of most states represents a much greater threat to global security, while the absence of international commitment to face these new challenges remains. The adoption of a specific system of protection aims not only at the formal recognition of a new category of persons, but at the global commitment to protecting people in this condition; thus promoting internalisation of such future commitments under domestic laws in the states and stimulating a coordinated action among the actors involved in the issues of refugees, migration, environment and human rights, including the preventive aspect.

Recognising the link between global environmental degradation, forced migration and security is also important for the development of instruments and policies to prevent conflicts, improve actions to promote international peace and security, and prevent worst scenarios in the near future. Thus, an international instrument to contemplate this emerging challenge must be developed independently, but be connected to these systems, incorporating principles, standards and mechanisms that can be adapted to meet the complexity of this new demand.

The responsibility for protection and assistance should then be shared among the affected states and the entire international community with the adoption of a global agreement – founded on coexistence, cooperation and solidarity – that points to a significant structural change in international law.

## *References*

CCDP/Convention for Persons Displaced by Climate Change, undated, *Frequently Asked Questions*, available at http://www.ccdpconvention.com/documents/CCDPConventionFAQs.pdf, last accessed 18 August 2012.

CRIDEAU/Centre de recherche interdisciplinaire en droit de l'environnement, de l'aménagement et de l'urbanisme, CRDP/Centre de recherche sur les droits de la personne, Université de Limoges & CIDCE/Centre International de Droit Comparé de l'Environnement, 2010, *Projet de Convencion Relative au Statut International des "Desplacés Environnementaux"*, Deuxième version, Montaigut, commune de St Yrieix la Perche, Limousin, France, 31 May 2010 (original version 2008), available at http://www.cidce.org/pdf/Projet%20de%20convention%20relative%20au%20statut%20international%20des%20d%C3%A9plac%C3%A9s%20environnementaux%20%28deuxi%C3%A8me%20version%29.pdf, last accessed 13 May 2012.

El-Hinnawi, Essam, 1985, *Environmental Refugees*, Nairobi, United Nations Environment Programme (UNEP).

Hodgkinson, David, Tess Burton, Heather Anderson & Lucy Young, 2009, Copenhagen, Climate Change 'Refugees' and the Need for a Global Agreement, *Public Policy*, 4 (2), 155–174.

IFRC/International Federation of Red Cross and Red Crescent Societies, 2012, *World Disasters Report – Focus on Forced Migration and Displacement*, Geneva, IFRC, available at: http://www.ifrcmedia.org/assets/pages/wdr2012/resources/1216800-WDR-2012-EN-FULL.pdf, last accessed 24 April 2012.

Jacobson, Jodi L., 1988, Environmental Refugees: A Yardstick of Habitability, Worldwatch Paper 86, November, Washington, Worldwatch Institute.

Leão, Márcia B.C., 2010, *Direitos Humanos e Meio Ambiente: Mudanças Climáticas, "Refugiados" Ambientais e Direito Internacional*, available at http://www.nima.puc-rio.br/aprodab/artigos/clima_e_refugiados_ambientais_marcia_brandao_carneiro_leao.pdf, last accessed 24 August 2012.

Prieur, Michel, 2010, Le Conseil de l'Europe, les Catastrophes et les Droits de l'homme *VertigO – La Revue Électronique en Sciences de l'environnement*, numéro hors série (8), Octobre 2010, available at http://vertigo.revues.org/10270, last accessed 13 May 2012.

Ramos, Erika P., 2011, *Refugiados Ambientais: Em Busca de Reconhecimento pelo Direito Internacional/Environmental Refugees: In Search of Recognition by International Law*, PhD Thesis, Faculty of Law, University of São Paulo (USP), Brazil, available at http://www.acnur.org/t3/portugues/informacao-geral/catedra-sergio-vieira-de-mello/, last accessed 10 January 2013.

Republic of Maldives, Ministry of Environment, Energy and Water, 2006, First Meeting on Protocol on Environmental Refugees: Recognition of Environmental Refugees in the 1951 Convention and 1967 Protocol Relating to the Status of Refugees, Male, 14–15th August 2006.

Sachs, Jeffrey D., 2008, *Common Wealth: Economics for a Crowded Planet*, London, Allen Lane.

Soares, Guido F.S., 2001, *Direito Internacional do Meio Ambiente: Emergência, Obrigações e Responsabilidades,* São Paulo, Atlas.

Trindade, Antônio A.C., 1993, *Direitos Humanos e Meio Ambiente: Paralelo dos Sistemas de Proteção Internacional*, Porto Alegre, Sergio Antonio Fabris.

UNU-EHS/United Nations University Institute for Environment and Human Security, 2005, As Ranks of "Environmental Refugees" Swell Worldwide, Calls Grow for Better Definition, Recognition, Support. World Day for Disaster Reduction (press release), available at http://www.ehs.unu.edu/article/read/130, last accessed 20 November 2012.

Wijnberg, Harry, 2007, *Environmentally-induced Forced Migration,* Dissertation (summary), Faculty of Social Sciences, Erasmus University, Rotterdam.

25

# Enhancing Adaptation Options and Managing Human Mobility in the Context of Climate Change: Role of the United Nations Framework Convention on Climate Change

*Koko Warner*

*Abstract*

This article analyses the potential role of the United Nations Framework Convention on Climate Change (UNFCCC) in addressing human mobility issues and climate change. It sets the stage by laying out research findings about types of migration related to climate stressors. Such research explores how changing climatic patterns interact with livelihood and food security systems in ways that contribute to different forms of mobility – or to lack of mobility ('trapped populations'). Against this background, the contribution then examines the first mention of human mobility in paragraph 14(f) of the Cancun Adaptation Framework agreed at the Sixteenth Conference of the Parties (COP16) in 2011, and subsequent treatment in policy areas such as paragraph 7a(iii) in the Doha Climate Gateway decision on loss and damage at COP18 in 2012. With these policy developments, the article explores the possible roles of the UNFCCC in dealing with migration, displacement and relocation associated with climate change. Future climate policy regarding human mobility could have relevance for arenas like the Adaptation Committee, National Adaptation Planning Processes, the Green Climate Fund, and other areas, such as loss and damage in the UNFCCC process. Other policy arenas dealing with human mobility in the context of climate change may also be influenced by how the UNFCCC processes deal with the issue.

## *A. Introduction*

This contribution discusses the potential role of the United Nations Framework Convention on Climate Change (UNFCCC) in addressing some of the governance, legal and institutional issues arising with human mobility in the context of global anthropogenic environmental change. It analyses the applicability of the UNFCCC in respect of addressing migration, displacement and relocation associated with climate change. The article examines the suitability and efficacy of using the UNFCCC to address movement associated with climate change, with a view to informing the development of guiding principles and effective practices to address crisis migration.

To provide a research basis for discussions about types of crisis migration, section B of the article first draws on new research findings particularly relevant to scenarios where changing climatic patterns stress livelihoods and contribute to migration, e.g. changes in weather patterns that require people to relocate for shorter or longer periods in an effort to manage climate-related livelihood stressors. Section B also addresses a basic question of whether migration and other forms of human mobility can be considered adaptation to climate change. Reference is made to new findings that help address the question *Under what circumstances do households use migration as a risk management strategy when facing climatic stressors?*, and looks at four profiles of households along an adaptation continuum. The annex presents data and household characteristics from the study results shared in Section B. Section C examines the framing of human migration and displacement and its first-time-ever appearance in an outcome of a Conference of the Parties – in the Cancun Adaptation Framework of December 2010. The article then analyses paragraph 14(f) on migration and displacement in the Cancun Adaptation Framework, and the range of possible activities which may take shape in the future, both within the emerging climate adaptation regime. This may include the Adaptation Committee, National Adaptation Planning Processes, the Green Climate Fund, and other areas like loss and damage in the UNFCCC process (Section D). Section E examines how the topic is emerging in other areas such as loss and damage, notably in the Doha Climate Gateway decision at the Eighteenth Conference of the Parties (COP18).[1] Other policy arenas dealing with human mobility in the context of climate change may also be influenced by how the UNFCCC processes deal with the issue. The final section of this article examines gaps and draws conclusions (Section F).

1 Para. 7a(iii).

### B. *Climate Change and Migration: Emerging Understanding in the Context of Adaptation*

Climate change is likely to worsen the situation in parts of the world that already experience high levels of stressors to livelihood and food security, among other societal impacts of global environmental change. The consequences of the greater variability of climatic factors, such as rainfall conditions, affect the livelihoods and safety of vulnerable people. Less predictable seasons, more erratic rainfall, unseasonable events, or the loss of transitional seasons have significant repercussions for millions of people regarding food security, livelihoods, and the migration decisions of vulnerable households. In order to make informed decisions about adaptation planning, development, and a transition to a more climate-resilient future, policymakers and development actors need a better understanding of how migration decisions are spurred by the linkages among changes in the climate, household livelihood and food security profiles, particularly in 'acute' situations of climate stress like too little rain, rain at the wrong times for the planting season, or too much rain.

Since at least the mid-1980s, scientists have linked environmental change to human mobility.[2] Early debates emerged around future projections and predictions of the number of 'environmental migrants'.[3] More recently, conceptual and empirical work has examined broad relationships between environmental factors and human mobility in different situations.[4] These studies have identified broad patterns as a point of departure for further, more nuanced work on the interactions of climatic and socio-economic factors.[5] Research since that time has determined that environmental factors do play

---

2 El-Hinnawy (1985) introduced the first definition of *environmental migrants* in a United Nations Environmental Programme report. His definition has been refined and made more comprehensive by other authors and institutions, such as the International Organization for Migration in 2007.

3 See e.g. Brown (2008), Christian Aid (2007) and Myers (2005), who attempt to assign estimate numbers on current and future environmentally induced migration.

4 Jäger et al. (2009) synthesised the results of the Environmental Change and Forced Migration Scenarios Project (EACH-FOR, www.each-for.eu, last accessed 14 May 2013) – the first global survey of its kind employing fieldwork to investigate environmental change and migration in 23 case studies. Warner et al. (2009) brought EACH-FOR's results to policymakers, particularly in the UNFCCC process.

5 Afifi (2011); Brown (2008); Gunvor (2010); Hugo (2008); Laczko & Aghazarm (2009); Martin, P. (2010); Martin, S. (2010); Morrissey (2009); Tacoli (2009).

a role in human mobility[6] and emphasises that some people who are more exposed to environmental stressors – particularly farmers, herders, pastoralists, fishermen and others who rely on natural resources and the weather for their livelihoods – may be the least able to move very far away, if at all.[7] In the decades ahead, these potentially 'limited-mobility' populations could face deteriorating habitability of their traditional homelands, with fewer options for moving to more favourable places in safety and dignity. The implications of climate change for a wider scope of issues related to population movement in the medium and longer term have driven a quest for a better understanding of the circumstances under which climatic factors affect human decisions about whether to leave, where to go, when to leave, and when to return.

Research findings were first formally reported to climate negotiators in a submission to the UNFCCC in August 2008 in the Accra Session in Ghana of the Ad Hoc Working Group on Long-term Cooperative Action.[8] Furthermore, the Fifth Assessment Report of the Intergovernmental Panel on Climate Change[9] (IPCC) will include a chapter on human security, which will particularly deal with migration and conflict among others.[10] The topic will also be addressed in several other chapters in the IPCC report as a cross-cutting issue (particularly in regional chapters). Emerging empirical evidence considers whether migration and other forms of human mobility can be regarded as adaptation to climate change. These research efforts contribute to policy discussions on the topic, and emerging policy responses nationally, regionally and internationally.

---

6 Jäger et al. (2009); Warner et al. (2009, 2011).

7 See Betts (2010); Black et al. (2011).

8 Third Session of the UNFCCC Ad Hoc Working Group on Long-Term Cooperative Action under the Convention (AWG-LCA 3), Accra, Ghana, 21–27 August 2008, FCCC/AWGLCA/2008/MISC.3; Submission entitled Climate Change and Migration: Impacts, Vulnerability, and Adaptation Options made on 18 August 2008 by the United Nations University to the Third Session of the UNFCCC Ad Hoc Working Group on Long-Term Cooperative Action under the Convention (AWG-LCA 3), Accra, Ghana, 21–27 August 2008.

9 For more information on the IPCC's Fifth Assessment Report (AR5), see http://www.ipcc.ch, last accessed 24 May 2013.

10 Contribution of Working Group II to the IPCC's Fifth Assessment Report, Chapter 12.

### I. *Migration as Adaptation or as Failure to Adapt? Four Migration Profiles along the Adaptation Spectrum*

This section draws on new findings from the project entitled "Where the Rain Falls: Understanding Relationships between Changing Rainfall Variability, Food and Livelihood Security, and Human Mobility", undertaken by the United Nations University Institute for Environment and Human Security, and CARE.[11] The Rainfalls work is supported by the AXA Group and the John D. and Catherine T. MacArthur Foundation. Findings emanate from Bangladesh, Guatemala, Ghana, India, Peru, Tanzania, Thailand and Vietnam. The case studies offer insights about current relationships between rainfall-dependent livelihoods and food security, and the circumstances under which households currently use migration to manage the risks of impacts on household consumption and income.[12]

- Rural people surveyed overwhelmingly perceive climatic changes occurring today in terms of *rainfall variability*. These perceptions shape household risk management decisions. The most common changes reported relate to the timing, quality, quantity and overall predictability of rainfall, including delayed onset and shorter rainy seasons; reduced number of rainy days per year; increased frequency of heavy rainfall events; and more frequent prolonged dry spells during rainy seasons. In many

---

11 Warner et al. (2012), hereinafter *Rainfalls*. The Rainfalls research explores the interrelationships among rainfall variability, food and livelihood security, and human mobility in a diverse set of research sites in eight countries. While climate change affects nearly all aspects of food security – from production and availability to the stability of food supplies, access to food, and food utilisation (Schmidhuber & Tubiello 2007) – the Rainfalls research focused on linkages between shifting rainfall patterns and food production and the stability of food supplies (Jennings & Magrath 2009). The central focus of the Rainfalls initiative was to explore the circumstances under which households in eight case study sites in Africa, Asia and Latin America use migration as a risk management strategy when faced with rainfall variability and food and livelihood insecurity. See www.wheretherainfalls.org, last accessed 13 May 2013.

12 The data presented in this section was gathered during the execution of a project to assess the circumstances under which households use migration as a risk management strategy when facing rainfall variability and food and livelihood insecurity in Bangladesh, Guatemala, Ghana, India, Peru, Tanzania, Thailand and Vietnam. The research for Rainfalls was undertaken by the UN University Institute for Environment and Human Security, in partnership with CARE France. Funding came from the AXA Group and the John D. and Catherine T. MacArthur Foundation.

cases, these perceived changes correlate with an analysis of local meteorological data over the last several decades.

- The largely agriculture-based households in the research sites overwhelmingly report that rainfall variability is already negatively affecting production and contributing to *food and livelihood insecurity*. Levels of food insecurity varied significantly across the eight sites, depending on factors such as the total amount and seasonality of rainfall; the degree of agricultural intensification; the extent of livelihood diversification; and the access of poor households to social safety nets and other support services.
- *Migration*, which was common in the research sites, was observed to have the following characteristics: almost entirely within national borders; predominantly male, but with growing participation by women in a number of countries; largely by individual household members (with India as the exception, where entire nuclear families moved together); largely driven by livelihood-related needs (household income) in most countries, but with a growing number of migrants seeking improved skills sets (e.g. through education) in countries like Peru, Thailand and Vietnam; and a mix of rural–rural and rural–urban, with more productive agricultural areas (Bangladesh, Ghana, Tanzania), nearby urban centres (India, Peru), mining areas (Ghana), and industrial estates (Thailand, Vietnam) being the most common destinations.
- Households manage climatic risks, such as changes in rainfall variability, with migration. Migration – seasonal, temporal and permanent – plays an important part in the struggle of many families to deal with rainfall variability and food and livelihood insecurity. Migration was found to have increased in recent decades in a number of the research sites. Rainfall was observed to have a more direct relationship with household migration decisions in research sites where the dependence on rain-fed agriculture (often with a single harvest per year) was high and local livelihood diversification options were low. Pressure on rainfall-dependent livelihoods is likely to grow as a driver of long-term mobility in the coming decades if vulnerable households are not assisted in building more climate-resilient livelihoods, in situ.
- Households with more diverse assets and access to a variety of adaptation, livelihood diversification or risk management options – through social networks, education, or community or government support programmes – can use migration in ways that enhance resilience. Those households which have the least access to such options – few or no

livelihood diversification opportunities, no land, little education – use (usually) internal migration during the hunger season as a survival strategy in an overall setting of erosive coping measures, which leave or trap such households at the margins of decent existence.

A preliminary analysis of the household survey data was used to generate four broad profiles. These profiles relate to the use of migration in response to rainfall variability and food and livelihood insecurity. The first profile is most commonly found in countries that have been able to provide alternative livelihoods and food security to most people. This group uses migration in ways which improve their resilience, such as investing in education, health and climate-resilient livelihood opportunities. These households use migration as one of a variety of adaptation strategies, moving seasonally or temporally, often to non-agricultural jobs in cities or internationally. The second profile often occurs in countries with less food security and fewer options for diversifying livelihoods. This group uses migration to survive, but not flourish. They move seasonally within their countries to find work, often to other rural areas as agricultural labourers. The third profile occurs where food security is even more tenuous and where adaptation options are fewer, or are not pursued vigorously. This group uses migration as a means of gaining security, in what can be seen as an erosive coping strategy which can become part of a negative cycle in wider crisis situations. This group often moves during the hunger season to other rural areas in their region in search of food, or they work to buy food for their families. The fourth profile appear to be 'trapped populations' that struggle to survive in their areas of origin and cannot easily use migration to adapt to the negative impacts of rainfall stressors.

Each of these four profiles was visible across all the research sites, but some countries manifested clusters of households with dominant patterns. The profiles represent a spectrum, with households within a profile being closer to one or the other of the profiles on either side. They are, thus, not mutually exclusive, and serve as a point of departure for further research and work to refine key explanatory variables regarding forms of human mobility related to crisis and climate change.

## II. Modelling Results and Future Scenarios

This section relates the four household profiles to an agent-based modelling approach applied in the Tanzania case to explore the scenarios in terms of which rainfall variability and food security have the potential to become significant drivers of human mobility in particular regions of the world in the next two to three decades.

In order to understand the potential for rainfall to become a significant driver of human mobility in the future, it is important to identify the range of impacts that likely scenarios may have on migration flows. By investigating the impact of rainfall variability on household- and community-level factors such as food and livelihood security, the influence of such variability on the decisions made by individual migrants can be further understood. Using the Rainfalls case study sites as examples of locations where changes in rainfall might contribute to increased food insecurity and human mobility, a process of future-oriented simulation and analysis provides a valuable opportunity to understand the circumstances under which rainfall variability might become a significant driver of migration.

The Rainfalls Agent-based Migration Model (RABMM) represents vulnerability and migration decision-making at two levels of agent analysis: the household and the individual, both of which can be generated from the household survey data collected in each case study location. The RABMM is designed to represent the degree of vulnerability of households to rainfall-variability-induced changes in livelihood and food security, and the subsequent impact of these on the migration of household members. The research identified a range of impacts that likely scenarios may have on migration flows, and showed that rainfall changes have the potential to become a significant driver of human mobility in the future.

### ***Tanzania Results: Migration from 2014–2040 under drier, wetter and extremely dry/wet rainfall scenarios***

Using the conceptual framework described above, the Tanzania RABMM outputs the number of migrants originating from contented and vulnerable households across the case study villages.[13] Figure 1 shows modelling results under the same scenarios for migrants from *Resilient* households.

13 Warner et al. (2012).

***Figure 1: Five-year Moving Averaged Normalised Difference in the Rate of RABMM-modelled Contented Migration****

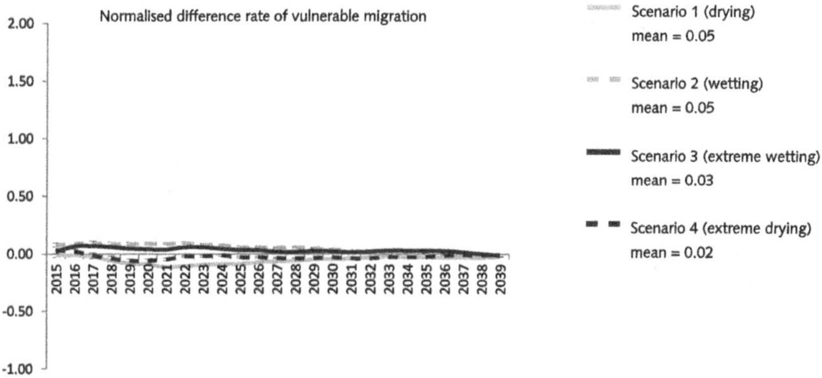


* Error bars indicate the envelope of changes modelled under five member ensembles.

Source: Warner et al. (2012:110)

The results of the modelling for contented migration shown in Figure 1 show a much lower level of sensitivity to changes in rainfall than is the case for vulnerable migration. The mean annual normalised rate of contented migration under the scenario is 0.05 – only 5% greater than that seen under the 'average' scenario. In contrast, Figure 2 shows the modelling results for *vulnerable* households in Tanzania's Same district.

***Figure 2: Five-year Moving Averaged Normalised Difference in the rate of RABMM-modelled Vulnerable Migration****

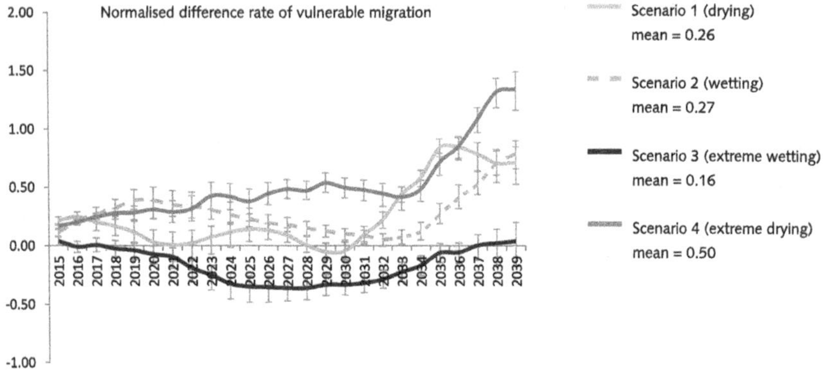


* Error bars indicate the envelope of changes modelled under five-member ensembles.

Source: Warner et al. (2012:108)

The agent-based modelling results from Rainfalls are pertinent to discussions of crisis migration in the context of climate change: households using migration to build resilience (*Contented Migration*) show a much lower level of sensitivity to changes in climatic patterns than is the case for *Vulnerable Migration*. Vulnerable households have a higher sensitivity to different rainfall scenarios and feel an imminent need to change their situation through migration. Changes in rainfall patterns can impact food and livelihood security in the future and have the potential to increase the vulnerability of many households worldwide.

The two graphics illustrate the key finding of the modelling exercise in Tanzania and the main message from the Rainfalls study: resilient households use migration in ways that appear to reduce their sensitivity to climate stressors over time (the first graphic), while vulnerable households use migration in ways that either does not affect their climate sensitivity over time, or may exacerbate it through related actions such as selling land or productive assets, migration-interrupting skill-building, and education in children (as seen in the India case). When such vulnerable households face scenarios of changing rainfall variability, particularly of extreme drying, migration rises notably over time.

Case study and modelling results illustrate the circumstances under which migration decisions occur, showing that both *Contented* and *Vulnerable* households use migration, but in markedly different ways that either enhance resilience or reinforce a downward spiral of vulnerability to climatic and other stressors. These findings point towards the key importance of the types and quality of adaptation measures chosen by countries. Many of these activities happen under the umbrella of economic and social development. Increasingly, a newer set of (ideally) complementary activities are emerging under the umbrella of climate adaptation efforts at the national level – spurred by international discussions in the 'climate negotiations' of the UNFCCC. We now turn to examining the potential of UNFCCC discussions to provide guidance for adaptation activities that include the management of human mobility.

### *C. The UNFCCC and Human Mobility in the Context of Climate Change*

To understand the treatment of human mobility in the climate policy arena, one must understand a larger discussion around impacts of climate change, framed largely by a discourse between the Alliance of Small Island States (AOSIS) and industrialised countries. The period from the early 1990s to the early 2000s was marked by an emphasis on mitigation – the collective reduction of greenhouse gas (GHG) emissions linked to changes in global temperature increases. This period saw the creation of the Kyoto Protocol, carbon markets, the Clean Development Mechanism, and other measures. By the mid-2000s, and certainly with the publication of the Intergovernmental Panel on Climate Change (IPCC) Fourth Assessment Report in 2007, the process reflected an emerging scientific realisation that emissions targets are too low to prevent anthropogenic interference with the earth's atmosphere. Hence, it is also necessary to discuss adaptation and issues around negative impacts of climatic change on human society.

The position of the AOSIS was underpinned by the idea that states that had experienced loss and damage related to climate change could seek assistance to rehabilitate their societies (ideally to pre-climate-change status). AOSIS had articulated this proposal since the early 1990s, framing it as a kind of 'assurance' against a wide range of climate change impacts. The early focus was on cautioning high-emitting countries about the consequences of not curbing their emissions (e.g. 'the polluter pays' principle). AOSIS and other allies have emphasised that sea-level rise (which can lead to displacement) could drastically affect the functionality of societies in low-lying countries. A range of possible outcomes, including population movement, were framed as indicators of severe challenges to in situ adaptation.

Avoidance, according to this view, was the only acceptable approach, and some parties (least-developed countries, AOSIS, and other vulnerable countries) championed the 1.5°C goal.

A second strand of discussion was introduced around the time of the 2007 IPCC Fourth Assessment Report and the release of the Stern Review:[14] scientists and policymakers began to concur that some impacts of climate change may already be manifest, and that adaptation was, therefore, a necessary complement to mitigation in order to cushion the blow to society from some of the expected impacts of climate change. By 2007, the 2007 IPCC Report and other scientific and policy discussions had firmly laid the case for the need for mitigation to be accompanied by adaptation in the UNFCCC process. This contributed to discussions about the need for coherence and coordination of adaptation activities, appropriate finance, and planning activities that would help countries (particularly those most vulnerable to the negative impacts of climate change) to adapt.

Thus, these two parts of the dialogue under UNFCCC discussions – mitigation and adaptation – fundamentally shaped how human mobility became couched within adaptation, and have contributed to thinking about issues like governance, funding and management of human mobility. Perhaps of greater long-term significance, the recognition of human mobility within the UNFCCC process has helped spur United Nations (UN), regional, and national discussions of finding stable trajectories for societal transformations within changing climate regimes. Such discussions include security and notions of borders, population shifts in particular regions, and moving from current adaptive practice to those practices which will be appropriate in the future.

### D. *Cancun Adaptation Framework and Potential Roles in Addressing Human Mobility*

At the Conference of the Parties in Cancun, Mexico (COP16), the draft text containing several key elements for the operationalisation and funding of adaptation – notably the Cancun Adaptation Framework – was accepted,. This Framework outlined key areas that would qualify for adaptation sup-

14 Stern (2006).

port, including the first-ever reference to adaptation and human mobility in an internationally agreed climate policy. Paragraph 14(f) reads as follows:

> 14. Invites all Parties to enhance action on adaptation under the Cancun Adaptation Framework, taking into account their common but differentiated responsibilities and respective capabilities, and specific national and regional development priorities, objectives and circumstances, by undertaking, inter alia, the following:
>
> ...
>
> (f) Measures to enhance understanding, coordination and cooperation with regard to climate change induced displacement, migration and planned relocation, where appropriate, at national, regional and international levels.

The Cancun talks also articulated elements necessary for implementation of such activities as the Adaptation Committee, National Adaptation Planning processes, the Green Climate Fund, and the Work Programme on Loss and Damage.

### *I. Significance of Paragraph 14(f) for the Management of Human Mobility*

The framing of human mobility in the climate negotiations is important for several reasons:

- In the context of the UNFCCC, mobility is acknowledged as having a link to climate change and is framed as an issue *to be managed.* It provides a stepping stone on a 'technical' level (rather than as political dialogue) for transitions between immediate-term use of existing approaches to necessary longer-term paradigm changes about population shifts, governance of borders and mobility, livelihood viability, planning in certain regions, etc.
- Paragraph 14(f) couches human mobility within the realm of adaptation to climate change and subtly introduces the thought that adaptation may require longer-term societal transformations. This suggests that adaptation may be understood not only as incremental changes in the way people live in certain locations, but also more broadly to include "transformational adaptation" that could include new locations.[15]
- Paragraph 14(f) frames human mobility as part of a wider range of measures that can be funded under the emerging climate finance regime to

15 Kates et al. (2012).

assist vulnerable countries to adjust to current and expected climate changes. Depending on how the states parties articulate their adaptation needs, human-mobility-related activities will be eligible for climate finance, i.e. managing migration, preventing or reducing displacement, and – where appropriate – undertaking planned relocation. Other than the UNFCCC, no other forum internationally or regionally has created a space in which a range of issues and possible activities is recognised and linked to the upcoming climate finance regime.

- Paragraph 14(f) has significance for *implementation*. As the institutional arrangements for adaptation continue to be shaped, human mobility (and the other areas mentioned in the Cancun Adaptation Framework) will expand from a topic for discussion into a topic for policy and operations. This will have meaning for development cooperation (particularly around livelihoods), humanitarian and disaster-risk reduction work, urban and rural planning, etc.
- Finally, Paragraph 14(f) provides an opportunity to further articulate *policy options* at appropriate levels (subnational, national, regional, international) and along the spectrum of human mobility. The work of the Adaptation Committee has now advanced to a draft, three-year programme, including the development of guiding principles for adaptation and efforts to coordinate and increase policy coherence for items included in the Cancun Adaptation Framework. The decision will be made in Autumn 2012 regarding the location and implementation of the Green Climate Fund as a vehicle for funding activities outlined in the Cancun Adaptation Framework. The National Adaptation Planning processes are moving forward, and nations are in the process of integrating climate policy into wider national planning efforts. The COP18 in Doha reached a decision on loss and damage, which included continuing work to understand, enhance coordination, and facilitate action on loss and damage as well as a mandate to establish institutional arrangements at COP19 in Warsaw. Such an arrangement could assess, address and coordinate issues that may extend beyond (certain current definitions of) adaptation, including migration, displacement and relocation. Furthermore, the presence of human mobility in one policy forum (UNFCCC) has and will continue to influence discussions in other arenas, including the UN Security Council, the Global Forum on Migration, the high-level dialogue on migration, and regional forums.

These last two points are important because, arguably, few other arenas emphasise discussion, action/planning and financial resources for implementation like the UNFCCC does. As described in other literature,[16] the existing institutional arrangements to manage voluntary migration and mobility related to natural disasters are full of gaps. Few coordination or planning mechanisms are in place to address relocation related to environmental or climate change: most are related to development projects. Paragraph 14(f) provides initial inroads into these areas for the future.

### *II. Potential Adaptation Actions Related to Paragraph 14(f)*

Questions now arise regarding the activities and modalities for implementation which could emanate from the inclusion of migration and displacement in the Cancun Adaptation Framework. This section explores what some of the likely combinations of measures, types of movement and levels may emerge in the short term. It also explores what kinds of entities might be involved in future interventions, and how these might be funded.

The discussion outlines kinds of activities (enhanced understanding), types of human mobility (displacement, migration, planned relocation), and levels of addressing the issue (national, regional and international). Figure 3 represents a matrix of the text in paragraph 14(f). Figure 4 displays an assessment of possible types of measures that could emerge – and have already partially emerged in 2011 – in relation to paragraph 14(f).

16 Warner (2010).

***Figure 3: Structure of Paragraph 14(f): Type of Human Mobility and Action, and Levels of Action***

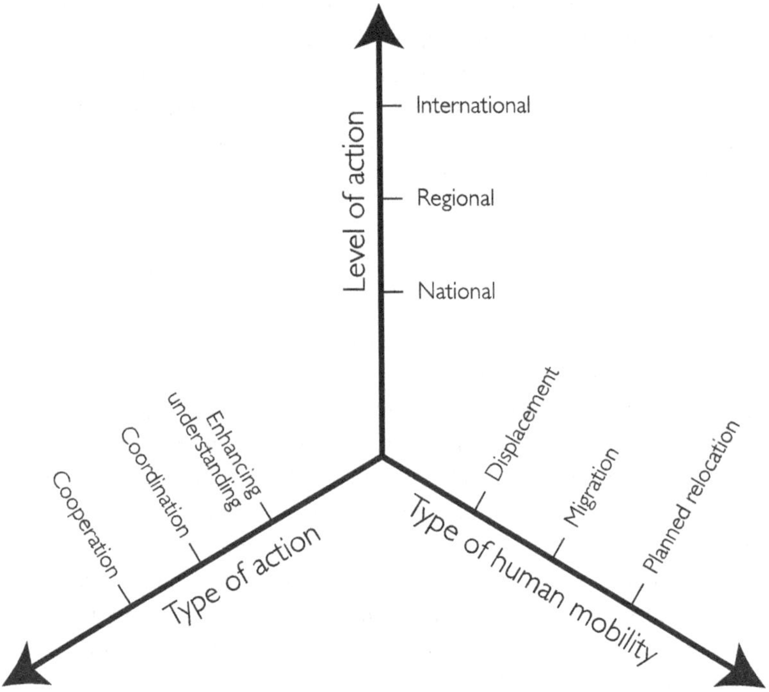


Source: Warner et al. in Foresight (2011)

***Figure 4: Possible Types of Measures that Could Emerge in Relation to Paragraph 14(f)***

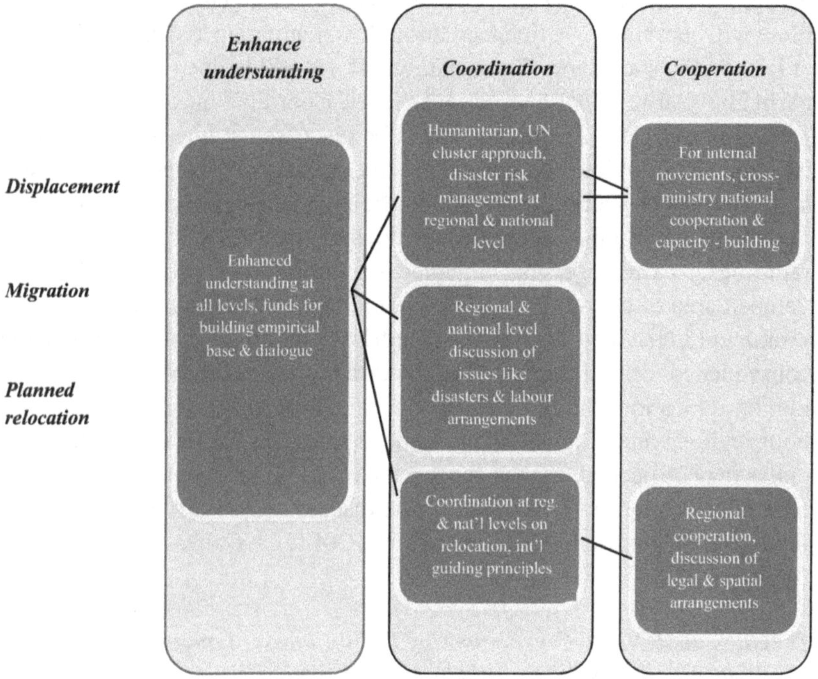


Paragraph 14(f) is likely to result in enhanced understanding at national, regional and international levels. If the current pattern continues, it is likely that funding for building the empirical base via research, case studies, etc. will come from bilateral sources, as individual countries call for specific studies and dialogue at conferences, meetings, etc. This has already happened, e.g. the Nansen Conference hosted by Norway in June 2011, which focused on climate-induced displacement, as well as several research projects supported by various Asian and European governments. The latter projects looked mostly at migration, but also at displacement. Regional and international dialogue on research findings is also expected, particularly with the IPCC's Fifth Assessment Report due in 2014, which will feature a chapter reviewing migration and displacement in the context of adaptation to climate change, and several additional chapters presenting scientific findings on human mobility in different regions.

It is expected that coordination efforts on displacement will continue along current humanitarian organisation lines through the UN cluster approach, and under the auspices of disaster risk management. It is likely that these will continue to be financed through humanitarian assistance channels, at least in the short term. Coordination efforts will happen at all levels, but particularly at regional and national level. Coordination on voluntary migration is less clear, but may begin to be discussed at international and regional level. Individual countries may choose to address whether tools such as temporary protection status should be broadened or altered to include a variety of environmental processes beyond natural disasters – typically, rapid-onset extreme events. Planned relocation is possibly the least-developed area of coordination at this point, but is likely to become far more prominent in the medium term as countries begin to think through potential consequences of mitigation and adaptation projects which may require population relocation. These kinds of coordination measures will be needed at both regional and national level, while at international level, guiding principles may be needed, such as those now available for development-project-related relocation. International expert discussions on the topic, including the two Bellagio roundtables in 2010 and 2011, have already begun.

### *E. Gaps the UNFCCC Process May Help Address: Longer-term Thinking about Human Mobility and Climate Change*

Now that migration, displacement and planned relocation have been highlighted in the UNFCCC climate negotiations, governments increasingly want to know more about the potential impacts of climate change and human mobility in order to prepare their own appropriate legal, institutional and governance approaches. Research suggests that the complex forms of migration and displacement will mix internal and cross-border movements. These movements will raise policy-relevant questions when people cannot return to their places of origin because of environmental factors that include sea-level rise, desertification and water issues. Parties to the UNFCCC have a role to play in minimising pressure on vulnerable populations and providing adaptation options, as well as guiding the management of migration, displacement and planned relocation in harmony with the protection of and respect for the dignity and safety of those involved.

*I. Leading up to 2015 and milestones in development, humanitarian and climate policy arenas*

The next few years will provide opportunities to fill knowledge gaps and support decision-makers with more and better-quality information about the role of environmental factors in the combination of issues that affect human migration, displacement and planned relocation.

Emerging dialogue around human mobility in the context of climate change focuses on climate variability and the use of existing tools. There is a risk that emerging issues related to human mobility and climate change may introduce needs that are not addressed by existing tools and institutions. One of the potential challenges to the position of countries that emphasise using current institutional frameworks is that these frameworks are already insufficient and may become more stressed in the future. A few examples of policy frameworks addressing this issue are available, such as temporary protection status in the United States and Europe, or principles and soft laws for protecting people who have been displaced by environmental events. Yet beyond humanitarian approaches for rapid-onset extreme events, there are significant governance gaps. Complex and slow-onset events could pose a major challenge to legal and governance frameworks, in part because responsibility and temporal limits are difficult to assign. Moreover, various institutions that deal with different issues related to the impacts of climate change may have a tendency to operate in 'silos', and may approach issues such as climate change within narrow, sectoral perspectives.

Paragraph 14(f) of the Cancun Adaptation Framework, the emerging work of the Adaptation Committee, National Adaptation Planning processes, the Green Climate Fund and the Work Programme on Loss and Damage provide spaces where some of these potential risks can be addressed.

*II. Between 2015 and 2020*

Policy spheres like the UNFCCC could be useful spaces for enhancing understanding, building dialogue, and facilitating regional cooperation and coordination at the policy and operational level around human migration, displacement and planned relocation. The key will be to align the appetite and needs of the governments with a range of appropriate – and politically feasible – measures. Given sensitivities of governments about the causes and societal consequences of climate change, calls for complex arrangements or measures that require significant 'political capital' to achieve may face little success. For example, in the current political environment, recommendations for commitments for expended or new protection that touch on existing arrangements for the protection of refugees could face resistance.

The development of a guiding framework around climate change and related human mobility, based on the positive experience with the guiding principles for internally displaced persons in the late 1990s, could emerge to help states prepare for the expected impacts of climate change on migration and displacement. Such a framework could be a constructive input from relevant stakeholders to bodies such as the Adaptation Committee, which will help provide coherence and coordination across other areas, i.e. activities of the Adaptation Fund and Green Climate Fund, National Adaptation Planning processes, and the Work Programme on Loss and Damage.

In the medium to longer term, when human mobility related to climatic change is expected to become more apparent, operational cooperation will be needed at both regional and national level to manage flows of people. Where movements – displacement, migration, relocation – are internal, cross-ministry national cooperation and capacity-building may be needed. These kinds of activities may be funded through existing bilateral channels or potentially through the emerging climate finance architecture. Where movements occur in border areas, regional cooperation may be necessary. Examples of regional labour migration agreements may be models for the future, but will require several years to design and implement.

### *III. Beyond 2020*

The need for large-scale, unplanned human mobility may be ameliorated to some degree through effective mitigation and adaptation measures, particularly in the areas of sustainable agriculture and rural livelihood diversification. It will become increasingly important to ensure that poorer countries and communities become institutionally and operationally equipped to support widespread adaptation (including livelihood diversification) to manage climatic risks and shifts in population distribution (including various types of mobility). Measures should be implemented which ease tensions that could arise around food security, resource availability, and issues around national borders. The period before 2020 will be a time of setting trajectories and laying the groundwork for new or adjusted institutional forms to deal with a broader range of climatic impacts, including human mobility.

In the medium and longer term, the humanitarian response could be overwhelmed by growing disaster-related displacement. Disaster risk reduction and measures to avoid loss and damage may not keep pace with the incremental and potentially permanent changes associated with desertification, sea level rise, ocean acidification, loss of geologic and other freshwater sources, etc. which can add pressure to human mobility. Such scenarios underscore the need for new thinking about managing and planning for the impacts of climate change on human mobility, ranging from migration to displacement to relocation. It will be necessary to address longer-term developments in human mobility, and it will become increasingly important to develop approaches that consider shifts in the baseline situation of many regions. Current risk management approaches, many of which include mobility, may be insufficient or inappropriate in a changed climate situation in the future. It will be important to incorporate long-term time horizons (or 'climate foresight') as opposed to simple 'impact/vulnerability' mapping (which results in providing short-term 'coping' strategies) in adaptation planning.

In scenarios of the world beyond 2°C, the impacts of climate change – combined with other megatrends such as the world population growth, changes in technology, and other unforeseen shifts in society – could require a new approach or forum for particular discussions including on migration, displacement and planned relocation.

There is a need for longer-term planning mechanisms related to human mobility which may be difficult to attain in the context of voluntary, non-binding international cooperation. It would be useful to include 'transformational' adaptation strategies – as opposed to merely 'improved coping' strategies – for current climatic conditions in specific locations. For example, typical community-based adaptive activities in coastal Bangladesh include providing assistance to vulnerable communities in low-lying areas, raising house plinths to keep the houses above flood level, and harvesting rainwater to ensure clean drinking water is available and to offer protection against the intrusion of salinity into surface and groundwater supplies.

A longer-term strategy would include empowering, training and building the skills of younger generations – including children – in those communities to be in a position to adjust not only to variability, but also to change. This may include development-related resilience-building, such as enabling young people to get climate-appropriate and/or better-paying jobs in nearby towns over the next decade. Such resilience-building would include social, financial and environmental aspects. In this way, the younger generation would be able to take their families with them if survival in their current location became increasingly more difficult. This kind of longer-term, 'empowered relocation' strategy, as opposed to 'forced migration' or even 'planned relocation', would ideally be part of a participatory process about when, how, where and who would move within affected households and communities. The Adaptation Committee, National Adaptation Planning processes, and managers of climate finance could take such longer-term perspectives into account when recommending or funding adaptation-related activities – in this case, those that relate to mobility.

## *F. Conclusions*

The existing UNFCCC agreed-upon language around human mobility in the Cancun Adaptation Framework and the Doha Climate Gateway Decision provide milestones and points of departure for this journey. In the future, appropriate frameworks, policies and governance structures are needed to address human population movements. Such developments would enhance the development of effective measures that would enable governments to manage climate-change-related human mobility in proactive ways that safeguard the security, dignity and living standards of migrants, displaced and relocated people – as well as those 'left behind'.

## *References*

Afifi, Tamer, 2011, Economic or Environmental Migration? The Push Factors in Niger, *International Migration (International Organization of Migration Special Issue)* 49, s1, e95–e124.

Betts, Alexander, 2010, Survival Migration: A New Protection Framework, *Global Governance* 16 (3), 361–382.

Black, Richard, Stephen R.G. Bennett, Sandy M. Thomas & John R. Beddington, 2011, Climate Change: Migration as Adaptation, *Nature* 478 (7370), 447–449.

Brown, Oli, 2008, Migration and Climate Change, IOM Research Series No. 31, Geneva, International Organization for Migration.

Christian Aid, 2007, *Human Tide: The Real Migration Crisis*, available at http://www.christianaid.org.uk/Images/human_tide3__tcm15-23335.pdf, last accessed 10 December 2012.

El-Hinnawy, Essam, 1985, *Environmental Refugees*, Nairobi, United Nations Environment Programme.

Foresight, 2011, *Migration and Global Environmental Change*, Final Project Report, London, The Government Office for Science.

Hugo, Graeme, 2008, *Migration, Development and Environment*, Geneva, International Organization for Migration.

IPCC/Intergovernmental Panel on Climate Change, 2007, *Fourth Assessment Report, Climate Change 2007 (AR4)*, available at http://www.ipcc.ch/publications_and_data/publications_and_data_reports.shtml, last accessed 24 May 2013.

Jäger, Jill, Johannes Frühmann, Sigrid Grünberger & Andreas Vag (Eds), 2009, *Environmental Change and Forced Migration Scenarios Project Synthesis Report, Deliverable D.3.4 for the European Commission*, Gipuzkoa, Universidad del Pais Vasco.

Jennings, Steve & John Magrath, 2009, *What Happened to the Seasons?, Oxfam Research Report*, London, Oxfam.

Jónsson, Gunvor, 2010, The Environmental Factor in Migration Dynamics – A Review of African Case Studies, International Migration Institute Working Paper 21, Oxford, International Migration Institute.

Kates, Robert W., William R. Travis & Thomas J. Wilbanks, 2012, Transformational adaptation when incremental adaptations to climate change are insufficient, *Proceedings of the National Academy of Sciences of the United States of America* 109, 7156–7161.

Laczko, Frank & Christine Aghazarm (Eds), 2009, *Migration, Environment and Climate Change: Assessing the Evidence*, Geneva, International Organization for Migration.

Martin, Philip, 2010, *Climate Change, Agricultural Development, and Migration: Background Paper for the Transatlantic Study Team on Climate Change and Migration*, Washington, D.C., Georgetown University and United Nations University Institute for Environment and Human Security.

Martin, Susan F., 2010, *Climate Change and International Migration: Background Paper for the Transatlantic Study Team on Climate Change and Migration*, Washington, D.C., Georgetown University and United Nations University Institute for Environment and Human Security.

Morrissey, James, 2009, Environmental Change and Forced Migration: A State of the Art Review, Refugee Studies Centre Background Paper, Oxford, Refugee Studies Centre.

Myers, Norman, 2005, Environmental Refugees: An Emergent Security Issue, conference paper presented at the Thirteenth Economic Forum, Prague, 23–27 May.

Schmidhuber, Josef & Francesco N. Tubiello, 2007, Global Food Security under Climate Change, *Proceedings of the National Academy of Science of the United States of America* 104, 19703–19708.

Stern, Nicholas, 2006, *The Economics of Climate Change, The Stern Review*, Cambridge, Cambridge University Press.

Tacoli, Cecilia, 2009, Crisis or Adaptation? Migration and Climate Change in a Context of High Mobility, in: Guzmán, José Miguel, George Martine, Gordon McGranahan, Daniel Schensul & Cecilia Tacoli (Eds), *Population Dynamics and Climate Change*, New York/London, United Nations Population Fund & International Institute for Environment and Development, 104–118.

Warner, Koko, Charles Erhart, Alex de Sherbin, Susana Adamo & Tricia Chai-Onn, 2009, *In Search of Shelter: Mapping the Effects of Climate Change on Human Migration and Displacement*, Geneva, United Nations University, Institute for Environment and Human Security.

Warner, Koko et al. 2011, Migration and Displacement in the Context of Adaptation to Climate Change: Developments in the UNFCCC Climate Negotiations and Potential for Future Action, in: Foresight, *Migration and Global Environmental Change*, Final Project Report, London, The Government Office for Science.

Warner, Koko, Tamer Afifi, Kevin Henry, Tonya Rawe, Chris Smith & Alex de Sherbinin, 2012, *Where the Rain Falls: Climate Change, Food and Livelihood Security, and Migration*, Global Policy Report of the Where the Rain Falls Project, Bonn, United Nations University Institute for Environment and Human Security & CARE.

# 26
# Adaptation to Climate Change under Changing Urban Patterns: The Climatic Perspective of Migration

*María Máñez Costa, Kathleen Schwerdtner Máñez & Sainab Husain Paragay*

*Abstract*

Climate change in human history is having the same effect as the opening of Pandora's box in Greek mythology. Although some still argue that the consequences of climate change might be trivial or harmless, the reality shows that at least some of these consequences are already severe and far-reaching. This is particularly true for climate-induced migration and urbanisation, a two-directional problem: urban areas are already accounting for 75% of global emissions, but the impacts of climate change are increasingly bringing people to urbanised areas, where they contribute to even greater emissions. In this article on climate-induced migration and urbanisation, we take a closer look at the city of Makassar in South Sulawesi, Indonesia. The city and its surroundings – including some 70 small coral islands in the Makassar Strait – are affected by impacts of climate change, especially water scarcity. Makassar's population is growing due to migration flows from nearby rural areas, and the city struggles to meet the needs of its inhabitants. Because political attention and adaptation planning is often focused on megacities, peripheral cities like Makassar as well as medium-sized and small towns might soon become hotspots of climate-change-related urbanisation.

## *A. Introduction*

If one thinks of an environmental or climate migrant, the vision of a poor person in a dry landscape or in an endangered seascape might come to mind. Reality looks different. Environmentally or climate-induced migration is enlarging urban populations faster than is currently recognised. The global urban population already accounts for half of humanity. For the first time in history, the world is facing a reverse situation: there are more people living in urban than in rural areas. In 2011, urban populations were already rising to 31% in Indian cities, for example. Projections for 2025 predict increases of up to 42.5%. The role that climate change is playing in this respect is currently hardly acknowledged in the political debate on urbanisation, yet it is a crucial role: climate-induced migration serves to enlarge human numbers in cities, and cities are the main contributors to carbon dioxide emissions. At present, cities account for 75% of global emissions. Furthermore, this is not the only challenge we face that is caused by expanding urban areas. Domestic material consumption, sanitation, infrastructure, and racial issues are among the others. All the challenges together will create severe problems for future urban sustainability, and will increase the potential for conflict.

Today, the main focus of attention regarding adaptation plans for countries affected by climate change is on rural development, food security and energy. What has been ignored, or at least is less frequently acknowledged, is that major adaptation measures will be needed in and around urban areas because they are the main destinations for climate and environmental migrants. The lack of adaptation research on the interface between climate change, urbanisation and migration is mainly the result of restrictions implied by disciplinary studies. In this chapter, we aim to combine three research fields – environmental history, natural resources management and migration studies – in an example of current urbanisation trends related to climate change perceptions and induced actions.

## *B. Opening Pandora's Box*

In many parts of the world, climate change is supposed to be acting as the hand that has opened Pandora's box. The impacts expected from this release are only partly known, but they are already causing difficult situations in many regions. There are numerous reports on these impacts and their negative effects on human lives. Human migration is one of the impacts of climate change, yet neither the consequences nor the possible solutions to this challenge are part of the ongoing political discourse on climate change. Little policy attention has been focused on the complex and multidirectional relationships between climate change and migration.[1]

1 Hugo (2008).

Environmental history provides us with many examples of human migration connected to climate change. Temporary or permanent migration has always been a way of adapting to changing climatic conditions. In many cases, it has been climate change that has driven the economic, social and environmental development of societies – and, therefore, their futures. The Viking colonisation of Greenland was only possible during a warm period, which enabled migration to formerly unsuitable areas. A couple of hundred years later, during a cold period which also caused the fall of the Yuan Dynasty in China, those Viking colonies were abandoned. Thus, as has been already stated, during its course, humanity has experienced many migration flows attached to climate change. But what we are currently experiencing is likely to be staggering, and will surpass any historical antecedent.[2] The main difference between historical events and the present is that, today, the highest percentages of migrants end up in cities or in urban peripheries. All trends in the developing world predict a further increase in urban populations. Statistics show that already around 50 million people have migrated to urban areas for climatic and/or environmental reasons.[3] Rapid urban growth is critically surpassing the capacity of most cities to provide adequate services for their inhabitants.[4] The face of the earth has changed, with around 3 billion people living in cities. Almost 400 cities already contain 1 million people or more. Most of these cities are in the developing world. The main concentrations of energy and domestic material consumption are also found in and around urban areas. This explains the high emission levels of cities, mentioned previously. But how do we understand climate-induced migration and link it to the effects on urbanisation and to adaptation needs? How do we make a clear distinction regarding reasons for migration, and link those reasons to climate change?

## C. Climate-induced Migration and Urbanisation

In recent decades, natural disasters, extreme weather events and the loss of environmental services have contributed to increasing resource scarcity and led to environmental conflicts which have forced millions of people to migrate. Most migration routes end up in cities. Predictions for 2050 estimate

2 Warner et al. (2009).
3 Warner et al. (2010).
4 Cohen (2006).

some 200 million migrants ending up in and around urban areas. Additionally, climate change projections for the same time show that cities will be increasingly exposed to major impacts, such as temperature rise, heatwaves, sea-level rise, and an increase in storm severity. Cities and urban areas will not only face a tremendous increase of climate migrants, but will also have to deal with the effects of climate change that will reduce their resilience and ability to respond to these climatic events. An acknowledgement of climate migrants and the recognition of the rapid increase in their numbers, mostly in urban areas, will facilitate the preparation of adaptation policies to address this challenge.

If we are going to use the many research results and observations at our disposal to start adequate policy processes, we need to develop standards that allow for a uniform treatment of climate-related migration. This requires agreeing, firstly, on a proper definition of *climate-induced migration*. On the international level, *migration* is currently defined by the 1951 United Nations (UN) Convention Relating to the Status of Refugees[5] as well as by the associated 1967 Protocol Relating to the Status of Refugees.[6] Additionally, in 2002 recommendations were set by the UN Special Rapporteur on the Human Rights of Internally Displaced Persons to recognise migrants and their rights.[7] However, these definitions are not legally binding at the national level. Also, they have considerable gaps, e.g. none of the documents mention climate- or environment-related migration, or provide advice on how countries can deal with this issue on a national level. The Inter-agency Standing Committee[8] mentions four scenarios which may lead to environmental and climate migration:

5 The 1951 Convention is the key legal document in defining who is a *refugee*, their rights and the legal obligations of states; see http://www.unhcr.org/3b66c2aa10.html, last accessed April 2013.

6 The 1967 Protocol to the Convention of 1951 removed geographical and temporal restrictions on the definition of refugee; see http://www.unhcr.org/3b66c2aa10.html, last accessed April 2013.

7 Report E/CN.4/2003/85 of the Special Rapporteur on Migrants; see http://www.unhcr.org/refworld/topic,4565c22541,4565c25f509,45377ac50,0,UNCHR,,.html, last accessed April 2013.

8 The Inter-Agency Standing Committee (IASC) is a unique inter-agency forum for coordination, policy development and decision-making involving the key UN and non-UN humanitarian partners. The IASC was established in June 1992 in response to UN General Assembly Resolution 46/182 on the strengthening of humanitarian assistance.

- Hydro-meteorological catastrophes
- Environmental degradation and/or long-standing catastrophes
- Land loss caused by sea-level rise, and
- Conflicts caused by resource scarcity.

It was not until 2010 that countries agreed to sign the Cancún Agreement in which they recognised climate-induced migration as a fact that also has to be acknowledged in their national climate policies. As climate migration is often caused by poor adaptation strategies, the Cancún Agreement sets the political framework for addressing the causes of migration.

Since the beginning of the 1990s, researchers have provided serious attention to the study of connections between environmental change, security issues and migration.[9] At the same time, a discussion has evolved about people who could be forced to leave their homes as a result of environmental destruction.[10] Current research also focuses strongly on defining *environment- and climate-induced migration*,[11] based on what has already been identified as major issues with regard to migration and the governance of waves of climate-induced migration.[12] Surprisingly, there is hardly any research on the linkages between urbanisation and climate-induced migration.[13] On a general level, the results of current scientific research are not percolating into proper adaptation policies that might be able to tackle the challenge of increasing migration associated with climate change and other environmental issues. The policy instruments available at present are not sufficient to support climate migrants and to respond to their needs. Such instruments also lack an adequate and useful definition of *climate-induced migration*. Bilsborrow and DeLargy are scholars working on a proper definition of this concept.[14] They suggested as early as 1991 that a cluster of causes exacerbated by climate change induce people to migrate:[15]

9 Homer-Dixon (1991, 1994); Bächler et al. (1996); Bächler (1998); Percival & Homer-Dixon (1998). For a historical perspective, see McLeman (2011).

10 Richter (1998); Scheffran (1994); Wöhlcke (1994).

11 Fernando et al. (2009); Gemenne (2010, 2011); Grote & Warner (2010); Máñez & Scheffran (2011); Renaud et al. (2007); Warner et al. (2009, 2010).

12 Mayer (2010, 2011).

13 DePaul (2012).

14 Bilsborrow & DeLargy (1991).

15 (ibid.).

- The reduction of income
- The increased risk of income reduction in the future, and
- The reduction of environmental health.

We might add to those causes the population trends in many areas of the world, the economic opportunities in cities, the diaspora networks, and the increasingly changing climatic patterns. The International Organisation for Migration assumes 192 million migrants globally. Several predictions show that this number will increase considerably. Direct causes will be environmental change and weather extremes such as droughts. Indirectly, economic problems and conflicts resulting from climate change impacts will also take their toll. Estimates of future numbers of climate migrants differ widely in the scientific literature. While Myers[16] expects some 200 million climate migrants, the office of the UN General Secretary assumes between 50 and 350 million.[17] The German Advisory Council on Global Change predicts that 10–25% of all future migrations will be caused by climate change and its impacts.[18] However, Brown[19] and Jakobeit and Methmann[20] criticise this prediction for being too high. Leaving numbers and projections on migrants aside, and irrespective of the uncertainties attached to those calculations, reality shows that the problems of climate- and environment-related migration have increasingly developed into one of the main political challenges of the 21st century.[21] In 2008, approximately 208 million people worldwide were affected by different types of natural disasters. Around 36 million of them migrated, some 20 million because of climate-related incidents.[22] It is already possible today to identify the specific areas in the world where climate-induced migration will be a major issue. Cities, mostly the small and medium-sized ones, will become the focal points for climate change impacts and growing urbanisation. This is expected to result in growing urban vulnerability to environmental and social change.[23]

16 Myers (2002).

17 UN Report of the Secretary-General on Climate change and its Possible Security Implications, 11 September 2009, Document A/64/350.

18 WGBU (2007).

19 Brown (2011).

20 Jakobeit & Methmann (2007).

21 Morton et al. (2008).

22 OCHA et al. (2009).

23 DePaul (2012).

It is sometimes difficult to differentiate between climate-induced and other reasons for migration. Black et al. make such distinctions by introducing a framework of five drivers – environmental, economic, political, demographic and social – that contribute to migration, with climate change being an external factor affecting all five.[24]

### *D. Climate Urbanisation Trends in Asia and the Pacific*

Environment-induced migration within countries, especially from rural to urban areas, has become significant. In 2010, more than 30 million people in Asia and the Pacific were displaced by environmental disasters such as storms, floods or droughts. Many climate migrants returned home after the extreme events, but others found a new life in urban areas. This enlarged population numbers in cities in Asia, for example. Furthermore, not only big cities experienced such population growth: increasingly, small and medium-sized cities do so too. In China, for instance, it is expected that, by 2030, more than 1 billion people will live in cities, with 221 cities having more than 1 million inhabitants.[25]

In the next few decades, climate change in Asia is expected to contribute to the increase in the frequency of extreme coastal weather events and to a significant sea-level rise. At the same time, the region's population – currently around 4 billion – will continue to increase. These developments will result in growing numbers of people on the move, for reasons that include environmental factors.[26] Countries and populations of Asia and the Pacific will be affected by climate change in different ways, leading to various migration scenarios. Cross-border migration is likely to increase. Already, the region is home to the most important source of international migrants worldwide.

#### *An Example – Sulawesi, Indonesia*

Environment-induced migration in Indonesia is not a recent phenomenon. The largest archipelagic country of the world is situated in a disaster-prone

24 Black et al. (2011).
25 Roberts (2008).
26 ADB (2012).

region: earthquakes, volcanic eruptions, floods and other extreme events have had a strong influence on the Indonesian people throughout time. For example, several million people were displaced as a result of the devastating tsunami in December 2004. For those who have lost their livelihoods, migration to urban areas has always been an option.

The city of Makassar is the capital of the province of South Sulawesi on the island of Sulawesi. This multi-ethnic city is settled by Makassar, Bugis, Mandar and Toraja, to name only the major ethnic groups. Due to its regional importance as having been the biggest city in the outer islands for centuries, Makassar has been characterised as a "peripheral metropolis".[27] Makassar is the centre of migration on Sulawesi. Its population grew from 15,000 in the early 19th Century to 85,000 in 1930, mainly driven by in-migration from the rural areas of South Sulawesi. Over the last few decades, the population of Makassar has increased very rapidly: by 5.5% between 1971 and 1980, and by 2.92% between 1980 and 1990. In 2011, 1,352,136 people lived in Makassar, making it the fourth biggest city in Indonesia. Given that thousands of people live there without being registered, it is likely that this number is actually higher.[28]

Following the 2005 Indonesia Intercensal Population Survey,[29] Makassar stood out in terms of in-migration as a major enclave in the eastern islands, with 331,000 long-term and 82,000 short-term migrants of rural origin. Historically, in-migration had largely resulted from political developments and/or economic considerations, but more careful scrutiny reveals that many of these migration flows were also driven by environmental change. Given that Indonesia is influenced by the monsoon, climate variations also play an important role in this respect. During the inter-monsoon months, approximately 10,000 peasants temporarily move to Makassar to work as rickshaw drivers.[30] In the other seasons, they return to their villages to work in the agricultural sector. These temporal migrants are adding to the steadily growing population of the city, and represent a further challenge as regards the city's functioning and provision of services.

However, permanent in-migration to Makassar will also increase in future. Some of these migrants might come from the Spermonde Archipelago, some 70 very small coral atolls situated in the Makassar Strait just in front

27 Antweiler (2002:231).
28 (ibid.).
29 Survei Penduduk Antar Sensus (SUPAS).
30 Antweiler (2002:231).

of the city. Spermonde's population is growing, and most of the tiny islands are already very densely settled. On the other hand, climate change is already having considerable impacts on the archipelago. Figure 1 reflects the perceptions of island inhabitants on climate change. Changes in season variability, increased wind velocity, higher waves, beach erosion, and temperature increases are increasingly putting people's lives under pressure and inducing them to migrate.[31] Additionally, the scarcity of fresh water has increased significantly, with several islands now importing their supplies from Makassar.[32] Makassar itself is struggling to meet the demand for a supply of clean water; and the problem is expected to increase over the next ten years. The majority of Spermonde's inhabitants are trying to stay where they are, but doing so is not a long-term solution. People who have lost their houses due to erosion are moving towards the centre of the island, or live with their relatives. Some people move between islands, but space is obviously limited. Only those with sufficient financial resources buy land and houses in Makassar, which will enable them to migrate there. However, over the next few decades, people from the Spermonde Archipelago will have to migrate to the mainland. Given their family ties and social connections, many of them might end up in Makassar.

31 Husain (2012).
32 Schwerdtner Máñez et al. (2012).

***Figure 1: Responses with Regard to Perceived Climate Change***

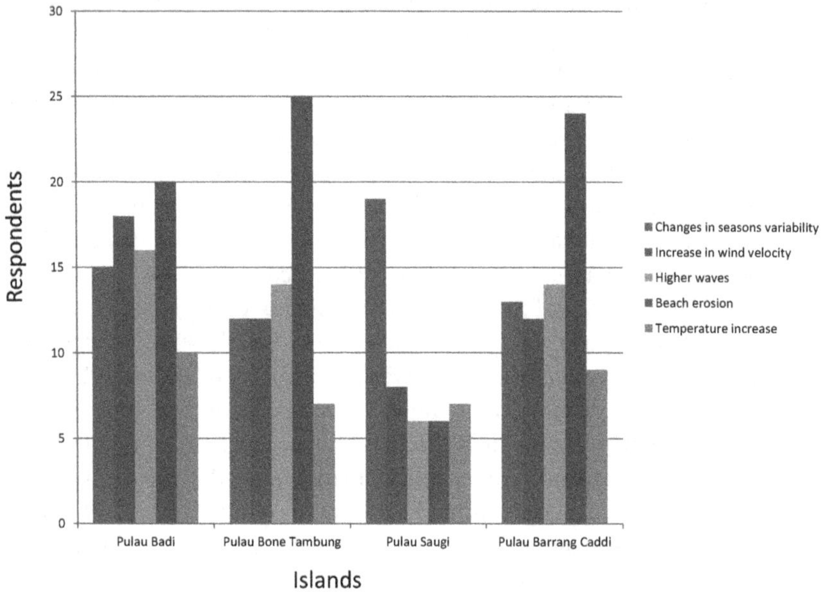


*E. Conclusions and Policy Recommendations*

Climate change will influence the capability of urban areas to adapt to expected and unexpected changes. Many growing cities are located in coastal areas and will be heavily impacted by storm surges, sea-level rise, heatwaves, and more intense floods and droughts. This situation will increase the need for cities to find new ways of adapting to gradually adverse effects within the environmental and social realms. One can also not forget the carbon dioxide budget of cities nowadays: if city sizes increase, one might expect a concomitant increase in carbon dioxide, which might have negative impacts in disaster-prone areas, as many cities are.

Climate-induced migration to urban areas is expected to become a major part of the broader global migration dynamics. In Asia, a region already subject to extensive migration, these dynamics will cause many urban areas, mainly medium-scale and small cities, to grow to previously unforeseen sizes. Climate-induced migration does not have to be seen as a threat; however, since urban areas are the migrant's principal destination, new ways of adaptation to this situation need to be found. When rural environmental or climate migrants settle in urban areas, they mainly settle in slums around the cities. In the face of climate change, these areas – known as hotspots of migration – will be the most vulnerable due, among other things, to the high density of their inhabitants, poor sanitation, and limited water and power supplies. Furthermore, landslides are a problem in mountain regions associated with illegal or uncontrolled urban migration, and might be exacerbated in the near future. Therefore, governments need to look at increasing the capacity of their cities to manage larger populations and to serve these populations appropriately.[33] The planning associated with urban infrastructure and transportation to meet the needs of the increasing population is also vital. In many cases, the reality shows that poor planning has caused severe transportation problems, sewage problems when local flooding occurs, and problems of water supply, due to the increase in demand coupled with depleting aquifers. The city of Makassar is just one typical example of this.

However, like any type of migration, the climate-induced form is also an opportunity. People who migrate into urban areas may profit from more economic possibilities, better education and improved health services. Because many migrants support family members back home, their rural relatives become better off and increase their resilience to disaster and crisis. Migrants also bring to rural areas the new knowledge that they have acquired in the urban areas. Therefore, migration can provide benefits to those who migrate as well as to those who stay behind. Thus, rural–urban migration has been shown to be an effective solution for reducing rural poverty. For example, Indonesia has experienced rapid urbanisation with a concomitant rapid decline in rural poverty over the last few decades.

Climate change is already a challenge to many cities, especially in the developing world. Climate-induced migration might become one of the major issues that governments have to consider in their national adaptation policies. If planners and decision-makers do not react properly and timeously, many urban areas will become disaster hotspots. Actions are urgently needed in order to link urban planning with disaster management and dealing with increasing migrant populations. This is true not only for Asia, but also for other parts of the world, especially Africa, where urban areas are growing as fast as those in Asia. Governments will have to deal with the costs of disasters if they do not start taking into account the rapid development of urban areas due to increasing numbers of climate-associated migrants. In

33 ADB (2012).

their planning, governments should also include investments in sustainable infrastructure and basic services in migrant destination cities. One suggestion, as a first step, would be for governments to develop hotspot maps – maps showing vulnerable areas – so that projections of future development become visible and, on that basis, adaptive management plans can be developed.

## *References*

ADB/Asian Development Bank, 2012, *Addressing Climate Change and Migration in Asia and the Pacific*, available at http://www.adb.org/publications/addressing-climate-change-and-migration-asia-and-pacific, last accessed December 2012.

Antweiler, Christoph, 2002, Rationalities in Makassar: Cognition and Mobility in a Regional Metropolis in the Indonesian Periphery, in: Nas, Peter J.M. (Ed.), *The Indonesian Town Revisited*, Münster, LIT Verlag, 230–258.

Bächler, Günther, 1998, Why Environmental Transformation Causes Violence: A Synthesis, *Environmental Change and Security Project Report* 4, 24–44.

Bächler, Günther & Kurt R. Spillmann, 1996, *Environmental Degradation as a Cause of War: Regional and Country Studies of Research Fellows*, Zürich, Rüegger Verlag.

Bilsborrow, Richard E. & Pamela DeLargy, 1991, Land Use, Migration, and Natural Resource Deterioration in the Third World: The Cases of Guatemala and Sudan, *Population and Development Review, Supplement to Vol. 16*, 125–147.

Black, Richard, W. Neil Adger, Nigel Arnell, Stefan Dercon, Andrew Geddes & David Thomas, 2011, The Effect of Environmental Change on Human Migration, *Global Environmental Change* 21 (4), 3–11.

Brown, Oli, 2011, *Migration, Climate Change and the Environment – A Complex Nexus*, Geneva, International Organization for Migration.

Cohen, Barney, 2006, Urbanization in Developing Countries: Current Trends, Future Projections, and Key Challenges for Sustainability, *Technology in Society* 28, 63–80.

DePaul, Miles, 2012, Climate Change, Migration, and Megacities: Addressing the Dual Stresses of Mass Urbanization and Climate Vulnerability, *Paterson Review of International Affairs*, 12, 145–162.

Fernando, Nishara, Koko Warner & Jörn Birkmann, 2009, Migration and Natural Hazards: Is Relocation a Secondary Disaster or an Opportunity for Vulnerability Reduction?, in: Afifi, Tamer & Jill Jäger (Eds), *Environment, Forced Migration, and Social Vulnerability*, Bonn/Geneva, United Nations University Institute for Environment and Human Security & International Organization for Migration, 145–156.

Gemenne, François, 2010, Migrations et Environnement, Numéro Spécial de la Revue, *Hommes et Migrations* 1284, 6–9.

Gemenne, François, 2011, Climate-induced Population Displacements in a 4°C+ World, *Philosophical Transactions of the Royal Society*, A, 369, 182–195.

Grote, Ulrike & Koko Warner, 2010, Environmental Change and Forced Migration: Evidence from Sub-Saharan Africa, *International Journal of Global Warming* 2 (1), 17–47.

Homer-Dixon, Thomas F., 1991, On the Threshold: Environmental Changes as Causes of Acute Conflict, *International Security* 16 (2), 76–116.

Homer-Dixon, Thomas F., 1994, Environmental Scarcities and Violent Conflict: Evidence from Cases, *International Security* 19, 5–40.

Hugo, Graeme, 2008, Migration, Development and Environment, *IOM Report*, Geneva, International Organization for Migration.

Husain, Sainab, 2012, *Faktor Faktor Yang Mempengaruhi Resiliensi Masyarakat Terhadap Dampak Perubahan Iklim, di Kepulauan Spermonde*, Sulawesi Selatan, University of Makassar.

Jakobeit, Cord & Chris Methmann, 2007, *Klimaflüchtlinge*, available at http://www.greenpeace.de/fileadmin/gpd/user_upload/themen/klima/klimafluechtlinge_endv.PDF, last accessed January 2013.

Máñez, María & Jürgen Scheffran, 2011, *Die Flucht vor dem Klimawandel*, available at http://www.climate-service-center.de/imperia/md/content/csc/Migration___Klimawandel_pdf.pdf, last accessed April 2013.

Mayer, B., 2010, Governing International Climate Change-induced Migration, in: Elliott, L. (Ed.), *Climate Change, Migration and Human Security in Southeast Asia*, Singapore, Nanyang Technical University Press.

Mayer, Benoit, 2011, Fraternity, Responsibility and Sustainability: The International Legal Protection of Climate (or Environmental) Migrants at the Crossroads, Earth System Governance Working Paper No. 14, Lund, Earth System Governance.

McLeman, Robert, 2011, Climate Change, Migration, and Critical International Security Considerations, Migration Research Series Report, No. 42, Geneva, International Organization for Migration.

McLeman, Robert & Barry Smit, 2006, Migration as an Adaptation to Climate Change, *Climatic Change* 76 (1–2), 31–53.

Morton, Andrew, Philippe Boncour & Frank Lankzo, 2008, Human Security Policy Challenges, *Forced Migration Review* 31, 5–7.

Myers, Norman, 2002, Environmental Refugees: A Growing Phenomenon of the 21st Century, *Philosophical Transactions of the Royal Society: Biological Science* 357, 609–611.

OCHA, IDMC & NRC/United Nations Office for the Coordination of Humanitarian Affairs, Internal Displacement Monitoring Centre and Norwegian Refugee Council, 2009, Monitoring Disaster Displacement in the Context of Climate Change – Findings of a Study by the United Nations Office for the Coordination of Humanitarian Affairs (OCHA), the Internal Displacement Monitoring Centre (IDMC) and the Norwegian Refugee Council (NRC), available at http://www.internaldisplacement.org/8025708F004BE3B1/(httpInfoFiles)/12E8C7224C2A6A9EC125763900315AD4/$file/monitoring-disasterdisplacement.pdf, last accessed December 2012.

Percival, Val & Thomas F. Homer-Dixon, 1998, Environmental Scarcity and Violent Conflict: The Case of South Africa, *Journal of Peace Research* 35, 279–298.

Renaud, Fabrice, Janos J. Bogardi, Olivia Dunn & Koko Warner, 2007, Control, Adapt or Flee: How to Face Environmental Migration?, *InterSecTions No. 5*, Bonn, United Nations University Institute for Environment and Human Security.

Richter, Roland, 1998, Umweltflüchtlinge in Afrika, in: Scheffran, Jürgen & Wolfgang R. Vogt (Eds), *Kampf um die Natur: Umweltzerstörung und die Lösung ökologischer Konflikte*, Darmstadt, Wissenschaftliche Buchgesellschaft, 42–73.

Roberts, Dexter, 2008, China prepares for urban revolution, *Bloomberg Businessweek*, 13 November 2008, available at http://www.businessweek.com/stories/2008-11-13/china-prepares-for-urban-revolutionbusinessweek-business-news-stock-market-and-financial-advice, last accessed December 2012.

Scheffran, Jürgen, 1994, Kriegs- und Umweltflüchtlinge, in: Böhme, Gernot, Rabindra N. Chakraborty & Frank Weiler (Eds), *Migration und Ausländerfeinlichkeit*, Darmstadt, Wissenschaftliche Buchgesellschaft, 23–38.

Schwerdtner Máñez, Kathleen, Sainab Husain, Sebastian C.A. Ferse & María Máñez Costa, 2012, Water Scarcity in the Spermonde Archipelago, Sulawesi, Indonesia: Past, Present and Future, *Environmental Science and Policy* 23, 74–84.

Warner, Koko, Charles Ehrhart, Alex de Sherbinin, Susana B. Adamo & Tricia Chai-Onn, 2009, *In Search of Shelter: Mapping the Effects of Climate Change on Human Migration and Displacement*, London, CARE International.

Warner, Koko, Mohammed Hamza, Tony Oliver-Smith, Fabrice Renaud & Alex Julca, 2010, Climate Change, Environmental Degradation and Migration, *Natural Hazards* 55 (3), 689–715.

WGBU/Wissenschaftliche Beirat der Bundesregierung Globale Umweltveränderungen/German Advisory Council on Global Change, 2007, *World in Transition: Climate Change as a Security Risk*, Berlin, WBGU.

Wöhlcke, Manfred, 1994, Bevölkerungswachstum, ökologische Degradierung, Unterentwicklung, Migration, *Comparativ – Zeitschrift für Globalgeschichte und Vergleichende Gesellschaftsforschung* 2, 78–92.

27

# Climate-change-induced Movement of Persons in Africa: Human Rights Responses to Aspects of Human Security*

*Oliver C. Ruppel & Sanita van Wyk*

## *Abstract*

This article reviews the possible and expected effects of climate change on certain components of human security in Africa. Due to multiple stresses, Africa is one of the continents most vulnerable to climate change and climate variability. Climate change and climate variability not only have the potential to impose additional pressures on human security and to overwhelm adaptive capacities of societies, climate change is also deemed to influence a diverse array of conflicts. According to the Intergovernmental Panel on Climate Change, one of the greatest impacts of climate change might be on human mobility. To this end, the chapter focuses on the socio-political and legal aspects of climate-change-induced movement of environmental migrants, refugees and displaced persons, and potential human rights responses thereto.

## *A. Introduction*

The purpose of the concept *human security* is to protect the vital core of all human lives in ways that enhance human freedoms and human fulfilment.[1] Human security has already been addressed by the founders of the United Nations (UN), yet an important milestone of the development of today's notion of the concept was the 1994 United Nations Development Programme (UNDP) Human Development Report,[2] which defines *human security* as "a

* This article is an updated version of Ruppel & van Wyk (2011a).
1 CHS (2003).
2 UNDP (1994).

concern with human life and dignity".[3] The Report also emphasises four vital characteristics of human security:[4]

- Human security is a universal concern
- All the components of human security are interdependent
- It is easier to ensure human security through prevention than intervention, and
- Human security is people-centred.

Seven broad interdependent components of human security have been identified. They are –[5]

- economic security
- food security
- health security
- environmental security
- personal security
- community security, and
- political security.

However, this list is not exhaustive or mutually exclusive in its representation of the existing risks to human security.

The concept of human security is established on two pillars. The first, the "freedom from fear" factor, focuses on protecting individuals from violent conflicts and from the denial of civil liberties, and ensures freedom of expression and belief. The second pillar is the "freedom from want" factor, which emphasises satisfying the individual's basic need for food, shelter and clothing.[6] In addressing the root problem of insecurity in Africa, violence, poverty and inequality – be it social or economic – play a core role.[7] A human security approach focusing on people as the prime referents of security is increasingly being integrated into policymaking and jurisprudence.[8]

It has been stated that "in no other continent are threats to human security more dire and the absence of protection infrastructure more conspicuous,

---

3 (ibid.:22).
4 (ibid.:22–23).
5 King & Murray (2002:585–610).
6 Kumssa & Jones (2010:453–461); UNDP (1994:24).
7 (ibid.).
8 Abass (2010:3).

than in Africa."[9] Undoubtedly, this assessment also applies to climate-change-related threats to human security. In the following section, possible and expected effects of climate change on the various components of human security in Africa will thus be outlined briefly, before making some references to potential human rights responses.

## *B. Climate Change and Human Security*

Despite Africa's relatively low contribution to the world's total greenhouse gas (GHG) emissions, the continent is one of the most vulnerable to climate change and climate variability.[10]

### *I. Environmental Security*

Climate change and variability have the potential to impose additional pressures on human security along with many socio-economic factors, and overwhelm the adaptive capacities of societies in many world regions. The most direct link between climate change and threats to human security is probably the aspect of environmental security, where access to clean water is considered to be one of the greatest environmental threats.[11]

### *II. Economic Security*

As another aspect of human security, the effects of climate change on economic security are manifold. Economic security requires an assured basic income for individuals, usually from productive and remunerative work or, as a last resort, from a publicly financed safety net.[12] The impacts of global

---

9 (ibid.:10).

10 Boko et al. (2007:433–467). While *climate change* "refers to a change in the state of the climate that can be identified ... by changes in the mean and/or the variability of its properties, and that persists for an extended period, typically decades or longer", *climate variability* "refers to variations in the mean state and other statistics ... of the climate on all spatial and temporal scales beyond that of individual weather events"; see IPCC (2007:Annex II, Glossary).

11 Bantekas (2010:43).

12 UNDP (1994).

warming on the agricultural sector in Africa are considered to be a direct and profound threat. Again, water scarcity has a direct impact on many economic development initiatives on the agricultural sector, which is one of the most important in African economies.

### *III. Food Security*

Climate change has economic impacts on crop and livestock farming systems; warmer and drier climates adversely affect net farm revenues translating into a worsening food security situation on the continent.[13] The ultimate damages of climate change may significantly affect economic growth.[14] Although agricultural productivity might increase in the short run in some regions of Africa because of global warming, many African countries and regions are likely to be severely affected by climate change and climate variability.[15] Increasing temperatures and declining precipitation in Africa resulting from climate change are likely to reduce yields for primary crops in the next two decades; these changes will have a substantial impact on food security, although the precise extent and nature of that impact cannot yet be determined.[16] Periods of droughts and floods will have an impact on the availability of and access to food.[17] It is predicted that the impacts of climate change such as sea-level rise, droughts, heatwaves, floods and rainfall variation could, by 2080, push another 600 million people into the abyss of malnutrition and increase the number of people facing water scarcity by 1.8 billion.[18]

### *IV. Health Security*

Health security aims to guarantee a minimum protection from diseases and unhealthy lifestyles. Climate change is considered to be a serious global

13 See e.g. Nhemachena et al. (2010).
14 Lecocq & Shalizi (2007).
15 Boko et al. (2007:435).
16 (ibid.).
17 Ziervogel et al. (2006).
18 UNDP (2008).

health threat in the 21st Century.[19] Africa is particularly vulnerable as threats to health security are usually greater for poor people in rural areas, particularly women and children, due to malnutrition and insufficient access to health services, clean water and other basic necessities. Major diseases could expand their coverage as a result of global warming. For example, an additional 220–400 million people could be exposed to malaria – a disease that already claims around 1 million lives annually.[20] Other health security threats in Africa include the spread of infectious diseases such as cholera – which is influenced by both global and regional climatic variability,[21] dengue fever,[22] and meningitis.[23]

### *C. Human Security, Conflict and Environmental Migration*

While *personal security* aims to protect people from physical violence by states or individuals, *community security* is concerned with protecting people from the loss of traditional relationships and values and from sectarian and ethnic violence. *Political security* addresses the question as to whether or not people live in a society that honours their basic human rights. All the aforementioned factors are relevant when it comes to the issues of violent conflicts and migration. Conflict, climate-change-induced disasters, and human mobility are all interlinked.[24] Climate change is deemed to influence, to a greater or lesser degree, a diverse array of conflicts including war, terrorism, and diplomatic and trade-related disputes.[25]

Conflict is deemed to become an indirect impact of climate change.[26] In other words, climate change impacts the natural environment, and the state of the natural environment may cause future conflict. In attempting to demarcate those that are affected by the impacts of climate change, it is necessary to focus on the link between the reasons for environmental migration caused by specific climate change impacts.[27] Thus, a person who migrates

19 Costello et al. (2009).
20 UNDP (2008).
21 De Magny et al. (2007:20).
22 Jansen & Beebe (2010:272–279).
23 Cuevas et al. (2007:A12–A17).
24 Kolmannskog (2010:104, 108).
25 Homer-Dixon (1991:77).
26 Biermann & Boas (2010:64).
27 (ibid.).

because his/her environment was directly impacted by climate change can be considered an *environmental migrant*. But a person who migrates because conflict has erupted in his/her environment – perhaps due to an environment affected by climate change – should not be considered an environmental migrant because the direct reason of migration concerns the conflict and not the state of the environment.

Indeed, the relationship between conflict and environmental migration is complex and troublesome. Resettling environmental migrants may place an extra or unbearable burden on their new environment.[28] This may cause friction between the environmental migrants and the local inhabitants of the land and lead to conflict in their new settlement area.[29] This can set in motion another migratory cycle, as conflict can damage the natural environment to such an extent that another group of people decides to migrate.[30] Conflict-ridden zones in general also reflect a lack of state control or effective governance. This, in turn, implies ineffective environmental management, which also contributes to degradation of the environment – potentially leading to environmental migration.[31] In addition to exacerbating environmental degradation, conflicts also make it difficult for environmental migrants to leave such zones peacefully.[32]

Climate change, climate-related environmental variability and conflict have attracted much attention and debate. While there seems to be consensus about the environment being only one of several interconnected causes of conflict rather than the decisive factor,[33] the changing climate arguably increases the risk of civil unrest in Africa. This is because conflicts are more likely in regions with more vegetation such as Africa (possibly resulting from vegetation recovery after population has been displaced out of conflict zones), and that increased levels of malnutrition, which run high in Africa, are related to armed conflicts.[34] Another view emphasises the role of arable land and renewable resources such as fresh water. It is argued that, as a long-term trend, population growth and resource scarcities result in violent com-

28 Warner et al. (2010:710).
29 (ibid.).
30 (ibid.).
31 Kolmannskog (2010:109).
32 (ibid.).
33 (ibid.:103–119).
34 Rowhani et al. (2011:207–222).

petition,[35] while short-term causes may trigger the outbreak of conflict.[36] Therefore, distributional conflicts can arise out of the degradation of natural resources as a result of overexploitation and the effects of climate change.

In determining with certainty whether climate change will be a cause of conflict, one first needs to ask what social changes are initiated by a changing climate; and secondly, one needs to know what type of conflict is likely to arise from a specific type of social change.[37] Potential social changes are numerous and complex, and are dependent on political and economic factors.[38] In addition to social, political and economic factors, environmental change is can lead to conflict and migration. In fact, migration is probably the oldest way of dealing with environmental change.[39] Historically, people have always set out to inhabit environments that would allow not only survival, but also a stable human existence.[40] But when people are faced with environmental change, they may have other options as well: they could choose to adapt to the change and, in so doing, might lessen the impact of such a change, or they could do nothing and, in effect, accept a lower quality of life.[41]

### *D. Climate-change-induced Movement of Persons*

In its First Assessment Report, the Intergovernmental Panel on Climate Change (IPCC) stated that one of the greatest impacts of climate change might be on human mobility.[42] Burgeoning human populations and the accelerated degradation of natural resources are expected to swell the numbers of migrants as well, both internally and across borders.[43] Increased environmental migration due to the effects of climate change is considered a new phenomenon, unprecedented in its scale and scope,[44] and is closely related to the concept of human security. In addition to low-lying islands, coastal

35 Homer-Dixon (1994:5–40).
36 Hendrix & Glaser (2007:695–715).
37 Homer-Dixon (1991:87).
38 (ibid.).
39 Kolmannskog (2008).
40 Warner et al. (2010:691).
41 (ibid.:690).
42 IPCC (1990:Chapter 5, 9).
43 Gleditsch et al. (2007:3).
44 Warner et al. (2010:692).

and deltaic regions, various parts of Africa are expected to be affected by climate-change-induced migration.[45]

Climate change impacts the frequency and severity of extreme or sudden weather events.[46] Such events become disastrous when a community is vulnerable to its effects.[47] A community's vulnerability is dependent on its exposure to the event and its ability to adapt or recover after the event has occurred.[48] Africa's low adaptive capacity[49] makes it particularly vulnerable to suffering disastrous consequences of climate change. A global study conducted in 2009[50] reveals that, in 2008, at least 36 million people were newly displaced by sudden-onset natural disasters. Of the 36 million, over 20 million were displaced by sudden-onset climate-related disasters. Further estimates put the number of displaced people in Africa as having increased almost 700,000 in 2008 to 1.1 million in 2009, and 1.7 million by 2010.[51]

It is predicted that the effect of climate change on future forced migrations will be determined by, among other things, the following factors:[52]

- The amount of GHG emissions
- Population growth
- Population distribution
- Evolution of climate change, and
- The adaptive capabilities of communities on various levels.

It should, however, be noted that estimates on migration flows resulting from climate change remain speculative, as migration drivers are usually not monocausal but influenced by multiple factors.[53] Only by understanding the environment in the broader social, economic and political context can one evaluate the role that the environment plays in migration behaviour.[54]

The impact of climate change on food security and water stress in rural areas of developing countries is expected to accelerate rural–urban migration

45 Gemenne (2011:182–195).
46 (ibid.).
47 Brown (2008:9, 24).
48 (ibid.).
49 Boko et al. (2007:435).
50 OCAH & IDMC (2009).
51 IDMC (2011:16).
52 Brown (2008:27).
53 Smith et al. (2011:180).
54 Lonergan (1998:8).

movements.[55] In sub-Saharan Africa, climate change is already considered an important determinant of urban growth.[56] In 2009, almost 40% of Africa's total population of 1 billion lived in urban areas; it is estimated that Africa's population will become 50% urban by 2030, rising to 60% by 2050.[57] African inland cities are exposed to higher ambient temperatures and more frequent heatwaves, with a concomitant potential risk of water shortages, damage to infrastructure (e.g. transportation infrastructure including roads and railways), and desiccating vegetation. Declining rainfall, droughts and floods have the potential of rendering agricultural lands in Africa unproductive or making rural settlements uninhabitable; this affects the livelihoods of rural residents, forcing them to migrate to the urban areas.[58]

Africa counts 37 cities with populations above 1 million, half of which are within low-elevation coastal zones,[59] which are particularly vulnerable to extreme weather events caused by climate change. These zones are likely to experience storm surges, a rise in sea level, increased flooding, (semi-)permanent inundation, coastal erosion, landslides, and the increase of water-borne diseases, which may all have devastating effects on human settlements – especially if no measures are taken to ensure risk reduction in terms of urban planning, land-use management, and the quality of housing and infrastructure.[60] Depending on their location and nature of construction, buildings and supporting infrastructure are vulnerable to flooding and other extreme weather events which increase the likelihood of landslides and building subsidence, especially on clay soils. Such threats require enhanced construction and infrastructural standards as a first barrier of resistance and protection, which include raising building foundations, strengthening roads, and increasing storm-water drainage capacity.[61] In this regard, the high risk for low-lying urban slums has to be pointed out. Although the proportion of urban slum dwellers is decreasing, informal settlements remain one of the major threats to African urban stability and, by extension, to overall political stability.[62]

---

55 Brown (2008:32).
56 Barrios et al. (2006:357–371).
57 UN-Habitat & UNEP (2010).
58 Hope (2011:37–56).
59 Mosha (2011:69–102).
60 (ibid.).
61 UN-Habitat & UNEP (2010).
62 (ibid.).

## *E. Migrants, Refugees and Internally Displaced Persons*

Not only is there a range of terms referring to environmental migrants, their use is inconsistent.[63] This contribution explores some of this terminology. The various legal options in dealing with environmental migrants are mentioned to illustrate that dealing with this problem on a terminological scale is problematic. This is a large and urgent issue to address, as the label and corresponding definition awarded to environmental migrants affects the world community's obligations under international law. Not only is a sound and accurate definition needed, but this problem should also be practically addressed by countries already struggling to adhere to existing international obligations. The way in which nations deal with environmental migrants is a litmus test of the way they deal with human rights. Safeguarding human dignity should be identified as a top priority in protecting such migrants.[64] In fact, the International Bill of Rights should apply, meaning that citizens and non-citizens alike should be able to claim rights in terms of –

- the Universal Declaration of Human Rights
- the International Covenant on Civil and Political Rights (ICCPR)
- the International Covenant on Economic, Social and Cultural Rights (ICESCR)
- the First Optional Protocol to the ICCPR, and
- the Second Optional Protocol to the ICCPR.

Most African countries have acceded to both the ICCPR[65] and the ICESCR.[66] On 10 December 2008, the UN General Assembly adopted, by consensus, the Optional Protocol to the ICESCR. The Optional Protocol provides a mechanism through which persons can petition the UN Committee on Economic, Social and Cultural Rights about violations of their rights. The Optional Protocol was opened for signing on 24 September 2009. Both the ICCPR and the ICESCR call on states parties to take legislative or other measures to effect the rights contained in them. Most of the rights and freedoms recognised in the ICCPR are also entrenched in national constitutions'

63 Warner et al. (2010:689–715).

64 OHCHR [n.d].

65 As at the time of writing, only one of the 54 fully recognised states in Africa is not a party to the ICCPR, namely South Sudan.

66 Of the 54 fully recognised states in Africa, Botswana and South Sudan are not parties to the ICESCR as at the time of writing.

Bill of Rights. This may include the right to dignity, the right to life, the right to health, the right of access to water, the right to legal representation, the guarantee against torture and other cruel or inhuman treatment or punishment, and protection against discrimination on any ground. Both the ICESCR and the ICCPR provide protection for specific rights and freedoms; both recognise the right of peoples to self-determination; both prohibit all forms of discrimination in the exercise of human rights; and both have the force of law for those countries that have ratified them. States have obligations under international human rights law to respect, to protect and to fulfil human rights. In the context of climate change, such obligations specifically include ensuring that policies aimed at limiting the effects of climate change are implemented effectively and in ways that do not overburden or discriminate against specific vulnerable groups, e.g. women, children, the disabled, and indigenous peoples.[67]

Since environmental migrants – just as migrants in general – experience a higher risk of falling into poverty and social exclusion than the general population, they can also be seen as a vulnerable group. Thus, environmental migrants de facto need a high level of protection under international human rights law. However, one problem encountered by using the term *environmental migrant* is that the term *migrant* may imply a voluntary decision to move – the pull of the new destination being stronger than the push of the old.[68] Some of the 'push factors' of environmental migration include sudden natural disasters, slow-onset natural disasters, environmental conflicts, environmental destruction due to conflict, environmental conservation, development projects, and industrial accidents.[69] Sudden natural disasters include floods, storms, hurricanes and typhoons.[70] Slow-onset natural disasters include sea-level rise, degradation of agricultural land, desertification, increasing water stress and food insecurity.[71] Climate change particularly impacts slow-onset natural disasters such as sea-level rise and water stress.[72]

The term *climate change refugee* (or, in short, *climate refugee*) is often used instead of the term *environmental migrant* or *environmentally dis-*

67 Ruppel (2010); Ruppel & Ruppel-Schlichting (2011).
68 Brown (2008:14).
69 Kolmannskog (2008).
70 Brown (2008:18).
71 (ibid.).
72 Gemenne (2011:182–195).

*placed person* to convey the urgency of the situation.[73] But the term *climate refugee* can also refer to a subcategory of *environmental migrants* – migrants whose relocation can be directly linked to climate change. There are also subcategories of climate refugees: they are categorised according to whether the migration is voluntary or forced, or anticipated or unexpected, as well as according to the role that environmental degradation plays in their decision to migrate.[74] However, it is difficult to determine the level of force or anticipation involved in the decision to migrate, and adaptive capabilities are influenced by additional factors,[75] such as the affected person's financial capacity and his/her available personal or official support networks at the time.[76] Those who are forced to stay behind because they do not have the financial resources or appropriate networks in place to ensure successful migration may fall through the gaps of an already limited international law regime.[77]

In trying to categorise environmental migrants (and not necessarily the subcategory of climate refugees), the International Organization for Migration proposes a useful definition:[78]

> Environmental migrants are persons or groups of persons who, for compelling reasons of sudden or progressive changes in the environment that adversely affect their lives or living conditions, are obliged to leave their habitual homes, or choose to do so, either temporarily or permanently, and who move either within their country or abroad.

The United Nations Environment Programme defines environmental refugees as –[79]

> … people who have been forced to leave their traditional habitat, temporarily or permanently, because of a marked environmental disruption (natural and/or triggered by people) that jeopardised their existence and/or seriously affected the quality of their life.

---

73 Brown (2008:13). For other terms for *environmentally induced migrants* and the environmental factors that influence the meanings and use of these terms, see Warner et al. (2010:697). For normative gaps and possible approaches relating to the protection of people crossing borders in the context of climate change, see Kälin & Schrepfer (2012).

74 Biermann & Boas (2010:65).

75 (ibid.).

76 Kolmannskog (2010:106).

77 Ruppel (2011d).

78 IOM (2007).

79 Biermann & Boas (2010:62).

However, there remains a need for a clear definition of climate refugee as the term creates conflict of a legal nature when it comes to whether or not climate refugees can legally be classified as refugees under international refugee law. But regardless of this, states should ensure that all persons – both citizens and non-citizens – classified as refugees or environmental migrants or persons fleeing from armed conflict or poverty will receive safeguarding of their basic human dignity and respect for their fundamental human rights. This assurance is perhaps the best point of departure in trying to expand the definition of refugee, and the UN Office of the High Commissioner for Human Rights also highlights this point.[80]

International refugee law applies to any person defined as refugee by the Geneva Refugee Convention of 1951,[81] namely as a person with a –[82]

> … wellfounded fear of being persecuted for reasons of race, religion, nationality, membership of a particular social group or political opinion, is outside the country of his nationality and is unable or, owing to such fear, is unwilling to avail himself of the protection of that country; or who, not having a nationality and being outside the country of his former habitual residence as a result of such events, is unable or, owing to such fear, is unwilling to return to it.

There are numerous complications in attempting to classify climate refugees as refugees under international refugee law. The definition of the term refugee, as set out under international law, does not (explicitly) provide for environmental reasons in leaving the country as being sufficient grounds to qualify for refugee status.[83] The fear of persecution has to be present and, unless nature could be seen as a persecutor, environmental reasons technically do not suffice.[84] In addition, the term refugee relies on the fact that an international border was crossed: a person moving within a country is not referred to as a refugee, but as an *internally displaced person*.[85] Moreover, the term refugee implies that the person is unable to return to the country from which s/he fled due to a well-founded fear of persecution,[86] while in some cases a climate refugee will be unable to return to the place from where

80 OHCHR [n.d.].

81 Nowak (2003:39).

82 Article 1.A(2), 1951 Convention Relating to the Status of Refugees, hereinafter the *Geneva Refugee Convention*.

83 Brown (2008:11–15).

84 Warner et al. (2010:695).

85 Brown (2008:13–14).

86 (ibid.).

s/he migrated for reasons other than a well-founded fear of persecution. Such a reason may be a rising sea level.[87]

Other instruments, such as the 1969 Organisation of African Unity Convention Governing the Specific Aspects of Refugee Problems in Africa[88] and the 1984 Cartagena Declaration on Refugees[89] have defined refugee in broader terms than the 1951 Geneva Refugee Convention. The 1969 Convention applies the term refugee to a person who is "compelled to leave his place of habitual residence in order to seek refuge in another place outside his country of origin or Nationality."[90]

According to Conclusion No. 3 of the 1984 Cartagena Declaration on Refugees, the definition or concept of a refugee –

> … recommended for use in the region is one which, in addition to containing the elements of the 1951 Convention and the 1967 Protocol, includes among refugees persons who have fled their country because their lives, safety or freedom have been threatened by generalized violence, foreign aggression, internal conflicts, massive violation of human rights or other circumstances which have seriously disturbed public order.

Although certain environmental events caused or worsened by climate change may comprise events which have "seriously disturbed public order", these two instruments were not originally intended to protect climate refugees.[91]

Even though the term refugee is debatable, Biermann and Boas advocate the use of the term climate refugees.[92] They argue that using a term other than climate refugee would not be appropriate as it detracts from the "legitimacy and urgency".[93] However, neither author supports the idea that refugee protection hinges on whether or not a person has crossed an international border (in the case of climate refugees).[94] They propose that climate refugees be defined as –[95]

---

87 (ibid.).
88 Article 1.2.
89 Section III.3. This declaration concerns refugees from Central America, Mexico, and Panama.
90 Article 1 of the 1969 Organisation of African Unity Convention Governing the Specific Aspects of Refugee Problems in Africa.
91 Biermann & Boas (2010:73).
92 (ibid.:66).
93 (ibid.:67).
94 (ibid.).
95 (ibid.).

> … people who have to leave their habitats, immediately or in the near future, because of sudden or gradual alterations in their natural environment related to at least one of the three impacts of climate change: sea-level rise, extreme weather events, and drought and water scarcity.

They add that a good definition of climate refugees should address three aspects, namely –[96]

- the cause of migration, i.e. the type of environmental impact that would create climate refugees
- the type of migration, i.e. whether it is forced or voluntary, and whether it is permanent or temporary, and
- appropriate terminology, i.e. whether or not the term refugee is appropriate.

There are many other proposed academic definitions for climate refugee.[97] Most of these do not specify whether the person needs to cross an international border, but they differ widely in stating whether or not the migration needs be temporary or permanent.[98]

Some also contend that climate refugees should be afforded the same protection as refugees under international refugee law.[99] Legally, this might be the simplest route; but there are numerous political and economic reasons why countries do not wish to extend 'traditional refugee protection' to climate refugees. By including climate refugees as refugees under current international refugee law, countries would have to deal with 20 times more refugees than they already do.[100] A potential problem with this definition lies in its implementation: it remains questionable whether developing countries will be able to monitor environmental changes as well as the concomitant human migration in a consistent and effective manner. Biermann and Boas suggest the creation of a sui generis regime designed specifically to

96 (ibid.:63).

97 For examples of a definition for *environmental refugee*, see Warner et al. (2010:694–695).

98 (ibid.:694).

99 (ibid.:695).

100 Biermann & Boas (2010:74).

protect climate refugees.[101] But even with the successful establishment of a "tailor-made"[102] regime, effective implementation of new legal and policy solutions will still present a significant stumbling block due to the enormous potential costs involved. Whether or not climate refugees are added to the traditional refugee group or divided into their own sui generis group is not that important if the ultimate goal is to protect the human rights of both groups – traditional refugees and climate refugees. Indeed, the biggest hurdle for countries in respect of defining climate refugees as a designated group lies in the fact that no country really wants to take on the responsibility in dealing with climate refugees; unfortunately for them, there is no definition or grouping of climate refugee that will make the problem disappear.

The International Organization for Migration and the UN Office of the High Commissioner for Refugees use the term *environmentally displaced persons* when referring to climate refugees.[103] The term *distress migration* can also be considered in a context where displacement is not forced, or the entire family does not have to migrate.[104] An environmental migrant (or climate refugee) not crossing the borders of his/her country is known as an *internally displaced person*, and internally displaced persons are not protected under international refugee law.[105] Clearly, this shows that climate refugees – who move within their own country as well as across borders – should be treated differently to internally displaced persons if they are to receive appropriate protection. Persons labelled as *internally displaced* do not seem to receive the same amount of protection and care that refugees and persons displaced due to conflict and sudden-onset disasters do.[106] The responsibility of protecting internally displaced persons lies with national

101 (ibid.:76–77). Various principles should form the basis of such a sui generis regime, according to Biermann & Boas. Briefly, these principles concern relocating and resettling climate refugees in a planned and voluntary manner; granting the majority of climate refugees the status of permanent immigrants to a new country; designing the regime for groups of people rather than for individual cases; making it a priority to support national governments in coping with an influx of climate refugees; and obliging developed countries to bear the bulk of the substantial financial burden in relocating climate refugees.

102 Kolmannskog (2010:117).

103 Biermann & Boas (2010:66).

104 Kolmannskog (2010:114).

105 Nowak (2003:39).

106 Kolmannskog (2010:105).

governments,[107] and, depending on the situation on the ground, this responsibility is often shirked.[108] However, the UN Human Rights Council recently expressed concern over the impact that climate change has on the problem of internally displaced persons.[109] To this end, the Council declared that "natural disasters are a cause of internal displacement", that such disasters were worsened by climate change, and that urgent "disaster risk reduction" was needed.[110]

The African Union Convention for the Protection and Assistance of Internally Displaced Persons in Africa[111] was adopted in Kampala on 23 October 2009. It is the first regional legal instrument in the world containing legal obligations for states with regard to the protection and assistance of internally displaced persons. The Kampala Convention defines internally displaced persons as –

> ... persons or groups of persons who have been forced or obliged to flee or to leave their homes or places of habitual residence, in particular as a result of or in order to avoid the effects of armed conflict, situations of generalized violence, violations of human rights or natural or human-made disasters, and who have not crossed an internationally recognized State border.

The Kampala Convention explicitly recognises its relevance for climate-change-induced displacement, stating in Article 5 that "States Parties shall take measures to protect and assist persons who have been internally displaced due to natural or human[-]made disasters, including climate change." However, the Kampala Convention applies to all situations of internal displacement regardless of its causes (Article 15), which makes sense as migration drivers in general are usually not monocausal but influenced by multiple factors.

Also, several obligations are imposed on states parties by the Kampala Convention. It addresses the need to prevent displacement from happening, e.g. by establishing early warning systems and adopting disaster preparedness and management measures to prevent displacement caused by natural

107 Biermann & Boas (2010:74).

108 Kolmannskog (2010:114).

109 IISD (2012).

110 UN General Assembly, Human Rights Council Doc. A/HRC/20/L.14 (29 June 2012).

111 Hereafter *Kampala Convention*; text available at http://www.au.int/en/sites/default/files/african_union_convention_for_the_protection_and_assistance_of_internally_displaced_persons_in_africa_(kampala_convention).pdf, last accessed 24 February 2012.

disasters. The Convention also requires states parties to protect people against displacement resulting from conflict and violence, discriminatory policies, or human rights violations. It further requires that displacement should neither be used as a method of warfare nor as collective punishment, and that forced evacuations should only take place for reasons of health and safety. If people are displaced, the Kampala Convention provides that they be protected and assisted (Article 4). According to Article 5, states parties are obliged to assess the needs and vulnerabilities of internally displaced persons and of host communities in cooperation with international organisations or agencies, and to provide adequate assistance, if need be, with support from relevant local and international agencies. Another objective of the Kampala Convention is to provide for durable solutions with regard to internally displaced persons, who have the right to make a free and informed choice on whether to return to their original homes, integrate into their new area, or relocate elsewhere in the country (Article 11). Furthermore, states parties are responsible for establishing an effective legal framework to provide just and fair compensation and other forms of reparations for damage incurred as a result of displacement (Article 12).

So far, the Kampala Convention has 39 signatories; 17 countries[112] have ratified it and it entered into force on 6 December 2012.

### *F. Human Rights Responses*

The impacts of climate change on human rights have been recognised explicitly by the African Commission on Human and Peoples' Rights (ACHPR). In 2009, the ACHPR called on the Assembly of Heads of State and Government to take all necessary measures to ensure that the African Commission is included in the African Union's negotiating team on climate change.[113] Yet, all the efforts made to date to place rights at the centre of any future climate change regime have not focused on human rights. In fact, most international human rights instruments were drafted before the emer-

112 See http://au.int/en/sites/default/files/Convention%20on%20IDPs%20-%20displa ced..._0.pdf, last accessed 23 May 2013.

113 ACHPR Resolution 153 on Climate Change and Human Rights and the Need to Study its Impacts in Africa, available at http://www.achpr.org/sessions/46th/resol utions/153/, last accessed 6 May 2013.

gence of climate change as a common concern.[114] Therefore, the specific rights potentially affected by climate change, such as rights to food, water, shelter, and health, or rights associated with gender, children and indigenous peoples need to be addressed in the context of climate change. To this end, in 2009, the Human Rights Council adopted Resolution 10/4 recognising the effects of climate change on the enjoyment of human rights.[115]

One could argue that human rights could inform approaches to climate change and human security. This dimension includes arguments based on states' human rights obligations under a variety of international law instruments. These obligations range from the integration of human rights into country strategies in terms of priority entitlements to be protected from the impacts of climate change (e.g. the right to health, housing, water, or food), or more procedural human rights that are relevant to the design and implementation of policies related to climate change (e.g. the right to information, participation, or access to decision-making).[116] In terms of this view, human rights obligations may provide a legal baseline for how climate change is tackled, and what needs to be protected from its impacts.[117]

The social impacts of climate change endanger human security in Africa and increase the vulnerability of specific groups and populations.[118] This vulnerability has also become a key element in human rights discussions. Rights and responsibilities need to be distributed with greater fairness among communities – globally, regionally and domestically. This entails ensuring that poor and marginalised communities in developing countries do not suffer a disproportionate burden associated with climate change.[119]

The interrelationship between human rights and economic development has become closer over the past few years due to increasing discussions in the world community on the issue. This interconnection can be seen as a two-way relationship insofar as economic development is obliged to respect human rights in a democratic society. Conversely, human rights can be given more effect through economic and sustainable development, as one outcome

114 Ruppel (2011c:220ff.).
115 UN General Assembly 2009 A/HRC/10/L.11, available at http://www2.ohchr.org/english/bodies/hrcouncil/docs/10session/A.HRC.10.L.11.pdf, last accessed 13 September 2011.
116 Ruppel (2011b:281).
117 McInerney-Lankford (2009:431–437).
118 Ruppel (2011a).
119 Ruppel (2011b).

of such development is the increasing availability of resources, resulting in the reduction of poverty and a higher standard of living.[120] In this light, the issue of climate-change-induced human movement of persons in Africa prompts significant questions about justice and distribution. There is an acute need for intelligent collective action focusing on human security and the human suffering that climate change will cause in future.[121] On the one hand, as a matter of law, the human rights of individuals need to be viewed in terms of state obligations: it is principally the state that is responsible for human rights fulfilment.[122] On the other hand, the assignment of such responsibility to the state alone seems inadequate in the context of climate change and human security. The specific rights potentially affected by climate change, such as rights to food, water, shelter and health, or rights associated with gender, children and indigenous peoples, need to be addressed in a supra-national context. In 2009, the Human Rights Council noted the effects of climate change on the enjoyment of human rights, and reaffirmed the potential that human rights obligations and commitments have "to inform and strengthen international *and* national policy-making in the area of climate change".[123]

In the 2009 Resolution, the Council welcomed the exchange of information between the Office of the High Commissioner for Human Rights and the UNFCCC Secretariat, stating, among other things, that climate change and human rights were governed by international regimes that had evolved separately, with different premises underlying the legal frameworks of multilateral environmental agreements such as the UNFCCC and human rights treaties.[124]

What is also remarkable is the emphasis in Outcome of the Work of the Ad Hoc Working Group on Long-term Cooperative Action under the UNFCCC[125] on a human-rights-oriented approach to deal with all issues relating

---

120 Passage largely taken from Ruppel (2009:279).

121 Ruppel & Van Wyk (2011b).

122 Ruppel (2011c).

123 UN General Assembly 2009 A/HRC/10/L.11, available at http://www2.ohchr.org/english/bodies/hrcouncil/docs/10session/A.HRC.10.L.11.pdf, last accessed 13 September 2011 (emphasis added).

124 McInerey-Lankford (2009:431–437).

125 Decision 1/CP.16 The Cancun Agreements: Outcome of the Work of the Ad Hoc Working Group on Long-term Cooperative Action under the Convention, available at http://unfccc.int/resource/docs/2010/cop16/eng/07a01.pdf#page=2, last accessed 6 May 2013.

to climate change. By "recognizing that climate change represents an urgent and potentially irreversible threat to human societies and the planet, and thus requires to be urgently addressed by all Parties …" and

> … noting resolution 10/4 of the United Nations Human Rights Council on human rights and climate change, which recognizes that the adverse effects of climate change have a range of direct and indirect implications for the effective enjoyment of human rights and that the effects of climate change will be felt most acutely by those segments of the population that are already vulnerable owing to geography, gender, age, indigenous or minority status, or disability …[,]

the Conference of the Parties "… emphasizes that Parties should, in all climate change related actions, fully respect human rights."

The intersections of human rights and climate change should be taken into account in all future climate change and human rights policy and legislation. Moreover, there may be complementarity identifiable in the principles to be found in both the UNFCCC and ICESCR regimes, such as the duty of co-operation, doing no harm, or equity. Human rights are relevant to the design and implementation of responses to the impacts of climate change. Arguably, human rights could inform approaches to climate change in policy and legal terms. It may be possible to identify ways in which addressing climate change can help realise human rights, and how realising such rights can help ensure a greater capacity to adapt to climate change, underscoring the core compatibility of aims and outcomes between addressing climate change and realising human rights.[126]

In the aforementioned context, the special and differential responsibility of industrialised countries remains particularly important with regard to climate migrants, climate refugees and the climate-change-induced internal displacement of persons.[127] This also seems in line with the key principles provided for in Article 3 of the UNFCCC. Today's accumulated GHG emissions originate mainly from over 150 years of carbon-based industrial activity in developed states. Therefore, the UNFCCC recognises that all countries have a common responsibility to tackle climate change, but places a heavier burden on industrialised states as fulfilment of their respective historic accountability for the causes of climate change. The key principles in Article 3 are reflected in the obligations established for developed and developing countries, including those relating to mitigation, adaptation, tech-

126 (ibid.).

127 Kiss & Shelton (2004:12ff.).

nology transfer, finance, and communication of information relating to the UNFCCC. Furthermore, the UNFCCC provides for countries in special situations, including particularly vulnerable countries, least developed countries, countries undergoing transition to a market economy.[128] For instance, Article 4(4) of the UNFCCC states the following: "[T]he developed country parties … shall assist the developing country parties that are particularly vulnerable to the adverse effects of climate change in meeting costs of adaptation to those adverse effects."

The human security of certain groups in Africa is being threatened by climate change, and it needs to be taken into account that climate change has largely emerged from actions undertaken elsewhere in the world. The impacts of climate change on (national and international) human displacement and movement, caused by acts of omission and commission by the industrialised world, violate or are deemed to violate the fundamental human rights of Africans (e.g. the rights to health, life, physical integrity, determination and security, and the freedoms of movement and residence). Since Africans are protected by international human rights regimes, the logical answer to such human rights violations would be to hold the largest contributors to GHG emissions liable for their impacts.[129] Even in the absence of a binding human rights regime that explicitly establishes such human rights protection with regard to the impacts of climate change, the responsibility of developed countries is not limited to mitigating climate change, but to help the people in Africa and other developing countries adapt to and cope with the growing injustices that are caused by the changing climate. After all, human rights are not static, and their recognition and implementation is a never-ending process. Moreover, it is commonly known that human rights are usually formulated in response to acts or regimes of injustice as they become responsive to such acts or regimes. In the context of the changing climate, human security and environmental migration, environmental human rights can no longer be seen in isolation from other human (and humanitarian) rights. As with all human rights, environmental human rights are Janus-faced, "simultaneously embracing morality and the law",[130] and call for positive conceptualisation by national and international legislature.

128 See e.g. Articles 4(4), 4(6), 4(8), 4(9) and 4(10).
129 Wold et al. (2009:432).
130 See Mushkat (2009:11).

## *G. Conclusion*

This contribution has been written from an African perspective – where violence, poverty and inequality often hinder the realisation of human rights. Against this background, climate change exacerbates the challenges faced by the most vulnerable groups of society. This article raised the question of whether or not it is possible to respond to these challenges by employing human rights mechanisms. The basic conclusion reached is that it can be done, but that it requires certain gaps to be closed and for rights and responsibilities to be assigned with greater fairness in future. This, in turn, would work to ensure in particular that poor and marginalised communities in Africa do not suffer a disproportionate burden associated with climate change.

Climate-change-induced disasters and conflict related to the impacts of climate change are linked to human mobility. The relationship between environmental migration and conflict is complex and further research needs to be done in terms of interwoven causes and effects. Both conflict and environmental migration are caused by a multitude of factors, and both trigger a domino effect of consequences for humankind and nature. Environmental migration presents a substantial set of problems – theoretically and procedurally. The terminology in referring to environmental migrants is often confused and intertwined with the terminology usually used to refer to political refugees or displaced persons; and while attempts are being made to seek a specific solution, the situation for the people who have to move away from their homes remains unresolved. What needs to be addressed is the current critical lack of consistent terminology and legal instruments dealing with problems specific to environmental migration, whether across borders and within countries, distinctly related to climate change. Looking beyond the detail, it remains imperative to protect the basic human rights of all vulnerable people in the interim, until more sophisticated and inclusive legal (and political) solutions can be formulated and implemented.

It is an unfortunate surprise to discover that the current climate change regime still does not adequately consider appropriate human-rights-focused ways in which to deal with the massive effects of climate change such as environmental migration. It has not been a priority of the international climate change regime to deal with corresponding human rights infringements along the way – notable exceptions being the 2009 Human Rights Council Resolution 10/4, the 2009 African Commission on Human and Peoples' Rights Resolution ACHPR/Res. 153 XLV09, and the 2012 Report by the

Special Rapporteur on the Human Rights of Migrants.[131] The latter considers, among other things, how international law approaches the matter of climate-induced migration, including some deficiencies in the currently existing categories. In developing appropriate responses to this complex issue, the Special Rapporteur also takes note of the political engagement that will be required on the issue.

## *References*

Abass, Ademola, 2010, An Introduction to Protecting Human Security in Africa, in: Abass, Ademola (Ed.), *Protecting Human Security in Africa*, Oxford, Oxford University Press, 1–20.

Bantekas, Ilias, 2010, Environmental Security in Africa, in: Abass, Ademola (Ed.), *Protecting Human Security in Africa,* Oxford, Oxford University Press, 43–62.

Barrios, Salvador, Luisito Bertinelli & Eric Strobl, 2006, Climatic Change and Rural–urban Migration: The Case of Sub-Saharan Africa, *Journal of Urban Economics* 60, 357–371.

Biermann, Frank & Ingrid Boas, 2010, Preparing for a Warmer World: Towards a Global Governance System to Protect Climate Refugees, *Global Environmental Politics* 10, 1, 60–88.

Boko, Michel, Isabelle Niang, Anthony Nyong, Coleen Vogel, Andrew Githeko, Mahmoud Medany, Balgis Osman-Elasha, Ramadjita Tabo & Pius Yanda, 2007, Africa, in: Parry, Martin L., Osvaldo F. Canziani, Jean Palutikof, Paul J. van der Linden & Clair E. Hanson (Eds), *Climate Change 2007: Impacts, Adaptation and Vulnerability. Contribution of Working Group II to the Fourth Assessment Report of the Intergovernmental Panel on Climate Change*, Cambridge, Cambridge University Press, 433–467.

Brown, Oli, 2008, Migration and Climate Change, *IOM Migration Research Series No. 31*, Geneva, International Organization for Migration, available at http://www.iisd.org/pdf/2008/migration_climate.pdf, last accessed 7 May 2011.

CHS/Commission on Human Security, 2003, *Human Security Now*, New York, CHS.

Costello, Anthony, Mustafa Abbas, Adriana Allen, Sarah Ball, Sarah Bell, Richard Bellamy, Sharon Friel, Nora Groce, Anne Johnson, Maria Kett, Maria Lee, Caren Levy, Mark Maslin, David McCoy, Bill McGuire, Hugh Montgomery, David Napier, Christina Pagel, Jinesh Patel, Jose Antonio Puppim de Oliveira, Nanneke Redclift, Hannah Rees, Daniel Rogger, Joanne Scott, Judith Stephenson, John Twigg, Jonathan Wolff & Craig Patterson, 2009, Managing the Health Effects of Climate Change, *The*

131 UN A/67/299, available at http://daccess-dds-ny.un.org/doc/UNDOC/GEN/N12/460/71/PDF/N1246071.pdf?OpenElement, last accessed 7 May 2013.

*Lancet* 373, 1693–1733, available at http://www.ucl.ac.uk/global-health/ucl-lancet-climate-change.pdf, last accessed 13 September 2011.

Cuevas, Luis E., Isabelle Jeanne, Anna Molesworth, Michael Bell, Emily C. Savory, Stephen J. Connor & Madeleine C. Thomson, 2007, Risk Mapping and Early Warning Systems for the Control of Meningitis in Africa, *Vaccine* 25, A12–A17.

De Magny, Guillaume, Jean-François Guégan, Michel Petit & Bernard Cazelles, 2007, Regional-scale Climate-variability Synchrony of Cholera Epidemics in West Africa, *BMC Infectious Diseases* 7, 20, available at http://www.biomedcentral.com/content/pdf/1471-2334-7-20.pdf, last accessed 7 May 2013.

Gemenne, François, 2011, Climate-induced Population Displacements in a 4°C+ World, *Philosophical Transactions of the Royal Society A, Mathematical, Physical and Engineering Sciences* 369, 182–195.

Gleditsch, Nils P., Ragnhild Nordås & Idean Salehyan, 2007, *Climate Change and Conflict: The Migration Link*, New York, International Peace Academy.

Hendrix, Cullen S. & Sarah M. Glaser, 2007, Trends and Triggers: Climate, Climate Change and Civil Conflict in Sub-Saharan Africa, *Political Geography* 26, 695–715.

Homer-Dixon, Thomas F., 1991, Environmental Changes as Causes of Acute Conflict, *International Security* 16 (2), 76–116.

Homer-Dixon, Thomas F., 1994, Environmental Scarcities and Violent Conflict: Evidence from Cases, *International Security* 19 (1), 5–40.

Hope, Kempe, 2011, Climate Change in the Context of Urban Development in Africa, in: Yuen, Belinda & Asfaw Kumssa (Eds), *Climate Change and Sustainable Urban Development in Africa and Asia*, Dordrecht, Springer, 37–56.

IDMC/Internal Displacement Monitoring Centre, 2011, *Displacement Due to Natural Hazard-induced Disasters, Global Estimates for 2009–2010*, Geneva, IDMC.

IISD/International Institute for Sustainable Development, 2012, Human Rights Council Links Internal Displacement with Climate Change, *IISD News*, 6 July 2012, available at http://climate-l.iisd.org/news/human-rights-council-links-internal-displacement-with-climate-change/145266/, last accessed 7 May 2013.

IOM/International Organization for Migration, 2007, Discussion Note: Migration and the Environment, MC/INF/288, available at http://www.iom.int/jahia/webdav/shared/shared/mainsite/about_iom/en/council/94/MC_INF_288.pdf, last accessed 07 May 2013.

IPCC/Intergovernmental Panel on Climate Change, 1990, *Climate Change – The IPCC Impacts Assessment*, Final Report of Working Group II of the IPCC, Tegart, William J., Gordon W. Sheldon & D. Collin Griffiths (Eds), Canberra, Australian Government Publishing Services.

IPCC/Intergovernmental Panel on Climate Change, 2007, *Climate Change 2007: Synthesis Report. Contribution of Working Groups I, II and III to the Fourth Assessment Report of the Intergovernmental Panel on Climate Change*, Core Writing Team: Pachauri, Rajendra K. & Andy Reisinger (Eds), Geneva, IPCC, available at http://www.ipcc.ch/publications_and_data/publications_ipcc_fourth_assessment_report_synthesis_report.htm, last accessed 13 January 2013.

Jansen, Cassie C. & Nigel W. Beebe, 2010, The Dengue Vector *Aedes aegypti*: What Comes Next, *Microbes and Infection* 12, 272–279.

Kälin, Walter & Nina Schrepfer, 2012, Protecting People Crossing Borders in the Context of Climate Change: Normative Gaps and Possible Approaches, *Legal and Protection Policy Research Series*, Geneva, United Nations High Commissioner for Refugees, available at http://www.unhcr.org/protect, last accessed 14 February 2013.

King, Gary & Chistopher J.L. Murray, 2002, Rethinking Human Security, *Political Science Quarterly* 116, 585–610, available at http://gking.harvard.edu/files/gking/files/hs.pdf, last accessed 6 May 2013.

Kiss, Alexandre & Dinah Shelton, 2004, *International Environmental Law*, New York, Transnational Publishers Inc.

Kolmannskog, Vikram O., 2008, *Future Floods of Refugees: A Comment on Climate Change, Conflict and Forced Migration*, available at http://www.nrc.no/arch/_img/9268480.pdf, last accessed 13 September 2011.

Kolmannskog, Vikram O., 2010, Climate Change, Human Mobility, and Protection: Initial Evidence from Africa, *Refugee Survey Quarterly* 29 (3), 103–119.

Kumssa, Asfaw & John F. Jones, 2010, Climate Change and Human Security in Africa, *Journal of Sustainable Development & World Ecology* 17 (6), 453–461.

Lecocq, Frank & Zmarak Shalizi, 2007, *How Might Climate Change Affect Economic Growth in Developing Countries? A Review of the Growth Literature with a Climate Lens*, available at http://www-wds.worldbank.org/servlet/WDSContentServer/WDSP/IB/2007/08/13/000158349_20070813145822/Rendered/PDF/wps4315.pdf, last accessed 13 September 2011.

Lonergan, Steve, 1998, The Role of Environmental Degradation in Population Displacement, *Environmental Change and Security Report* 4, 5–15.

McInerney-Lankford, Siobhan, 2009, Climate Change and Human Rights: An Introduction to Legal Issues, *Harvard Environmental Law Review* 33, 431–437.

Mosha, Aloysius, 2011, Climate Change in the Context of Urban Development in Africa, in: Yuen, Belinda & Asfaw Kumssa (Eds), *Climate Change and Sustainable Urban Development in Africa and Asia*, Dordrecht, Springer, 69–102.

Myers, Norman & Jennifer Kent, 1995, *Environmental Exodus: An Emergent Crisis in the Global Arena*, Washington, DC, Climate Institute.

Nhemachena, Charles, Rashid Hassan & James Chakwizira, 2010, *Economic Impacts of Climate Change on Agriculture and Implications for Food Security in Southern Africa*, available at http://climsec.prio.no/papers/Economic%20Impacts%20of%20Climate%20Change%20on%20Agriculture%20and%20Implications%20for%20Food%20Security%20in%20Southern%20Africa.pdf, last accessed 20 August 2011.

Nowak, Manfred, 2003, *Introduction to the International Human Rights Regime*, Leiden, Martinus Nijhoff Publishers.

OCAH/United Nations Office for the Coordination of Humanitarian Affairs & IDMC/Internal Displacement Monitoring Centre, 2009, *Monitoring Disaster Displacement in the Context of Climate Change*, Geneva, OCAH & IDMC.

OHCHR/Office of the High Commissioner for Human Rights, [n.d.], Fact Sheet 20, available at http://www.ohchr.org/Documents/Publications/FactSheet20en.pdf, last accessed 23 January 2012.

Rowhani, Pedram, Olivier Degomme, Debarati Guha-Sapir & Eric F. Lambin, 2011, Malnutrition and Conflict in East Africa: The Impacts of Resource Variability on Human Security, *Climatic Change* 105, 207–222.

Ruppel, Oliver C., 2009, Regional Economic Communities and Human Rights in East and Southern Africa, in: Bösl, Anton & Joseph Diescho (Eds), *Human Rights in Africa*, Windhoek, Macmillan, 275–318.

Ruppel, Oliver C., 2010, Climate Change and Human Vulnerability: Law and Policy, Unpublished paper presented at an expert workshop entitled Identifying Priorities for Climate Change, organised by the Human Rights and Documentation Centre of the University of Namibia in cooperation with the Tyndall Centre for Climate Change, University of Oxford, held at the Safari Court Conference Centre, Windhoek, 9 March 2010.

Ruppel, Oliver C., 2011a, Climate Change and Human Vulnerability in Africa, in: Ruppel, Oliver C. & Katharina Ruppel-Schlichting (Eds), *Environmental Law and Policy in Namibia*, Windhoek/Essen, Orumbonde Press Windhoek & Welwitschia Verlag Dr. A. Eckl, 308–315.

Ruppel, Oliver C., 2011b, Environmental Justice: Advocacy, Litigation and Mediation, in: Ruppel, Oliver C. & Katharina Ruppel-Schlichting (Eds), *Environmental Law and Policy in Namibia*, Windhoek/Essen, Orumbonde Press Windhoek & Welwitschia Verlag Dr. A. Eckl, 281–288.

Ruppel, Oliver C., 2011c, Human Rights and the Environment, in: Ruppel, Oliver C. & Katharina Ruppel-Schlichting (Eds), *Environmental Law and Policy in Namibia*, Windhoek/Essen, Orumbonde Press Windhoek & Welwitschia Verlag Dr .A. Eckl, 219–238.

Ruppel, Oliver C., 2011d, The International, African and Regional Institutional, Legal and Policy Framework of Climate Change, in: African Ministerial Conference on the Environment (AMCEN) (Eds), *Addressing Climate Change Challenges in Africa: A Practical Guide Towards Sustainable Development*, Nairobi, AMCEN, 169–202, available at www.unep.org/roa/amcen/docs/publications/guidebook_ClimateChange.pdf, last accessed 5 August 2012.

Ruppel, Oliver C. & Katharina Ruppel-Schlichting (Eds), 2011, *Environmental Law and Policy in Namibia*, Windhoek/Essen, Orumbonde Press Windhoek & Welwitschia Verlag Dr A. Eckl.

Ruppel, Oliver C. & Sanita Van Wyk, 2011a, Climate Change Induced Movement of Persons in Africa: Human Rights Responses to Aspects of Human Security?, *South African Journal of Environmental Law and Policy* 18 (1), 49–64.

Ruppel, Oliver C. & Sanita van Wyk, 2011b, It is a Problem that Can't Wait, *The Star*, 12 December 2011, available at http://www.iol.co.za/the-star/it-s-a-problem-that-can-t-wait-1.1196378, last accessed 7 May 2013.

Smith, Christopher, Dominic R. Kniveton, Sharon Wood & Richard Black, 2011, Climate Change and Migration: A Modelling Approach, in: Williams, Charles J.R. and Dominic R. Kniveton (Eds), *African Climate and Climate Change – Physical, Social and Political Perspectives,* Dordrecht, Springer, 179–202.

UNDP/United Nations Development Programme, 1994, *Human Development Report*, New York, UNDP.

UNDP/United Nations Development Programme, 2008, *Fighting Climate Change: Human Solidarity in a Divided World*, New York, UNDP.

UN-Habitat & UNEP/United Nations Habitat & United Nations Environment Programme, 2010, *The State of African Cities 2010: Governance, Inequalities and Urban Land Markets*, Nairobi, UN-Habitat & UNEP.

Warner, Koko, Mohamed Hamza, Anthony Oliver-Smith, Fabrice Renaud & Alex Julca, 2010, Climate Change, Environmental Degradation and Migration, *Natural Hazards* 55, 689–715.

Wold, Chris, David Hunter & Melissa Powers, 2009, *Climate Change Law and Policy*, Newark, NJ, Matthew Bender & Company, Inc. & LexisNexis.

Ziervogel, Gina, Anthony Nyong, Balgis Osman, Cecilia Conde, Sergio Cortés & Tom Downing, 2006, Climate Variability and Change: Implications for Household Food Security, *AIACC* [Assessments of Impacts and Adaptations to Climate Change] *Working Paper No. 20*, available at http://www.aiaccproject.org/working_papers/Working%20Papers/AIACC_WP_20_Ziervogel.pdf, last accessed 21 August 2011.

# PART VII: CLIMATE CHANGE AND THE LOSS AND DAMAGE DEBATE

# 28
# Framing the Loss and Damage Debate: A Thought Starter by the Loss and Damage in Vulnerable Countries Initiative*

*Sönke Kreft, Koko Warner, Sven Harmeling & Erin Roberts*

*Abstract*

*Loss and damage* refers to the negative effects of climate variability and climate change that people have not been able to cope with or adapt to. Loss and damage is already a significant – and in some places growing – consequence of an inadequate ability to adapt to changes in climate patterns across the world. Potential future loss and damage depends on emissions, vulnerability, and exposure variables of the impacted human (or natural) system. Today, loss and damage arising from climate change impacts is mostly a local problem, with changes in extreme weather events and slow-onset impacts. Future loss and damage is potentially of inconceivable magnitude – especially considering non-economic values and the interconnectivity leading to cascading, transnational effects. Addressing loss and damage is important because it will affect how society manages the negative impacts of climate change while pursuing other goals, such as resilient and low-emission development. The potential impacts of unmitigated anthropogenic climate change have significant implications for the current social organisation. Future loss and damage can be limited through the mitigation and adaptation choices that are made today. Mitigation ambitions will largely influence the degree to which loss and damage is averted, particularly from around 2030

* This article has been prepared in the context of the Loss and Damage in Vulnerable Countries Initiative, which is part of the Climate Development Knowledge Network. Responsibility for the content lies solely with the authors. The text is an output from a project funded by the United Kingdom's Department for International Development (DFID) for the benefit of developing countries. However, the views expressed and information contained in the text are not necessarily those of or endorsed by DFID or the members of the Climate and Development Knowledge Network, which can accept no responsibility or liability for such views, completeness or accuracy of the information, or for any reliance placed on them.

onwards. Until 2030, decisions that affect the level, scale and efficacy of adaptation will affect the ability of societies to adjust to manifestations of climate change such as alterations in climatic variability (e.g. shifts in seasonality of rainfall, heat waves, and the magnitude and frequency of extreme weather events). An implicit decision not to take ambitious mitigation action on a global scale, and/or decisions not to invest in and actively drive adaptation, could lead to loss and damage which exceeds the ability of all levels of society to manage climate-change-induced phenomena.

Loss and damage discussions under the United Nations Framework Convention on Climate Change (UNFCCC) have emerged as a distinct thematic area since the Cancun Agreements at the Sixteenth Conference of the Parties (COP16) in Mexico in 2010, and today decision-makers are grappling with both the current and future policy steps that need to be taken in order to understand and address loss and damage. Immediate steps will include pursuing the Doha Climate Gateway package – born in the final hours of the COP18 climate negotiations – to establish institutional arrangements to address loss and damage associated with the impacts of climate change. This article outlines initial thoughts by the Loss and Damage in Vulnerable Countries Initiative[1] to provide some conceptual and framing input into the loss and damage negotiations[2] under the UNFCCC. Given both the early stage of these discussions and the complexity of the issues of loss and damage, a spectrum of relevant scientific and policy perspectives and areas of expertise are presented to inform ongoing dialogue.

## A. *Introduction: What is Loss and Damage Associated with Climate Change Impacts?*

The authors view the phrase "loss and damage associated with the adverse effects of climate change" from the Cancun (COP16) Decision as the starting point for any definition of the theme. Paragraph 25 of 1/CP.16 states the following: "Recognises the need to strengthen international cooperation and expertise in order to understand and reduce loss and damage associated with the adverse effects of climate change, including impacts related to extreme

1 See www.lossanddamage.net, last accessed 13 April 2013.
2 COP16 launched a work programme to develop its recommendation entitled "Approaches to Address Loss and Damage Associated with Climate Change Impacts" for consideration at COP18 in 2012 in Doha.

weather events and slow onset events." *Slow-onset events* are further clarified by a footnote therein as "including sea level rise, increasing temperatures, ocean acidification, glacial retreat and related impacts, salinisation, land and forest degradation, loss of biodiversity and desertification."

*Loss and damage* includes the full range of climate-change-related impacts from (changes in) extreme events to slow-onset processes and combinations thereof. For example, the *process* of glacial melting can lead to the harmful *event* of glacier lake outburst floods. Addressing loss and damage requires an understanding of the kinds of events and processes that are associated with the adverse impacts of climate change.[3] Loss and damage impacts fall along a continuum, ranging from *events* associated with variability around current climatic norms (e.g. weather-related natural hazards) to *processes* associated with anticipated changes in climatic norms in different parts of the world. Loss and damage encompasses both incurred loss and damage, as well as future loss and damage.

### B. Working Definition of Loss and Damage

Although the impacts of climate change have been discussed for at least two decades since the UNFCCC's adoption in 1992 – and in other arenas for an even longer period, widely agreed-upon definitions of loss and damage related to those climate change impacts do not yet exist. This section aims to address that gap by offering a working definition that is meant to support discussion and further conceptual framing.

---

3 Although the terms *extreme weather events* (usually discrete temporal events) and *slow-onset climatic processes* (non-discrete continuous processes) are used throughout this article, the literature review also acknowledges that, for practitioners, this distinction is not as clear-cut. The climate stimuli above interact with each other in complex ways, and also interact with human systems in ways that drive loss and damage.

***Working Definition of Loss and Damage***

*Loss and damage* represents the actual and/or potential manifestations of climate impacts that negatively affect human and natural systems.

*Damage* could be seen as negative impacts that can be repaired or restored (such as windstorm damage to the roof of a building, or damage to a coastal mangrove forest from coastal surges which affect villages).

*Loss* could be characterised as negative impacts that cannot be repaired or restored (such as loss of geologic freshwater sources related to glacial melt or desertification, or loss of culture or heritage associated with potential population redistribution away from areas that become less habitable over time with climate change).

This broad working definition includes some further caveats:

- **Multiple temporal and spatial scales:** Loss and damage encapsulates historic and present (occurring and observed) manifestations of climate impacts as well as those that will occur in the future. Potential future loss and damage, by definition, relies on assumptions regarding parameters such as emissions, vulnerability, and exposure variables of the impacted human (or natural) system. Today, loss and damage arising from climate change impacts is mostly a local problem, with changes in extreme events and slow-onset impacts. Future loss and damage is potentially of inconceivable magnitude – especially considering non-economic values, and the interconnectivity leading to cascading, transnational effects. The concept of tipping points in climate, natural and societal systems – a moment where profound and potentially irreversible system changes occurs – is an important factor in weighing potential loss and damage.
- **Human and natural systems:** Loss and damage refers to the negative impacts of climate change on human systems, which are, in turn, often affected by impacts on natural systems. For example, sea-level rise and glacial melt result from climate change stimuli, and these shifts in natural systems in turn result in loss and damage to human systems such as habitable land or fresh water. Additionally, characteristics of human systems, such as development policies and poverty, affect the dependency of human systems on natural systems. Yet this connectedness does not change the fact that climate change impacts drive loss and damage, which occurs through the 'path' of natural system shifts and their effects on human systems.

- **Negative impacts:** Loss and damage is an undesirable phenomenon associated with climate change impacts, and does not include the impacts from managing climate change itself; the latter is discussed within the rubric of the UNFCCC under the policy forum of response measures.

### *C. Addressing Loss and Damage: Why it Matters Now*

Due to the uncertainty and volatility associated with them, extreme weather events already impose loss and damage which is difficult to deal with by the most vulnerable communities. In the future, even greater loss and damage is expected from the impacts of changing norms of extreme weather, distinct slow-onset climatic processes, and interaction between the two.

Addressing loss and damage is important because it will affect how society manages the negative impacts of climate change while pursuing other goals, such as resilient and low-emission development. Geologic records indicate that profound shifts in earth systems and life forms have accompanied climatic changes in the past. In what has relatively recently come to be termed the *Anthropocene*,[4] human interaction with our natural environment has led to patterns of loss and damage that are relevant for society. The potential impacts of unmitigated anthropogenic climate change have significant implications for the current organisation of society. For example, sea-level rise could redefine the borders of some countries; desertification and glacial melt could shape the habitability of large areas of the world where people rely on arable land and fresh water for survival; and temperature change could affect plant fertility and biodiversity. Failure to address loss and damage in a timely way could leave humankind unprepared to manage and adjust to these negative climate change impacts.

Success in addressing loss and damage would mean that the impacts of climate change could be somewhat contained or reduced while shifting gradually to new forms of organisation that will enable humans to continue living in balance with new states of climate in the future.

---

4 Popularised by the atmospheric chemist and Nobel laureate Paul Crutzen in 2000; see http://www.smithsonianmag.com/science-nature/What-is-the-Anthropocene-and-Are-We-in-It-183828201.html, last accessed 29 March 2013.

## D. Decision Pathways and Consequences of Loss and Damage

Fully addressing loss and damage involves two components. Firstly, potential future loss and damage could be avoided through appropriate mitigation and adaptation activities. The second component entails tackling loss and damage when it occurs, both today and in the future, through a range of mechanisms. Figure 1 helps illustrate this idea.

***Figure 1: Conceptual Framing for the Loss and Damage Debate***

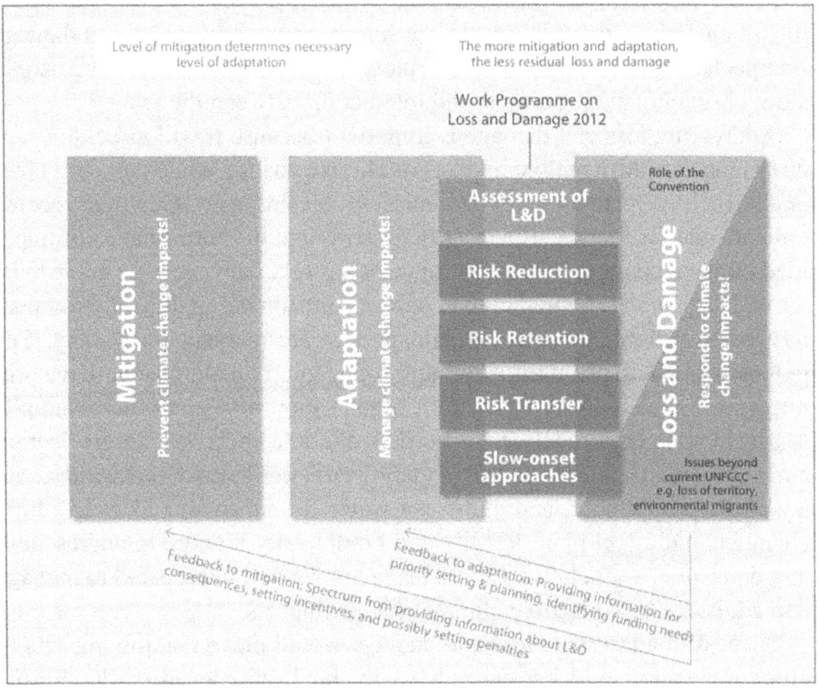


The frontiers of future loss and damage can be limited through the mitigation and adaptation choices that are made today. Climate change impacts are driven by the level of greenhouse gases (GHGs) in the atmosphere. Negative climate change impacts that lead to loss and damage also influence the ability of human systems to adapt to changes in climate. Present choices about mitigation and adaptation determine not only current, but also, and especially, future loss and damage potential – while acknowledging significant uncertainty in the decision-making context.

Mitigation ambitions will have the greatest influence on the degree to which loss and damage is avoided, particularly from around 2030 onwards. Until 2030, decisions that affect the level, scale and efficacy of adaptation will also affect the ability of societies to adjust to manifestations of climate change, including alterations in climatic variability such as shifts in seasonality of rainfall, heat waves, and the magnitude and frequency of extreme weather events. The most effective approach towards addressing loss and damage in the long term – in respect of avoiding future loss and damage, and minimising impacts in the short and medium terms – is to enhance both mitigation and adaptation measures.

An implicit decision not to take ambitious mitigation action on a global scale and/or make decisions not to invest in and actively drive adaptation could lead to loss and damage which exceeds the ability of all levels of society to manage climate-change-induced phenomena. The global community – or, more specifically, governments of more than 190 countries that are states parties to the UNFCCC – have agreed on the objective to limit the increase in global warming to below 2°C above pre-industrial levels.[5] Warming above this level can, therefore, be regarded as "dangerous climate change", which Article 2 of the UNFCCC expressly seeks to avoid. Should mitigation efforts fail to keep GHG concentrations below the equivalent of a '2°C world', the implications for loss and damage could be profound in terms of the availability of resources on which humankind depends, i.e. water, food, shelter, livelihoods, etc.

### *E. Loss and Damage as an Equity and Climate Justice Issue*

As noted previously, the magnitude of 'residual' loss and damage – negative climate change impacts that remain and demand to be addressed – depends on how effective mitigation and adaptation efforts are. However, as a result of both historical and current GHG emissions, some degree of climate change impacts is already 'locked in'. Thus, even after the best possible mitigation and adaptation action, societies worldwide will still face some residual loss and damage.

---

5 The majority of UNFCCC states parties even endorsed a global temperature goal of below 1.5°C. Parties therefore agreed that the review decided in Cancun should periodically revisit the adequacy of the goal in light of achieving the UNFCCC's ultimate objective.

Addressing loss and damage is not only of common concern for humankind,[6] but also an issue of climate justice. The element of (in)justice has a spatial and temporal dimension. The potential spatial distribution of negative consequences related to loss and damage –particularly in respect of intangible elements which currently elude quantification, such as social, cultural and psychological loss and damage – will burden those countries which have historically contributed least to global GHG emissions and which have the most limited capacities to deal with the consequences of loss and damage. Without adequate action, communities in these countries will experience loss and damage with significant consequences – both nationally and globally. The temporal dimension of loss and damage lies in the fact that future generations could be left with significantly different and possibly constrained opportunities if we collectively fail to raise ambition around mitigation and adaptation today, and miss the opportunity to design approaches to address loss and damage in the long term.

## F. *What Needs to be Done Next to Address Loss and Damage and Move the Discussions Forward?*

The impacts of loss and damage associated with climate-related stressors such as weather extremes and long-term climatological shifts can impair socio-economic development and reinforce cycles of poverty across the globe. Planning 'only' for the extreme climate-related events of today due to a static understanding of climate change impacts could leave countries without enough resources tomorrow. By contrast, planning for approaches to address loss and damage associated with both current climate variability and long-term shifts in climate patterns are needed. This holistic approach will help smooth development pathways as well as cushion the expected negative impacts of loss and damage in the future.

In today's world, there are challenges associated with creating strategies to address loss and damage. Faced with financial crises, political strife, population growth and a multitude of other challenges, decision-makers may be tempted to postpone considering approaches to address loss and damage related to climate change impacts. Sceptics (see text box) question the evidence on linkages between loss and damage (from disasters) and climate

6 Article 2, UNFCCC.

change, and implicitly suggest waiting to address the issue until more evidence is available.

However, in spite of these challenges, international and national policy forums as well as communities of policy, science and practice have many tools to help them begin to address loss and damage. Tapping into and jump-starting action of these different communities and processes should be an essential next step for the UNFCCC process, as the discussion on loss and damage becomes more mature and probably more institutionalised.

***Sceptics claim loss and damage related to extreme events cannot yet be attributed to climate change. Would it be prudent to postpone the discussion until more conclusive evidence is found?***

The findings of the Intergovernmental Panel on Climate Change (IPCC)[7] have suggested uncertainty today about the relationship between climate change and long-term trends in normalised losses from weather-related extreme events. These findings have led some critics to focus on the current inability of science to definitively address the attribution of loss and damage from weather extremes to climate change; however, this critique is misleading. The IPCC findings reflect a lack of longer-term evidence and gaps in research rather than conclusive, positive evidence that there is no link between extreme weather events and loss and damage.

Furthermore, the inconclusive IPCC findings, related to the attribution of disaster losses to climate change, highlight the potential pitfalls of focusing only on extreme events to inform decision-making about the wider spectrum of policy that may be needed to address current and future negative climate change impacts. In time, science may develop to the state where the consequences of various manifestations of climate change may be attributable to anthropogenic activities. Yet, it is likely that, by the time science can conclusively establish those relationships, loss and damage related to those impacts will already have occurred. At that point, a number of the windows of opportunity for shaping policies to anticipate, reduce, plan for, and manage negative climate change impacts (ranging from extreme weather events to slow-onset changes like sea-level rise) will have closed.

Attribution is a difficult issue. Article 1 of the UNFCCC defines *climate change* as "a change of climate which is attributed directly or indirectly to human activity that alters the composition of the global atmosphere and

7 IPCC (2012).

which is in addition to natural climate variability observed over comparable time periods."

Following this definition, approaches to address loss and damage only deal with the anthropogenic component of changing climate norms. However, extreme events are often the starting point for actions by governments and communities. It is often not feasible to conduct activities that distinguish the climate change component of an extreme event from existing weather variability. Therefore, the authors agree with the path of the UNFCCC work programme on approaches to address loss and damage. The first step is to engage in an option-based approach that includes risk reduction, risk retention and risk transfer, and which starts from existing experience – especially as regards managing loss and damage around existing climate variability – and, from that, derive the action necessary at UNFCCC level. At the same time, slow-onset processes – an area were experience is still sparse but growing – should always feature specifically in the discussion to avoid a 'status quo bias'.

Policy discussions on loss and damage are important today because a 'science and evidence only' approach will not sufficiently anticipate and inform society about decision pathways and consequences related to the negative impacts of climate change. Relying solely on questions of attribution truncates discussions and prevents full consideration of a range of options to address loss and damage today.

## *G. The COP18 Outcome on Loss and Damage*

In the climate change negotiations at COP18 in Doha in December 2012, the Subsidiary Body for Implementation (SBI)[8] considered progress made on the implementation of the work programme on loss and damage,[9] and noted the achievements in respect of understanding loss and damage as well as the gaps in such understanding. In the final hours of COP18, governments reached a decision – incorporated as part of the Doha Climate Gateway package – to establish institutional arrangements to address loss and damage

8 The SBI supports the work of the COP and the CMP through the assessment and review of the effective implementation of the Convention and its Kyoto Protocol; see http://unfccc.int/bodies/body/6406.php, last accessed 11 May 2013.

9 In accordance with Decision 1/CP.16 at COP16 and Decision 7/CP.17 at COP17.

associated with the impacts of climate change.[10] The decision acknowledged that further work was needed to advance understanding on the topic, including how loss and damage affected vulnerable segments of the population and how the implementation of approaches could benefit vulnerable people,[11] and how to develop appropriate approaches like risk reduction, risk sharing and risk transfer tools.[12]

The key points of the decision on loss and damage are as follows:

- Paragraph 5: An explication of the role of the UNFCCC in an institutional arrangement
- Paragraph 9: To establish institutional arrangements to address loss and damage by COP19, and
- Paragraphs 7 and 10: To define elements of the work by COP19 to help define the functions and modalities of an institutional arrangement to address loss and damage.

Paragraph 5 discusses the UNFCCC's role in implementing approaches to address loss and damage associated with the adverse effects of climate change. The implications of paragraph 5 are that a decision at COP19 and further related decisions should consider the UNFCCC's role in designing options (functions and modalities). According to the Doha Gateway Decision text, the UNFCCC could enhance knowledge and understanding of comprehensive risk management approaches; strengthen dialogue, coordination, coherence, and synergies; and enhance action and financial, technological and capacity-building support.

Paragraph 7 issues an invitation to pursue additional work to enhance understanding. These areas are also referred to for work in 2013, and may be undertaken voluntarily in addition to the work mandated to the Secretariat. They are as follows:

10 Paragraph 9 of the Doha Draft Decision reads as follows: "[d]ecides to establish, at its nineteenth session, institutional arrangements, such as an international mechanism, including functions and modalities, elaborated in accordance with the role of the Convention as defined in paragraph 5 above, to address loss and damage associated with the impacts of climate change in developing countries that are particularly vulnerable to the adverse effects of climate change".

11 Decision 3/CP.18 (2012:para. 7iii).

12 (ibid.:para. 7iv).

- Enhancing understanding of slow-onset events, non-economic losses, impacts on particularly vulnerable groups, impacts on climate-resilient development, and impacts on human mobility
- Strengthening the collection and management of relevant data for assessing loss and damage
- Enhancing coordination, synergies and linkages among organisations to enable development and support of approaches to address loss and damage
- Strengthening regional collaboration, centres and networks
- Enhancing capacity-building at national and regional levels, and
- Strengthening institutional arrangements at national regional and international levels.

For COP19 scheduled for December 2013, paragraph 9 lays out a "decision to establish institutional arrangements, such as an international mechanism, including functions and modalities, elaborated in accordance with the role of the Convention to address loss and damage."

Thus, key work needs to involve exploring the functions and modalities of different options for institutional arrangements to address loss and damage. These ideas should consider the role of the UNFCCC, as elaborated in paragraph 5.

Paragraph 10 requests the Secretariat to carry out the following interim activities under the SBI Work Programme on Loss and Damage prior to COP19. The Secretariat is mandated to carry out three activities in 2013 (before COP19), namely –

- hold and prepare a report on an experts meeting to consider future needs, including capacity needs associated with possible approaches to address slow onset
- prepare a technical paper on non-economic losses, and
- prepare a technical paper on gaps in existing institutional arrangements in and outside the UNFCCC to address loss and damage, including slow onset.

The SBI Work Programme on Loss and Damage is requested –

- to consider the gaps analysis technical paper[13] in developing institutional arrangements, and

13 Decision 3/CP.18, para. 10(c).

- according to paragraph 7, to prepare a technical paper to suggest further activities by the June SBs in Bonn.

## *H. Future Potential Directions for Loss and Damage Policy*

The UNFCCC mandate to pursue further work on loss and damage[14] and the decision to establish institutional arrangements suggest a longer-term commitment towards understanding and addressing the expected loss and damage related to climate change. While the UNFCCC is a key policy forum on these topics, other opportunities arise in the context of different policy processes that will mark major milestones in 2015, namely –

- target for the next international climate agreement (UNFCCC)
- renewal of the Hyogo Framework for Action on disaster risk reduction, and
- reports on achievements of the United Nations Millennium Development (and Sustainable Development) Goals.

Furthermore, embedding considerations of loss and damage in these and other processes will be vital in determining how those climate change consequences which cannot or cannot feasibly be dealt with at different levels are addressed beyond the 2015–2020 period. The ultimate objective for the loss and damage discussions will be to anchor a consolidated response to loss and damage in the post-2015 development and climate context in order to contribute to an accelerated paradigm change towards climate-resilient development.

## *Reference*

IPCC/Intergovernmental Panel on Climate Change, 2012, *Managing the Risks of Extreme Events and Disasters to Advance Climate Change Adaptation – A Special Report of Working Groups I and II of the Intergovernmental Panel on Climate Change*, Field, Christopher B., Vicente Barros, Thomas F. Stocker, Dahe Qin, David J. Dokken, Kristie L. Ebi, Michael D. Mastrandrea, Katharine J. Mach, Gian-Kasper Plattner, Simon K. Allen, Melinda Tignor and Pauline M. Midgley (Eds), Cambridge, Cambridge University Press.

14 From 1/CP.16.

29

# Legal and Policy Responses to Loss and Damage Associated with Climate Change

*M. Hafijul Islam Khan*


*Abstract*

Due to inadequate mitigation and adaptation efforts, loss and damage associated with climate change is now a reality. Some recent studies reveal the empirical evidence on loss and damage resulting from the adverse impacts of climate change. Therefore, the adverse impacts of climate change have imposed additional challenges for the global legal community to address loss of life, property, traditional livelihoods, values, culture, heritage and territory and damages including ecology. A plaintiff and defendant can be identified persuasively in respect of a claim related to loss or damage resulting from climate change, but in a specific case, the legal community will find it difficult to pin such loss or damage on climate change. Empirical data can lend a hand in this regard, but the problem lies in choosing the appropriate legal avenues to address the claim.

Loss and damage associated with the adverse impacts of climate change are a major emerging challenge for the global community: they demand contemporary legal and policy frameworks with specific substantive and procedural mechanisms. Mitigation and adaptation can prevent and reduce loss and damage, but a specific regulatory regime is needed in order to deal with residual loss and damage resulting from climate change. Hence, states parties to the United Nations Framework Convention on Climate Change (UNFCCC) decided to establish the required institutional arrangements such as international mechanisms to address loss and damage associated with the impacts of climate change. Following efforts to conceptualise loss and damage associated with climate change, this paper looks at substantive and procedural mechanisms for addressing loss and damage within the contexts of conventional international law and the emerging legal regime of climate change, with the ultimate objective of exploring potential legal and policy responses to loss and damage.

## A. Introduction

Climate change and its adverse impacts and vulnerabilities are now a reality.[1] Vulnerability as the consequence of climate change will be most severe for the developing world, in other words, those areas which are both least responsible for climate change and least able to deal with its effects.[2] Thus, the most vulnerable are the people living in least developed countries (LDCs), Small Island Developing States (SIDSs), and African countries. Although past and current global emissions of greenhouse gases (GHGs) originated in developed countries,[3] LDCs like Bangladesh face the vulnerabilities of climate change disproportionately. As such, the very nature of climate change exacerbates the inequities associated with proportional contribution to the causes, and suffering from the consequences.[4] The adverse impacts of climate change, including increased frequency and intensity of disasters and slow-onset processes like sea-level rise and saline water intrusion, have continued to devastate the lives and livelihoods of millions of people and inflict huge economic losses, particularly in developing countries.

In accordance with figures from the international disaster database of the Centre for Research on the Epidemiology of Disasters, the number of disasters in first seven years of the 21st Century doubled in comparison with 1987–1997. Developing countries, where more than 95% of deaths from natural disasters in the past 25 years have occurred, bear the brunt of this increase. According to the global reinsurance company Munich Re, direct economic losses (averaging US$100 billion per annum in the first decade of this century in relation to national income were more than double in low-income countries, compared with their high-income counterparts. On average, 250 million people are affected by disasters annually – up by more than 30% in just a decade.[5] Moreover, the United Nations Office for the Coordination of Humanitarian Affairs (UNOCHA) estimated that, in 2008, over 20 million people were displaced by disasters.[6] Bangladesh, an LDC, faced two

1 IPCC (2007:5): "Warming of the climate system is unequivocal, as is now evident from observations of increases in global average air and ocean temperatures, widespread melting of snow and ice, and rising global average sea level".

2 Stern (2006); IPCC (2007).

3 Preamble, United Nations Framework Convention on Climate Change (UNFCCC).

4 Khan (2010).

5 Zakieldeen & Warner (2012).

6 UNOCHA (2009).

consecutive floods in 2007, which caused economic damage in the amount of approximately US$1.1 billion, followed by Cyclone Sidr in November 2007, which killed 3,500 people and led to estimated economic damage of US$1.7 billion.[7] Moreover, in 2009, Cyclone Aila hit Bangladesh's western border with India and caused the initial displacement of 201,982 people, with a further 60,000 people having since migrated to other areas of the country in search of employment.[8]

Climate change impact and vulnerability, particularly current extreme weather events, bring up the serious legal question of liability for the damage caused, based on proportional contribution to climate change. The disproportionate contribution to the cause of climate change shifts the burden to the industrialised countries to take the entire responsibility for the adverse impacts and vulnerabilities of such change, in accordance with causal liability. In the absence of the required response to climate change, advocacy groups, public authorities, communities and individuals are coming up before the judiciary to seek compensation for loss and damage resulting from climate change, and for judicial direction to compel those entities responsible to act in response to climate change through the required mitigation and adaptation measures. Decisions also have begun to emerge through different judicial forums as a means of compelling decision-makers to address the issues for future action and to make those responsible liable for the harm caused to the climatic system.[9]

In the context of climate science, the relative contributions by different states towards climate change can be estimated based on the cumulative contribution and, as such, each state should be liable proportionally. Thus, each state's proportion of liability can be estimated by its cumulative contribution in relation to other states' cumulative contributions. Taking into account this simple legal equivalence, although one can convincingly establish substantive arguments to apportion liability and to compensate for climate-induced loss and damage based on the rules of customary international law, there are often no certain procedural means to pursue this legitimate claim further.[10] In response to demands for broad, system-changing solutions to the climate crisis and to address the loss and damage associated with climate change, contemporary legal and policy frameworks with spe-

7 Khan et al. (2012).
8 ActionAid et al. (2009).
9 Khan (2010).
10 (ibid.).

cific substantive and procedural mechanisms are required to deal with this shortfall.

Very recently, states parties to the United Nations Framework Convention on Climate Change (UNFCCC) decided to establish the institutional arrangements required, such as an international mechanism to address loss and damage associated with the impacts of climate change.[11] The international climate regime, which began in 1992 with the adoption of the UNFCCC, is still struggling to set up governance mechanisms for mitigation and adaptation with respect to climate change. However, even if adequate mitigation measures are taken now, given the levels of GHGs that have already been released into the atmosphere, some climate change impacts and the associated loss and damage are inevitable. While adaptation measures can reduce loss and damage resulting from climate change to some extent, there will be a certain unavoidable degree of loss and damage, for which a separate framework is needed. Clearly, an agreed framework within the UNFCCC process for addressing loss and damage is still a long way off.[12]

Nonetheless, UNFCCC states parties from developing countries are enthusiastically negotiating the establishment of an international mechanism to this end. Indeed, the inclusion of an "international mechanism" in the Doha Decision on loss and damage[13] marks an important window of opportunity for the further development of such mechanisms. Against this backdrop, this paper explores the legal avenues for addressing loss and damage associated with climate change within the context of customary international law and the UNFCCC. The first part of this paper attempts to conceptualise the issue of loss and damage associated with such adverse impacts, and provides legal arguments for a liability regime whose ultimate objective is exploring the potential legal and policy frameworks. The second part of the paper examines the scope and limitations of conventional international law in this context. The concluding section scrutinises the development of the UNFCCC process, and proposes some policy guidance for establishing national, regional and international mechanisms to deal with loss and damage associated with the adverse impacts of climate change.

11 Decision 3/CP.18, para. 9.
12 Al Faruque & Khan (2012).
13 Decision 3/CP.18, para. 9.

## B. Loss and Damage Associated with Climate Change and the Liability Regime

An appropriate conceptualisation of *loss* and *damage* associated with climate change will provide the necessary guidance for identifying the entities responsible for such change, including the private sector and for developing the liability regime. So, conceptualisation of loss and damage is the prerequisite for structuring the required legal and policy frameworks. The first step in framing this discussion is to properly identify the issues and challenges related to loss and damage associated with climate change. Under the UNFCCC, the issue of loss and damage was discussed from the beginning, but state parties only finally agreed to establish a work programme on loss and damage in 2010 and recognised the complexity of the subject matter.[14] The 2010 Cancun Decision at COP16[15] distinguished the need to strengthen international cooperation and expertise in order to understand and reduce loss and damage associated with climate change.[16] Under the agreed work programme, a series of expert workshops organised by the UNFCCC Secretariat throughout 2012 and the 2012 Doha Decision at COP18[17] noted the importance of enhancing knowledge and understanding of the comprehensive risk management approaches to address loss and damage associated with the adverse effects of climate change, including slow-onset impacts. COP18 also reflects an agreement on comprehensive, inclusive and strategic responses needed in order to address loss and damage associated with the adverse effects of climate change, taking into account regional, national and local capacity, context and circumstances, and the involvement of relevant stakeholders.[18] Moreover, COP18 invited all parties to identify options and design and implement country-driven risk management strategies and approaches, including risk reduction, and risk transfer and risk-sharing mechanisms.

Thus, in order to understand loss and damage associated with climate change, the context of a highly vulnerable country like Bangladesh can provide some food for thought. Bangladesh experiences frequent natural disasters such as floods, tropical cyclones, storm surges and droughts, which cause loss of lives and livelihoods and damage to infrastructure and econo-

---

14 Decision 1/CP.16, para. 26.

15 At the Sixteenth Conference of the Parties to the UNFCCC (COP16).

16 (ibid.:para. 25).

17 At the Eighteenth Conference of the Parties to the UNFCCC (COP18).

18 Draft Decision-/CP.18, para.'s 2 and 5.

mic assets.[19] The frequency and intensity of these natural hazards has already increased in Bangladesh;[20] hence, the context of the frequency and intensity of these events might be considered as climatic impacts. Although it remains a challenge to try to segregate climate-induced hazards, in general, the increased frequency and intensity of natural disasters such as floods, droughts, cyclones and associated storm surges, heat stress and other extreme hydro-meteorological events can be considered as sudden-onset events associated with climate impact on Bangladesh. On the other hand, a rise in sea level and the salination being detected in coastal regions can be considered as slow-onset events or processes. In terms of slow-onset processes, the Cancun Decision also listed impacts such as rising sea levels, increasing temperatures, ocean acidification, glacial retreat and related impacts, salination, land and forest degradation, loss of biodiversity, and desertification.[21] Therefore, the rise in sea level as a slow-onset process and extreme weather events as rapid-onset events are considered in this article to understand loss and damage within the geographical context of Bangladesh.

The 2009 Bangladesh Climate Change Strategy and Action Plan recognises that the rise in sea level would lead to the submergence of low-lying coastal areas, the intrusion of saline water from coastal rivers into groundwater aquifers – reducing the availability of fresh water and damaging the Sundarban's mangrove forest – and drainage congestion inside coastal polders, which will adversely affect agriculture.[22] It is worth mentioning that the average land elevation is about 7.62 m above mean sea level, whereas for coastal and offshore islands it is about 1.5 m above mean sea level. As such, major portions of the waterways are under tidal influence.[23] Therefore, increasing rates of sea-level rise would cause permanent inundation, drainage congestion, salinity intrusion and frequent storm surge inundation.[24] As a result, a rise in sea level and the resulting salination would adversely affect the coastal agrarian economy and will force communities to migrate to search for alternative livelihoods due to a loss of territory and traditional livelihoods. Nevertheless, the experience of recent cyclones and storm

19 GPRB (2009:para 3).
20 UNDP (2011).
21 Decision 1/CP.16, para. 25.
22 GPRB (2009:para. 21).
23 Mondal (2009).
24 (ibid.).

surges would provide considerable gestures for demonstrations of actual and potential loss and damage associated with sea-level rise in Bangladesh.

Cyclone Aila, a sudden-onset event which hit coastal areas of Bangladesh in May 2009, caused the death of 193 people, damaged infrastructure, houses, institutions, cultivated land and crops,[25] and displaced over 100,000 people.[26] Aila affected an area of about 1,200 km and the livelihoods of millions of people were simply destroyed in the affected areas. Particularly due to storm surge, most of the areas including agricultural land were submerged by salty water and, as a result, saline water intrusion and being waterlogged for a long time brought about a loss of crop productivity. Consequently, initially displaced people could not return home due to a loss of their traditional livelihoods. Moreover, the slow-onset process of salination caused further harm to livelihoods and prevented displaced people from returning home. These displaced people migrated to urban areas and other countries such as India in search of alternative means of making a living. This gradually increased the number of forced migrants, when some of them failed to adapt to ecological changes. Finally, about 123,000 people migrated due to Aila-related impacts, and an additional 23,000 migrated at a later stage due to failed efforts at ecological restoration, i.e. desalination of soil in which to grow crops.[27]

In a recent study, which consulted people from Aila-affected areas, 81% of respondents reported high salinity levels in their soil, compared with just 2% 20 years ago. One adaptation that farmers had employed was to plant saline-tolerant varieties of rice. This worked until 2009, when Aila hit and caused a sudden and drastic increase of the salt content in the soil. Almost all of the farmers lost their complete harvest that year. Two years later, rice yields were still extremely poor. From 2009 to 2011, the total loss in respect of rice harvests was US$1.9 million in only four villages surveyed. These findings exemplify a case where seemingly successful measures to adapt to slow-onset processes are not strong enough to avoid loss and damage when the situation is aggravated by an extreme weather event.[28] Although empirical data suggest that extreme weather events such as Aila have grown in frequency and intensity, scientific experts are divided on how to quantify the extent to which climate change has contributed to the destruction caused

25 Mehedi et al. (2010).
26 McAdam & Saul (2010:239).
27 Mehedi et al. (2010).
28 Rabbani (2012).

by Aila.[29] However, if not at a macro scale, a few micro-scale disasters and related data can establish the causal relationship between intensity and frequency of extreme weather events and climate-induced loss and damage. Some recent research studies[30] reveal the evidence of the plight of the coastal fishermen in Bangladesh under a climate-change-induced rise in sea surface temperature (SST) which is devastating the lives and livelihoods of these fishermen.

One such study, titled "Livelihood of Coastal Fishermen in Peril: In Search of Early Evidence of Climate Change Induced Adverse Impacts in Bangladesh",[31] reveals that increasing SST fulfils one of the major preconditions of the formation of an increased number of depressions and low-pressure systems in the Bay of Bengal. Since the SST of the Bay of Bengal has been unusually high, one finds a scientific link between rising SST with increasing episodes of rough sea conditions, the latter having serious livelihood implications on especially the impoverished fishers of Bangladesh. With increasing SST, they can hardly survive one unusual year: how would they be able to sustain their livelihoods for generations to come?[32] Another recent study assessed the relevant data on tropical storms in the Bay of Bengal, including cyclones and depressions, during the period 1985–2009:[33] the evidence also revealed an increasing frequency of cyclones. Furthermore, the Bangladesh Climate Change Strategy and Action Plan 2009 acknowledged that rough weather along the coast might prevail for longer durations in future, adversely impacting on fishermen's livelihoods.[34]

A media report published in October 2010 provides further related data on the frequency of rough sea events in the Bay of Bengal, stating that 10

29 McAdam & Saul (2010:239).

30 Ahmed & Neelormi (2007/2009); Chowdhury et al. (2012).

31 Ahmed & Neelormi (2007/2009).

32 (ibid.). Using data on the frequency of rough sea events in the Bay of Bengal in 2007, the study argued that the year had been unusually rough. Of 22 incidences of low pressure and depressions in the Bay of Bengal, 12 had occurred during July and mid-November, the peak of the fishing season along the south-eastern coastal region. The apparent high energy in the sea affected the entire coastal zone by bringing in unusually high tides and frequent rough seas. The latter effect was so pronounced that the Port Authority issued a total of 89 warnings through the year, 12 of which were issued during July and mid-November. Moreover, the latter were higher than level 3 or above, which marks a potentially dangerous situation.

33 Chowdhury et al. (2012).

34 GPRB (2009:39).

cyclones and 50 incidents of formations of low pressure and depressions in the Bay of Bengal occurred from mid-2010 to 2012, as a result of which about 30 million people were affected.[35] This report stated that, this was because of the rough weather influenced by a depression in the Bay of Bengal. In reference to the incidence of a depression on 11 October 2010, the said report stated further that over 100 fishers were feared missing as 10 fishing trawlers had not returned to land, while 7,000 trawlers were kept near the coast for safety.[36] Unfortunately, coastal fishers go missing quite regularly due to frequent depressions in the Bay of Bengal. Moreover, because of increased storms due to climate change, coastal fishers cannot go fishing for several days and, hence, they lose their traditional livelihoods. Storms also cause damage to property and other coastal fishers' assets. The disruption of settlements and the reduction of livelihood opportunities can also cause displacement.

The latter 2010 newspaper report and the aforementioned scientific arguments offer proof of the causal link between increased SST and global warming, which causes harm to the lives and livelihoods of millions of coastal residents in Bangladesh. The Fourth Assessment Report by the Intergovernmental Panel on Climate Change (IPCC) also recognises that "altered frequencies and intensities of extreme weather, together with sea-level rise, are expected to have mostly adverse effects on natural and human systems."[37]

Therefore, the increased frequency and intensity of rough sea events in Bangladesh, which cause the loss of lives and livelihoods as well as damage to property and other assets owned by coastal fishers, provide some important contexts of loss and damage associated with climate change. The above contextual analysis on sea-level rise and saline water intrusion, as well as the implication of micro-level disasters such as frequent rough sea events, provides factual evidence of actual loss and damage. Thus, a particular climate-related incident or a series of incidents from a specific country can provide some insights for conceptualising the loss and damage associated with climate change. However, it is necessary to consider country by country and case by case. Certainly, some similarities may be found across countries

35 *BD News*, 12 October 2010, available online at http://bdnews24.com/bangladesh/2005/09/19/over-200-trawlers-with-over-3500-people-aboard-missing-in-bay-of-bengal.-1st-ld, last accessed 15 January 2013.

36 (ibid.).

37 IPPC (2007:58).

and across cases that will be useful for influencing policymakers at the global level to adopt a common definition of *loss and damage associated with climate change*; however, providing a definition is a very difficult task – especially taking into account the complexity of the issues related to loss and damage. However, certain studies and experts have recently provided a few broader definitional outlines of *loss* and *damage* associated with climate change; these should also be useful to discuss in order to explore potential mechanisms for addressing such loss and damage.

## C. *Definition of Loss and Damage*

The Loss and Damage in Vulnerable Countries Initiative[38] provided a working definition of *loss* and *damage* related to climate change, and stated that *loss* and *damage* represent the actual and/or potential manifestations of climate impacts that negatively affect human and natural systems. It further considered *damage* as "negative impacts that can be repaired or restored (such as windstorm damage to the roof of a building, or damage to a coastal mangrove forest from coastal surges which affect villages)."[39]

On the other hand, *loss* is characterised as negative impacts that cannot be repaired or restored, such as loss of geologic freshwater sources related to glacial melt or desertification, or loss of culture or heritage associated with potential population redistribution away from areas that become less habitable over time with climate change.[40]

Distinguished adaptation specialist, Saleemul Huq, also recently provided some views on loss and damage along these lines:[41]

38 The Loss and Damage in Vulnerable Countries Initiative is a project initiated in order to move forward the debate on loss and damage for the benefit of LDCs and other vulnerable states parties, while the Government of the People's Republic of Bangladesh requested assistance from the Climate and Development Knowledge Network (CDKN) to help build a common understanding around loss and damage and provide insight into what it entailed for vulnerable countries. CDKN has appointed a consortium of organisations, which includes Germanwatch, the UN University Institute for Environmental and Human Security, the International Centre for Climate Change and Development, and the Munich Climate Insurance Initiative.

39 Germanwatch (2012).

40 (ibid.).

41 Huq (2012).

> It is still not clear what the difference is between the two terms, "loss" and "damage." One way of thinking of this difference is to consider "loss" to mean the "complete loss" of something (e.g. human life or biodiversity, or land that goes under water, etc). These losses are in fact irrecoverable.
> "Damage," in contrast, can be considered to refer to "partial loss" or "partial damage," such as to infrastructure and human livelihoods, which can be repaired.
> These distinctions are of course not watertight compartments, as there will still be some overlaps between loss and damage, but it is worth keeping these terms separate in this context.

Doreen Stabinsky and Juan P. Hoffmaister also provided a brief definition with an overview of approaches to loss and damage, as follows:[42]

> [T]he phrase 'loss and damage' refers broadly to the entire range of damage and permanent loss "associated with climate change impacts in developing countries that are particularly vulnerable to the adverse effects of climate change"[43] that can no longer be avoided through mitigation nor can be avoided through adaptation. There are multiple approaches to address those damages and losses, some which may have synergies with adaptation efforts, while others will require taking action through new arrangements and stand-alone approaches.

Roda Verheyen, an eminent legal expert, provided a definition of *loss and damage* in the following way:[44]

> [i]n legal terms, loss and damage are not separate concepts. Rather, loss is a specific term to describe a particular type of damage such as loss of earnings or loss of office. Damage is a legal concept equated with "tort" or "liability", which often leads to a claim for damages, with monetary or in kind compensation as a remedy, but it is also the generic term for harm afflicted to a legal entity or person or other systems (e.g. a particular ecosystem) which may give rise to a legal claim.

There are some common elements found in the aforementioned definitional views, such as that *damage* can be repaired or restored, but *loss* is considered irrecoverable damage, i.e. complete loss that can no longer be avoided through mitigation or adaptation. With regard to approaches to address such damage and loss, one suggestion is to have synergies with adaptation efforts, while others will require taking action through new arrangements and stand-alone approaches. In terms of legal definitions, *damage* equates with "tort" or "liability" and pleads for a claim for damage, with monetary forms of

42 Stabinsky & Hoffmaister (2012).
43 (ibid.), but no source indicated in the original for this citation.
44 Verheyen (2012).

compensation as a remedy. Also, remedial measures can be offered for ecological harm and monetary compensation can be awarded for loss of infrastructure and property. However, the resettlement and rehabilitation of forced migrants is a complex consequence of damage or loss, and questions also remain on non-economic losses. These elements or definitions can be scrutinised, taking into account the facts from Bangladesh discussed above, to conceptualise the loss and damage associated with climate change and compensation and remedy in response to such adverse consequences.

The contextual analysis on intensified cyclones such as Aila and frequent rough sea events in the Bay of Bengal provides evidence of actual loss and damage, including loss of life and property, ecological damage, loss of traditional livelihoods, displacement and migration, and loss of territory, values and culture. Moreover, forced migration means people lose their freedom to choose a profession, and they face challenges with new lodgings, drinking water, food, sanitation, security, and so forth. In terms of the loss of property, monetary compensation can be awarded; in terms of ecological damage and loss of livelihood, remedial measures can be useful. Also, displacement and migration can be dealt with by way of appropriate resettlement/relocation and rehabilitation measures, with the greatest challenge being relocation to another country. Then the question arises as to how death and non-economic losses such as the loss of territory, values, heritage and culture are to be compensated. However, we need to think of who will compensate, who will be compensated, and what the compensation mechanisms would be. Thus, we now turn to a discussion on a liability regime.

### *D. Climate Justice and a Liability Regime*

In its Preamble, the UNFCCC recognises that the largest historical and current global emissions of GHGs originated in industrialised countries,[45] which is the cause of anthropogenic climate change. Hence, the ultimate objective of the UNFCCC, as set forth under Article 2, is to stabilise GHG concentrations in the atmosphere at a level that would prevent dangerous

45 The UNFCCC also notes that the largest share of historic and current global emissions of GHGs originated in developed countries, that per capita emissions in developing countries are still relatively low, and that the share of global emissions originating in developing countries will grow to meet their social and development needs.

anthropogenic interference with the climate system. The UNFCCC also called for meeting such targets within a time frame sufficient to allow ecosystems to adapt naturally to climate change. Thus, Article 4(2) compels developed nations to take measures related to mitigation and adaptation. Moreover, Article 4(4) requires developed countries to assist their developing country counterparts – who are particularly vulnerable to the adverse effects of climate change – in meeting the cost of adapting to such adverse effects. Furthermore, the UNFCCC states parties adopted the Kyoto Protocol with its legally binding commitments for a reduction in GHG emissions, particularly for developed countries, and also to finance adaptation.[46]

While there is a struggle to extend the Kyoto Protocol for a second commitment period, states parties to the UNFCCC started a process to develop another instrument. The latter is to be adopted by 2015 and implemented by 2020.[47] Therefore, the UNFCCC climate regime is still exploring the framework for mitigation and adaptation measures along with the relevant finance, technology and capacity-building. However, while the global community is exploring the mechanisms for mitigation and adaptation, loss and damage resulting from climate change has become a reality.

The findings of an analysis on loss and damage in LDCs and other vulnerable countries today suggests that communities are observing and experiencing changes in climate stresses, in both extreme weather events and slow-onset climatic changes.[48] Research reveals that communities are experiencing significant loss and damage to quality of life, livelihoods, food and livelihood security, as well as secondary loss and damage in the form of stress on the social fabric – essential to adaptive capacity and resilience.[49]

---

46 In 1995, at the First Conference of the Parties (COP1) in Berlin, the states parties agreed on legally binding commitments, in acknowledgment of the inadequacy of voluntary commitments under the UNFCCC to reduce GHGs. In accordance with the Berlin Mandate, therefore, states parties to the UNFCCC initiated further negotiations for legally binding instruments. This led to the Kyoto Protocol being adopted at the Third Conference of the Parties (COP3) in Kyoto, Japan, in 1997.

47 Decision 1/CP.17.

48 See the Loss and Damage in Vulnerable Countries Initiative, available at http://www.lossanddamage.net/empirical-research, last accessed 17 January 2013: "Case studies in Africa, Asia and Oceania illustrate the effects of climate change beyond adaptation. The case studies look at several climate threats, such as drought, flooding, changing rainfall patterns, cyclones and sea-level rise. The case studies are conducted by the UN University, Institute for Environment and Human Security, in cooperation with local research institutions in Least Developed Countries".

49 Government of the Gambia (2012).

The current low ambition levels of emission reductions are taking us to a 4°C warmer world, and it would be a question of survival of millions of people in the world. Consequently, actual and potential loss and damage associated with climate change raises the question of liability and immediate response to loss and damage based on causal liability.

The context of climate justice, taking into account the legal commitments under the UNFCCC and the Kyoto Protocol as well as by way of historical responsibility, shifts the burden causally to developed country states parties to take the entire responsibility for climate change. Legal commitments oblige developed countries to reduce GHG emissions and to facilitate the implementation of adaptation measures to counteract the adverse impacts of climate change. From the adaptation perspective, developed countries should take on the responsibility of reacting to the consequences and preventing further deterioration. Technological and financial resources should be provided, based on proportional contributions to climate change and the relevant state's respective capacity.[50]

It is worth mentioning that, in its Preamble, the UNFCCC recognises states parties' responsibility to ensure that activities within their jurisdiction or control do not cause damage to the environment of other states or to areas beyond the limits of national jurisdiction.[51] Therefore, in response to unavoidable loss and damage resulting from climate impacts, the states responsible for atmospheric pollution should also provide compensation and remedial measures. In the context of climate science, the relative contributions by different states to climate change can be estimated based on their cumulative contributions and, as such, each state should be liable proportionally. The breach of an international obligation can be derived from international treaties or customary law, and may be committed through an act of commission or omission.[52] Treaties that are relevant exclusively to loss and damage associated with climate change are the UNFCCC and the Kyoto Protocol. Thus, state liability and attributed wrong can be identified under

50 Khan (2010).

51 The Preamble to the UNFCCC declares the following: "States have, in accordance with the Charter of the United Nations and the principles of international law, the sovereign right to exploit their own resources pursuant to their own environmental and developmental policies, and the responsibility to ensure that activities within their jurisdiction or control do not cause damage to the environment of other States or of areas beyond the limits of national jurisdiction".

52 ILC (2001).

the UNFCCC. However, the literature predominantly considers that the states parties' primary obligations under the UNFCCC are too vague, and that the compliance system under the UNFCCC and the Kyoto Protocol is too weak to exclude the application of general international law on state responsibility.[53]

The compliance mechanism of the UNFCCC suggests a consultative process. The Kyoto Protocol entails reporting, monitoring and compliance within its own mechanism, and any binding requirements demand amendment to the Protocol's provisions. As such, the UNFCCC and the Kyoto Protocol prefer self-governing dispute settlement mechanisms and bar member states from seeking legal remedy outside the UNFCCC process. Thus, even if, in terms of climate change, contentious state liability and attributed wrongful acts are present in the UNFCCC regime, the UNFCCC does not provide the procedural means to lodge claims for climate-induced loss and damage. The notion of *climate justice* presented a challenge to the global legal community to protect the rights violated by atmospheric pollution. The current literature, however, predominantly suggests that a violation of international law could be based on the so-called No-harm Rule.[54]

### *E. No-harm Rule and Customary International Law*

A widely recognised principle of customary international law is the No-harm Rule, which obliges a state to prevent damage and to minimise the risk of damage to other states. This principle was first applied in the *Trail Smelter* case.[55] The basis of the case was that a Canadian smelter's sulphur dioxide emissions had caused air pollution damages across the border in the United States (US). The arbitral tribunal decided that the government of Canada had to pay the US compensation for the damage that the smelter had caused along the Columbia River Valley in the US. The no-harm principle employed in *Trail Smelter* case was subsequently confirmed by different decisions of international courts and tribunals.

In the 1949 *Corfu Channel* case, the International Court of Justice (ICJ) observed that there were "general and well-recognised principles" of inter-

53 Schwarte & Byrne (2010).

54 (ibid.).

55 *United States v Canada*, United Nations, Reports of International Arbitral Awards, Vol. III, 1906, 1982.

national law concerning "every State's obligation not to allow knowingly its territory to be used for acts contrary to the rights of other States."[56]

A 1996 advisory opinion of the ICJ on the legality of the threat or use of nuclear weapons stated that –[57]

> … the existence of the general obligation of states to ensure that activities within their jurisdiction and control respect the environment of other states or of areas beyond national control is now part of the corpus of international law relating to the environment.

The No-harm Rule was also restated and accepted by both parties – Hungary and Slovakia – in the *Gabčikovo* case decided by the ICJ in 1997.[58]

The No-harm Rule has also been incorporated into international law and policy documents. Principle 21 of the Stockholm Declaration, for example, provides as follows:

> [S]tates have, in accordance with the Charter of the United Nations and the principles of international law, the sovereign right to exploit their own resources pursuant to their own environmental policies, and the responsibility to ensure that activities within their jurisdiction or control do not cause damage to the environment of other states or of areas beyond the limits of national jurisdiction.

This principle is also included in the 1992 Convention on Biological Diversity,[59] the 1985 Vienna Convention for the Protection of the Ozone Layer,[60] and UN General Assembly Resolution 2996.[61] It has also been reiterated in the UNFCCC.[62]

### *F. No-harm Rule and State Responsibility*

A state's failure to comply with the No-harm Rule is an internationally wrongful act that gives rise to an obligation to take responsibility. A state's breach of obligations not to cause damage, to prevent harm, or to minimise the risk of harm occurring, would constitute an internationally wrongful act, which entails the international responsibility of that state. Atmospheric pol-

56 1949 ICJ Rep. 4.
57 (ibid.:241, para. 29).
58 The *Gabčikovo* case (1997 ICJ Rep. 7, in particular at 41).
59 Article 3.
60 Preamble, para. 2.
61 XXVII, 15 December 1972.
62 Preamble, para. 9.

lution clearly falls within the purview of the No-harm Rule of international law. However, international law will not support a conclusion that a state emitting GHGs and, thus, contributing to global climate change should be held responsible for damage occurring per se simply because it has emitted such gases.

A state's behaviour has to be found contrary to a specific standard of care. Once this duty of care is defined, if a state fails to take proportionate measures to minimise the risk of foreseeable damage, the No-harm Rule is breached.[63] The problem with damage from climate change is that it is diffuse and hard to trace back to any particular state's actions. The general rule under international law, however, appears to be that states that are jointly responsible for a wrongful act are jointly and separately liable. There exists a relatively clear estimate of different countries' relative contributions to the tons of GHGs emitted globally. It has, therefore, been suggested that, because of the cumulative causation of climate change, each actor should only be held responsible for its share of the overall wrong.[64]

Bangladesh, in the context of climate vulnerability and related loss and damage, can convincingly establish substantive arguments under public international law that one or more states are responsible for wrongful acts based on causation and liability. While the substantive law may provide a clear basis for the claim of the climate victim community of Bangladesh, there are often no procedural means to pursue it further and enforce compliance under public international law. The ICJ is the UN's principal judicial organ, and has been described as the guardian of the international legal community as a whole. It may hear contentious disputes concerning an alleged breach of an international obligation if – and to the extent that – the states concerned have accepted such obligation.[65]

There is no governing authority that automatically addresses the legality of an act or situation at international level. This reflects the fundamental principle of international relations that states are sovereign and free to choose the methods of resolving their disputes. In practice, political pressure and diplomatic negotiations remain the primary tools in the international arena to influence state conduct.[66] However, the international community has committed itself to increasing efforts to develop international law on liability

---

63 Schwarte & Byrne (2010).

64 (ibid.).

65 Article 36, Statute of International Court of Justice.

66 Schwarte & Byrne (2010).

and compensation for the victims of pollution damage. In 1972, states committed to develop international law on liability and compensation for environmental damage.[67]

Moreover, in 1992, through the Rio Declaration at the Earth Summit, states parties agreed to increase their efforts in this regard:[68]

> States shall develop national law regarding liability and compensation for the victims of pollution and other environmental damage. States shall also cooperate in an expeditious and more determined manner to develop further international law regarding liability and compensation for adverse effects of environmental damage caused by activities within their jurisdiction or control to areas beyond their jurisdiction.

Therefore, in the context of climate-change-induced loss and damage, the global community should develop contemporary and comprehensive legal and policy frameworks with appropriate institutional arrangements under the UNFCCC with compensation approaches based on the broadly accepted 'The Polluter Pays' Principle.

### *G. UNFCCC Processes in Response to Loss and Damage*

The preceding section shows there is a sound legal basis under customary international law for individual cases brought by states seeking compensation for loss and damage associated with climate change. Such individual cases should not, however, be the path of choice. International law is based on the notion of cooperation and the avoidance of adjudication, where possible, in favour of diplomatic solutions. Cumbersome individual cases should not be necessary, given that the climate regime is based on the notion of cooperation and good faith. International law scholars have also expressed the view that states even have a legal duty to provide negotiated solutions where environmental damage is expected to occur, so that prompt and adequate compensation can be obtained in practice.[69]

Against this backdrop, one better understands the position of the Alliance of Small Island States (AOSIS) and the idea that states harmed by loss and

67 Stockholm Declaration on the United Nations Conference on the Human Environment, Principle 22, 11 ILM 1416 (1972).

68 Rio Declaration on Environment and Development, United Nations Conference on Environment and Development, Principle 13, 31 ILM 874 (1992).

69 Schwarte & Byrne (2010).

damage related to climate change should seek compensation to rehabilitate their societies (ideally to pre-anthropogenic climate change conditions).[70] In connection with the negotiation for adopting the UNFCCC in 1991, AOSIS proposed the establishment of an International Climate Fund to counter the adverse consequences of climate change, and a separate International Insurance Pool to provide financial insurance against the consequences of sea-level rise. Revenue was to be drawn from mandatory sources and, in particular, from developed countries.[71]

AOSIS and the LDCs have raised this issue of compensation and rehabilitation in oral interventions at a number of international negotiating sessions. Bangladesh, on behalf of the LDCs, called for compensation for damages caused by climate change at the Eleventh Conference of the Parties to the UNFCCC (COP11) in Montreal in 2005.[72] AOSIS argued that[73] –

> [w]here adaptation cannot fully address the impacts of climate change on countries and their communities, impacted countries are justified in seeking compensation from those countries most responsible for the greenhouse gas emissions that have led to those impacts.

The spectre of liability and possibly needing to pay unspecified amounts of money to compensate 'sinking island states' or other countries facing a range of catastrophic climate-related impacts made this area of negotiation controversial for many industrialised countries. However, despite the calls for compensation approaches in the climate change negotiations, the issue of loss and damage associated with climate change damage remains to be squarely addressed and placed under ongoing adaptation framework negotiation.

## *H. The Bali Action Plan and Loss and Damage*

In 2007, under the Bali Action Plan (BAP), states parties agreed to enhance action on adaptation, and loss and damage associated with climate change

70 Zakieldeen & Warner (2012).
71 AOSIS (2007).
72 *Earth Bulletin*, available at http://www.iisd.ca/vol12/enb12291e.html, last accessed 14 January 2013.
73 AOSIS (2005).

were addressed within an adaptation framework. BAP expressly agreed on two options:[74]

- Consideration of "disaster risk reduction strategies and means to address loss and damage associated with climate change impacts in developing countries that are particularly vulnerable to the adverse effects of climate change", and
- "[R]isk management and risk reduction strategies, including risk sharing and transfer mechanisms such as insurance".

Although BAP contained an entire section on (disaster) risk management as well as loss and damage associated with climate change, any association or mention of compensation or liability for such negative effects caused discomfort for industrialised countries.[75]

After BAP formed an Ad Hoc Working Group on Long-term Cooperative Action (AWG–LCA), the issue of loss and damage was advanced further under adaptation negotiations. AOSIS also continues its stance on the compensation approach under the AWG–LCA process, but has proposed some additional elements under the multi-window mechanism in addressing loss and damage. In 2008, AOSIS proposed a mechanism for risk reduction, management and sharing to be established with the following components:[76]

- A *risk management and prevention component* to promote risk assessment and risk management tools and strategies at all levels, with a view to facilitating and supporting the implementation of risk reduction and risk management measures
- An *insurance component* to address climate-related extreme weather events, and risks to crop production, food security and livelihood, and
- A *rehabilitation and compensation component* to address progressive negative impacts that result in loss and damage.

The AOSIS proposal's three-pronged approach clearly sets out how different challenges of loss and damage will be tackled, and is the most comprehensive and far-reaching proposal to date in respect of moving the discussion forward. Some industrialised states parties that were uncomfortable with the elements of rehabilitation and compensation attempted to avoid the discus-

74 Decision 1/CP.13, para. 1c(iii).
75 Zakieldeen & Warner (2012:1).
76 AOSIS (2008).

sions – particularly on compensation for loss and damage. Parties wary of 'compensation' may have wanted to manoeuvre the issue of loss and damage out of the process; however, they needed to build consensus with the mass of countries that are anticipated to experience loss and damage in the future, and preferred to address only risk management, insurance and related capacity-building.[77] A compromise was found at COP16 in Cancun in 2010, and a work programme on loss and damage was established.

### *I. Work Programme on Loss and Damage*

The Cancun Agreements (COP16) recognised the need to strengthen international cooperation and expertise in order to understand and reduce loss and damage associated with the adverse effects of climate change. The decision of the COP16 also considered the loss and damage associated with adverse impacts related to extreme weather and slow-onset events. Sea-level rise, increasing temperatures, ocean acidification, glacial retreat and its related impacts, salination, land and forest degradation, loss of biodiversity, and decortication are identified as slow-onset events associated with climate change. The same COP16 decision established a work programme to consider the approaches required to address loss and damage associated with climate change impacts in developing countries that are particularly vulnerable to such adverse effects.[78] At the COP17 in Durban, states parties further decided to request the Subsidiary Body for Implementation (SBI)[79] to continue the implementation of this work programme.[80] The COP17 decision also provided the required guidance to advance the work programme through a set of activities related to agreed thematic areas, as follows:[81]

- Assessing the risk of loss and damage associated with the adverse effects of climate change, and current knowledge

77 Zakieldeen & Warner (2012).

78 Decision 1/CP.16, paras 25, 26.

79 The SBI is one of two permanent subsidiary bodies to the UNFCCC established by the Conference of the Parties (COP) and the COP serving as the Meeting of the Parties (CMP) to the Kyoto Protocol, through the assessment and review of the effective implementation of the UNFCCC and the said Protocol.

80 Decision 7/CP.17, para. 1.

81 (ibid.:paras 8–11).

- A range of approaches to address loss and damage associated with the adverse effects of climate change, including impacts related to extreme weather events and slow-onset events, taking into consideration experience at all levels, and
- The role of the UNFCCC in enhancing the implementation of approaches to address loss and damage associated with the adverse effects of climate change.

Decision 7/CP.17 also appreciated the need to explore a range of possible approaches and potential mechanisms, including an international mechanism, to address loss and damage, with a view to making recommendations on loss and damage to the Eighteenth Conference of the Parties (COP18) for consideration, including elaborating the elements set out in Decision 1/CP. 16, paragraph 28(a–d).[82]

### *J. Work Programme on Loss and Damage and Related Activities*

In accordance with mandate of Decision 7/CP.17, a technical paper was prepared by the Secretariat before the expert meeting on Thematic Area 1, namely Assessing the Risk of Loss and Damage. This document, titled "Current Knowledge on Relevant Methodologies and Data Requirements as well as Lessons Learned and Gaps Identified at Different Levels, in Assessing the Risk of Loss and Damage Associated with the Adverse Effects of Climate Change",[83] assessed 18 selected approaches, methods and tools in terms of their data and information requirements, strengths, weaknesses, lessons learned, gaps at different levels, and relevance for social and environmental impacts; the document also discussed capacity needs for applying risk assessment methods in developing countries. Moreover, the Secretariat prepared the notes of the expert meeting held in Tokyo, Japan, from 26 to 28 March 2012 on assessing the risk of loss and damage associated with the adverse effects of climate change.[84] In accordance with these notes, the key issues identified by the experts are as follows:

---

82 (ibid.:para. 5).
83 FCCC/TP/2012/1.
84 FCCC/SBI/2012/INF.3.

- The data and information requirements for assessing impacts and climate risks
- Methods and tools for risk assessment, including their requirements, strengths and weaknesses
- Capacity needs for applying risk assessment methods on the ground, and
- Linking risk assessment with decision-making.

At its Thirty-sixth Session, the SBI considered the progress made on the implementation of the work programme on loss and damage. In accordance with Decision 1/CP.16, paragraphs 26–29 noted a number of points relevant to assessing the risk of loss and damage associated with the adverse effects of climate change and the current knowledge on the same, including the following:[85]

(a) The assessment of climate-related risk is complex, involving the consideration of hazards, exposure and vulnerability, and takes into account underlying risk drivers;
(b) A range of approaches, methods and tools are available to assess the risk of loss and damage associated with the adverse effects of climate change. The selection of appropriate approaches, methods and tools depends upon regional, national and local capacity, contexts and circumstances and involves the engagement of all relevant stakeholders; …

The SBI recalled Decision 7/CP.17, and requested the Secretariat to organise four expert meetings – three at regional level and one for SIDs – to be held before the SBI's Thirty-seventh Session.[86] The SBI also provided the required guidance to organise these workshops.[87]

In accordance with the COP17 Decision, and following the guidance provided by the conclusion adopted at the SBI's Thirty-Sixth Session, as discussed above, the UNFCCC Secretariat organised a further three regional expert meetings in Africa, Asia, and Latin America and the Caribbean, to address issues related to a range of approaches for addressing loss and damage associated with the adverse effects of climate change, including impacts related to extreme weather and slow-onset events, and taking into consideration experience at all levels. Another expert meeting with the identical brief was organised for SIDs. Furthermore, the Secretariat produced a literature

85 FCCC/SBI/2012/L.12, para. 3.
86 (ibid.:para. 5).
87 (ibid.:para. 7).

review on the topics in the context of Thematic Area 2 of the work programme on loss and damage, which also guided the workshop participants in identifying gaps in and the scope of existing approaches to loss and damage at the regional level.[88]

The Secretariat compiled the experts workshop report and published it in accordance with the mandates of Decision 7/CP.17.[89] The report includes an overview of the issues discussed at the meetings, including gaps, needs and challenges, as well as region-specific issues related to the impacts of climate change, and possible areas for further action in addressing loss and damage associated with the adverse effects of climate change at different levels.[90] This report identified the necessity of a new approach to address loss and damage, in combination with prevention, reduction, retention and sharing mechanisms. The report also finds –[91]

> … the need for the further clarification of the operational aspects of the international mechanism proposed by the Alliance of Small Island States was expressed, such as how it would interact with other levels and institutions, including its linkages to capacities and corresponding structures required to be implemented at the national level in order to benefit from the opportunities that such a mechanism will provide.

Moreover, the report identified the necessity of a better understanding of the role of national governments in creating enabling environments for minimising loss and damage associated with the adverse effects of climate change and for developing a global architecture as well as a multi-institutional approach to loss and damage.[92] The experts workshop report provided some of the essential information to the negotiators and influenced the decision-making process at the Eighteenth Conference of the Parties (COP18) in Doha.

In addition, the Secretariat compiled all the views and information from states parties and relevant organisations on the possible elements to be included in the recommendations on loss and damage in accordance with Decision 1/CP.16. Nauru, on behalf of AOSIS, recommended adopting a decision at COP18 in Doha to establish an international mechanism to address loss and damage with three mutually reinforcing components to address loss

88 FCCC/SBI/2012/INF.14.
89 FCCC/SBI/2012/29.
90 (ibid.:Abstract).
91 (ibid.:para. 80).
92 (ibid.:para. 38, 39).

and damage in line with their previous submission in 2008 (discussed above).[93] The LDC group submitted their recommendation based on some recent research studies[94] revealing the reality of loss and damage resulting from climate change, and provided the projection for potential loss and damage in the context of a 2–4°C increase in temperature. LDCs proposed establishing an international mechanism to address loss and damage which would work as an umbrella for activities required on different levels and would perform the key functions required for an adequate response to loss and damage. LDCs suggested the COP as the central oversight body of the mechanism for providing the political direction, and for developing key guidance on the elaboration and operation of the mechanism and its elements.[95] At the UNFCCC informal pre-session meeting held on 24 November 2012 prior to the COP18, states parties exchanged further views on the possible recommendations on loss and damage associated with the adverse effects of climate change.[96]

### *K. Doha Decision and the Way Forward*

The Doha Decision recognised that comprehensive, inclusive and strategic responses were needed to address loss and damage associated with the adverse effects of climate change, and expressed appreciation of the progress made not only in the implementation, but also the importance of the continuation, of the work programme to address climate-change-induced loss and damage through a range of approaches. It was also agreed that the UNFCCC's role in promoting the implementation of approaches to address loss and damage included –

- enhancing knowledge and understanding of comprehensive risk management approaches
- strengthening dialogue among relevant stakeholders, and

93 AOSIS (2012); FCCC/SBI/2012/MISC.14, Add.1 and Add.2, *Submission by Nauru on behalf of AOSIS*, 5.

94 Government of the Gambia (2012).

95 FCCC/SBI/2012/MISC.14, Add.1 and Add.2, Submission by The Gambia on Behalf of the Least Developed Countries (LDCs).

96 Available at http://unfccc.int/adaptation/cancun_adaptation_framework/loss_and_damage/items/7157.php, last accessed 13 January 2013.

- enhancing action and support, including finance, technology and capacity-building.

The Doha Decision recognised the UNFCCC's important and fundamental role in addressing loss and damage through promoting leadership, collaboration and cooperation at national, regional and international levels.[97] The Doha Decision invites all states parties to enhance the actions on addressing loss and damage by, inter alia, –[98]

- designing and implementing country-driven risk management strategies and approaches
- implementing comprehensive climate risk management approaches and approaches including risk reduction, risk transfer, and risk-sharing mechanisms, and
- promoting an enabling environment that would encourage investment and the involvement of relevant stakeholders in climate risk management.

The Doha Decision declares that a range of approaches, methods and tools is available to assess the risk of and to respond to loss and damage associated with the adverse effects of climate change, and that their selection depends upon regional, national and local capacity, context and circumstances, and involves the engagement of all relevant stakeholders.[99] As such, the Decision requests developed country parties to provide developing country parties with finance, technology and capacity-building in order to respond adequately to loss and damage associated with the adverse effects of climate change.

The Doha Decision acknowledges the necessity of strengthening institutional arrangements at national, regional and international levels to address loss and damage associated with the adverse effects of climate change. The Decision also resolves to establish institutional arrangements, such as an international mechanism, to address loss and damage in developing countries that are particularly vulnerable to the adverse effects of climate change. The said Decision mandates the establishment of such institutional arrange-

97 Decision 3/CP.18, Preamble, and para.'s 4 and 5.

98 (ibid.:para. 6). See also Summary of the Doha Climate Change Conference, *Earth Negotiations Bulletin*, http://www.iisd.ca/vol12/enb12567e.html, last accessed 13 January 2013.

99 Decision 3/CP.18, para. 2).

ments and their functions and modalities in accordance with the UNFCCC's role as defined in paragraph 5 of the same decision.[100] However, challenges remain to develop the required institutional mechanisms and their related functions and modalities by COP19.

Before COP19, states parties will work through one official SBI session in June, 2013, one experts meeting, and a technical paper on gaps in existing institutional arrangements within and outside of the UNFCCC as agreed interim activities under the work programme leading up to COP19. Nonetheless, states parties to the UNFCCC may utilise the scope of paragraph 12 of the Doha Decision, which requests the SBI, at its June 2013 session, to elaborate activities under the work programme on loss and damage, taking into account the provisions contained in paragraph 7 of the Doha Decision. The latter paragraph includes the notion of strengthening institutional arrangements at national, regional and international levels in order to address loss and damage in order to further implement the work programme.[101] Therefore, while states parties will work on the functions and modalities of institutional arrangements such as an international mechanism, they also need to take into account the important aspects of national and regional institutional arrangements and their functions and linkages with international mechanisms. An international mechanism is expected to be established at COP19 with micro-level institutional arrangements, so that a bottom-up approach can assess and redress the loss and damage associated with adverse impacts of climate change. AOSIS and LDCs negotiated with the same spirit at COP18 in order to resolve to establish an international mechanism to address loss and damage, and COP18 indeed mandates the establishment of such mechanism. Thus, further coordinated and collective efforts are required in order to develop the required governance mechanism through the vehicle of the work programme on loss and damage under the UNFCCC.

## *L. Concluding Remarks*

The prerequisite of addressing a particular issue like loss and damage associated with climate change is to assess and quantify each case, taking into account the geographical context. Understanding and successfully assessing

---

100 (ibid.:para. 10).
101 (ibid.:para. 7 and 12).

a particular case of climate-change-induced loss and damage would also suggest the required approaches for dealing with such loss and damage. Thus, the identification, measurement and characterisation of loss and damage are primary requirements for developing local, national, regional and international policy and legal frameworks. Assessment efforts at a micro level demand a particular role of a particular state in collaboration with regional and international efforts to deal with climate-change-related loss and damage.

Against this backdrop, this paper initially concentrated on the conceptualisation of *loss* and *damage* associated with climate change in the context of a particular territory, and considered two cases from Bangladesh that dealt with extreme weather events and slow-onset processes of climate change. The case of the extreme weather event, namely Cyclone Aila, has divided scientific experts in respect of how to quantify the extent to which climate change contributed to Aila's path of destruction.[102] Nonetheless, the destruction wrought by Aila provided a portrait of actual and potential loss and damage related to climate change. On the other hand, increased SST is identified as an impact of climate change with the empirical data, which also provided some unique features of loss and damage.

The factual evidence on existing loss and damage from both cases includes loss of lives; loss of property; ecological damage and loss of traditional livelihoods; displacement and migration; and loss of territory, values, culture and heritage. Moreover, people who were forced to migrate lost their freedom to choose a profession, and also faced challenges with new lodgings, drinking water, food, sanitation, security and so forth. The discussion on definitions revealed that *damage* can be repaired or restored, but *loss* is considered irrecoverable damage, i.e. complete loss that can no longer be avoided through mitigation or adaptation. With regard to approaches to address such damage and loss, adaptation efforts need to be synergised, while others will require taking action through new arrangements and stand-alone approaches. In terms of a legal definition, *loss and damage* equates with "tort" or "liability", and pleads for a claim for such negative impacts, with the monetary compensation as a remedy. Also, remedial measures can be offered for ecological harm, and monetary compensation can be awarded for lost infrastructure and property. However, complex legal and administrative

102 McAdam & Saul (2010).

procedures are needed for resettlement and rehabilitation for forced migrants as well as for issues related to non-economic losses.

In accordance with the legal direction of the definitional views, if Bangladesh can put forward a climate-change-induced loss and damage claim under a climate legal regime and/or under customary international law, it can convincingly establish substantive arguments under both regimes that one or more states are responsible for wrongful acts based on causation and liability. While the substantive law may provide a clear basis for such a claim, there are often no procedural means to pursue them further due to adequate procedural mechanisms under public international law. There is no governing authority that automatically addresses the legality of an act or situation at international level. This reflects the fundamental principle of international relations that states are sovereign and free to choose their own methods of resolving their disputes. In practice, political pressure and diplomatic negotiations remain the primary tools in the international arena to influence state conduct.[103]

Taking into account the gaps and constraints of public international laws involved with climate-change-related loss and damage, broad, system-changing solutions to the climate crisis are called for. Thus, a contemporary legal and policy framework with specific substantive and procedural mechanisms are required. UNFCCC states parties from developing countries are enthusiastically negotiating the establishment of an international mechanism to address loss and damage, and the inclusion of an "international mechanism" in the Doha Decision on loss and damage marks an important window of opportunity for further development of procedural mechanisms in this respect.

Careful proactive policy can minimise the risks of potential loss and damage and can maximise the extent to which community resilience copes with climatic hazards. Mitigation is primary in this respect, with secondary focus being placed on adequate adaptation measures that can prevent and reduce the loss and damage related to climate change. It is, of course, the reactive legal response that is also needed to redress the unavoidable climate-change-induced loss and damage with compensation and remedial approaches. In terms of resettlement/relocation and rehabilitation, adaptation can also be a proactive measure, but it needs to take into account the non-economic consequences like loss of territory. Therefore, for the purposes of regulatory

103 Khan (2010).

responses to loss and damage, policy and legal frameworks will be reactionary and anticipatory. Reactionary measures will be taken in case of unavoidable climate-change-induced loss and damage, whereas anticipatory measures are taken for planning ahead for avoidable loss and damage.[104]

Therefore, it is imperative to establish an autonomous international mechanism with micro-level institutional arrangements, so that a bottom-up approach can assess and redress the real loss and damage.[105] A compensation fund can be launched at international level to meet the financial needs of these institutions (including micro-level institutions at national level) to deliver their functions with executive authority. A quasi-judicial authority, such as an independent dispute settlement body, can be formed to respond by way of compensation and remedial measures in dealing with the claims of loss and damage cases. Certainly, in this regard, a fundamental role needs to be played by the UNFCCC in collaboration with other relevant actors.

While the international climate regime began in 1992 with the adoption of the UNFCCC, it is still struggling to set up governance mechanisms for mitigation and adaptation to climate change, and the establishment of an autonomous international mechanism under UNFCCC will take time. However, the basic foundation needs to be built by 2015, when a new legal instrument will be adopted (Durban Platform). Therefore, LDCs, SIDSs and other vulnerable states such as African countries will have to work together to advance the work programme on loss and damage towards a legally binding instrument to deal with loss and damage. At the same time, some of the research initiatives such as the Loss and Damage in Vulnerable Countries Initiative are needed to generate knowledge and information and to build capacity of the negotiators to act effectively to develop international and national regulatory mechanisms on loss and damage. Without waiting for the development of international mechanisms, efforts can also be taken immediately by LDCs in particular to develop national mechanisms to address loss and damage, which could also provide bottom-up support to developing regional and international mechanisms.

104 Khan (2011a).
105 Khan (2011b).

## References

ActionAid, Concern WorldWide, DanChurchAid, MuslimAid, Islamic Relief, Oxfam-GB & Save the Children-UK, 2010, *In-depth Recovery Needs Assessment of Cyclone Aila Affected Areas*, Conducted by International Agencies Currently Involved in Aila Response Programme funded by ECHO (Humanitarian Aid and Civil Protection department of the European Commission), available at http://www.lcgbangladesh.org/DERweb/Needs%20Assessment/Reports/In-depth%20Recovery%20Needs%20Assessment%20of%20Cyclone%20Aila%20Affected%20Areas_ECHO%20funded.pdf, last accessed 15 January 2013.

Ahmed, Ahsan Uddin & Sharmind Neelormi, 2007 (reprinted 2009), *Livelihoods of Coastal Fishermen in Peril: In Search of Early Evidence of Climate Change Induced Adverse Impacts in Bangladesh*, Dhaka, Campaign for Sustainable Rural Livelihoods.

Al Faruque, Abdullah & M. Hafijul Islam Khan, 2012, *Loss and Damage Associated with Climate Change: The legal and Institutional Context in Bangladesh: Draft Report under Loss and Damage in Vulnerable Countries Initiative*, supported by the Climate and Development Knowledge Network, Executive Summary available at http://www.lossanddamage.net/download/6836.pdf, last accessed 13 May 2013.

AOSIS/Alliance of Small Island States, 2007, Submission from AOSIS to the Dialogue on Long-term Cooperative Action to Address Climate Change, *Working Paper No. 14*, available at http://unfccc.int/files/meetings/dialogue/application/pdf/wp14-aosis.pdf, last accessed 14 January 2013.

AOSIS/Alliance of Small Island States, 2008, *Proposal to the AWG–LCA: Multi-window Mechanism to Address Loss and Damage from Climate Change Impacts*, available at http://unfccc.int/files/kyoto_protocol/application/pdf/aosisinsurance061208.pdf, last accessed 12 April 2013.

Chowdhury, Sayedur R., M. Shahadat Hossain, Md Shamsuddoha & S.M. Munjurul Hannan Khan, 2012, *Coastal Fishers' Livelihood in Peril: Sea Surface Temperature and Tropical Cyclones in Bangladesh*, Dhaka, Centre for Participatory Research and Development.

*Earth Bulletin*, available at http://www.iisd.ca/vol12/enb12291e.html, last accessed 14 January 2013.

Germanwatch, 2012, *Framing the Loss and Damage Debate*, available at http://germanwatch.org/en/download/6673.pdf, last accessed 17 January 2013.

Government of the Gambia, 2012, *Submission by the Gambia on Behalf of the Least Developed Countries Group on Loss and Damage*, available at http://unfccc.int/files/adaptation/application/pdf/submission_by_the_gambia_on_behalf_of_the_least_developed_countries_on_loss_and_damage.pdf, last accessed 6 January 2013.

GPRB/Government of the People's Republic of Bangladesh, 2009, *Bangladesh Climate Change Strategy and Action Plan*, Dhaka, Ministry of Environment and Forests.

Huq, Saleemul, 2012, Loss and Damage from Climate Change: After Adaptation, *The Daily Star*, 10 September, available at http://www.thedailystar.net/newDesign/news-details.php?nid=249053, last accessed 12 January 2013.

ILC/International Law Commission, 2001, Draft Articles on Responsibility of States for Internationally Wrongful Acts, in: ILC (Eds), *Report of the ILC on the Work of its 53rd Session,* New York, ILC.

IPCC/Intergovernmental Panel on Climate Change, 2007, *Climate Change 2007: Synthesis Report. Contribution of Working Groups I, II and III to the Fourth Assessment Report of the Intergovernmental Panel on Climate Change*, Core Writing Team: Pachauri, Rajendra K. & Andy Reisinger (Eds), Geneva, IPCC, available at http://www.ipcc.ch/publications_and_data/publications_ipcc_fourth_assessment_report_synthesis_report.htm, last accessed 13 January 2013.

Khan, M. Hafijul Islam, 2010, *Exploring the Legal Avenues to Address the Plight of Climate Victims*, available at http://www.thedailystar.net/forum/2010/November/exploring.htm, last accessed 15 January 2013.

Khan, M. Hafijul Islam, 2011, Policy Responses for Adaptation to Climate Change, *The Daily Star*, 12 March 2011, available at http://www.thedailystar.net/newDesign/news-details.php?nid=177354, last accessed 10 January 2013.

Khan, M. Hafijul Islam, 2011, Post Cancun: Prospects of Climate Regulatory Regime, *The Daily Star*, 15 January 2011, available at http://www.thedailystar.net/newDesign/news-details.php?nid=170065, last accessed 10 January 2013.

Khan, S.M. Munjurul Hannan, Saleemul Huq & Md Shamsuddoha, 2012, The Bangladesh National Climate Funds: A Brief History and Description of the Bangladesh Climate Change Trust Fund and the Bangladesh Climate Change Resilience Fund, *LDC (Least Developed Countries) Paper Series*, available at http://ldcclimate.files.wordpress.com/2012/05/bangladeshnationalfund.pdf, last accessed 1 May 2013.

McAdam, Jane & Ben Saul, 2010, *Displacement with Dignity: International Law and Policy Responses to Climate Change: Migration and Security in Bangladesh*, Kensington, University of New South Wales.

Mehedi, Hasan, Anup K. Nag & Sayeeda Farhana, 2010, *Climate Induced Displacement – Case Study of Cyclone Aila in the Southwest Coastal Region of Bangladesh*, Khulna, Humanitywatch.

Mohal, Nasreen, Zahirul Haque Khan & Nazibur Rahman, 2009, *Impact of Sea Level Rise on Coastal Rivers of Bangladesh*, Dhaka, Coast, Port and Estuary Division, Institute of Water Modelling, available at http://archive.riversymposium.com/2006/index.php?element=06MOHALNasreen, last accessed 4 December 2012.

Mondal, Md. Abdul Matin, 2009, *Sea Level Rise along the Coast of Bangladesh*, Dhaka, Bangladesh Inland Water Transport Authority, Ministry of Shipping, available at http://www.gloss-sealevel.org/publications/documents/bangladesh_2001.pdf, last accessed 18 January 2013.

Rabbani, Gulam, 2012, Loss and Damage on Rice Production and Drinking Water due to Salinity Intrusion in Coastal Bangladesh, *CDKN (Climate and Development Knowledge Network) Loss and Damage in Vulnerability Countries Initiative: Bangladesh Case Study Report, First Draft*, Bonn, United Nations University Institute for Environment and Human Security.

SBI/Subsidiary Body for Implementation, 2012a, *Literature Review on the Topics in the Context of Thematic Area 2 of the Work Programme on Loss and Damage: A Range of Approaches to Address Loss and Damage Associated with the Adverse Effects of Climate Change*, FCCC/SBI/2012/INF.14, available at http://unfccc.int/resource/docs/2012/sbi/eng/inf14.pdf, last accessed 10 January 2013.

SBI/Subsidiary Body for Implementation, 2012b, *Report on the Regional Expert Meetings on a Range of Approaches to Address Loss and Damage Associated with the Adverse Effects of Climate Change, including Impacts Related to Extreme Weather Events and Slow Onset Events*, FCCC/SBI/2012/29, 19 November 2012, available at http://unfccc.int/resource/docs/2012/sbi/eng/29.pdf, last accessed 5 January 2013.

Schwarte, Christoph & Ruth Byrne, 2010, *International Climate Change Litigation and the Negotiation Process: Working Paper*, London, Foundation for International Environmental Law and Development.

Stabinsky, Doreen & Juan P. Hoffmaister, 2012, Loss and Damage: Some Key Issues and Considerations, *Briefing Paper No. 1*, Penang, Third World Network Bookshop, available at http://www.twnside.org.sg/title2/climate/info.service/2012/climate20120705/BP_LossDamage_Latin-America.pdf, last accessed 4 January 2013.

Stern, Nicholas, 2006, *Stern Review on the Economics of Climate Change*, Cambridge, Cambridge University Press.

UN/United Nations, 1982, *Reports of International Arbitral Awards*, Vol. III, 1906 New York, UN.

UNDP/United Nations Development Programme, 2011, *Project Facts – Bangladesh*, available at http://www.undp.org/content/dam/undp/library/crisis%20prevention/disaster/asia_pacific/erf_bangladesh_feb2011.pdf, last accessed 10 January 2013.

UNOCHA/United Nations Office for the Coordination of Humanitarian Affairs, 2009, *Natural Disasters and Forced Displacement in the Context of Climate Change*, New York/Geneva, UNOCHA.

Verheyen, Roda, 2012, *Tackling Loss and Damage – A New Role for the Climate Regime?*, Bonn, The Loss and Damage in Vulnerable Countries Initiative, available at http://www.loss-and-damage.net/download/6877.pdf, last accessed 2 January 2013.

Verheyen, Roda & Peter Roderick, 2008, Beyond Adaptation: The Legal Duty to Pay Compensation for Climate Change Damage, *WWF-UK Climate Change Programme Discussion Paper*, Surrey, World Wildlife Fund – United Kingdom.

Zakieldeen, Sumaya Ahmed & Koko Warner, 2012, Loss and Damage in the Context of Adaptation to Climate Change, *European Capacity Building Initiative Background Paper*, available at http://www.oxfordclimatepolicy.org/publications/documents/LossandDamage.pdf, last accessed 12 January 2013.

# 30
# Insurance Solutions in the Context of Climate-Change-Related Loss and Damage: Needs, Gaps and Roles of the UNFCCC in Addressing Loss and Damage

*Koko Warner, Sönke Kreft, Michael Zissener, Peter Höppe, Christoph Bals, Thomas Loster, Joanne Linnerooth-Bayer, Silvio Tschudi, Eugene Gurenko, Armin Haas, Simon Young, Paul Kovacs, Andrew Dlugolecki & Aaron Oxley*

*Abstract*

The burden of loss and damage – the actual and/or potential manifestation of climate change impacts that negatively affect human and natural systems – is not evenly distributed across the world because of differing exposures, vulnerabilities and coping capabilities. As the risks often fall more heavily on those least able to reduce or recover from them, there is a need for assistance for the most vulnerable people and countries. In 2012, at the United Nations Climate Change Conference in Doha (COP18), the Doha Gateway Package entailed a decision to establish an institutional arrangement, including functions and modalities, to address loss and damage. This article outlines the potential roles of insurance in the contexts of adaptation and loss and damage, highlighting a set of recommendations to the United Nations Framework Convention on Climate Change (UNFCCC). Insurance-related approaches are designed for managing loss and damage caused by events which cannot be foreseen, where and when they occur. This contribution also offers insights into design principles – objectives and functions – that could guide a range of approaches, including insurance. It suggests that the UNFCCC can foster long-term commitment to risk transfer in order to enable sustainable solutions and partnerships, and makes a case for an international climate risk insurance facility that could be part of a wider coordination function of a loss and damage mechanism operationalised through a series of regional risk-management platforms, including risk insurance pools, which collaborate and coordinate on the management of loss and damage. Such a facility would help diversify risks of loss and damage from extreme weather events, lower the costs of managing these risks, and ensure more timely and targeted delivery of support when catastrophes strike.

## *A. Executive Summary*

### *I. Challenges of Addressing Loss and Damage caused by Extreme Weather Events*

The burden of loss and damage – the actual and/or potential manifestation of climate change impacts that negatively affect human and natural systems – is not evenly distributed across the world because of differing exposures, vulnerabilities and coping capabilities. As the risks often fall more heavily on those least able to reduce or recover from them, there is a need for assistance for the most vulnerable people and countries. All countries will require pathways that lead to development that is more climate-resilient in the face of potentially growing weather extremes and incremental, profound shifts in natural systems, such as sea-level rise and desertification driven by climate change.

The challenge of addressing both the impacts of weather extremes and incremental change is daunting, yet there is a great need to manage loss and damage, today and in the future, by avoiding, reducing and sharing the risks imposed by climate change.

Proactive planning and management of climate-related stressors have to become a central part of decision-making now and in the future because patterns of loss and damage related to climate change threaten to derail climate-resilient development in many parts of the world. Delays in action will worsen the plight of developing countries in particular.

### *II. Strategies for Managing Weather Extremes*

Strategies are needed to manage unexpected shocks from weather extremes. These strategies should complement and facilitate the design of strategies to address longer-term incremental loss and damage associated with climate change. Risk assessment as required by insurance approaches can help identify climate stressors and thresholds. Insurance can help manage loss and damage from weather extremes in ways that bolster rather than diminish efforts to achieve climate-resilient development.

Insurance-related approaches are designed for managing loss and damage caused by events which cannot be foreseen, where and when they occur. Prudently employing a combination of insurance-like approaches/solutions with risk-reduction measures such as early warning, education, infrastructure strengthening, and maintenance and livelihood strengthening, creates a space of reduced societal disruption when extreme weather events occur. Approaches that manage unexpected extremes can create a buffer for developing countries, i.e. by providing financial liquidity through fast payouts immediately after a loss event, and can help the international community in planning more accurately with respect to issues such as financial requirements for adaptation as well as for managing loss and damage.

The UNFCCC could establish a global climate risk insurance facility coordinated internationally but operationalised through a series of regional risk-management platforms which could receive funding from sources like the Green Climate Fund. Such a climate risk insurance facility could incentivise loss reduction and resilience-building, create more certainty in investing and other decision-making, and facilitate the provision of timely finance to prepare for and recover from extreme weather events.

Insurance-related approaches, in combination with a wide range of others at local, national, regional and international levels, can contribute towards creating a space of certainty within which it would be safe to make investments in climate-resilient development. Thus, insurance-related approaches should be part of a comprehensive strategy to manage climate-related stressors now and in the future.

In the recent past, a wide variety of insurance and other risk-transfer mechanisms have been introduced at different scales in emerging markets. Combining private insurance with insurance supported in a public–private arrangement with other forms of social protection at the local level can help people in the low-income bracket to better absorb shocks. Including risk-transfer mechanisms in national budgets can contribute to climate-resilient development. At regional and international levels, countries can create insurance pools that build on solidarity concepts to share and transfer loss and damage resulting from extreme weather events.

As the hazard situation for the most vulnerable people in developing countries is, in many instances, increasing due to processes they have not caused themselves, in the interest of fairness, countries that have contributed to a larger share of human-induced climate change should consider supporting the risk-management activities of the most vulnerable countries.

## *III. A Unique Role for the UNFCCC*

The UNFCCC has a unique role to play in facilitating short- and long-term strategies to address loss and damage. The UNFCCC should include a global climate risk insurance facility in its decision on loss and damage. This facility, operationalised through regional risk-management platforms, could fulfil three functions in order not only to address loss and damage, but also to complement adaptation and mitigation efforts, as follows:

1. ***Assess loss and damage:*** The climate risk insurance facility can provide guidelines for assessing loss and damage. Technical assistance may involve pooling technical expertise, coordinating data repositories, and encouraging collaborative worldwide networks and coherence across information frameworks – such as adequate standards for data-gathering, open-source remote sensing, and other information needed to assess risk exposures – that are sensitive to vulnerable people and groups.
2. ***Facilitate regional and international dialogue to advance policy coherence and regulations*** on insurance-related measures that address loss and damage at local, national and regional level. Such dialogue should improve conditions for regulators and decision-makers in developing countries to develop appropriate local, national and regional financial risk-management approaches, including insurance. Policy coherence should enhance resilience-building and risk reduction through links to adaptation and national development planning processes.
3. ***Operationalise a global risk insurance facility through regional risk management to address loss and damage,*** including regional risk insurance pools, which could, in the longer term, become part of a future global system for managing weather extremes. This operationalisation would include appropriate financial and other support. These regional platforms could provide technical assistance to facilitate appropriate combinations of insurance measures which could, together with other tools, address the impacts of extreme weather events.
4. ***Enable systematic capacity development for risk-management tools and expertise within governments and civil society,*** particularly through the use of country or sectoral risk officers: Capacity development could include participatory design processes so that approaches to address loss and damage, including insurance, complement and strengthen social safety networks and other resilience-building measures.

**Box 1**

The UNFCCC can foster long-term commitment to risk transfer in order to enable sustainable solutions and partnerships. A global approach to risk transfer, embedded in a coherent strategy to manage the negative impacts of climate change, can be a sustainable solution to parts of the loss and damage spectrum. An international climate-risk insurance facility will help better diversify risks of loss and damage from extreme weather events, lower the costs of managing these risks, and ensure more timely and targeted delivery of support when catastrophes strike. This could be part of a wider coordination function of a loss-and-damage mechanism, which could be operationalised through a series of regional risk-management platforms, including risk insurance pools, which could collaborate and coordinate on the management of loss and damage.

## *B. Introduction*

The Cancun Adaptation Framework recognises –[1]

> … the need to strengthen international cooperation and expertise to understand and reduce loss and damage associated with the adverse effects of climate change, including impacts related to extreme weather events and slow onset events.

The Framework invites views and information on possible approaches to address loss and damage, including a climate risk insurance facility:[2]

- Options for risk management and reduction; risk sharing and transfer mechanisms such as insurance, including options for microinsurance; and resilience building, including through economic diversification, and[3]
- Approaches for addressing rehabilitation measures associated with slow onset events.[4]

The Cancun Adaptation Framework asked the Subsidiary Body for Implementation (SBI) to make recommendations on loss and damage to the Con-

1 Report of the Conference of the Parties on its Sixteenth Session, held in Cancun from 29 November to 10 December 2010, para.'s 25–29, FCCC/CP/2010/7/Add.
2 (ibid.:para. 28(a)).
3 (ibid.:para. 28(b)).
4 (ibid.:para. 28(c)).

ference of the Parties (COP) for consideration at COP18,[5] as well as to strengthen international cooperation and expertise in order to understand and reduce loss and damage associated with the adverse effects of climate change, including impacts related to extreme weather events and slow-onset events.[6]

The Munich Climate Insurance Initiative (MCII)[7] has written the current contribution in response to the invitation to engage stakeholders with relevant specialised expertise in order to share their views on exploring approaches to address loss and damage.[8] In particular, the MCII's submission is a response to the invitation to explore a "[p]ossible development of a climate risk insurance facility to address impacts associated with severe weather events."[9] The submission further addresses "[o]ptions for risk management and reduction; risk sharing and transfer mechanisms such as insurance, including options for microinsurance; and resilience building, including through economic diversification."[10]

This contribution addresses issues related to managing loss and damage associated with extreme weather events. It explores the potential roles of a range of insurance-related approaches which transfer risk in the context of loss and damage, including social safety nets, solidarity and catastrophe funds, insurance pools, microinsurance, catastrophe bonds, and insurance linked to sectoral or community risk-management programmes. Although beyond the scope of this contribution, it is clear that a wider spectrum of approaches needs to be employed across the full scope of loss and damage,

5 (ibid.:para. 29).
6 (ibid.:para. 25).
7 The Munich Climate Insurance Initiative (MCII) was launched in April 2005 in response to the growing realisation that insurance-related solutions can play a role in adaptation to climate change, as advocated in the United Nations Framework Convention on Climate Change and the Kyoto Protocol associated with it. The MCII brings together insurers, experts on climate change and adaptation, non-governmental organisations, and policy researchers who are intent on finding solutions to the risks posed by climate change. The MCII provides a forum and gathering point for insurance-related expertise on climate change impact issues. The MCII is hosted at the United Nations University Institute for Environment and Human Security (UNU-EHS) in Bonn, Germany; www.climate-insurance.org, info@climate-insurance.org.
8 Decision 1/CP.16, para. 28(d).
9 Report of the Conference of the Parties on its Sixteenth Session, held in Cancun from 29 November to 10 December 2010, para.'s 25–29, FCCC/CP/2010/7/Add.
10 (ibid.:para. 28(b)).

particularly for slow incremental changes that also cause significant long-term loss and damage.

**Box 2**

**The MCII Submission in the Context of UNFCCC Discussions on Loss and Damage Related to Insurance**

- This contribution responds to the invitation[11] to give a submission to COP18 on the possible elements to be included in the recommendations on loss and damage, under the SBI Work Programme on Loss and Damage. This submission addresses some of the questions related to the use of insurance in the context of loss and damage, as follows:[12]
- The cost-effectiveness of various approaches, and at what level various tools are employed (local, national, regional and global)
- The resources required for the successful implementation of various tools, including budget, technical capacity for implementation, data and infrastructure
- Lessons learnt from existing efforts within both the public and private sectors, considering elements of design, limitations, challenges and best practices
- Links and synergies between risk reduction and other instruments such as risk transfer, and how comprehensive risk-management portfolios or tool kits can be designed, and
- Tailoring risk-management approaches to national contexts, and ways to evaluate which tools might be most appropriate for the particular risks and circumstances of a country.

This submission provides further insights into design principles that could guide a range of approaches, including an international mechanism.[13]

## *I. The Burden of Loss and Damage Today*

Since 1980, a general upward trend has been recorded as regards frequency of weather-related loss events. This trend is detectable in both rich and poor

11 See Decision 7/CP.17, para.'s 1–9; available at http://unfccc.int/files/meetings/durban_nov_2011/decisions/application/pdf/cop17_loss_damage.pdf, last accessed 14 May 2013.

12 (ibid.:para. 2 and Annex 2).

13 (ibid.:para. 5).

countries. The average annual weather-related disaster losses in the last five years (2007–2011) in countries with 'low' and 'lower-middle' economies have reached US$1.3 billion and US$6.8 billion, respectively. Data from 1980 onwards reveal that over 80% of people killed due to these weather-related disasters lived in developing countries.

In Figures 1 and 2, the annual numbers of weather-related loss events and their relative changes are shown for countries, broken down into the four income groups defined by the World Bank (starting point in 1980 = 100%).

***Figure 1: Annual Numbers of Weather-related Loss Events Globally in Countries with Different Economies (1980–2011)***

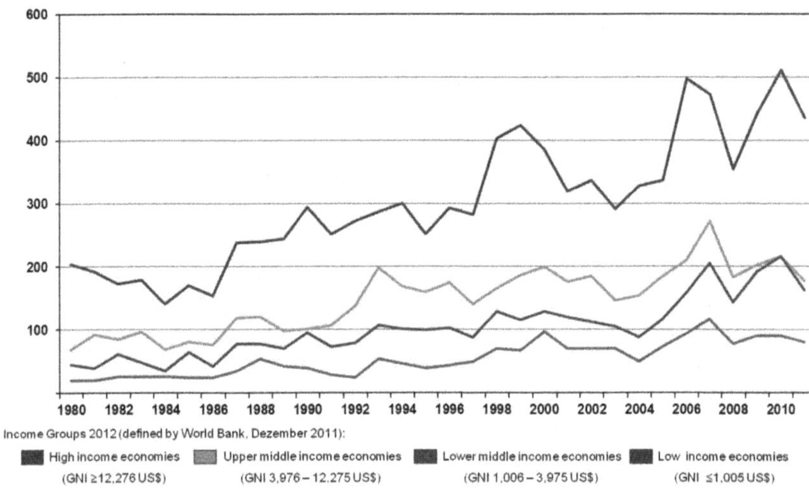


GNI = gross national income

Source: Munich Re, Geo Risks Research, NatCatSERVICE, 2012

Figure 2 shows that the countries with the lowest-income economies show not only the lowest number of events, but also the largest increase from 1980 to 2010. The relative number of loss events has increased by a factor of six in those countries with the lowest-income economies while, in the richest countries, loss events has increased by a factor of three, i.e. half as much. To what extent this difference is due to increasing wealth in developing nations, or to more frequent extreme weather events, is an open question. In terms of managing future risks, we recommend that the possibility of changing weather patterns impacting developing countries severely in decades to come should be taken seriously.

***Figure 2: Relative Trends of Annual Numbers of Weather-related Loss Events Globally in Countries with Different Economies (1980–2011)***

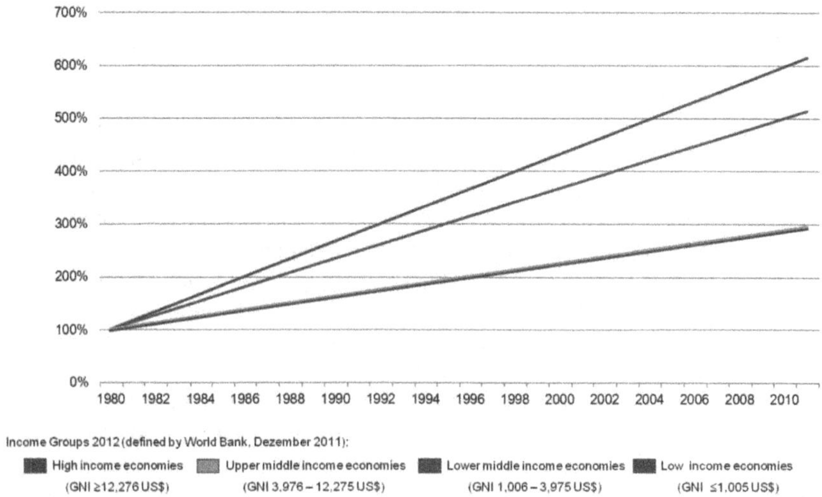


GNI = gross national income

Source: Munich Re, Geo Risks Research, NatCatSERVICE, 2012

## *II. Loss and Damage Tomorrow: Avoiding the Worst-case Scenario*

Managing loss and damage involves avoiding the potential for loss and damage in the future through appropriate mitigation and adaptation. It also involves preparing for and addressing actual loss and damage when it occurs, today and in the future.

Choices about mitigation and adaptation will be the main factor determining the degree of climate change and, thus, will have an influence on the magnitude of loss and damage, particularly from around 2030 onwards, when measures will have to be taken to adapt to the unavoidable changes that will have taken place, since global warming is a given until 2030. Decisions that affect the level, scale and efficacy of adaptation will affect the ability of societies to adjust to manifestations of changes in climatic variability, e.g. shifts in seasonality of rainfall, heatwaves, and magnitude and frequency of extreme weather events. The preeminent approach to loss and damage in the medium and longer term – in respect of avoiding future loss and damage, and minimising impacts in the short and medium term – lies in our choices about mitigation and adaptation.

**Box 3**

**What Does a 4°C World Mean in the Context of Loss and Damage?**
At COP16 in Cancun in December 2011, states parties agreed "to hold the increase in global average temperature below 2°C above pre-industrial levels". In 2011, a United Nations Environment Programme (UNEP) report anticipated a gap in 2020 between expected emissions and the global emissions consistent with the 2°C target, even if pledges were to be implemented fully.[14] One year after COP16, a follow-up report concluded that even with the full implementation of the current Cancun pledges, "the planet is heading to a temperature rise of at least 3.5°C, but that could be even more if the 2020 pledges are not met".[15]
Even this might be an optimistic scenario, however. According to the global carbon budget in 2010, growth rates of global emissions are not decreasing but increasing. In a worst-case scenario, where no action is taken to dampen the rise in greenhouse gas (GHG) emissions, "temperatures would most likely rise by more than 5°C by the end of the century".[16]
This has at least two consequences for all climate insurance concepts:

- The question of insurability has to be discussed for each of these different risk levels. For a 5° world, the risk of regional or continental scale might become unmanageable or at least be very different to manage in different parts of the world.
- Moral hazard has a second face in the climate-related insurance debate. The traditional understanding is that a badly designed insurance scheme can give an incentive for maladaptation, along the lines of, "I'm insured; I don't have to prepare for a possible disaster." Now, a second wrong incentive signal by insurance also has to be taken into account. If polluters don't contribute to the premium, the insurance scheme could send the signal, "I don't have to reduce emissions; others pay for the damage."

14 UNEP (2011).
15 See http://climateactiontracker.org/countries.html, last accessed 20 May 2013.
16 Pope (2008).

The consequences for the design and context of climate insurance instruments are as follows:

- Risk reduction, GHG reduction, disaster preparedness and loss prevention – all of which can be incentivised with insurance and cannot stand alone as solutions to the climate change challenge.
- In the interest of equity, countries with large per capita emissions of GHGs could contribute to insurance premiums. To avoid the disincentives this might create for loss prevention (by lowering the price of the risk), financial support could target the administrative and capital costs ('load') of the premium.

An implicit decision not to take ambitious mitigation action on a global scale and/or decisions not to invest in and actively drive adaptation could lead to loss and damage which exceeds the ability of human society to manage such loss or damage – at all scales.[17]

## C. *What Role can Insurance Play in the Context of Loss and Damage?*

This section outlines the key functions that insurance can play at the individual, community, national, regional and international level in the context of loss and damage. Section D revisits this discussion by asking what the UNFCCC can do to harness these functions, possibly in the form of a climate risk insurance facility operationalised through regional risk-management platforms that address climate-change-related loss and damage.

It should, however, be emphasised that insurance is not a universal remedy for all types of loss and damage resulting from climate change. As Figure 3 shows, insurance options can support adaptation and risk resilience for extreme weather events, but such options are not appropriate for many, usually slower-onset, climate-induced impacts.

Figure 3 also illustrates that insurance is not appropriate or even generally feasible for slowly developing and foreseeable events, or for processes that happen with high certainty under different climate change scenarios. The losses from long-term, foreseeable risks, such as sea-level rise, desertification and the loss of glaciers and other cryospheric water sources, are estimated to be substantial in the future.[18] Even for weather-related events, in-

17 See e.g. Stern (2007).
18 IPCC (2012:9); also see Parry et al. (2007:23–78).

surance would be an ill-advised solution for disastrous events that occur with very high frequency, such as recurrent flooding. Resilience-building and prevention of loss and damage in such instances may be cost-effective ways to address these risks.

Nonetheless, insurance is a feasible adaptation measure to address extreme weather events, including insurance for households (e.g. microinsurance), farms (e.g. index-based crop insurance) and governments (e.g. sovereign insurance). As we discuss in this contribution, insurance arrangements at these scales might be usefully supported by regional and global risk-management facilities.

### *I. Insurance as Adaptation*

By spreading losses among people and across time, insurance reduces the catastrophic impact of disasters and enables a timely recovery. Insurance is an adaptation measure when it reduces the *burden* of loss and damage, if not the average loss.[19]

In addition to providing timely capital after a disaster, as illustrated in Figure 3, insurance can and should be linked to risk-reducing, preventive activities.[20] Prudently employing a combination of insurance measures with risk reduction – including early warning, education, infrastructure strengthening, and land-use regulations – can greatly reduce the immediate losses and long-term development setbacks from disasters.[21] In addition, by creating a secure investment environment, insurance instruments can enable productive risk-taking on the part of individuals and governments, and in this way mitigate disaster-induced poverty traps.

Insurance, however, is not affordable to many in the most vulnerable countries; nor is it always advisable.[22] In Box 4 we discuss the principles that guide the MCII proposals for assisting vulnerable communities and governments to pool and reduce their losses from extreme weather events.

19 Linnerooth-Bayer et al. (2010a).
20 Warner et al. (2009).
21 ClimateWise (2010); Warner et al. (2010).
22 Linnerooth-Bayer et al. (2010b).

**Box 4**

**MCII Principles for Weather-Related Insurance Targeted at the Most Vulnerable**

Insurance solutions, as proposed by the MCII, should serve the interests of the most vulnerable people, communities and countries. The following principles suggest how insurance can be guided to fulfil this mission:

- ***Intelligent mix:*** Prevention and insurance should be closely linked with an ex-ante climate risk-management strategy that places priority on preventing human and economic losses. Action can be guided by a risk-layering approach. Cost-effective risk reduction is the first priority for limiting loss and damage. The costs of preventing low-impact, frequent events are typically much lower than the losses that would occur without investments in prevention measures. Alternatively, prevention measures for high-impact, low-frequency events can be far costlier with respect to the losses prevented. For this high layer of risk, insurance and other risk-transfer mechanisms may be more appropriate.
- ***Economic efficiency and risk-based premiums:*** By pricing risk, insurance can provide an important price signal to incentivise risk-reducing behaviour. For example, high insurance premiums will discourage people from locating in high-risk areas. Care should be taken, therefore, not to significantly distort insurance prices or market competition while addressing affordability and accessibility needs.
- ***Solidarity and responsibility:*** While risk-based pricing promotes loss reduction, an equally important principle relates to solidarity and the allocation of responsibility for climate change impacts. The loss burden can be far more severe in vulnerable developing countries and, within these countries, among poor households and communities. Since these communities have contributed little to climate change, it is incumbent on countries with high per-capita emissions of GHGs to take a share of responsibility. Pilot projects are demonstrating that market-based insurance can be a viable option for providing security to the poor, but generally not without donor support. Combined with other forms of social protection, premium support for the poorest will be an important feature of any insurance approach for vulnerable people and countries. This can take many forms, including direct financial support that minimally distorts incentives, capital support for local insurers (thus lowering premiums), technical assistance, and education programmes.
- ***Subsidiarity principle*:** Decisions should be made as close as possible to their point of application and to where the need is manifest. Transparency and accountability are important criteria for the creation of insurance programmes. International finance may best be allocated on a strategic basis and not involve international micromanagement at the project level.

***Figure 3: Tree of Options for Managing Climate-change-related Loss and Damage***

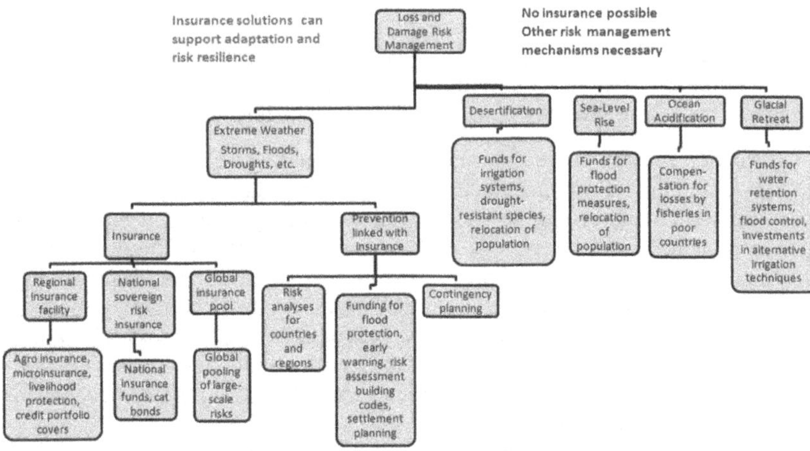


Source: Warner et al. (2012)

## *II. Assessing Loss and Damage Potential*

Assessment of loss and damage is a prerequisite for identifying needs and policy priorities and is a core function of insurance approaches. Risk assessment frequently serves to bring attention to the hazard potential, the exposure, and the vulnerability, and in this way it can raise awareness and expose new options for managing the risks involved. Publicly collected and open source data and risk assessments, as well as open source hazard modelling, can contribute meaningfully to national and regional risk-management and investment decisions. Insurance risk assessment can also facilitate regional and international data analysis, such as establishing data standards, comparability, methods, and data repositories.

## *III. Incentivising Loss Reduction and Resilience-building Activities*

Countries can define nationally appropriate risk-reduction priorities and identify and make plans for reducing weather-related risks. The principles of climate-resilient development – including principles from the Hyogo Framework[23] – can guide these actions. Such activities include –

- mapping risks and avoiding settlement in high-risk zones
- building hazard-resistant infrastructure and houses
- protecting and developing hazard buffers (forests, reefs, mangroves, etc.)
- improving early warning and response systems
- building institutions and developing policies and plans, and
- developing a culture of prevention and resilience.

Many of these measures will be cost-effective for low-impact events, but not for very extreme disasters. This suggests a layered approach to risk management, as discussed in Box 4. Applying loss-avoiding measures can reduce insurance premiums in many contexts, e.g. building hazard-resilient structures. In this way, insurance sends a signal to households, businesses and governments to reduce risks. Besides reduced premiums to reward risk reduction, additional design elements can be incorporated into insurance contracts. Ongoing participation/renewal of insurance coverage with public or international support could be dependent on evidence that participating vulnerable countries are making tangible progress in implementing their loss-reduction plans.

### *IV. Reducing Financial Repercussions of Volatility and Create more Certainty in Decision-making*

The volatility in economies and social systems caused by weather extremes is a challenge for social and economic development. Insurance can help create a space of certainty within which investments and planning can be undertaken. This certainty, in turn, can help create an environment that is more conducive to climate-resilient investments in sectors like tourism and agriculture, which are typically heavily exposed to climatic stressors, as well as in job creation and market development. Moreover, insurance can provide the safety net essential for making productive yet high-risk investments. As an example, a microinsurance scheme in Malawi enabled farmers to receive loans for purchasing hybrid seeds that increased their productivity fivefold.[24]

23 UNISDR (2005).
24 Suarez et al. (2008).

Apart from illustrating the costs of insurance, Figure 4[25] indicates its main benefits as well as the complementary nature of risk transfer with risk-reduction and risk-retention approaches. Insurers operating in developing countries have high start-up and transaction expenses, which can greatly limit affordability and constrain insurance penetration. Moreover, because disasters can affect whole communities or regions (covariant risks), insurers need to be prepared to meet large claims all at once. Their costs as regards the requisite back-up capital, diversification or reinsurance[26] to cover covariant claims can add greatly to business expenses and raise the premium far above the client's expected losses. Without government or donor support, private insurance is not easily affordable by households or small- and medium-scale enterprises in highly exposed and vulnerable countries, where the opportunity costs of private risk-financing instruments can be prohibitively high in terms of meeting other human needs.

## *V. Determining Whether Risk Transfer Can Help Ease Climatic Stressors and Related Poverty*[27]

Risk is ever-present in the lives of the poor. When a crisis occurs, the poor often resort to a variety of coping strategies such as reducing food consumption, selling assets, asking family or friends for help, changing livelihoods, moving away, taking children out of school, and/or borrowing from moneylenders or microfinance institutions. Selling productive assets or borrowing from moneylenders who charge high interest rates can jeopardise the economic basis of a household. Few of these households have access to formal insurance services. The result is that their trajectory out of poverty follows a zigzag route: advances reflect times of asset-building and income growth; declines are the result of shocks and economic stresses that often push expenditure beyond current income (Figure 4). The role of microinsurance, like any effective risk-management instrument, is to temper these downturns, which are major impediments to escaping poverty.

25 ECA (2009).
26 Insurance for insurers.
27 See also Churchill (2006).

***Figure 4: Impacts of Shocks on Household Income and Assets***

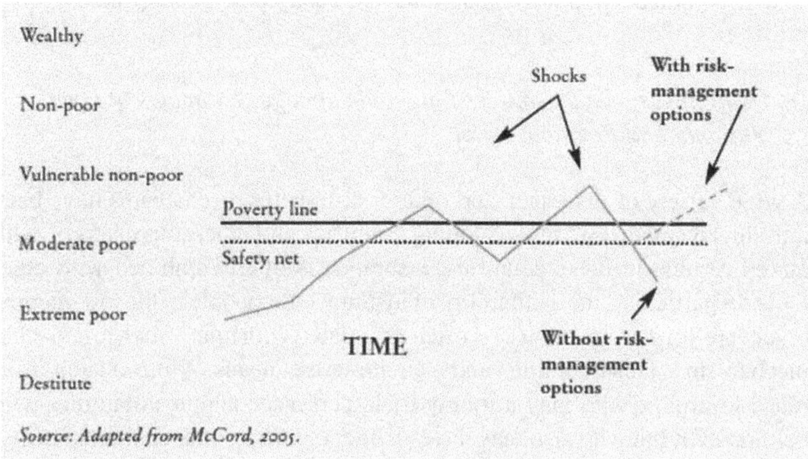


Source: Churchill (2006)

## *VI. Providing Timely Finance to Cover Loss and Damage*

As previously mentioned, there are numerous roles that insurance can play at the individual, community, national, regional and international level in the context of loss and damage, i.e. –

- providing security against the wholesale loss of assets, livelihoods and even lives in the post-disaster period
- ensuring reliable and dignified post-disaster relief
- setting powerful incentives for prevention
- providing certainty for weather-affected public and private investments, and
- spurring economic development and easing disaster-related poverty.

A major advantage of insurance over post-disaster financing options, including aid, loans and family assistance, is its timeliness and reliability. In comparison with (usually) ad hoc disaster assistance, insured clients have a 'right' to post-disaster compensation. Index-based contracts, which require no inspections for claim settlements, can, in principle, provide payouts immediately following the triggering event. Timely payouts, in turn, enable households to purchase food and other necessities without resorting to selling household assets, which could trap them in poverty. Timely payouts also help governments avoid fiscal deficits and costly post-disaster loans.

### D. *Using Insurance to Address Loss and Damage: Examples at Local, National and Regional Level*

A wide variety of insurance and other risk-transfer mechanisms have been introduced since 2002 in developing countries and emerging markets, with mixed results. In these countries, insurance is often combined with other tools. In particular, the availability of insurance for people in the low-income bracket (e.g. microinsurance) is often associated with microfinance and other mechanisms. This coupling can be an attractive means of introducing insurance to groups who may not only be underserved and/or unfamiliar with risk transfer, but who also may have an understanding and need for security. Combined products can reduce the costs of insurance to consumers, and enhance access to financial resources so as to minimise effective losses. Organised groups in particular, such as trusts, self-help groups and mutual, understand risk for their community and, therefore, develop an awareness of security and safety. Insurance can be linked to effective disaster risk management (DRM), as is shown from the example in Box 5.[28]

**Box 5**

**Early Warning Community Disaster Teams and Risk Transfer in Sofala, Mozambique**

A people-centred early warning project in central Mozambique is based on an impressively simple structure. A number of villagers have been nominated for the job of measuring daily precipitation levels at strategic points in the Búzi and Save River basins. Water levels along the rivers are also monitored using straightforward gauges. If there is particularly heavy rainfall, or the water level becomes critical, this information is passed on by radio. Should reports reaching the central coordination point indicate widespread heavy rainfall, the alarm is raised. Local disaster-prevention teams have been formed in a number of villages along these rivers. The system includes younger citizens and women in order to reinforce the part they play in the village community and in society.

28 The Munich Re Foundation is developing this approach, with partners, for Mozambique.

In the Mozambique case, early warning and insurance can reduce risk in a low-cost way. Money that has to be spent on post-disaster recovery by the Mozambican government and donors after an extreme weather event is split between two Funds:

- *The Standard Recovery Fund:* This Fund is used in the usual manner to support recovery, i.e. serving affected communities and people to repair damage, and
- *The Fast-Track Recovery Fund:* This Fund is paid out quickly and serves much faster recovery in case a disaster strikes. Communities receive funds much more speedily, and loss assessment can be managed more easily because risk awareness and management skills are in place (see also the bullet points below).

There are preconditions to these funds, however. Communities can only make use of the Fast-Track Recovery Fund if they take part in a tailor-made DRM programme, e.g. an awareness-raising programme at community level (capacity-building) and/or adopt a DRM strategy (e.g. appropriate land-use planning and evacuation plans).

If one links this approach to private-sector insurance, leveraging can be very effective. Through insurance mechanisms, countries can get the following:

- Professional risk assessment by private-sector risk specialists
- Tailor-made products and effective administration (existing professionalism)
- Sustainable solutions (since insurers will look for economic sustainability), and
- A real public–private partnership.

### *I. General Remarks: Innovations and Partnerships in Using Insurance*

Innovations in using insurance together with other tools to address loss and damage should be tailored to the level where needs are manifest, i.e. there should be a mix of private-sector, public-sector, and public–private partnership (PPP) solutions. The public-sector and PPP solutions may differ significantly from standard private-sector insurance solutions. There is scope for much innovation in providing for the needs of affected communities, countries and regions, as the examples below illustrate:

- ***Private-sector solutions for well-off households and governments:*** In some cases, countries may choose to share a layer of risk with the private insurance market for assets such as public infrastructure (sovereign insurance). Frequently, the private-sector reinsurance markets are involved in covering some portion of the largest risks a country or sector may face from extreme weather events. Private-sector solutions can be 'traditional' indemnity products, for which insurance payouts are made proportionate to the loss, or 'parametric' (or index) products, which establish parameters or triggers for extreme events to determine insurance payout levels. In the latter case, no loss adjustment (which, as a rule, is very time-consuming) is needed, and payout levels are agreed to in advance for the particular trigger levels. However, parametric products bear significant basis risk, i.e. the potential mismatch between the defined trigger level such as wind speed or amount of precipitation and the actual occurrence of loss. However, the rapid money flows in parametric products make them very attractive to all stakeholders. About 40% of the weather-related damage in developed countries is covered by private-sector insurance, with strong differences occurring from country to country. This includes most of the loss and damage to homes and businesses as a result of severe wind, wildfire, winter storms and – in some countries – floods. Most of the loss and damage not covered by insurance in developed countries involves damage to public infrastructure and, again, in some countries, flood damage to public and private assets.

- ***Public-sector solutions to protect people in the low-income bracket and their policy priorities:*** Pure market solutions are not always desirable or appropriate. Some in the low-income bracket are not in a position to pay private market prices, may not have access to insurance markets for a variety of reasons, or may not demand the standard products on offer. When private-sector markets for insurance are not fully developed – which is the case in most developing countries – public-sector risk-transfer solutions sometimes appear.[29] Such solutions can have higher transaction costs than private-sector solutions because market infrastructure and expertise, a developed client base, and a degree of standardisation may not be in place.

29 Melecky & Raddatz (2011).

As the following examples show, public-sector solutions are often innovative:

- ***They are designed to overcome barriers and link to broader social goals:*** Public-sector risk-transfer schemes sometimes evince new ways of thinking in their design. These ways of thinking aim to overcome some of the barriers of private-sector insurance. Public-sector insurance is often designed to link public programmes to existing social protection schemes (e.g. Ethiopia, Honduras and Nicaragua), employing early warning and disaster-risk reduction tools in combination with insurance (e.g. the Caribbean, Mongolia, Tanzania and Vietnam).
- ***They provide services that complement risk transfer for the low-income sector:*** Publicly supported insurance approaches sometimes provide services that are not always available in private-sector product lines, such as helping people in the low-income bracket access credit, offering support to protect livelihoods and not only to cover assets, and employing agricultural extension officers to educate people about good risk-management practices for extreme weather events.
- ***They offer public support to enable participation by the low-income sector:*** Public-sector insurance programmes use public resources to develop approaches, support premium payments and make payouts. In some programmes, publicly funded insurance payouts occur in a form that is valuable to the target group. This may be by way of seeds and other agricultural products for farmers in the low-income bracket, rapid cash payouts to poor households immediately after an extreme event, or benefits to sectors like tourism or agriculture to help them recover quickly after an extreme event.

It should be noted that a weakness of publicly funded insurance schemes is that they can be destabilised through changes in government priorities, lack of sufficient funding, and insufficient support to sectors or community level clients.

Insurance-related measures can be driven by the public sector and employed to promote a spectrum of public priorities. Some examples include the following:

- ***Protect priority sectors and households from climatic stressors:*** Some public programmes protect jobs and livelihoods in activities like agriculture and tourism through, for example, loan protection, targeted support programmes and livelihood protection.

- ***Reliable provision of public services:*** In the Caribbean, a regional risk insurance pool improves the governments' ability to keep basic public services functioning in the aftermath of a major catastrophic event. The Caribbean Catastrophe Risk Insurance Facility (CCRIF)[30] is a sovereign insurance pool designed to make rapid payouts to member governments after hurricanes or earthquakes. Since 2005, the CCRIF has paid money to the governments of Dominica, Haiti, St Lucia, and the Turks and Caicos Islands.
- ***Early identification of threats and resource provision to address them:*** In Africa, a new regional risk insurance pool is being developed to help governments quickly identify emerging drought situations and accumulate resources to avoid famine. Africa Risk Capacity, the pan-African contingency planning and food security insurance pool, requires member governments to have drought-risk and food-security plans in place, and provides payouts to help them purchase and stockpile grain in a timely way to prevent famine.

PPPs can offer the market sustainability of private-sector approaches and the flexibility and innovation of public-sector approaches. *Subsidiarity* means that each partner will have clearly defined, distinct roles to play, and decisions need to be made not only where the need is manifest, but also as close to their point of application as possible. For instance, the public sector may undertake data collection and needs assessment, and may shape the regulatory framework for insurance-related approaches. The public sector may also work with private-sector actors to design tools that meet the targeted needs, and may, under appropriate circumstances, provide some financing to support programme costs, such as those which groups in the low-income bracket cannot afford to pay. The private sector, on the other hand, can help implement the approaches over time, ideally ensuring that such approaches are effective and affordable, and comply with consumer protection and technical standards, such as premiums being sufficient to cover the risk insured. Strong commitment over a longer period is needed when creating sustainable solutions.

30 CCRIF (2010).

**Box 6**

**Caribbean Adaptation and Insurance for People in the Low-income Bracket**

Studies of low-income groups in the Caribbean have shown a relatively high demand for weather risk insurance-related solutions.[31] A new multi-country approach is linking livelihood protection with other ex-ante tools to provide timely and unbureaucratic recovery aid following excessive wind and rainfall events. However, these approaches have thus far experienced difficulties in reaching out to a larger proportion of the vulnerable population due to a shortage of information on local weather risks, insufficient risk-management and risk-transfer experience on the part of the initiators, insurance illiteracy on the part of stakeholders and potential clients, and the lack of a clearly viable reinsurance concept.

The Climate Risk Adaptation and Insurance in the Caribbean Programme, developed by MCII, bundles an early warning system with risk-reduction information and insurance to protect the livelihoods of low-income groups in Grenada, Jamaica and St Lucia, which will be expanded after 2014. Germany's Federal Ministry for the Environment provides funding for the Programme. The approach features two insurance products: the first protects the livelihoods of people in the low-income bracket, i.e. a livelihood protection policy, while the second protects loan portfolios exposed to weather risks, i.e. loan portfolio cover. These products were developed collaboratively with the respective Ministries of Agriculture and Tourism, local stakeholder groups, the local private sector, and the Programme partners – MCII, Munich Re, MicroEnsure, and the Caribbean Catastrophe Risk Insurance Facility (CCRIF).

The approach facilitates access to new market segments. Its partners include a company that specialises in matching local needs with tailored risk-management products; a regional facility (CCRIF) that has access to governments, an understanding of the regulatory environment and the ability to serve as a regional risk aggregator; and a reinsurer with expertise in modelling, product structuring, and international practice and policy. The regional-level approach allows underserved, low-income groups to gain protection from weather risks. It also fosters the development of local enterprise.[32]

---

31 Lashley & Warner (2012).

32 For more information, see www.climate-insurance.org, last accessed 20 December 2012.

### *II. Local: Building Resilience with Local Insurance and Safety Nets – Helping People in the Low-income Bracket Absorb Shocks and Temper Downturns*

Evidence of local-level insurance approaches to manage extreme weather events suggests that safety nets can be enhanced when linked to or designed to have some insurance-like properties. The role of insurance-related approaches at the local level, like any effective risk-management instrument, helps people in the low-income bracket to better absorb shocks and to temper downturns, which are major impediments to escaping poverty. Many examples and pilot projects exist which demonstrate the combination of insurance mechanisms with livelihood protection, social safety nets, and prevention measures on the local level. A promising example is Horn of Africa Risk Transfer for Adaptation (HARITA) in Ethiopia (Box 7).

**Box 7**

**HARITA, Ethiopia**[33]

In Ethiopia, 85% of the population rely on smallholder, non-irrigation farming for their livelihood. The people are, therefore, highly vulnerable to drought-related risks. Initially targeting teff farmers in the village of Adi Ha, an index insurance product was developed which allows farmers to pay their premiums either in cash or in kind by contributing labour to projects that increase the community's resilience to climate change. Farmer participation is ensured by a management team of five village members. Financial literacy workshops are given. To overcome data limitations and to reduce basis risk, new techniques such as satellite data or simulation models are being explored. This clearly demonstrates how insurance, besides addressing monetary issues, improves research and minimises risks. Horn of Africa Risk Transfer for Adaptation (HARITA) is embedded in an important government initiative, namely the Productive Safety Net Programme (PSNP),[34] which integrates insurance with both risk reduction and credit. It allows very vulnerable farmers, even the poorest of the poor, to pay their premiums through risk-reducing labour, such as helping to plant, compost or plant for protection. Thus, farmers benefit even when there is

33 For more information, see Oxfam America, available at http://www.oxfamamerica.org/publications/harita-quarterly-report-jan-mar-2011, last accessed 20 December 2012.

34 The PSNP is the Ethiopian Government's conditional cash-transfer programme that serves around 8 million chronically food-insecure households.

no payout because these risk-reduction activities improve yields and help minimise vulnerability to drought.[35]
Resilience-building activities for smallholders participating in HARITA include –

- learning to make and use compost, which is critical for rebuilding soil nutrients and improving soil moisture retention
- constructing small-scale water-harvesting structures on farmland
- planting nitrogen-fixing trees and grasses to promote soil regeneration and water conservation, and
- learning how to clean teff seeds before sowing them in order to boost productivity.

Through HARITA, farmers enrolled in PSNP have the option to work extra days beyond those required for their normal government payments, but instead of earning cash or food for this additional labour, they earn an insurance certificate which protects them against deficit rainfall.
The HARITA project started in 2008 and was developed by institutions such as Oxfam America, Swiss Re, the International Research Institute for Climate and Society (IRI), and the Relief Society of Tigray (REST). The risk carriers are Nyala Insurance in Ethiopia and the global reinsurer, Swiss Re.
In 2011, a payout was triggered and 1,810 farmers received US$17,392. Although this amount may sound low on average, it helped the affected poor to a large degree.

Several gaps need to be overcome in order to improve the links between and among programmes aimed at improving the resilience of low-income groups at the local level by way of risk transfer. Two such gaps are mentioned in the following paragraphs, and some additional gaps will be discussed in Section F.

35 The insurance-for-work model also allows insurance and credit to stand as independent components. In most index insurance pilots, farmers have been required to take insurance and credit as a package. Under HARITA, however, farmers may choose whether or not to bundle the two. The independence of credit and risk transfer means that farmers do not lose access to insurance once they have repaid their loans, and farmers who do not want a loan can still obtain insurance.

### *1. Basic Financial Infrastructure and Regulatory Environment*

Many insurance schemes at the local level are started without the benefit of basic foundational requirements. This implies that pilot, local-level approaches often face almost insurmountable obstacles. A financial infrastructure is essential for well-functioning risk-transfer systems, especially for low-income communities. Clients have to know – ideally in advance – what risks they wish to 'insure away', what the cost of such risks is, and how they will collect their payments. Basic financial infrastructure such as savings accounts, affordable and accessible credit, and other features needed to manage financial transactions is lacking when it comes to managing shocks and building resilience; this implies that insurance providers have to build not only new relationships with clients, but also a new technical infrastructure for premium payments. In addition, providers of risk-transfer solutions need to have a relationship with the appropriate regulatory authority to ensure consumers are protected and that adequate financial infrastructure is in place.

### *2. Education about Weather-related Extremes and Risk-transfer Functions*

Insurance solutions for low-income communities are often driven by microfinance organisations, community groups, cooperatives, trusts, associations, self-help groups and other grass-roots organisations. Insurance knowledge is not always available in such organisations. Even if a microfinance organisation knows how to manage large numbers of microfinance clients successfully, it may not have the necessary knowledge to assess risks and adequately price them. Support from technical assistance providers, or cooperation between an insurance organisation and, for example, a microfinance organisation, can help to overcome this knowledge gap. Understanding the concept of *insurance* is crucial – how it works and what it can and cannot do for the provider and the client. Significant investment in customer education is necessary, therefore, to reduce insurance illiteracy for providers, consumers, government officials and donors. This is another area where rules and regulations are needed: providers of risk transfer are required to have a sound understanding of the tools and the underlying technical issues, and should know how to educate and protect consumers at the local level.

### *III. National: Combining Risk Transfer and Measures to Protect National Development Priorities*

Retaining and transferring the appropriate risk layers can contribute to achieving climate-resilient development. For example, in a World Bank comparative study of countries with different insurance market penetration, the post-catastrophe patterns of economic growth were evaluated.[36] The results, summarised in Figure 5, show the mean and possible ranges of a weather-related, catastrophe-triggered trend deviation on gross domestic product (GDP) development. The solid lines mark the mean developments, while the dotted (for countries with high insurance market penetration) and dashed (for countries with low insurance market penetration) lines mark the range.

***Figure 5: Comparison of GDP after a Weather-related Loss Event in Countries with High and Low Insurance Market Penetration***

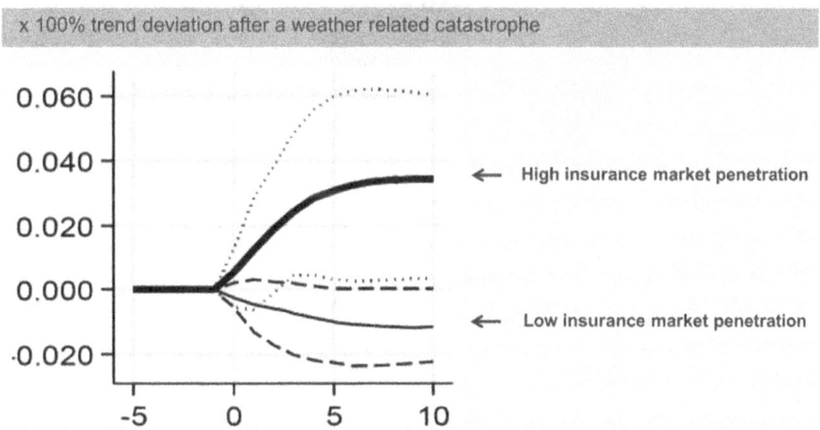


Source: Melecky & Radatz (2011)

The study shows that, after an extreme weather-related event, countries with high insurance market penetration reveal a positive GDP trend deviation, with sustainable additional growth generated. In contrast, countries with low insurance market penetration suffer from a negative GDP deviation, which, if not compensated for by other growth factors, can lead to long-term re-

36 Melecky & Radatz (2011).

ductions in GDP, which further inhibits development. If several such extreme weather-related events occur in short succession within a few years of each other, they will drive poor countries even further into the poverty trap. Studies such as this illustrate the potential which insurance-related approaches – public, private or PPP – have to increase the resilience of countries in respect of extreme weather-related events. Most developed countries already benefit from the shock-absorbing function of public and private insurance measures as well as from PPP risk-transfer arrangements. The map in Figure 6 shows the distribution of insurance penetration worldwide.

***Figure 6: Insurance Penetration Worldwide Since 2012***

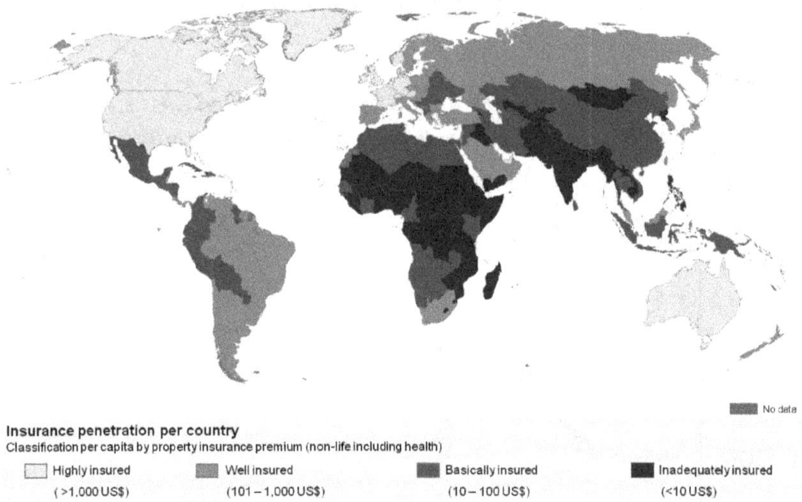


Source: Munich Re (2012)

**Box 8**

**An example from the private sector**
Insurance companies anticipate, and pre-fund, loss events with accumulated capital and the purchase of reinsurance. As a result, the use of insurance supports an earlier and fuller recovery for society from a loss and damage event. Damage claims are paid promptly, so homeowners and businesses can quickly return to a state similar to that which existed before the loss event. Moreover, in developed countries, consumers have high confidence in the role of insurance, which is bolstered by regulation and by experience with previous loss and damage events. Insurance-related

approaches can help to make economic activity more resilient to climate-related loss and damage, such as in the agricultural and tourism sectors in many developing countries, by protecting the livelihoods of people in the low-income bracket and by providing coverage for business interruption from extreme weather events.

Reliable data is essential in order to give a price to risk, to come up with options to manage that risk (including insurance), and to adequately assess the potential loss and damage from extreme weather events. However, countries interested in exploring risk-transfer solutions frequently have to deal with inhomogeneous, inadequate or inappropriate data. Historical data are often not available for longer time periods, and are only occasionally in digital format. Many countries struggle to establish sufficient networks of weather stations, making the assessment of weather-related risks difficult. Data-gathering and quality assurance of the data often requires time and resources to improve such information, e.g. through interviews, or by transferring historical data from written documents to electronic databases. Nonetheless, some databases do exist regarding loss and damage from weather-related extremes, such as those from reinsurers.[37] The compilation of meaningful and useful data on loss and damage, especially for developing countries, remains a premier obstacle to developing more comprehensive approaches – not only insurance – to address loss and damage. Where insurance exists or is being built up, data-gathering and processing exist too, and the interest to collect better data is systemic. Thus, insurance can address many of the problems described above.

### *IV. Regional and International: Combining Risk Transfer with Regional Risk Capacity and Forecasting*

A trend is emerging whereby countries in a region create insurance pools to share and transfer loss and damage from extreme weather events. An underlying principle of insurance is the diversification of risk, i.e. reducing the likelihood that an insurance scheme will be overwhelmed by the same types of stressors (a single event can cause simultaneous losses to many insured

37 Munich Re NatCatService, or the Swiss Re sigma, or the Emergency Events Database (EM-DAT) of the World Health Organization Collaborating Centre for Research on the Epidemiology of Disasters (CRED) (Munich Re 2012).

assets) or the same group of insured needing a payout all at the same time (such as a community, where most households are affected by the same stressor). A multi-country or multi-regional approach can prove viable where local and national pooling arrangements may not be feasible for statistically dependent (covariant) risks that cannot be sufficiently diversified. For this reason, primary insurers, individuals and governments, particularly in small countries, do and may need to rely on risk-sharing and transfer instruments that diversify their risks regionally and even globally.

### *1. Light governance structures for risk pools*

For regional- and international-level insurance approaches, examples such as the CCRIF show that light governance structures for risk pools are able to contribute to regional risk-management efforts and make rapid payouts in the case of extreme events. Such institutional models can be designed to have transparent governance structures, allow private-sector engagement, and serve as conduits for international adaptation funding. As with lower-level risks pooled at a national level and then transferred at a regional level, insurance pools at the regional level would need a fund of last resort to provide a reinsurance function for very rare catastrophic events. A fund of last resort – or global climate-risk insurance pool – would be important because this is a level at which large private-sector entities may not engage due to the capital requirements involved to cover the risks. At this level, most of the money paid in premiums for the highest level of risks relate to the costs of keeping capital. International support, such as in a global climate risk pool, could ensure the necessary cover for regions and countries following a catastrophic event.

**Box 9**

**Africa Risk Capacity: An Approach Linking Contingency Planning and Insurance for Food Security and Drought in Africa**

African countries regularly experience drought, which often turns to famine if timely assistance is not available. For many people, traditional ex-post humanitarian aid often comes too late to avoid loss of life and property. Today, luckily, organisations such as the World Food Programme support victims of drought. Often, the support comes late due to time-consuming processes that include support requests, verification, confirmation, claim assessments, and payout). With ex-ante mechanisms, e.g. money flows after no rain in April because there will be known effects on yield in September, people can be served even before the crisis materialises. Establishing a contingency fund or resources that can be made available automatically if an extreme drought, flood or cyclone occurs in a vulnerable area ensures a more timely and reliable response. As extreme weather events do not happen in the same year across the continent, pan-African solidarity was deemed financially effective when a disaster risk pool was created. Such a facility will provide participating member countries with readily available resources in the event of severe droughts, with additional hazards to be incorporated later.

The Africa Risk Capacity (ARC) is one of several tools that governments can use to eliminate delays in disaster response due to a lack of predictable funding, and to limit reallocation of government resources from planned development activities in times of crisis. In advance of joining the ARC, each participating country needs to create a contingency plan to identify how ARC funds will be used to assist those affected.

The ARC's capacity-building programme will not only enable governments to make informed decisions on their participation in the ARC's financial services, but will also, significantly, enable meaningful, risk-informed fiscal management of natural disaster risk for African governments, with enhanced national capacity to respond to these predictable catastrophes.

The ARC aims to provide parametric funding for approved contingency plans for events of a frequency of 1:5 or greater, up to an initial maximum of US$30 million per season.

The ARC supports national disaster risk managers in identifying realistic contingency plans maximising the value of early and reliable funding for events greater than roughly 1:5. At less frequent but more severe risks, roughly above 1:5, contingency funding makes sense for two reasons: firstly, investments are unlikely to create resilience for events less frequent than 1:5 in a reasonable time frame; secondly, the potential for pooling, as shown in ARC's dynamic financial analysis, reduces cost.

Source: www.africanriskcapacity.org, last accessed 12 December 2012

### 2. *Payouts*

There are many different ways to differentiate a payout from a (regional) climate insurance pool. It could be a proportional payout to all weather-related losses, or a payout of 100% of the losses of a percentile (e.g. 30%) of the most extreme losses. In the latter case, a regional analysis on the return periods of losses can be made, and the payout calibrated regionally.

After the 2010 earthquake calamity in Haiti, the CCRIF – designed to address hurricane and earthquake risk in the Caribbean – paid out almost US $8 million within two weeks of the disaster. Experts estimate, though, that the amount could have been as high as US$100 million, or a 40:1 ratio, had the government chosen that particular premium-to-payout ratio. In this instance, the insurance provided a rapid payout in a crisis situation when liquidity was greatly needed. This is a notable feature of the CCRIF, which was originally envisaged as a mechanism to assist governments by providing short-term liquidity during the 'funding gap' – the hiatus between the immediate flow of response goods and services after a major disaster and the launch of long-term rebuilding programmes.[38]

### E. *Considerations on the UNFCCC's Role in Insurance Approaches to Address Loss and Damage*

This section calls attention to gaps that can best be filled through regional and international action, supported by UNFCCC guidance. It outlines regional-level and international elements that may become part of a COP19 decision on arrangements to address loss and damage. These elements are required to address needs or gaps arising from loss and damage due to failure to achieve the UNFCCC objective, particularly those that cannot be adequately addressed at the national level.

**Box 10**

**Recommendation**

It is recommended that the international community consider the following:

- A risk-layering approach to addressing loss and damage, which can increase efficiency and value added by targeting support differently for infrequently occurring, high-consequence risks versus frequently occurring, low-consequence risks, and
- The establishment of a climate risk insurance facility operationalised as a network of international and regional risk-management and transfer platforms embedded in wider efforts to address loss and damage, and in coordination with adaptation and mitigation efforts. The rationale for

38 CCRIF (2010).

coordinated international and regional platforms is they can serve multiple functions, including pooling and transferring risk more cost-effectively than if such functions were carried out at community or national level.

Principles underlying the design of such an approach should include the following:

- ***Ex-ante approach, emphasising assessment, planning and decision support:*** The UNFCCC can play a role in helping support purposeful rather than ad hoc responses to negative impacts of climate change. The UNFCCC can also help to ensure threats are identified, and can bring this information to decision-making and planning to address loss and damage.
- ***Risk layering/subsidiarity:*** The UNFCCC has a special role to play in facilitating strategies to address loss and damage. Following the principle of subsidiarity, efforts to address the spectrum of loss and damage – ranging from extreme weather and other kinds of climatic variability to incremental but profound climate change – may best be designed and implemented on various levels. These include country and local levels, under the jurisdiction of nation states, or on a regional and international scale. Implementation of risk-transfer approaches should be embedded in wider programmes designed to reduce loss and damage and enhance the ability of societies to adjust to the negative impacts of climate change. Such approaches should address the needs and engage the participation of key stakeholders as close as possible to the level where the needs are manifest.
- ***Finance and other means of supporting implementation:*** The international community can play a role in helping to overcome some of the current obstacles. These obstacles include a lack of meaningful back-up mechanisms, i.e. reinsurance, the lack of technical and financial capacity and expertise, and the quality and availability of loss and exposure-related data. Playing this role would mean that countries could employ risk-transfer solutions from a broader tool set for promoting climate-resilient growth and adaptation, and for dampening the negative impacts of climate-change-related loss and damage.[39]

39 Cummins (2008).

### *I. Functions of a Climate Risk Insurance Facility, Coordinated Internationally and Operationalised Regionally*

The functions outlined below have a transboundary nature and will, therefore, be particularly useful if implemented at a regional or international level rather than in compartmentalised national contexts.

A climate risk insurance facility could have capacities that include, but are not limited to, the objectives and functions shown in Table 1, and explained thereafter.

***Table 1: Possible Roles of the UNFCCC in Facilitating Insurance to Address Loss and Damage***

| No. | Objective | Function |
|---|---|---|
| 1 | Provide loss and damage potential assessments that support decision-making and facilitate management of weather-related risks | Guide and enable assessments of loss and damage potential for extreme weather events |
| 2 | Provide timely finance to cover loss and damage to reduce the financial repercussions of volatility related to extreme weather events | Operationalise climate risk insurance, including finance mechanisms and other means of implementation |
| 3 | Incentivise loss reduction and embed risk transfer into wider resilience-building efforts | Ensure policy coherence and appropriate use of risk-transfer tools in a wider context of climate risk management |

#### *1. Objective 1: Provide Loss and Damage Potential Assessments that Support Decision-making and Facilitate Management of Weather-related Risks*

*Function 1: Guide and Enable Assessments of Loss and Damage Potential for Extreme Weather Events*

The UNFCCC process can help to fulfil this function, inter alia, in the following ways:

- ***Provide guidance on assessment methods and data-collection standards*** for risk transfer that could benefit wider efforts in the assessment of loss and damage. This could be done by supporting 'open source' projects,[40] where risk-assessment approaches are made available for a defined audience, e.g. political decision-makers and the insurance industry. On the other hand, guidelines and methods could also be spread by publication and presentation, i.e. knowledge transfer.

- ***Support development of standardised hazard maps***, e.g. maps providing information on river flood zones, extreme precipitation estimations and wind-speed zones. This could include support for establishing regional/international catastrophe loss indices.[41] Technical assistance may also involve pooling technical expertise as well as collaborative worldwide networks.
- ***Coordinate data repositories and encourage coherence across information frameworks***, such as adequate standards for data-gathering and open source assessment methods, including remote sensing, open source risk models, and other information needed to assess risk exposures, which are sensitive to vulnerable people and groups.
- ***Offer systematic capacity-building*** for tools that, in combination, can be appropriately used to manage and reduce loss and damage potential. This involves *technical assistance* to facilitate dialogue between countries on experiences regarding design and implementation of packages of different tools, foundational requirements, and outcomes of appropriate combinations of insurance measures with other tools, to address the impacts of extreme weather events.

National governments, with the engagement of relevant public- and private-sector actors, can help to fulfil this function, inter alia, by –

- obtaining reliable sources of information about managing, reducing and transferring risks
- investing in systematic and reliable risk-exposure data
- understanding the risks of greatest concern by identifying key risks and vulnerabilities, and estimating exposure
- putting a price on risks and adaptation options, and

---

40 Similar to the Global Earthquake Model; see http://www.globalquakemodel.org/ last accessed 20 May 2013.

41 Akin to the Property Claim Services in the United States or Perils AG in Europe; see http://www.iso.com/Products/Property-Claim-Services/Property-Claim-Services-PCS-info-on-losses-from-catastrophes.html and http://www.perils.org/, both last accessed 20 December 2012.

- helping to evaluate the relative merits of specific adaptation interventions for national implementation, e.g. by a cost–benefit analysis.

## 2. *Objective 2: Provide Timely Finance to Cover Loss and Damage to Reduce Repercussions of Volatility Related to Extreme Weather Events*

The regional risk-management and transfer platforms that form the climate risk insurance facility can have a distributive function, help regions to absorb and manage higher layers of financial loss and damage, and help to capitalise risk-management approaches at lower risk layers that are tailored to local and national contexts. The regional platforms would help to manage and limit financial losses which may be incurred from possible yet uncertain loss events.

*Function 2: Operationalise Climate Risk Insurance, including Finance Mechanisms and Other Means of Implementation*

The UNFCCC process can help to fulfil this function, inter alia, by –

- ***setting up an international risk-management and transfer platform (or a network of regional ones)*** that covers catastrophic layers of risk. This may include seed funds for national and regional risk-reduction and risk-transfer initiatives.
- ***supporting an evaluation of different roles of finance to support approaches under the UNFCCC***, particularly areas for facilitating, providing platforms, considerations of price support, and investments in elements necessary for the functioning of appropriate risk-transfer approaches.
- ***channelling commitment by the donor community to provide expertise, capacity- building and financial support*** to innovative mechanisms for addressing the financial aspects of loss and damage associated with extreme weather events. It is essential that innovative risk-transfer mechanisms are designed in a way that meets the needs and priorities of vulnerable people and those in the low-income bracket.
- ***Planning and implementing packages of tools to reduce risk and enhance resilience in regional cooperation:*** Such packages of tools should help create the context within which decisions can be taken with greater certainty.

National governments, with the engagement of relevant public- and private-sector actors, can help to fulfil Function 2, inter alia, by –

- acting on lessons learnt about regional public-private partnerships
- designing and implementing measures to avoid loss and damage, and transfer risk which cannot be avoided, and
- using risk reduction as a criterion for participation in insurance schemes.

### 3. *Objective 3: Incentivise Loss Reduction and Embed Risk Transfer into Wider Resilience-building Efforts*

*Function 3: Ensure Policy Coherence and Appropriate Use of Risk-transfer Tools in a Wider Context of Climate Risk Management*

The UNFCCC process can help fulfil this function, inter alia, by –

- providing guidance on purposeful, planned approaches to loss and damage
- providing guidance on technical measures and design elements of risk transfer to incentivise loss-reduction and resilience-building activities for beneficiaries of the international mechanism
- fostering a better understanding of the value addition and the scalability of a package of tools, of how they work together, and of the cost savings of jointly implementing approaches, including innovative risk-financing mechanisms
- facilitating regional and international dialogue to advance policy coherence and regulations on insurance-related measures at local and national level to address loss and damage. Such dialogue should improve conditions for regulators and decision-makers in developing countries to devise appropriate regional and national financial risk-management tools, including insurance. Policy coherence should enhance consumer protection, links to resilience-building and risk reduction, and links to adaptation and national development planning processes, and
- coordinating, where appropriate, with bodies on technical matters related to assessments, such as the International Association of Insurance Supervisors. Such a process could ensure the compiling, open access and standardisation of data.

National governments, with the engagement of relevant public- and private-sector actors, can help to fulfil this function, inter alia, by conducting risk-reduction activities and providing an enabling environment for risk management, insurance, governance, etc.

### II. Some Cost Figures

Estimating costs for a global coverage for developing countries is a challenging task because the (technical) premium costs are individual, and depend heavily on regional and international settings. Nevertheless, there are first estimates of capital costs and costs of maintaining regional risk-sharing facilities.[42]

A global extreme risk fund, possibly like the one proposed by the MCII,[43] could need US$10 billion in initial capitalisation and would be maintained at that level. Young[44] estimates the initial capitalisation needs for regionally organised risk-pooling solutions at US$5–10 billion over five years, and ongoing premium support costs of US$2–5 billion per year for multiple, regional, risk-sharing facilities covering extreme weather risk at both local and national levels. Additional funds would be required to provide technical support, alongside other adaptation initiatives, and for capitalisation of a global risk fund of last resort to cover the most extreme events (perhaps an additional US$10 billion). Investment return on the latter could cover technical support in the long term.[45]

### III. Accompanying Activities in the Emerging Institutional Set-up of Adaptation and Mitigation

The UNFCCC, through the Cancun Decisions, has already achieved major advances on the issue of adaptation. Several elements that are under way towards their operationalisation have to play synergetic roles for advancing a climate-insurance approach.

42 Young (2009).
43 MCII (2008).
44 Young (2009).
45 (ibid.).

### *1. National Adaptation Programmes of Action*

States parties agreed to operationalise the National Adaptation Programmes of Action (NAPAs) as mandated by the Cancun Adaptation Framework. This includes a medium- to long-term strategic approach for least-developed countries (LDCs) as regards how to manage adaptation at the national level. The developed modalities and guidelines should also be applied by other developing countries.

NAPAs will be accompanied by concrete investment activities. The Cancun Adaptation Framework already offers guidance on eligible adaptation activities. Countries should consider embracing a risk-layering approach, and should include elements of a climate-insurance approach in their concrete activities.

There is no immediate mention in the NAPAs concept regarding loss and damage. However, many approaches to be discussed under the loss and damage work programme, such as assessment of loss and damage and relevant decision-making tools, also have a high relevance for medium- to long-term adaptation planning. In elaborating the work programme on loss and damage, therefore, states parties should link the programme with the NAPAs concept and possibly include the concept in the review of the guidelines to be conducted by the LDCs Expert Group.

### *2. The Green Climate Fund*

At COP17 in Durban in 2011, states parties succeeded in operationalising the Green Climate Fund. The decision includes an annex on the governing instrument, which lays out the fundamental structures and procedures of the Fund. Part of this decision was to fund adaptation, which is likely to be interpreted as funding eligible activities under paragraph 14 of the Cancun Adaptation Framework. Up to now, however, loss and damage has not been considered an eligible activity for funding.

Nonetheless, possible loss-and-damage-related activities might well be eligible for funding. Such activities include –

- impact
- vulnerability and adaptation assessments
- climate-change-related disaster risk-reduction strategies
- risk assessment and management
- sharing and transfer mechanisms

- enhancing understanding, coordination and cooperation with regard to climate-change-induced displacement
- strengthening data, and
- improving climate-related research and systematic observation.

In the medium and long term, funding of risk-transfer mechanisms for developing countries to address loss and damage should generally also be financed and capitalised by, among other international sources,[46] the Green Climate Fund. The regional facilities can be a conduit for distribution of payments, other appropriate forms of support, etc.

### 3. *Adaptation Committee*

In Durban, states parties also operationalised the Adaptation Committee.[47] This Committee will serve as the major advisory body on adaptation under the UNFCCC; it will also extract lessons learnt, make recommendations to states parties, and provide general coherence. The Committee should, therefore, work not only on the general guidance on risk-transfer solutions as part of such adaptation, but also on the loss and damage portfolio.

## *F. Outlook*

The impacts of loss and damage associated with climate-related stressors including weather extremes and long-term climatological shifts can impair socio-economic development and reinforce cycles of poverty across the globe. Building the management capacity for dealing with today's extreme climate-related events will provide the basis for dealing with both current climate variability and long-term shifts in climate patterns. This comprehensive approach will help both to smooth development pathways, and cushion the expected negative impacts of loss and damage in the future.

In today's world, creating strategies to address loss and damage is challenging. Faced with financial crises, political strife, population growth, and

---

46 Some countries take the position that national funding should not compete with funding for regional purposes. Therefore, international funding sources are one option, but more discussion is needed to ensure that national and regional priorities are addressed.

47 Decision 2/CP.17.

a multitude of other hurdles, decision-makers may be tempted to postpone considering having to determine suitable approaches to dealing with loss and damage related to the impact of climate change. In spite of these challenges, international and national policy forums, as well as communities of policy, science and practice, have many tools to help them begin to address loss and damage. Jump-starting or tapping into activities by different communities and processes should be an essential next step for the UNFCCC process, as the discussions on loss and damage mature and become, in all probability, more institutionalised.

## *References*

CCRIF/Caribbean Catastrophe Risk Insurance Facility, 2010, *Annual Report 2009–2010*, George Town, CCRIF.

Churchill, Craig, 2006, *Protecting the Poor – A Microinsurance Compendium*, Vol. I, Munich/Geneva, International Labour Organization.

Climate Action Tracker, 2012, Emissions Gap Looks Set to Increase if Government Action Doesn't Step Up, available at http://climateactiontracker.org/news/126/Emissions-gap-looks-set-to-increase-if-government-action-doesnt-step-up.html, last accessed 10 December 2012.

ClimateWise, The Geneva Association, Munich Climate Insurance Initiative & United Nations Environment Programme Finance Initiative, 2010, *Global Insurance Industry Statement on Adapting to Climate Change in Developing Countries*, Cambridge/Geneva/Munich/Nairobi, ClimateWise, The Geneva Association, Munich Climate Insurance Initiative and United Nations Environment Programme Finance Initiative, available at http://www.unepfi.org/fileadmin/documents/insurance_climatechange_statement.pdf, last accessed 20 May 2013.

Cummins, David & Oliver Mahul, 2008, *Catastrophe Risk Financing in Developing Countries: Principles for Public Intervention*, Washington, DC, World Bank Publications.

ECA/Economics of Climate Adaptation Working Group, 2009, *Shaping Climate-resilient Development: A Framework for Decision-making, Report of the Economics of Climate Adaptation Working Group*, Grand Cayman, ECA [Climate Works Foundation, European Commission, Global Environmental Facility, McKinsey & Company, Rockefeller Foundation and Swiss Re].

IPCC/Intergovernmental Panel on Climate Change, 2007, *Climate Change 2007: Synthesis Report. Contribution of Working Groups I, II and III to the Fourth Assessment Report of the Intergovernmental Panel on Climate Change*, Core Writing Team: Pachauri, Rajendra K. & Andy Reisinger (Eds), Geneva, IPCC, available at http://www.ipcc.ch/publications_and_data/publications_ipcc_fourth_assessment_report_synthesis_report.htm, last accessed 13 January 2013.

IPCC/Intergovernmental Panel on Climate Change, 2012, *Managing the Risks of Extreme Events and Disasters to Advance Climate Change Adaptation. A Special Report of Working Groups I and II of the Intergovernmental Panel on Climate Change*, Field, Christopher B., Vicente Barros, Thomas F. Stocker, Dahe Qin, David J. Dokken, Kristie L. Ebi, Michael D. Mastrandrea, Katharine J. Mach, Gian-Kasper Plattner, Simon K. Allen, Melinda Tignor & Pauline M. Midgley (Eds), Cambridge, Cambridge University Press.

Lashley, Jonathan & Koko Warner, 2012, *Weather-related Insurance and Risk Management: A Demand Study in the Caribbean*, Bonn, Deutsche Gesellschaft für Internationale Zusammenarbeit.

Linnerooth-Bayer, Joanne, Christoph Bals & Reinhard Mechler, 2010, Insurance as Part of a Climate Adaptation Strategy, in: Hulme, Mike & Henry Neufeldt (Eds), *Making Climate Change Work for Us: European Perspectives on Adaptation and Mitigation Strategies*, Cambridge, Cambridge University Press.

Linnerooth-Bayer, Joanne, Koko Warner, Christoph Bals, Peter Höppe, Ian Burton, Thomas Loster & Armin Haas, 2010, Insurance Mechanisms to Help Developing Countries Respond to Climate Change, Special Issue on Climate Change and its Economic Impact on Insurance, *The Geneva Papers on Risk and Insurance – Issues and Practice*, Geneva, The Geneva Association.

MCII/Munich Climate Insurance Initiative, 2008, *Insurance Instruments for Adapting to Climate Risks: A Proposal for the Bali Action Plan*, Submission by the Munich Climate Insurance Initiative (MCII), 30 September 2008, Fourth session of the Ad Hoc Working Group on Long-term Cooperative Action under the Convention (AWG-LCA 3), Poznan, 1–13 December 2008.

Melecky, Martin & Claudio Raddatz, 2011, *How Do Governments Respond After Catastrophes? Natural-disaster Shocks and the Fiscal Stance*, Geneva, World Bank.

Munich Re, 2012, *Geo Risks Research*, Munich, NatCatSERVICE.

Parry, Martin, Osvaldo Canziani, Jean Palutikof, Paul van der Linden & Clair Hanson, 2007, Technical Summary, in: Parry, Martin, Osvaldo Canziani, Jean Palutikof, Paul van der Linden & Clair Hanson (Eds), *Climate Change 2007: Impacts, Adaptation and Vulnerability. Contribution of Working Group II to the Fourth Assessment Report of the Intergovernmental Panel on Climate Change*, Cambridge, Cambridge University Press, 23–78.

Pope, Vicky, 2008, Up to 6°C, *World Energy Outlook* (November), Paris, International Energy Agency.

Stern, Nicholas, 2006, *The Economics of Climate Change: The Stern Review*, Cambridge, Cambridge University Press.

Stern, Nicholas, 2007, *The Stern Review: The Economics of Climate Change*, Cambridge, Cambridge University Press.

Suarez, Pablo, Joanne Linnerooth-Bayer & Reinhard Mechler (Eds), 2008, *Feasibility of Risk Financing Schemes for Climate Adaptation: The Case of Malawi*, Report prepared for the World Bank Development Economics Research Group, Laxenburg, International Institute for Applied Systems Analysis.

UNEP/United Nations Environment Programme, 2010, *The Emission Gap Report: Are the Copenhagen Accord Pledges Sufficient to Limit Global Warming to 2°C or 1.5°C?: A Preliminary Assessment*, available at http://www.unep.org/publications/ebooks/emissionsgapreport/chapter1.asp?c=1.6, last accessed 3 January 2013.

UNEP/United Nations Environment Programme, 2011, *Bridging the Emissions Gap: A UNEP Synthesis Report*, available at http://www.unep.org/publications/ebooks/bridgingemissionsgap/, last accessed 3 January 2013.

UNISDR/United Nations International Strategy for Disaster Reduction, 2005, *Hyogo Framework for Action (HFA): Building the Resilience of Nations and Communities to Disasters*, Geneva, UNISDR.

Warner, Koko, Thomas Loster, Michael Zissener, Sönke Kreft, Joanne Linnerooth-Bayer, Christoph Bals, Peter Hoeppe, Eugene Gurenko, Ian Burton & Armin Haas, 2009, Vulnerable Countries and People: How Disaster Risk Reduction and Insurance Can Help Manage the Risks of Climate Change, UNU-EHS Policy Brief, Bonn, United Nations University Institute for Environment and Human Security.

Warner, Koko, Michael Zissener, Sönke Kreft, Peter Hoeppe, Christoph Bals, Joanne Linnerooth-Bayer, Armin Haas, Eugene Gurenko, Thomas Loster & Ian Burton, 2010, Solutions for Vulnerable Countries and People: Designing and Implementing DRR [Disaster Risk Relief] and Insurance for a COP16 Outcome on Adaptation, UNU-EHS Policy Brief, Bonn, United Nations University Institute for Environment and Human Security.

Young, Simon, 2009, *Cost Estimates for Multi-regional Risk-sharing Pools*, Castries, Caribbean Risk Managers Ltd.

# Climate Change

## International Law and Global Governance

**Climate Change: International Law and Global Governance**

**Volume I: Legal Responses and Global Responsibility**

Edited by Prof. Dr. Oliver C. Ruppel,
Prof. Dr. Christian Roschmann and
Dr. Katharina Ruppel-Schlichting

With a foreword by Prof. Dr. Hans-Gert Poettering

*2013, 1.001 pp., hc., € 168.00*
*ISBN 978-3-8329-7797-9*

*www.nomos-shop.de/19846*

**Volume I** assesses the most pressing impacts of climate change on various international law regimes. The main focus lies on

- international climate change law as a new international law discipline
- climate change and human rights
- climate change, international trade and investment law
- the law of the sea and sea level rise
- judicial review and international climate change litigation
- other subjects such as mitigation regulation, natural resource management and climate-engineering.

This publication was produced in cooperation with the

Konrad
Adenauer
Stiftung

Nomos

# Annex
# Contents Volume I
# Climate Change: International Law and Global Governance
# Volume I: Legal Responses and Global Responsibility


FOREWORD 9

PREFACE 11

ACKNOWLEDGEMENTS 13

THE EDITORS 15

THE CONTRIBUTORS 17

**PART I: INTERNATIONAL CLIMATE CHANGE LAW – A NEW LEGAL DISCIPLINE?**

1 Intersections of Law and Cooperative Global Climate Governance – Challenges in the Anthropocene 29
*Oliver C. Ruppel*

2 Whose Climate, which Ethics? On the Foundations of Climate Change Law 95
*Johan Hattingh*

3 Legal Strategies to Come to Grips with Climate Change 121
*Jaap Spier*

4 Climate Change Law: Objectives, Instruments and Structures of a New Area of Law 153
*Hans-Joachim Koch*

**PART II: CLIMATE CHANGE AND HUMAN RIGHTS**

5 Climate Change and Human Rights 203
*Christian Roschmann*

6 Economic, Social and Cultural Rights and Climate Change 243
*Ariranga G. Pillay*

7 Climate Change Adaptation and Human Rights: An Equitable View 261
*Margaux J. Hall & David C. Weiss*

8 Climate Change and Human Rights: What Follows for Corporate Human Rights Responsibility? 299
*Stefanie Ricarda Roos*

9 Climate Change and Gender Justice: International Policy and Legal Responses 323
*Patricia Kameri-Mbote*

10 Climate Change and Children's Rights: An International Law Perspective 349
*Katharina Ruppel-Schlichting, Sonia Human & Oliver C. Ruppel*

**PART III: CLIMATE CHANGE, TRADE, INVESTMENT AND REGIONAL INTEGRATION**

11 Climate, Trade and Investment Law in the Global Green Economy 381
*Markus W. Gehring & Jarrod Hepburn*

12 Two Stories about EU Climate Change Law and Policy 419
*Navraj Singh Ghaleigh*

13 The Emissions Trading System in the Context of Climate Change: China's Response 463
*Tianbao Qin*

14 Climate Change, Human Security and Regional Integration: The Example of the Southern African Development Community 505
*Oliver C. Ruppel & Katharina Ruppel-Schlichting*

**PART IV: CLIMATE CHANGE, THE LAW OF THE SEA AND SEA-LEVEL RISE**

15 Climate Change Challenges and the Law of the Sea Responses 549
*Guifang (Julia) Xue*

16 Holding Back the Waves? Sea-level Rise and Maritime Claims 593
*Clive Schofield*

17 Disappearing States: Harnessing International Law to Preserve Cultures and Society 615
*Gregory E. Wannier & Michael B. Gerrard*

18 Climate Change and Small Island Claims in the Pacific 657
*Yukari Takamura*

**PART V: JUDICIAL REVIEW AND INTERNATIONAL CLIMATE CHANGE LITIGATION**

19 Some Perspectives on Global Governance, Judicial Review and Climate Change 687
*Hennie A. Strydom*

20 Climate Change, Global Governance and International Criminal Justice 711
*Gerhard Kemp*

21 Climate Change Litigation: A Global Tendency 741
*Noriko Okubo*

22 International Climate Change Cases 759
*Roda Verheyen & Cathrin Zengerling*

23 Public Interest Litigation and Climate Change – An Example from Kenya 805
*Collins Odote*

24 Injunctions against Climate Change? 831
*Ulrich Magnus*

25 Climate Change and Liability: An Overview of Legal Issues 859
*Ina Ebert*

**PART VI: INTERNATIONAL CLIMATE CHANGE LAW AND CROSSCUTTING ISSUES**

26 Limitations of Risk Law 869
*Ivo Appel*

27 International Climate Law and Mining Regulation – Perspectives from Developing Countries 899
*Yemi Oke*

28 Legal and Regulatory Aspects of Carbon Capture and Storage: A Developed and Developing Country Perspective 933
*Jan Glazewski*

29 Climate Change Mitigation and Adaptation: What is the Role of Intellectual Property and Traditional Knowledge? 957
*Eliamani I. Laltaika & Joy Faida*

30 Climate Engineering and International Law: Final Exit or the End of Humanity? 979
*Gerd Winter*

Annex
Contents Volume II
Climate Change: International Law and Global Governance
Volume II: Policy, Diplomacy and Governance in a Changing Environment 1013